Neural-Immune Interactions in Brain Function and Alcohol Related Disorders

Changhai Cui • Lindsey Grandison
Antonio Noronha
Editors

Neural-Immune Interactions in Brain Function and Alcohol Related Disorders

Springer

Editors
Changhai Cui
Division of Neuroscience and Behavior
National Institute on Alcohol Abuse
and Alcoholism
National Institutes of Health
Bethesda, MD, USA

Lindsey Grandison
Division of Neuroscience and Behavior
National Institute on Alcohol Abuse
and Alcoholism
National Institutes of Health
Bethesda, MD, USA

Antonio Noronha
Division of Neuroscience and Behavior
National Institute on Alcohol Abuse
and Alcoholism
National Institutes of Health
Bethesda, MD, USA

ISBN 978-1-4614-4728-3 ISBN 978-1-4614-4729-0 (eBook)
DOI 10.1007/978-1-4614-4729-0
Springer New York Heidelberg Dordrecht London

Library of Congress Control Number: 2012947993

Printed on acid-free paper

Springer is part of Springer Science+Business Media (www.springer.com)

Preface

Recent studies have provided clear evidence on the role of neural–immune interactions in normal brain function and neuropathological conditions. Neuroimmune factors, which play an essential role in neuroinflammatory response, have been implicated in the regulation of neuronal function and plasticity. Thus, neural–immune interactions provide a new framework for understanding the role of the neuroimmune system in normal brain function, neurodevelopment, and a variety of neurological disorders. These advances have a far-reaching impact on many areas of neuroscience, including alcohol research. Studies using human alcoholic brains, gene knockout mice, and gene expression profiling have established a clear link between alcoholism and an altered neuroimmune profile. This book integrates emerging knowledge on neural–immune interactions with key discoveries in alcohol research and provides a comprehensive overview of neural–immune interactions in brain function and behavior associated with alcohol use disorders.

This book addresses this rapidly developing area of research. It contains three main sections. Section I focuses on the basics of the neuroimmune system and the neuroimmunological alterations that are directly relevant to alcoholism. It integrates the traditional role of the neuroimmune system with emerging novel functions of neuroimmune factors in the brain. This section highlights recent advances in microglia function, neuron–glia communication, as well as neural–immune interactions in synaptic function, stress, mood disorder, and neuroAIDS. It also provides a chapter on the clinical detection of brain–immune interactions using neuroimaging approaches. Section I sets the foundation for understanding alcohol and neuroimmune interactions, which are primarily presented in the second section. Section II provides a detailed overview of the impact of alcohol exposure on the neuroimmune system as well as the contribution of neural–immune interactions to alcohol use disorders. It highlights several key research discoveries in this area in the context of neuroimmune signaling, neurodevelopment, neurotransmission, alcohol drinking behavior, aging, HPA axis, and neuroAIDS. Section II also discusses the therapeutic potential of targeting neuroimmune modulation for the treatment of alcoholism and substance use. Section III provides a summary and future research directions by the book editors. Collectively, these three sections review exciting advances in neuroimmune research and summarize current understanding of alcohol and neuroimmune interactions.

While this book focuses on neural–immune interactions in areas directly related to alcohol use disorders, it is not intended to be all inclusive. Several areas, including sleep disorders, pain, and cholinergic anti-inflammatory pathways, are not covered as independent chapters but briefly mentioned in the text. The close relevance of these topics to neural–immune interactions and alcohol use disorders warrants future discussion and more research efforts.

This book will appeal to scientists, students, and educators who are interested in cross disciplinary topics, particularly those in the fields of alcohol use disorders, addiction, neuroimmunology, neuroimmune pharmacology, neuropsychopharmacology, and neuroAIDS. Readers will benefit from cutting-edge insights provided by outstanding, active researchers in the fields of neural–immune interactions and alcohol use disorders. We would like to thank all the scientists who have contributed chapters to this book and those who have provided constructive comments and suggestions for the manuscripts.

Bethesda, MD, USA

Changhai Cui
Lindsey Grandison
Antonio Noronha

Contents

Section II Alcohol and Neuroimmune Interactions

Section III Summary

Section I
Neural-Immune Interactions

Chapter 1
An Introduction to CNS-Resident Microglia: Definitions, Assays, and Functional Roles in Health and Disease

Deirdre S. Davis and Monica J. Carson

1.1 Introduction and Overview of Topics

Microglia are the resident immune cells of the brain. As such, their activation is associated with a range of conditions in the healthy and inflamed CNS. In the healthy brain, microglia maintain homeostasis by continuously sampling their environment [1]. When CNS infection and/or inflammation occurs, microglia respond appropriately by changing their activation state. Recent studies illustrate that microglial activation is not an "all or nothing" event. Instead, like peripheral macrophages, microglia exhibit a range of activation states associated with pathogen defense, wound healing, and tissue repair functions (Fig. 1.1) [2, 3]. In addition, although microglia can change their morphology in response to specific stimuli, morphology is now well demonstrated to be a poor discriminator between different types of activation (Fig. 1.1) [2–4].

Though microglial activation has been observed in the developing brain, the type of activation and the functions associated with the particular activation state(s) have not been closely examined. Studies by Hristova and others have clearly demonstrated that microglia in subcortical white matter areas of the brain convert from an activated phenotype during the early postnatal period to a resting or homeostatic phenotype observed during the second postnatal week. Though they also identified that several markers of microglial activation changed with healthy postnatal brain development, whether microglia adopt a specific type of activation state during this period is unknown.

Systemic inflammation experienced during the perinatal period and subsequent microglial activation have recently been associated with the development and/or

D.S. Davis • M.J. Carson, Ph.D. (✉)
Division of Biomedical Sciences, Center for Glial-Neuronal Interactions,
University of California, Riverside, 900 University Avenue, Riverside, CA 92521-0121, USA
e-mail: monica.carson@microglia.org

C. Cui et al. (eds.), *Neural-Immune Interactions in Brain Function and Alcohol Related Disorders*, DOI 10.1007/978-1-4614-4729-0_1,

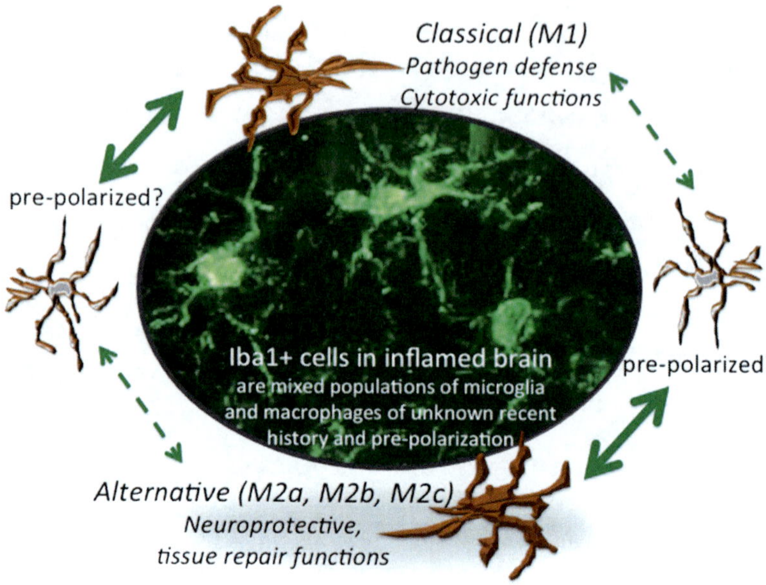

Fig. 1.1 Morphology defines neither cell type nor activation state. Morphology does not reveal whether a cell expressing macrophage markers is of CNS origin or recently of blood-derived origin. Similarly, microglia can be primed by environmental and/or developmental cues to be biased toward cytotoxic (M1/classical activation states) or tissue repair (M2/alternative activation states) without overt changes to their morphology. This figure depicts an inflammatory foci in a murine model of CNS autoimmunity in which both CNS-resident microglia and CNS-infiltrating macrophages participate and both M1 and M2 polarized cells are present

pathogenesis of several neurodevelopmental disorders [5–7]. Moreover, epidemiological, immunohistochemical, and imaging techniques have shown that microglia are activated in the brains of patients with autism, schizophrenia, and cerebral palsy even very early in disease progression [6–9]. These findings strengthen the hypothesis that microglial activation in the brains of these patients is associated with the pathogenesis of these disorders. However, because microglial functions in the postnatal developing brain are not well defined, the potential roles that microglia may have in these and other neurodevelopmental disorders are unknown.

Interestingly, studies have also shown that microglia become activated in the brain with healthy aging [10, 11]. This activation is also associated with increased expression of pro-inflammatory molecules such as B7.2, IFNγ, IL-1β, and IL-6 [10, 12]. There is currently some debate as to whether anti-inflammatory molecules are upregulated in the aged brain as hippocampal IL-10 mRNA expression has been shown to decrease in the aged brain while cortical IL-10 protein expressed by microglia has been shown to increase [10, 13]. Currently, it is unknown what causes microglia to become activated in the healthy aged brain. It is also unknown whether microglia become activated with age or whether the increased expression of inflammatory molecules serves as an indicator of aged or senescent microglia. However, because systemic inflammation causes greater induction of pro-inflammatory molecules in the brains of aged mice than in the brains of young adult mice, this demonstrates that microglia do become primed in the aged brain.

Specialized glial cells called microglia serve as the resident immunocompetent cells of the central nervous system (CNS). Though many describe microglia as having a "resting" state in the healthy brain, this is a misnomer. In the healthy unmanipulated CNS, microglial processes are highly mobile and constantly survey their microenvironment [1]. In response to trauma, inflammation, or infection, microglia become activated and peripheral immune cells are recruited to the CNS by microglia and other CNS-resident cells. Both microglial morphology and the collection of receptors that they express change with their activation state. In the adult CNS, the type and magnitude of microglial activation and macrophage influx triggered by inflammation are better characterized than in the developing CNS. However, many researchers use in vitro methods, including microglia from mixed glial cultures or immortalized microglial cell lines to study microglia in the "adult" brain. Microglial phenotype in the intact brain is highly dependent on the summation of cues provided by cells in the microenvironment. Though in vitro methods help us understand how these cells may react to a specific stimulus in isolation, experimental findings from in vitro studies may differ significantly from what occurs in vivo. Why? For the simple reason that one of the primary characteristics of all macrophages including microglia is their phenotype is highly plastic and is determined by the summation of all the cues from their environment.

1.2 CNS Environment: Neurons and Glia

The CNS is a complex organ system consisting of several cell types. The main cell types of the brain and spinal cord are the neurons and glia. Glia can be further divided into the macroglia including astrocytes and oligodendrocytes which are of neuroectodermal origin and microglia which are of mesenchymal origin.

Though neurons form the major communication networks that allow the brain to control all other organ systems in the body, all glia perform a diverse array of functions that serve to maintain homeostasis in the CNS. Specifically, glia provide active regulation of the network's operations, ongoing maintenance to deal with normal wear and tear, as well as active defense and repair following injury or pathogen attack [14–17]. To perform the appropriate functions, glial cells constantly communicate with neurons. Specifically, neurons, macroglia, and microglia are equipped with molecules that allow them to communicate with each other and with immune cells from the periphery that are recruited to the CNS following inflammation (see Sect. 1.4) [4, 17, 18]. Though neuroimmunologists are becoming increasingly aware of how peripheral immune cells affect the mature CNS, we have yet to fully explore whether these cells exert distinct effects on the developing CNS.

1.2.1 What Are Microglia?

Microglia are commonly referred to as the tissue macrophage of the CNS. As these cells express most macrophage markers, their origin has served as the source of debate for some time. To date, two hypotheses exist on when microglia populate the CNS [19–21]. The first and most commonly accepted view is that microglia appear in the CNS in the nonvascularized embryonic period, and these cells originate from progenitors in the yolk sac [20, 21]. The second hypothesis postulates a second wave of microglia may originate from circulating monocytes that populate the brain during the neonatal and/or very early postnatal period [19–21]. Regardless of whether there are one or two waves of microglial progenitors, several studies have established that during the period of early CNS development and in the adult CNS, microglia exist in specific subpopulations [22, 23]. It is currently unclear how the functions of these specific subpopulations differ.

Evidence for the earliest wave of microglia into presumptive CNS was obtained using flow cytometric and lineage studies. These studies illustrate that embryonic and adult mouse microglia arise from primitive myeloid precursors in the extra embryonic yolk sac [24]. Using parabiotic mice, Ginhoux et al. demonstrated that though 30% of monocytes and tissue macrophages were donor derived at 1 and 12 months following parabiosis, only 5% of microglia were donor derived at these time points. These findings illustrate that CNS-resident microglia are rarely, if ever replenished by blood, derived cells from young adulthood onward. This finding also illustrates that experiments using bone marrow chimeric mice and sublethal irradiation may present inaccurate results as Ginhoux et al. [24] also observed that 10–20% of parenchymal microglia in the recipient mouse were of donor origin in their own bone marrow chimeric studies. Irradiation is known to cause changes to the blood–brain barrier (BBB), and thus the higher percentage of donor derived microglia in bone marrow chimeric recipients may occur as a result of BBB damage [24, 25]. Ginhoux and colleagues also used lineage tracing studies to depict that primitive myeloid progenitors that arise before embryonic day 7.5 contribute significantly to adult microglial homeostasis in the healthy brain. These same progenitors gave rise to very few circulating leukocytes in the adult [24]. Contrary to previous studies, Ginhoux et al. reported that myeloid progenitors require blood vessels to migrate from the yolk sac to the brain between embryonic days 8.5 and 9.5.

As glial cells, microglia perform various functions that facilitate proper function of the brain and spinal cord. However, microglia are also immune cells and they are able to detect when CNS immunity has been compromised and respond appropriately by changing their morphology and the collection of receptors and molecules that they express. Though their origin has served as a major debate since their discovery, many now agree that microglia are myeloid-derived cells that can be detected in the CNS during early prenatal development [25, 26]. In the healthy brain, microglia have a stellate morphology and can be found throughout the brain and spinal cord [27]. When microglia become fully activated, they resemble blood-derived inflammatory macrophages as their processes shorten and their cell bodies become more round.

1.2.2 How Do We Study Microglia?

Microglia are highly plastic cells and their phenotype depends greatly on the summation of their external cues provided by cells within their CNS microenvironment [27, 28]. In order to effectively study microglia in the lab, in vitro and in vivo methods have been devised. However, as we are not yet able to study microglia in the fully intact and living brain, both types of experimental methods have pros and cons that must be carefully considered before studying these cells.

1.2.3 In Vitro Studies

In the brain microglia are surrounded by neurons, astrocytes, oligodendrocytes all in a three-dimensional matrix. However, to study specific molecular mechanisms of microglial activation, in vitro methods are often used due to the difficulty of obtaining sufficient cells from adult tissue. However, when performing in vitro studies, one must remember that in the intact brain, microglia are in constant three-dimensional contact with multiple CNS cell types. Moreover, in vivo microglial activation can be caused by the loss of these cues which do not exist in most in vitro culture systems [25, 28].

The in vitro methods that are most often used to study microglia are (1) immortalized microglial cell lines including the BV-2 and N9 cell lines, (2) primary microglia generated from embryonic and neonatal mixed glial cultures, (3) primary microglia isolated from CNS tissue and analyzed ex vivo, and (4) organotypic slice cultures isolated from the living and intact brain [25]. As mentioned above, microglial phenotype is highly dependent on cues from their environment [25, 28]. When these cues are removed and microglia are cultured in isolation, studies have shown that microglial physiology may differ greatly from their in vivo physiology.

CD45, also called leukocyte common antigen, is a type I transmembrane protein tyrosine phosphatase that is expressed by all nucleated hematopoietic cells and their precursors [29, 30]. Studies have shown that CD45 is involved in the negative regulation of cellular activation [30]. While all other cells of hematopoietic origin increase their expression of CD45 from the neonatal period through adulthood, microglia do not [14, 31, 32]. In the intact brain, activated microglia upregulate CD45 [14, 27, 31]. However, the CD45 levels of activated microglia remain lower than that expressed by peripheral macrophage populations [14, 27, 31]. Curiously, unactivated microglia cultured in the absence of neurons express higher CD45 levels than that expressed by activated microglia in vivo [14, 25, 27]. This clearly depicts that microglia cultured in the absence of CNS cells do not necessarily mirror microglial phenotype and/or physiology in the intact brain.

Organotypic brain slice cultures are also a widely used setup that can be used to study microglia. In many ways, slice cultures may prove to be more beneficial to specific studies because, unlike immortalized cell lines or primary microglial cultures, slice cultures allow researchers to study microglia in a brain slice that retains not only the cellular diversity of the CNS but also the tissue structure [25, 33]. Slice cultures also allow researchers to treat the slice with pharmacological agents, toxins, pathogens, and cells to observe how cells within the slice react [25]. Though slice cultures can be beneficial for studies, there are several reasons why they may be problematic in microglial studies. When the slice itself is cut from the whole brain, "wounds" are made leading to microglial and astrocyte activation [25]. Moreover, extensive and ongoing cell death occurs in these slices, potentially skewing results and observations [33]. Depending on the type of experiment planned, slice cultures do not allow for the infiltration of blood-derived immune cells following pathology.

Isolation and analysis of microglia (without culture) from the intact murine brain can also serve as a useful method to study microglia. These cells develop and mature in the intact murine brain and thus are more likely to resemble in vivo microglia than cultured cells. However, depending on the types of studies planned, a flow cytometer or a real-time PCR (RT-PCR) machine may be necessary. Though the numbers of microglia isolated from the murine brain may be sufficient for RT-PCR analysis, it is often difficult to collect adequate numbers for high-throughput assays [25].

Though the various in vitro methods used to study microglia can be readily helpful in understanding their physiology, the results should be analyzed with extreme caution. More specifically, in order to truly validate results using these in vitro methods, researchers may want to run studies using several of these methods to create valid conclusions. Not only will this help to solidify the results, but it will also provide researchers with confidence in the usefulness of these methods.

1.2.4 In Vivo Studies

In vivo studies are useful because they allow us to observe microglia in their intact microenvironment. Though informative methods have been created to study these cells in the living brain, methods have not been devised that allow for a clear understanding of their physiology using noninvasive techniques. The main methods used to study these cells in vivo are (1) bone marrow chimeric studies, though the cells are analyzed in postmortem tissues, and (2) imaging techniques [25].

In bone marrow chimeric studies, mice generally receive sublethal irradiation to kill off their bone marrow, and then they receive bone marrow from a donor mouse to replenish their peripheral immune compartment [25]. Interestingly, these studies show that though recipient mice receive donor bone marrow, the majority of microglia remain recipient derived [24]. Generally, these studies involve genetically altering the donor bone marrow in order to trace the cells in the recipient [25]. Bone marrow chimeric studies can be incredibly useful in studies looking at the degree of influx of peripheral immune cells into the brain, and also the role that these peripheral cells have in several CNS pathologies. However, studies have shown that the irradiation itself can damage the vasculature and BBB, indicating that the numbers of peripheral immune cells detected in the CNS may occur as a result of the irradiation and not the healthy or pathological physiology [24, 25]. Moreover, these studies are typically analyzed in postmortem tissue, making it difficult to observe the interactions between the donor and recipienT cells in the live brain.

Recently, powerful imaging techniques have been used to observe microglia in the intact brain. Two-photon microscopy is a powerful tool that has been used by researchers studying microglial physiology and motility in the healthy and inflamed brain [1, 34]. Typically, those using two-photon microscopy use explants of the tissue of interest and remove the organ entirely from the animal or intravital methods where the animal is immobilized and the tissue is exposed [35]. Most reported microglial studies have used preparations in which the skull is removed or thinned [25, 35]. Moreover, to visualize microglia, transgenic mice with fluorescent microglia are typically used [1, 34]. Two major studies performed by Nimmerjahn and colleagues and Davalos and colleagues used this powerful technique and found that microglia are not "resting" in the healthy brain. Instead, in the healthy brain, microglia constantly extend their processes to survey their microenvironment every few hours [1]. Though two-photon microscopy has become a very powerful tool to visualize these cells in intact tissue, this technique also has limitations. A major caveat for intravital setups of two-photon microscopy is that motion artifacts can occur because of movements generated from breathing or involuntary muscle contractions [35].

1.3 What Are Microglial Functions in the Healthy CNS?

For some time, microglial functions have been studied in the context of pathology caused by trauma or a specific disease. This focus is potentially due to the fact that microglia change their morphology and the types of genes that are expressed when they become activated. Because of the visible changes microglia undergo during their activation, many have regarded microglia in the healthy brain to be in a "resting" state. Recently, several labs have used two-photon imaging of the cerebral cortex of live mice and show that in the healthy brain, microglial processes are highly motile and constantly sample their environment [1]. Because microglial processes monitor their microenvironment extensively, Nimmerjahn et al. suggest that the entire brain parenchyma is surveyed by these "homeostatic" microglia once every few hours. In these studies, it was observed that microglia also phagocytose tissue samples which are then transported to the soma [1].

It seems we are just beginning to understand the roles that microglia play in the healthy CNS. Specifically, studies have shown that a receptor expressed by microglia may directly or indirectly affect cognitive and other important functions in the healthy CNS [36–38]. For example, Nasu–Hakola disease is a rare recessive human genetic disorder that causes early-onset cognitive dementia, bone cysts, and death in the fifth decade of life [39, 40]. Studies have shown that this disease is caused by loss of function mutations in the triggering receptor expressed on myeloid cells 2 (TREM2) or in the TREM2 signaling pathway regulated by the adaptor molecule DNAX-activating protein of molecular mass 12 kDa (DAP12) [40, 41]. DAP12 is a transmembrane adaptor molecule expressed by several immune cells that signals via an immunoreceptor tyrosine-based activation motif (ITAM) [40, 42]. Though DAP12 is required for TREM2 signaling, because Nasu–Hakola disease has obvious effects on cognition, a process most often thought to be regulated by neurons, many believed that the cognitive symptoms were caused by a defect in neurons. Interestingly, we and others have found that TREM2 is expressed by subsets of microglia in the healthy murine brain [23, 40].

Overexpression and knockdown studies have shown that TREM2 expression promotes microglial phagocytosis of cellular debris and apoptotic cells, decreases microglial motility, inhibits microglial and macrophage cytokine expression, and limits the severity of a mouse model of multiple sclerosis [43–46]. Though there is much that is not understood, these data demonstrate that lack of a microglial protein can lead to the development of a disease with cognitive manifestations.

More direct demonstrations of the ability of microglia to modulate cognition are provided by two additional studies using murine models of Rett syndrome and obsessive–compulsive disorder (OCD) [36, 38]. In the first study, Capecchi and colleagues demonstrated that the absence of HOXb8 in microglia was sufficient to cause OCD-like over-grooming behaviors and restoring HOXb8 expression only in microglia was sufficient to prevent OCD grooming behavior [36]. Similar types of observations were made in murine model of Rett syndrome [38]. Rett syndrome is a neurodevelopmental syndrome with multiple clinical symptoms that places it partially under the autism spectrum of neurodevelopmental disorders. Rett syndrome is caused by lack of function mutations in the ubiquitously expressed gene, MECP2. To date, most of the focus on pathogenesis has focused on neuronal dysfunction. However, recently Dericki et al. demonstrated that by reconstituting, MECP2 function in the microglial compartment was sufficient to arrest and even partially restore behavioral Rett syndrome pathology.

In the murine models of Nasu–Hakola and OCD, the mechanisms by which microglia lead to changes in neuronal function are not defined. Therefore, it is striking that the study using the Rett mouse model was able to demonstrate that inhibiting microglial phagocytosis was sufficient to block the restorative functions of microglia expressing wild-type MECP2 [38]. This data is also consistent with new hypotheses of microglial function in synaptic plasticity [37]. Specifically, multiple studies have now implicated microglia in editing or phagocytosing non-active or weakly active synapses as part of the natural process of synaptic maturation [37]. As yet the mechanisms of this action are still being defined, and different mechanisms may be used in different brain regions or during different periods of development/regeneration. However, in the development of the visual system, microglial editing of unwanted synapses appears dependent on the C1q pathway [37].

1.3.1 What Are Microglial Roles in the Developing CNS?

While there is a current lack of understanding of the roles microglia have in the healthy mature brain, there is perhaps an equal lack in knowledge regarding microglial functions in the developing brain. It has been well documented by several labs that microglial morphology and markers of microglial activation change as the brain develops [47]. Hristova and colleagues performed informative immunohistochemical studies showing that periventricular microglia express high levels of several integrin subunits and B7.2, a co-stimulatory molecule, at birth that decreased by the second postnatal week. This clearly depicts that microglial activation state changes during the first postnatal week in rodents, corresponding to the last trimester in humans. A curious observation from this study is that though both cortical and subcortical microglia would presumably be exposed to cellular apoptosis during this postnatal period, which could influence microglial activation status, only subcortical microglia displayed an activated phenotype. Are the microglia in the subcortical regions more active or performing different functions than the cortical microglia in the developing brain?

The developing brain experiences a great deal of proliferation followed by programmed cell death in order to shape the final form and connections of the mature CNS [48, 49]. A readily accepted hypothesis is that microglia are involved in the removal of apoptotic neurons in the developing brain. Using time-lapse recordings of cultured ex vivo cells and immunohistochemistry, Parnaik et al. demonstrate that professional phagocytes of the brain can indeed engulf apoptotic cells in the postnatal developing brain. Though Parnaik et al. clearly demonstrate that microglia from brain cultures can phagocytose experimentally induced apoptotic neurons, they fail to clearly explain how microglia are capable of clearing such substantial numbers of apoptotic neurons during this critical postnatal period of brain development. Another study using silver impregnated tissues and immunohistochemistry also demonstrated that apoptotic cells in the cerebral cortex were engulfed by phagocytes in the brain during the first postnatal week of life [50]. It was not clear whether the transitory phagocytes were CNS-resident microglia or blood-derived macrophages.

1.3.2 What Are the Types of Microglial/Macrophage Activation?

As previously stated, for much of the last century, microglia were thought to be maintained in an "off" state in the healthy CNS and turned "on" in response to injury or infection. However, current research also reveals that microglial activation is not as simple as it was once thought to be. Though microglial activation has mostly been discussed in the context of CNS pathology, microglia are highly plastic and can display a range of activation states that are not always associated with pathogen defense functions [3, 51]. Now that it has become common knowledge that microglial activation is not synonymous with cytotoxicity, researchers must take a closer look at the activation states of these cells in order to accurately define their roles in the developing, damaged, diseased, or aging CNS. Based on characterizations of peripheral macrophage activation states, microglia are now realized to have a spectrum of activation states described by two nomenclatures: classical versus alternative or M1 versus M2 activation spectrums (Fig. 1.1) [2].

The activation state that most researchers are familiar with is the classical (also termed M1) activation states of microglia and macrophages. Very detailed studies have shown that cytokines that are produced by activated T helper 1 (Th1) cells of the immune system, namely, TNFα and/or IFNγ, cause microglia to convert from "homeostatic" cells into classically activated cells that are able to defend the host from invading pathogens [3, 52–54]. These signals in turn cause the microglia to produce factors such as IL-1 and IL-6 and reactive oxygen species that primarily serve to destroy foreign pathogens by amplifying the immune response, recruiting immune cells to the site of inflammation, and activating immune and nonimmune cells [3, 55]. However, if excess pro-inflammatory molecules are produced and/or not cleared in a timely fashion, these molecules can also damage host tissues [3, 55]. Because of the potential cytotoxic responses, the functions of classically activated macrophages must be carefully regulated.

As mentioned, classically activated phagocytes will increase their production of inducible nitric oxide synthase (iNOS) in order to produce reactive oxygen species to defend the host from pathogens. Normally, classically activated phagocytes will also increase their expression of molecules associated with antigen presentation, including MHC class II, and molecules that will allow them to effectively activate T cells, namely, co-stimulatory molecules such as B7.2 and CD40.

Though the removal of invading pathogens and/or debris is an essential function performed by microglia and macrophages, at times, they are required to repair tissue that has been damaged by a foreign organism or by an immune response. To perform these functions, microglia and macrophages must become alternatively activated. IL-4 and IL-13 can be produced by cells associated with innate and/or adaptive immunity, including basophils, masT cells, eosinophils, NKT cells, CD4+ T helper 2 (Th2) cells, and CD8+ T cells [56–60]. These factors have been shown to promote arginase activity in microglia and macrophages which can then convert arginine into ornithine [61]. Ornithine is then converted by ornithine aminotransferase into proline, which is important in collagen production [3, 61, 62]. Ornithine can also be converted by ornithine decarboxylase into polyamines which are involved in cellular proliferation [3, 61, 62]. Both products allow alternatively activated cells to promote wound healing and tissue repair in part by building up the extracellular matrix.

When phagocytes become alternatively activated (also termed M2 activation), it is generally accepted that these cells upregulate a number of specific molecules that serve to resolve parasitic infections, build up the extracellular matrix, and in general mediate tissue repair. Consistent with their repair functions, alternatively activated cells can express a variety of classic neuroprotective growth factors including IGF-1.

In addition, as discussed before, arginase 1 is an enzyme that is consistently induced by alternatively activated microglia and macrophages, and it competes with iNOS for L-arginine in order to produce factors that promote tissue repair [3, 61, 62]. Much has been discussed in the literature that the arginase 1 versus iNOS expression defines the choice point between alternative versus classical activation. However, it is important to note that while this is true of murine cells, arginase 1 expression does not appear to be a feature of human alternatively activated microglia/macrophages!

There is a vast amount of literature that also states that alternatively activated microglia and macrophages induce mannose receptor (CD206) and Ym1 and found in inflammatory zone 1 (Fizz1) [51, 63]. Mannose receptor is a phagocytic receptor that mediates the binding and ingestion of microorganisms containing surface mannose residues [51]. The functions of Ym1 and Fizz1 are not completely understood, and their roles in alternative activation have not been discerned. However, these two molecules are often upregulated by alternatively activated phagocytes in vitro, by IL-4 stimulation and in vivo during Th2-mediated immune responses.

Ym1 is a member of the chitinase family that was originally described as a chemotactic factor for eosinophils that was produced by CD8+ T cells [64, 65]. Fizz1 is a member of a family of secreted cysteine-rich molecules [63, 66]. It is known that Fizz1 expression is induced in alternatively activated microglia and macrophages. Studies have suggested that Fizz1 promotes wound healing and fibrosis in Th2-mediated immune responses [67]. Interestingly, a recent study using Fizz1 knockout mice and *Schistosoma mansoni* infection showed that Fizz1 acts as a negative regulator of helminth-induced Th2-mediated immune responses [68]. In this study, they demonstrated that Fizz1 knockout mice infected with *S. mansoni* had increased inflammation in the lung and accelerated fibrosis in the liver [68]. Though these studies were focused on peripheral organs, their findings would suggest that though Fizz1 is induced following IL-4 stimulation, it functions to regulate the degree of the immune response.

An important first step has been made by recognizing that microglia and macrophages have the potential to adopt a variety of activation states. Yet in vivo, we must also acknowledge that these cells very rarely adopt a single fully polarized phenotype. Despite this caveat, the ability to characterize the activation states of microglial or macrophage populations in terms of classical versus alternative activation states is providing a useful measure to discriminate between distinct types of activation programs in resolving and chronic neuroinflammatory disorders.

1.3.3 Can Microglia Be Distinguished from Blood Macrophages?

There are a variety of macrophage-like cells that can be found within the CNS. These include parenchymal microglia, perivascular microglia, and acutely infiltrating blood-derived macrophages. Though these populations have been recognized for some time, no conclusive method that will allow one to distinguish microglia from CNS-infiltrating macrophages in tissue sections has been identified. However, methods have been created that allow researchers to separate the two populations using single-cell brain suspensions and flow cytometry [32]. Moreover, recent studies may have identified markers that may one day be used to detect bone marrow-derived macrophages in the brain [69].

In the brain, microglia and macrophages express CD11b, an integrin that promotes migration, adhesion, and phagocytosis [70]. CD45 is expressed by all nucleated hematopoietic cells and their precursors [29]. Using single-cell brain suspensions and flow cytometry, Sedgwick et al. demonstrated that microglia can be distinguished from CNS-infiltrating leukocytes based on the levels of CD45 that are expressed by each population. Specifically, if one were to label myeloid cells with antibodies against CD11b and CD45, microglia would be the CD11b+CD45lo population while CNS-infiltrating blood-derived macrophages would be the CD11b+CD45hi population [31].

For some time, it was thought that the main role of homeobox (Hox) genes was to establish body plans by providing positional information along the anteroposterior axis of the embryo [36, 69]. Recent studies, however, demonstrate that Hox genes may also play roles in the development of specific organs, such as the kidney [71]. Thorough investigations performed by Chen et al. demonstrate that Hoxb8 mutant mice show behavioral deficits including excessive and pathological grooming that is characteristic of human patients with the OCD trichotillomania [69]. Instead of finding Hoxb8 expressed by cells in a neural circuit controlling grooming, Chen et al. established that Hoxb8 was expressed by bone marrow-derived microglia in the brain. They came to this conclusion because (1) only a subpopulation of microglia in the brain, specifically 40%, were Hoxb8+ and (2) in newborn mice these Hoxb8+ cells were found in the choroid plexus, meninges, and ventricular lining eventually populating cortical and forebrain regions by postnatal day 14 [69]. This migration strongly suggests that the Hoxb8+ microglia are bone marrow derived. Though these studies were performed in the naïve brain, it would be interesting to determine whether blood-derived macrophages that enter the CNS during pathology were also Hoxb8+. This would help to elucidate whether Hoxb8 could be used as a marker to distinguish yolk sac-derived microglia from bone marrow-derived macrophages in the brain tissue sections.

1.3.4 Do Microglia Have Different Functions than Macrophages?

Microglia and macrophages express many of the same markers including F4/80, Iba-1, and CD11b [15]. During an immune response, both cell types can induce molecules such as MHC class II, B7.2, and CD40 that allow them to communicate with and activate T cells in the presence of antigen [15]. Because of these similarities, many have believed that these cell types are very similar, if not the same cell.

However, studies such as those performed by Chen and colleagues (discussed in the previous section) in relation to Hoxb8 expression, and also studies by Ford et al. showing that microglia express lower levels of CD45 than peripheral macrophages, clearly demonstrate that in the CNS, microglia may exhibit different functions than bone marrow-derived macrophages.

In seminal studies performed by Hickey and Kimura, bone marrow chimeric rats were used to demonstrate that antigen-specific interactions between CNS-resident microglia and myelin-specific T cells were not required to initiate or sustain autoimmune responses. In these studies, experimental autoimmune encephalomyelitis (EAE), an experimental model of multiple sclerosis in rodents, was used. Specifically, these studies and others have nicely demonstrated that perivascular macrophages and dendritic cells are capable of initiating and sustaining EAE [72, 73]. Moreover, bone marrow chimeras lacking antigen-presenting cells (APC) in the periphery (macrophages, dendritic cells) were not able to induce EAE, indicating that CNS-resident microglia were in some manner insufficient to induce this autoimmune response [73].

Differences between CNS microglia and macrophages have also been established in a model of facial axotomy. Used as a model to study the mechanisms of survival of the facial motoneuron cell body and regeneration of the axon, the facial axotomy model involves transecting the axon of the facial motoneuron at the site where it exits the CNS to the periphery [74]. This transection permanently prevents reconnection of the axon to the cell body while the neuron cell body survives [74]. Previous studies using the facial axotomy model showed that activated CD4+ T cells promote facial motoneuron survival [75]. Studies by Byram and colleagues support this study by demonstrating that bone marrow-derived peripheral APCs (macrophages, dendritic cells) are required to initiate CD4+ T cell activation but cannot sufficiently sustain a protective T cell response [76]. Furthermore, CNS-resident microglia are required to reactivate CD4+ T cells in order to sustain survival of the facial motoneuron [76].

Together, this data demonstrates that microglia and macrophages have certain similarities with respect to their functions and the repertoire of molecules they express in the naïve and inflamed brain. However, their roles in different CNS pathologies differ greatly.

1.4 How Do Microglia Detect Changes in Their Microenvironment?

In the CNS, microglia are resident innate immune cells and are generally the first responders to inflammation or injury. As such, mechanisms must be in place that allow them to detect when the CNS has been damaged or invaded by pathogens. Research has shown that several cell types throughout the body and within the CNS are capable of detecting a number of evolutionarily conserved motifs on pathogens as well as molecules produced by CNS intrinsic cells following tissue damage.

These molecules are collectively referred to as pathogen-associated molecular patterns (PAMPs) and danger-associated molecular patterns (DAMPs), respectively. Moreover, by virtue of several receptor–ligand interactions, microglia are incredibly sensitive to minute changes in the well-being of specific cell types.

1.4.1 Recognition of Conserved Pathogenic Signals

Toll-like receptors (TLRs) are a class of pattern recognition receptors expressed by immune and nonimmune cells that can detect PAMPs [77, 78]. Though the Toll pathway was initially defined as a family of genes involved in the dorsal–ventral patterning of the *Drosophila embryo*, in mammals, TLRs have been shown to be involved in pathogen detection and in the initiation of immune responses [78–80]. Both extracellular and intracellular TLRs have been found, and they are characterized by a cytosolic motif termed the Toll/IL-1 receptor domain (TIR) [78, 81]. Though studies have shown that astrocytes (TLRs 2, 4, 5, and 9) and neurons (TLRs 2, 6, and 8) express some TLRs, it has also been shown that microglia express the mRNA for all currently known murine TLRs (TLR 1–9) [82–88]. This allows microglia to defend the brain from a wide range of pathogens. Specifically, studies have shown that microglia can detect bacterial lipopeptides, double-stranded RNA, lipopolysaccharides from gram-negative bacteria, bacterial flagellin, single-stranded RNA, and unmethylated CpG dinucleotides [85, 89]. There is some debate as to whether astrocytes express TLR2 in vivo, as labs have had difficulty reproducing the data using in situ hybridization [89]. These differences may be due to species differences used in experiments or even the route of PAMP administration [89].

1.4.2 Recognition of Tissue Integrity

Microglia recognize tissue integrity by integrating both the presence and absence of signals. For example, microglia express multiple receptors to detect products of necrotic and apoptotic cells such as purinergic receptor and phosphatidylserine receptors. Conversely, stressed, infected, necrotic, or pre-apoptotic neurons and glia also trigger microglia activation by ceasing to produce "I am well" signals.

An important example of such an "I am happy" signal is the interaction between CD200 and CD200 receptor (CD200R). CD200 is a membrane glycoprotein of the immunoglobulin superfamily that is expressed by endothelial cells and neurons in the brain [90, 91]. CD200R is expressed by myeloid cells in the brain including microglia and macrophages [91]. Studies have shown that in the healthy CNS when CD200 is expressed, CD200R expressing microglia display a ramified morphology, characteristic of their "homeostatic" activation status. Using CD200 knockout (CD200KO) mice, Hoek and colleagues clearly demonstrate that even in the absence of any other inflammatory signal, microglia in the unmanipulated CD200 KO mice exhibit several features characteristic of activated microglia including shorter glial processes and elevated CD11b and CD45 expression. This clearly indicates that the CD200–CD200R interaction between neurons and microglia is required to prevent microglial activation in the noninflamed brain. Moreover, to determine whether CD200–CD200R interactions also regulate the degree of microglial activation in specific disease processes, Hoek and colleagues examined microglial activation and the timeline of disease onset in CD200 KO mice with EAE. Interestingly, Hoek et al. found that in CD200 KO mice with myelin oligodendrocyte glycoprotein (MOG)-induced EAE, the onset of EAE occurred 3 days before the onset in wild-type C57BL/6 mice.

Because EAE requires activation of autoreactive T cells, some have speculated as to whether these studies demonstrate a faulty T cell compartment, mainly the hyperactivation of T cells. Hoek et al. have performed various in vitro and in vivo experiments and have not observed T cell dysregulation in CD200R knockout mice, strengthening the hypothesis that the problem lies within the myeloid compartment. Together, these results demonstrate that CD200–CD200R interactions are required in the brain to maintain homeostatic functions of microglia and prevent unwanted microglial activation.

CX3CL1–CX3CR1 interactions have also been shown to regulate microglial activation in the brain and spinal cord. The chemokine CX3CL1 (fractalkine) is expressed in various organs in the body including the brain, heart, kidney, lung, and small intestine [92]. In the healthy brain, the chemokine CX3CL1 is expressed by neurons while the receptor is expressed by myeloid cells including microglia and macrophages [18, 93]. However, in vitro studies have also concluded that TNFα and IFNγ stimulation can cause human astrocytes to express CX3CL1 [94]. Though CX3CL1 exists as a membrane form, it can be cleaved to produce a soluble form as well. Parallel to these findings, studies have shown that the soluble form of fractalkine can be used as a chemoattractant for several leukocytes while the membrane-bound form can facilitate adhesion of these cells [94, 95]. Several essential functions of this ligand–receptor pair have been established. However, the role that this chemokine and receptor pair may play in neuronal–microglial communication has not been explored until recently.

Various in vitro studies have demonstrated that the CX3CL1–CX3CR1 interaction is neuroprotective. Using microglial cultures prepared from newborn rats, Zujovic et al. demonstrated that CX3CL1 lowered LPS-induced TNFα release in a dose-dependent manner. Interestingly, Zujovic and colleagues also demonstrated that when a neutralizing anti-CX3CL1 antibody was added to neuronal cultures prior to the addition of microglia, neuronal death was detected (following the addition of microglia). This would illustrate that CX3CL1 expression by neurons is required to prevent neurotoxic microglial activation in the noninflamed brain. However, these results have not been replicated in vivo. What both of these results reveal is that the CX3CL1–CX3CR1 interaction does allow neurons and microglia to communicate and that this "communication" leads to a neuroprotective microglial response. Though these in vitro studies were incredibly useful and informative, many qualities of in vitro studies, particularly with microglial cells, do not model what actually occurs in vivo.

To determine the roles of the CX3CL1–CX3CR1 interaction in the intact brain, Cardona et al. induced systemic inflammation in CX3CR1$^{+/GFP}$ mice (mice with functional CX3CR1 and green microglia) and CX3CR1$^{GFP/GFP}$ mice (mice that lack functional CX3CR1 but have green microglia) and examined microglial morphology and the degree of neuronal death in the hippocampus of the brain. Interestingly, Cardona and colleagues found that in LPS-injected CX3CR1$^{+/GFP}$ mice, microglia maintained a ramified morphology, indicative of a "homeostatic" state. On the contrary, microglia from LPS-injected CX3CR1$^{GFP/GFP}$ mice had shorter and thicker processes and a larger cell body, indicative of activated microglia [96]. These studies also nicely demonstrated that systemic inflammation in CX3CR1$^{GFP/GFP}$ mice caused a significant number of hippocampal neurons to become apoptotic, whereas no apoptotic neurons were detected in the CX3CR1$^{+/GFP}$ mice [96]. These studies clearly show that the threshold for microglial activation is perhaps lowered in mice lacking functional CX3CR1 and that the consequences of microglial hyperactivation to systemic inflammation in the absence of CX3CR1 are neuronal damage and death.

Another somewhat surprising neuron–microglia signaling is one also shared with B cells. As discussed in a previous section, microglia constitutively express CD45, a protein tyrosine phosphatase that is also the receptor for CD22 [30]. While CD22 has long been known to be expressed by B cells, Tan and colleagues discovered that neurons constitutively expressed and secreted CD22 as well [128]. To investigate whether neurons regulate microglial activation state through CD45, primary murine microglial cultures were treated with CD40 ligand (CD40L) and a cross-linking antibody for CD45 [30]. These studies demonstrated that activation of CD45 led to a significant decrease in the amount of TNFα produced by CD40 ligation. Tan and colleagues further illustrated the fact that CD45 expression by microglia is required to regulate the levels of TNFα produced following CD40 ligation by exposing primary microglial cultures from CD45-deficient mice to CD40L. As one may predict, the microglia from this experiment produced significantly higher levels of TNFα than CD45 expressing primary cultures.

These studies nicely illustrate that a molecule commonly used to identify microglial activation and distinguish them from CNS-infiltrating macrophages in brain cell suspensions is also used by neurons to limit the degree of microglial activation. Moreover, it appears that microglial activation state in the healthy and inflamed brain is tightly regulated by a variety of ligands that are expressed by neurons. The redundancy in receptor–ligand interactions between these two cell types that specifically control microglial activation status strongly suggests that dysregulated microglial activation is not tolerated in the CNS. From another viewpoint, these interactions also allow microglia to monitor neuronal health.

1.5 What Is Systemic Inflammation and How Does It Affect the CNS?

Viral and bacterial infections that occur outside of the CNS can cause immune responses within the brain and spinal cord. When viruses, bacteria, or parasites invade the body, PAMPs, or in the case of tissue injury, DAMPS, are detected by cells of the innate immune system. Innate immune cells in the periphery including macrophages and neutrophils then produce inflammatory cytokines such as IL-1β, IL-6, and TNFα that can circulate in the blood [97]. It is well established from a variety of studies that systemic inflammation can lead to activation of microglia and astrocytes and changes in neurons. However, the methods by which these cells become activated are still being studied. In order for peripheral immune signals to affect cells within the brain, they must either (1) enter circumventricular (CVO) organs and diffuse directly from the blood into the brain parenchyma (though this is debatable), (2) activate endothelial cells forming the blood vessels in the brain that can in turn activate perivascular microglia, (3) be directly transported across the BBB, or (4) activate the vagus nerve which can then communicate with neuronal populations within the brainstem [97–99].

Specific regions within the brain called CVOs are unique within the brain in that they lack a complete BBB. The blood vessels of most regions in the brain provide an extra level of protection for the CNS from a variety of molecules within the blood because the endothelial cells form tight junctions between one another. CVOs have fenestrated capillaries, and thus, many believe that cytokines can freely diffuse from the blood into the brain parenchyma and interact with immune cells such as the perivascular microglia in these regions [97]. However, other studies have established that following systemic inflammation induced by LPS, perivascular phagocytic cells upregulate IL-1β in CVOs, and this expression gradually shifts from the CVOs to adjacent brain nuclei [100]. Furthermore, Konsman and colleagues nicely demonstrate that the IL-1β is bioactive since iNOS mRNA can be detected at the interface between the CVOs and the adjacent neural nuclei. This study nicely illustrates that systemic inflammation can lead to activation of cells and the production of inflammatory cytokines in CVOs that can then propagate the signal into the brain parenchyma.

Several studies have also demonstrated that brain cells become activated following systemic inflammation due to endothelial cell production of inflammatory molecules. Specifically, it is hypothesized that peripheral cytokines produced in response to systemic inflammation lead to the production of the rate limiting enzyme for prostanoid biosynthesis, cyclooxygenase-2 (COX-2), and membrane-associated prostaglandin synthase-1 (mPGES-1) by endothelial cells in the brain [99, 101, 102]. Depending on the dose and agent used to induce systemic inflammation, others have reported that perivascular microglia also induce COX-2 expression [103]. These in turn lead to the production of prostaglandin E2 in the brain which has been shown to activate neurons [99, 104]. Several studies have demonstrated that systemic inflammation leads to production of circulating IL-1 and that interaction of IL-1 with its receptor on brain endothelial cells and some neurons can lead to the production of COX-2 [102, 103]. Ching and colleagues demonstrated that knockdown of endothelial cell IL-1R1 prevented COX-2 induction, hypothalamic neuronal activation and fever induced by intravenous (IV), and intracerebroventricular (ICV)-injected IL-1. Curiously, endothelial-specific knockdown of IL-1R1 suppressed intraperitoneal (IP) IL-1-induced fever but not neuronal activation in the hypothalamus [102]. These results suggest that the IL-1 injected into the peritoneal cavity led to the activation of CNS neurons using another mechanism besides endothelial cell activation.

Several studies have illustrated that some cytokines are actively transported across the BBB. Although several cytokines are produced in the brain, many have also observed that cytokines that circulate in the blood can have effects on the CNS. For example, when cancer patients receive IV administration of TNFα for its antitumor properties, they often experience several side effects that involve the CNS including headache, memory loss, and acute blindness which are dose dependent and resolve following therapy [105, 106]. These studies strongly suggest that the cytokines somehow cross the BBB to act directly on the brain. Following these studies, it was shown that TNFα and other cytokines, such as IL-1α, were able to exert direct effects on the CNS because they are transported across the BBB. Studies by Gutierrez et al. demonstrate that following IV injection, radiolabeled murine TNFα (mTNFα) was found in its intact form in brain tissue. By administering an IV injection of unlabelled mTNFα along with the radiolabeled form, Gutierrez and colleagues demonstrated that TNFα was transported into the CNS via a saturable mechanism. Further studies from this same group found that the TNFα receptors on endothelial cells were required to transport the cytokine across the BBB [107]. Specifically, in the absence of p55 and p75, the two TNFα receptors, influx of TNFα into the brain was completely abolished following IV administration [107]. This suggests that specific cytokines bind directly to a receptor and are then transported, via this receptor, into the brain. TNFα is not the only cytokine that has been shown to cross the BBB. In similar studies, it has also been shown that cytokines such as IL-1β cross the BBB and that the type II IL-1 receptor plays a role in its transport [108, 109].

So far, mainly BBB-dependent mechanisms of transferring peripheral immune signals to the brain have been discussed. However, studies have also nicely depicted that BBB-independent mechanisms exist that transmit peripheral immune signals to the brain. Specifically, many have suggested that peripheral immune signals can be transmitted to the brain via sensory nerves. In 1994, Bluthe and colleagues demonstrated that vagotomy could block LPS-induced changes in social investigative behavior. Moreover, several studies have also demonstrated that the intact vagal nerve is essential for a number of physiological changes that occur following systemic immune challenge including hyperalgesia, fever, and increased levels of plasma corticosteroids [99, 110]. Multiple research groups have implicated many different neurotransmitter-specific pathways of immune regulation most notably the cholinergic, PCAP, and VIP pathways [111, 112]. Though many of these studies present convincing data, it seems that the route through which the inflammatory inducing agent is administered, the inflammatory inducing agent itself and the dose given can lead to debatable and contrasting results.

It is clear that peripheral immune molecules can either be transmitted themselves or transmit the ensuing inflammatory response to the CNS. The routes through which this occurs have been widely studied and discussed here. What is not yet well understood is (1) what the physiological criteria must be in order to determine which route is used in what instance or (2) whether all routes are used following all peripheral immune challenges (though given the data this seems unlikely).

1.5.1 How Is Systemic Inflammation Correlated with the Development and/or Pathogenesis of Neurodevelopmental Disorders?

Systemic inflammation, such as colds and flu viruses, is commonly experienced by adults and children alike. In the adult, in addition to well-known sickness behaviors such as fever, loss of appetite, and reduced activity, peripheral inflammation causes activation of CNS-resident microglia and the acute influx of blood-derived macrophages into the brain and spinal cord [31, 97, 113]. Though newborns, infants, and toddlers also experience bouts of systemic inflammation, their brains and spinal cords are relatively immature. This adds an extra level of complexity in the comprehension of how peripheral inflammation experienced during periods of brain development can affect the ensuing development of the CNS. Several epidemiological studies have strongly suggested that systemic inflammation experienced during the perinatal period may play a role in the development and/or pathogenesis of several neurodevelopmental disorders including cerebral palsy, autism, and schizophrenia. These studies loosely speculate that systemic inflammation experienced during the perinatal period causes glial activation and the production of inflammatory molecules that can affect developing oligodendrocytes and white matter and lead to the loss of neurons [5, 114]. Though a great deal of research has been devoted to understanding neurodevelopmental disorders, the roles that microglia may have in these disorders are still relatively unknown.

Cerebral palsy (CP) is a neurodevelopmental disorder beginning in early childhood and persisting throughout the life of the individual [115]. This group of disorders affects movement and posture and is believed to be the result of a disturbance in the developing fetal or infant brain [115, 116]. Clinical and epidemiological studies strongly suggest that the "disturbance" that leads to the development of periventricular leukomalacia (PVL), which is believed to precede many cases of CP, and CP itself, is systemic inflammation experienced during the perinatal period [115–117].

Systemic inflammation occurring in the pregnant mother, fetus, or neonate is correlated with the increased incidence of PVL and CP [118]. Haynes and colleagues clearly observed microglial and astrocyte activation and apoptotic oligodendrocytes in the periventricular white matter of brain tissue from deceased infants with PVL. Moreover, they also nicely illustrated that the microglia, astrocytes, and possibly infiltrating macrophages expressed oxidative stress markers and markers of protein nitration [9]. Experimental studies have shown that when mice or rats receive ICV, IV, intrauterine, or maternal exposure to LPS, gliosis and elevated levels of TNFα, IL-1β, and IL-6 in the blood or in brain tissue are commonly observed [117]. These findings clearly demonstrate that microglia and other glial cells have intricate roles in the pathogenesis and perhaps the development of PVL and CP. However, their exact functions in this disorder and whether they exhibit phenotypes that promote and/or prevent the progression of the disorder are not well understood.

Studies have also demonstrated that systemic inflammation experienced during the perinatal period may contribute, in part, to the development of autism. Autism is a complex neurodevelopmental disorder characterized by impairments in social interaction, deficits in verbal and nonverbal communication, and repetitive and stereotyped patterns of behavior [114, 119]. Though the phenotypic heterogeneity of autism leads many to believe that a variety of factors contribute to the development of this disorder, including environmental and genetic, there is a growing consensus that the immune system is involved in its development and/or pathogenesis as well [119]. Many have demonstrated that autistic patients have aberrant immunological responses in the peripheral compartment and abnormal cytoarchitectural organization of the cerebral cortex and cerebellum [120, 121]. However, Vargas and colleagues published some of the first studies clearly illustrating glial activation and abnormal cytokine profiles in brain tissue from deceased autistic patients.

In postmortem autistic brains, Vargas and colleagues found microglial and astrocyte activation in cortical regions, in white matter, and in the cerebellum in areas with widespread Purkinje and granule cell loss. They also found elevated expression of monocyte chemoattractant protein-1 (MCP-1) and IL-6 in postmortem brain tissue and cerebrospinal fluid (CSF) from living autistic patients and demonstrated that activated astrocytes were the main source of both of these inflammatory molecules in the brain [114]. Moreover, they also observed elevated TGF-β expression in postmortem brain tissue but not in CSF [114]. Interestingly, Vargas and colleagues found microglial nodules, or collections of activated microglia, in the postmortem autistic brains. This illustrates that there is persistent or chronic activation of microglia, as is seen in chronic neurodegenerative disorders such as Alzheimer's disease and Parkinson's disease, and is also observed in the brains of patients with neurodevelopmental disorders such as autism.

Though these studies elegantly depicted that glial cells are involved in the pathogenesis of autism, this study has also left many questions unanswered. During development, does a specific environmental or genetic insult occur that causes microglia and astrocytes to become activated which then leads to abnormalities in the formation of the brain? Do abnormalities in brain development occur that then lead to microglial and astrocyte activation? Many researchers are undoubtedly attempting to answer these questions. However, with these results, we now know that glial cells are somehow involved in autism progression and can only speculate on their specific roles.

Schizophrenia is a chronic, severe, and disabling psychotic disorder believed to result, in part, from inflammation experienced during the prenatal or perinatal period [8, 122]. The etiology of schizophrenia is currently unknown. However, many speculate that like autism, genetic, environmental, and immunological factors contribute to its development [8]. In the schizophrenic brain, researchers have found elevated levels of the neurotransmitter dopamine and abnormalities in the prefrontal cortex, striatum, and limbic system [123, 124]. In the periphery, studies show that schizophrenic patients have abnormal cytokine levels in the blood [125, 126]. These studies suggest that schizophrenic patients have abnormal levels of inflammatory molecules present in the blood and defects in specific regions of the brain. What is currently not well understood is whether immune cells of the brain are affected by the circulating inflammatory molecules and also how brain immune cells are involved in the pathogenesis of this disease.

Epidemiological studies strongly suggest that infections experienced during childhood can lead to the development of schizophrenia in the adult [127]. Using various imaging techniques, researchers have also demonstrated that activated glial cells, including microglia and astrocytes, could be detected in the brains of patients with schizophrenic psychosis, specifically in the hippocampus [6, 8]. Though these results strongly suggest that microglia may have a role in the pathogenesis of this disease, fundamental experiments have not yet been performed to illustrate when these cells become activated in the disorder, what inflammatory molecules they produce, and whether they promote pathology or become activated to protect the brain from further pathology.

1.6 Conclusion: Do Long-Lived Microglia Serve as an "Immunologic Memory" of CNS Damage and Infection?

Like all macrophages, microglia are highly plastic in their ability to acquire a broad range of phenotypes (classical, cytotoxic pathogen defense functions to alternative, neuroprotective, tissue repair functions). Here we discussed multiple methods by which the state of neuronal function drives microglial phenotype toward activation or homeostasis. Because microglia are distinct from most other macrophages in being long-lived, microglia have the potential to generate "immune memory" within the CNS. Depending on the nature of prior insults, microglia may become partially polarized toward either classical or alternative activation states. Thus, it is tempting to speculate that exposure to pathogens or toxins that alter microglial phenotype directly (or indirectly by altering neuronal/glial function) may have long-lasting consequences on either increased or decreased risk of neurologic disorders associated with dysregulated neuroinflammatory responses.

References

1. Nimmerjahn A, Kirchhoff F, Helmchen F (2005) Resting microglial cells are highly dynamic surveillants of brain parenchyma in vivo. Science (New York) 308:1314–1318
2. Colton CA et al (2006) Expression profiles for macrophage alternative activation genes in AD and in mouse models of AD. J Neuroinflammation 3:27
3. Mosser DM, Edwards JP (2008) Exploring the full spectrum of macrophage activation. Nat Rev Immunol 8:958–969
4. D'Mello C, Le T, Swain MG (2009) Cerebral microglia recruit monocytes into the brain in response to tumor necrosis factor alpha signaling during peripheral organ inflammation. J Neurosci 29:2089–2102
5. Chew LJ, Takanohashi A, Bell M (2006) Microglia and inflammation: impact on developmental brain injuries. Ment Retard Dev Disabil Res Rev 12:105–112
6. Doorduin J, deVries EFJ, Willemsen ATM, Dierckx RA, Klein HC (2008) Neuroinflammation in schizophrenic patients: a positron emission tomography study with [11C]-(R)-PK11195. Neuroimage 41:T109

7. Pardo CA, Vargas DL, Zimmerman AW (2005) Immunity, neuroglia and neuroinflammation in autism. Int Rev Psychiatry (Abingdon) 17:485–495
8. Hickie IB, Banati R, Stewart CH, Stewart CH, Lloyd AR (2009) Are common childhood or adolescent infections risk factors for schizophrenia and other psychotic disorders? Med J Aust 190:S17–S21
9. Haynes R et al (2003) Nitrosative and oxidative injury to premyelinating oligodendrocytes in periventricular leukomalacia. J Neuropathol Exp Neurol 62:441–450
10. Frank M et al (2006) mRNA upregulation of MHC II and pivotal pro-inflammatory genes in normal brain aging. Neurobiol Aging 27:717–722
11. Dilger RN, Johnson RW (2008) Aging, microglial cell priming, and the discordant central inflammatory response to signals from the peripheral immune system. J Leukoc Biol 84:932–939
12. Godbout J et al (2005) Exaggerated neuroinflammation and sickness behavior in aged mice following activation of the peripheral innate immune system. FASEB J 19:1329–1331
13. Henry C, Huang Y, Wynne A, Godbout J (2009) Peripheral lipopolysaccharide (LPS) challenge promotes microglial hyperactivity in aged mice that is associated with exaggerated induction of both pro-inflammatory IL-1β and anti-inflammatory IL-10 cytokines. Brain Behav Immun 23:309–317
14. Carson MJ, Reilly CR, Sutcliffe JG, Lo D (1998) Mature microglia resemble immature antigen-presenting cells. Glia 22:72–85
15. Melchior B, Puntambekar SS, Carson MJ (2006) Microglia and the control of autoreactive T cell responses. Neurochem Int 49:145–153
16. Carson M, Doose J, Melchior B, Schmid C, Ploix C (2006) CNS immune privilege: hiding in plain sight. Immunol Rev 213:48–65
17. Carson M, Thrash J, Walter B (2006) The cellular response in neuroinflammation: the role of leukocytes, microglia and astrocytes in neuronal death and survival. Clin Neurosci Res 6:237–245
18. Biber K, Neumann H, Inoue K, Boddeke HWGM (2007) Neuronal "on" and "off" signals control microglia. Trends Neurosci 30:596–602
19. Andjelkovic AV, Nikolic B, Pachter JS, Zecevic N (1998) Macrophages/microglial cells in human central nervous system during development: an immunohistochemical study. Brain Res 814:13–25
20. Lichanska AM et al (1999) Differentiation of the mononuclear phagocyte system during mouse embryogenesis: the role of transcription factor PU.1. Blood 94:127–138
21. Soulet D, Rivest S (2008) Bone-marrow-derived microglia: myth or reality? Curr Opin Pharmacol 8:508–518
22. Hutchins K, Dickson D, Rashbaum W, Lyman W (1990) Localization of morphologically distinct microglial populations in the developing human fetal brain: implications for ontogeny. Brain Res Dev Brain Res 55:95–102
23. Schmid CD et al (2002) Heterogeneous expression of the triggering receptor expressed on myeloid cells-2 on adult murine microglia. J Neurochem 83:1309–1320
24. Ginhoux F et al (2010) Fate mapping analysis reveals that adult microglia derive from primitive macrophages. Science (New York) 330:841–845
25. Carson MJ, Crane J, Xie AX, Silva GA (2009) Modeling CNS microglia: the quest to identify predictive models. Drug Discov Today 5:19–25
26. Ransohoff RM, Cardona AE (2010) The myeloid cells of the central nervous system parenchyma. Nature 468:253–262
27. Carson M et al (2007) A rose by any other name? The potential consequences of microglial heterogeneity during CNS health and disease. Neurotherapeutics 4:571–579
28. Carson MJ, Thrash JC, Walter B (2009) The cellular response in neuroinflammation: the role of leukocytes, microglia and astrocytes in neuronal death and survival. Clin Neurosci 6:237–245
29. Thomas M (1989) The leukocyte common antigen family. Annu Rev Immunol 7:339–369

30. Tan J, Town T, Mullan M (2000) CD45 inhibits CD40L-induced microglial activation via negative regulation of the Src/p44/42 MAPK pathway. J Biol Chem 275:37224–37231
31. Sedgwick JD et al (1991) Isolation and direct characterization of resident microglial cells from the normal and inflamed central nervous system. Proc Natl Acad Sci USA 88: 7438–7442
32. Ford AL, Goodsall AL, Hickey WF, Sedgwick JD (1995) Normal adult ramified microglia separated from other central nervous system macrophages by flow cytometric sorting. Phenotypic differences defined and direct ex vivo antigen presentation to myelin basic protein-reactive CD4+ T cells compared. J Immunol 154:4309–4321
33. Kovács R, Papageorgiou I, Heinemann U (2011) Slice cultures as a model to study neurovascular coupling and blood brain barrier in vitro. Cardiovasc Psychiatry Neurol 2011:646958
34. Davalos D et al (2008) Stable in vivo imaging of densely populated glia, axons and blood vessels in the mouse spinal cord using two-photon microscopy. J Neurosci Methods 169:1–7
35. Germain R, Miller M, Dustin M, Nussenzweig M (2006) Dynamic imaging of the immune system: progress, pitfalls and promise. Nat Rev Immunol 6:497–507
36. Greer J, Capecchi M (2002) Hoxb8 is required for normal grooming behavior in mice. Neuron 33:23–34
37. Tremblay M, Stevens B, Sierra A, Wake H, Bessis A, Nimmerjahn A (2011) The role of microglia in the healthy brain. J Neurosci 31:16064–16069
38. Derecki NC, Cronk JC, Lu Z, Xu E, Abbott SBG, Guyenet PG, Kipnis J (2012) Wild type microglia arrest pathology in mouse model of Rett syndrome. Nature 484:105–109
39. Verloes A et al (1997) Nasu–Hakola syndrome: polycystic lipomembranous osteodysplasia with sclerosing leucoencephalopathy and presenile dementia. J Med Genet 34:753–757
40. Thrash JC, Torbett BE, Carson MJ (2009) Developmental regulation of TREM2 and DAP12 expression in the murine CNS: implications for Nasu–Hakola disease. Neurochem Res 34:38–45
41. Paloneva J et al (2000) Loss-of-function mutations in TYROBP (DAP12) result in a presenile dementia with bone cysts. Nat Genet 25:357–361
42. Paloneva J et al (2003) DAP12/TREM2 deficiency results in impaired osteoclast differentiation and osteoporotic features. J Exp Med 198:669–675
43. Takahashi K, Prinz M, Stagi M, Chechneva O, Neumann H (2007) TREM2-transduced myeloid precursors mediate nervous tissue debris clearance and facilitate recovery in an animal model of multiple sclerosis. PLoS Med 4:e124. doi:10.1371/journal.pmed.0040124
44. Melchior B et al (2010) Dual induction of TREM2 and tolerance related transcript, Tmem176b, in amyloid transgenic mice: implications for vaccine-based therapies for Alzheimer's disease. ASN Neuro 2:e00037. doi:10.1042/AN20100010
45. Bouchon A, Facchetti F, Weigand MA, Colonna M (2001) TREM-1 amplifies inflammation and is a crucial mediator of septic shock. Nature 410:1103–1107
46. Piccio L et al (2007) Blockade of TREM-2 exacerbates experimental autoimmune encephalomyelitis. Eur J Immunol 37:1290–1301
47. Hristova M et al (2010) Activation and deactivation of periventricular white matter phagocytes during postnatal mouse development. Glia 58:11–28
48. Blaschke AJ, Staley K, Chun J (1996) Widespread programmed cell death in proliferative and postmitotic regions of the fetal cerebral cortex. Development 122:1165–1174
49. Cheng X-S et al (2011) Neuronal apoptosis in the developing cerebellum. Anat Histol Embryol 40:21–27
50. Ferrer I, Bernet E, Soriano E, del Rio T, Fonseca M (1990) Naturally occurring cell death in the cerebral cortex of the rat and removal of dead cells by transitory phagocytes. Neuroscience 39:451–458
51. Stein BM, Keshav S, Harris N, Gordon S (1992) Interleukin 4 potently enhances murine macrophage mannose receptor activity: a marker of alternative immunologic macrophage activation. J Exp Med 176:287–292

52. Buchmeier N, Schreiber R (1985) Requirement of endogenous interferon-gamma production for resolution of *Listeria monocytogenes* infection. Proc Natl Acad Sci USA 82:7404–7408
53. Nathan BYCF, Murray HW, Wlebe IE, Rubin BY, Berish Y (1983) Identification of interferon-gamma as the lymphokine that activates human macrophage oxidative metabolism and antimicrobial activity. J Exp Med 158:670–689
54. Martinez FO, Helming L, Gordon S (2009) Alternative activation of macrophages: an immunologic functional perspective. Annu Rev Immunol 27:451–483
55. Agarwal S, Piesco N, Johns L, Riccelli A (1995) Differential expression of IL-1 beta, TNF-alpha, IL-6, and IL-8 in human monocytes in response to lipopolysaccharides from different microbes. J Dent Res 74:1057–1065
56. Seder R et al (1991) Mouse splenic and bone marrow cell populations that express high-affinity Fcε receptors and produce interleukin 4 are highly enriched in basophils. Proc Natl Acad Sci USA 88:2835–2839
57. Moqbel R et al (1995) Identification of messenger RNA for IL-4 in human eosinophils with granule localization and release of the translated product. J Immunol 155:4939–4947
58. Yoshimoto T, Bendelac A, Watson C, Hu-Li J, Paul W (1995) Role of NK1.1+ T cells in a TH2 response and in immunoglobulin E production. Science 270:1845–1847
59. Demeure C et al (1995) Human naïve CD4 T cells produce interleukin-4 priming and acquire a Th2 phenotype upon repetitive stimulations in neutral conditions. Eur J Immunol 25:2722–2725
60. Seder R et al (1992) CD8+ T cells can be primed in vitro to produce IL-4. J Immunol 148:1652–1656
61. Wu G, Morris SJ (1998) Arginine metabolism: nitric oxide and beyond. Biochem J 336:1–17
62. Kreider T, Anthony RM, Urban JF, Gause WC (2007) Alternatively activated macrophages in helminth infections. Curr Opin Immunol 19:448–453
63. Nair M, Cochrane D, Allen J (2003) Macrophages in chronic type 2 inflammation have a novel phenotype characterized by the abundant expression of Ym1 and Fizz1 that can be partly replicated in vitro. Immunol Lett 85:173–180
64. Owhashi M, Arita H, Niwa A (1998) Production of eosinophil chemotactic factor by CD8+ T cells in *Toxocara canis*-infected mice. Parasitol Res 84:136–138
65. Owhashi M, Arita H, Hayai N (2000) Identification of a novel eosinophil chemotactic cytokine (ECF-L) as a chitinase family protein. J Biol Chem 275:1279–1286
66. Steppan CM et al (2001) A family of tissue-specific resistin-like molecules. Proc Natl Acad Sci USA 98:502–506
67. Mora A et al (2006) Activation of alveolar macrophages via the alternative pathway in herpesvirus induced lung fibrosis. Am J Respir Cell Mol Biol 35:466–473
68. Pesce JT et al (2009) Retnla (relmalpha/fizz1) suppresses helminth-induced Th2-type immunity. PLoS Pathog 5:e1000393
69. Chen S-K et al (2010) Hematopoietic origin of pathological grooming in Hoxb8 mutant mice. Cell 141:775–785
70. Akiyama H, McGeer P (1990) Brain microglia constitutively express beta-2 integrins. J Neuroimmunol 30:81–93
71. Wellik D, Hawkes P, Capecchi M (2002) Hox11 paralogous genes are essential for metanephric kidney induction. Genes Dev 16:1423–1432
72. Hickey W, Kimura H (1988) Perivascular microglial cells of the CNS are bone marrow-derived and present antigen in vivo. Science 239:290–292
73. McMahon E, Bailey S, Castenada C, Waldner H, Miller S (2005) Epitope spreading initiates in the CNS in two mouse models of multiple sclerosis. Nat Med 11:335–339
74. Sanders V, Jones K (2006) Role of immunity in recovery from a peripheral nerve injury. J Neuroimmune Pharmacol 1:11–19
75. Serpe C, Coers S, Sanders V, Jones K (2003) CD4+ T, but not CD8+ or B, lymphocytes mediate facial motoneuron survival after facial nerve transection. Brain Behav Immun 17:393–402
76. Byram S et al (2004) CD4-positive T cell-mediated neuroprotection requires dual compartment antigen presentation. J Neurosci 24:4333–4339

77. Bsibsi M, Ravid R, Gveri CD, van Noort J (2002) Broad expression of Toll-like receptors in the human central nervous system. J Neuropathol Exp Neurol 61:1013–1021
78. Crack PJ, Bray PJ (2007) Toll-like receptors in the brain and their potential roles in neuropathology. Immunol Cell Biol 85:476–480
79. Kimbrell D, Beutler B (2001) The evolution and genetics of innate immunity. Nat Rev Genet 2:256–267
80. Valanne S, Wang J, Rämet M (2011) The drosophila toll signaling pathway. J Immunol 186:649–656
81. Bowie A, O'Neill L (2000) The interleukin-1 receptor/Toll-like receptor superfamily: signal generators for pro-inflammatory interleukins and microbial products. J Leukoc Biol 67:508–514
82. Bowman C, Rasley A, Tranguch S, Marriott I (2003) Cultured astrocytes express Toll like receptors for bacterial products. Glia 43:281–291
83. Kurt-Jones E et al (2004) Herpes simplex virus 1 interaction with Toll-like receptor 2 contributes to lethal encephalitis. Proc Natl Acad Sci USA 101:1315–1320
84. Mishra B, Mishra PT, Teale J (2006) Expression and distribution of Toll-like receptors in the brain during murine neurocysticercosis. J Neuroimmunol 181:46–56
85. Kielian T, Mayes P, Kielian M (2002) Characterization of microglial responses to *Staphylococcus aureus*: effects on cytokine, costimulatory molecule, and Toll like receptor expression. J Neuroimmunol 130:86–99
86. Rasley A, Anguita J, Marriott I (2002) *Borrelia burgdorferi* induces inflammatory mediator production by murine microglia. J Neuroimmunol 130:22–31
87. Dalpke A et al (2002) Immunostimulatory CpG DNA activates murine microglia. J Immunol 168:4854–4863
88. Olson JK, Miller SD (2004) Microglia initiate central nervous system innate and adaptive immune responses through multiple TLRs. J Immunol 173:3916–3924
89. Mitchell J, Paul-Clark M, Clarke G, McMaster S, Cartwright N (2007) Critical role of Toll-like receptors and nucleotide oligomerisation domain in the regulation of health and disease. J Endocrinol 193:323–330
90. Wright GJ et al (2000) Lymphoid/neuronal cell surface OX2 glycoprotein recognizes a novel receptor on macrophages implicated in the control of their function. Immunity 13:233–242
91. Masocha W (2009) Systemic lipopolysaccharide (LPS)-induced microglial activation results in different temporal reduction of CD200 and CD200 receptor gene expression in the brain. J Neuroimmunol 214:78–82
92. Mizuno T, Kawanokuchi J, Numata K, Suzumura A (2003) Production and neuroprotective functions of fractalkine in the central nervous system. Brain Res 979:65–70
93. Zujovic V, Benavides J, Vigé X, Carter C, Taupin V (2000) Fractalkine modulates TNFα secretion and neurotoxicity induced by microglial activation. Glia 29:305–315
94. Yoshida H et al (2001) Synergistic stimulation, by tumor necrosis factor-alpha and interferon-gamma, of fractalkine expression in human astrocytes. Neurosci Lett 303:132–136
95. Bazan J et al (1997) A new class of membrane-bound chemokine with a CX3C motif. Nature 385:640–644
96. Cardona AE et al (2006) Control of microglial neurotoxicity by the fractalkine receptor. Nat Neurosci 9:917–924
97. Perry VH (2004) The influence of systemic inflammation on inflammation in the brain: implications for chronic neurodegenerative disease. Brain Behav Immun 18:407–413
98. Dantzer R (2001) Cytokine-induced sickness behavior: mechanisms and implications. Ann N Y Acad Sci 933:222–234
99. Quan N (2008) Immune to brain signaling: how important are the blood–brain barrier independent pathways. Mol Neurobiol 37:142–152
100. Konsman J, Kelley K, Dantzer R (1999) Temporal and spatial relationships between lipopolysaccharide-induced expression of fos, interleukin-1β and inducible nitric oxide synthase in rat brain. Neuroscience 89:535–548
101. Engblom D et al (2003) Microsomal prostaglandin E synthase-1 is the central switch during immune-induced pyresis. Nat Neurosci 6:1137–1138

102. Ching S et al (2007) Endothelial-specific knockdown of interleukin-1 (IL-1) type 1 receptor differentially alters CNS responses to IL-1 depending on its route of administration. J Neurosci 27:10476–10486
103. Schiltz JC, Sawchenko PE (2002) Distinct brain vascular cell types manifest inducible cyclooxygenase expression as a function of the strength and nature of immune insults. J Neurosci 22:5606–5618
104. Andreasson K (2010) Emerging roles of PGE2 receptors in models of neurological disease. Prostaglandins Other Lipid Mediat 91:104–112
105. Steinmetz T et al (1988) Phase I study of 24-hour continuous intravenous infusion of recombinant human tumor necrosis factor. J Biol Response Mod 7:417–423
106. Triozzi P, Kinney P, Rinehart J (1990) Central nervous system toxicity of biological response modifiers. Ann N Y Acad Sci 594:347–354
107. Pan W, Kastin A (2002) TNFα transport across the blood–brain barrier is abolished in receptor knockout mice. Exp Neurol 174:193–200
108. Banks WA, Kastin AJ, Ehrensing CA (1994) Blood-borne interleukin-lla is transported across the endothelial blood-spinal cord barrier of mice. J Physiol 479:257–264
109. Skinner R, Gibson R, Rothwell N, Pinteaux E, Penny J (2009) Transport of interleukin-1 across cerebromicrovascular endothelial cells. Br J Pharmacol 156:1115–1123
110. Maier S, Goehler L, Fleshner M, Watkins L (1998) The role of the vagus nerve in cytokine to brain communication. Ann N Y Acad Sci 840:289–300
111. Pavlo VA, Parrish WR, Rosas-Ballina M, Ochani M, Puerta M, Ochani K, Chavan S, Al-Abed Y, Tracey KJ (2009) Brain acetylcholinesterase activity controls systemic cytokine levels through the cholinergic anti-inflammatory pathway. Brain Behav Immun 23:41–45
112. Carson MJ, Lo DD (2007) Perspective is everything: an irreverent discussion of CNS-immune system interactions as viewed from different scientific traditions. Brain Behav Immun 21:367–373
113. Perry V, Anderson P (1992) The inflammatory response in the CNS. Neuropathol Appl Neurobiol 18:454–459
114. Vargas DL, Nascimbene C, Krishnan C, Zimmerman AW, Pardo CA (2005) Neuroglial activation and neuroinflammation in the brain of patients with autism. Ann Neurol 57:67–81
115. Bax M et al (2005) Proposed definition and classification of cerebral palsy. Dev Med Child Neurol 27:571–576
116. Saadani-Makki F et al (2008) Intrauterine administration of endotoxin leads to motor deficits in a rabbit model: a link between prenatal infection and cerebral palsy. Am J Obstet Gynecol 199:651.e1–651.e7
117. Wang X, Rousset C, Hagberg H, Mallard C (2006) Lipopolysaccharide-induced inflammation and perinatal brain injury. Semin Fetal Neonatal Med 11:343–353
118. Dammann O, O'Shea TM (2008) Cytokines and perinatal brain damage. Clin Perinatol 35:643–663
119. Ashwood P, Van de Water J (2004) A review of autism and the immune response. Clin Dev Immunol 11:165–174
120. Gupta S, Aggarwal S, Rashanravan B, Lee T (1998) Th1-and Th2-like cytokines in CD4+ and CD8+ T cells in autism. J Neuroimmunol 85:106–109
121. Kemper T, Bauman M (1998) Neuropathology of infantile autism. J Neuropathol Exp Neurol 57:645–652
122. Meyer U, Feldon J, Dammann O (2011) Schizophrenia and autism: both shared and disorder-specific pathogenesis via perinatal inflammation? Pediatr Res 69:26R–33R
123. Breier A et al (1992) Brain morphology and schizophrenia. A magnetic resonance imaging study of limbic, prefrontal cortex, and caudate structures. Arch Gen Psychiatry 49:921–926
124. Seeman P, Kapur S (2000) Schizophrenia: more dopamine, more D2 receptors. Proc Natl Acad Sci USA 97:7673–7675
125. O'Donnell M et al (1996) Increased production of interleukin-2 (IL-2) but not soluble interleukin-2 receptors (sIL-2R) in unmedicated patients with schizophrenia and schizophreniform disorder. Psychiatry Res 65:171–178

126. Potvin S et al (2008) Inflammatory cytokine alterations in schizophrenia: a systematic quantitative review. Biol Psychiatry 63:801–808
127. Rantakallio P, Jones P, Moring J, Von Wendt L (1997) Association between central nervous system infections during childhood and adult onset schizophrenia and other psychoses: a 28-year follow-up. Int J Epidemiol 26:837–843
128. Mott RT et al (2004) Neuronal expression of CD22: novel mechanism for inhibiting microglial proinflammatory cytokine production. Glia 46:369–379

Chapter 2
Astrocyte–Neuron Communications

Sally R. McIver, Mathilde Faideau, and Philip G. Haydon

2.1 Introduction

When astrocytes were first visualized by Virchow in 1846, he characterized them as a type of "glue" filling in the interstitial space. The term "astrocyte" first appeared in 1893 when improvements in histological techniques made it possible to distinguish individual cell morphology within this cerebral "glue" [1]. The importance of these cells, though not well understood, was appreciated by the fact that astrocytes occupy a substantial amount of space in the brain, representing up to 50% of cerebral volume [2]. Interestingly, the ratio of astrocytes to neurons varies among species and according to the relative complexity of the brain [3], increasing proportionally with the complexity of the neural network. This is perhaps one of the first pieces of evidence hinting at a role for astrocytes in the integration of neuronal activity. With the advancement of staining techniques came a greater appreciation for the unique structure of these cells which subsequently provided great insight into their diverse functions. Astrocytes have multiple primary processes and fine branching processes which are able to expand and contract, allowing them to dynamically contact both synapses and microvasculature. In addition, by forming independent microdomains, with little or no overlap with neighboring astrocytes, astrocytes are able to effectively modulate communication between neuronal networks and glial–vascular coupling. For example, the end feet of astrocytes contact blood vessels and modulate blood flow via Ca^{2+}-dependent release of vasoactive agents, effectively regulating neuronal access to nutrients required to sustain metabolic demand. Similarly, astrocyte morphology can change in response to their environment. Hormonally responsive astrocytes in the arcuate nucleus of the adult female rat respond to estradiol with dramatic changes in their morphology, including an increased coverage of neuronal perikarya, impacting synaptic communication [4]. These changes in

S.R. McIver • M. Faideau • P.G. Haydon, Ph.D. (✉)
Department of Neuroscience, Tufts University School of Medicine,
136 Harrison Avenue, Boston, MA 02111, USA
e-mail: philip.haydon@tufts.edu

C. Cui et al. (eds.), *Neural-Immune Interactions in Brain Function and Alcohol Related Disorders*, DOI 10.1007/978-1-4614-4729-0_2,

morphology have been recently correlated with changes in glutamate–glutamine cycling, indicating functional plasticity of neuronal–glial communication in the normal adult brain [5]. Another important function of astrocytes is their role in the "tripartite synapse," or the communication between the astrocytic process and the pre- and postsynaptic terminals [6]. Despite the complex morphology and numerous ramifications of astrocytes, it is still rather surprising to consider that a single astrocyte residing in area CA1 of the rat hippocampus can contact up to 140,000 synapses [7].

Much of what is widely known about the function of astrocytes has been discovered using in vitro and in situ preparations, providing valuable information on how these glial cells help maintain extracellular homeostasis by buffering potassium and reuptaking glutamate, for example. However, the dynamic nature of astrocyte physiology is undoubtedly best appreciated in vivo. The recent progress in molecular genetic techniques has allowed for astrocyte-specific expression of reporter proteins, such as GFP and various transgenes of interest, and contributed to the surge in research investigating the multifaceted role of these cells as key players in brain function. It is now widely accepted that astrocytes are in direct communication with neurons, modulating neuronal function at the synaptic and network levels, ultimately providing a significant impact on physiological and pathological conditions. This chapter begins by describing different types of glia–neuron communication at the synaptic and network levels. It then aims to link these glia–neuron signaling mechanisms to what is currently known regarding how glia–neuron communication, or perturbations thereof, contribute to healthy brain physiology as well as neuroinflammation and other related neurological conditions such as epilepsy, Alzheimer's disease, amyotrophic lateral sclerosis (ALS), and addiction.

2.2 Gliotransmission

According to the definition of "classical neurotransmitter," the signaling molecule must be released from the presynaptic terminal. This definition reflects the historically neuron-centric considerations of brain function. As neuroscience has progressed toward a greater encompassment of glial biology, it is now widely accepted that astrocytes are also capable of transmitter release through a process called "gliotransmission." Similar to neurons, astrocytes contain the cellular mechanics necessary for transmitter packaging and release, which can occur either via lysosomal or small vesicle fusion [8–11]. Vesicle fusion and exocytosis of gliotransmitters is calcium-dependent and involves formation of the SNARE complex, similar to neurons. Early demonstrations of gliotransmission, particularly those using calcium imaging, were primarily performed in cell culture, and though invaluable to the progress in current understanding of glial biology, these studies posed limitations that would later impact interpretations of astrocyte function [12–16]. Thanks to recent advances in technology and innovation, many of these early studies are being confirmed in situ and in vivo. This section summarizes the discoveries and current understanding of gliotransmission as a key player in glia–neuron communication.

2.2.1 Glutamate

Glutamate is the major excitatory neurotransmitter in the nervous system and is critically involved in many functions, including cerebral development and motor and cognitive functions. It has been recognized as a neurotransmitter since the 1950s [17], but its role as a gliotransmitter, which was first demonstrated in the 1990s in vitro [18–21], is currently under debate [22]. Using astrocyte–neuron cocultures and photostimulation of astrocytes, Parpura et al. [20] was among the first to demonstrate that an increase in intracellular calcium in astrocytes causes an increase in intracellular calcium in neighboring neurons in vitro. This effect was blocked with administration of a broad-spectrum glutamate receptor antagonist (D-glutamylglycine), suggesting a calcium-dependent mechanism of glutamate release by astrocytes [20]. Subsequent studies using rat and mouse hippocampal and cortical slices employed either pharmacological or cell-specific photolytic elevation of Ca^{2+} and substantiated the relevance of Ca^{2+}-dependent glutamate release from astrocytes as a mechanism of gliotransmission that could modulate either neuronal excitability or synaptic transmission and plasticity [23–25].

Further evidence supporting the capacity of astrocytes to release glutamate was later provided by biochemical and structural evidence demonstrating that VGLUT transporters and SNARE complex proteins are expressed in astrocyte processes in vitro and in acutely isolated brain slices, showing that these cells are potentially capable of vesicular glutamate release [10, 11, 26, 27]. Indeed, in purified astrocyte cultures, calcium-induced glutamate release depends on proteins involved in vesicular function, including the SNARE element synaptobrevin 2 and the calcium-binding protein synaptotagmin IV [11, 28].

Astrocytic glutamate can be synthesized de novo in astrocytes through conversion of alpha-ketoglutarate by glutamate dehydrogenase (GDH), which is mainly expressed in astrocytes [29–34]. Astrocytic glutamate also comes from the high-capacity uptake of synaptic glutamate by the astrocytic transporters GLAST and GLT-1 and the Cl-dependent glutamate/cysteine exchanger [35, 36]. Following uptake, glutamate is quickly metabolized into glutamine by the glutamine–glutamate cycle via a reaction mediated by glutamine synthetase [36].

Multiple factors have been shown to stimulate glutamate release from astrocytes in vitro, from the classic neurotransmitters, including glutamate itself, to inflammatory molecules such as TNF-alpha [37], ATP [38, 39], prostaglandins, and CXCL12 [40, 41]. In some cases, multiple convergent signaling cascades are required to produce elevated calcium-induced release, suggesting that astrocytes are capable of integrating incoming signals to produce a response. For example, basal concentrations of TNF-alpha are required for P2Y1R-evoked and Ca^{2+}-dependent glutamate release from astrocytes [41, 42], whereas higher concentrations of TNF-alpha can directly increase the glutamate release, independently of P2Y1R in granule cells of the mouse hippocampus [41].

The release of glutamate from astrocytes has been shown to modulate synaptic activity in multiple brain regions in situ including the hippocampus, thalamus, nucleus accumbens, and olfactory bulb. In many cases, it produces these effects by activating neuronal extrasynaptic NMDA and metabotropic glutamate receptors to modulate neuronal activity [23, 43–49].

These early studies were critical in demonstrating that astrocytes not only communicate with neurons but are capable of modulating neuronal activity via calcium-dependent release of transmitters. However, further studies using slice preparations and in vivo models have become increasingly necessary for understanding the functional impact of astrocytic glutamate release.

Although numerous studies have shown the potential for glutamate to be released from astrocytes and signal directly to neurons, efforts to directly monitor glutamate-mediated gliotransmission by many methods, including microdialysis and biosensor detection, have been challenged by the fact that astrocytes are endowed with high-affinity glutamate transporters that rapidly and avidly remove glutamate from the extracellular space. Of course, the inability to detect does not mean that signaling does not occur. Because of the difficulties inherent in direct measurements, more indirect approaches have been used to infer the functional relevance of glutamate release.

Recently, studies from one laboratory used molecular-genetic approaches to alter intracellular calcium signaling in astrocytes and concluded that calcium-dependent glutamate-mediated gliotransmission does not occur in the hippocampus, contrary to other reports [13]. Specifically, the authors used two lines of transgenic mice, one in which intracellular calcium levels could be selectively increased in astrocytes by pharmacological activation of ectopic expression of a receptor (MrgA1R+) and one in which the IP_3 signaling pathway was selectively impaired in astrocytes ($IP_3R2^{-/-}$), causing reduced intracellular calcium. Based on whole cell recordings of CA1 pyramidal neurons in hippocampal slices, the authors failed to detect a difference in evoked or spontaneous NMDA-mediated EPSCs, as well as LTP induction or maintenance, in either of the transgenic mouse lines compared to WT [13]. Results from this study directly challenged previous results from several independent laboratories, fueling a debate concerning the importance of glutamate-mediated gliotransmission. A crucial consideration, however, is that "absence of evidence is not evidence of absence." The contradictions in results are likely to reflect variations in experimental approaches, all of which pose limitations [50]. For example, cell culture studies undoubtedly are restricted by the lack of intact physiology; slice preparations maintain network connections, but the electrophysiological approaches that are generally employed in these studies do not always abide by consistent paradigms and on some occasions may depend upon a narrow range of physiological metrics (such as measurements of basal neuronal or synaptic currents) that may obscure functionally relevant responses. Finally, in vivo studies using astrocyte-specific manipulations in transgenic mice minimize the need for pharmacological approaches but should be regarded with careful criticism as well, since in many cases the transgenes expressed may function by introducing new cellular pathways that are not inherent in astrocytes (rather than introducing mutations in preexisting astrocyte pathways), or may be expressed constitutively, allowing for potential developmental compensations to impact function. These difficulties necessitate a combined approach employing multiple independent measurements to assess the role of glutamatergic gliotransmission in the regulation of neuronal function.

In contrast to the reports indicating a lack of impact of disrupting astrocytic calcium on some forms of activity, other studies employed a multifaceted approach and demonstrated that glutamatergic gliotransmission does play a role in modulating synaptic plasticity in the hippocampus. For example, Navarette et al. recently showed that in vivo calcium signals in astrocytes are required for cholinergic-induced LTP in the hippocampus [51]. Using in vivo calcium imaging and electrophysiology, they demonstrated that somatosensory stimulation (via tail pinch) or electrical stimulation of cholinergic activity evoked increases in intracellular calcium in hippocampal astrocytes along with subsequent cholinergic LTP, through a pathway that required activation of muscarinic receptors and, in the case of LTP, of metabotropic glutamate receptors (mGluRs). To further examine astrocytic mechanisms contributing to cholinergic LTP, the authors use a combination of calcium imaging, electrophysiology, and calcium uncaging in hippocampal slice preparations. Consistent with the in vivo results, these experiments showed that cholinergic LTP required calcium-dependent glutamate release from astrocytes—LTP was abolished when calcium chelators were applied to astrocytes in rat hippocampus and in hippocampal slices prepared from mice in which IP_3-mediated calcium signaling in astrocytes is transgenically knocked out (IP_3R2–/–). Interestingly, using simultaneous calcium uncaging in astrocytes and a minimal stimulation paradigm, they show that astrocytic calcium elevations induce release of glutamate that activates presynaptic mGluRs and that LTP requires a tight coordination of these events with postsynaptic depolarization. Thus, by using similar techniques (including the same IP_3R2–/– as used in the aforementioned study) but also by broadening the scope of physiologically relevant paradigms for monitoring synaptic activity, this group provided rigorous evidence showing that calcium-dependent release of glutamate from astrocytes modulates cholinergic LTP in the hippocampus [51].

Thus, the pendulum is swinging back in favor of substantial evidence supporting calcium-dependent glutamate-mediated gliotransmission. However, in the future, it will be important to use cell-specific recombination of floxed alleles of genes to understand the precise physiological and behavioral role of this form of glial–neuronal interaction.

2.2.2 D-Serine

D-serine has been generally considered to be selectively expressed in astrocytes, along with serine racemase, which converts L-serine into D-serine [52, 53]. D-serine is an endogenous co-agonist of NMDA receptors and binds to the "glycine-binding site" on the NR1 subunit, a subunit which is present on all NMDA receptor assemblies [54–56]. In vitro studies of astrocytes demonstrated that D-serine release is Ca^{2+}-dependent and occurs via SNARE-mediated exocytosis following AMPA receptor activation [57]. Evidence from in vitro cortical astrocyte cultures and in vivo hippocampal astrocytes studies show that D-serine and glutamate are contained in similar vesicular organelles called synaptic-like microvesicles (SLMVs) in astrocytes [58, 59].

In contrast to early studies showing astrocyte specificity of D-serine, the recent development of new antibodies suggests that neurons express more serine racemase than astrocytes [60–64]. Serine racemase expression has been observed in pyramidal neurons of the cortex and hippocampus and GABAergic medium-spiny neurons of the striatum, but was not detected in glial cells [62, 63]. More recently, Ding et al. produced new antibodies against serine racemase and found immunoreactivity in mouse cortical neurons and to a lower extent in oligodendrocytes and astrocytes of the corpus callosum [64]. Similar to what was reported in astrocyte cultures, AMPA receptor stimulation on cortical neuronal cultures leads to D-serine release; however this seems to occur independently of exocytosis [61]. Neuronal D-serine release can be differentiated from astrocytic D-serine release using application of veratridine in cortical slices, which enhances depolarization and subsequent neuronal release of D-serine through Asc-1 transporters [61]. The precise mechanism of D-serine signaling pathways is still rather unclear, though Wolosker et al. [60] proposed the existence of a neuron–astrocyte "serine shuttle," where neurons and astrocytes provide sources of D-serine and L-serine, respectively. In this scenario, astrocytic L-serine produced from glucose would be used by neurons to produce D-serine, which would then accumulate in astrocytes [61].

The action of D-serine on the glutamatergic synapses of the supraoptic nucleus of the hypothalamus is finely modulated by the distance between the astrocytic processes and the postsynaptic element. During lactation, the astrocytic processes retract from the synapse, which reduces the D-serine concentration at the synapse and thereby induces long-term depression (LTD) [49]. The contribution of D-serine in synaptic plasticity also has been demonstrated within the hippocampus and cortex [65, 66]. For example, impairing intracellular Ca^{2+} within a single astrocyte prevents LTP induction at Schaffer collateral/CA1 synapses, and LTP can be rescued with subsequent application of D-serine [65]. Furthermore, LTP is also suppressed following treatment with a serine racemase inhibitor, which decreases astrocytic release of D-serine. Astrocytic D-serine was also shown to be involved in LTP formation in layers V/VI of the prefrontal cortex, where inhibition of astrocytic metabolic activity with fluoroacetate reduced LTP [66].

Since several lines of evidence indicate localization of serine racemase expression in neurons as well as astrocytes, the relative contribution of astrocytic and neuronal D-serine was investigated using cell-selective serine racemase KO mice [67]. These studies showed that neuronal absence of serine racemase significantly reduced LTP in Schaffer collateral/CA1, whereas astrocytic loss of serine racemase did not. This study does not necessarily contradict original reports of D-serine as a gliotransmitter but suggests that in normal conditions neuronal D-serine predominantly impacts synaptic activity. Future studies using these mice or other cell-specific approaches may determine whether astrocytic D-serine plays a more important role in synaptic function under conditions of neuronal dysfunction or metabolic dysregulation.

2.2.3 GABA

Transient expression of the GABA-synthesizing enzyme, glutamic acid decarboxylase (GAD), has been observed in astrocytes during development [68, 69]. GAD67

(67 kDa form of GAD) and the GABA transaminase, GABA-T, have also been reported to be expressed in human astrocytes [70]. However, the potential role of GABA release in astrocyte–neuron signaling has been unclear until recently, although an astrocytic inhibitory modulation on neuronal activity was suggested in the past [71, 72]. Kozlov et al. demonstrated that mitral cells of the olfactory bulb present synchronous slow outward currents, which are dependent on GABA release from astrocytes [48]. Since then, studies have confirmed a role for astrocytic GABAergic signaling in tonic inhibition, modulating a continuous current which was found to be dependent on extrasynaptic GABA-A receptors that control the neuronal excitability [73] and which has been shown to be involved in sleep, memory, epilepsy, and alcohol vulnerability [74–77].

Tonic inhibition mediated by astrocytic GABA appears to be released by the anion channel Best1, as was shown recently in the cerebellum [78]. Best1 is activated by intracellular Ca^{2+} and by changes in cell volume and is also permeable to glutamate. Although Best1 is expressed in astrocytes of the CA1 layer in the hippocampus, Yoon et al. [79] observed a very low tonic inhibition correlated with low levels of GABA in astrocytes [79]. Bergmann glia of the cerebellum provide a higher tonic inhibition due to their higher concentration of intracellular GABA [79].

In addition to providing a source of tonic inhibition to neurons, astrocytic GABA signaling has anti-inflammatory properties. GABA receptors expressed on astrocytes and microglia (GABA-A and GABA-B) inhibit the NF-κB and p38-MAP kinase pathways and the release of TNF-alpha and IL-6 under inflammatory stimulation [70]. In return, activation of GABA-A and GABA-B on astrocytes increases their release of GABA in a Ca^{2+}-dependent manner [80]. GABA transporters GAT1, GAT2, and GAT3 are able to release GABA from astrocytes via a Ca^{2+}-independent pathway when intracellular GABA concentration is too high [80], although these mechanisms are less clear.

2.2.4 Adenosine Triphosphate

Adenosine triphosphate (ATP) is critical for metabolic cellular processes but can also act as a transmitter. Astrocytes are equipped with machinery for ATP release, as was appreciated by visualization of ATP-containing vesicles in cultured astrocytes [39, 81]. Early studies using astrocyte cultures demonstrated that calcium wave propagation was mediated by astrocyte release of ATP [82–84]. Astrocytic ATP signaling was thought to propagate calcium waves via chemical coupling [85], providing a mechanism for rapid intercellular communication across broad domains. ATP also proved important for astrocyte cross talk with neurons. Zhang et al. elegantly demonstrated a functional impact of glia–neuron signaling by showing that ATP release from astrocytes tonically suppresses glutamatergic synapses via activation of presynaptic purinergic receptors in vitro, an effect which was dependent on synaptic glutamate release and subsequent activation of glutamate receptors on astrocytes [84]. Similar findings were reported in studies of CA1 synapses in hippocampal slices, where adenosine, a metabolite of

ATP, was found to modulate synaptic depression [84]. A key role for purinergic gliotransmission was further established through the development of transgenic mice in which SNARE-dependent exocytosis was selectively blocked in astrocytes [86]. Specifically, this study demonstrated that astrocytic ATP and its metabolite, adenosine, act to tonically suppress synaptic activity via activation of A1 receptors.

ATP release from astrocytes can occur through activation of purinergic P2Y and P2X receptors and impacts physiological and pathological responses [87–89]. Indeed, activation of P2Y1 and P2Y2 receptors on astrocytes has been shown to be critical for intercellular calcium signaling [90], the source of which is thought to be ATP.

Purinergic signaling from astrocytes extends beyond ATP release. Once released, ATP can be rapidly hydrolyzed into adenosine, which can act on adenosine A1 and A2a receptors. Activation of presynaptic A1 receptors tonically inhibits glutamate release, whereas activation of A2a receptors upregulates synaptic transmission, and there is evidence for dynamic astrocytic regulation of synaptic activity via both A1 and A2a receptors [86, 91, 92]. Several lines of evidence point to a critical role for this astrocytic source of adenosine in modulation of synaptic plasticity, discussed in the subsequent sections, memory consolidation and sleep homeostasis.

2.2.5 TNF-Alpha

TNF-alpha is a well-known proinflammatory cytokine, but the studies of Beattie and Stellwagen in 2002 and 2006 suggest that it can also act as a gliotransmitter [93, 94]. When applied to neurons in vitro, TNF-alpha taken from conditional media of astrocyte cultures increases the amount of AMPAR expression at the membrane surface, thereby increasing neuronal sensitivity to glutamate [93]. In response to prolonged blockade of activity (via TTX treatment), TNF-alpha treatment still increases AMPAR surface expression [94], suggesting that TNF-alpha is secreted by astrocytes and modulates neuronal activity by increasing surface AMPAR expression and increasing the synaptic strength.

2.3 Synaptic Plasticity

As part of the tripartite synapse, astrocytes are uniquely positioned to monitor synaptic activity. Signaling at the synapse can activate receptors on astrocytes, and astrocytes are readily equipped to modulate the uptake and release of neuroactive signaling molecules, including glutamate, D-serine, ATP, and TNF-alpha, subsequently regulating activation of pre-, post-, and extrasynaptic receptors. Communication at the tripartite synapse, therefore, can be both static and dynamic, modulating synaptic plasticity and network activity through a variety of mechanisms, many of which are calcium-dependent.

2.3.1 Intracellular Calcium Signaling

There has been conflicting evidence regarding the role of calcium-dependent gliotransmitter release on LTP. Several different methods have been used to assess this, and it is likely that a difference in techniques accounts for the discrepancies. For example, several studies have shown that calcium clamping astrocytes impairs LTP in astrocytes [95]. However, results from Agulhon et al. [13] stimulated quite a bit of controversy by demonstrating a lack of an effect on LTP in two transgenic mouse lines with an induced impairment in calcium signaling specifically in astrocytes [13]. These authors failed to detect a difference in LTP induction and maintenance in hippocampal slices isolated from transgenic mice with astrocyte-specific impairment in the IP_3 signaling pathway. However, a recent study using the same IP_3 knockout mouse showed that calcium-dependent glutamate-mediated gliotransmission does indeed contribute to cholinergic LTP in the hippocampus [51]. Despite conflicting evidence, there are several additional sophisticated approaches that demonstrate the impact of intracellular calcium signaling on synaptic transmission (for review [96]). For example, visualization of high spatial and temporal resolution using two-photon laser scanning microscopy revealed two types of local calcium dynamics in individual hippocampal astrocytes [97]. "Focal" events were characterized by small increases in calcium $[Ca^{2+}]_i$ that were highly localized to a small subregion and accounted for the majority of calcium events in astrocytic processes. "Expanded" events were detected simultaneously in contiguous segments of astrocytic processes and were marked by large calcium increases $[Ca^{2+}]_i$. The expanded events showed a higher degree of dependence on neuronal activity, whereas the focal events only slightly decreased in frequency when neuronal activity was blocked and therefore appeared to primarily reflect spontaneous synaptic release. Furthermore, these calcium events were dependent on IP_3 receptor signaling and purinergic receptor activation and participated in bidirectional communication, such that activation of the purinergic receptor, P2Y1R, stimulated IP_3-mediated release of intracellular calcium causing transmitter release at excitatory synapses.

Using combined techniques of whole cell recordings, calcium imaging in astrocytes and glutamate imaging in hippocampal slices, it is possible to measure simultaneous responses in neurons and astrocytes, providing solid evidence for astrocytic glutamate release. Panatier et al. [92] visualized calcium dynamics localized within "compartments" of astrocyte processes using real-time confocal imaging and found that stimulation of single synapses induced calcium events confined to a local astrocytic compartment and occurring simultaneously with postsynaptic glutamatergic transmission [92]. These events were dependent on mGluR5 activation of astrocytes, which induced calcium signaling and subsequent release of an astrocytic source of adenosine acting on presynaptic A2a receptors to modulate basal excitatory synaptic transmission.

Recent studies have also discovered that astrocytic calcium signaling can occur in microdomains of the astrocytic processes [98], which further impacts the ability of these cells to communicate discretely at the synapse.

2.3.2 *Astrocyte Signaling Molecules and Synaptic Plasticity*

Astrocyte–neuron signaling also occurs locally through interactions between signaling molecules and ligand receptors. These interactions are often bidirectional, impacting both astrocyte and neuron physiology and, in many cases, synaptic plasticity.

2.3.2.1 Ephrin

Ephrin ligands and their receptors are widely recognized for their role in axon guidance and synapse development, but recent studies show that they are also involved in synaptic plasticity in the adult brain. For example, recent [99, 100] studies indicate that ephrin-mediated signaling among astrocytes and synapses impacts LTP in the hippocampus. Astrocytic processes express ephrin-A3, whose binding partner, EphA4, is expressed neuronally. LTP is attenuated in mice lacking either ephrin-A3 or EphA4 in the CA1 region of the hippocampus. Furthermore, reduced synaptic plasticity in these knockout mice was associated with increased glutamate uptake via a selective increase in expression of the astrocytic glutamate transporters GLT1 and GLAST, subsequently reducing the availability of synaptic glutamate [99].

2.3.2.2 Endocannabinoid Signaling

Endocannabinoids and their endogenous receptor CB1 have widespread effects on neuronal activity, including synaptic plasticity, mainly by reducing presynaptic release of neurotransmitters. For example, astrocytes were recently shown to express functional CB1 receptors that, upon activation, regulate glutamate release from hippocampal astrocytes in a calcium-dependent manner [101]. Further studies revealed that activation of CB1 receptors modulates synaptic plasticity in a bidirectional manner. Specifically, activation of presynaptic CB1 receptors depresses synaptic activity, whereas activation of astrocytic CB1 receptors potentiates synaptic plasticity via calcium-dependent release of glutamate and subsequent activation of presynaptic mGluR1 receptors [102]. These results suggest that endocannabinoid signaling may induce inhibitory effects locally at the synapse but may also potentiate synaptic transmission across a broader network via CB1R activation (and subsequently calcium signaling) on astrocytes. To investigate the specific role of astrocytic CB1R, Han et al. recently generated a conditional transgenic mouse lacking CB1R selectively in astrocytes, GABAergic, or glutamatergic neurons [103]. This study showed that CB1Rs expressed by astrocytes, but not by glutamatergic or GABAergic hippocampal neurons, are necessary for endocannabinoid-induced synaptic depression in hippocampal CA3/CA1 synapses and cannabinoid-induced impairment of spatial working memory [103]. These studies provide further evidence of astrocytic modulation of synaptic plasticity.

2.3.2.3 Aquaporin

Aquaporin 4 (APQ4) is a water-permeable channel that is mainly expressed by astrocytes in the CNS. It responds to changes in extracellular milieu and has been implicated in cerebral edema following injury [104]. Since APQ4 is colocalized with K^+ channels [104], it was hypothesized that it has the potential to alter excitability. Indeed, mice null for APQ4 show attenuated theta-burst stimulation (TBS) LTP and LTD, though baseline activity and short-term plasticity are intact [105]. Interestingly, LTP impairment in APQ4–/– mice was attributed to reduced BDNF, which is known to be required for TBS–LTP. How a lack of APQ4 modifies BDNF expression is unknown.

2.4 Network Modulation

While astrocytes are ideally situated to monitor synaptic activity at the tripartite synapse, they are also able to communicate long distances via the expression of gap junctions. Astrocytes have been shown to be organized in structurally nonoverlapping domains [106] and make contact with neurons and blood vessels. With its impressive volume, one astrocyte can contact tens of thousands of synapses [7, 106, 107], while one synapse is surrounded by only one astrocyte. This anatomical aspect is essential for the comprehension of their role in the plasticity and regulation of synaptic activity. Indeed, astrocytes are organized in a network, or syncytium, to communicate with one another via gap junctions, which are formed by connexins that are permeable to compounds with molecular weights lower than 1,000 Da [108]. The astrocytic syncytium can then be easily visualized via dye propagation through gap junctions, where it is confined within the network.

Though it is clear that astrocytic signaling modulates neuronal network activity, both through intercellular calcium signaling, gap junctions, and through lactate transport, the intricacies of the molecular mechanisms governing these pathways are still somewhat unclear.

2.4.1 Intercellular Calcium Signaling ("Calcium Waves")

Unlike neurons, astrocytes do not produce action potentials. Their membrane potential is very stable and has low resistance due to the high resting permeability to K^+ ions and because of the gap junction connectivity in an astrocytic syncytium. In early work, Nedergaard used astrocyte–neuron cocultures to show that astrocytes communicate to each other and to neurons through calcium signals, or "waves" that propagate via gap junctions [109]. Since then, two forms of excitation have been characterized in astrocytes based on "calcium waves": a spontaneous, autonomous excitation and an excitation which is dependent on chemical signals relayed within the neuronal circuit [107]. Until recently, calcium waves were mainly characterized in vitro or in acute brain slices and were found to be independent of neuronal activity and relatively slow and localized [110, 111]. There are discrepancies in the results of these earlier studies, most of which were done in vitro or in situ, and subsequently there are two proposed mechanisms of calcium wave propagation: one poses that it occurs via cytosolic transfer of IP_3 through gap junctions [112], whereas the other proposes that it occurs via ATP release and diffusion, activation of purinergic P2YRs on astrocytes, and then subsequent activation of IP_3 signaling [85]. It is important to realize that these two pathways are not mutually exclusive, and it is likely that maintenance of calcium wave propagation varies according to brain region and the related heterogeneity of astrocyte cell types. Models of these two pathways have been described [113]. Further studies using in vivo techniques are required to determine the relevance of calcium wave propagation and the signaling mechanisms that govern it.

As an example, recent improvements in calcium imaging techniques in vivo contributed to the confirmation that calcium waves, or "glissandi," propagate through astrocytes en masse in the hippocampus [114], occurring more frequently than sporadic calcium activity. Interestingly, glissandi were blocked by tetrodotoxin, demonstrating for the first time that calcium wave propagation is dependent on neuronal activity, relative to sporadic activity, which is TTX-insensitive, and therefore results from intrinsic intracellular calcium signaling in individual astrocytes. Calcium wave propagation was also found to be initiated by ATP application and blocked by inhibitors of gap junctions or ATP receptors, suggesting a role for both intra- and extracellular signaling molecules. Moreover, these large-scale calcium waves coincided with reduced flow of red blood cells through vessels within the same region, providing even more evidence that astrocytic signaling coordinates network activity on multiple levels.

2.4.2 Gap Junctions

When a large number of gap junctions were detected among astrocytes over three decades ago, many scientists were led to the conclusion that this "glial syncytium" functioned as more of a support system. However, more recent studies indicate that astrocytes communicate through gap junctions to integrate intra- and extracellular signals in response to changes in neurovasculature and neuronal metabolic demands, effectively integrating signals to modulate network activity. Contrary to early belief, astrocytes are uniquely structured to morphologically occupy discrete anatomical domains, without structural overlap, therefore necessitating an efficient mechanism for intercellular communication. This is achieved primarily by the formation of gap junctions, which are composed of connexins 30 and 43 (Cx30, Cx43) in astrocytes. Regulation of network activity by gap junction signaling in astrocytes can be easily appreciated in the somatosensory barrel cortex of rodents, where dye coupling of astrocytes shows strong coupling almost exclusively within a barrel [115]. Similarly, astrocytic networks also modulate sensory integration within olfactory glomeruli, which are highly structurally organized in functional units, mainly via activity-dependent generation of extracellular potassium and Cx30-mediated gap junction communication [116].

Studies from Cx30 and Cx43 knockout mice also highlight the impact of astrocytic regulation of hippocampal network activity. Hippocampal network activity is altered in these mice, with increased synaptic activity at CA1 pyramidal cells and increased basal excitatory synaptic transmission, which occludes LTP and enhances LTD [117]. This aberrant synaptic plasticity was attributed to decreased/slower glutamate uptake and slower K^+ clearance. Despite the finding that Cx30–/– and Cx43–/– astrocytes still uptake a significant amount of glutamate and potassium, they are uncoupled and therefore unable to redistribute it, causing cell swelling in response to neuronal activity. These results show the importance of astrocytic networks in maintaining extracellular homeostasis during basal synaptic transmission.

2.4.3 Energy Metabolism (Lactate Transport)

Neuronal activity, and in particular glutamatergic activity, requires a great quantity of energy resulting from the metabolism of glucose. Cerebral consumption of glucose accounts for 20 % of the total body glucose utilization [2] and is mainly utilized in support of excitatory neuronal activity; action potentials, reestablishment of ion gradients and postsynaptic responses all rely on glucose availability [118].

Astrocytes are ideally located to interact with both blood vessels, via end-feet contact, and neurons within the tripartite synapse. Here, astrocytes are critical regulators of brain energy metabolism as they provide energy to neurons during synaptic activity [119–121]. As proposed in the astrocyte–neuron lactate shuttle (ANLS) [122], the astrocytic energy supply to neurons depends on their glutamate and glucose uptake and their high glycolytic activity [119, 120]. Astrocytes uptake glutamate via Na^+-coupled glutamate transporters (GLAST and GLT-1) in response to glutamate release from neurons. The Na^+/K^+ pump then restores the Na^+ gradient, requiring ATP consumption, which in turn stimulates glucose uptake by astrocytes via the transporter GLUT-1 [123] and glycolysis [122, 124–127]. The glycolytic product, lactate, is then transported via the astrocytic MCT transporters (MCT1 and MCT4) to the neuronal MCT2. In addition to their direct utilization of glucose, neurons can rapidly use the lactate oxidative energy substrate [128–131].

An interesting study performed by Rouach et al. [132] showed that within the hippocampus, an entire astrocyte syncytium, supported by the connexins Cx43 and Cx30, is able to restore neuronal activity under conditions of glucose deprivation due to lactate provided by the astrocytes and its ability to diffuse through the syncytium [132]. This study showed that the fluorescent derivative of glucose 2-[*N*-(7-nitrobenz-2-oxa-1,3-diazol-4-yl)amino]-2-deoxyglucose (2-NBDG), injected in a perivascular astrocyte of a mouse hippocampal slice, can diffuse in a score of minutes in a network of astrocytes (60–90 astrocytes). This diffusion was shown to be dependent on the gap junction proteins, Cx43 and Cx30, and increased during neuronal activity. Moreover, it suggests that diffusion through the astrocytic network, including the perivascular astrocyte, is capable of shifting differentially according to energy demands. Specifically, this study showed that the diffusion of 2-NBDG propagated toward the electrically stimulated area, which was deprived of glucose, and subsequently restored neuronal activity. This restoration was shown to be dependent on lactate [132]. Thus, the ANLS is more than the response of one individual astrocyte to synaptic activity but is also a dynamic response to an entire network, with the ability to adapt its size to the intensity of the energetic demand. This metabolic coupling between the astrocytes, the neurons, and of course the blood capillaries is a perfect illustration of the complex integration of signaling in the cerebral parenchyma [133].

2.5 Impact of Glia–Neuron Communication on Behavior

The previous sections were designed to introduce astrocyte–neuron signaling as it occurs at the synaptic and network levels. The recent surge in technology and interest focused on the potential contribution of astrocyte signaling to brain function has challenged traditional views of neuroscience, which are primarily neuron-centric, promoting the idea that neuronal activity is the basis of cognition and behavior. Indeed, there are currently several lines of evidence that highlight the impact of astrocyte signaling on behavior.

2.5.1 Sleep Homeostasis

Ramon y Cajal was among the first neuroscientists to appreciate the unique structure of astrocytes, predicting the importance of these glial cells in brain function, namely, sleep–wake states [134]. Despite the availability of only static micrographs of astrocytes, Cajal imagined a very dynamic structure and function of astrocyte processes, based on their morphology and relative proximity to neurons. Indeed, he theorized that astrocytes retract their processes during wakefulness, allowing neurons to directly communicate, and then induce sleep by extending their processes to the synapse, subsequently interrupting synaptic communication [134]. Over a century later, through the advancements in techniques applied to glial biology research, aspects of his precocious insight have been confirmed. Using a transgenic mouse line in which SNARE-mediated transmitter release is selectively attenuated in astrocytes (called "dnSNARE" mice [86]), it was recently shown that gliotransmission regulates sleep pressure by providing a tonic source of adenosine acting on adenosine A1R [91]. Specifically, 6-h sleep deprivation effectively increased sleep pressure (measured by the intensity of low-frequency SWA component of NREM sleep in the cortex) in WT, but not in dnSNARE mice. This effect on sleep pressure was demonstrated to be dependent on A1R activation, as the A1R antagonist CPT mimicked the dnSNARE phenotype when administered to WT mice. Interestingly, the cognitive impairments associated with sleep deprivation, while present in WT mice, were markedly attenuated in dnSNARE mice, suggesting that an astrocytic source of adenosine regulates the accumulation of sleep pressure that is associated with memory impairments following sleep loss. Extending on these findings, a recent study examined the late-phase LTP (L-LTP) in hippocampal slices of dnSNARE or WT mice after sleep deprivation and found that sleep loss does not impair L-LTP in hippocampus of dnSNARE mice, as it does in WT mice, an effect which was shown to be dependent on A1R activation [135]. Similarly, spatial object memory, a hippocampal-dependent function, was not impaired following sleep deprivation in dnSNARE mice or CPT-treated mice, as it was in WT and vehicle controls. These studies consistently demonstrate the impact of an astrocytic source of adenosine acting on synaptic A1Rs on sleep homeostasis and memory impairments following sleep loss.

Studies linking genetic influences to sleep disorders also point to a role of adenosine, which may likely include astrocytic-derived adenosine. Specifically, a functional polymorphism in adenosine deaminase (ADA), the enzyme responsible for the breakdown of adenosine into inosine, was linked to high sleep pressure in otherwise healthy individuals [136, 137]. These results are consistent with animal models showing the importance of ADA as a regulator of sleep homeostasis [138, 139]. In light of aforementioned studies of dnSNARE mice, and given that ADA is thought to be more abundantly expressed in astrocytes than neurons [138, 140], adenosine signaling in astrocytes may be a critical regulator of sleep homeostasis.

2.5.2 *Breathing*

Recent evidence has shown a crucial role for astrocyte–neuron signaling in regulating respiration. In an elegant study by Gourine et al., astrocytes in the ventral medulla oblongata were genetically induced to selectively express a calcium indicator to examine whether astrocytes respond to respiratory pH changes via calcium signaling [141]. Indeed, artificially induced decreases in pH stimulated immediate increases in astrocytic calcium, particularly in astrocytes adjacent to blood vessels, independent of neuronal activation. pH-sensitive calcium signaling was found to partially involve gap junctions but was primarily dependent on activation of ionotropic ATP receptors. Using optogenetically controlled calcium channels in astrocytes, they next showed that calcium signaling in astrocytes induced release of ATP which subsequently depolarized neighboring neurons. The functional importance of this signaling was then remarkably demonstrated by optogenetic stimulation of calcium influx in astrocytes that robustly induced respiratory activity in anesthetized rats [141].

Astrocytes also couple with respiratory neurons and regulate rhythmic firing in the pre-Botzinger complex in the midbrain. The burst firing of action potentials causes an abundance of extracellular K^+ and glutamate, which initiate astrocytic responses characterized by rhythmic inward currents carried by inward-rectifying K^+ channels and electrogenic uptake of glutamate via GLT-1 transporter [142].

2.5.3 Circadian Regulation

Thanks to the amenability of genetic manipulations in Drosophila, Suh and Jackson discovered that glial cells synthesize (and secrete) biogenic amines necessary for circadian regulation of locomotor behaviors [143]. More recently, several different lines of transgenic mice were used to assay circadian release of ATP from cultured astrocytes [144] and showed that it is dependent on clock genes. Using cultures isolated from dnSNARE mice, they show that circadian release of ATP does not rely on SNARE-dependent transmitter release. Using another line of transgenic mice in which IP_3 signaling (and therefore intracellular Ca^{2+}) is selectively upregulated in astrocytes, they showed that IP_3-dependent calcium signaling contributes to the amplitude of rhythmic ATP release. In light of evidence showing astrocytic SNARE-dependent modulation of sleep homeostasis [91], it is interesting to consider the idea that two mechanisms of ATP release from astrocytes independently regulate two components of sleep and wake: circadian rhythms and sleep pressure.

2.5.4 Memory

Astrocytic modulation of network activity via lactate transport was recently demonstrated in an elegant study by Suzuki et al. [145], where it was shown to have an impact on plasticity and related memory impairments [145]. Specifically, they showed that inhibition of glycolysis in the hippocampus impaired long-term memory but had no effect on acquisition ("training") or short-term memory. Administration of lactate restored long-term memory, suggesting that there is rapid glycogenesis during training and, subsequently, lactate release that is important for long-term memory consolidation. Impairment of lactate release via the knockdown of the astrocytic lactate transporters MCT1 or MCT4 significantly impaired long-term memory (but not acquisition or short-term memory), an effect which could be rescued by injection of lactate, but not glucose. Furthermore, long-term memory was also impaired when the neuronal lactate transporter MCT2 was knocked down, but neither lactate nor glucose rescued it, suggesting lactate release by MCT1 and MCT4 from astrocytes and import by MCT2 into neurons is necessary for long-term memory consolidation.

Mice lacking expression of the receptor for the cytokine IL1 (which is highly, but not exclusively, expressed in astrocytes) also exhibit impairments in hippocampal-dependent memory and LTP [146–148]. However, when neural stem cells (NSC) from wild-type mice were implanted into the hippocampus of IL1−/− mice and allowed to differentiate into astrocytes, spatial memory was recovered [149], suggesting that astrocytic IL1 signaling plays an important role in hippocampal-dependent memory. Similarly, LTP is impaired and learning and memory deficits manifest in transgenic mice with astrocyte-targeted deletion of nuclear factor-kappa B (NF-κB), a transcription factor originally recognized for its involvement in inflammatory and immune responses, but shown to play a role at the synapse [150–153]. Interestingly, these deficits were associated with reduced expression of the metabotropic glutamate receptor mGluR5 and PSD-95 in the hippocampus and cortex and importantly were exclusively pronounced in female mice [154], implicating a role for sexually dimorphic pathways in astrocytes.

2.6 Glia–Neuron Dysfunction in Disease and Neuroinflammation

Astrocytes have long been recognized as key players in neuroinflammatory conditions. Notably, along with microglia, astrocytes respond to injury by becoming "reactive," a somewhat controversial term used to describe the morphological and histochemical profile of these cells under injury conditions. Indisputable changes occurring in "reactive" astrocytes include hypertrophic morphology, upregulation of the cytoskeletal protein GFAP, proliferation, loss of nonoverlapping domains, and ultimately glial scar formation. The functional roles of reactive astrocytes appear to be both neuroprotective and neurotoxic. For example, selective ablation of reactive astrocytes increases neuronal death under several injury conditions (e.g., [155]), and inducible, transgenic blockade of STAT3, a signaling molecule for several cytokines, impairs astrogliosis, increases inflammation, and impedes functional recovery after spinal cord injury [156, 157]. On the other hand, blockade of other inflammatory pathways in astrocytes, such as NF-kappa B, has been shown to be neuroprotective in models of spinal cord injury, ischemia, and experimental autoimmune encephalopathy [158–160]. To date, there are several reviews describing the role of astrocytes in neuroinflammation associated with neurodegenerative diseases (e.g., see [160–164]).

Collectively, these and other studies suggest that numerous signaling pathways are likely to be involved in astrocytic response to inflammation.

2.6.1 *Epilepsy*

Multiple osmotic functions of astrocytes are impaired in epilepsy, including ionic balance (K^+ buffering), energy metabolism, vascular coupling, and glutamate uptake (for review, see [165]). In general, these astrocytic changes were thought to occur secondary to neuronal dysfunction. However, recent evidence suggests that astrocyte signaling contributes to epileptogenesis and seizure activity through changes in glutamate transport and release, energy metabolism, calcium signaling, and adenosine.

Regions affected by epilepsy in human brains show high extracellular levels of glutamate and the associated excitotoxic neuronal death. Both decreases and increases in expression of astrocytic glutamate transporters have been observed, so how these transporters are involved in aberrant levels of extracellular glutamate is not clear [166–169]. Similarly, animal models usually exhibit an impairment in glutamate transport function [170–176]. Shortly after status epilepticus, an increased coupling and decreased efficiency of glutamate uptake has been observed in the hippocampus of a kainate-induced status epilepticus rat model [177], supporting a role for astrocytic glutamate transport in epileptogenesis.

The glutamine–glutamate cycle in astrocytes modulates levels of glutamate and GABA produced by neurons. This cycle has been shown to be impaired in sclerotic hippocampus [178, 179]. Epileptiform activity is dependent on neuronal glutamine uptake [180], and reactive astrocytes have a decreased expression of glutamine synthetase, ultimately resulting in a decreased GABA inhibition [181]. In this manner, it is possible that astrocytes contribute to the hyperexcitability of epileptiform activity.

Furthermore, dysfunctional energy metabolism in astrocytes was recently shown to be involved in the latent phase of epileptogenesis, which marks the transition from status epilepticus to seizure behaviors. As described recently in Alvestad et al., impairment of glutamate–glutamine cycle and astrocytic GABA production from glucose have been observed specifically in the latent phase of a kainate rat model of epilepsy [182]. Thus, changes in metabolism of both glutamate and GABA may contribute to epileptogenesis.

Astrocytic calcium signaling may also contribute to epileptiform activity. For example, seizure activity induces an upregulation of metabotropic glutamate receptor expression in astrocytes and increases intracellular Ca^{2+} [183, 184]. As described previously, Ca^{2+} signaling in astrocytes increases Ca^{2+}-dependent gliotransmission. Though glutamate release from astrocytes does not generate epileptiform activity [185], it is possible that it contributes to sustained hyperexcitability during seizures and subsequent excitotoxic neuronal death. Intracellular calcium signaling in astrocytes and extrasynaptic (potentially astrocytic) sources of glutamate have also been shown to contribute to aberrant depolarization during interictal events [186], suggesting that an astrocytic source of glutamate contributes to epileptiform activity. Indeed, inhibition of astrocytic metabotropic glutamate receptors (mGluR5) after status epilepticus delays neuronal death, as does application of a Ca^{2+} chelator [187]. To more closely examine the role of astrocytes on epileptiform activity, Gomez-Gonzalez et al. monitored astrocytic Ca^{2+} signaling in parallel with neuronal epileptiform activity in entorhinal cortical slices from rat brain. Intracellular astrocytic Ca^{2+} signaling increased during the development of a focal ictal event; selective inhibition of intracellular astrocytic Ca^{2+} near the focal ictal event reduced the intensity of this event. Thus, these results showed that astrocytes can play a role in the initiation of ictal activity during seizures [188].

Adenosine, a by-product of the gliotransmitter ATP, typically counteracts neuronal hyperexcitability through its action on presynaptic A1 receptors. Along the same line, overexpression of adenosine kinase (ADK), which metabolizes adenosine into ADP and therefore reduces adenosine levels, is sufficient to induce epileptiform activity [189]. Immunohistology from brains of temporal lobe epilepsy patients reveals enhanced ADK expression in the hippocampus [190], and inhibition of ADK expression is able to prevent seizures in mouse models of epilepsy [191].

In addition to local signaling, the astrocytic syncytium has been shown to be involved in seizure development, as deletion of the gap junction proteins, Cx30 and Cx43, increases the potency of epileptiform activity [192]. Furthermore, astrocytic domains were shown to be disrupted in three different mouse models of epilepsy. Diolistic labeling in cortical slices showed that the domains normally occupied by independent astrocytes begin to overlap following seizure events, an effect which was lessened by treatment with antiepileptic drugs [193]. Interestingly, the loss of astrocytic domains coincided with changes in dendritic morphology, including an increase in spine density, though it is yet unknown whether the changes in astrocyte morphology precede those in dendrites, or vice versa.

2.6.2 *Neurodegenerative Diseases*

Neurodegenerative diseases are characterized by neuronal degeneration and necrotic cell death and in later stages are often associated with severe cognitive and/or motor deficits. Though initially thought to result from a primary impairment in neuronal function, accumulating evidence suggests a role for dysfunction in astrocyte–neuron signaling as a critical contributor to pathology for most neurodegenerative diseases, including Alzheimer's disease (AD), Parkinson's disease (PD), Huntington's disease (HD), and ALS [164, 194–199]. AD, PD, and ALS primarily occur sporadically, with the few inherited forms resulting from abnormal mutation of specific genes, while HD is inherited through acquisition of the well-known huntingtin gene mutation. Reactive astrocytes are hallmark features of pathology in these diseases and correlate with the intensity of neuronal dysfunction and death [198, 200, 201]. This section will focus on AD and ALS, for which there is substantial evidence showing the role of astrocytes in neurodegeneration.

Accumulation of amyloid plaques is a hallmark pathology of AD and is generally associated with neighboring reactive astrocytes (e.g., [202]). Though astrocytes accumulate around amyloid plaques, they are not able to degrade amyloid β (Aβ) as they do in culture [203]. Rather, these reactive astrocytes, along with microglia, release proinflammatory cytokines, including interleukin 1β, interleukin 6, and TNFα [204], though whether this response is neuroprotective or toxic is not well understood. Inhibitors of proinflammatory cytokines were found to attenuate both synaptic and cognitive deficits in mice treated with Aβ infusion, suggesting that these inflammatory responses of astrocytes have detrimental downstream effects [205]. However, the relevance of these findings to clinical evidence is questionable [206].

Notably, an abnormally increased level of TNFα is a characteristic feature of neuroinflammation in AD patients [207], and there is convincing evidence highlighting a role for astrocytic release of TNFα in aberrant Ca^{2+} signaling and extracellular glutamate levels in a mouse model of AD [208, 209]. Specifically, TNFα-stimulated, Ca^{2+}-dependent release of glutamate was decreased in hippocampal slices isolated from aged AD mice [209], a finding which is somewhat incongruous with studies showing increased Ca^{2+} signaling in astrocytes (which should increase glutamate gliotransmission) in AD mouse models [208, 210, 211]. The altered levels of glutamate transporter and receptor expression that is characteristically pronounced in AD models [212, 213] may partially account for these results (see for review [164]). Given the established role of astrocytes in modulating synaptic activity, as described in previous sections, it is likely that aberrations in astrocyte signaling contribute to the progression of impairments in synaptic transmission associated with cognitive decline in AD. The gliotransmitter ATP has recently been proposed as a neuroprotective agent in AD. A study on primary astrocyte cultures observed an increased release of ATP after application of the peptide Aβ42 in the medium [214]. ATP application on neuronal cultures prevented Aβ42 toxicity via P2-type purinergic receptors and prevented Aβ42-induced impairment of LTP on acute hippocampal slices.

ALS is characterized by a progressive paralysis due to motor neuron death. The etiology of ALS is mainly sporadic, but 10 % of the patients present a dominant negative mutation, and among them 25 % have a mutation of the antioxidant enzyme superoxide dismutase 1 gene (SOD1) [215]. Changes in microglia, macrophages, and astrocytes have been documented in transgenic mice that express mutant SOD1 ubiquitously. Immunohistochemical examination of the CNS from these mice revealed that microglia activation was pronounced in the motor cortex, motor nuclei of the brainstem, corticospinal tract, and ventral horn of the spinal cord, and preceded the onset of motor neuron degeneration, which suggests the potential to serve as a preclinical indicator of ALS [216, 217]. In fact, positron emission tomography (PET) has been used to examine microglia activation in patients with ALS, revealing an association between microgliosis and upper (but not lower) motor neuron damage, again suggesting that microglia responses may occur prior to motor neuron degeneration in humans [218]. Macrophage activation is prominent in the sciatic nerve in SOD1 mutant mice, occurring throughout the disease onset and progression [219]. Astrogliosis appears in the dorsal and ventral horn of the spinal cord and is not restricted to motor areas of the brain, extending through both cortical gray and subcortical white matter in SOD1 mutant mice [220–222]. The activation of astrocytes occurs later than that of microglia, in time with the onset of motor neuron degeneration.

Given the changes that occur in microglia and astrocytes in SOD1 model of ALS, transgenic mice with a specific manipulation of mutant SOD1 expression in Cd11b-expressing microglia or GFAP-expressing astrocytes were created to examine more closely how these cells contribute to the initiation and progression of ALS [194, 223]. Transgenic mice expressing mutant SOD1 do not display features of neurodegeneration when transgene expression is restricted to motoneurons [224]. This observation suggests that neurodegeneration in ALS is not cell-autonomous, but rather it also results from dysfunction in nonneuronal cells. Specific deletion of mutant SOD1 in Cd11b-expressing cells, including microglia and macrophages, or in GFAP-expressing astrocytes does not change the disease onset but slows the disease progression and increases the survival of mutant SOD1 mice [194, 223, 225]. When mutant SOD1 expression is absent in microglia only, astrogliosis and microgliosis persist [223]. When mutant SOD1 is absent in astrocytes only, astrogliosis, but not microgliosis, is delayed compared to mice with ubiquitous expression of mutant SOD1 [225]. This suggests that the microglia response depends in part on factors released by astrocytes expressing mutant SOD1 [225]. In conclusion, neuronal mutant SOD1 expression is necessary for the onset of SOD1-related ALS but expression in nonneuronal cells is involved in the progression of the disease. More particularly, both astrocytes and microglia are important in the late progression phase of the disease and in survival [223, 225].

Several lines of evidence indicate that astrocytes are actively involved in ALS pathogenesis, particularly due to an impairment in glutamate uptake. For example, brain regions affected by the sporadic form of the ALS show a 95 % loss of the astrocytic glutamate transporter EAAT2 and an increased level of glutamate in CSF [226, 227].

The impairment in energy metabolism observed in ALS also appears to involve astrocytes. For example, mutant SOD1 expression in astrocytes decreases lactate efflux to motor neurons in vitro, and this loss of metabolic support leads to a decrease in neuronal survival [228].

Similar to AD pathology, inflammation is also a characteristic feature of ALS pathogenesis [229]. Depletion of SOD1 in microglia or astrocytes leads to secretion of cytokines [229], and treatment with anti-inflammatory drugs slows down the progression of the disease in transgenic mice [230, 231].

Collectively, this evidence suggests that neuron–astrocyte interactions have a strong impact on the progression of neurodegenerative diseases, and therefore, astrocytes may be a novel therapeutic target for treatment of these diseases.

2.6.3 *Drugs of Abuse*

To date, there are limited studies investigating the putative contribution of glial cells to drug and alcohol abuse and addiction. However, there is some evidence from both human and mouse models that suggest astrocytes may play a role in drug and alcohol behavior. For example, there are marked changes in the immunohistochemical profile and morphology of astrocytes during chronic morphine treatment, such as increased GFAP expression and reduced expression of GLT-1 and GLAST [232, 233], suggesting an alteration in glutamate clearance. Similar changes occur following alcohol exposure, with the degree (and direction) of GFAP expression dependent on the duration of alcohol exposure [234, 235]. Interestingly, the density of GFAP-positive astrocytes is decreased in the prefrontal cortex of alcohol-preferring rats compared to non-preferring rats [236, 237], which may be either coincidental or meaningful to the same finding in prefrontal cortex of depressed patients, who also show decreased GFAP protein and mRNA expression in prefrontal cortex, and have a high comorbidity with alcoholism [238–242]. The link between astrocyte dysfunction and neuronal signaling is not yet clear in models of addiction; however, there is evidence from genetic screens and manipulations to suggest that it involves glutamate signaling. Notably, a single nucleotide mutation in the EAAT2 (GLT-1) gene has been linked to impulsive behaviors in alcoholics [243]. Furthermore, mutations in the circadian clock gene, *Per2*, lead to increased alcohol consumption in mice, coincident with decreased expression of another glial glutamate transporter, EAAT1 [244]. This phenotype is consistent with reports of associations between *Per2* mutations and increased alcohol intake in humans.

Both astrocytes and microglia have been recognized as key players in neuroinflammation and, as stated above, are often characterized as "reactive" based on changes in morphology and immunohistochemical profile [245]. However, dysfunction in glia–neuron signaling during neuroinflammation has only recently received attention as a key contributor to neuropathological conditions. Toll-like receptor (TLR) type 4, which is responsible for immediate detection and response to insults to the immune system, is mainly expressed in glial cells [246] and is receiving increasing attention for its role in pathology associated with neurodegenerative diseases and chronic alcohol exposure [247–250]. For example, astrocyte and microglia activation and associated impairments in memory and anxiety behaviors that typically occur following chronic alcohol exposure were attenuated in mice lacking expression of TLR4 [248]. Though TLR4 is expressed on microglia and astrocytes, and upregulated in response to a neuroimmune challenge (including chronic alcohol exposure), a recent study using mixed glial cultures with or without microglia depletion showed that the astrocytic response to lipopolysaccharide, an activator of TLR4, was completely dependent on the presence of microglia [251]. This study suggests that activation of TLR4 on microglia precedes and induces subsequent activation of astrocytes in vitro. Further investigation is required to examine the extent to which microglia TLR4 signaling impacts astrocyte response in vivo.

2.7 Summary

Our understanding of astrocyte–neuron signaling mechanisms is ever-increasing. It is now widely appreciated that astrocytes' contribution to brain function extends beyond the basic extracellular "housekeeping" tasks to include modulation of neuronal communication at the synaptic, network, and behavioral levels. As a component of the tripartite synapse, astrocytes are capable of fine-tuning synaptic transmission via calcium-dependent gliotransmitter release, dynamically regulating the physiological range required for synaptic plasticity. Intercellular calcium signaling, or calcium waves, among astrocytes helps to coordinate neuronal activity and neurovascular resources to satisfy the metabolic demands of network activity. Recent advances in molecular genetic techniques, including transgenic mice with astrocyte-specific knockout, overexpression or mutation of genes, and in vivo application of optogenetics, especially when used in combination with rigorous electrophysiological, real-time imaging, and behavioral assays, are allowing glial biologists to more directly probe the functions of astrocytes. From studies published so far, there is now mounting evidence showing direct involvement of astrocytes in several behaviors, including sleep, circadian rhythms, memory, and respiration. It is also becoming increasingly clear, both in experimental models and clinical cases, that astrocytes play a role in neuropathological conditions, including neurodegeneration, neuroinflammation, and addiction. Any remaining debate in regard to the significance of astrocyte–neuron communication should serve as an impetus to set higher and more specified experimental standards, incorporating in vivo techniques with acute slice preparations and pharmacology, and importantly sampling a range of electrophysiological parameters which may be more physiologically relevant to conditions in which astrocytes modulate synaptic activity. In summary, the understanding that astrocytes contribute to overall brain function, but do so by continuously monitoring and tuning network activity rather than directly relaying it (as neurons do), is becoming an essential component to fundamental neuroscience and may be a key to the progression toward therapeutics for neurological conditions.

2.8 Conclusion

Over the past decade, discoveries in glial biology have brought forth a new era in brain research, moving beyond a neuron-centric approach to encompass the increasingly defined impact of glial signaling. Glia not only play an active role in modulating synaptic activity but, through intercellular communication, serve to integrate information at the network level. In this manner, glia–neuron signaling has a much more significant impact on both physiological and pathological conditions than what was initially thought. For example, glial cell involvement in the onset and/or progression of pathology in many neurodegenerative diseases is now widely accepted. Though the role of astrocytes in drug and alcohol abuse is only beginning to be understood, progress made thus far has provided a greater potential for insight into mechanisms of addiction. Astrocytes express proteins known to be impacted by drugs and alcohol, but the signaling pathways that contribute to addiction are not known. Given the ever-increasing development of techniques for probing glial function, and what is currently known regarding the role of astrocytes in neurodegenerative diseases and addiction, the future holds the possibility for identifying novel diagnostic techniques and therapeutics.

References

1. Kettenmann H, Verkhratsky A (2008) Neuroglia: the 150 years after. Trends Neurosci 31(12):653–659
2. Magistretti PJ, Pellerin L (1999) Astrocytes couple synaptic activity to glucose utilization in the brain. News Physiol Sci 14:177–182
3. Banaclocha MA (2007) Neuromagnetic dialogue between neuronal minicolumns and astroglial network: a new approach for memory and cerebral computation. Brain Res Bull 73(1–3):21–27
4. Garcia-Segura LM, Chowen JA, Duenas M, Torres-Aleman I, Naftolin F (1994) Gonadal steroids as promoters of neuro-glial plasticity. Psychoneuroendocrinology 19(5–7):445–453
5. Blutstein T, Baab PJ, Zielke HR, Mong JA (2009) Hormonal modulation of amino acid neurotransmitter metabolism in the arcuate nucleus of the adult female rat: a novel action of estradiol. Endocrinology 150(7):3237–3244
6. Haydon PG, Blendy J, Moss SJ, Rob JF (2009) Astrocytic control of synaptic transmission and plasticity: a target for drugs of abuse? Neuropharmacology 56(suppl 1):83–90
7. Bushong EA, Martone ME, Jones YZ, Ellisman MH (2002) Protoplasmic astrocytes in CA1 stratum radiatum occupy separate anatomical domains. J Neurosci 22(1):183–192
8. Li D, Ropert N, Koulakoff A, Giaume C, Oheim M (2008) Lysosomes are the major vesicular compartment undergoing Ca2+-regulated exocytosis from cortical astrocytes. J Neurosci 28(30):7648–7658
9. Liu T, Sun L, Xiong Y, Shang S, Guo N, Teng S et al (2011) Calcium triggers exocytosis from two types of organelles in a single astrocyte. J Neurosci 31(29):10593–10601
10. Montana V, Malarkey EB, Verderio C, Matteoli M, Parpura V (2006) Vesicular transmitter release from astrocytes. Glia 54(7):700–715
11. Zhang Q, Pangrsic T, Kreft M, Krzan M, Li N, Sul JY et al (2004) Fusion-related release of glutamate from astrocytes. J Biol Chem 279(13):12724–12733
12. Agulhon C, Petravicz J, McMullen AB, Sweger EJ, Minton SK, Taves SR et al (2008) What is the role of astrocyte calcium in neurophysiology? Neuron 59(6):932–946

13. Agulhon C, Fiacco TA, McCarthy KD (2010) Hippocampal short- and long-term plasticity are not modulated by astrocyte Ca2+ signaling. Science 327(5970):1250–1254
14. Fiacco TA, Agulhon C, Taves SR, Petravicz J, Casper KB, Dong X et al (2007) Selective stimulation of astrocyte calcium in situ does not affect neuronal excitatory synaptic activity. Neuron 54(4):611–626
15. Fiacco TA, Agulhon C, McCarthy KD (2009) Sorting out astrocyte physiology from pharmacology. Annu Rev Pharmacol Toxicol 49:151–174
16. Petravicz J, Fiacco TA, McCarthy KD (2008) Loss of IP3 receptor-dependent Ca2+ increases in hippocampal astrocytes does not affect baseline CA1 pyramidal neuron synaptic activity. J Neurosci 28(19):4967–4973
17. Watkins JC (2000) L-glutamate as a central neurotransmitter: looking back. Biochem Soc Trans 28(4):297–309
18. Bezzi P, Carmignoto G, Pasti L, Vesce S, Rossi D, Rizzini BL et al (1998) Prostaglandins stimulate calcium-dependent glutamate release in astrocytes. Nature 391(6664):281–285
19. Jeftinija SD, Jeftinija KV, Stefanovic G, Liu F (1996) Neuroligand-evoked calcium-dependent release of excitatory amino acids from cultured astrocytes. J Neurochem 66(2):676–684
20. Parpura V, Basarsky TA, Liu F, Jeftinija K, Jeftinija S, Haydon PG (1994) Glutamate-mediated astrocyte-neuron signalling. Nature 369(6483):744–747
21. Parpura V, Fang Y, Basarsky T, Jahn R, Haydon PG (1995) Expression of synaptobrevin II, cellubrevin and syntaxin but not SNAP-25 in cultured astrocytes. FEBS Lett 377(3):489–492
22. Hamilton NB, Attwell D (2010) Do astrocytes really exocytose neurotransmitters? Nat Rev Neurosci 11(4):227–238
23. Fellin T, Pascual O, Gobbo S, Pozzan T, Haydon PG, Carmignoto G (2004) Neuronal synchrony mediated by astrocytic glutamate through activation of extrasynaptic NMDA receptors. Neuron 43(5):729–743
24. Pasti L, Volterra A, Pozzan T, Carmignoto G (1997) Intracellular calcium oscillations in astrocytes: a highly plastic, bidirectional form of communication between neurons and astrocytes in situ. J Neurosci 17(20):7817–7830
25. Santello M, Volterra A (2009) Synaptic modulation by astrocytes via Ca2+-dependent glutamate release. Neuroscience 158(1):253–259
26. Montana V, Ni Y, Sunjara V, Hua X, Parpura V (2004) Vesicular glutamate transporter-dependent glutamate release from astrocytes. J Neurosci 24(11):2633–2642
27. Ormel L, Stensrud MJ, Bergersen LH, Gundersen V (2012) VGLUT1 is localized in astrocytic processes in several brain regions. Glia 60(2):229–238
28. Zhang Q, Fukuda M, Van Bockstaele E, Pascual O, Haydon PG (2004) Synaptotagmin IV regulates glial glutamate release. Proc Natl Acad Sci U S A 101(25):9441–9446
29. Aoki C, Milner TA, Sheu KF, Blass JP, Pickel VM (1987) Regional distribution of astrocytes with intense immunoreactivity for glutamate dehydrogenase in rat brain: implications for neuron-glia interactions in glutamate transmission. J Neurosci 7(7):2214–2231
30. Kaneko T, Shigemoto R, Mizuno N (1988) Metabolism of glutamate and ammonia in astrocyte: an immunocytochemical study. Brain Res 457(1):160–164
31. Madl JE, Clements JR, Beitz AJ, Wenthold RJ, Larson AA (1988) Immunocytochemical localization of glutamate dehydrogenase in mitochondria of the cerebellum: an ultrastructural study using a monoclonal antibody. Brain Res 452(1–2):396–402
32. Wenthold RJ, Altschuler RA, Skaggs KK, Reeks KA (1987) Immunocytochemical characterization of glutamate dehydrogenase in the cerebellum of the rat. J Neurochem 48(2):636–643
33. Rothe F, Brosz M, Storm-Mathisen J (1995) Quantitative ultrastructural localization of glutamate dehydrogenase in the rat cerebellar cortex. Neuroscience 64(4):iii–xvi
34. Rothe F, Brosz M, Storm-Mathisen J (1994) Quantitative ultrastructural localization of glutamate dehydrogenase in the rat cerebellar cortex. Neuroscience 62(4):1133–1146
35. Anderson CM, Swanson RA (2000) Astrocyte glutamate transport: review of properties, regulation, and physiological functions. Glia 32(1):1–14

36. Danbolt NC (2001) Glutamate uptake. Prog Neurobiol 65(1):1–105
37. Bezzi P, Domercq M, Brambilla L, Galli R, Schols D, De Clercq E et al (2001) CXCR4-activated astrocyte glutamate release via TNFalpha: amplification by microglia triggers neurotoxicity. Nat Neurosci 4(7):702–710
38. Bal-Price A, Moneer Z, Brown GC (2002) Nitric oxide induces rapid, calcium-dependent release of vesicular glutamate and ATP from cultured rat astrocytes. Glia 40(3):312–323
39. Coco S, Calegari F, Pravettoni E, Pozzi D, Taverna E, Rosa P et al (2003) Storage and release of ATP from astrocytes in culture. J Biol Chem 278(2):1354–1362
40. Cali C, Bezzi P (2010) CXCR4-mediated glutamate exocytosis from astrocytes. J Neuroimmunol 224(1–2):13–21
41. Santello M, Bezzi P, Volterra A (2011) TNFalpha controls glutamatergic gliotransmission in the hippocampal dentate gyrus. Neuron 69(5):988–1001
42. Domercq M, Brambilla L, Pilati E, Marchaland J, Volterra A, Bezzi P (2006) P2Y1 receptor-evoked glutamate exocytosis from astrocytes: control by tumor necrosis factor-alpha and prostaglandins. J Biol Chem 281(41):30684–30696
43. Angulo MC, Kozlov AS, Charpak S, Audinat E (2004) Glutamate released from glial cells synchronizes neuronal activity in the hippocampus. J Neurosci 24(31):6920–6927
44. Araque A, Sanzgiri RP, Parpura V, Haydon PG (1999) Astrocyte-induced modulation of synaptic transmission. Can J Physiol Pharmacol 77(9):699–706
45. Fellin T, D'Ascenzo M, Haydon PG (2007) Astrocytes control neuronal excitability in the nucleus accumbens. Sci World J 7:89–97
46. Haydon PG, Carmignoto G (2006) Astrocyte control of synaptic transmission and neurovascular coupling. Physiol Rev 86(3):1009–1031
47. Jourdain P, Bergersen LH, Bhaukaurally K, Bezzi P, Santello M, Domercq M et al (2007) Glutamate exocytosis from astrocytes controls synaptic strength. Nat Neurosci 10(3):331–339
48. Kozlov AS, Angulo MC, Audinat E, Charpak S (2006) Target cell-specific modulation of neuronal activity by astrocytes. Proc Natl Acad Sci USA 103(26):10058–10063
49. Panatier A, Theodosis DT, Mothet JP, Touquet B, Pollegioni L, Poulain DA et al (2006) Glia-derived D-serine controls NMDA receptor activity and synaptic memory. Cell 125(4):775–784
50. Nedergaard M, Verkhratsky A (2012) Artifact versus reality – how astrocytes contribute to synaptic events? Glia 60(7):1013–1023
51. Navarrete M, Perea G, de Sevilla DF, Gomez-Gonzalo M, Nunez A, Martin ED et al (2012) Astrocytes mediate in vivo cholinergic-induced synaptic plasticity. PLoS Biol 10(2):e1001259
52. Williams SM, Diaz CM, Macnab LT, Sullivan RKP, Pow DV (2006) Immunocytochemical analysis of D-serine distribution in the mammalian brain reveals novel anatomical compartmentalizations in glia and neurons. Glia 53(4):401–411
53. Wolosker H, Blackshaw S, Snyder SH (1999) Serine racemase: A glial enzyme synthesizing D-serine to regulate glutamate-N-methyl-D-aspartate neurotransmission. Proc Natl Acad Sci USA 96(23):13409–13414
54. Johnson JW, Ascher P (1987) Glycine potentiates the Nmda response in cultured mouse-brain neurons. Nature 325(6104):529–531
55. Schell MJ, Molliver ME, Snyder SH (1995) D-Serine, an endogenous synaptic modulator – localization to astrocytes and glutamate-stimulated release. Proc Natl Acad Sci USA 92(9):3948–3952
56. Mothet JP, Parent AT, Wolosker H, Brady RO, Linden DJ, Ferris CD et al (2000) D-Serine is an endogenous ligand for the glycine site of the N-methyl-D-aspartate receptor. Proc Natl Acad Sci USA 97(9):4926–4931
57. Mothet JP, Pollegioni L, Ouanounou G, Martineau M, Fossier P, Baux G (2005) Glutamate receptor activation triggers a calcium-dependent and SNARE protein-dependent release of the gliotransmitter D-serine. Proc Natl Acad Sci USA 102(15):5606–5611
58. Bergersen LH, Morland C, Ormel L, Rinholm JE, Larsson M, Wold JF et al (2011) Immunogold detection of L-glutamate and D-serine in small synaptic-like microvesicles in adult hippocampal astrocytes. Cereb Cortex 22(7):1690–1697

59. Martineau M, Galli T, Baux G, Mothet JP (2008) Confocal imaging and tracking of the exocytotic routes for D-serine-mediated gliotransmission. Glia 56(12):1271–1284
60. Wolosker H (2011) Serine racemase and the serine shuttle between neurons and astrocytes. Biochim Biophys Acta-Proteins Proteomics 1814(11):1558–1566
61. Rosenberg D, Kartvelishvily E, Shleper M, Klinker CMC, Bowser MT, Wolosker H (2010) Neuronal release of D-serine: a physiological pathway controlling extracellular D-serine concentration. FASEB J 24(8):2951–2961
62. Yoshikawa M, Takayasu N, Hashimoto A, Sato Y, Tamaki R, Tsukamoto H et al (2007) The serine racemase mRNA is predominantly expressed in rat brain neurons. Arch Histol Cytol 70(2):127–134
63. Miya K, Inoue R, Takata Y, Abe M, Natsume R, Sakimura K et al (2008) Serine racemase is predominantly localized in neurons in mouse brain. J Comp Neurol 510(6):641–654
64. Ding XH, Ma N, Nagahama M, Yamada K, Semba R (2011) Localization of D-serine and serine racemase in neurons and neuroglias in mouse brain. Neurol Sci 32(2):263–267
65. Henneberger C, Papouin T, Oliet SHR, Rusakov DA (2010) Long-term potentiation depends on release of D-serine from astrocytes. Nature 463(7278):232–U120
66. Fossat P, Turpin FR, Sacchi S, Dulong J, Shi T, Rivet JM et al (2011) Glial D-serine gates NMDA receptors at excitatory synapses in prefrontal cortex. Cereb Cortex 22(3):595–606
67. Benneyworth MA, Li Y, Basu AC, Bolshakov VY, Coyle JT (2012) Cell selective conditional null mutations of serine racemase demonstrate a predominate localization in cortical glutamatergic neurons. Cell Mol Neurobiol 32(4):613–624
68. Lake N (1992) Taurine, Gaba and Gfap immunoreactivity in the developing and adult-rat optic-nerve. Brain Res 596(1–2):124–132
69. Ochi S, Lim JY, Rand MN, During MJ, Sakatani K, Kocsis JD (1993) Transient presence of Gaba in astrocytes of the developing optic-nerve. Glia 9(3):188–198
70. Lee M, Schwab C, Mcgeer PL (2011) Astrocytes are gabaergic cells that modulate microglial activity. Glia 59(1):152–165
71. Sakatani K, Hassan AZ, Ching W (1991) Age-dependent extrasynaptic modulation of axonal conduction by exogenous and endogenous Gaba in the rat optic-nerve. Exp Neurol 114(3): 307–314
72. Kang J, Jiang L, Goldman SA, Nedergaard M (1998) Astrocyte-mediated potentiation of inhibitory synaptic transmission. Nat Neurosci 1(8):683–692
73. Farrant M, Nusser Z (2005) Variations on an inhibitory theme: phasic and tonic activation of GABA(A) receptors. Nat Rev Neurosci 6(3):215–229
74. Jia F, Pignataro L, Schofield CM, Yue M, Harrison NL, Goldstein PA (2005) An extrasynaptic GABA(A) receptor mediates tonic inhibition in thalamic VB neurons. J Neurophysiol 94(6):4491–4501
75. Cope DW, Hughes SW, Crunelli V (2005) GABA(A) receptor-mediated tonic inhibition in thalamic neurons. J Neurosci 25(50):11553–11563
76. Martin LJ, Zurek AA, MacDonald JF, Roder JC, Jackson MF, Orser BA (2010) Alpha 5GABA(A) receptor activity sets the threshold for long-term potentiation and constrains hippocampus-dependent memory. J Neurosci 30(15):5269–5282
77. Fleming RL, Acheson SK, Moore SD, Wilson WA, Swartzwelder HS (2011) GABA transport modulates the ethanol sensitivity of tonic inhibition in the rat dentate gyrus. Alcohol 45(6): 577–583
78. Lee S, Yoon BE, Berglund K, Oh SJ, Park H, Shin HS et al (2010) Channel-mediated tonic GABA release from glia. Science 330(6005):790–796
79. Yoon BE, Jo S, Woo J, Lee JH, Kim T, Kim D et al (2011) The amount of astrocytic GABA positively correlates with the degree of tonic inhibition in hippocampal CA1 and cerebellum. Mol Brain 4(1):42
80. Lee M, Mcgeer EG, Mcgeer PL (2011) Mechanisms of GABA release from human astrocytes. Glia 59(11):1600–1611
81. Maienschein V, Marxen M, Volknandt W, Zimmermann H (1999) A plethora of presynaptic proteins associated with ATP-storing organelles in cultured astrocytes. Glia 26(3):233–244

82. Cotrina ML, Lin JH, Nedergaard M (1998) Cytoskeletal assembly and ATP release regulate astrocytic calcium signaling. J Neurosci 18(21):8794–8804
83. Wang Z, Haydon PG, Yeung ES (2000) Direct observation of calcium-independent intercellular ATP signaling in astrocytes. Anal Chem 72(9):2001–2007
84. Zhang JM, Wang HK, Ye CQ, Ge W, Chen Y, Jiang ZL et al (2003) ATP released by astrocytes mediates glutamatergic activity-dependent heterosynaptic suppression. Neuron 40(5):971–982
85. Guthrie PB, Knappenberger J, Segal M, Bennett MV, Charles AC, Kater SB (1999) ATP released from astrocytes mediates glial calcium waves. J Neurosci 19(2):520–528
86. Pascual O, Casper KB, Kubera C, Zhang J, Revilla-Sanchez R, Sul JY et al (2005) Astrocytic purinergic signaling coordinates synaptic networks. Science 310(5745):113–116
87. Illes P, Verkhratsky A, Burnstock G, Franke H (2011) P2X receptors and their roles in astroglia in the central and peripheral nervous system. Neuroscientist
88. Peterson TS, Camden JM, Wang Y, Seye CI, Wood WG, Sun GY et al (2010) P2Y2 nucleotide receptor-mediated responses in brain cells. Mol Neurobiol 41(2–3):356–366
89. Tozaki-Saitoh H, Tsuda M, Inoue K (2011) Role of purinergic receptors in CNS function and neuroprotection. Adv Pharmacol 61:495–528
90. James G, Butt AM (2002) P2Y and P2X purinoceptor mediated Ca2+ signalling in glial cell pathology in the central nervous system. Eur J Pharmacol 447(2–3):247–260
91. Halassa MM, Florian C, Fellin T, Munoz JR, Lee SY, Abel T et al (2009) Astrocytic modulation of sleep homeostasis and cognitive consequences of sleep loss. Neuron 61(2):213–219
92. Panatier A, Vallee J, Haber M, Murai KK, Lacaille JC, Robitaille R (2011) Astrocytes are endogenous regulators of basal transmission at central synapses. Cell 146(5):785–798
93. Beattie EC, Stellwagen D, Morishita W, Bresnahan JC, Ha BK, von Zastrow M et al (2002) Control of synaptic strength by glial TNF alpha. Science 295(5563):2282–2285
94. Stellwagen D, Malenka RC (2006) Synaptic scaling mediated by glial TNF-alpha. Nature 440(7087):1054–1059
95. Henneberger C, Rusakov DA (2010) Synaptic plasticity and Ca(2+) signalling in astrocytes. Neuron Glia Biol 6(3):141–146
96. Parpura V, Grubisic V, Verkhratsky A (2011) Ca(2+) sources for the exocytotic release of glutamate from astrocytes. Biochim Biophys Acta 1813(5):984–991
97. Di Castro MA, Chuquet J, Liaudet N, Bhaukaurally K, Santello M, Bouvier D et al (2011) Local Ca2+ detection and modulation of synaptic release by astrocytes. Nat Neurosci 14(10):1276–1284
98. Shigetomi E, Kracun S, Khakh BS (2010) Monitoring astrocyte calcium microdomains with improved membrane targeted GCaMP reporters. Neuron Glia Biol 6(3):183–191
99. Filosa A, Paixao S, Honsek SD, Carmona MA, Becker L, Feddersen B et al (2009) Neuron-glia communication via EphA4/ephrin-A3 modulates LTP through glial glutamate transport. Nat Neurosci 12(10):1285–1292
100. Murai KK, Pasquale EB (2011) Eph receptors and ephrins in neuron-astrocyte communication at synapses. Glia 59(11):1567–1578
101. Navarrete M, Araque A (2008) Endocannabinoids mediate neuron-astrocyte communication. Neuron 57(6):883–893
102. Navarrete M, Araque A (2010) Endocannabinoids potentiate synaptic transmission through stimulation of astrocytes. Neuron 68(1):113–126
103. Han J, Kesner P, Metna-Laurent M, Duan T, Xu L, Georges F et al (2012) Acute cannabinoids impair working memory through astroglial CB(1) receptor modulation of hippocampal LTD. Cell 148(5):1039–1050
104. Saadoun S, Papadopoulos MC (2010) Aquaporin-4 in brain and spinal cord oedema. Neuroscience 168(4):1036–1046
105. Skucas VA, Mathews IB, Yang J, Cheng Q, Treister A, Duffy AM et al (2011) Impairment of select forms of spatial memory and neurotrophin-dependent synaptic plasticity by deletion of glial aquaporin-4. J Neurosci 31(17):6392–6397
106. Halassa MM, Fellin T, Takano H, Dong JH, Haydon PG (2007) Synaptic islands defined by the territory of a single astrocyte. J Neurosci 27(24):6473–6477

107. Volterra A, Meldolesi J (2005) Astrocytes, from brain glue to communication elements: the revolution continues. Nat Rev Neurosci 6(8):626–640
108. Rose CR, Ransom BR (1997) Gap junctions equalize intracellular Na+ concentration in astrocytes. Glia 20(4):299–307
109. Nedergaard M (1994) Direct signaling from astrocytes to neurons in cultures of mammalian brain cells. Science 263(5154):1768–1771
110. Cornell-Bell AH, Finkbeiner SM, Cooper MS, Smith SJ (1990) Glutamate induces calcium waves in cultured astrocytes: long-range glial signaling. Science 247(4941):470–473
111. Scemes E, Giaume C (2006) Astrocyte calcium waves: what they are and what they do. Glia 54(7):716–725
112. Giaume C, Venance L (1998) Intercellular calcium signaling and gap junctional communication in astrocytes. Glia 24(1):50–64
113. Goldberg M, De Pitta M, Volman V, Berry H, Ben Jacob E (2010) Nonlinear gap junctions enable long-distance propagation of pulsating calcium waves in astrocyte networks. PLoS Comput Biol 6(8). pii:e1000909
114. Kuga N, Sasaki T, Takahara Y, Matsuki N, Ikegaya Y (2011) Large-scale calcium waves traveling through astrocytic networks in vivo. J Neurosci 31(7):2607–2614
115. Houades V, Koulakoff A, Ezan P, Seif I, Giaume C (2008) Gap junction-mediated astrocytic networks in the mouse barrel cortex. J Neurosci 28(20):5207–5217
116. Roux L, Benchenane K, Rothstein JD, Bonvento G, Giaume C (2011) Plasticity of astroglial networks in olfactory glomeruli. Proc Natl Acad Sci USA 108(45):18442–18446
117. Pannasch U, Vargova L, Reingruber J, Ezan P, Holcman D, Giaume C et al (2011) Astroglial networks scale synaptic activity and plasticity. Proc Natl Acad Sci USA 108(20):8467–8472
118. Attwell D, Laughlin SB (2001) An energy budget for signaling in the grey matter of the brain. J Cereb Blood Flow Metab 21(10):1133–1145
119. Magistretti PJ, Sorg O, Naichen Y, Pellerin L, de Rham S, Martin JL (1994) Regulation of astrocyte energy metabolism by neurotransmitters. Ren Physiol Biochem 17(3–4):168–171
120. Pellerin L, Magistretti PJ (2011) Sweet sixteen for ANLS. J Cereb Blood Flow Metab 32(7):1152–1166
121. Sokoloff L, Takahashi S, Gotoh J, Driscoll BF, Law MJ (1996) Contribution of astroglia to functionally activated energy metabolism. Dev Neurosci 18(5–6):344–352
122. Pellerin L, Bouzier-Sore AK, Aubert A, Serres S, Merle M, Costalat R et al (2007) Activity-dependent regulation of energy metabolism by astrocytes: an update. Glia 55(12):1251–1262
123. Kacem K, Lacombe P, Seylaz J, Bonvento G (1998) Structural organization of the perivascular astrocyte endfeet and their relationship with the endothelial glucose transporter: a confocal microscopy study. Glia 23(1):1–10
124. Herard AS, Dubois A, Escartin C, Tanaka K, Delzescaux T, Hantraye P et al (2005) Decreased metabolic response to visual stimulation in the superior colliculus of mice lacking the glial glutamate transporter GLT-1. Eur J Neurosci 22(7):1807–1811
125. Pellerin L, Magistretti PJ (1994) Glutamate uptake into astrocytes stimulates aerobic glycolysis: a mechanism coupling neuronal activity to glucose utilization. Proc Natl Acad Sci USA 91(22):10625–10629
126. Schousboe A, Sickmann HM, Walls AB, Bak LK, Waagepetersen HS (2010) Functional importance of the astrocytic glycogen-shunt and glycolysis for maintenance of an intact intra/extracellular glutamate gradient. Neurotox Res 18(1):94–99
127. Voutsinos-Porche B, Bonvento G, Tanaka K, Steiner P, Welker E, Chatton JY et al (2003) Glial glutamate transporters mediate a functional metabolic crosstalk between neurons and astrocytes in the mouse developing cortex. Neuron 37(2):275–286
128. Bouzier-Sore AK, Merle M, Magistretti PJ, Pellerin L (2002) Feeding active neurons: (re) emergence of a nursing role for astrocytes. J Physiol Paris 96(3–4):273–282
129. Bouzier-Sore AK, Serres S, Canioni P, Merle M (2003) Lactate involvement in neuron-glia metabolic interaction: (13)C-NMR spectroscopy contribution. Biochimie 85(9):841–848
130. Itoh Y, Esaki T, Shimoji K, Cook M, Law MJ, Kaufman E et al (2003) Dichloroacetate effects on glucose and lactate oxidation by neurons and astroglia in vitro and on glucose utilization by brain in vivo. Proc Natl Acad Sci USA 100(8):4879–4884

131. Ivanov A, Mukhtarov M, Bregestovski P, Zilberter Y (2011) Lactate effectively covers energy demands during neuronal network activity in neonatal hippocampal slices. Front Neuroenerg 3:2
132. Rouach N, Koulakoff A, Abudara V, Willecke K, Giaume C (2008) Astroglial metabolic networks sustain hippocampal synaptic transmission. Science 322(5907):1551–1555
133. Giaume C, Koulakoff A, Roux L, Holcman D, Rouach N (2010) Astroglial networks: a step further in neuroglial and gliovascular interactions. Nat Rev Neurosci 11(2):87–99
134. Garcia-Marin V, Garcia-Lopez P, Freire M (2007) Cajal's contributions to glia research. Trends Neurosci 30(9):479–487
135. Florian C, Vecsey CG, Halassa MM, Haydon PG, Abel T (2011) Astrocyte-derived adenosine and A1 receptor activity contribute to sleep loss-induced deficits in hippocampal synaptic plasticity and memory in mice. J Neurosci 31(19):6956–6962
136. Bachmann V, Klaus F, Bodenmann S, Schafer N, Brugger P, Huber S et al (2011) Functional ADA polymorphism increases sleep depth and reduces vigilant attention in humans. Cereb Cortex 22(4):962–970
137. Retey JV, Adam M, Honegger E, Khatami R, Luhmann UF, Jung HH et al (2005) A functional genetic variation of adenosine deaminase affects the duration and intensity of deep sleep in humans. Proc Natl Acad Sci USA 102(43):15676–15681
138. Okada T, Mochizuki T, Huang ZL, Eguchi N, Sugita Y, Urade Y et al (2003) Dominant localization of adenosine deaminase in leptomeninges and involvement of the enzyme in sleep. Biochem Biophys Res Commun 312(1):29–34
139. Franken P, Chollet D, Tafti M (2001) The homeostatic regulation of sleep need is under genetic control. J Neurosci 21(8):2610–2621
140. Fredholm BB, Chen JF, Cunha RA, Svenningsson P, Vaugeois JM (2005) Adenosine and brain function. Int Rev Neurobiol 63:191–270
141. Gourine AV, Kasymov V, Marina N, Tang F, Figueiredo MF, Lane S et al (2010) Astrocytes control breathing through pH-dependent release of ATP. Science 329(5991):571–575
142. Schnell C, Fresemann J, Hulsmann S (2011) Determinants of functional coupling between astrocytes and respiratory neurons in the pre-Botzinger complex. PLoS One 6(10):e26309
143. Suh J, Jackson FR (2007) Drosophila ebony activity is required in glia for the circadian regulation of locomotor activity. Neuron 55(3):435–447
144. Marpegan L, Swanstrom AE, Chung K, Simon T, Haydon PG, Khan SK et al (2011) Circadian regulation of ATP release in astrocytes. J Neurosci 31(23):8342–8350
145. Suzuki A, Stern SA, Bozdagi O, Huntley GW, Walker RH, Magistretti PJ et al (2011) Astrocyte-neuron lactate transport is required for long-term memory formation. Cell 144(5):810–823
146. Avital A, Goshen I, Kamsler A, Segal M, Iverfeldt K, Richter-Levin G et al (2003) Impaired interleukin-1 signaling is associated with deficits in hippocampal memory processes and neural plasticity. Hippocampus 13(7):826–834
147. Ban EM, Sarlieve LL, Haour FG (1993) Interleukin-1 binding sites on astrocytes. Neuroscience 52(3):725–733
148. Goshen I, Kreisel T, Ounallah-Saad H, Renbaum P, Zalzstein Y, Ben Hur T et al (2007) A dual role for interleukin-1 in hippocampal-dependent memory processes. Psychoneuroendocrinology 32(8–10):1106–1115
149. Menachem-Zidon O, Avital A, Ben Menahem Y, Goshen I, Kreisel T, Shmueli EM et al (2011) Astrocytes support hippocampal-dependent memory and long-term potentiation via interleukin-1 signaling. Brain Behav Immun 25(5):1008–1016
150. Albensi BC, Mattson MP (2000) Evidence for the involvement of TNF and NF-kappaB in hippocampal synaptic plasticity. Synapse 35(2):151–159
151. Meffert MK, Chang JM, Wiltgen BJ, Fanselow MS, Baltimore D (2003) NF-kappa B functions in synaptic signaling and behavior. Nat Neurosci 6(10):1072–1078
152. O'Mahony A, Raber J, Montano M, Foehr E, Han V, Lu SM et al (2006) NF-kappaB/Rel regulates inhibitory and excitatory neuronal function and synaptic plasticity. Mol Cell Biol 26(19):7283–7298

153. Yu Z, Cheng G, Wen X, Wu GD, Lee WT, Pleasure D (2002) Tumor necrosis factor alpha increases neuronal vulnerability to excitotoxic necrosis by inducing expression of the AMPA-glutamate receptor subunit GluR1 via an acid sphingomyelinase- and NF-kappaB-dependent mechanism. Neurobiol Dis 11(1):199–213
154. Bracchi-Ricard V, Brambilla R, Levenson J, Hu WH, Bramwell A, Sweatt JD et al (2008) Astroglial nuclear factor-kappaB regulates learning and memory and synaptic plasticity in female mice. J Neurochem 104(3):611–623
155. Sofroniew MV, Bush TG, Blumauer N, Lawrence K, Mucke L, Johnson MH (1999) Genetically-targeted and conditionally-regulated ablation of astroglial cells in the central, enteric and peripheral nervous systems in adult transgenic mice. Brain Res 835(1):91–95
156. Herrmann JE, Imura T, Song B, Qi J, Ao Y, Nguyen TK et al (2008) STAT3 is a critical regulator of astrogliosis and scar formation after spinal cord injury. J Neurosci 28(28):7231–7243
157. Okada S, Nakamura M, Katoh H, Miyao T, Shimazaki T, Ishii K et al (2006) Conditional ablation of Stat3 or Socs3 discloses a dual role for reactive astrocytes after spinal cord injury. Nat Med 12(7):829–834
158. Brambilla R, Bracchi-Ricard V, Hu WH, Frydel B, Bramwell A, Karmally S et al (2005) Inhibition of astroglial nuclear factor kappaB reduces inflammation and improves functional recovery after spinal cord injury. J Exp Med 202(1):145–156
159. Brambilla R, Hurtado A, Persaud T, Esham K, Pearse DD, Oudega M et al (2009) Transgenic inhibition of astroglial NF-kappa B leads to increased axonal sparing and sprouting following spinal cord injury. J Neurochem 110(2):765–778
160. Dvoriantchikova G, Barakat D, Brambilla R, Agudelo C, Hernandez E, Bethea JR et al (2009) Inactivation of astroglial NF-kappa B promotes survival of retinal neurons following ischemic injury. Eur J Neurosci 30(2):175–185
161. Allaman I, Belanger M, Magistretti PJ (2011) Astrocyte-neuron metabolic relationships: for better and for worse. Trends Neurosci 34(2):76–87
162. Chung YC, Ko HW, Bok E, Park ES, Huh SH, Nam JH et al (2010) The role of neuroinflammation on the pathogenesis of Parkinson's disease. BMB Rep 43(4):225–232
163. Li C, Zhao R, Gao K, Wei Z, Yin MY, Lau LT et al (2011) Astrocytes: implications for neuroinflammatory pathogenesis of Alzheimer's disease. Curr Alzheimer Res 8(1):67–80
164. Rossi D, Volterra A (2009) Astrocytic dysfunction: insights on the role in neurodegeneration. Brain Res Bull 80(4–5):224–232
165. de Lanerolle NC, Lee TS, Spencer DD (2010) Astrocytes and epilepsy. Neurotherapeutics 7(4):424–438
166. Hoogland G, Spierenburg HA, van Veelen CW, van Rijen PC, van Huffelen AC, de Graan PN (2004) Synaptosomal glutamate and GABA transport in patients with temporal lobe epilepsy. J Neurosci Res 76(6):881–890
167. Tessler S, Danbolt NC, Faull RL, Storm-Mathisen J, Emson PC (1999) Expression of the glutamate transporters in human temporal lobe epilepsy. Neuroscience 88(4):1083–1091
168. Mathern GW, Mendoza D, Lozada A, Pretorius JK, Dehnes Y, Danbolt NC et al (1999) Hippocampal GABA and glutamate transporter immunoreactivity in patients with temporal lobe epilepsy. Neurology 52(3):453–472
169. Proper EA, Hoogland G, Kappen SM, Jansen GH, Rensen MG, Schrama LH et al (2002) Distribution of glutamate transporters in the hippocampus of patients with pharmaco-resistant temporal lobe epilepsy. Brain 125(Pt 1):32–43
170. Moreira JD, de Siqueira LV, Lague VM, Porciuncula LO, Vinade L, Souza DO (2011) Short-term alterations in hippocampal glutamate transport system caused by one-single neonatal seizure episode: implications on behavioral performance in adulthood. Neurochem Int 59(2):217–223
171. Binder DK, Steinhauser C (2006) Functional changes in astroglial cells in epilepsy. Glia 54(5):358–368
172. Zhang G, Raol YS, Hsu FC, Brooks-Kayal AR (2004) Long-term alterations in glutamate receptor and transporter expression following early-life seizures are associated with increased seizure susceptibility. J Neurochem 88(1):91–101

173. Ueda Y, Doi T, Tokumaru J, Yokoyama H, Nakajima A, Mitsuyama Y et al (2001) Collapse of extracellular glutamate regulation during epileptogenesis: down-regulation and functional failure of glutamate transporter function in rats with chronic seizures induced by kainic acid. J Neurochem 76(3):892–900
174. Tanaka K, Watase K, Manabe T, Yamada K, Watanabe M, Takahashi K et al (1997) Epilepsy and exacerbation of brain injury in mice lacking the glutamate transporter GLT-1. Science 276(5319):1699–1702
175. Simantov R, Crispino M, Hoe W, Broutman G, Tocco G, Rothstein JD et al (1999) Changes in expression of neuronal and glial glutamate transporters in rat hippocampus following kainate-induced seizure activity. Brain Res Mol Brain Res 65(1):112–123
176. Miller HP, Levey AI, Rothstein JD, Tzingounis AV, Conn PJ (1997) Alterations in glutamate transporter protein levels in kindling-induced epilepsy. J Neurochem 68(4):1564–1570
177. Takahashi DK, Vargas JR, Wilcox KS (2010) Increased coupling and altered glutamate transport currents in astrocytes following kainic-acid-induced status epilepticus. Neurobiol Dis 40(3):573–585
178. Petroff OA, Errante LD, Rothman DL, Kim JH, Spencer DD (2002) Glutamate-glutamine cycling in the epileptic human hippocampus. Epilepsia 43(7):703–710
179. Eid T, Thomas MJ, Spencer DD, Runden-Pran E, Lai JC, Malthankar GV et al (2004) Loss of glutamine synthetase in the human epileptogenic hippocampus: possible mechanism for raised extracellular glutamate in mesial temporal lobe epilepsy. Lancet 363(9402):28–37
180. Tani H, Dulla CG, Huguenard JR, Reimer RJ (2010) Glutamine is required for persistent epileptiform activity in the disinhibited neocortical brain slice. J Neurosci 30(4):1288–1300
181. Ortinski PI, Dong J, Mungenast A, Yue C, Takano H, Watson DJ et al (2010) Selective induction of astrocytic gliosis generates deficits in neuronal inhibition. Nat Neurosci 13(5):584–591
182. Alvestad S, Hammer J, Qu H, Haberg A, Ottersen OP, Sonnewald U (2011) Reduced astrocytic contribution to the turnover of glutamate, glutamine, and GABA characterizes the latent phase in the kainate model of temporal lobe epilepsy. J Cereb Blood Flow Metab 31(8): 1675–1686
183. Aronica E, Gorter JA, Jansen GH, van Veelen CW, van Rijen PC, Ramkema M et al (2003) Expression and cell distribution of group I and group II metabotropic glutamate receptor subtypes in taylor-type focal cortical dysplasia. Epilepsia 44(6):785–795
184. Steinhauser C, Seifert G (2002) Glial membrane channels and receptors in epilepsy: impact for generation and spread of seizure activity. Eur J Pharmacol 447(2–3):227–237
185. Fellin T, Gomez-Gonzalo M, Gobbo S, Carmignoto G, Haydon PG (2006) Astrocytic glutamate is not necessary for the generation of epileptiform neuronal activity in hippocampal slices. J Neurosci 26(36):9312–9322
186. Tian GF, Azmi H, Takano T, Xu Q, Peng W, Lin J et al (2005) An astrocytic basis of epilepsy. Nat Med 11(9):973–981
187. Ding S, Fellin T, Zhu Y, Lee SY, Auberson YP, Meaney DF et al (2007) Enhanced astrocytic Ca2+ signals contribute to neuronal excitotoxicity after status epilepticus. J Neurosci 27(40):10674–10684
188. Gomez-Gonzalo M, Losi G, Chiavegato A, Zonta M, Cammarota M, Brondi M et al (2010) An excitatory loop with astrocytes contributes to drive neurons to seizure threshold. PLoS Biol 8(4):e1000352
189. Li T, Lan JQ, Boison D (2008) Uncoupling of astrogliosis from epileptogenesis in adenosine kinase (ADK) transgenic mice. Neuron Glia Biol 4(2):91–99
190. Aronica E, Zurolo E, Iyer A, de Groot M, Anink J, Carbonell C et al (2011) Upregulation of adenosine kinase in astrocytes in experimental and human temporal lobe epilepsy. Epilepsia 52(9):1645–1655
191. Theofilas P, Brar S, Stewart KA, Shen HY, Sandau US, Poulsen D et al (2011) Adenosine kinase as a target for therapeutic antisense strategies in epilepsy. Epilepsia 52(3):589–601
192. Wallraff A, Kohling R, Heinemann U, Theis M, Willecke K, Steinhauser C (2006) The impact of astrocytic gap junctional coupling on potassium buffering in the hippocampus. J Neurosci 26(20):5438–5447

193. Oberheim NA, Tian GF, Han X, Peng W, Takano T, Ransom B et al (2008) Loss of astrocytic domain organization in the epileptic brain. J Neurosci 28(13):3264–3276
194. Clement AM, Nguyen MD, Roberts EA, Garcia ML, Boillee S, Rule M et al (2003) Wild-type nonneuronal cells extend survival of SOD1 mutant motor neurons in ALS mice. Science 302(5642):113–117
195. Faideau M, Kim J, Cormier K, Gilmore R, Welch M, Auregan G et al (2010) In vivo expression of polyglutamine-expanded huntingtin by mouse striatal astrocytes impairs glutamate transport: a correlation with Huntington's disease subjects. Hum Mol Genet 19(15):3053–3067
196. Halliday GM, Stevens CH (2011) Glia: initiators and progressors of pathology in Parkinson's disease. Mov Disord 26(1):6–17
197. Maragakis NJ, Rothstein JD (2006) Mechanisms of disease: astrocytes in neurodegenerative disease. Nat Clin Pract Neurol 2(12):679–689
198. Salmina AB (2009) Neuron-glia interactions as therapeutic targets in neurodegeneration. J Alzheimers Dis 16(3):485–502
199. Mallajosyula JK, Kaur D, Chinta SJ, Rajagopalan S, Rane A, Nicholls DG et al (2008) MAO-B elevation in mouse brain astrocytes results in Parkinson's pathology. PLoS One 3(2):e1616
200. Escartin C, Bonvento G (2008) Targeted activation of astrocytes: a potential neuroprotective strategy. Mol Neurobiol 38(3):231–241
201. Serrano-Pozo A, Mielke ML, Gomez-Isla T, Betensky RA, Growdon JH, Frosch MP et al (2011) Reactive glia not only associates with plaques but also parallels tangles in Alzheimer's disease. Am J Pathol 179(3):1373–1384
202. Wyss-Coray T (2006) Inflammation in Alzheimer disease: driving force, bystander or beneficial response? Nat Med 12(9):1005–1015
203. Wyss-Coray T, Loike JD, Brionne TC, Lu E, Anankov R, Yan F et al (2003) Adult mouse astrocytes degrade amyloid-beta in vitro and in situ. Nat Med 9(4):453–457
204. Mcgeer EG, Mcgeer PL (2010) Neuroinflammation in Alzheimer's disease and mild cognitive impairment: a field in its infancy. J Alzheimers Dis 19(1):355–361
205. Ralay RH, Craft JM, Hu W, Guo L, Wing LK, Van Eldik LJ et al (2006) Glia as a therapeutic target: selective suppression of human amyloid-beta-induced upregulation of brain proinflammatory cytokine production attenuates neurodegeneration. J Neurosci 26(2):662–670
206. Leuba G, Savioz A, Vernay A, Carnal B, Kraftsik R, Tardif E et al (2008) Differential changes in synaptic proteins in the Alzheimer frontal cortex with marked increase in PSD-95 postsynaptic protein. J Alzheimers Dis 15(1):139–151
207. Llano DA, Li J, Waring JF, Ellis T, Devanarayan V, Witte DG et al (2011) Cerebrospinal fluid cytokine dynamics differ between alzheimer disease patients and elderly controls. Alzheimer Dis Assoc Disord
208. Kuchibhotla KV, Lattarulo CR, Hyman BT, Bacskai BJ (2009) Synchronous hyperactivity and intercellular calcium waves in astrocytes in Alzheimer mice. Science 323(5918):1211–1215
209. Rossi D, Brambilla L, Valori CF, Crugnola A, Giaccone G, Capobianco R et al (2005) Defective tumor necrosis factor-alpha-dependent control of astrocyte glutamate release in a transgenic mouse model of Alzheimer disease. J Biol Chem 280(51):42088–42096
210. Stix B, Reiser G (1998) Beta-amyloid peptide 25-35 regulates basal and hormone-stimulated Ca2+ levels in cultured rat astrocytes. Neurosci Lett 243(1–3):121–124
211. Haughey NJ, Mattson MP (2003) Alzheimer's amyloid beta-peptide enhances ATP/gap junction-mediated calcium-wave propagation in astrocytes. Neuromolecular Med 3(3):173–180
212. Li S, Mallory M, Alford M, Tanaka S, Masliah E (1997) Glutamate transporter alterations in Alzheimer disease are possibly associated with abnormal APP expression. J Neuropathol Exp Neurol 56(8):901–911
213. Jacob CP, Koutsilieri E, Bartl J, Neuen-Jacob E, Arzberger T, Zander N et al (2007) Alterations in expression of glutamatergic transporters and receptors in sporadic Alzheimer's disease. J Alzheimers Dis 11(1):97–116
214. Jung ES, An K, Seok HH, Kim JH, Mook-Jung I (2012) Astrocyte-originated ATP protects abeta1-42-induced impairment of synaptic plasticity. J Neurosci 32(9):3081–3087

215. Rosen DR, Siddique T, Patterson D, Figlewicz DA, Sapp P, Hentati A et al (1993) Mutations in Cu/Zn superoxide dismutase gene are associated with familial amyotrophic lateral sclerosis. Nature 362(6415):59–62
216. Hall ED, Oostveen JA, Gurney ME (1998) Relationship of microglial and astrocytic activation to disease onset and progression in a transgenic model of familial ALS. Glia 23(3):249–256
217. Kawamata T, Akiyama H, Yamada T, Mcgeer PL (1992) Immunologic reactions in amyotrophic lateral sclerosis brain and spinal cord tissue. Am J Pathol 140(3):691–707
218. Turner MR, Cagnin A, Turkheimer FE, Miller CC, Shaw CE, Brooks DJ et al (2004) Evidence of widespread cerebral microglial activation in amyotrophic lateral sclerosis: an [11C](R)-PK11195 positron emission tomography study. Neurobiol Dis 15(3):601–609
219. Chiu IM, Phatnani H, Kuligowski M, Tapia JC, Carrasco MA, Zhang M et al (2009) Activation of innate and humoral immunity in the peripheral nervous system of ALS transgenic mice. Proc Natl Acad Sci USA 106(49):20960–20965
220. Nagy D, Kato T, Kushner PD (1994) Reactive astrocytes are widespread in the cortical gray matter of amyotrophic lateral sclerosis. J Neurosci Res 38(3):336–347
221. Kushner PD, Stephenson DT, Wright S (1991) Reactive astrogliosis is widespread in the subcortical white matter of amyotrophic lateral sclerosis brain. J Neuropathol Exp Neurol 50(3):263–277
222. Schiffer D, Cordera S, Cavalla P, Migheli A (1996) Reactive astrogliosis of the spinal cord in amyotrophic lateral sclerosis. J Neurol Sci 139(suppl):27–33
223. Boillee S, Yamanaka K, Lobsiger CS, Copeland NG, Jenkins NA, Kassiotis G et al (2006) Onset and progression in inherited ALS determined by motor neurons and microglia. Science 312(5778):1389–1392
224. Pramatarova A, Laganiere J, Roussel J, Brisebois K, Rouleau GA (2001) Neuron-specific expression of mutant superoxide dismutase 1 in transgenic mice does not lead to motor impairment. J Neurosci 21(10):3369–3374
225. Yamanaka K, Chun SJ, Boillee S, Fujimori-Tonou N, Yamashita H, Gutmann DH et al (2008) Astrocytes as determinants of disease progression in inherited amyotrophic lateral sclerosis. Nat Neurosci 11(3):251–253
226. Bristol LA, Rothstein JD (1996) Glutamate transporter gene expression in amyotrophic lateral sclerosis motor cortex. Ann Neurol 39(5):676–679
227. Sasaki S, Komori T, Iwata M (2000) Excitatory amino acid transporter 1 and 2 immunoreactivity in the spinal cord in amyotrophic lateral sclerosis. Acta Neuropathol 100(2):138–144
228. Ferraiuolo L, Higginbottom A, Heath PR, Barber S, Greenald D, Kirby J et al (2011) Dysregulation of astrocyte-motoneuron cross-talk in mutant superoxide dismutase 1-related amyotrophic lateral sclerosis. Brain 134(Pt 9):2627–2641
229. Philips T, Robberecht W (2011) Neuroinflammation in amyotrophic lateral sclerosis: role of glial activation in motor neuron disease. Lancet Neurol 10(3):253–263
230. West M, Mhatre M, Ceballos A, Floyd RA, Grammas P, Gabbita SP et al (2004) The arachidonic acid 5-lipoxygenase inhibitor nordihydroguaiaretic acid inhibits tumor necrosis factor alpha activation of microglia and extends survival of G93A-SOD1 transgenic mice. J Neurochem 91(1):133–143
231. Kiaei M, Petri S, Kipiani K, Gardian G, Choi DK, Chen J et al (2006) Thalidomide and lenalidomide extend survival in a transgenic mouse model of amyotrophic lateral sclerosis. J Neurosci 26(9):2467–2473
232. Beitner-Johnson D, Guitart X, Nestler EJ (1993) Glial fibrillary acidic protein and the mesolimbic dopamine system: regulation by chronic morphine and Lewis-Fischer strain differences in the rat ventral tegmental area. J Neurochem 61(5):1766–1773
233. Bruce-Keller AJ, Turchan-Cholewo J, Smart EJ, Geurin T, Chauhan A, Reid R et al (2008) Morphine causes rapid increases in glial activation and neuronal injury in the striatum of inducible HIV-1 Tat transgenic mice. Glia 56(13):1414–1427
234. Evrard SG, Duhalde-Vega M, Tagliaferro P, Mirochnic S, Caltana LR, Brusco A (2006) A low chronic ethanol exposure induces morphological changes in the adolescent rat brain that are

not fully recovered even after a long abstinence: an immunohistochemical study. Exp Neurol 200(2):438–459
235. Lewohl JM, Wixey J, Harper CG, Dodd PR (2005) Expression of MBP, PLP, MAG, CNP, and GFAP in the Human Alcoholic Brain. Alcohol Clin Exp Res 29(9):1698–1705
236. Miguel-Hidalgo JJ, Overholser JC, Meltzer HY, Stockmeier CA, Rajkowska G (2006) Reduced glial and neuronal packing density in the orbitofrontal cortex in alcohol dependence and its relationship with suicide and duration of alcohol dependence. Alcohol Clin Exp Res 30(11):1845–1855
237. Miguel-Hidalgo JJ (2006) Withdrawal from free-choice ethanol consumption results in increased packing density of glutamine synthetase-immunoreactive astrocytes in the prelimbic cortex of alcohol-preferring rats. Alcohol Alcohol 41(4):379–385
238. Miguel-Hidalgo JJ, Waltzer R, Whittom AA, Austin MC, Rajkowska G, Stockmeier CA (2010) Glial and glutamatergic markers in depression, alcoholism, and their comorbidity. J Affect Disord 127(1–3):230–240
239. Rajkowska G, Miguel-Hidalgo JJ, Wei J, Dilley G, Pittman SD, Meltzer HY et al (1999) Morphometric evidence for neuronal and glial prefrontal cell pathology in major depression. Biol Psychiatry 45(9):1085–1098
240. Johnston-Wilson NL, Sims CD, Hofmann JP, Anderson L, Shore AD, Torrey EF et al (2000) Disease-specific alterations in frontal cortex brain proteins in schizophrenia, bipolar disorder, and major depressive disorder. The Stanley Neuropathology Consortium. Mol Psychiatry 5(2):142–149
241. Miguel-Hidalgo JJ, Baucom C, Dilley G, Overholser JC, Meltzer HY, Stockmeier CA et al (2000) Glial fibrillary acidic protein immunoreactivity in the prefrontal cortex distinguishes younger from older adults in major depressive disorder. Biol Psychiatry 48(8):861–873
242. Si X, Miguel-Hidalgo JJ, O'Dwyer G, Stockmeier CA, Rajkowska G (2004) Age-dependent reductions in the level of glial fibrillary acidic protein in the prefrontal cortex in major depression. Neuropsychopharmacology 29(11):2088–2096
243. Sander T, Ostapowicz A, Samochowiec J, Smolka M, Winterer G, Schmidt LG (2000) Genetic variation of the glutamate transporter EAAT2 gene and vulnerability to alcohol dependence. Psychiatr Genet 10(3):103–107
244. Spanagel R, Pendyala G, Abarca C, Zghoul T, Sanchis-Segura C, Magnone MC et al (2005) The clock gene Per2 influences the glutamatergic system and modulates alcohol consumption. Nat Med 11(1):35–42
245. Farina C, Aloisi F, Meinl E (2007) Astrocytes are active players in cerebral innate immunity. Trends Immunol 28(3):138–145
246. Okun E, Griffioen KJ, Lathia JD, Tang SC, Mattson MP, Arumugam TV (2009) Toll-like receptors in neurodegeneration. Brain Res Rev 59(2):278–292
247. Alfonso-Loeches S, Pascual-Lucas M, Blanco AM, Sanchez-Vera I, Guerri C (2010) Pivotal role of TLR4 receptors in alcohol-induced neuroinflammation and brain damage. J Neurosci 30(24):8285–8295
248. Pascual M, Balino P, Alfonso-Loeches S, Aragon CM, Guerri C (2011) Impact of TLR4 on behavioral and cognitive dysfunctions associated with alcohol-induced neuroinflammatory damage. Brain Behav Immun 25(suppl 1):S80–S91
249. Waak J, Weber SS, Waldenmaier A, Gorner K, Alunni-Fabbroni M, Schell H et al (2009) Regulation of astrocyte inflammatory responses by the Parkinson's disease-associated gene DJ-1. FASEB J 23(8):2478–2489
250. Salaria S, Badkoobehi H, Rockenstein E, Crews L, Chana G, Masliah E et al (2007) Toll-like receptor pathway gene expression is associated with human immunodeficiency virus-associated neurodegeneration. J Neurovirol 13(6):496–503
251. Holm TH, Draeby D, Owens T (2012) Microglia are required for astroglial Toll-like receptor 4 response and for optimal TLR2 and TLR3 response. Glia 60(4):630–638

Chapter 3
Neuroimmune Modulation of Synaptic Function

Horia Pribiag and David Stellwagen

3.1 Introduction

It is becoming increasingly clear that molecules classically associated with the peripheral immune system participate in the modulation of synaptic function. These molecules are from both the innate and adaptive immune systems, including cytokines, class I major histocompatibility complex (MHCI), and complement cascade components, and are used constitutively by the nervous system as signaling molecules that regulate developmental synapse formation and several forms of synaptic plasticity. In addition, the upregulation of immune molecules associated with neuroinflammatory pathological states such as stroke, trauma, and neurodegenerative disease is likely to disrupt normal signaling events and result in maladaptive modulation of neurotransmission. This chapter will present and discuss current evidence implicating neuroimmune molecules in the constitutive regulation of central nervous system (CNS) synapse function, under both normal and pathological conditions.

3.2 Neuroimmune Molecules Constitutively Expressed in the CNS

Formerly, the CNS was viewed as immune privileged, whereby the blood–brain barrier kept the brain largely devoid of immune cells under healthy conditions, and thus similarly devoid of immune signaling molecules, which were not thought to be

H. Pribiag • D. Stellwagen (✉)
Centre for Research in Neuroscience, McGill University, Montreal General Hospital, L7 217 1650 Cedar Av, Montreal, QC H3G 1A4, Canada
e-mail: david.stellwagen@mcgill.ca

C. Cui et al. (eds.), *Neural-Immune Interactions in Brain Function and Alcohol Related Disorders*, DOI 10.1007/978-1-4614-4729-0_3,

expressed in neuronal tissue. However, there is now general agreement that many immune signaling molecules are expressed in healthy nervous tissue, albeit at significantly lower levels than seen in the immune system, and that these proteins regulate normal neuronal functions [1–3]. Conversely, a few proteins initially identified in the nervous system, such as DSCAM and semaphorins, have been found to have a role in the immune system. This chapter will focus on classic immune molecules having significant and distinct neuronal functions—cytokines, class I MHC, and members of the complement cascade.

Cytokines are primarily expressed by glia, particularly by astrocytes and microglia, although neuronal expression has been observed in some studies [3, 4]. Cytokine receptors are ubiquitously expressed by all cell types in the nervous system [5]. All major cytokines are expressed in the CNS, including tumor necrosis factor-α (TNFα), interleukin 1α (IL-1α), IL-1β, IL-4, IL-6, IL-10, IL-12, and IL-18 [6, 7]. The pro-inflammatory cytokines TNFα, IL-1, and IL-6 are elevated in most inflammatory states [8]; however, they are also constitutively expressed at low levels in the normal brain [9]. This chapter will focus on these three pro-inflammatory cytokines because they have the clearest role in modulating synaptic function; however, future work may yet indentify fundament roles for the other cytokines in the nervous system.

TNFα and TNFα receptors (TNFRs) are members of the TNF- and TNFR-related superfamily of proteins that have various functions in host defense, inflammation, apoptosis, autoimmunity, and organogenesis [10]. TNFα is a 26-kDa single-pass transmembrane protein, cleaved at the membrane by TNFα-converting enzyme (TACE, also known as ADAM17) to generate a 17-kDa ectodomain that is soluble in the extracellular milieu [11, 12]. It is active as a homomeric trimer in both the transmembrane and soluble forms [13, 14] and signals through two receptors: TNFR1 and TNFR2 [15]. Under normal conditions, TNFRs are constitutively expressed throughout the CNS [16, 17] by neurons, astrocytes, microglia, and oligodendrocytes [17–19]. Similarly, in the normal brain, TNFα mRNA is detectable at low levels [20], and in vitro glial preparations constitutively secrete TNFα [21]. Expression of TNFα and TNFRs is strongly upregulated in response to ischemia, head trauma, infections, stroke, and other pathological brain insults [22, 23]. Of note, elevated expression of TNFα in response to endotoxins [24, 25] or during prolonged action potential blockade in vitro [26] is of glial origin. Taken together, these findings indicate that capacity for TNFα signaling exists in both normal and pathological CNS states.

Like TNFα, IL-1β is a pro-inflammatory cytokine playing major roles in inflammatory and autoimmune diseases [8, 27]. IL-1 represents a family of cytokines which includes IL-1α, IL-1β, IL-18, and other members [27]; however, most evidence regarding regulation of synaptic strength through neurotransmitter receptor trafficking focuses on IL-1β. IL-1β is synthesized as a 31-kDa protein, which is cleaved by IL-1β-converting enzyme (ICE) to generate a mature 17-kDa active protein that is released into the extracellular space [27]. IL-1β can bind to three different receptors: IL-1RI, IL-1RII, and IL-1 receptor accessory protein (IL-1RAcP) [28]. IL-1RII is considered a "decoy" receptor owing to its lack of enzymatic activity and presumably acts to buffer excessive concentrations of IL-1 ligands [29]. Upon binding of IL-1β to IL-1RI, IL-1RAcP is recruited, resulting in the initiation of various signaling cascades [27]. The presence of the naturally occurring receptor antagonist IL-1Ra—which binds IL-1RI with affinity similar to IL-1 ligands but does not induce an intracellular response [30]—further increases the complexity of the IL-1 signaling system. Components of the IL-1 signaling system are constitutively expressed in the CNS under normal physiological conditions: IL-1R1 and IL-1RII are expressed throughout the brain, in both neurons and glial cells [31, 32]; endogenous IL-1β bioactivity is present in various brain regions including the cortex and hippocampus [33], and IL-1β is immunolocalized sporadically to glial cells [34]; IL-1Ra mRNA is also present in the cerebral cortex [35]. Expression of IL-1β is greatly elevated during CNS injury and neurodegenerative disease [36], primarily in microglia [37, 38] and astrocytes [38, 39].

IL-6 belongs to a group of structurally related cytokines—referred to as either the IL-6 family, the gp-130 family, or the neuropoietic family—which includes IL-6, IL-11, IL-27, leukemia inhibitory factor (LIF), ciliary neurotrophic factor (CNTF), and several other members [40]. This family of cytokines is implicated in a broad range of physiological functions including infection, pregnancy, and bone, muscle, and cardiovascular functions; however, it has also been implicated in developing and mature brain functions such as neurogenesis, stem cell fate determination, synaptic plasticity, and the regulation of stress, feeding, sleep, and depressive behaviors [40]. In the context of regulation of synaptic plasticity and ion channel trafficking, most studies implicate IL-6 itself. IL-6 is classified as having both pro-inflammatory (e.g., participates in the acute phase reaction and in inflammation) and anti-inflammatory actions (stimulation of intestinal cell proliferation, inhibition of apoptosis) [41]. Following binding to the IL-6 receptor (IL-6R), IL-6 signals through the ubiquitously expressed membrane glycoprotein gp130 as do most members of the IL-6 family [42, 43]. Both IL-6 and IL-6R mRNAs are detectable at low levels in various brain regions including the hippocampal formation, cerebral cortex, hypothalamus, piriform cortex, cerebellum, and olfactory bulb [41, 43]; in vitro, it is produced by both glial cells and neurons [41]. IL-6 expression in the CNS is induced dramatically in response to infectious diseases [41], acute brain and spinal cord injury [44, 45], stroke [46, 47], and seizures [48, 49].

MHCI are antigen-presenting molecules in the immune system, although it is unclear if they present peptides in the nervous system [2]. The complex consists of polymorphic and polygenic alpha heavy chain coupled with an invariant β2-microglobulin (β2m) light chain. Peptide loading, a prerequisite for MHCI surface expression in antigen-presenting cells, requires the transporter associated with antigen processing 1 (TAP1). The polygenic nature of MHCI makes it difficult to eliminate genetically, but deletion of elements of the signaling complex is simpler. MHCI surface expression can be greatly reduced by removing the invariant β2m subunit and by deleting TAP1 [50, 51]. In the immune system, MHCI binds to the T-cell receptor complex, which does not appear to be expressed in the nervous system [52]. MHCI also binds to a variety of natural killer cell receptors, which are expressed on neurons, including the paired immunoglobulin-like receptor B (PirB)

[53], Ly49 [54], and the killer cell-immunoglobulin receptor (KIR) [55]. Based on available data, other neuronal receptors for MHCI are likely to exist but have not been identified. The associated effector protein CD3ζ also appears to be a component of the neuronal receptor(s) for MHCI [56, 57], although it is unknown if it associates with all the aforementioned receptors or just a subset. DAP12/KARAP encodes another transmembrane adaptor polypeptide structurally and functionally related to CD3ζ [58–60] and associated with natural killer cell receptors. While neurons clearly express both MHCI and a variety of receptors, glial expression of most of these proteins is uncertain, though likely. CD3ζ is expressed in neurons [56], but DAP12 is only expressed in microglia and not in neurons, astrocytes, or oligodendrocytes [58]. At the neuromuscular junction, motor neuron axons express MHCI, and Schwann cells (but not muscle fibers) express the PirB receptor [61].

In the classic complement cascade, C1q initiates the cascade by binding to a target and recruiting the other elements of the C1 complex. This then cleaves C2 and C4 into active fragments, which ultimately leads to cleavage of C3 and formation of the effector complex. C3 cleavage and activation can also occur through nonclassical pathways, but as of yet there is no evidence that this occurs in the nervous system. C1q is expressed by astrocytes, as well as by astrocyte-stimulated neurons, while C3 is primarily expressed by microglia [62]. The signaling cascades downstream of C3 remain unclear.

Neuronal pentraxins are another class of immune-related molecules that regulate synapse function. Although neuronal pentraxins are specific to the nervous system, they have homology to pentraxins, which are secreted members of the innate immune system and mark cells for degradation and phagocytosis. There are two neuronal pentraxins, NP1 and NP2 (which are also called Narp). Neuronal pentraxins are secreted but can form heteromultimeric complexes tethered to the membrane by neuronal pentraxin receptor (NPR) [63]. NP1, NP2, and NPR are broadly expressed in the hippocampus, cerebral cortex, retina, and cerebellum but are likely absent from midbrain and thalamic structures [64–66].

3.3 Neuroimmune Molecules Contribute to Developmental Synapse Formation and Refinement

3.3.1 Synapse Formation

Immune signaling molecules may affect a number of basic neuronal processes, starting with early development and initial synapse formation. MHCI is present throughout development, including during the period of synapse formation, and is localized both pre- and postsynaptically [67]. In dissociated cortical cultures, MHCI overexpression negatively regulates synapse formation, while decreasing surface MHCI levels increases synapses density [68]. Similarly, synapse density is increased in β2m knockout mice, which also have reduced surface expression of MHCI [68].

Curiously, no change in total content of synaptically localized proteins is observed in hippocampal cultures from β2m/TAP knockout mice [69], although the cultures do have increased miniature excitatory postsynaptic current (mEPSC) frequency, which is consistent with an increase in synapse density [70]. Additionally, treating cortical cultures with anti-H2-K^b and H2-D^b MHCI antibodies decreases the number of synapsin puncta [54], which suggests that MHCI signaling should increase synapse number. Whether this represents a difference in preparations or manipulations, or that measuring synaptic protein content is a poor measure of functional synapse number is unclear. Indirect measures could be complicated by changes in synaptic strength induced by these molecules (see below).

There is limited evidence that cytokines can also impact synapse formation. A component of the IL-1 receptor complex, IL-1RAcP, appears synaptogenic as knockdown of it in cortical culture reduces synapse number, while overexpression increases it [71]. Furthermore, knockout mice have reduced cortical spine density in vivo [71]. However, IL-1 does not appear to be the relevant ligand for synaptogenesis; rather, protein tyrosine phosphatase δ (PTP) is the likely activator of the signaling complex. These data suggest that cytokine receptor complexes are capable of regulating synaptogenesis and indicate that it is possible that cytokines themselves may regulate this process.

Neuronal pentraxins are also immune-related molecules that can regulate synapse formation. Both neuronal pentraxins (NP1 and NP2/Narp) appear to be synaptogenic, with overexpression increasing the number of excitatory synapses [63, 72] and dominant-negative expression decreasing synapse number [73]. By virtue of their sequence and structural homology to peripheral pentraxins, neuronal pentraxins may employ clustering mechanisms similar to those of the peripheral immune-related pentraxins. The clustering of synaptic components, including AMPA-type glutamate receptors, may trigger the stabilization or strengthening of nascent excitatory contacts. Indeed, cultured retinal ganglion cells (RGCs) from NP1/NP2 double-knockout mice display delayed functional maturation of excitatory synapses, despite no changes in the number of synaptic contacts [74]. Neuronal pentraxins may modulate synapse formation, but are clearly not the prime initiators, as the phenotype in the NP1/NP2 knockout mouse is relatively subtle [74].

3.3.2 Synaptic Refinement

Activity-dependent refinement of neuronal connectivity has been best characterized in the developing visual system (reviewed in [75, 76]). In the mammalian visual system, RGCs project to their thalamic target, the lateral geniculate nucleus (LGN). In binocular vision, RGC projections from the two eyes subserving the same visual field project to the same LGN. These projections from the two eyes are initially overlapping in the LGN but slowly segregate into eye-specific layers in a manner dependent on spontaneous retinal activity [77, 78]. A similar process occurs subsequently in the binocular zone of primary visual cortex of highly binocular mammals (such as carnivores or primates), where inputs from the eye-specific layers of the LGN are initially intermingled in layer 4 but subsequently segregate into ocular dominance columns, again through an activity-mediated process [79]. Traditional immune molecules appear to regulate these processes.

In the LGN, the activity-dependent segregation of eye-specific domains occurs over the first two postnatal weeks in rodents [77, 78]. Neuronal expression of MHCI was first observed in a screen for activity-regulated genes in the developing LGN [56]. Double-knockout mice of β2m/TAP1 have enlarged projections in the LGN from the ipsilateral eye, indicating a reduction in axonal refinement [80]. A similar phenotype is observed in knockout mice for CD3ζ [80] and for mice lacking both the H2-K^b and H2-D^b MHCI genes [81]. PirB does not appear to be the MHCI receptor in the LGN, as the PirB knockout mouse has no phenotype in the LGN [53]. It is unclear if the LGN phenotype is due to a change in plasticity in the LGN itself [80], or due to a change in the patterns of retinal activity which drive the axonal refinement [57], or both. Conditional knockouts may help address this issue, allowing tissue-specific deletions. Such control would also help define which cells need to express the ligand (MHCI) or the receptor complex, helping to define the direction of signaling.

The complement cascade also regulates axonal refinement in the LGN. There is reduced segregation of eye-specific inputs in animals deficient in either C1q or C3, and LGN neurons remain innervated by multiple RGCs [62]. Initially, LGN cells receive inputs from several ganglion cells, but this is slowly refined to 1–2 dominant inputs through activity-dependent competition. The C1q and C3 phenotype suggests a reduction in the normal process of synapse elimination, although the mechanistic details remain unclear. It also suggests a role for microglia, the primary cells expressing C3, in this process. Neuronal pentraxins also contribute to LGN refinement, as NP1/NP2 double-knockout mice have reduced refinement and segregation of retinal axons, although again it is uncertain if this is due to pentraxins in the LGN itself or due to subtle changes in retinal activity patterns [74]. This deficit is striking in early development but largely resolved by adulthood [74]. It remains to be determined if MHCI, neuronal pentraxins, and the complement signaling pathway interact, or if they independently regulate distinct aspects of refinement. Peripheral pentraxins are known to bind directly to C1q and can initiate or inhibit the cascade [82]. It is interesting that the LGN phenotypes are similar (though not identical), but it is unknown if they are additive. It is appealing to put these molecules in the same pathway, targeting synapses for removal (and thus allowing refinement). It is unclear which of these factors are instructive in this process and which are permissive. The complement cascade would presumably be the end process for synapse removal (perhaps by causing the phagocytosis of synaptic elements) and would therefore be unlikely to be instructive and determine the synapses to eliminate. Indeed, while C1q knockout mice retain inputs from higher numbers of RGCs, there is still only a single dominant input [62], suggesting that activity-dependent competition occurs but is unable to complete the process of synapse elimination.

There is no established role for cytokines in retinogeniculate axonal refinement. TNFα knockout mice have no obvious defect in this projection (D. Stellwagen, unpublished data), but it has never been examined in detail. However, chronic treatment with TNFα increases tectal interconnectivity in *Xenopus* tadpoles [83], which suggests a lack of refinement in these connections. This could be due to an artificial stabilization of nascent synapses, however, and does not establish a role for TNFα in normal tectal development and refinement. TNFα does regulate morphological refinement of hippocampal neurons, but this could be due to alterations in synaptic maturation [84].

Immune molecules also contribute to the developmental plasticity observed in the primary visual cortex. Mice unfortunately do not have ocular dominance columns; instead, the thalamocortical inputs from the ipsilateral eye-specific area of the LGN define a discrete binocular zone in layer 4 of the primary visual cortex. Manipulation of visual experience during a critical period alters the extent of this zone, with closing or removing the contralateral eye leading to an expansion of the ipsilateral input area. Genetic deletion of H2-K^b and H2-D^b MHCI genes [81] or of receptor PirB [53] enhances this plasticity—the deprivation-induced expansion of the ipsilateral eye inputs is greater in mice deficient in MHCI signaling. These data suggest that MHCI restricts the changes in thalamocortical axonal arbors induced by alterations in visual experience. There is also refinement of the layer 4 inputs to layer 2/3 cells, which is sensitive to a slightly later critical period. Layer 2/3 neurons are to varying degrees binocular, but generally preferentially respond to the contralateral eye. Monocular deprivation during a critical period of heightened sensitivity in early postnatal life profoundly alters visual function by causing a shift of responsiveness toward the open eye [85]. This shift is comprised of two components: an initial loss of responsiveness to the closed eye and a subsequent gain of responsiveness to the open eye [86]. TNFα knockout mice display the normal loss of responsiveness to the closed eye but are completely deficient in the gain of responsiveness to the open eye [87]. The phenotype may be explained by the types of plasticity involved in the individual components (see below). This deficit is phenocopied in mice cortically infused with a pharmacological inhibitor of TNFα signaling, indicating that the deficiency is not due to a developmental defect but rather due to a defect in plasticity [87]. Curiously, adult ocular dominance plasticity, which has a smaller and slower shift in responsiveness than that observed during the critical period and consists almost exclusively of a gain of responsiveness to the open eye, is normal in TNFα knockout mice [88]. Again, it is unknown if the MHCI and TNFα pathways interact in the visual cortex, or if MHCI regulates layer 2/3 plasticity, or if TNFα regulates layer 4 plasticity. Nor has a role for the complement cascade been found in this system, although C1q knockout mice do have enhanced cortical interconnectivity [89]. Much work remains to be done to delineate the roles of these pathways in refinement. Clearly, the regulation of refinement will be system dependent. For example, MHCI does not regulate refinement in all systems, as refinement of the cerebellar climbing fiber axons is unaffected by alteration in MHCI signaling [90, 91]. In contrast, refinement of the number of motor neurons contacting a muscle does depend on MHCI signaling, as H2-K^b and H2-D^b knockout mice have more multiply innervated muscle fibers [92]. It should be noted that PirB knockout mice have an extended critical period for experience-dependent plasticity of the visual cortex [53]. However, PirB, in addition to being an MHCI receptor, is also a receptor for the myelin inhibitory protein Nogo and can complex with the Nogo receptor (NgR) [93]. Furthermore, the extension of the critical period mirrors the phenotype seen in the NogoA knockout mice and the NgR knockout mice, both of which do not have any phenotype related to monocular deprivation. Lastly, the critical period extension is not phenocopied in any of the other MHCI mutant mice. These data indicate that PirB participates in multiple signaling pathways in the same tissue and suggest that other neuroimmune receptor complexes may also have similarly multifaceted roles.

3.4 Regulation of Neurotransmitter Release by Neuroimmune Molecules

Regulation of synapse function may occur presynaptically, by regulating the probability of neurotransmitter release, or postsynaptically, typically by regulating the availability of neurotransmitter receptors. Several neuroimmune molecules have been implicated in regulation of neurotransmitter release. IL-6 treatment reduces glutamate release from cerebrocortical synaptosomes, possibly by inhibiting MAPK/ERK-induced phosphorylation of synapsin I [94]. Recently, interferon-γ has been shown to decrease release probability at GABAergic synapses onto CA1 neurons [95]. Acute treatment of cultured hippocampal neurons with TNFα increases mESPC frequency [21, 96], suggesting an increase in release probability. It is unclear if this is a direct effect as cytokines such as TNFα may indirectly regulate transmitter release by controlling the release of potentiating factors from nearby glia (see below) [97]. The MHCI signaling system may also regulate basal synaptic transmission, although this regulation is more likely presynaptic than postsynaptic. Cultures deficient in β2M/TAP exhibit higher mEPSC frequency, stronger vesicular glutamate transporter (vGluT) staining, and more synaptic vesicles per bouton [70]. MHCI signaling in vivo regulates neurotransmitter release probability at cerebellar synapses [90, 98]. CD3ε knockout mice display an enhanced paired-pulse facilitation ratio at parallel fiber synapses onto Purkinje cells, signifying a lower probability of neurotransmitter release compared to wild-type mice [98]. This is associated with poor rotarod behavioral task performance, suggesting an impairment in cerebellar function [98]. Conversely, double-knockout mice for the MHCI molecules $H2\text{-}K^b$ and $H2\text{-}D^b$ show reduced paired-pulse depression at climbing fiber synapses onto Purkinje cells, suggesting enhanced glutamate release compared to wild-type mice; these findings are associated with an improved performance on the rotarod behavioral task [90]. There is no evidence for changes in postsynaptic receptor content with MHCI deficiency. Mice deficient in DAP12, which associates with some MHCI receptors, do have a reduction in AMPA-type glutamate receptor levels in the postsynaptic density, changes in AMPA receptor rectification, and changes in

AMPA/NMDA receptor current ratios [58, 59], indicating changes in postsynaptic function. However, MHCI is unlikely to be acting as the ligand at a DAP12-associated receptor due to the discrepancy between the MHCI and DAP12 loss of function phenotypes. Taken together, these studies implicate several neuroimmune molecules in the modulation of neurotransmitter release at CNS synapses; however, further studies are needed in order to better characterize such effects.

3.5 Pro-inflammatory Cytokines Regulate Neurotransmitter Receptor Trafficking (Fig. 3.1)

Synaptic strength at excitatory and inhibitory synapses is in large part determined by the cell-surface abundance of ligand-gated ion channels clustered at the postsynaptic specialization. AMPA receptors (AMPARs) are the primary mediators of fast

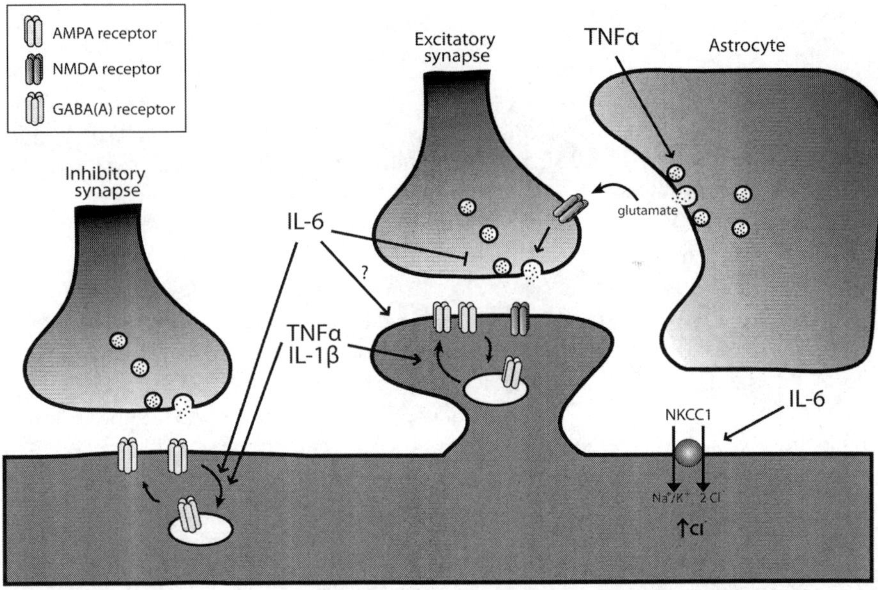

Fig. 3.1 Regulation of synapse function by pro-inflammatory cytokines. TNFα and IL-1β acutely regulate both excitatory and inhibitory synapse function by inserting AMPARs and removing GABA(A)R from the synapse. IL-6 reduces glutamate release at excitatory synapses, and evidence suggests that it may also regulate receptor trafficking, particularly at inhibitory synapses. In dorsal root ganglion neurons, IL-6 enhances the function of the Na-K-Cl cotransporter NKCC1, thereby increasing intracellular chloride concentration. At perforant path synapses onto granule cells in the hippocampus, TNFα potentiates excitatory neurotransmission by controlling the docking of glutamate-controlling vesicles in astrocytes; the glutamate released acts on presynaptic NMDARs to increase neurotransmitter release probability

excitatory neurotransmission in the brain [99]. They are glutamate-gated channels permeable to monovalent cations and have a tetrameric dimer of dimers composition of subunits GluA1 through GluA4 [100]. NMDA receptors (NMDARs) primarily act as molecular switches controlling the induction of excitatory synaptic plasticity [99], although they can also contribute to basal synaptic transmission particularly during early development [101]. NMDA receptors are also glutamate-gated ion channels and consist of a heterotetrameric structure of GluN1 through GluN3 subunits [102, 103]. $GABA_A$ receptors (GABA(A)Rs) are the primary mediators of fast inhibitory neurotransmission in the brain [104]. They are GABA-gated chloride-permeable channels consisting of a pentameric assembly of 19 distinct subunits [105]. Cell-surface trafficking of all three types of receptors is modulated by proinflammatory cytokines (TNFα, IL-1β, and IL-6) on a rapid time scale. All three are known to exert strong effects on synaptic strength through trafficking of ligand-gated ion channels.

3.5.1 TNFα Regulates AMPAR and GABA(A)R Trafficking

Application of exogenous TNFα to mature dissociated hippocampal neuron cultures or acute hippocampal slices induces a rapid (within 15–20 min) increase in excitatory synaptic strength as indicated by an increase in average mEPSC amplitude and frequency [21, 26, 106]. Similarly, increases in mEPSC, sEPSC (spontaneous excitatory postsynaptic current recorded in the absence of TTX), and AMPA-induced whole-cell current amplitudes are observed in spinal cord lamina II neurons [107] and lumbar motoneurons [108]. Cell-surface immunostaining experiments and electrophysiological recordings using polyamine block indicate that the increase in synaptic strength is mediated by a rapid exocytosis of GluA1-containing AMPARs that lack or have lower stoichiometric amounts of the GluA2 subunit [106]. Consistent with increased surface expression of GluA2-lacking AMPARs—which are permeable to Ca^{2+}—the neuronal response to TNFα exposure includes an enhancement of AMPA-mediated Ca^{2+} rises [109], suggesting a plausible mechanism for neuroinflammatory disease progression through excitotoxicity [110, 111]. Signaling occurs through TNFR1, but not TNFR2, activation [106, 112]. Interestingly, genetic deletion of TNFR1 or application of TNFR1 neutralizing antibody to dissociated cortical/hippocampal cultures decreases surface GluA1 expression [106, 112] and reduces mEPSC amplitude and frequency [112], suggesting that TNFα is used constitutively to regulate synaptic AMPAR expression. Downstream of TNFR1 activation, the intracellular signaling cascade involves activation of PI3 kinase [106]; other components of this pathway—specific to the rapid surface delivery of AMPARs—remain unidentified.

The effects of TNFα on NMDAR trafficking are ambiguous. Several lines of evidence from studies of cortical/hippocampal neurons indicate that NMDARs are not trafficked in response to acute exposure to TNFα: synaptic localization of GluN1 and whole-cell currents in response to NMDA application are unaltered [21, 112]; furthermore, the AMPAR to NMDAR current ratio is increased in response to TNFα [26], suggesting that as synaptic AMPAR content increases, NMDAR content does not change significantly. Conversely, a few studies document an increase in NMDAR function using electrophysiological measurements [107, 108], albeit at spinal cord synapses. Only one report describes a decrease in NMDAR function in response to TNFα [113].

The plasticity of inhibitory synapses via trafficking of GABA(A)Rs provides a significant point of regulation for action potential firing frequency and for synchronization of neuronal activity oscillations [114, 115]. TNFα induces a rapid (within 20 min) decrease in surface expression of GABA(A)R β subunits, an increase in internalized β subunits, and a decrease in miniature inhibitory postsynaptic current (mIPSC) amplitude on CA1 pyramidal neurons in acute hippocampal slices [106]. This downregulation of phasic inhibition, in combination with an increase in excitatory synaptic transmission, increases the ratio of excitatory to inhibitory synaptic transmission in acute hippocampal slices [106], which is likely to result in altered network properties in intact preparations. Similar effects are observed in the spinal cord. In spinal cord lamina II neurons (a critical site for nociceptive transmission to the brain), exogenous TNFα application induces a decrease in the frequency of mIPSCs and sIPSCs (spontaneous inhibitory postsynaptic currents recorded in the absence of TTX) through activation of TNFR1 (but not TNFR2), p38 MAPK, and inhibition of hyperpolarization-activated cation currents (Ih) [17]. It remains unknown whether TNFα can regulate tonic GABA(A)R currents, which have been implicated in numerous developmental processes and in adult dentate gyrus neurogenesis [115, 116].

Several noteworthy questions remain unexplored, including the effects of chronic TNFα exposure on synapse function, the persistent effects on synapse function following a brief exposure to TNFα, whether anti-inflammatory cytokines (such as IL-10) can interfere with the receptor trafficking function of pro-inflammatory cytokines like TNFα, and the role of the associated Ca^{2+} influx in regulating synapse function on longer time scales, perhaps through transcriptional activation.

3.5.2 IL-1β Regulates Excitatory and Inhibitory Neurotransmission

The effects of IL-1β on excitatory neurotransmission parallel those of TNFα. Acute exposure to IL-1β also rapidly increases the surface expression of GluA1 on hippocampal neurons, although to a lesser extent than TNFα [106]. Similarly, application of IL-1β to acute spinal cord slice lamina II neurons increases the frequency and amplitude of sEPSCs and potentiates AMPA-induced currents [107]. Treatment of acute hippocampal slices with IL-18—a member of the IL-1 family—also increases mEPSC amplitude and frequency in CA1 pyramidal neurons [117]. More compellingly than the data for TNFα, evidence suggests that IL-1β upregulates NMDAR abundance and/or function. Following IL-1β treatment, NMDA-induced currents are potentiated in lamina II neurons [107], and in hippocampal neurons, NMDA-induced Ca^{2+} currents are enhanced through a mechanism involving Src family kinases [118, 119] and enhanced GluN2B binding to the postsynaptic anchoring protein PSD-95 [120]. IL-1β activates p38 MAPK in neurons and activates both p38 MAPK and NF-κB in astrocytes [119], pointing to potential signaling components of trafficking mechanisms. The response of GABA(A)Rs to IL-1β remains largely unexplored, particularly in neocortical neurons; however, in spinal cord lamina II neurons, sIPSC amplitude and frequency are significantly lower after IL-1β treatment, with both GABA- and glycine-induced currents being downregulated [107]. Many of the studies on the effects of IL-1β should be supplemented by cell-surface immunostaining or cell-surface biotinylation experiments in order to show that acute trafficking mechanisms underlie the electrophysiological observations.

3.5.3 IL-6 Regulates Excitatory and Inhibitory Neurotransmission

IL-6 has less consistent effects on synaptic function. Acute IL-6 application does not appear to regulate AMPA receptor trafficking [106]. To our knowledge, only one report suggests rapid ion channel trafficking effects of IL-6: a 2-min exposure to 10 ng/mL IL-6 reduces sIPSC frequency and GABA- and glycine-induced currents in spinal cord lamina II neurons [107]. However, on a chronic time scale, a 2-week exposure of cerebellar Purkinje neuron cultures to IL-6 leads to a reduction in action potential generation and enhanced AMPA-induced currents, while at lower doses of IL-6 increased calcium signals induced by DHPG (a group I mGluR agonist) [121] and AMPA [122]. Unlike TNFα and IL-1β, IL-6 appears to be neuroprotective against NMDA receptor-mediated excitotoxicity in the brain [123, 124], suggesting IL-6 may downregulate NMDA receptor function or surface expression. Along with TNFα and IL-1β, IL-6 will be discussed in further detail below for its role in regulation of long-term potentiation (LTP).

While this topic remains largely unexplored for cytokine-dependent regulation, one recent report implicates IL-6 in the regulation of intracellular Cl^- concentration [125]. Intracellular Cl^- concentration is a crucial regulator of inhibition—the effect of open GABA(A)Rs on membrane potential (i.e., either hyperpolarizing, depolarizing, or simply shunting) depends on the Cl^- gradient established by transporters [126]. Sciatic nerve axotomy induces elevation of IL-6 expression in adult dorsal root ganglion neurons, which consequently phosphorylates NKCC1 and increases intracellular Cl^- concentration [125].

The implications of this study for regulation of Cl^- concentration in neocortical neurons remain unknown.

3.5.4 Conclusion

Despite signaling through different receptor complexes, pro-inflammatory cytokines generally exert similar effects on neuronal synaptic strength: (1) increased excitatory synaptic strength through upregulation of cell-surface AMPARs, (2) enhanced calcium influx at the synapse—either through Ca^{2+}-permeable AMPARs or NMDARs, and (3) decreased inhibitory synaptic strength, likely via downregulation of surface GABA(A)Rs/glycine receptors or increased intracellular Cl^- concentration. The combined action of these effects is likely to alter network behavior by increasing overall neuron firing frequency. Furthermore, these effects constitute a form of synaptic plasticity that is distinct from Hebbian types of plasticity, such as LTP and LTD, which are input specific and dependent on correlated firing of pre- and postsynaptic sides. Rather, as discussed in further detail below, while pro-inflammatory cytokines can affect LTP and LTD induction, their effects on baseline synaptic strength resemble a form of homeostatic synaptic plasticity that compensates for chronic inactivity.

3.6 Modulation of Hebbian Plasticity by Class I MHC Proteins and Pro-inflammatory Cytokines

Many of the neuroimmune molecules implicated in synapse development and direct trafficking of ligand-gated ion channels also regulate long-term plasticity. LTP and long-term depression (LTD) are widely studied neurophysiological correlates of learning and memory in the mammalian brain [127, 128]. Resulting from patterns of coincident pre- and postsynaptic activity, LTP and LTD are classified as "Hebbian" forms of synaptic plasticity owing to two characteristics they display: input specificity (i.e., the strength of single pathways can be altered without affecting neighboring inputs to the same cell) [129] and associativity (i.e., a stimulus not strong enough to induce plasticity by itself will induce plasticity when paired with a stronger stimulus) [130]. LTP and LTD have been studied most commonly using transverse hippocampal slice preparations, for which numerous induction protocols have been documented (reviewed in [131, 132]). However, more broadly defined, several types of long-term plasticity have been characterized in the hippocampus, cortex, amygdala, and cerebellum, including NMDAR-dependent LTP and LTD, mossy fiber LTP, mGluR-dependent LTD, and endocannabinoid-mediated LTD (reviewed in [99, 133]). Several of these involve NMDAR-dependent Ca^{2+} influx and postsynaptic expression through trafficking of AMPARs (reviewed in [134]), while others involve regulation of neurotransmitter release by NMDAR-dependent or NMDAR-independent presynaptic mechanisms (reviewed in [99, 133, 135]).

3.6.1 Loss of Class I MHCs Enhances Long-Term Plasticity

Class I MHC proteins, in addition to their roles in synapse development and refinement, also regulate several forms of long-term plasticity. LTP and LTD have been examined using knockout and knock-in strategies, targeting components required for MHCI receptor signaling or MHCI surface expression [58, 59, 80, 136, 137]; more recently, MHCI molecules themselves have been knocked out [90]. At Schaffer collateral-CA1 synapses, both CD3ζ knockout mice and β2M/TAP1 knockout mice display enhanced NMDAR-dependent LTP and abolished LTD [80], while CD3ζ knockout mice have also been shown to display aberrant expression of long-lasting (likely transcription-dependent) LTP with subthreshold stimuli [136]. MHCI mRNAs are upregulated in mice constitutively expressing CREB [136], which suggests that MHCIs may actively participate in the process of memory formation. Enhanced Schaffer collateral-CA1 LTP is also observed in knock-in mice bearing a loss-of-function DAP12 gene [58, 59]. Unlike in CD3ζ knockout mice, which only display enhancement of NMDAR-dependent LTP [80], enhanced LTP in DAP12 knockout mice is at least partially independent of NMDARs [58]. Furthermore, while basal neurotransmission is unaltered in CD3ζ knockout mice [80], DAP12 knockout mice display an increased AMPAR to NMDAR current ratio [59]. These differences may be explained by differential expression profiles: CD3ζ is expressed in neurons [56], while DAP12 is only expressed in microglia, and not in neurons, astrocytes, or oligodendrocytes [58]. It is therefore conceivable that MHCIs make use of distinct neuron- and glial-mediated mechanisms to restrict LTP. MHCIs may also use different receptors to differentially affect structural plasticity versus activity-dependent plasticity: PirB knockout mice display normal LTP and LTD at Schaffer collateral-CA1 synapses [137] but enhanced ocular dominance plasticity in the visual cortex [53]. More recently, the use of knockout mice for the classical MHCI molecules H2-K^b and H2-D^b, both of which are expressed by Purkinje cells, has shown a lower threshold for induction of LTD at parallel fiber to PC synapses [90]. Thus, in agreement with their restrictive function on synaptogenesis [68], loss of MHCI molecules or MHCI receptor signaling components enhances long-term plasticity, suggesting a similarly restrictive role for MHCI in Hebbian plasticity.

3.6.2 Pro-inflammatory Cytokines Inhibit LTP

Numerous studies have studied the impact of cytokines on LTP induction. As elevated pro-inflammatory cytokine levels are associated with inflammatory disease states and "sickness behavior"—that is, decreased motor activity, social withdrawal, reduced food and water intake, increased slow-wave sleep, and altered cognition [138]—it is perhaps not surprising that most pro-inflammatory cytokines affect long-term plasticity.

In contrast to its potentiating effects on basal neurotransmission [106, 139], treatment with TNFα blocks the induction of perforant path-dentate gyrus LTP [140, 141]. Early-phase LTP blockade in the dentate gyrus requires signaling through TNFR1, p38 MAPK, mGluRI, and c-Jun N-terminal kinase (JNK) [141–143]. However, late-phase LTP (2–3 h post tetanus) blockade in the dentate gyrus is independent of p38 MAPK [142], suggesting that TNFα may use distinct pathways to control receptor trafficking-dependent (early-phase) LTP versus transcription-dependent (late-phase) LTP. In contrast, TNFα does not affect Schaffer collateral-CA1 LTP and LTD [26, 87]. While some reports have suggested that CA1 LTP is blocked by TNFα treatment [139] and that CA1 LTD is absent in TNFR knockout mice [144], recent studies have convincingly shown that, despite increasing surface levels of GluA1 and increasing the AMPAR to NMDAR current ratio, treatment with TNFα does not affect the induction of LTP or LTD at Schaffer collateral-CA1 synapses [26]; similarly, TNFα knockout and TNFR1/TNFR2 double-knockout mice display normal CA1 LTP and LTD [26] and normal LTP in $TNF\alpha^{-/-}$ visual cortex slices [87].

In contrast, IL-1β inhibits LTP broadly in the hippocampus: in areas CA1 [145], CA3 [146], and the dentate gyrus [140, 147]. Inhibition in the dentate gyrus is observed for both NMDAR-dependent [147] and NMDAR-independent [148] forms of LTP and is dependent on activation of p38 MAPK [148, 149] and JNK [143]. In apparent contradiction, however, induction of LTP in the hippocampus increases IL-1β gene expression, and IL-1β appears to be required for the maintenance phase of CA1 LTP [150]. IL-18—a member of the IL-1 superfamily—also impairs LTP induction in the dentate gyrus through mechanisms involving mGluRI and the classic inflammatory signaling components cyclooxygenase-2 (COX-2), peroxisome proliferator-activated receptor-gamma (PPARγ), and inducible nitric oxide synthase (iNOS) [151–153]. The effects of IL-1β on hippocampal LTD remain largely unexplored, with only one study suggesting that LTD induction is unaffected at Schaffer collateral-CA1 synapses [154]. Similarly, CA1 LTP and LTD are normal in IL-1α/IL-1β double-knockout mice [154].

Other cytokines also significantly inhibit LTP in the hippocampus. IL-6 inhibits LTP at Schaffer collateral-CA1 synapses [155, 156] and in the dentate gyrus [157]. At least in the CA1 region, this occurs through activation of tyrosine kinases and potentially through activation of STAT3 and MAPK/ERK [156]. However, similar to IL-1β, IL-6 gene expression is increased in the hippocampus following LTP induction, and blockade of IL-6 using an anti-IL-6 Ab sustains high levels of LTP at 90 min after tetanus [158]. Lastly, several other cytokines have been reported to inhibit LTP induction or maintenance, including IL-2 [159], IL-8 [160], and interferon [161, 162]. Given the varied and somewhat inconsistent results, it is likely that cytokines modulate various forms of Hebbian plasticity but are not part of the core mechanisms of this type of plasticity. The details of the modulation of plasticity induction thresholds or maintenance remain to be determined.

Neuronal pentraxins may also regulate certain forms of plasticity. Cleavage of the NPR complex is required for mGluR1/5-dependent LTD [163], although NMDAR-dependent LTP and LTD are normal in the NP1/NP2 knockout mouse [74].

3.7 Requirement of Neuroimmune Molecules for Expression of Homeostatic Synaptic Plasticity

3.7.1 Introduction to Homeostatic Synaptic Plasticity

Under normal physiological conditions, neuronal network activity is stably maintained within a functional range. For example, excitation and inhibition must be set to appropriate levels in order to prevent network activity from ending prematurely or increasing to epileptiform levels [164]. Furthermore, because "Hebbian" forms of plasticity such as LTP and LTD can lead to unconstrained positive feedback and eventual occlusion of further plasticity, mechanisms that conserve total synaptic strength are thought to exist [165, 166]. It is conceivable that numerous mechanisms contribute to generating network stability and conserving total synaptic strength, for example, through regulation of synapse development and refinement, intrinsic excitability, intracellular chloride concentration, neurotransmitter release probability, and metabotropic and ionotropic neurotransmitter-gated receptors. To date, the majority of experimental evidence has focused on mechanisms that regulate neurotransmitter release and postsynaptic efficacy through receptor trafficking (reviewed in [166, 167]). Collectively referred to as homeostatic synaptic plasticity, these negative feedback mechanisms are used by neurons to compensate for experimentally induced chronically elevated activity [typically induced pharmacologically by GABA(A)R blockade for 24–48 h] or chronic inactivity [typically induced by action potential blockade with tetrodotoxin (TTX) for 24–48 h] [168]. For example, prolonged activity elevation results in a compensatory decrease in mEPSC amplitude (referred to as "scaling down") [168–172] and an increase in mIPSC amplitude [173, 174] on excitatory principal neurons. Conversely, prolonged activity blockade results in a compensatory increase in mEPSC amplitude (referred to as "scaling up") [26, 168, 175–177] and a decrease in mIPSC amplitude [178, 179]. Homeostatic synaptic plasticity has been observed both globally (i.e., at all synapses, in a multiplicative manner called "synaptic scaling" [166, 168]) and locally (within dendritic regions or at individual synapses [180, 181]). Several forms of homeostatic synaptic plasticity have been characterized on the basis of whether action potentials alone are blocked (typically slower expression, i.e., after >24 h [26, 168, 181]) or whether both mEPSCs and action potentials are concurrently blocked (typically faster expression, as rapidly as within 3 h [181–183]). Distinct signaling mechanisms underlie each of these induction protocols (reviewed in [166, 167]). The neuroimmune molecules TNFα, MHCI, and Narp have been implicated in homeostatic plasticity observed in response to activity blockade (Fig. 3.2).

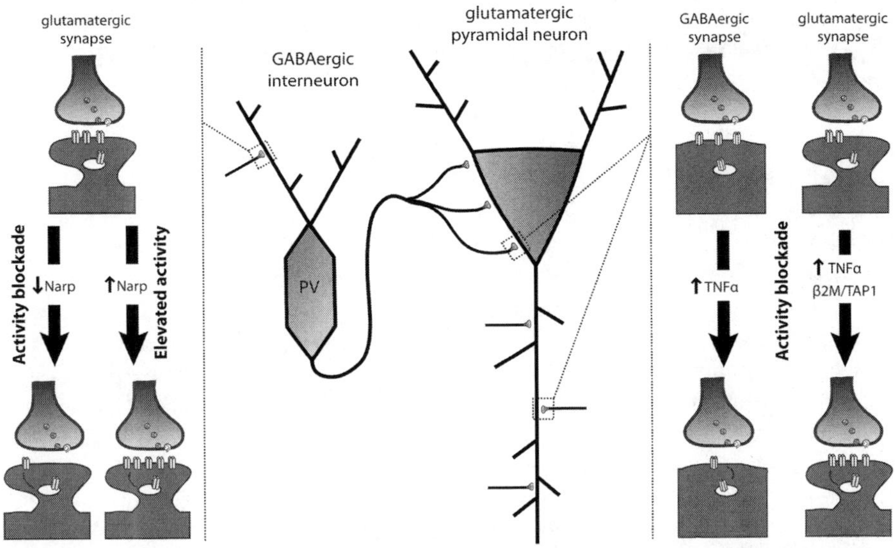

Fig. 3.2 Requirement of neuroimmune molecules for expression of homeostatic synaptic plasticity. *Right panel.* In response to activity blockade, GABAergic synapses onto pyramidal neurons are weakened through removal of postsynaptic GABA(A)Rs, while glutamatergic synapses onto pyramidal neurons are strengthened through addition of AMPARs. These homeostatic adaptations require soluble TNFα, which is elevated in response to activity blockade. At glutamatergic synapses, β2M/TAP 1 (necessary for MHCI surface expression) is also required for the homeostatic response to activity blockade. *Left panel.* At glutamatergic synapses onto parvalbumin-expressing GABAergic interneurons (PV), Narp is required for an inverse form of homeostatic plasticity: activity blockade weakens glutamatergic synapse while elevated activity strengthens them through trafficking of GluA4-containing AMPARs. Surface expression of Narp is bidirectionally regulated in the same manner as GluA4 surface expression, suggesting that Narp controls the level of AMPARs at PV excitatory synapses in proportion to the level of circuit activity

3.7.2 Requirement of TNFα for Inactivity-Induced Homeostatic Synaptic Plasticity

As discussed above in this chapter, exogenous application of TNFα to dissociated hippocampal neuron cultures or acute hippocampal slices concomitantly increases AMPAR-mediated currents while decreasing GABA(A)R-mediated currents [106]. The similarity of these findings to the compensatory response occurring following prolonged activity blockade [168, 178] prompted an investigation into the role of TNFα in homeostatic synaptic plasticity [26]. Indeed, interfering with TNFα signaling during prolonged TTX treatment by adding a soluble form of TNFR1 (sTNFR) blocks scaling up of excitatory synapses and scaling down of inhibitory synapses [26]. Similarly, dissociated hippocampal cultures and hippocampal or visual cortex organotypic slice cultures from TNFα knockout mice also do not display scaling up in response to prolonged treatment with TTX or CNQX/APV (AMPAR/NMDAR blockers) [26, 87]. The absence of scaling up in TNFα$^{-/-}$ cultures when induced either by prolonged action potential blockade alone or concomitant blockade of mEPSCs and action potentials suggests that TNFα is required for several forms of scaling up. However, TNFα is not involved in the process of scaling down, as TNFα$^{-/-}$ cultures maintain the ability to tune down excitatory synapses in response to prolonged treatment with picrotoxin [PTX, a GABA(A)R antagonist] [26]. Moreover, the TNFα is of glial origin, as TNFα$^{-/-}$ neurons grown on wild-type glia demonstrate synaptic scaling, while wild-type neurons grown on TNFα$^{-/-}$ glia do not [26]. Recent work has suggested that TNFα may act as a permissive rather than instructive cue for scaling up because a brief blockade with sTNFR following 24 h of TTX treatment cannot reverse scaling [184]. However, this interpretation is inconsistent with the previous finding that acute application of conditioned medium from TTX-treated cultures to naïve sister cultures rapidly increases mEPSC amplitudes—an effect prevented by pretreating the conditioned medium with sTNFR [26]. Further work is necessary to differentiate between the acute effects of TNFα on receptor trafficking versus the effects of prolonged exposure to TNFα, as the latter could also induce a persistent, transcription-dependent, scaled up state.

Studies have shown that the characteristics of homeostatic synaptic plasticity observed in vitro are also present in vivo for the visual cortex. Manipulations that decrease visual activity at developmental [185] and adult [186] stages result in the homeostatic increase in synaptic strength of the visual cortex. Similarly, the in vitro findings implicating TNFα in homeostatic synaptic plasticity have been extended in vivo using a model for developmental plasticity involving the readjustment of the visual system to a chronic decrease in activity [87]. As noted above, during monocular deprivation during the critical period, TNFα knockout mice display the normal loss of responsiveness to the closed eye but are completely deficient in the gain responsiveness to the open eye [87]. The failure to gain responsiveness to the open eye is consistent with a loss of homeostatic scaling up in response to deprivation of visual activity from the closed eye. In contrast to the findings in young animals, recent findings indicate that ocular dominance plasticity in the adult is independent of TNFα-mediated multiplicative scaling [88], which is consistent with previous findings of a distinct non-multiplicative form of homeostatic synaptic plasticity in the adult visual cortex [186].

3.7.3 Involvement of MHCI in Inactivity-Induced Homeostatic Synaptic Plasticity

A role for MHCI molecules in brain function was initially suggested by a screen for altered gene expression in response to action potential blockade in the developing LGN [56]. It is therefore not surprising that MHCI molecules play a role in homeostatic regulation of synapse function. Mature (14DIV) cultures of dissociated hippocampal neurons from β2M/TAP1 double-knockout mice fail to scale up mEPSC amplitudes in response to several days of TTX treatment [70]. Younger (8DIV) dissociated cortical cultures treated with TTX for 24 h display decreased levels of surface MHCI and an increase in synapse density that is blocked by MHCI overexpression [68]. Further studies are necessary to determine whether MHCI signaling plays a role in synaptic scaling down at excitatory synapses or in homeostatic synaptic plasticity at inhibitory synapses.

3.7.4 Narp-Dependent Homeostatic Synaptic Plasticity

Evidence suggests that homeostatic synaptic plasticity at inhibitory (GABAergic) interneurons either does not occur or occurs in the opposite direction to that of glutamatergic neurons [109, 168, 187]. This would either maintain the normal activity of interneurons, thus leaving the homeostatic response of principal neurons unaffected, or the inverted homeostatic response of interneurons would enhance the homeostatic response of principal neurons. Narp (neuronal activity-regulated pentraxin, also referred to as NP2) is one example of a protein involved in regulating homeostatic scaling of excitatory synapses on parvalbumin (PV)-expressing interneurons (the most abundant subtype of interneuron in the hippocampus [188]). Narp is enriched at excitatory synapses on PV interneurons, and its surface expression in vitro is bidirectionally controlled by chronic activity manipulation—48-h incubation with either TTX or bicuculline decreases or increases surface Narp, respectively [187]. The expression of the AMPAR subunit GluA4 parallels that of surface Narp in response to chronic activity manipulation and requires Narp. Indeed, compared to glutamatergic neurons, PV interneurons respond inversely to chronic activity manipulations by decreasing mEPSC amplitude in response to TTX treatment and increasing mEPSC amplitude in response to bicuculline treatment [187]; crucially, this inverse homeostatic response is absent in $Narp^{-/-}$ cultures. Narp knockout mice are hypersensitive to kindling-induced seizures, suggesting that Narp-dependent regulation of PV interneuron excitatory synapses is essential to the adaptive homeostatic response of the network as a whole in response to seizure development [187].

3.7.5 Conclusion

Accumulating evidence implicates the neuroimmune molecules TNFα, MHCI, and Narp in regulation of homeostatic synaptic plasticity. Despite belonging to three different classes of neuroimmune molecules, the expression of all three proteins is regulated by prolonged activity manipulations. There may be a common activity sensor that regulates the expression of one or several of these molecules, which could in turn regulate other components of the homeostatic plasticity response. For example, activity regulates *Narp* gene expression [63], glutamate regulates TNFα secretion [26], and TNFα is known to regulate MHCI expression in hippocampal neurons [189].

3.8 Control of Gliotransmission by TNFα

Gliotransmission describes the process whereby glial cells (primarily astrocytes) participate in the control of synaptic neurotransmission by releasing neuroactive molecules (gliotransmitters) such as glutamate, D-serine, ATP, adenosine, GABA, and others [190]. Typically, these molecules are released in response to Ca^{2+} transients in astrocytes in response to neuronal activity [191]. One example of gliotransmission occurs in the dentate gyrus molecular layer: activation of the perforant path synapses onto granule cells (PP–GC synapses) induces a Ca^{2+}-dependent astrocytic release of glutamate which acts on presynaptic NMDARs to induce potentiation of excitatory transmission at PP–GC synapses [192]. This potentiation of excitatory transmission is associated with an increase in synaptic release probability as assessed by mEPSC frequency [97]. Intriguingly, in $TNF\alpha^{-/-}$ slices, this phenomenon is absent despite a normal occurrence of astrocytic Ca^{2+} transients [97]. In fact, mechanistically, TNFα controls the release of glutamate from astrocytes by promoting efficient docking and synchronous exocytosis of glutamate-containing astrocytic vesicles. Thus, in the absence of TNFα, release of glutamate lacks temporal focus and likely does not reach the concentrations necessary to activate presynaptic NMDARs [97]. These findings highlight an additional role for normal brain levels of TNFα in regulating exocytosis of gliotransmitter vesicles, perhaps in analogous fashion to its ability to induce rapid surface expression of AMPARs [26, 106].

3.9 Neuroimmune Molecules Regulate Synapse Function in Pathological States

Dysregulation of CNS function occurs in a number of pathological states, including stroke, chronic pain, and common neurodegenerative diseases such as Alzheimer's, Parkinson's, Huntington's, amyotrophic lateral sclerosis (ALS), and glaucoma. Dysregulation of synapse function is a major component of many of these conditions, thus leading to the hypothesis that many of the neuroimmune molecules discussed above may also regulate synapse function in pathological CNS states.

Inflammatory responses to injury and prolonged CNS disease lead to the production of high levels of pro-inflammatory cytokines [193, 194]. Pro-inflammatory cytokines are prime candidates for the induction and maintenance of chronic pain, owing to their effects on AMPAR and GABA(A)R trafficking which result in increased excitatory synaptic transmission and decreased inhibitory transmission [106, 107, 117]. Furthermore, TNFα-induced delivery of calcium-permeable

AMPARs at the synapse [106, 109] constitutes an important mechanism underlying neuronal death due to excitotoxicity [110, 194] and spinal cord injury [111]. Similarly, an increase in synaptic GluA1 and a decrease of synaptic GluA2 in presymptomatic G93ASOD1 mice (a model of ALS) [195] may result in increased Ca^{2+}-permeable (GluA2-lacking) AMPARs at the synapse, which in turn may contribute to neuronal death in ALS. Conversely, several studies have shown a neuroprotective effect of TNFα and other pro-inflammatory cytokines in various models of excitotoxicity [196–198].

Significant evidence implicates the complement cascade component C1q in early stages of neurodegenerative disease. C1q is elevated in the early stages of Alzheimer's disease and is concentrated in amyloid beta (Aβ) plaques [199–202]. In addition, injections of C1q in the hippocampus and cortex of Alzheimer's disease mouse models increase fibrillar Aβ [203], which is consistent with findings that C1q inhibits the uptake of Aβ by microglia [204]. Taken together, these findings suggest that the presence of C1q contributes to Alzheimer's pathology by modulating the immune response to Aβ. This hypothesis is supported by in vivo evidence showing that absence of C1q in mouse models of Alzheimer's disease reduces neuropathology [205]. C1q may also contribute to the neurodegeneration of RGCs characteristic of glaucoma. C1q levels in the retina are upregulated in the early stages of a mouse model of glaucoma [62, 206, 207].

In conclusion, the activation of microglia and the expression of immune-related molecules are commonly associated with various neurodegenerative diseases (reviewed in [208]). However, the mechanistic roles played by immune molecules in development, maintenance, and resolution of CNS pathology remain largely unexplored.

3.10 Regulation of Cognitive Function by Cytokines

Much of this chapter has focused on the role of neuroimmune molecules in regulating synapse development and function. It is intriguing to consider how effects at the synaptic level may translate to the level of neural network activity and behavior. While an extensive review of this topic is beyond the scope of this chapter, a brief overview is offered here. As mentioned above, elevation of pro-inflammatory cytokines induces a variety of behavioral effects—including decreased motor activity, social withdrawal, reduced food and water intake, and increased slow-wave sleep—collectively termed "sickness behavior" [138]. It remains to be fully understood to what extent this behavioral phenotype relies on direct cytokine-mediated modulation of synaptic efficacy via effects on receptor trafficking and activity-dependent plasticity. Numerous cognitive-behavioral studies in rodents have investigated the involvement of the pro-inflammatory cytokines TNFα, IL-1, and IL-6 in learning and memory tasks (reviewed in [9]), finding either impairment or enhancement in cognitive tasks in the absence of the cytokine or its receptor(s). Many such studies would benefit from brain- or cell-type-specific conditional knockout strategies in order to differentiate between central and peripheral cytokine functions. Interestingly, pro-inflammatory cytokines may also regulate cognitive function by regulating adult neurogenesis: TNFα, IL-1β, and IL-6 limit adult hippocampal neurogenesis, suggesting a common regulatory mechanism. To date, similar roles for C1q, MHCI, neuronal pentraxins, and other neuroimmune molecules in cognitive function remain largely unexplored.

References

1. Boulanger LM (2009) Immune proteins in brain development and synaptic plasticity. Neuron 64:93–109
2. Shatz CJ (2009) MHC class I: an unexpected role in neuronal plasticity. Neuron 64:40–45
3. Gosselin D, Rivest S (2007) Role of IL-1 and TNF in the brain: twenty years of progress on a Dr. Jekyll/Mr. Hyde duality of the innate immune system. Brain Behav Immun 21:281–289
4. Rivest S (2009) Regulation of innate immune responses in the brain. Nat Rev Immunol 9:429–439
5. Cahoy JD et al (2008) A transcriptome database for astrocytes, neurons, and oligodendrocytes: a new resource for understanding brain development and function. J Neurosci 28:264–278
6. Zhang XM, Zhu J (2011) Kainic acid-induced neurotoxicity: targeting glial responses and glia-derived cytokines. Curr Neuropharmacol 9:388–398
7. Vitkovic L et al (2000) Cytokine signals propagate through the brain. Mol Psychiatry 5:604–615
8. Scheller J, Chalaris A, Schmidt-Arras D, Rose-John S (2011) The pro- and anti-inflammatory properties of the cytokine interleukin-6. Biochim Biophys Acta 1813:878–888
9. McAfoose J, Baune BT (2009) Evidence for a cytokine model of cognitive function. Neurosci Biobehav Rev 33:355–366
10. Locksley RM, Killeen N, Lenardo MJ (2001) The TNF and TNF receptor superfamilies: integrating mammalian biology. Cell 104:487–501
11. Pennica D et al (1984) Human tumour necrosis factor: precursor structure, expression and homology to lymphotoxin. Nature 312:724–729
12. Moss ML et al (1997) Cloning of a disintegrin metalloproteinase that processes precursor tumour-necrosis factor-alpha. Nature 385:733–736
13. Kriegler M, Perez C, DeFay K, Albert I, Lu SD (1988) A novel form of TNF/cachectin is a cell surface cytotoxic transmembrane protein: ramifications for the complex physiology of TNF. Cell 53:45–53
14. Smith RA, Baglioni C (1987) The active form of tumor necrosis factor is a trimer. J Biol Chem 262:6951–6954
15. MacEwan DJ (2002) TNF receptor subtype signalling: differences and cellular consequences. Cell Signal 14:477–492
16. Kinouchi K, Brown G, Pasternak G, Donner DB (1991) Identification and characterization of receptors for tumor necrosis factor-alpha in the brain. Biochem Biophys Res Commun 181:1532–1538
17. Zhang H, Nei H, Dougherty PM (2010) A p38 mitogen-activated protein kinase-dependent mechanism of disinhibition in spinal synaptic transmission induced by tumor necrosis factor-alpha. J Neurosci 30:12844–12855
18. Dopp JM, Mackenzie-Graham A, Otero GC, Merrill JE (1997) Differential expression, cytokine modulation, and specific functions of type-1 and type-2 tumor necrosis factor receptors in rat glia. J Neuroimmunol 75:104–112

19. Yang L, Lindholm K, Konishi Y, Li R, Shen Y (2002) Target depletion of distinct tumor necrosis factor receptor subtypes reveals hippocampal neuron death and survival through different signal transduction pathways. J Neurosci 22:3025–3032
20. Vitkovic L, Bockaert J, Jacque C (2000) “Inflammatory” cytokines: neuromodulators in normal brain? J Neurochem 74:457–471
21. Beattie EC et al (2002) Control of synaptic strength by glial TNFalpha. Science 295:2282–2285
22. Tonelli LH, Postolache TT (2005) Tumor necrosis factor alpha, interleukin-1 beta, interleukin-6 and major histocompatibility complex molecules in the normal brain and after peripheral immune challenge. Neurol Res 27:679–684
23. Wang CX, Shuaib A (2002) Involvement of inflammatory cytokines in central nervous system injury. Prog Neurobiol 67:161–172
24. Li Y, Ji A, Weihe E, Schafer MK (2004) Cell-specific expression and lipopolysaccharide-induced regulation of tumor necrosis factor alpha (TNFalpha) and TNF receptors in rat dorsal root ganglion. J Neurosci 24:9623–9631
25. Wang Y et al (2007) Lipopolysaccharide-induced upregulation of tumor necrosis factor-alpha (TNF-alpha) and TNF receptors in rat sciatic nerve. J Mol Neurosci 32:207–216
26. Stellwagen D, Malenka RC (2006) Synaptic scaling mediated by glial TNF-alpha. Nature 440:1054–1059
27. Braddock M, Quinn A (2004) Targeting IL-1 in inflammatory disease: new opportunities for therapeutic intervention. Nat Rev Drug Discov 3:330–339
28. O’Connor JJ, Coogan AN (1999) Actions of the pro-inflammatory cytokine IL-1 beta on central synaptic transmission. Exp Physiol 84:601–614
29. Bourke E et al (2003) IL-1 beta scavenging by the type II IL-1 decoy receptor in human neutrophils. J Immunol 170:5999–6005
30. Arend WP, Malyak M, Guthridge CJ, Gabay C (1998) Interleukin-1 receptor antagonist: role in biology. Annu Rev Immunol 16:27–55
31. French RA et al (1999) Expression and localization of p80 and p68 interleukin-1 receptor proteins in the brain of adult mice. J Neuroimmunol 93:194–202
32. Hammond EA et al (1999) The interleukin-1 type I receptor is expressed in human hypothalamus. Brain 122(Pt 9):1697–1707
33. Quan N, Zhang Z, Emery M, Bonsall R, Weiss JM (1996) Detection of interleukin-1 bioactivity in various brain regions of normal healthy rats. Neuroimmunomodulation 3:47–55
34. Huitinga I et al (2000) IL-1beta immunoreactive neurons in the human hypothalamus: reduced numbers in multiple sclerosis. J Neuroimmunol 107:8–20
35. Loddick SA et al (1997) Endogenous interleukin-1 receptor antagonist is neuroprotective. Biochem Biophys Res Commun 234:211–215
36. Loddick SA, Liu C, Takao T, Hashimoto K, De Souza EB (1998) Interleukin-1 receptors: cloning studies and role in central nervous system disorders. Brain Res Brain Res Rev 26:306–319
37. Giulian D, Baker TJ, Shih LC, Lachman LB (1986) Interleukin 1 of the central nervous system is produced by ameboid microglia. J Exp Med 164:594–604
38. Pearson VL, Rothwell NJ, Toulmond S (1999) Excitotoxic brain damage in the rat induces interleukin-1beta protein in microglia and astrocytes: correlation with the progression of cell death. Glia 25:311–323
39. Davies CA et al (1999) The progression and topographic distribution of interleukin-1beta expression after permanent middle cerebral artery occlusion in the rat. J Cereb Blood Flow Metab 19:87–98
40. Bauer S, Kerr BJ, Patterson PH (2007) The neuropoietic cytokine family in development, plasticity, disease and injury. Nat Rev Neurosci 8:221–232
41. Juttler E, Tarabin V, Schwaninger M (2002) Interleukin-6 (IL-6): a possible neuromodulator induced by neuronal activity. Neuroscientist 8:268–275
42. Heinrich PC et al (2003) Principles of interleukin (IL)-6-type cytokine signalling and its regulation. Biochem J 374:1–20

43. Gruol DL, Nelson TE (1997) Physiological and pathological roles of interleukin-6 in the central nervous system. Mol Neurobiol 15:307–339
44. Hans VH et al (1999) Experimental axonal injury triggers interleukin-6 mRNA, protein synthesis and release into cerebrospinal fluid. J Cereb Blood Flow Metab 19:184–194
45. Hayashi M, Ueyama T, Nemoto K, Tamaki T, Senba E (2000) Sequential mRNA expression for immediate early genes, cytokines, and neurotrophins in spinal cord injury. J Neurotrauma 17:203–218
46. Suzuki S et al (1999) Temporal profile and cellular localization of interleukin-6 protein after focal cerebral ischemia in rats. J Cereb Blood Flow Metab 19:1256–1262
47. Tarkowski E et al (1995) Early intrathecal production of interleukin-6 predicts the size of brain lesion in stroke. Stroke 26:1393–1398
48. De Simoni MG et al (2000) Inflammatory cytokines and related genes are induced in the rat hippocampus by limbic status epilepticus. Eur J Neurosci 12:2623–2633
49. Peltola J et al (2000) Interleukin-6 and interleukin-1 receptor antagonist in cerebrospinal fluid from patients with recent tonic-clonic seizures. Epilepsy Res 41:205–211
50. Dorfman JR, Zerrahn J, Coles MC, Raulet DH (1997) The basis for self-tolerance of natural killer cells in beta2-microglobulin- and TAP-1- mice. J Immunol 159:5219–5225
51. Tourne S et al (1996) Biosynthesis of major histocompatibility complex molecules and generation of T cells in Ii TAP1 double-mutant mice. Proc Natl Acad Sci USA 93:1464–1469
52. Syken J, Shatz CJ (2003) Expression of T cell receptor beta locus in central nervous system neurons. Proc Natl Acad Sci USA 100:13048–13053
53. Syken J, Grandpre T, Kanold PO, Shatz CJ (2006) PirB restricts ocular-dominance plasticity in visual cortex. Science 313:1795–1800
54. Zohar O et al (2008) Cutting edge: MHC class I-Ly49 interaction regulates neuronal function. J Immunol 180:6447–6451
55. Bryceson YT, Foster JA, Kuppusamy SP, Herkenham M, Long EO (2005) Expression of a killer cell receptor-like gene in plastic regions of the central nervous system. J Neuroimmunol 161:177–182
56. Corriveau RA, Huh GS, Shatz CJ (1998) Regulation of class I MHC gene expression in the developing and mature CNS by neural activity. Neuron 21:505–520
57. Xu HP et al (2010) The immune protein CD3zeta is required for normal development of neural circuits in the retina. Neuron 65:503–515
58. Roumier A et al (2004) Impaired synaptic function in the microglial KARAP/DAP12-deficient mouse. J Neurosci 24:11421–11428
59. Roumier A et al (2008) Prenatal activation of microglia induces delayed impairment of glutamatergic synaptic function. PLoS One 3:e2595
60. Tomasello E et al (2000) Combined natural killer cell and dendritic cell functional deficiency in KARAP/DAP12 loss-of-function mutant mice. Immunity 13:355–364
61. Cullheim S, Thams S (2010) Classic major histocompatibility complex class I molecules: new actors at the neuromuscular junction. Neuroscientist 16:600–607
62. Stevens B et al (2007) The classical complement cascade mediates CNS synapse elimination. Cell 131:1164–1178
63. Xu D et al (2003) Narp and NP1 form heterocomplexes that function in developmental and activity-dependent synaptic plasticity. Neuron 39:513–528
64. Dodds DC, Omeis IA, Cushman SJ, Helms JA, Perin MS (1997) Neuronal pentraxin receptor, a novel putative integral membrane pentraxin that interacts with neuronal pentraxin 1 and 2 and taipoxin-associated calcium-binding protein 49. J Biol Chem 272:21488–21494
65. Schlimgen AK, Helms JA, Vogel H, Perin MS (1995) Neuronal pentraxin, a secreted protein with homology to acute phase proteins of the immune system. Neuron 14:519–526
66. Tsui CC et al (1996) Narp, a novel member of the pentraxin family, promotes neurite outgrowth and is dynamically regulated by neuronal activity. J Neurosci 16:2463–2478
67. Needleman LA, Liu X-B, El-Sabeawy F, Jones EG, McAllister AK (2010) MHC class I molecules are present both pre- and postsynaptically in the visual cortex during postnatal development and in adulthood. Proc Natl Acad Sci USA 107:16999–17004

68. Glynn MW et al (2011) MHCI negatively regulates synapse density during the establishment of cortical connections. Nat Neurosci 14:442–451
69. Fourgeaud L et al (2010) MHC class I modulates NMDA receptor function and AMPA receptor trafficking. Proc Natl Acad Sci USA 107:22278–22283
70. Goddard CA, Butts DA, Shatz CJ (2007) Regulation of CNS synapses by neuronal MHC class I. Proc Natl Acad Sci USA 104:6828–6833
71. Yoshida T et al (2012) Interleukin-1 receptor accessory protein organizes neuronal synaptogenesis as a cell adhesion molecule. J Neurosci 32:2588–2600
72. O'Brien RJ et al (1999) Synaptic clustering of AMPA receptors by the extracellular immediate-early gene product Narp. Neuron 23:309–323
73. O'Brien R et al (2002) Synaptically targeted narp plays an essential role in the aggregation of AMPA receptors at excitatory synapses in cultured spinal neurons. J Neurosci 22:4487–4498
74. Bjartmar L et al (2006) Neuronal pentraxins mediate synaptic refinement in the developing visual system. J Neurosci 26:6269–6281
75. Leamey CA, Van Wart A, Sur M (2009) Intrinsic patterning and experience-dependent mechanisms that generate eye-specific projections and binocular circuits in the visual pathway. Curr Opin Neurobiol 19:181–187
76. Ruthazer ES, Aizenman CD (2010) Learning to see: patterned visual activity and the development of visual function. Trends Neurosci 33:183–192
77. Guido W (2008) Refinement of the retinogeniculate pathway. J Physiol 586:4357–4362
78. Feller MB (2009) Retinal waves are likely to instruct the formation of eye-specific retinogeniculate projections. Neural Dev 4:24
79. Katz LC, Shatz CJ (1996) Synaptic activity and the construction of cortical circuits. Science 274:1133–1138
80. Huh GS et al (2000) Functional requirement for class I MHC in CNS development and plasticity. Science 290:2155–2159
81. Datwani A et al (2009) Classical MHCI molecules regulate retinogeniculate refinement and limit ocular dominance plasticity. Neuron 64:463–470
82. Nauta AJ et al (2003) Biochemical and functional characterization of the interaction between pentraxin 3 and C1q. Eur J Immunol 33:465–473
83. Lee RH et al (2010) Neurodevelopmental effects of chronic exposure to elevated levels of pro-inflammatory cytokines in a developing visual system. Neural Dev 5:2
84. Golan H, Levav T, Mendelsohn A, Huleihel M (2004) Involvement of tumor necrosis factor alpha in hippocampal development and function. Cereb Cortex 14:97–105
85. Wiesel TN (1982) Postnatal development of the visual cortex and the influence of environment. Nature 299:583–591
86. Frenkel MY, Bear MF (2004) How monocular deprivation shifts ocular dominance in visual cortex of young mice. Neuron 44:917–923
87. Kaneko M, Stellwagen D, Malenka RC, Stryker MP (2008) Tumor necrosis factor-alpha mediates one component of competitive, experience-dependent plasticity in developing visual cortex. Neuron 58:673–680
88. Ranson A, Cheetham CEJ, Fox K, Sengpiel F (2012) Homeostatic plasticity mechanisms are required for juvenile, but not adult, ocular dominance plasticity. Proc Natl Acad Sci USA 109:1311–1316
89. Chu Y et al (2010) Enhanced synaptic connectivity and epilepsy in C1q knockout mice. Proc Natl Acad Sci USA 107:7975–7980
90. McConnell MJ, Huang YH, Datwani A, Shatz CJ (2009) H2-K(b) and H2-D(b) regulate cerebellar long-term depression and limit motor learning. Proc Natl Acad Sci USA 106:6784–6789
91. Letellier M, Willson ML, Gautheron V, Mariani J, Lohof AM (2008) Normal adult climbing fiber monoinnervation of cerebellar Purkinje cells in mice lacking MHC class I molecules. Dev Neurobiol 68:997–1006
92. Thams S et al (2009) Classical major histocompatibility complex class I molecules in motoneurons: new actors at the neuromuscular junction. J Neurosci 29:13503–13515

93. Atwal JK et al (2008) PirB is a functional receptor for myelin inhibitors of axonal regeneration. Science 322:967–970
94. D'Arcangelo G et al (2000) Interleukin-6 inhibits neurotransmitter release and the spread of excitation in the rat cerebral cortex. Eur J Neurosci 12:1241–1252
95. Zhu PJ et al (2011) Suppression of PKR promotes network excitability and enhanced cognition by interferon-gamma-mediated disinhibition. Cell 147:1384–1396
96. Grassi F et al (1994) TNF-alpha increases the frequency of spontaneous miniature synaptic currents in cultured rat hippocampal neurons. Brain Res 659:226–230
97. Santello M, Bezzi P, Volterra A (2011) TNFalpha controls glutamatergic gliotransmission in the hippocampal dentate gyrus. Neuron 69:988–1001
98. Nakamura K et al (2007) CD3 and immunoglobulin G Fc receptor regulate cerebellar functions. Mol Cell Biol 27:5128–5134
99. Malenka RC, Bear MF (2004) LTP and LTD: an embarrassment of riches. Neuron 44:5–21
100. Greger IH, Ziff EB, Penn AC (2007) Molecular determinants of AMPA receptor subunit assembly. Trends Neurosci 30:407–416
101. Daw NW, Stein PS, Fox K (1993) The role of NMDA receptors in information processing. Annu Rev Neurosci 16:207–222
102. Paoletti P, Neyton J (2007) NMDA receptor subunits: function and pharmacology. Curr Opin Pharmacol 7:39–47
103. Pachernegg S, Strutz-Seebohm N, Hollmann M (2012) GluN3 subunit-containing NMDA receptors: not just one-trick ponies. Trends Neurosci 35:240–249
104. Jacob TC, Moss SJ, Jurd R (2008) GABA(A) receptor trafficking and its role in the dynamic modulation of neuronal inhibition. Nat Rev Neurosci 9:331–343
105. Luscher B, Fuchs T, Kilpatrick CL (2011) GABAA receptor trafficking-mediated plasticity of inhibitory synapses. Neuron 70:385–409
106. Stellwagen D, Beattie EC, Seo JY, Malenka RC (2005) Differential regulation of AMPA receptor and GABA receptor trafficking by tumor necrosis factor-alpha. J Neurosci 25:3219–3228
107. Kawasaki Y, Zhang L, Cheng JK, Ji RR (2008) Cytokine mechanisms of central sensitization: distinct and overlapping role of interleukin-1beta, interleukin-6, and tumor necrosis factor-alpha in regulating synaptic and neuronal activity in the superficial spinal cord. J Neurosci 28:5189–5194
108. Han P, Whelan PJ (2010) Tumor necrosis factor alpha enhances glutamatergic transmission onto spinal motoneurons. J Neurotrauma 27:287–292
109. Ogoshi F et al (2005) Tumor necrosis-factor-alpha (TNF-alpha) induces rapid insertion of Ca^{2+}–permeable alpha-amino-3-hydroxyl-5-methyl-4-isoxazole-propionate (AMPA)/kainate (Ca-A/K) channels in a subset of hippocampal pyramidal neurons. Exp Neurol 193:384–393
110. Leonoudakis D, Zhao P, Beattie EC (2008) Rapid tumor necrosis factor alpha-induced exocytosis of glutamate receptor 2-lacking AMPA receptors to extrasynaptic plasma membrane potentiates excitotoxicity. J Neurosci 28:2119–2130
111. Ferguson AR et al (2008) Cell death after spinal cord injury is exacerbated by rapid TNF alpha-induced trafficking of GluR2-lacking AMPARs to the plasma membrane. J Neurosci 28:11391–11400
112. He P, Liu Q, Wu J, Shen Y (2012) Genetic deletion of TNF receptor suppresses excitatory synaptic transmission via reducing AMPA receptor synaptic localization in cortical neurons. FASEB J 26:334–345
113. Furukawa K, Mattson MP (1998) The transcription factor NF-kappaB mediates increases in calcium currents and decreases in NMDA- and AMPA/kainate-induced currents induced by tumor necrosis factor-alpha in hippocampal neurons. J Neurochem 70:1876–1886
114. Buzsaki G (2001) Hippocampal GABAergic interneurons: a physiological perspective. Neurochem Res 26:899–905
115. Farrant M, Nusser Z (2005) Variations on an inhibitory theme: phasic and tonic activation of GABA(A) receptors. Nat Rev Neurosci 6:215–229

116. Duveau V et al (2011) Spatiotemporal specificity of GABAA receptor-mediated regulation of adult hippocampal neurogenesis. Eur J Neurosci 34:362–373
117. Kanno T, Nagata T, Yamamoto S, Okamura H, Nishizaki T (2004) Interleukin-18 stimulates synaptically released glutamate and enhances postsynaptic AMPA receptor responses in the CA1 region of mouse hippocampal slices. Brain Res 1012:190–193
118. Viviani B et al (2003) Interleukin-1beta enhances NMDA receptor-mediated intracellular calcium increase through activation of the Src family of kinases. J Neurosci 23:8692–8700
119. Huang Y, Smith DE, Ibanez-Sandoval O, Sims JE, Friedman WJ (2011) Neuron-specific effects of interleukin-1beta are mediated by a novel isoform of the IL-1 receptor accessory protein. J Neurosci 31:18048–18059
120. Viviani B et al (2006) Interleukin-1 beta released by gp120 drives neural death through tyrosine phosphorylation and trafficking of NMDA receptors. J Biol Chem 281: 30212–30222
121. Nelson TE, Netzeband JG, Gruol DL (2004) Chronic interleukin-6 exposure alters metabotropic glutamate receptor-activated calcium signalling in cerebellar Purkinje neurons. Eur J Neurosci 20:2387–2400
122. Nelson TE, Ur CL, Gruol DL (2002) Chronic interleukin-6 exposure alters electrophysiological properties and calcium signaling in developing cerebellar purkinje neurons in culture. J Neurophysiol 88:475–486
123. Ali C et al (2000) Ischemia-induced interleukin-6 as a potential endogenous neuroprotective cytokine against NMDA receptor-mediated excitotoxicity in the brain. J Cereb Blood Flow Metab 20:956–966
124. Pizzi M et al (2004) Prevention of neuron and oligodendrocyte degeneration by interleukin-6 (IL-6) and IL-6 receptor/IL-6 fusion protein in organotypic hippocampal slices. Mol Cell Neurosci 25:301–311
125. Pieraut S et al (2011) An autocrine neuronal interleukin-6 loop mediates chloride accumulation and NKCC1 phosphorylation in axotomized sensory neurons. J Neurosci 31: 13516–13526
126. Staley K, Smith R (2001) A new form of feedback at the GABA(A) receptor. Nat Neurosci 4:674–676
127. Cooke SF, Bliss TV (2006) Plasticity in the human central nervous system. Brain 129:1659–1673
128. Kandel ER (2001) The molecular biology of memory storage: a dialogue between genes and synapses. Science 294:1030–1038
129. Barrionuevo G, Brown TH (1983) Associative long-term potentiation in hippocampal slices. Proc Natl Acad Sci USA 80:7347–7351
130. Levy WB, Steward O (1979) Synapses as associative memory elements in the hippocampal formation. Brain Res 175:233–245
131. Albensi BC, Oliver DR, Toupin J, Odero G (2007) Electrical stimulation protocols for hippocampal synaptic plasticity and neuronal hyper-excitability: are they effective or relevant? Exp Neurol 204:1–13
132. Vogt KE, Canepari M (2010) On the induction of postsynaptic granule cell-Purkinje neuron LTP and LTD. Cerebellum 9:284–290
133. Duguid I, Sjostrom PJ (2006) Novel presynaptic mechanisms for coincidence detection in synaptic plasticity. Curr Opin Neurobiol 16:312–322
134. Kessels HW, Malinow R (2009) Synaptic AMPA receptor plasticity and behavior. Neuron 61:340–350
135. Fioravante D, Regehr WG (2011) Short-term forms of presynaptic plasticity. Curr Opin Neurobiol 21:269–274
136. Barco A et al (2005) Gene expression profiling of facilitated L-LTP in VP16-CREB mice reveals that BDNF is critical for the maintenance of LTP and its synaptic capture. Neuron 48:123–137
137. Raiker SJ et al (2010) Oligodendrocyte-myelin glycoprotein and Nogo negatively regulate activity-dependent synaptic plasticity. J Neurosci 30:12432–12445

138. Dantzer R, O'Connor JC, Freund GG, Johnson RW, Kelley KW (2008) From inflammation to sickness and depression: when the immune system subjugates the brain. Nat Rev Neurosci 9:46–56
139. Tancredi V et al (1992) Tumor necrosis factor alters synaptic transmission in rat hippocampal slices. Neurosci Lett 146:176–178
140. Cunningham AJ, Murray CA, O'Neill LA, Lynch MA, O'Connor JJ (1996) Interleukin-1 beta (IL-1 beta) and tumour necrosis factor (TNF) inhibit long-term potentiation in the rat dentate gyrus in vitro. Neurosci Lett 203:17–20
141. Wang Q, Wu J, Rowan MJ, Anwyl R (2005) Beta-amyloid inhibition of long-term potentiation is mediated via tumor necrosis factor. Eur J Neurosci 22:2827–2832
142. Butler MP, O'Connor JJ, Moynagh PN (2004) Dissection of tumor-necrosis factor-alpha inhibition of long-term potentiation (LTP) reveals a p38 mitogen-activated protein kinase-dependent mechanism which maps to early-but not late-phase LTP. Neuroscience 124:319–326
143. Curran BP, Murray HJ, O'Connor JJ (2003) A role for c-Jun N-terminal kinase in the inhibition of long-term potentiation by interleukin-1beta and long-term depression in the rat dentate gyrus in vitro. Neuroscience 118:347–357
144. Albensi BC, Mattson MP (2000) Evidence for the involvement of TNF and NF-kappaB in hippocampal synaptic plasticity. Synapse 35:151–159
145. Bellinger FP, Madamba S, Siggins GR (1993) Interleukin 1 beta inhibits synaptic strength and long-term potentiation in the rat CA1 hippocampus. Brain Res 628:227–234
146. Katsuki H et al (1990) Interleukin-1 beta inhibits long-term potentiation in the CA3 region of mouse hippocampal slices. Eur J Pharmacol 181:323–326
147. Coogan A, O'Connor JJ (1997) Inhibition of NMDA receptor-mediated synaptic transmission in the rat dentate gyrus in vitro by IL-1 beta. Neuroreport 8:2107–2110
148. Coogan AN, O'Connor JJ (1999) Interleukin-1beta inhibits a tetraethylammonium-induced synaptic potentiation in the rat dentate gyrus in vitro. Eur J Pharmacol 374:197–206
149. Coogan AN, O'Neill LA, O'Connor JJ (1999) The P38 mitogen-activated protein kinase inhibitor SB203580 antagonizes the inhibitory effects of interleukin-1beta on long-term potentiation in the rat dentate gyrus in vitro. Neuroscience 93:57–69
150. Schneider H et al (1998) A neuromodulatory role of interleukin-1beta in the hippocampus. Proc Natl Acad Sci USA 95:7778–7783
151. Cumiskey D, Curran BP, Herron CE, O'Connor JJ (2007) A role for inflammatory mediators in the IL-18 mediated attenuation of LTP in the rat dentate gyrus. Neuropharmacology 52:1616–1623
152. Cumiskey D, Pickering M, O'Connor JJ (2007) Interleukin-18 mediated inhibition of LTP in the rat dentate gyrus is attenuated in the presence of mGluR antagonists. Neurosci Lett 412:206–210
153. Curran B, O'Connor JJ (2001) The pro-inflammatory cytokine interleukin-18 impairs long-term potentiation and NMDA receptor-mediated transmission in the rat hippocampus in vitro. Neuroscience 108:83–90
154. Ikegaya Y, Delcroix I, Iwakura Y, Matsuki N, Nishiyama N (2003) Interleukin-1beta abrogates long-term depression of hippocampal CA1 synaptic transmission. Synapse 47:54–57
155. Li AJ, Katafuchi T, Oda S, Hori T, Oomura Y (1997) Interleukin-6 inhibits long-term potentiation in rat hippocampal slices. Brain Res 748:30–38
156. Tancredi V et al (2000) The inhibitory effects of interleukin-6 on synaptic plasticity in the rat hippocampus are associated with an inhibition of mitogen-activated protein kinase ERK. J Neurochem 75:634–643
157. Bellinger FP, Madamba SG, Campbell IL, Siggins GR (1995) Reduced long-term potentiation in the dentate gyrus of transgenic mice with cerebral overexpression of interleukin-6. Neurosci Lett 198:95–98
158. Balschun D et al (2004) Interleukin-6: a cytokine to forget. FASEB J 18:1788–1790
159. Tancredi V, Zona C, Velotti F, Eusebi F, Santoni A (1990) Interleukin-2 suppresses established long-term potentiation and inhibits its induction in the rat hippocampus. Brain Res 525:149–151

160. Xiong H et al (2003) Inhibition of long-term potentiation by interleukin-8: implications for human immunodeficiency virus-1-associated dementia. J Neurosci Res 71:600–607
161. D'Arcangelo G et al (1991) Interferon inhibits synaptic potentiation in rat hippocampus. Brain Res 564:245–248
162. Mendoza-Fernandez V, Andrew RD, Barajas-Lopez C (2000) Interferon-alpha inhibits long-term potentiation and unmasks a long-term depression in the rat hippocampus. Brain Res 885:14–24
163. Cho RW et al (2008) mGluR1/5-dependent long-term depression requires the regulated ectodomain cleavage of neuronal pentraxin NPR by TACE. Neuron 57:858–871
164. Turrigiano GG, Nelson SB (2004) Homeostatic plasticity in the developing nervous system. Nat Rev Neurosci 5:97–107
165. Miller KD (1996) Synaptic economics: competition and cooperation in synaptic plasticity. Neuron 17:371–374
166. Turrigiano GG (2008) The self-tuning neuron: synaptic scaling of excitatory synapses. Cell 135:422–435
167. Pozo K, Goda Y (2010) Unraveling mechanisms of homeostatic synaptic plasticity. Neuron 66:337–351
168. Turrigiano GG, Leslie KR, Desai NS, Rutherford LC, Nelson SB (1998) Activity-dependent scaling of quantal amplitude in neocortical neurons. Nature 391:892–896
169. Leslie KR, Nelson SB, Turrigiano GG (2001) Postsynaptic depolarization scales quantal amplitude in cortical pyramidal neurons. J Neurosci 21:RC170
170. Pak DT, Sheng M (2003) Targeted protein degradation and synapse remodeling by an inducible protein kinase. Science 302:1368–1373
171. Seeburg DP, Feliu-Mojer M, Gaiottino J, Pak DT, Sheng M (2008) Critical role of CDK5 and Polo-like kinase 2 in homeostatic synaptic plasticity during elevated activity. Neuron 58:571–583
172. Seeburg DP, Sheng M (2008) Activity-induced Polo-like kinase 2 is required for homeostatic plasticity of hippocampal neurons during epileptiform activity. J Neurosci 28:6583–6591
173. Peng YR et al (2010) Postsynaptic spiking homeostatically induces cell-autonomous regulation of inhibitory inputs via retrograde signaling. J Neurosci 30:16220–16231
174. Rannals MD, Kapur J (2011) Homeostatic strengthening of inhibitory synapses is mediated by the accumulation of GABA(A) receptors. J Neurosci 31:17701–17712
175. Aoto J, Nam CI, Poon MM, Ting P, Chen L (2008) Synaptic signaling by all-trans retinoic acid in homeostatic synaptic plasticity. Neuron 60:308–320
176. Groth RD, Lindskog M, Thiagarajan TC, Li L, Tsien RW (2011) Beta Ca2+/CaM-dependent kinase type II triggers upregulation of GluA1 to coordinate adaptation to synaptic inactivity in hippocampal neurons. Proc Natl Acad Sci USA 108:828–833
177. Gainey MA, Hurvitz-Wolff JR, Lambo ME, Turrigiano GG (2009) Synaptic scaling requires the GluR2 subunit of the AMPA receptor. J Neurosci 29:6479–6489
178. Kilman V, van Rossum MC, Turrigiano GG (2002) Activity deprivation reduces miniature IPSC amplitude by decreasing the number of postsynaptic GABA(A) receptors clustered at neocortical synapses. J Neurosci 22:1328–1337
179. Saliba RS, Michels G, Jacob TC, Pangalos MN, Moss SJ (2007) Activity-dependent ubiquitination of GABA(A) receptors regulates their accumulation at synaptic sites. J Neurosci 27:13341–13351
180. Beique JC, Na Y, Kuhl D, Worley PF, Huganir RL (2011) Arc-dependent synapse-specific homeostatic plasticity. Proc Natl Acad Sci USA 108:816–821
181. Jakawich SK et al (2010) Local presynaptic activity gates homeostatic changes in presynaptic function driven by dendritic BDNF synthesis. Neuron 68:1143–1158
182. Sutton MA et al (2006) Miniature neurotransmission stabilizes synaptic function via tonic suppression of local dendritic protein synthesis. Cell 125:785–799
183. Wang HL, Zhang Z, Hintze M, Chen L (2011) Decrease in calcium concentration triggers neuronal retinoic acid synthesis during homeostatic synaptic plasticity. J Neurosci 31: 17764–17771

184. Steinmetz CC, Turrigiano GG (2010) Tumor necrosis factor-alpha signaling maintains the ability of cortical synapses to express synaptic scaling. J Neurosci 30:14685–14690
185. Desai NS, Cudmore RH, Nelson SB, Turrigiano GG (2002) Critical periods for experience-dependent synaptic scaling in visual cortex. Nat Neurosci 5:783–789
186. Goel A, Lee HK (2007) Persistence of experience-induced homeostatic synaptic plasticity through adulthood in superficial layers of mouse visual cortex. J Neurosci 27:6692–6700
187. Chang MC et al (2010) Narp regulates homeostatic scaling of excitatory synapses on parvalbumin-expressing interneurons. Nat Neurosci 13:1090–1097
188. Freund TF, Buzsaki G (1996) Interneurons of the hippocampus. Hippocampus 6:347–470
189. Neumann H, Schmidt H, Cavalie A, Jenne D, Wekerle H (1997) Major histocompatibility complex (MHC) class I gene expression in single neurons of the central nervous system: differential regulation by interferon (IFN)-gamma and tumor necrosis factor (TNF)-alpha. J Exp Med 185:305–316
190. Perea G, Navarrete M, Araque A (2009) Tripartite synapses: astrocytes process and control synaptic information. Trends Neurosci 32:421–431
191. Halassa MM, Fellin T, Haydon PG (2007) The tripartite synapse: roles for gliotransmission in health and disease. Trends Mol Med 13:54–63
192. Jourdain P et al (2007) Glutamate exocytosis from astrocytes controls synaptic strength. Nat Neurosci 10:331–339
193. Allan SM, Rothwell NJ (2001) Cytokines and acute neurodegeneration. Nat Rev Neurosci 2:734–744
194. Leonoudakis D, Braithwaite SP, Beattie MS, Beattie EC (2004) TNFalpha-induced AMPA-receptor trafficking in CNS neurons; relevance to excitotoxicity? Neuron Glia Biol 1:263–273
195. Zhao P, Ignacio S, Beattie EC, Abood ME (2008) Altered presymptomatic AMPA and cannabinoid receptor trafficking in motor neurons of ALS model mice: implications for excitotoxicity. Eur J Neurosci 27:572–579
196. Carlson NG, Bacchi A, Rogers SW, Gahring LC (1998) Nicotine blocks TNF-alpha-mediated neuroprotection to NMDA by an alpha-bungarotoxin-sensitive pathway. J Neurobiol 35:29–36
197. Carlson NG et al (1999) Inflammatory cytokines IL-1 alpha, IL-1 beta, IL-6, and TNF-alpha impart neuroprotection to an excitotoxin through distinct pathways. J Immunol 163:3963–3968
198. Liu XH, Xu H, Barks JD (1999) Tumor necrosis factor-a attenuates N-methyl-D-aspartate-mediated neurotoxicity in neonatal rat hippocampus. Brain Res 851:94–104
199. Schwab C, Steele JC, McGeer PL (1996) Neurofibrillary tangles of Guam parkinson-dementia are associated with reactive microglia and complement proteins. Brain Res 707:196–205
200. Rogers J et al (1992) Complement activation by beta-amyloid in Alzheimer disease. Proc Natl Acad Sci USA 89:10016–10020
201. Rogers J et al (1992) Complement activation and beta-amyloid-mediated neurotoxicity in Alzheimer's disease. Res Immunol 143:624–630
202. Rogers J, Luber-Narod J, Styren SD, Civin WH (1988) Expression of immune system-associated antigens by cells of the human central nervous system: relationship to the pathology of Alzheimer's disease. Neurobiol Aging 9:339–349
203. Boyett KW et al (2003) Increased fibrillar beta-amyloid in response to human clq injections into hippocampus and cortex of APP+PS1 transgenic mice. Neurochem Res 28:83–93
204. Webster SD et al (2000) Complement component C1q modulates the phagocytosis of Abeta by microglia. Exp Neurol 161:127–138
205. Fonseca MI, Zhou J, Botto M, Tenner AJ (2004) Absence of C1q leads to less neuropathology in transgenic mouse models of Alzheimer's disease. J Neurosci 24:6457–6465
206. Steele MR, Inman DM, Calkins DJ, Horner PJ, Vetter ML (2006) Microarray analysis of retinal gene expression in the DBA/2J model of glaucoma. Invest Ophthalmol Vis Sci 47:977–985
207. Stasi K et al (2006) Complement component 1Q (C1Q) upregulation in retina of murine, primate, and human glaucomatous eyes. Invest Ophthalmol Vis Sci 47:1024–1029
208. Ringheim GE, Conant K (2004) Neurodegenerative disease and the neuroimmune axis (Alzheimer's and Parkinson's disease, and viral infections). J Neuroimmunol 147:43–49

Chapter 4
Neuroimaging and Clinical Studies on Brain–Immune Interactions

Jonas Hannestad

4.1 Introduction

In this chapter I will review the following areas: (1) basic immunology and the pathways through which the brain and the immune system communicate, (2) experimental paradigms that can be used to activate the innate immune system in humans, (3) functional brain imaging and its use in combination with those paradigms, (4) symptoms induced by immune activation that are relevant to the core pathology of addiction (reward, anhedonia, and fatigue), (5) the use of medications to reduce the effects of immune activation on the brain, and (6) imaging of microglia, cells which may mediate some of the effect of immune activation on the brain.

4.2 Basic Innate Immunology

The *innate immune system* evolved to allow an organism to recognize and defend itself against pathogens. *Pathogen-associated molecular patterns* (PAMPs) are molecular "signatures" of whole classes of microorganisms that can be detected by receptors called *pathogen recognition receptors* (PRRs). Because of the survival advantage conferred by being able to defend against pathogens, the ability of an organism to detect the presence of pathogens arose very early in evolution. In plants, PRRs are expressed on individual cells; this allows each cell to recognize PAMPs and release antimicrobial chemicals [1, 2]. In the animal kingdom, specialized cells have developed to recognize PAMPs and fight pathogens; these cells constitute what is referred to as the innate immune system [3]. When a PAMP binds to a PRR

J. Hannestad, M.D., Ph.D (✉)
Department of Psychiatry, Yale University School of Medicine,
New Haven, CT, USA
e-mail: jonas.hannestad@yale.edu

C. Cui et al. (eds.), *Neural-Immune Interactions in Brain Function and Alcohol Related Disorders*, DOI 10.1007/978-1-4614-4729-0_4,

on an innate immune cell, this cell responds by producing inflammatory cytokines, chemokines, and type I interferons, molecules which not only have antimicrobial properties but which also engage the adaptive immune system. Importantly, cells of the innate immune system are not able to distinguish between an isolated PAMP and the actual pathogen with the same PAMP embedded in its structure. This phenomenon is useful in research because a PAMP can be administered to an organism in vivo to induce an innate immune response without the risk of infection. That is, PAMPs can be used as sterile immune stimuli to study the immune response. For instance, the PAMP *lipopolysaccharide* (LPS or endotoxin) is a part of the external membrane of Gram-negative bacteria. When LPS is isolated from bacteria and introduced into an organism, it will elicit an initial immune response that is almost identical to that elicited by introduction of actual Gram-negative bacteria; however, the resulting immune response is shorter-lived because there is no replication of pathogens and further release of LPS. Whether attached to the bacterial wall or not, when LPS is introduced into an organism, it binds to a receptor called CD14 on innate immune cells. This enables LPS interaction with another receptor, *Toll-like receptor 4* (TLR4), which ultimately leads to the transcription (through the transcription factor *nuclear factor kappa B*; NFκB) of genes that encode for inflammatory cytokines such as tumor necrosis factor alpha (TNFα), interleukin-1 beta (IL-1β), IL-6, IL-8, and IL-18. These inflammatory cytokines have local and systemic effects, as discussed below. In the case of IL-1β and IL-18, even though transcription of these genes is induced by binding of LPS to TLR4, the inactive precursor forms pro-IL-1β and pro-IL-18 require cleavage by the enzyme caspase-1 to become biologically active. *Caspase-1* is not induced by the TLR4 pathway directly, but can be induced by other signals from the pathogen or from intracellular crosstalk between various pathways [4]. Caspase-1-deficient mice, which cannot produce bioactive IL-1β or IL-18, have reduced inflammatory gene expression in the brain after systemic LPS administration [5], which highlights the role of IL-1β in the systemic and brain response to LPS. Another important inflammatory mediator that is transcribed during an innate immune response is *cyclooxygenase-2* (COX-2), an enzyme that converts membrane-bound arachidonic acid into eicosanoids, for example, prostaglandins and leukotrienes. Eicosanoids are responsible for many of the local effects of inflammation (pain, increased vascular permeability, neutrophil chemotaxis). The so-called *acute phase proteins* include some components of the complement system as well as *C-reactive protein*, which is often used as an index of systemic inflammation in human studies because of its association with increased risk of a variety of diseases. Lastly, an innate immune response may lead to transcription of inducible nitric oxide synthase (iNOS) and synthesis of nitric oxide, which has deleterious effects on neurons and brain function [6]. The inflammatory cytokines, chemokines, and eicosanoids produced during an innate immune response have local effects on vessel permeability and C-fibers, producing the cardinal signs of inflammation: *dolor, calor, rubor, and tumor*. If the PAMP stimulus is sufficiently strong (e.g., large doses of LPS or other PAMPs or the presence of large numbers of pathogens), these inflammatory mediators are released in higher quantities, which results in systemic effects (e.g., fever, malaise, autonomic changes). During an

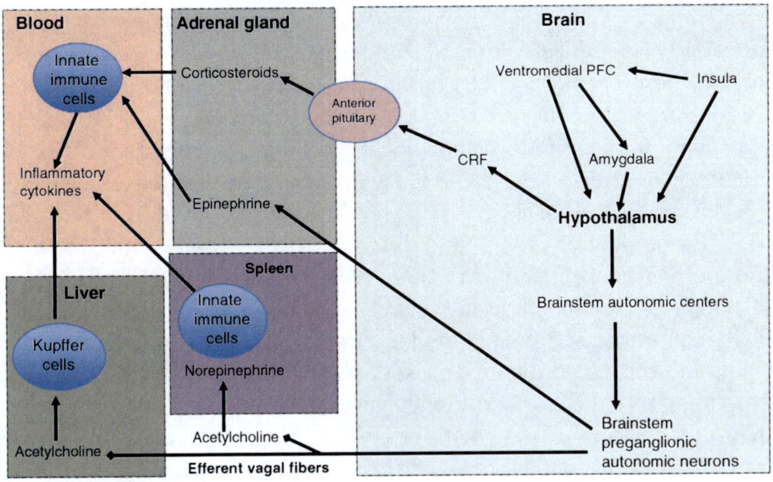

Fig. 4.1 Immunomodulatory pathways. The brain can inhibit systemic inflammation through the hypothalamic–pituitary–adrenal axis and the autonomic nervous system. Abbreviations: *CRF* corticotropin-releasing factor, *PFC* prefrontal cortex

innate immune response, so-called anti-inflammatory cytokines are also released. The role of these is believed to be a "brake" on the effects of the inflammatory mediators to prevent excessive damage from inflammation. There are also multiple intracellular "brakes" that keep the inflammatory response in check. This is important because some of these molecules have very potent effects, for example, TNFα and IL-1β. For instance, a deficiency of IL-1β receptor antagonist (IL1RA), which normally inhibits the effects of IL-1β, causes a disorder characterized by sterile multifocal osteomyelitis with periostosis [7]. For readers interested in more details about the innate immune system, basic immunobiology textbooks are a good source. Discussed next are the brain's immunomodulatory pathways, which are an important complement to the immune system's internal "brakes" (Fig. 4.1).

4.3 Brain–Immune Interactions

4.3.1 Brain Immunomodulatory Functions

It is well established that the brain and the immune system communicate. The brain can modulate immune function through neuroendocrine and autonomic pathways and, indirectly, through behavior (see Sect. 4.3.2). Of these *immunomodulatory* pathways, the most important one is the hypothalamic–pituitary–adrenal (HPA) axis, activation of which leads to release of cortisol, which has potent anti-inflammatory effects [8]. Efferent vagal fibers also exert anti-inflammatory effects through the release of acetylcholine, which can act directly on nicotinic receptors on immune cells [9] or by activating the splenic nerve, which in turn releases anti-inflammatory norepinephrine (NE) in the spleen [10]. NE inhibits innate immune cell activity through binding to α- and β-adrenergic receptors [11]. Other lines of evidence support the notion that catecholamines, particularly NE, inhibit inflammation: (1) monoamine depletion with reserpine resulted in an increase in LPS-induced production of TNFα [12], (2) NE reuptake inhibition reduced LPS-induced TNFα production [13], and (3) NE transporter-deficient mice produced less TNFα after LPS administration [13]. A recent study showed a novel pathway through which the sympathetic nervous system can modulate immune function, specifically cell traffic across the blood–brain barrier (BBB). In mouse experimental autoimmune encephalomyelitis, sensory neurons in the soleus muscle transmit a signal to lumbar spinal cord efferent sympathetic fibers, which produce IL-6-dependent upregulation of the chemokine CCL20 in blood vessels, which allows autoreactive T lymphocytes to access the central nervous system [14]. Interestingly, NE also appears to have effects on inflammation in the brain. For instance, stress-induced expression of IL-1β in the brain was inhibited by the β-adrenergic blocker propranolol [15], which suggests that stress causes release of NE, which binds to β-adrenergic receptors and induces IL-1β expression. On the other hand, lesions of the locus coeruleus, which depletes the cortex of NE, caused increased expression of IL-1β and IL-6 in the cortex after an inflammatory stimulus [16], indicating that NE can also have anti-inflammatory effects in the brain. In summary, the brain can modulate innate immune activity through the HPA axis and the parasympathetic nervous system. The effects of the sympathetic nervous system, and the effects of NE in the brain, can be either anti- or proinflammatory.

Some interventions suggest that the brain can have proinflammatory effects; however, it is unclear whether these are simply a result of an inhibition of the anti-inflammatory mechanisms described above. For instance, in rats injected systemically with LPS, prior inactivation of neurons in the anterior hypothalamus resulted in reduced peripheral levels of TNFα [17], suggesting that the anterior hypothalamus could have a proinflammatory effect. In general, the introduction of an immune stimulus (e.g., LPS) inside the brain (e.g., intraventricular or intraparenchymal) will cause sickness behavior (see Sect. 4.3.2) as well as increased release of inflammatory mediators systemically, especially in the liver. For instance, (1) adenoviral overexpression of IL-1β in the mouse brain led to increased systemic production of inflammatory cytokines [18], and (2) injection of LPS into the cerebral ventricles caused increased systemic IL-6 production [19]. Administration of IL-1β in the anterior hypothalamus causes sickness behavior, which can be blocked by peripheral neutralization of TNFα [20], suggesting that systemically released cytokines play a role in sickness behavior even when the immune stimulus is delivered directly into brain parenchyma. The presumed reason that inflammation increases systemically when an immune stimulus is introduced into the brain is because the brain needs to recruit the systemic immune system to fight a pathogen. That is, when, for example, LPS is introduced into the brain, the brain "thinks" that there are Gram-negative bacteria inside the brain and therefore "requests" help from the innate immune system outside the brain. For instance, administration of LPS or IL-1β into the brain results in leukocyte adhesion to blood vessels [21, 22]. In summary, the brain exerts some inhibition on the innate immune system through anti-inflammatory regulation by the HPA axis and autonomic nervous system; however, the presence of inflammatory mediators in brain parenchyma can have proinflammatory effects peripherally.

In a similar manner that immune stimuli introduced into the brain have systemic effects, "psychological" stress not only increases the expression of inflammatory mediators in the rodent brain parenchyma [23–25], in humans it has been shown that psychological stressors can increase peripheral inflammation [8, 26–28]. The effects of stress on immune parameters are likely mediated by the brain's immunomodulatory functions. Whether the effect of stress in increasing systemic inflammation in humans involves increased expression of inflammatory mediators inside the brain is not known. Given the effects of stress on systemic inflammation, it is possible that in depression or alcoholism, both of which are associated with elevated serum levels of TNFα and IL-6, this is due to the substantial stress associated with being in a depressive episode or heavy drinking. In a recent meta-analysis by our group, we did not find evidence that treatment of depression and resolution of depressive symptoms was associated with a normalization of elevated TNFα levels, whereas there was some indication that IL-6 levels may decrease [29]. This suggests that reducing the stress of being in a depressive episode with antidepressant medications does not, at least in the short term, reduce systemic inflammation.

4.3.2 Brain Immunosensory Functions

As described in Sect. 4.1, the brain can modulate immune function. Conversely, the immune system can modulate brain function, or, in a different conceptualization, the brain has *immunosensory* functions. The brain's ability to detect and respond to immune signals must be distinguished conceptually from the disruption of brain function that can occur in instances of severe systemic inflammation. For instance, sepsis is associated with high levels of inflammatory cytokines, which can disrupt the BBB, allowing leukocytes, inflammatory cytokines, and bacterial toxins to enter the brain parenchyma from the circulation [30]. Sepsis may therefore lead to significant disruption of brain function, including neuronal death [31]. Such severe effects on the brain may explain why sepsis in humans is associated with short-term cognitive impairment (delirium) [30] and long-term cognitive decline [32]. In milder forms of systemic inflammation, there are measurable effects on brain function in the absence of BBB disruption, neuronal death, or deleterious functional sequelae. Rodent studies have demonstrated that systemic inflammation causes increased expression of inflammatory mediators in the brain. This also occurs in human and nonhuman primates: systemic inflammatory stimuli cause increased CSF levels of IL-6 [33, 34] (Fig. 4.2).

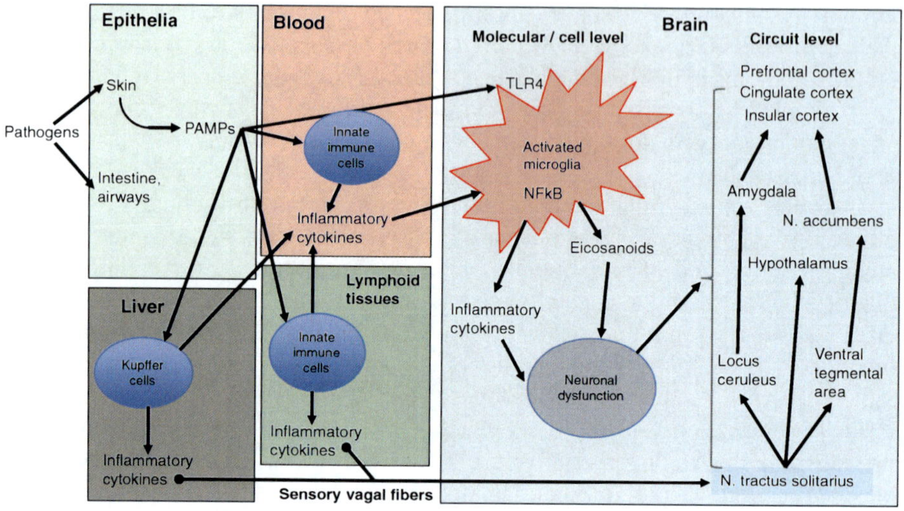

Fig. 4.2 Immunosensory pathways. The brain can detect systemic inflammatory signals, which affect brain function. This can occur through release of inflammatory mediators in brain parenchyma through activation of microglia and through activation of sensory vagal afferents. Abbreviations: *NFκB* nuclear factor kappa B, *PAMP* pathogen-associated molecular pattern, *TLR4* Toll-like receptor 4 (the lipopolysaccharide receptor). Lymphoid tissues: gut-associated lymphoid tissue, spleen, pulmonary lymphoid tissue, and lymph nodes

The effects on the brain of milder systemic inflammation include fever (mediated by the hypothalamus), activation of the HPA axis and the autonomic nervous system, and changes in behavior, emotions, and cognition [35]. The constellation of behavioral changes is often called *sickness behavior*, which in rodents includes behaviors such as reduced activity and reduced food intake [36]. Sickness behavior is phylogenetically old; for instance, locusts and caterpillars display reduced food intake during LPS-induced systemic inflammation [37, 38]. This phylogenetic conservation of the brain's ability to sense and respond to immune signals suggests that this is adaptive. That is, the brain has evolved to detect and respond to immune signals so that it can optimize the ability of the immune system to fight off the pathogen while minimizing inflammatory damage to tissues and organs [39]. In order for this to happen, the brain must receive and interpret information from the immune system. Experiments in rodents have helped elucidate the pathways through which this occurs [36, 40–42].

4.3.2.1 Signaling Through the Endothelium

The BBB is composed of tightly bound endothelial cells and perivascular astrocytes. Astrocytes promote the immune quiescence of BBB endothelial cells by decreasing the expression of proinflammatory mediators and the adhesion and migration of leukocytes [43]. When inflammatory mediators or PAMPs are present in the circulation, signaling to the brain involves mainly endothelial cells of the BBB. Endothelial cells have receptors for LPS and other PAMPs, and they also have receptors for inflammatory cytokines such as TNFα and IL-1β. When PAMPs/pathogens or inflammatory cytokines are present in the circulation and bind to these receptors, activation of intracellular pathways that lead to NFκB-dependent gene transcription occurs [44, 45]. The endothelium of the BBB appears to play an important role in mediating inflammatory signals from both the circulation and cerebrospinal fluid to the brain parenchyma [46]. When IL-1β is administered intracerebroventricularly (i.c.v.), the presence of IL-1β receptors on endothelial cells is required for activation of hypothalamic neurons and the occurrence of sickness behavior [22], which indicates that the endothelium is required for signaling both from blood and from CSF. The core sign of systemic inflammation, fever, requires signaling through the endothelium. In vitro, IL-1β induces COX-2 transcription in brain endothelial cells, and this results in prostaglandin E2 release, which leads to fever [47]. LPS-induced anhedonia is blocked by nonsteroidal anti-inflammatory agents, but expression of IL-1β and IL-6 in the hypothalamus is not [48], suggesting that eicosanoids are not involved in mediating the effects of endotoxin on brain expression of inflammatory cytokines.

4.3.2.2 Signaling Through the Afferent Vagus

When inflammatory mediators or PAMPs are present in viscera of the abdominal or thoracic cavities, signaling to the brain occurs through afferent vagal fibers [49–51]. The primary central projection area of the vagus nerve is the nucleus tractus solitarius, and secondary vagus nerve projection areas include parabrachial, paraventricular, and hypothalamic nuclei; the amygdala and extended amygdala (bed nucleus of the stria terminalis); and the insula. In addition to the mechanism involving the BBB endothelium described above, fever can occur when eicosanoids bind to vagal afferents [52].

4.3.2.3 Leukocyte Entry

Whether leukocytes can enter the brain parenchyma when there is no disruption of the BBB is controversial. In one study of hepatic inflammation in mice, monocytes from the blood entered the brain parenchyma, a process that was dependent on TNFα-induced expression of monocyte chemoattractant protein-1 in microglia [53]. Recently, a study found that T lymphocytes can be recruited into the central nervous system at the level of the lumbar spinal cord and that this is dependent on psoas muscle contraction [14].

In summary, systemic immune stimuli can be detected by the brain through various pathways, including signaling across the BBB, through vagal afferents, and potentially by direct entry of peripheral leukocytes into brain parenchyma. These signals affect the function of the brain.

4.3.3 Sickness Behavior

When inflammatory signals are detected by the brain, rodents display a constellation of behaviors which in some respects are similar to depression in humans. These include anhedonia, decreases in novelty-induced and social behaviors, reduced food intake, and sleep disturbance [54]. Depending on the dose of LPS used, other depressive-like behaviors (e.g., as measured in the tail-suspension or forced-swim tests) in rodents may continue after the acute sickness behavior ends [55]. Delayed effects from inflammatory stimuli may occur because increased brain and peripheral levels of cytokines can last for several weeks, especially after high doses of LPS [56, 57]. LPS administration in humans also produces depressive-like symptoms, for example, fatigue and anhedonia [58–60]. For instance, inflammatory cytokines released during viral infections are associated with changes in mood [61]. After experimental exposure of human subjects to rhinovirus or influenza virus, blood levels of inflammatory cytokines predicted a reduction in positive affect the following day [62]. In such instances, the infectious agent does not have access to the brain; therefore, the changes in mood are most likely due to the effect of inflammatory cytokines released by innate immune cells. Similarly, administration of sterile immune stimuli (e.g., vaccines, LPS) or cytokines such as interferon-alpha (IFNα) or IL-6 is associated with transient depressive symptoms in humans [59, 63–67].

In summary, the brain's immunosensory functions allow it to detect and monitor immune activity in the body. This information in turn causes the brain to respond, both through the immunomodulatory pathways described in Sect. 4.3.2 and through changes in behavior. Therefore, the relationship between the brain and the immune system is a loop in which the immune system provides information to the brain about immune "events" in the body, and the brain uses this information to modulate the function of the immune system. The impact on the brain, and the pathways through which this signaling occurs, depends on the degree and localization of the systemic inflammatory event. Whether dysregulation in these pathways plays a role in the pathogenesis of psychiatric disorders, including alcohol dependence, is a topic of great interest and controversy [35, 68]. Research in this area may lead to better understanding and improved treatments for a variety of neurobiological disorders, including alcoholism.

4.4 Clinical Experimental Models of Immune Activation

4.4.1 Interferon-Alpha Model

IFNα is used for the treatment of hepatitis C and melanoma. Up to 45% of patients treated with IFNα develop depressive symptoms [69], which in toto resemble idiopathic depression [66], respond to treatment with antidepressants [70, 71], and can be prevented by pretreatment with antidepressants [72]. This lends both face and predictive validity to this model. Fatigue occurs early after patients start treatment with IFNα, and fatigue predicts later emergence of other depressive symptoms such as depressed mood [66, 73]. Of note, IFNα-induced depressive symptoms are associated with increases in blood levels of IL-6 and TNFα, similar to levels found in depression [74, 75], and increases in CSF levels of IL-6 [76]. This suggests that the depressogenic effect of IFNα may be mediated by the increase in levels of these inflammatory cytokines. In patients with melanoma, after 4 weeks of treatment with IFNα, there was an increase in the Montgomery–Åsberg Depression Rating Scale (MADRS) total score from 2.7 to 12.7, which indicates mild depression [77]. This increase in MADRS total score was mostly due to increases in the lassitude item, which measures fatigue and motivation, and to reduced appetite, reduced sleep, and anhedonia [77], suggesting that IFNα treatment is associated especially with an increase in the neurovegetative symptoms of depression [66]. One limitation of the IFNα model is that, because the depressogenic effects of IFNα do not usually occur until 8–12 weeks of treatment [69], it cannot be used as an acute laboratory model. A second limitation, when IFNα is used for treatment of hepatitis C, is that the hepatitis C virus is neurotropic and activates microglia [78], which indicates that some of the behavioral effects of IFNα treatment in patients with hepatitis C (but not in patients with melanoma) may occur due to indirect effects of IFNα on viral load in neurons or microglia. Despite these real and theoretical drawbacks, the IFNα model has greatly expanded our knowledge of how the inflammation affects behavior and brain function.

4.4.2 *The Typhoid Vaccination Model*

Acute administration of the typhoid (*Salmonella typhi*) vaccine induces a small increase in IL-6 levels but no change in TNFα levels [64], which suggest that it is a milder immune stimulus than IFNα treatment. The mood effect of typhoid vaccination is also milder [79]: when measured with the total score on the Profile of Mood States [80], a scale that measures anxiety, depression, fatigue, "vigor," and "confusion," there was a five-point mean difference, which was significant with 30 subjects in a within-subject design [64]. Like IFNα treatment, typhoid vaccination also produces fatigue [81]. The typhoid vaccination model has been combined with functional neuroimaging to answer important questions about the brain's response to systemic inflammation (see Sect. 4.6).

4.4.3 *The Human LPS Model*

In rodents, the most commonly used immune stimulus is intraperitoneal (i.p.) or intracerebral administration of LPS. As described above, activation of the innate immune system with LPS leads to a constellation of behaviors similar to depression in humans [54, 55]. The human LPS model was reviewed in detail recently [35]. LPS administration in humans is a safe experimental procedure that has been used for several decades to study the systemic response to innate immune system activation [82]. In humans, LPS doses of 2–4 ng/kg body weight cause flu-like symptoms (fever, chills, myalgia, headache, nausea) and increases in blood levels of TNFα and IL-6, similar to those that occur in sepsis [83–85]. Lower doses of LPS (0.4–0.8 ng/kg) cause mild sickness symptoms, depressive symptoms (notably fatigue), and less pronounced increases in TNFα and IL-6 levels [59, 65, 86]. Although there is a clear dose–response relationship between LPS and blood levels of TNFα and IL-6, there are large individual differences in the immune response. There is also not a good correlation between blood levels of TNFα or IL-6 and the behavioral response. Total MADRS score increased after LPS administration, but this was mostly due to the lassitude item, and we did not find a significant effect on mood with either the MADRS mood item or the POMS depression subscale [58]. Although other groups found an increase in depressed mood after low-dose LPS administration, this effect was reduced when controlling for fatigue [60], which is consistent with our data, which show that the main effect of LPS is the induction of a state of fatigue and listlessness [87]. LPS administration does not appear to have an effect on cognition in young, healthy humans [88]. Women showed a more pronounced inflammatory response to LPS than men [84]. A recent study found that biofeedback can be used to reduce LPS-induced autonomic dysfunction, but this did not affect LPS-induced levels of inflammatory cytokines [89].

In summary, LPS, typhoid vaccination, or IFNα can be administered in human subjects to induce innate immune system activation, increased systemic levels of inflammatory cytokines, and various behavioral effects, most notably depressive-like symptoms. Next, we will discuss the use of functional imaging to measure brain activity during immune system activation.

4.5 Functional Brain Imaging

Functional neuroimaging refers to brain imaging modalities that are used to obtain an estimate of cellular (neuronal) activity. If a disease state or an experimental intervention is associated with changes in neuronal activity, one can use functional imaging to measure this and identify the brain regions involved. For instance, functional magnetic resonance imaging (fMRI) uses MR to measure changes in oxygenated blood in the brain. Changes in blood oxygenation correlate with blood flow; therefore, the *blood oxygen level dependent* (BOLD) signal indicates changes in blood flow. The assumption is that an increase in regional blood flow (which is under tight physiologic control) is indicative of increased metabolic demand and therefore increased cellular activity in a brain region, an assumption that is supported by experimental studies [90]. *Positron emission tomography* (PET) is an imaging modality that measures the amount of radioactivity emitted from a tissue or organ in vivo after the injection of a positron-emitting radiopharmaceutical. The radioactivity emitted is used to quantify a physiologic process. Radiopharmaceuticals are interchangeably called radioligands or *radiotracers*; the latter refers to the fact that the mass dose of radiopharmaceutical administered is very low. Such "tracer" amounts rarely have any measurable physiologic or biochemical effects and therefore do not interfere with the metabolic process that they are designed to measure, for example, glucose metabolism, oxygen uptake, or blood flow. PET imaging with ^{18}F-labeled 2-deoxy-D-glucose (FDG) is a commonly used functional PET modality. Deoxyglucose, like glucose, is taken up by cells in the brain and phosphorylated by hexokinase to FDG-6-PO_4. Unlike glucose-6-PO_4, FDG-6-PO_4 is not a substrate of phosphoglucose isomerase; the biological half-life of FDG is therefore very long, which means that the effective half-life is equivalent to the physical half-life of ^{18}F (110 min). In other words, FDG will emit positrons for several hours after it was administered to a subject, and this radioactivity will occur in the cells that took up FDG shortly after it was administered. The degree of FDG uptake is proportional to the metabolic activity of the cell at the time of FDG administration. Therefore, the radioactive signal detected from a certain brain region is an indication of the metabolic activity in that region at the time of FDG administration. A detailed discussion of the theoretical underpinnings and limitations of FDG-PET imaging can be found in [91].

The fact that the signal from FDG can be detected several hours after injection is an advantage of FDG-PET over fMRI because it allows us to use FDG-PET imaging to measure neuronal activity while the subject is not in the scanner. One can take advantage of this characteristic of FDG and design research studies in which a subject undergoes an experimental intervention that cannot be performed in the scanner but during which injection of FDG can be performed. This allows us to "capture" metabolic activity at the time of the experiment and measure it up to an hour later. Another factor in deciding to use FDG-PET over fMRI for functional imaging is whether it is important for the research question to measure absolute metabolism rather than relative changes in blood flow. With FDG-PET, provided that one has an appropriate input function (either through an arterial line or cardiac imaging), one can obtain precise estimates of absolute glucose utilization in brain regions of interest.

4.6 Functional Imaging of Brain Immunosensory Functions

In rodents, systemic LPS administration has been associated with increases in neuronal c-fos expression in various hypothalamic nuclei (including the paraventricular nucleus), in the central amygdala and in the bed nucleus of the stria terminalis (which is part of the extended amygdala), and in the nucleus tractus solitarius, locus coeruleus, parabrachial nucleus, and raphe nucleus [55, 92–95], while reduced neuronal c-fos expression has been found in the hippocampus, striatum, and ventral tegmental area [92]. Electrical activity increases in the insula and amygdala 2–3 h after peripheral administration of either LPS or staphylococcal enterotoxin in rats [96]. In humans, the brain regions that respond to systemic inflammation have been partially identified through the use of functional imaging in combination with various immune stimuli (summarized in Table 4.1); however, there are still many contradictory findings. Most importantly, a majority of studies in humans employed various behavioral paradigms that were done in combination with immune stimulation, whereas in rodents most studies measured LPS-induced c-fos expression with no behavioral intervention.

4.6.1 IFNα Treatment

In patients with hepatitis C treated with IFNα, compared to patients who received no treatment, fMRI showed increased activation in the dorsal anterior cingulate cortex during a task of visuospatial attention [97]. The authors attributed this increase to a need to exert greater mental effort to maintain performance. In patients with melanoma treated with IFNα, FDG-PET showed increased metabolism in basal ganglia and cerebellum and decreased metabolism in dorsolateral prefrontal cortex after 4 weeks of treatment, compared to baseline [77]. In these patients, the rating of "energy" on a visual analog scale decreased during treatment and correlated inversely with the change in glucose metabolism in the left putamen and left nucleus accumbens, while the "fatigue" rating on the visual analog scale did not correlate with any brain metabolism changes [77].

4.6.2 Typhoid Vaccination

In a cohort of male subjects who received typhoid vaccination on 1 day and placebo injection on the other day, fMRI showed no global effects of typhoid vaccination on blood flow, but it was associated with increased blood flow in the brainstem, thalamus, amygdala, cingulate, and insula during the presentation of various stimuli [81, 98, 99]. For instance, a greater IL-6 response to typhoid vaccination was associated with enhanced blood flow in the substantia nigra during a test of reaction time [98], suggesting that psychomotor slowing caused by systemic inflammation may be mediated by the substantia nigra. The subjective experience of fatigue induced by typhoid vaccination was predicted by blood flow changes within the mid/posterior insula and the left anterior cingulate [81]. After typhoid vaccination, reduced positive mood during an emotional face perception task was associated with increased blood flow in the subgenual anterior cingulate, and there was reduced connectivity of this region with the amygdala, medial prefrontal cortex, nucleus accumbens, and superior temporal sulcus [99]. Deterioration in total mood (see Sect. 4.4.2 for definition) produced by typhoid vaccination was associated with a marked *decrease* in activity in the amygdala.

Table 4.1 Functional imaging studies of the effects of immune activation on the brain in humans

Study	Design	Immune stimulus	Imaging modality/task	Brain regions
Capuron et al. (2005) [97]	Between-group Hepatitis C, IFNα treated, $n=10$ Hepatitis C, no treatment, $n=11$	IFNα 4 weeks vs. no treatment	fMRI Visuospatial attention	dACC ↑
Capuron et al. (2007) [77]	Within-subject Melanoma $n=12$	IFNα 4 weeks	FDG-PET None	Putamen ↑ Pallidum ↑ Accumbens ↑ Thalamus ↑ Cerebellum ↑ DLPFC ↓
Brydon et al. (2008)[a] [98]	Within-subject $n=16$ (men)	*Salmonella typhi* vaccination vs. placebo	fMRI Stroop task	Substantia nigra ↑
Harrison et al. (2009) (pp. 407–414)[b] [99]	Idem	Idem	fMRI Emotional face perception	Subgenual ACC ↑
Harrison et al. (2009) (pp. 415–422)[c,d] [81]	Idem	Idem	fMRI Stroop task	Cingulate, R ↑ Insula, B ↑ Periaqueductal gray,R ↑ Thalamus, R ↑ Amygdala, L ↑
Eisenberger et al. (2009)[e] [65]	Between-group $n=20$ LPS $n=16$ placebo	LPS vs. placebo	fMRI Social exclusion	Insula ↑ Perigenual ACC ↑ Posterior CC ↑ Caudate ↑ Dorsomedial PFC ↑ Temporal ↑ Occipital ↑ Cerebellum ↑

(continued)

Table 4.1 (continued)

Study	Design	Immune stimulus	Imaging modality/task	Brain regions
Eisenberger et al. (2010) [101]	Idem	Idem	fMRI Monetary reward	Ventral striatum, left ↓ Premotor, R ↓ Sup parietal, R ↓ Inf mid temp R ↓ Hippocampal gyrus, L ↓ Occipital, L ↓
Inagaki et al. (2012) [102]	Idem	Idem	fMRI Socially threatening images	Amygdala ↓
Hannestad et al. (2012) [100]	Within-subject $n=9$	LPS vs. placebo	FDG-PET None	Anterior cingulate ↓ Insula ↑ Putamen ↓ Thalamus ↑

[a]The IL-6 response to typhoid vaccination correlated with enhanced response to Stroop stimuli within the left substantia nigra

[b]Vaccination-induced mood deterioration correlated with enhanced activity in the subgenual anterior cingulate cortex (sACC) during emotional face processing

[c]Activity in the insula (an interoceptive brain region) predicted individual differences in vaccination-induced fatigue

[d]Vaccination-induced mood deterioration correlated with decreased activity in the amygdala

[e]Regions in which neural activation during social exclusion vs. inclusion correlated positively with increases in IL-6 levels in the endotoxin group ($n=20$)

4.6.3 LPS Administration

Using fMRI it was found that LPS administration, which induces robust systemic inflammation, was associated with reduced blood flow in the ventral striatum in response to monetary reward cues [101]. In a computerized social-rejection paradigm, the only between-groups difference in blood flow (social exclusion vs. inclusion) in subjects who received LPS was found in a region of the occipital cortex [65]. In women subjects, LPS administration was associated with increased blood flow in prefrontal cortex and anterior insula, and this increase in blood flow mediated the relationship between IL-6 levels and depressed mood in women [65]. Within the LPS group ($n = 20$) individual differences in IL-6 responses to LPS were associated with neural responses to social exclusion compared to social inclusion in various regions (see Table 4.1). In subjects who were presented with images, blood flow in the amygdala in response to threatening images, especially social ones (faces), was higher after LPS administration than after placebo, and this was associated with increased perceived social disconnection [102]. Our group recently used FDG-PET and showed that LPS administration, in the absence of any computerized task or mood-eliciting stimulus, was associated with a 2.5% decrease in glucose metabolism in the cingulate, which was mostly due to a 4.5% decrease in the right anterior cingulate [100]. In the insula there was a 3% increase in glucose metabolism, which was mostly due to a 6% increase in the right anterior insula. In secondary analyses, we found decreased metabolism in the left putamen and increased metabolism in the right ventrolateral thalamus. We did not find any changes in glucose metabolism in the amygdala induced by LPS alone.

4.6.4 Summary of the Response of Brain Regions to Inflammation in Humans

4.6.4.1 Amygdala

Typhoid vaccination was associated with a decrease in blood flow to the amygdala that was associated with a decrease in mood, while LPS administration did not cause any change in amygdala glucose metabolism on its own [100] or when combined with a social-exclusion paradigm [65]. When subjects received LPS and were presented with socially threatening visual stimuli, there was an increase in blood flow to the amygdala [102]. In rodents, studies have found that LPS causes increased neuronal c-fos expression [55, 92, 103] and increased electric activity [96] in the amygdala.

4.6.4.2 Striatum

Treatment with IFNα was associated with increased glucose metabolism in the putamen [77], while acute LPS administration was associated with decreased glucose metabolism in the putamen [100]. This discrepancy between the effect of IFNα and LPS on striatal neuronal activity may be because LPS induces acute, robust systemic inflammation while IFNα treatment induces chronic, low-grade systemic inflammation. When combined with a social-exclusion paradigm, LPS administration was associated with increased blood flow in the caudate [65], while when combined with a monetary reward paradigm, LPS administration was associated with decreased blood flow in the ventral striatum [101]. In rodents, one study found decreased neuronal c-fos expression in the dorsal striatum after LPS [92]. Another study found that acute LPS administration caused decreased syntaxin expression in the ventral striatum, while chronic LPS administration caused no change in syntaxin expression [104]; this is consistent with the difference between acute LPS administration and chronic IFNα treatment in humans described above.

4.6.4.3 Cingulate

Treatment with IFNα was associated with increased blood flow in the dorsal anterior cingulate [97]. Typhoid vaccination was associated with increased blood flow in the right cingulate and with increased blood flow to the subgenual anterior cingulate during presentation of emotional stimuli (faces) [99]. LPS administration combined with a social-exclusion task was associated with increased blood flow to the posterior cingulate, and the dorsal and perigenual anterior cingulate [65], whereas LPS administration in the absence of any stimulus was associated with reduced glucose metabolism in the right anterior cingulate [100].

4.6.4.4 Insula

Typhoid vaccination [81] and LPS administration [65] are associated with increased blood flow/glucose metabolism in the insula, especially the anterior insula. On the other hand, there are no reported effects of IFNα on neuronal activity in the human insula. This may suggest that the insula responds to acute, but not chronic, inflammatory stimuli. In rats, acute immune stimuli also increase electrical activity in the insula [96]. The similarity of the response of the insula to acute inflammatory stimuli in humans and rodents is consistent with its known role as one of the primary regions involved in interoception, including "sensing" immune events inside the body. The interoceptive information the insula receives is used to modulate peripheral responses to stress: through its projections to the hypothalamus, anterior cingulate, orbitofrontal cortex, and parabrachial nucleus [105], the insula participates in the modulation of autonomic and neuroendocrine activity [106, 107]. For instance, electric stimulation of the right insula causes increased sympathetic output [108].

4.6.4.5 Correlations Between Brain Activity and Systemic Inflammation

After LPS administration, we found a negative correlation between TNFα or IL-6 levels and glucose metabolism in the insula [100], while others found a positive correlation between IL-6 and blood flow to the insula [65]. After typhoid vaccination, there was no correlation between IL-6 and blood flow to the insula [81]. Several differences may explain these apparent contradictions, including the use of different imaging modalities (FDG-PET vs. fMRI), the timing of imaging relative to the immune stimulus, and whether imaging was performed during a computerized task or at rest.

In our cohort of subjects who received LPS, we found a positive correlation between levels of TNFα and IL-6 and social anhedonia. Interestingly, there was a positive correlation between social anhedonia and change in metabolism in the insula, that is, the more metabolism increased, the less LPS-induced social anhedonia the subject experienced, which suggests that the insula may have a "protective" role. Consistent with this, there was a negative correlation between levels of TNFα and IL-6 and metabolism in the insula, that is, subjects with higher TNFα and IL-6 levels had less LPS-induced increase in metabolism in the insula and more social anhedonia. Although these correlations may seem counterintuitive at first, they may indicate that increased metabolism in the insula has an inhibitory effect on the systemic immune response. As described in Sect. 4.3, the autonomic nervous system modulates systemic inflammatory responses [35], and the insula is a core brain region involved in this [109]. In subjects with higher LPS-induced metabolism in the insula, serum cytokine levels may have been lower because of an active inhibition of the systemic inflammatory response. Lower cytokine levels may, in turn, have caused less reduction in social interest. In subjects with lower LPS-induced metabolism in the insula, less inhibition of systemic cytokine levels may have allowed higher TNFα and IL-6 levels, which caused more social anhedonia. Because of the limited sample size, it was not possible to test this hypothesis statistically.

4.7 Reward, Anhedonia, and Fatigue

Reward, motivation, and fatigue are related neurobiological processes. Stimuli that elicit reward in the ventral striatum/nucleus accumbens are believed to be the main driver of approach behaviors. When there is a lack of ability to experience reward, there is a resultant lack of motivation. Anhedonia, the lack of ability to experience pleasure or reward, is a core symptom of depression [110]. Conversely, substances of abuse must cause increased reward in order to be abused; in the initial phases of substance use, reward drives the behavior to seek and continue to use the substance of abuse [111]. Our knowledge about the effect of inflammatory mediators on reward and motivation come largely from rodent studies, some of which will be reviewed briefly.

4.7.1 Sucrose Preference

In rodents, sucrose preference, that is, increased preference for sweetened water or milk, is inhibited by administration of LPS, IL-1β, or TNFα [48, 112, 113]. Nonsteroidal anti-inflammatory drugs block LPS-induced decreases in sucrose preference [48], suggesting that eicosanoids are involved in mediating the effect of LPS on sucrose preference. Sucrose consumption causes release of dopamine in the nucleus accumbens [114], which is involved in reward; however, sucrose preference is a complex behavior that also involves brain regions that control appetite and energy homeostasis such as the hypothalamus and brain stem [115, 116].

4.7.2 Intracranial Self-Stimulation

Reinforcing electrical stimulation of the forebrain bundle at the level of the posterior lateral hypothalamus is called *intracranial self-stimulation* (ICSS), which does not involve appetite regulation, and is therefore considered a better rodent model of anhedonia [117]. LPS administration inhibits ICSS, and this inhibition is associated with increased dopamine efflux in the nucleus accumbens [104, 118]. Studies have failed to show an effect of IL-1β and IL-6 on ICSS, although both cause sickness behavior [119] and IL-1β reduces sucrose preference [113], suggesting that (1) the effects of LPS on ICSS may involve other cytokines and/or (2) ICSS and sucrose preference are mediated by different brain circuits (as discussed in Sect. 4.7.1). Of note, stimulation of the ventral tegmental area, which supplies dopamine to the ventral striatum, counteracted LPS-induced sickness behavior [120], further supporting the notion that LPS-induced anhedonia may be at the core of other LPS-induced behavioral changes.

4.7.3 Exploratory and Social Behavior

In rodents, both LPS and IL-1β inhibit the exploration of novel environments and the engagement in social behaviors [54], suggesting that inflammation inhibits motivation. In mice presented with a new environment, concurrent LPS administration inhibited c-fos expression in motor and cingulate cortex, and nucleus accumbens and locus coeruleus [121]. In summary, LPS administration in rodents produces anhedonia as assessed by sucrose preference, ICSS, and exploratory and social behaviors. This is consistent with human studies of the effect of LPS on reward [101] and on social behavior [58, 60]. In substance abusers, if inflammatory stimuli produce anhedonia and reduce the ability to experience reward, it is possible that they increase the use of the substance of choice to override this reduction in reward. Supporting this, one recent study found that LPS administration increased alcohol drinking in mice [122], and another study found that inhibition of inflammation with minocycline reduced drinking [123]. This avenue of research has been insufficiently pursued. For instance, no published studies assessed whether inflammation increased cocaine or opioid self-administration in rodents, and no studies have measured this in human substance abusers.

4.7.4 *Fatigue and Psychomotor Slowing*

Fatigue is common in depression and in a host of medical diseases associated with inflammation [110, 124–127]. Fatigue can be induced in humans by administration of LPS, IFNα, and typhoid vaccination [58, 60, 77, 81, 128]. In cancer patients there is a correlation between fatigue and blood levels of IL-6, but not TNFα or IL-1β [129]. Fatigue and psychomotor slowing occur frequently during IFNα treatment [66, 69], which leads to increased IL-6 levels both peripherally and centrally [34, 130]. In humans, fatigue can be elicited directly by administration of IL-6 [67]. In rodents LPS-induced reductions in wheel-running are blunted by the administration of antibodies against IL-6 [131]. Taken together, these data suggest that IL-6 mediates the effects of illness, LPS, typhoid vaccination, and IFNα on fatigue, although IL-1β and TNFα may play a role in the central mediation of fatigue [132, 133].

4.8 Pharmacologic Studies of Brain–Immune Interactions

The largest gap in our understanding of how systemic inflammation produces various behavioral changes is at the levels of cells and molecular pathways inside the brain. As described above, the mediators released by the immune system and the brain regions affected by these mediators are fairly well characterized, especially in the rodent. Comparatively less is known about how the signal is transmitted across the BBB, and even less about the subsequent events inside the brain parenchyma, particularly how inflammatory cytokines or other mediators such as eicosanoids can affect neuronal function and thus produce the behavioral effects we observe. Recently, it was found that IL-1β induces COX-2 transcription in brain endothelial cells, which results in prostaglandin E2 release and lethargy [47]. Consistent with this, in depressed patients who had not remitted with reboxetine treatment, augmentation with a COX-2 inhibitor was superior to placebo [134]. One of the most studied potential mechanisms is the effect of inflammatory mediators on tryptophan metabolism and serotonin synthesis [135]. Inflammatory cytokines such as TNFα and IL-1β induce the expression of indoleamine 2,3-dioxygenase (IDO) in macrophages and microglia [135]. The potential effects of increased IDO are twofold. IDO converts tryptophan to kynurenic acid, leaving less tryptophan available for serotonin synthesis. This may have effects similar to tryptophan depletion, which induces depressive symptoms in susceptible individuals [136]. Inflammatory cytokines also induce microglia to produce more quinolinic acid than kynurenic acid [137]. Quinolinic acid is an *N*-methyl-D-aspartic acid receptor agonist and it can have neurotoxic effects [138, 139], either of which can contribute to depressive symptoms [140]. Treatment with IFNα activated IDO and central cytokine responses, which resulted in increased brain kynurenic and quinolinic acid, which correlated with depressive symptoms [76]. In summary, it appears that interference with serotonergic and glutamatergic neurotransmission may be important pathways through which systemic inflammation interferes with brain function.

4.8.1 Pretreatment with Antidepressants

Based on the potential role of serotonin in LPS-induced behavioral responses, we designed a study to determine whether pretreatment with the serotonin reuptake inhibitor citalopram could inhibit LPS-induced symptoms [58]. Subjects received 20 mg of citalopram daily for 5 days before each challenge. We found that the LPS-induced increase in total MADRS score was halved by pretreatment with citalopram, with a moderate effect size ($d=0.5$). Most of this effect was due to an inhibition of the increase in the lassitude item score, which measures fatigue and motivation. Citalopram pretreatment halved the LPS-induced increase in fatigue, with a large effect size ($d=0.9$). Citalopram pretreatment also reduced LPS-induced social anhedonia. These data suggest that citalopram treatment, which reduces serotonin turnover [141], can reduce the effects of systemic inflammation on fatigue and motivation in humans. In patients with Parkinson's disease, fatigue is associated with lower serotonin transporter levels in the striatum, cingulate, and thalamus [142], all regions in which glucose metabolism was affected by fatigue-inducing doses of LPS in humans [100]. Therefore, it is possible that citalopram reduces LPS-induced fatigue through inhibition of serotonin reuptake in the striatum, cingulate, or thalamus. In a rodent model of Parkinson's, the serotonin reuptake inhibitor paroxetine enhanced survival of dopaminergic neurons by reducing expression of IL-1β, TNFα, and iNOS in activated microglia [143], suggesting that SSRIs may have direct anti-inflammatory effects in the brain.

Most imaging studies of immune-induced fatigue implicate the nigrostriatal dopamine pathway, which facilitates movement and drives motivational behavior [144]. As discussed above, in IFNα-induced fatigue, there was increased activity in the nucleus accumbens and putamen [77], and in subjects who received a typhoid vaccine, psychomotor slowing correlated with activity in the substantia nigra [98]. In Parkinson's, fatigue is associated with reduced dopamine turnover in the caudate and insula [142]. Based on these studies implicating dopamine in fatigue, we designed a follow-up study to determine whether pretreatment with the dopamine and norepinephrine reuptake inhibitor bupropion pretreatment (75 mg twice a day for 7 days) could reduce LPS-induced fatigue. Surprisingly, bupropion treatment does not appear to have any effect on LPS-induced fatigue (Hannestad, unpublished). This may be because the dopamine reuptake effect of bupropion is comparatively less strong than the NE reuptake effect.

4.9 PET Imaging of Microglia

4.9.1 Basics of Receptor PET Imaging

PET imaging can be used functionally (as discussed in Sect. 4.5), but more importantly it can be used to measure the availability of a target molecule, for example, a neurotransmitter receptor or transporter, or molecule that is found in a specific cell type. PET radiotracers are synthetic or semisynthetic molecules that are designed to bind to a specific target molecule and to have certain chemical and biological properties (e.g., ability to cross the BBB). In a radiotracer molecule, a positron-emitting isotope has been substituted for an original nuclide (e.g., ^{11}C replaces ^{12}C). When this positron emitter decays in a tissue, the positron travels a short distance (usually <1 mm) before losing energy and interacting with an electron. This interaction leads to the annihilation of both the positron and the electron and gives rise to two photons traveling in opposite (approximately 180°) direction. Because these annihilation photons have high (gamma spectrum) energy, many will escape the body and be detected by the PET camera. The PET camera has scintillation detectors, which are crystal materials that produce visible-light photons (scintillations) when the annihilation photon is absorbed. A photomultiplier tube converts the energy of these photons into an electrical pulse; the pulses are amplified and sorted by a pulse-height analyzer. The photomultiplier tube only "accepts" photons within a certain energy window (usually 350–650 keV) to avoid scattered photons (i.e., photons that "collide" with other atoms and change direction). All the counts from all the detector pairs over the course of a PET scan are combined through a process called computed tomography, an algorithm that computes cross-sectional images that estimate the actual concentration of radioactivity in the tissue. This allows the event count for each pixel to be determined, which in turn is used to construct a three-dimensional image of radioactivity in the organ or tissue. The amount of radioactivity is an indication of how much radiotracer is present in a brain region. This in turn depends on how much of the target molecule is "available" for binding to the radiotracer. Therefore, PET imaging is used to estimate the amount (density) of the target molecule in a brain region. The PET image can be co-registered to a magnetic resonance image of the same brain, and the higher resolution facilitates region-of-interest analysis.

Although several measures of radiotracer binding are used in the literature, increasingly the field is moving toward a consensus nomenclature [145]. A detailed description of measures of binding is beyond the scope of this chapter; however, one frequently used measure is *volume of distribution* (V_T), which refers to the ratio of how many ml of blood contain the same amount of radiotracer as 1 cm^3 of brain tissue. For instance, $V_T = 20$ ml/cm^3 indicates that 20 ml of blood has the same amount of radiotracer as 1 cm^3 of brain, that is, the concentration in the brain is 20 times higher than in plasma. (Even though ml and cm^3 are both volumes and V_T is therefore technically unitless, the units of V_T are maintained.) The reason the concentration is higher in the brain is because the radiotracer binds to molecules in the brain and less free radiotracer is available to diffuse back into the blood. However, V_T does not indicate whether the radiotracer in the brain is specifically bound. Within the brain, binding may occur to the target molecule, to other molecules (nonspecific binding), and not at all (free tissue radiotracer); therefore, V_T indicates the total amount of radiotracer in the tissue (specifically bound, nonspecifically bound, and free).

4.9.2 Microglia

Microglia are the brain's innate immune cells, and they are involved in a variety of physiologic and pathologic processes [146]. Under physiologic conditions, resting microglia are tightly regulated by interactions with neurons [147]. When provided with molecular signals that indicate infection (e.g., PAMPs), tissue damage, or inflammation, microglia become activated; activation of microglia is a key event in the pathophysiology of brain ischemia, trauma, and infection, and in neurodegenerative, autoimmune, and possibly psychiatric disorders [146, 148, 149]. The signals that activate microglia can be local or systemic. The susceptibility of microglia to become activated depends on their baseline state; primed microglia are activated more easily; priming can occur by local signals or systemic signals [150]. Therefore, systemic inflammation and local signals interact and can lead to activation of microglia. For instance, microglia can be primed by a pathologic process in the brain, and a subsequent systemic inflammatory event may cause activation of these primed microglia. Conversely, systemic inflammatory signals may prime microglia, and these may then be activated by signals from the brain parenchyma. Although activated microglia can serve important repair functions, activated microglia can also worsen brain function and the course of brain pathologies through the release of substances that have the potential to cause neuronal dysfunction and damage, for example, inflammatory cytokines, eicosanoids, and reactive oxygen species [140, 151, 152]. The influence of systemic inflammatory signals on microglia is a mechanism through which systemic inflammation can affect brain diseases [153]. For instance, systemic LPS administration in rodents leads to systemic release of inflammatory cytokines, which signal through receptors on brain endothelial cells and activate microglia [31, 154–157]. Activation of microglia may therefore be a key pathogenic event in brain dysfunction associated with various degrees of systemic inflammation, from severe (e.g., septic encephalopathy) to mild (sickness behavior and fatigue in autoimmune disorders and cancers). In depression, there is mild systemic inflammation [158], and this may contribute to depressive symptoms through effects on microglia [140]. The effects of systemic inflammation on the diseased brain, for example, in neurodegenerative or autoimmune diseases, are much more pronounced than effects on a healthy brain [153, 159]. For instance, systemic infection can cause acute and long-term cognitive impairment in susceptible subjects, that is, subjects with mild cognitive impairment or dementia [32, 160–162]. This may be mediated by activation of microglia because activated microglia have been found in postmortem samples of patients with sepsis [163]. Milder forms of systemic inflammation (e.g., urinary tract infections) can also cause brain dysfunction (delirium) in, for example, Alzheimer's disease, and it is believed that repeated systemic inflammation can actually worsen the course of the Alzheimer's disease itself [159]. When systemic infection occurs concurrently with a cerebrovascular accident, the ischemic lesion is larger [164]. Therefore, systemic inflammation may have an impact on the progression and severity of a variety of brain diseases through activation of microglia.

Once stimulated, microglia can stay activated for several months and continue to express inflammatory mediators [56, 152, 165]. Activation of microglia by endotoxin or inflammatory cytokines leads to further release inflammatory cytokines and other potentially neurotoxic substances [152], while exposure to anti-inflammatory cytokines induces a neuroprotective microglial phenotype [166–168]. Because of the pivotal role that microglia play in a variety of neurodegenerative, neuroinflammatory, and ischemic disorders of the brain, the ability to measure activation of microglia in vivo with PET imaging is an area of great interest.

4.9.3 Translocator Protein

The translocator protein (TSPO) is an 18-kDa mitochondrial protein that is expressed in steroid-synthesizing cells, in which its role is to allow transport of cholesterol across the mitochondrial membranes. The expression of TSPO is very low in healthy brain tissue, but it increases in pathologies associated with microglial activation including stroke, trauma, infection, and autoimmune and neurodegenerative disorders [169, 170]; therefore, PET imaging of TSPO is used to measure microglial activation in various disease states [171]. Several PET radiotracers that bind to TSPO are available [172], including [^{11}C]PBR28 [173, 174]. The 20-min half-life of ^{11}C permits multiple measurements of microglial activation within a short time period, which allows for a variety of research design options. Other ways of imaging inflammation in the brain are being developed, including radiotracers that bind to COX-2 [175].

4.9.4 LPS-Induced Microglial Activation

Our group used PET and [^{11}C]PBR28 to measure binding to TSPO at various times after endotoxin administration in baboons, an example of which is illustrated in Fig. 4.3. The baboon has a response to LPS that is physiologically and immunologically very similar to humans [176]. Our hypothesis was that LPS administration would produce microglial activation and significantly increased [^{11}C]PBR28 binding compared to baseline. We found a $29 \pm 16\%$ increase in binding at 1 h (range 12–47%) and a $62 \pm 34\%$ increase in binding at 4 h (range 36–101%). There

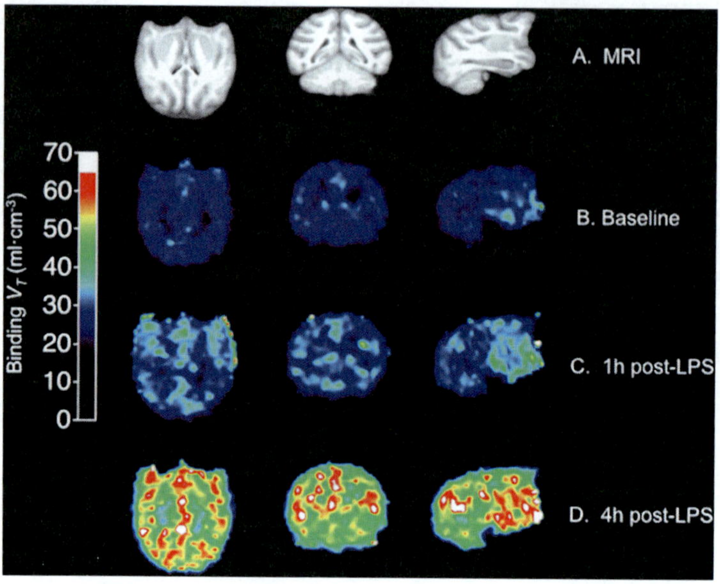

Fig. 4.3 Increase in [^{11}C]PBR28 binding after LPS administration in a baboon that had a PET scan at baseline and 1 and 4 h post-LPS. (*A*) MRI from this baboon. (*B*) [^{11}C]PBR28 binding at baseline. (*C*) [^{11}C]PBR28 binding 1 h post-LPS. (*D*) [^{11}C]PBR28 binding 4 h post-LPS. Abbreviations: *LPS* lipopolysaccharide, *MRI* magnetic resonance imaging, V_T total volume of distribution (a measure of binding)

was a positive correlation between IL-1β levels at 2 h and the increase in [^{11}C]PBR28 binding from baseline and between IL-6 levels at 3 h and the increase in [^{11}C]PBR28 binding from baseline to 4 h. There was a trend positive correlation between TNFα levels at 3 h and the increase in [^{11}C]PBR28 binding from baseline to 4 h. LPS is eliminated rapidly from the circulation after intravenous administration in baboons [177]. It is therefore likely that the effect of LPS on TSPO expression in the brain is mediated by LPS-induced inflammatory cytokines [177]. Immunohistochemistry indicated that the increase in TSPO expression measured with PET was not largely due to infiltrating monocytes or perivascular macrophages. This is consistent with data from humans, in whom a comparable dose of LPS does not cause a measurable disruption of the BBB [88]. The individual variation in the LPS-induced increase in [^{11}C]PBR28 binding (range: 12–47% at 1 h and 36–101% at 4 h) may point to individual differences in the susceptibility of microglia to a systemic inflammatory stimulus. The variability in microglial response within a group of animals of a similar age is consistent with data from young, healthy human subjects, in whom LPS-induced systemic inflammation can have highly variable effects on behavior [58] and on brain glucose metabolism [100]. It is notable that this dose of LPS causes an increase in microglial expression of TSPO in baboons, while the equivalent dose of LPS in healthy human subjects does not cause disruption of the BBB or any detectable changes in cognition [88]. This suggests that systemic inflammation can have subtle effects on microglia, which do not result in measurable cognitive or other behavioral effects. Whether the effects of LPS on noncognitive brain functions such as motivation and fatigue are mediated by microglial activation remains to be determined. Because microglial activation has been found in human alcoholics [178], and because rodent studies suggest that alcohol consumption leads to activation of microglia [154, 179], PET imaging with [^{11}C]PBR28 or similar ligands should be tested in alcoholics.

4.10 Clinical Relevance of Neuroimmune Dysregulation

Inflammation is a basic immunologic process that is a core feature of a host of diseases, both "medical" and neurobiological. Whether inflammation contributes to the pathophysiology of a disease, or whether it is simply a by-product of the core disease process, can be difficult to determine. For instance, it is well established that inflammation is a core pathophysiologic process in atherosclerosis [180]; however, in many other disorders, especially neurobiological ones, the evidence for this is less clear. In some psychiatric disorders (e.g., depression and alcohol dependence), there are systemic immune abnormalities; however, these may be a consequence of, rather than a contributing factor to, the core disease process. On the other hand, from the data discussed above, it is clear that systemic inflammation affects the brain and behavior. The susceptibility of the brain to the effects of systemic inflammation depends on whether the brain is healthy or diseased. Mild systemic inflammation (e.g., urinary tract infections) does not have any noticeable effects in most healthy persons, but it can cause brain dysfunction (delirium) in susceptible individuals such as those with mild cognitive impairment [159]. Similarly, when systemic inflammation occurs concurrently with a cerebrovascular accident, the ischemic lesion is larger [164]. In multiple sclerosis, systemic inflammation is associated with faster progression [153].

4.10.1 Depression

In depression, immunologic abnormalities have been described for decades [181, 182], but it is still unclear whether these abnormalities play a role in the pathogenesis of depressive symptoms. Serum levels of TNFα, IL-6, and perhaps IL-1β are elevated in depression [158, 183, 184]. The causes for this could include the following: (1) depression leads to a dysfunction in the brain's immunomodulatory functions and an increase in systemic TNFα and IL-6 levels, (2) there is a primary deficit in the innate immune system, or (3) genes that predispose to depression also predispose to increased inflammatory activity in the innate immune system [185]. Treatment with antidepressant medications that results in a reduction in symptoms

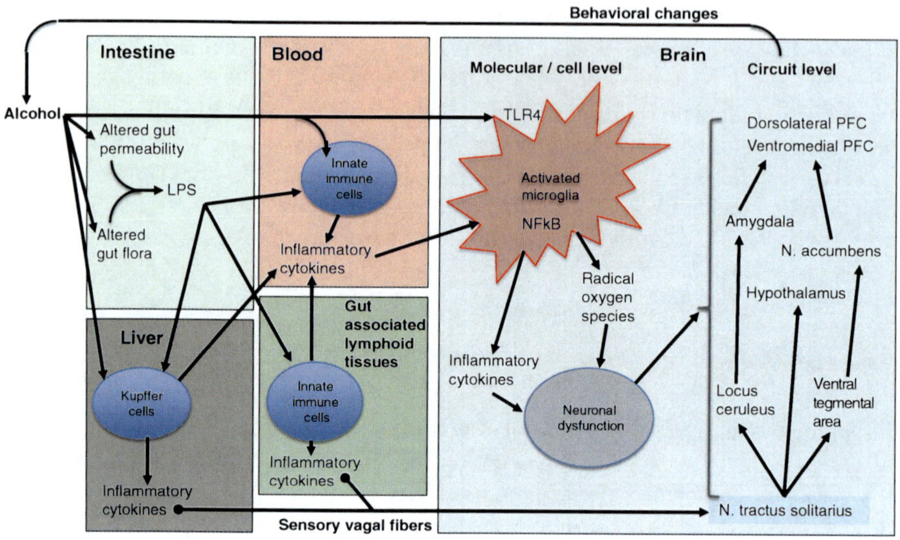

Fig. 4.4 Immune-brain pathways in alcohol dependence. (1) Potential pathways through which alcohol may increase systemic inflammation, (2) pathways through which systemic inflammation can activate microglia, (3) pathways through which microglia can affect neuronal function in regions involved in core behaviors of alcohol dependence, (4) pathways through which systemic inflammation can activate sensory vagal afferents, and (5) pathways through which vagal information modulates activity in the same brain regions. Abbreviations: *LPS* lipopolysaccharide, *NFκB* nuclear factor kappa B, *TLR4* Toll-like receptor 4 (the receptor for LPS)

does not reduce elevated TNFα or IL-6 levels [29]. This may indicate that (1) elevated TNFα and IL-6 levels do not contribute to depressive symptoms in depression, or (2) antidepressant medications "protect" the brain from the depressogenic effects of these cytokines, or (3) the dysfunction in the brain's immunomodulatory functions that occurs in depression requires more time to normalize than the time that is required for antidepressants to exert their antidepressant effect.

4.10.2 Alcoholism

Systemic immune abnormalities have been found in alcohol dependence [186]. Blood levels of TNFα are about twofold higher than in nonalcoholic subjects [187–190]; whether these normalize with cessation of alcohol use is not clear. Excessive alcohol intake is associated with higher production of inflammatory cytokines by monocytes [191]. Potential mechanisms for the increase in systemic inflammation in alcoholism, illustrated in Fig. 4.4, include the following: (1) effects of alcohol on gut permeability and flora, with leakage of LPS endotoxin and other PAMPs that stimulate the immune system [192], (2) effects of stress on the peripheral immune system [8], and (3) direct effects of alcohol on peripheral immune cells. Studies in vitro on the direct effect of alcohol on immune cells are inconclusive; some found an inhibitory effect [193–196] while others found a proinflammatory effect [197, 198]. It appears that acute alcohol exposure inhibits LPS-induced cytokine production, while chronic alcohol exposure increases inflammatory cytokines [199]. There are several mechanisms through which alcohol can directly affect immune cells: (1) interference with TLR4 [179, 200], which mediates the proinflammatory effects of LPS; (2) interaction with the transcription factor NFκB, which is involved in transcription of genes that encode for inflammatory mediators [201]; and (3) signaling through the gamma-aminobutyric acid receptor [202]. Regardless of the mechanisms involved, it is well established that systemic inflammation has important effects on the brain. It is therefore likely that systemic inflammation associated with chronic heavy alcohol drinking has similar effects on the brain. It was shown that peripheral blockade of TNFα in alcoholics reverses electroencephalographic abnormalities in sleep [203], which demonstrates that blocking systemic inflammatory cytokines can have beneficial effects on brain function.

4.10.2.1 Alcohol and Gut Permeability

The brain modulates gut motility, secretions, blood flow, and immunity through autonomic and endocrine pathways. The brain also receives information from the gut, including information about gut microbes. Neuroendocrine cells in the gut epithelium relay information about luminal microbes to nerve fibers [204], and M-cells sample microbial products and expose immune cells to these molecules [205]. The normal functioning of these pathways (brain-to-gut, gut-to-brain, and gut-to-immune-to-brain) is essential for maintaining health and optimal functioning of gut and immune function [204–206]. Heavy alcohol use has profound effects on the gut. It promotes overgrowth of LPS-containing Gram-negative bacteria [207], it alters the composition of immune cell subsets in gut-associated lymphoid tissues [208], and it increases gut permeability [209, 210], allowing influx of endotoxin which leads to secretion of TNFα and IL-6 by Kupffer cells and blood monocytes [192] (Fig. 4.4). Increased gut permeability may occur through effects of alcohol on tight junction proteins [211]. As chronic alcohol exposure sensitizes monocytes to the effects of LPS [199], heavy alcohol use may have a synergistic effect on inflammation by allowing influx of LPS from the gut and at the same time sensitizing monocytes to its effects. In summary, heavy, chronic alcohol intake may increase levels of inflammatory cytokines through direct effects on immune cells, effects on gut permeability (allowing influx of LPS and other PAMPs), and by disrupting the brain's immunomodulatory functions. Figure 4.4 illustrates some of the complex gut–immune–brain pathways that may be affected in alcohol abuse.

4.10.2.2 Alcohol and the HPA Axis

Corticotropin-releasing factor (CRF) from the hypothalamus activates the HPA axis, but CRF also affects brain structures such as the amygdala, ventral striatum, and locus coeruleus, effects which appear to play an important role in the dysphoric state that accompanies alcohol withdrawal, ultimately contributing to continued alcohol intake through negative reinforcement [212]. Consistent with this, both the negative state that accompanies alcohol withdrawal and alcohol drinking can be blocked by inhibition of CRF [111]. Interestingly, the most potent inducers of CRF release are inflammatory cytokines [213, 214], which may be the mechanism through which LPS or TNFα can sensitize to alcohol withdrawal anxiety-like behaviors [215] and increase drinking [122]. It is thus possible that elevated levels of circulating cytokines in alcohol dependence increase brain CRF release and contribute to negative reinforcement.

4.11 Conclusions and Future Directions

The brain has immunomodulatory functions, which may be disrupted in neurobiological disorders such as alcoholism, contributing to systemic inflammation. The brain has immunosensory functions, which may allow systemic inflammation to influence mood, cognition, and behavior, including functions such as reward and motivation, which are at the core of alcoholism. The study of the brain's immunosensory and immunomodulatory functions, using pharmacological interventions and/or neuroimaging in clinical populations (e.g., alcoholism), is only in its infancy. Potentially promising research avenues include (1) whether immune activation by, for example, LPS is associated with alcohol craving and increased alcohol consumption in alcoholics; (2) whether drugs that reduce the effects of LPS-induced systemic inflammation, either peripherally or in the brain, can reduce the increase in craving and/or drinking induced by LPS; (3) whether drugs that inhibit microglial activation specifically reduce craving/drinking at baseline or after LPS administration; and (4) whether drugs that tighten intestinal tight junctions can reduce systemic inflammation, liver damage, and craving/drinking in alcoholics.

Acknowledgments The author would like to thank Richard E. Carson, PhD; Kelly Cosgrove, PhD; John Krystal, MD; Marc Laruelle, MD; Robert Malison, MD; Evan Morris, PhD; Stephanie O'Malley, PhD; and Julie Staley, PhD for mentoring, research collaborations, and thoughtful discussions on various topics discussed herein. The author would also like to thank the staff of the Clinical Neuroscience Research Unit of the Connecticut Mental Health Center and the staff of the Yale PET Center. This work was made possible by support from NIDA (K12DA00167), Department of Defense (PT073595), the Society for Nuclear Medicine, NARSAD, the Nancy Taylor Foundation for Chronic Disease, and the Yale Center for Clinical Investigation (UL1 RR024139 from the National Center for Research Resources, a component of the National Institutes of Health, and NIH Roadmap for Medical Research).

References

1. Maekawa T, Kufer TA, Schulze-Lefert P (2011) NLR functions in plant and animal immune systems: so far and yet so close. Nat Immunol 12:817–826
2. Boller T, Felix G (2009) A renaissance of elicitors: perception of microbe-associated molecular patterns and danger signals by pattern-recognition receptors. Annu Rev Plant Biol 60:379–406
3. Kumar H, Kawai T, Akira S (2011) Pathogen recognition by the innate immune system. Int Rev Immunol 30:16–34
4. Barker BR, Taxman DJ, Ting JP (2011) Cross-regulation between the IL-1beta/IL-18 processing inflammasome and other inflammatory cytokines. Curr Opin Immunol 23:591–597
5. Mastronardi C et al (2007) Caspase 1 deficiency reduces inflammation-induced brain transcription. Proc Natl Acad Sci USA 104:7205–7210
6. Weberpals M et al (2009) NOS2 gene deficiency protects from sepsis-induced long-term cognitive deficits. J Neurosci 29:14177–14184
7. Dinarello CA (2009) Immunological and inflammatory functions of the interleukin-1 family. Annu Rev Immunol 27:519–550
8. Bierhaus A et al (2003) A mechanism converting psychosocial stress into mononuclear cell activation. Proc Natl Acad Sci USA 100:1920–1925
9. Pavlov VA, Tracey KJ (2005) The cholinergic anti-inflammatory pathway. Brain Behav Immun 19:493–499
10. Vida G, Pena G, Deitch EA, Ulloa L (2011) Alpha7-cholinergic receptor mediates vagal induction of splenic norepinephrine. J Immunol 186:4340–4346
11. Gosain A, Muthu K, Gamelli RL, DiPietro LA (2007) Norepinephrine suppresses wound macrophage phagocytic efficiency through alpha- and beta-adrenoreceptor dependent pathways. Surgery 142:170–179
12. Szelenyi J, Selmeczy Z (2002) Immunomodulatory effect of antidepressants. Curr Opin Pharmacol 2:428–432
13. Selmeczy Z, Szelenyi J, Vizi ES (2003) Intact noradrenaline transporter is needed for the sympathetic fine-tuning of cytokine balance. Eur J Pharmacol 469:175–181
14. Arima Y et al (2012) Regional neural activation defines a gateway for autoreactive T cells to cross the blood–brain barrier. Cell 148:447–457
15. Blandino P Jr, Barnum CJ, Deak T (2006) The involvement of norepinephrine and microglia in hypothalamic and splenic IL-1beta responses to stress. J Neuroimmunol 173:87–95
16. Heneka MT et al (2002) Noradrenergic depletion potentiates beta -amyloid-induced cortical inflammation: implications for Alzheimer's disease. J Neurosci 22:2434–2442
17. Yilmaz MS, Millington WR, Feleder C (2008) The preoptic anterior hypothalamic area mediates initiation of the hypotensive response induced by LPS in male rats. Shock 29:232–237
18. Campbell SJ et al (2007) Overexpression of IL-1beta by adenoviral-mediated gene transfer in the rat brain causes a prolonged hepatic chemokine response, axonal injury and the suppression of spontaneous behaviour. Neurobiol Dis 27:151–163
19. Huang Y, Henry CJ, Dantzer R, Johnson RW, Godbout JP (2008) Exaggerated sickness behavior and brain proinflammatory cytokine expression in aged mice in response to intracerebroventricular lipopolysaccharide. Neurobiol Aging 29:1744–1753
20. Jiang Y, Deacon R, Anthony DC, Campbell SJ (2008) Inhibition of peripheral TNF can block the malaise associated with CNS inflammatory diseases. Neurobiol Dis 32:125–132
21. Zhou H, Lapointe BM, Clark SR, Zbytnuik L, Kubes P (2006) A requirement for microglial TLR4 in leukocyte recruitment into brain in response to lipopolysaccharide. J Immunol 177:8103–8110
22. Ching S et al (2007) Endothelial-specific knockdown of interleukin-1 (IL-1) type 1 receptor differentially alters CNS responses to IL-1 depending on its route of administration. J Neurosci 27:10476–10486

23. Koo JW, Duman RS (2008) IL-1beta is an essential mediator of the antineurogenic and anhedonic effects of stress. Proc Natl Acad Sci USA 105:751–756
24. Koo JW, Russo SJ, Ferguson D, Nestler EJ, Duman RS (2010) Nuclear factor-kappaB is a critical mediator of stress-impaired neurogenesis and depressive behavior. Proc Natl Acad Sci USA 107:2669–2674
25. Anisman H, Merali Z, Hayley S (2008) Neurotransmitter, peptide and cytokine processes in relation to depressive disorder: comorbidity between depression and neurodegenerative disorders. Prog Neurobiol 85:1–74
26. Dickerson SS, Kemeny ME, Aziz N, Kim KH, Fahey JL (2004) Immunological effects of induced shame and guilt. Psychosom Med 66:124–131
27. Gundersen Y, Opstad PK, Reistad T, Thrane I, Vaagenes P (2006) Seven days' around the clock exhaustive physical exertion combined with energy depletion and sleep deprivation primes circulating leukocytes. Eur J Appl Physiol 97:151–157
28. Pace TW et al (2006) Increased stress-induced inflammatory responses in male patients with major depression and increased early life stress. Am J Psychiatry 163:1630–1633
29. Hannestad J, Dellagioia N, Bloch M (2011) The effect of antidepressant medication treatment on serum levels of inflammatory cytokines: a meta-analysis. Neuropsychopharmacology 36:2452–2459
30. Sharshar T, Polito A, Checinski A, Stevens RD (2010) Septic-associated encephalopathy – everything starts at a microlevel. Crit Care 14:199
31. Semmler A et al (2008) Sepsis causes neuroinflammation and concomitant decrease of cerebral metabolism. J Neuroinflammation 5:38
32. Iwashyna TJ, Ely EW, Smith DM, Langa KM (2010) Long-term cognitive impairment and functional disability among survivors of severe sepsis. JAMA 304:1787–1794
33. Reyes TM, Coe CL (1996) Interleukin-1 beta differentially affects interleukin-6 and soluble interleukin-6 receptor in the blood and central nervous system of the monkey. J Neuroimmunol 66:135–141
34. Raison CL et al (2009) Activation of central nervous system inflammatory pathways by interferon-alpha: relationship to monoamines and depression. Biol Psychiatry 65:296–303
35. DellaGioia N, Hannestad J (2010) A critical review of human endotoxin administration as an experimental paradigm of depression. Neurosci Biobehav Rev 34:130–143
36. Dunn AJ, Swiergiel AH, de Beaurepaire R (2005) Cytokines as mediators of depression: what can we learn from animal studies? Neurosci Biobehav Rev 29:891–909
37. Adamo SA, Fidler TL, Forestell CA (2007) Illness-induced anorexia and its possible function in the caterpillar, *Manduca sexta*. Brain Behav Immun 21:292–300
38. Goldsworthy G (2010) Locusts as model organisms in which to study immunogen-induced anorectic behaviour. J Insect Physiol 56:991–997
39. Raison CL, Capuron L, Miller AH (2006) Cytokines sing the blues: inflammation and the pathogenesis of depression. Trends Immunol 27:24–31
40. De La Garza R 2nd (2005) Endotoxin- or pro-inflammatory cytokine-induced sickness behavior as an animal model of depression: focus on anhedonia. Neurosci Biobehav Rev 29:761–770
41. Pecchi E, Dallaporta M, Jean A, Thirion S, Troadec JD (2009) Prostaglandins and sickness behavior: old story, new insights. Physiol Behav 97:279–292
42. Anisman H (2009) Cascading effects of stressors and inflammatory immune system activation: implications for major depressive disorder. J Psychiatry Neurosci 34:4–20
43. Alvarez JI et al (2011) The Hedgehog pathway promotes blood–brain barrier integrity and CNS immune quiescence. Science 334:1727–1731
44. Dauphinee SM, Karsan A (2006) Lipopolysaccharide signaling in endothelial cells. Lab Invest 86:9–22
45. Magder S, Neculcea J, Neculcea V, Sladek R (2006) Lipopolysaccharide and TNF-alpha produce very similar changes in gene expression in human endothelial cells. J Vasc Res 43:447–461

46. Gosselin D, Rivest S (2008) MyD88 signaling in brain endothelial cells is essential for the neuronal activity and glucocorticoid release during systemic inflammation. Mol Psychiatry 13:480–497
47. Ridder DA et al (2011) TAK1 in brain endothelial cells mediates fever and lethargy. J Exp Med 208:2615–2623
48. De La Garza R 2nd, Asnis GM, Fabrizio KR, Pedrosa E (2005) Acute diclofenac treatment attenuates lipopolysaccharide-induced alterations to basic reward behavior and HPA axis activation in rats. Psychopharmacology (Berl) 179:356–365
49. Kapas L, Hansen MK, Chang HY, Krueger JM (1998) Vagotomy attenuates but does not prevent the somnogenic and febrile effects of lipopolysaccharide in rats. Am J Physiol 274:R406–R411
50. Opp MR, Toth LA (1998) Somnogenic and pyrogenic effects of interleukin-1beta and lipopolysaccharide in intact and vagotomized rats. Life Sci 62:923–936
51. Konsman JP, Luheshi GN, Bluthe RM, Dantzer R (2000) The vagus nerve mediates behavioural depression, but not fever, in response to peripheral immune signals; a functional anatomical analysis. Eur J Neurosci 12:4434–4446
52. Blatteis CM (2007) The onset of fever: new insights into its mechanism. Prog Brain Res 162:3–14
53. D'Mello C, Le T, Swain MG (2009) Cerebral microglia recruit monocytes into the brain in response to tumor necrosis factor alpha signaling during peripheral organ inflammation. J Neurosci 29:2089–2102
54. Larson SJ, Dunn AJ (2001) Behavioral effects of cytokines. Brain Behav Immun 15:371–387
55. Frenois F et al (2007) Lipopolysaccharide induces delayed FosB/DeltaFosB immunostaining within the mouse extended amygdala, hippocampus and hypothalamus, that parallel the expression of depressive-like behavior. Psychoneuroendocrinology 32:516–531
56. Qin L et al (2007) Systemic LPS causes chronic neuroinflammation and progressive neurodegeneration. Glia 55:453–462
57. Moreau M et al (2008) Inoculation of Bacillus Calmette-Guerin to mice induces an acute episode of sickness behavior followed by chronic depressive-like behavior. Brain Behav Immun 22:1087–1095
58. Hannestad J, DellaGioia N, Ortiz N, Pittman B, Bhagwagar Z (2011) Citalopram reduces endotoxin-induced fatigue. Brain Behav Immun 25:256–259
59. Reichenberg A et al (2001) Cytokine-associated emotional and cognitive disturbances in humans. Arch Gen Psychiatry 58:445–452
60. Eisenberger NI, Inagaki TK, Mashal NM, Irwin MR (2010) Inflammation and social experience: an inflammatory challenge induces feelings of social disconnection in addition to depressed mood. Brain Behav Immun 24:558–563
61. Bucks RS et al (2008) Selective effects of upper respiratory tract infection on cognition, mood and emotion processing: a prospective study. Brain Behav Immun 22:399–407
62. Janicki-Deverts D, Cohen S, Doyle WJ, Turner RB, Treanor JJ (2007) Infection-induced proinflammatory cytokines are associated with decreases in positive affect, but not increases in negative affect. Brain Behav Immun 21:301–307
63. Hermann DM et al (1998) Endotoxin-induced changes in sleep and sleepiness during the day. Psychoneuroendocrinology 23:427–437
64. Wright CE, Strike PC, Brydon L, Steptoe A (2005) Acute inflammation and negative mood: mediation by cytokine activation. Brain Behav Immun 19:345–350
65. Eisenberger NI, Inagaki TK, Rameson LT, Mashal NM, Irwin MR (2009) An fMRI study of cytokine-induced depressed mood and social pain: the role of sex differences. Neuroimage 47:881–890
66. Capuron L et al (2009) Does cytokine-induced depression differ from idiopathic major depression in medically healthy individuals? J Affect Disord 119:181–185

67. Spath-Schwalbe E et al (1998) Acute effects of recombinant human interleukin-6 on endocrine and central nervous sleep functions in healthy men. J Clin Endocrinol Metab 83:1573–1579
68. Crews FT, Zou J, Qin L (2011) Induction of innate immune genes in brain create the neurobiology of addiction. Brain Behav Immun 25(Suppl 1):S4–S12
69. Asnis GM, De La Garza R 2nd (2006) Interferon-induced depression in chronic hepatitis C: a review of its prevalence, risk factors, biology, and treatment approaches. J Clin Gastroenterol 40:322–335
70. Kraus MR et al (2008) Therapy of interferon-induced depression in chronic hepatitis C with citalopram: a randomized, double-blind, placebo-controlled study. Gut 57:531–536
71. Musselman DL et al (2001) Paroxetine for the prevention of depression induced by high-dose interferon alfa. N Engl J Med 344:961–966
72. Raison CL et al (2007) Paroxetine for prevention of depressive symptoms induced by interferon-alpha and ribavirin for hepatitis C. Aliment Pharmacol Ther 25:1163–1174
73. Majer M et al (2008) IFN-alpha-induced motor slowing is associated with increased depression and fatigue in patients with chronic hepatitis C. Brain Behav Immun 22:870–880
74. Taylor JL, Grossberg SE (1998) The effects of interferon-alpha on the production and action of other cytokines. Semin Oncol 25:23–29
75. Wichers MC et al (2007) Interferon-alpha-induced depressive symptoms are related to changes in the cytokine network but not to cortisol. J Psychosom Res 62:207–214
76. Raison CL et al (2010) CSF concentrations of brain tryptophan and kynurenines during immune stimulation with IFN-alpha: relationship to CNS immune responses and depression. Mol Psychiatry 15:393–403
77. Capuron L et al (2007) Basal ganglia hypermetabolism and symptoms of fatigue during interferon-alpha therapy. Neuropsychopharmacology 32:2384–2392
78. Grover VP et al (2012) Cerebral microglial activation in patients with hepatitis C: in vivo evidence of neuroinflammation. J Viral Hepat 19:e89–e96
79. Strike PC, Wardle J, Steptoe A (2004) Mild acute inflammatory stimulation induces transient negative mood. J Psychosom Res 57:189–194
80. Norcross JC, Guadagnoli E, Prochaska JO (1984) Factor structure of the profile of mood states (POMS): two partial replications. J Clin Psychol 40:1270–1277
81. Harrison NA et al (2009) Neural origins of human sickness in interoceptive responses to inflammation. Biol Psychiatry 66:415–422
82. Lowry SF (2005) Human endotoxemia: a model for mechanistic insight and therapeutic targeting. Shock 24(Suppl 1):94–100
83. Michie HR et al (1988) Detection of circulating tumor necrosis factor after endotoxin administration. N Engl J Med 318:1481–1486
84. van Eijk LT et al (2007) Gender differences in the innate immune response and vascular reactivity following the administration of endotoxin to human volunteers. Crit Care Med 35:1464–1469
85. van der Poll T et al (1997) Effect of a recombinant dimeric tumor necrosis factor receptor on inflammatory responses to intravenous endotoxin in normal humans. Blood 89:3727–3734
86. Vedder H et al (2007) Immune-endocrine host response to endotoxin in major depression. J Psychiatr Res 41:280–289
87. Hannestad J, Dellagioia N, Ortiz N, Pittman B, Bhagwagar Z (2011) Citalopram reduces endotoxin-induced fatigue. Brain Behav Immun 25:256–259
88. van den Boogaard M et al (2010) Endotoxemia-induced inflammation and the effect on the human brain. Crit Care 14:R81
89. Lehrer P et al (2010) Voluntarily produced increases in heart rate variability modulate autonomic effects of endotoxin induced systemic inflammation: an exploratory study. Appl Psychophysiol Biofeedback 35:303–315
90. Lee JH et al (2010) Global and local fMRI signals driven by neurons defined optogenetically by type and wiring. Nature 465:788–792

91. Wienhard K (2002) Measurement of glucose consumption using [(18)F]fluorodeoxyglucose. Methods 27:218–225
92. Gaykema RP, Goehler LE (2011) Ascending caudal medullary catecholamine pathways drive sickness-induced deficits in exploratory behavior: brain substrates for fatigue? Brain Behav Immun 25:443–460
93. Dallaporta M et al (2007) c-Fos immunoreactivity induced by intraperitoneal LPS administration is reduced in the brain of mice lacking the microsomal prostaglandin E synthase-1 (mPGES-1). Brain Behav Immun 21:1109–1121
94. Sugama S, Takenouchi T, Fujita M, Conti B, Hashimoto M (2009) Differential microglial activation between acute stress and lipopolysaccharide treatment. J Neuroimmunol 207:24–31
95. Hollis JH, Evans AK, Bruce KP, Lightman SL, Lowry CA (2006) Lipopolysaccharide has indomethacin-sensitive actions on Fos expression in topographically organized subpopulations of serotonergic neurons. Brain Behav Immun 20:569–577
96. Doenlen R et al (2011) Electrical activity in rat cortico-limbic structures after single or repeated administration of lipopolysaccharide or staphylococcal enterotoxin B. Proc Biol Sci 278:1864–1872
97. Capuron L et al (2005) Anterior cingulate activation and error processing during interferon-alpha treatment. Biol Psychiatry 58:190–196
98. Brydon L, Harrison NA, Walker C, Steptoe A, Critchley HD (2008) Peripheral inflammation is associated with altered substantia nigra activity and psychomotor slowing in humans. Biol Psychiatry 63:1022–1029
99. Harrison NA et al (2009) Inflammation causes mood changes through alterations in subgenual cingulate activity and mesolimbic connectivity. Biol Psychiatry 66:407–414
100. Hannestad J et al (2012) Glucose metabolism in the insula and cingulate is affected by systemic inflammation in humans. J Nucl Med 53:601–607
101. Eisenberger NI et al (2010) Inflammation-induced anhedonia: endotoxin reduces ventral striatum responses to reward. Biol Psychiatry 68:748–754
102. Inagaki TK, Muscatell KA, Irwin MR, Cole SW, Eisenberger NI (2012) Inflammation selectively enhances amygdala activity to socially threatening images. Neuroimage 59:3222–3226
103. Rossi-George A, Urbach D, Colas D, Goldfarb Y, Kusnecov AW (2005) Neuronal, endocrine, and anorexic responses to the T-cell superantigen staphylococcal enterotoxin A: dependence on tumor necrosis factor-alpha. J Neurosci 25:5314–5322
104. Barr AM et al (2003) Tolerance to the anhedonic effects of lipopolysaccharide is associated with changes in syntaxin immunoreactivity in the nucleus accumbens. Int J Neuropsychopharmacol 6:23–34
105. Craig AD (2009) Emotional moments across time: a possible neural basis for time perception in the anterior insula. Philos Trans R Soc Lond B Biol Sci 364:1933–1942
106. Ongur D, Price JL (2000) The organization of networks within the orbital and medial prefrontal cortex of rats, monkeys and humans. Cereb Cortex 10:206–219
107. Jones CL, Ward J, Critchley HD (2010) The neuropsychological impact of insular cortex lesions. J Neurol Neurosurg Psychiatry 81:611–618
108. Oppenheimer SM, Gelb A, Girvin JP, Hachinski VC (1992) Cardiovascular effects of human insular cortex stimulation. Neurology 42:1727–1732
109. Medford N, Critchley HD (2010) Conjoint activity of anterior insular and anterior cingulate cortex: awareness and response. Brain Struct Funct 214:535–549
110. Kennedy SH (2008) Core symptoms of major depressive disorder: relevance to diagnosis and treatment. Dialogues Clin Neurosci 10:271–277
111. Koob GF (2009) Neurobiological substrates for the dark side of compulsivity in addiction. Neuropharmacology 56(Suppl 1):18–31
112. Weil ZM, Bowers SL, Pyter LM, Nelson RJ (2006) Social interactions alter proinflammatory cytokine gene expression and behavior following endotoxin administration. Brain Behav Immun 20:72–79

113. Brebner K, Hayley S, Zacharko R, Merali Z, Anisman H (2000) Synergistic effects of interleukin-1beta, interleukin-6, and tumor necrosis factor-alpha: central monoamine, corticosterone, and behavioral variations. Neuropsychopharmacology 22:566–580
114. de Araujo IE et al (2008) Food reward in the absence of taste receptor signaling. Neuron 57:930–941
115. Ahima RS, Antwi DA (2008) Brain regulation of appetite and satiety. Endocrinol Metab Clin North Am 37:811–823
116. Adamson TW, Corll C, Svec F, Porter J (2010) Role of the perifornical hypothalamic monoamine neurotransmitter systems in anorectic effects of endotoxin. Neuroendocrinology 91:48–55
117. Liebman JM (1983) Discriminating between reward and performance: a critical review of intracranial self-stimulation methodology. Neurosci Biobehav Rev 7:45–72
118. Borowski T, Kokkinidis L, Merali Z, Anisman H (1998) Lipopolysaccharide, central in vivo biogenic amine variations, and anhedonia. Neuroreport 9:3797–3802
119. Anisman H, Kokkinidis L, Borowski T, Merali Z (1998) Differential effects of interleukin (IL)-1beta, IL-2 and IL-6 on responding for rewarding lateral hypothalamic stimulation. Brain Res 779:177–187
120. Kentner AC et al (2008) The effects of rewarding ventral tegmental area stimulation and environmental enrichment on lipopolysaccharide-induced sickness behavior and cytokine expression in female rats. Brain Res 1217:50–61
121. Stone EA, Lehmann ML, Lin Y, Quartermain D (2006) Depressive behavior in mice due to immune stimulation is accompanied by reduced neural activity in brain regions involved in positively motivated behavior. Biol Psychiatry 60:803–811
122. Blednov YA et al (2011) Activation of inflammatory signaling by lipopolysaccharide produces a prolonged increase of voluntary alcohol intake in mice. Brain Behav Immun 25(Suppl 1):S92–S105
123. Agrawal RG, Hewetson A, George CM, Syapin PJ, Bergeson SE (2011) Minocycline reduces ethanol drinking. Brain Behav Immun 25(Suppl 1):S165–S169
124. Huffman JC, Smith FA, Quinn DK, Fricchione GL (2006) Post-MI psychiatric syndromes: six unanswered questions. Harv Rev Psychiatry 14:305–318
125. Bower JE (2007) Cancer-related fatigue: links with inflammation in cancer patients and survivors. Brain Behav Immun 21:863–871
126. Meyers CA, Albitar M, Estey E (2005) Cognitive impairment, fatigue, and cytokine levels in patients with acute myelogenous leukemia or myelodysplastic syndrome. Cancer 104:788–793
127. Davis MC et al (2008) Chronic stress and regulation of cellular markers of inflammation in rheumatoid arthritis: implications for fatigue. Brain Behav Immun 22:24–32
128. Bahador M, Cross AS (2007) From therapy to experimental model: a hundred years of endotoxin administration to human subjects. J Endotoxin Res 13:251–279
129. Schubert C, Hong S, Natarajan L, Mills PJ, Dimsdale JE (2007) The association between fatigue and inflammatory marker levels in cancer patients: a quantitative review. Brain Behav Immun 21:413–427
130. Loftis JM, Huckans M, Ruimy S, Hinrichs DJ, Hauser P (2008) Depressive symptoms in patients with chronic hepatitis C are correlated with elevated plasma levels of interleukin-1beta and tumor necrosis factor-alpha. Neurosci Lett 430:264–268
131. Harden LM, du Plessis I, Poole S, Laburn HP (2006) Interleukin-6 and leptin mediate lipopolysaccharide-induced fever and sickness behavior. Physiol Behav 89:146–155
132. Harden LM, du Plessis I, Poole S, Laburn HP (2008) Interleukin (IL)-6 and IL-1 beta act synergistically within the brain to induce sickness behavior and fever in rats. Brain Behav Immun 22:838–849
133. Netea MG et al (2007) Increased voluntary exercise in mice deficient for tumour necrosis factor-alpha and lymphotoxin-alpha. Eur J Clin Invest 37:737–741
134. Muller N et al (2006) The cyclooxygenase-2 inhibitor celecoxib has therapeutic effects in major depression: results of a double-blind, randomized, placebo controlled, add-on pilot study to reboxetine. Mol Psychiatry 11:680–684

135. Muller N, Schwarz MJ (2007) The immune-mediated alteration of serotonin and glutamate: towards an integrated view of depression. Mol Psychiatry 12:988–1000
136. Neumeister A (2003) Tryptophan depletion, serotonin, and depression: where do we stand? Psychopharmacol Bull 37:99–115
137. Saito K, Crowley JS, Markey SP, Heyes MP (1993) A mechanism for increased quinolinic acid formation following acute systemic immune stimulation. J Biol Chem 268:15496–15503
138. Myint AM, Leonard BE, Steinbusch HW, Kim YK (2005) Th1, Th2, and Th3 cytokine alterations in major depression. J Affect Disord 88:167–173
139. Myint AM et al (2007) Kynurenine pathway in major depression: evidence of impaired neuroprotection. J Affect Disord 98:143–151
140. McNally L, Bhagwagar Z, Hannestad J (2008) Inflammation, glutamate, and glia in depression: a literature review. CNS Spectr 13:501–510
141. Bosker FJ et al (2010) Biochemical and behavioral effects of long-term citalopram administration and discontinuation in rats: role of serotonin synthesis. Neurochem Int 57:948–957
142. Pavese N, Metta V, Bose SK, Chaudhuri KR, Brooks DJ (2010) Fatigue in Parkinson's disease is linked to striatal and limbic serotonergic dysfunction. Brain 133:3434–3443
143. Chung YC, Kim SR, Jin BK (2010) Paroxetine prevents loss of nigrostriatal dopaminergic neurons by inhibiting brain inflammation and oxidative stress in an experimental model of Parkinson's disease. J Immunol 185:1230–1237
144. Cools R (2008) Role of dopamine in the motivational and cognitive control of behavior. Neuroscientist 14:381–395
145. Innis RB et al (2007) Consensus nomenclature for in vivo imaging of reversibly binding radioligands. J Cereb Blood Flow Metab 27:1533–1539
146. Tambuyzer BR, Ponsaerts P, Nouwen EJ (2009) Microglia: gatekeepers of central nervous system immunology. J Leukoc Biol 85:352–370
147. Ponomarev ED, Veremeyko T, Barteneva N, Krichevsky AM, Weiner HL (2011) MicroRNA-124 promotes microglia quiescence and suppresses EAE by deactivating macrophages via the C/EBP-alpha-PU.1 pathway. Nat Med 17:64–70
148. Chan WY, Kohsaka S, Rezaie P (2007) The origin and cell lineage of microglia: new concepts. Brain Res Rev 53:344–354
149. Chen SK et al (2010) Hematopoietic origin of pathological grooming in Hoxb8 mutant mice. Cell 141:775–785
150. Ramaglia V et al (2012) C3-dependent mechanism of microglial priming relevant to multiple sclerosis. Proc Natl Acad Sci USA 109:965–970
151. van Gool WA, van de Beek D, Eikelenboom P (2010) Systemic infection and delirium: when cytokines and acetylcholine collide. Lancet 375:773–775
152. Block ML, Zecca L, Hong JS (2007) Microglia-mediated neurotoxicity: uncovering the molecular mechanisms. Nat Rev Neurosci 8:57–69
153. Teeling JL, Perry VH (2009) Systemic infection and inflammation in acute CNS injury and chronic neurodegeneration: underlying mechanisms. Neuroscience 158:1062–1073
154. Qin L et al (2008) Increased systemic and brain cytokine production and neuroinflammation by endotoxin following ethanol treatment. J Neuroinflammation 5:10
155. Chakravarty S, Herkenham M (2005) Toll-like receptor 4 on nonhematopoietic cells sustains CNS inflammation during endotoxemia, independent of systemic cytokines. J Neurosci 25:1788–1796
156. Masocha W (2009) Systemic lipopolysaccharide (LPS)-induced microglial activation results in different temporal reduction of CD200 and CD200 receptor gene expression in the brain. J Neuroimmunol 214:78–82
157. Terrando N et al (2010) The impact of IL-1 modulation on the development of lipopolysaccharide-induced cognitive dysfunction. Crit Care 14:R88
158. Dowlati Y et al (2010) A meta-analysis of cytokines in major depression. Biol Psychiatry 67:446–457

159. Cunningham C (2011) Systemic inflammation and delirium: important co-factors in the progression of dementia. Biochem Soc Trans 39:945–953
160. Davies NW, Sharief MK, Howard RS (2006) Infection-associated encephalopathies: their investigation, diagnosis, and treatment. J Neurol 253:833–845
161. Girard TD et al (2010) Delirium as a predictor of long-term cognitive impairment in survivors of critical illness. Crit Care Med 38:1513–1520
162. Godbout JP et al (2005) Exaggerated neuroinflammation and sickness behavior in aged mice following activation of the peripheral innate immune system. FASEB J 19:1329–1331
163. Lemstra AW et al (2007) Microglia activation in sepsis: a case–control study. J Neuroinflammation 4:4
164. Denes A, Humphreys N, Lane TE, Grencis R, Rothwell N (2010) Chronic systemic infection exacerbates ischemic brain damage via a CCL5 (regulated on activation, normal T-cell expressed and secreted)-mediated proinflammatory response in mice. J Neurosci 30: 10086–10095
165. Maeda J et al (2007) Phase-dependent roles of reactive microglia and astrocytes in nervous system injury as delineated by imaging of peripheral benzodiazepine receptor. Brain Res 1157:100–111
166. Li L, Lu J, Tay SS, Moochhala SM, He BP (2007) The function of microglia, either neuroprotection or neurotoxicity, is determined by the equilibrium among factors released from activated microglia in vitro. Brain Res 1159:8–17
167. Hanisch UK, Kettenmann H (2007) Microglia: active sensor and versatile effector cells in the normal and pathologic brain. Nat Neurosci 10:1387–1394
168. Pinteaux E, Rothwell NJ, Boutin H (2006) Neuroprotective actions of endogenous interleukin-1 receptor antagonist (IL-1ra) are mediated by glia. Glia 53:551–556
169. Cosenza-Nashat M et al (2009) Expression of the translocator protein of 18 kDa by microglia, macrophages and astrocytes based on immunohistochemical localization in abnormal human brain. Neuropathol Appl Neurobiol 35:306–328
170. Batarseh A, Papadopoulos V (2010) Regulation of translocator protein 18 kDa (TSPO) expression in health and disease states. Mol Cell Endocrinol 327:1–12
171. Venneti S, Lopresti BJ, Wiley CA (2006) The peripheral benzodiazepine receptor (translocator protein 18 kDa) in microglia: from pathology to imaging. Prog Neurobiol 80:308–322
172. Rupprecht R et al (2010) Translocator protein (18 kDa) (TSPO) as a therapeutic target for neurological and psychiatric disorders. Nat Rev Drug Discov 9:971–988
173. Brown AK et al (2007) Radiation dosimetry and biodistribution in monkey and man of 11C-PBR28: a PET radioligand to image inflammation. J Nucl Med 48:2072–2079
174. Imaizumi M et al (2008) Brain and whole-body imaging in nonhuman primates of [(11)C] PBR28, a promising PET radioligand for peripheral benzodiazepine receptors. Neuroimage 39:1289–1298
175. Shukuri M et al (2011) In vivo expression of cyclooxygenase-1 in activated microglia and macrophages during neuroinflammation visualized by PET with 11C-ketoprofen methyl ester. J Nucl Med 52:1094–1101
176. Redl H, Bahrami S (2005) Large animal models: baboons for trauma, shock, and sepsis studies. Shock 24(Suppl 1):88–93
177. Haudek SB et al (2003) Lipopolysaccharide dose response in baboons. Shock 20:431–436
178. He J, Crews FT (2008) Increased MCP-1 and microglia in various regions of the human alcoholic brain. Exp Neurol 210:349–358
179. Alfonso-Loeches S, Pascual-Lucas M, Blanco AM, Sanchez-Vera I, Guerri C (2010) Pivotal role of TLR4 receptors in alcohol-induced neuroinflammation and brain damage. J Neurosci 30:8285–8295
180. Weber C, Noels H (2011) Atherosclerosis: current pathogenesis and therapeutic options. Nat Med 17:1410–1422
181. Irwin MR, Miller AH (2007) Depressive disorders and immunity: 20 years of progress and discovery. Brain Behav Immun 21:374–383

182. Miller AH, Maletic V, Raison CL (2009) Inflammation and its discontents: the role of cytokines in the pathophysiology of major depression. Biol Psychiatry 65:732–741
183. Hamer M, Molloy GJ, de Oliveira C, Demakakos P (2009) Persistent depressive symptomatology and inflammation: to what extent do health behaviours and weight control mediate this relationship? Brain Behav Immun 23:413–418
184. Howren MB, Lamkin DM, Suls J (2009) Associations of depression with C-reactive protein, IL-1, and IL-6: a meta-analysis. Psychosom Med 71:171–186
185. Raison CL, Miller AH (2012) The evolutionary significance of depression in Pathogen Host Defense (PATHOS-D). Mol Psychiatry, Jan 31. doi: 10.1038/mp.2012.2. [Epub ahead of print]
186. Kelley KW, Dantzer R (2011) Alcoholism and inflammation: neuroimmunology of behavioral and mood disorders. Brain Behav Immun 25(Suppl 1):S13–S20
187. Kiefer F, Jahn H, Schick M, Wiedemann K (2002) Alcohol intake, tumour necrosis factor-alpha, leptin and craving: factors of a possibly vicious circle? Alcohol Alcohol 37:401–404
188. Szabo G, Mandrekar P (2009) A recent perspective on alcohol, immunity, and host defense. Alcohol Clin Exp Res 33:220–232
189. Gonzalez-Quintela A et al (2008) Serum TNF-alpha levels in relation to alcohol consumption and common TNF gene polymorphisms. Alcohol 42:513–518
190. Garcia-Valdecasas-Campelo E et al (2007) Brain atrophy in alcoholics: relationship with alcohol intake; liver disease; nutritional status, and inflammation. Alcohol Alcohol 42:533–538
191. Laso FJ, Vaquero JM, Almeida J, Marcos M, Orfao A (2007) Production of inflammatory cytokines by peripheral blood monocytes in chronic alcoholism: relationship with ethanol intake and liver disease. Cytometry B Clin Cytom 72:408–415
192. Keshavarzian A et al (2009) Evidence that chronic alcohol exposure promotes intestinal oxidative stress, intestinal hyperpermeability and endotoxemia prior to development of alcoholic steatohepatitis in rats. J Hepatol 50:538–547
193. Walker JE Jr et al (2009) Ethanol exposure impairs LPS-induced pulmonary LIX expression: alveolar epithelial cell dysfunction as a consequence of acute intoxication. Alcohol Clin Exp Res 33:357–365
194. Wandler A, Bruun JM, Nielsen MP, Richelsen B (2008) Ethanol exerts anti-inflammatory effects in human adipose tissue in vitro. Mol Cell Endocrinol 296:26–31
195. Andrade MC, Martins-Filho OA, Coelho-Neto J, Mesquita ON, Faria AM (2008) The long-term impaired macrophages functions are already observed early after high-dose ethanol administration. Scand J Immunol 68:306–314
196. Dai Q, Pruett SB (2006) Different effects of acute and chronic ethanol on LPS-induced cytokine production and TLR4 receptor behavior in mouse peritoneal macrophages. J Immunotoxicol 3:217–225
197. Mandrekar P, Catalano D, Jeliazkova V, Kodys K (2008) Alcohol exposure regulates heat shock transcription factor binding and heat shock proteins 70 and 90 in monocytes and macrophages: implication for TNF-alpha regulation. J Leukoc Biol 84:1335–1345
198. Gobejishvili L, Barve S, Joshi-Barve S, McClain C (2008) Enhanced PDE4B expression augments LPS-inducible TNF expression in ethanol-primed monocytes: relevance to alcoholic liver disease. Am J Physiol Gastrointest Liver Physiol 295:G718–G724
199. Mandrekar P, Bala S, Catalano D, Kodys K, Szabo G (2009) The opposite effects of acute and chronic alcohol on lipopolysaccharide-induced inflammation are linked to IRAK-M in human monocytes. J Immunol 183:1320–1327
200. Szabo G, Dolganiuc A, Dai Q, Pruett SB (2007) TLR4, ethanol, and lipid rafts: a new mechanism of ethanol action with implications for other receptor-mediated effects. J Immunol 178:1243–1249
201. Yakovleva T, Bazov I, Watanabe H, Hauser KF, Bakalkin G (2011) Transcriptional control of maladaptive and protective responses in alcoholics: a role of the NF-kappaB system. Brain Behav Immun 25(Suppl 1):S29–S38

202. Bhat R et al (2010) Inhibitory role for GABA in autoimmune inflammation. Proc Natl Acad Sci USA 107:2580–2585
203. Irwin MR, Olmstead R, Valladares EM, Breen EC, Ehlers CL (2009) Tumor necrosis factor antagonism normalizes rapid eye movement sleep in alcohol dependence. Biol Psychiatry 66:191–195
204. Rhee SH, Pothoulakis C, Mayer EA (2009) Principles and clinical implications of the brain-gut-enteric microbiota axis. Nat Rev Gastroenterol Hepatol 6:306–314
205. Abt MC, Artis D (2009) The intestinal microbiota in health and disease: the influence of microbial products on immune cell homeostasis. Curr Opin Gastroenterol 25:496–502
206. Clarke TB et al (2010) Recognition of peptidoglycan from the microbiota by Nod1 enhances systemic innate immunity. Nat Med 16:228–231
207. Mutlu E et al (2009) Intestinal dysbiosis: a possible mechanism of alcohol-induced endotoxemia and alcoholic steatohepatitis in rats. Alcohol Clin Exp Res 33:1836–1846
208. Poonia B, Nelson S, Bagby GJ, Veazey RS (2006) Intestinal lymphocyte subsets and turnover are affected by chronic alcohol consumption: implications for SIV/HIV infection. J Acquir Immune Defic Syndr 41:537–547
209. Farhadi A et al (2008) Susceptibility to gut leakiness: a possible mechanism for endotoxaemia in non-alcoholic steatohepatitis. Liver Int 28:1026–1033
210. Cariello R et al (2010) Intestinal permeability in patients with chronic liver diseases: its relationship with the aetiology and the entity of liver damage. Dig Liver Dis 42:200–204
211. Tang Y et al (2008) Effect of alcohol on miR-212 expression in intestinal epithelial cells and its potential role in alcoholic liver disease. Alcohol Clin Exp Res 32:355–364
212. Bale TL et al (2000) Mice deficient for corticotropin-releasing hormone receptor-2 display anxiety-like behaviour and are hypersensitive to stress. Nat Genet 24:410–414
213. Bienkowski MS, Rinaman L (2008) Noradrenergic inputs to the paraventricular hypothalamus contribute to hypothalamic–pituitary–adrenal axis and central Fos activation in rats after acute systemic endotoxin exposure. Neuroscience 156:1093–1102
214. Hermann GE, Rogers RC (2009) TNF activates astrocytes and catecholaminergic neurons in the solitary nucleus: implications for autonomic control. Brain Res 1273:72–82
215. Breese GR et al (2008) Repeated lipopolysaccharide (LPS) or cytokine treatments sensitize ethanol withdrawal-induced anxiety-like behavior. Neuropsychopharmacology 33:867–876

Chapter 5
Mechanisms of Stress-Dependent Neuroinflammation and Their Implications for Understanding Consequences of Alcohol Exposure

Terrence Deak, Anny Okrainets, and Tamara L. Doremus-Fitzwater

5.1 Introduction

Central inflammatory processes evoked by immunological challenges are responsible for directing a whole host of neural, hormonal and behavioral reactions in order to defend against infection and initiate recuperative processes. In particular, activation of brain cytokines appear especially important to the induction of sickness behaviors that are meant to help coordinate recovery during infection. More recently, research has shown that exposure to stressors may also activate peripheral and brain immune processes, with activation/inhibition of central cytokines being a key component of this response. Specifically, dynamic changes in central cytokines and other key inflammatory signaling pathways have been demonstrated to ultimately result in the expression of sickness behaviors as well, suggesting that illness- and stress-related neural consequences may share common mechanisms. Limited studies have now begun to show that alcohol is yet another exogenous stimulus that may also influence central cytokines. In addition to modulating behavioral responses incurred by alcohol exposure, early changes in cytokines may portend the development of long-term neuropathological consequences associated with chronic alcohol exposure. The goal of this chapter, therefore, is to review the literature surrounding stress-related neuroinflammation and alcohol-related changes in cytokines as a means for understanding the complex interaction between stress-responsive systems, neuroinflammatory processes, and their interactions with alcohol exposure.

Subsequently, the influence of alcohol exposure on the expression of cytokines will be addressed. Acute or chronic alcohol administration and its consequent withdrawal have been shown to alter cytokines, as well as to modulate the immune response to an

T. Deak, Ph.D. (✉) • A. Okrainets • T.L. Doremus-Fitzwater
Behavioral Neuroscience Program, Department of Psychology, Binghamton University, State University of New York, PO Box 6000, Binghamton, NY 13902-6000, USA
e-mail: tdeak@binghamton.edu

C. Cui et al. (eds.), *Neural-Immune Interactions in Brain Function and Alcohol Related Disorders*, DOI 10.1007/978-1-4614-4729-0_5,

immunological challenge. Though limited thus far, what literature is available would suggest that alcohol exposure has the capacity to significantly change both peripheral and central cytokines that could, in turn, ultimately impact behavior, stress-related processes, and (under long-term exposure circumstances) alcohol-associated brain/tissue damage. Alcohol-induced cytokine alterations may also interact with other processes more traditionally associated with stress challenges, such as the sympathetic nervous system (SNS) and hypothalamic–pituitary–adrenal (HPA) axis, which could either positively or negatively, respectively, feed back onto cytokine signaling.

Before proceeding, however, it is necessary to acknowledge from the outset a few critical distinctions. For instance, the field of stress research readily accepts the view that immunological challenges produced by antigen, pathogen, or direct injection of cytokines represent one distinct form of a "stress challenge" (e.g., [10]). This issue is well established by empirical studies demonstrating that immune activation leads to activation of the HPA axis (e.g., see [11, 12]) and SNS activity (e.g., see [13]) in much the same way as exposure to psychological and/or physical stress challenges that are otherwise devoid of antigenic or pathogenic components. However, because the impact of antigen or pathogen exposure on cytokine expression in brain is largely to be expected, use of the term "stress" in the present context will specifically exclude challenges that are, by their very nature, immunologically based. In this regard, much of what we describe as stress-related neuroinflammation really falls into the category of what the field is now referring to as "sterile inflammation," which simply refers to induction of inflammatory processes in the absence of antigen/pathogen/wound exposure.

The second semantic issue we would like to address from the outset is what is meant by the term "neuroinflammation" or "inflammatory processes in brain." This is a critical issue because, at some point, one would like to extrapolate from isolated changes in a specific cytokine (such as interleukin-1; IL-1), cellular process (e.g., morphological changes in microglia), or physiological condition (e.g., hyperthermia) toward these broader constructs. Though we will seek to avoid use of terms like this that tend to amalgamate specific features of the inflammatory response into broader statements, it will be occasionally necessary to make such extrapolations in order to synthesize the diverse range of inflammatory changes examined in brain after stress into some meaningful portrait of the relationship between stress, neuroinflammation, and larger issues surrounding central nervous system (CNS) dysfunction/function.

5.2 Stress and Neuroinflammation

5.2.1 Cytokines in the CNS Coordinate the Expression of Sickness Behaviors

Cytokines are a classification of small proteins, which are important for regulation of cell signaling processes within the immune system. Comprised of chemokines, interleukins, and lymphokines, these molecules are critical for initiation of the immune response to infection or damage and are secreted by several cell types within the CNS, including microglia, perivascular and meningeal macrophages [14], astrocytes [15], and even neurons [16]. Additionally, cytokine receptors have been identified on virtually all cell types within the CNS, and there is evidence to suggest that genes coding for cytokines are expressed in both neurons and glia in the brain under nonpathological conditions [17].

Following infection, it has been consistently shown that one especially important consequence of immune activation is expression of sickness behaviors. Studies have demonstrated that acute illness induced by LPS administration or other immunogens produces a constellation of classic behavioral alterations including reduced social and sexual interaction, decreased exploration in a novel environment, reduced overall activity, hyperalgesia, and suppressed food and water intake [18–25]. Collectively, these behavioral responses to acute illness are referred to as "sickness behaviors" [21] and, in combination with the physiological responses to infection (e.g., fever), are thought to act in an adaptive manner to promote efficient recovery from infection [21, 26]. Increased central expression of immune-derived factors, including the proinflammatory cytokines IL-1, tumor necrosis factor (TNF-α), and interleukin-6 (IL-6), appears to coordinate the manifestation of sickness behaviors, with IL-1 playing a particularly prominent role in their initiation. For instance, many sickness behaviors can be produced by central administration of IL-1 [24, 27, 28], whereas inhibition of central IL-1 activity blocks many of the sickness responses normally observed following peripheral immune activation [25]. At a mechanistic level, cytokine signals originating in peripheral organs such as the liver, gut, and spleen are communicated to the CNS through a variety of pathways (for a review, see [29]), ultimately leading to increased expression of proinflammatory cytokines in the CNS. Interestingly, both neurons and glia express cytokines under varying conditions [16, 30], and the hypothalamus appears to be one crucial CNS structure where proinflammatory cytokines precipitate the behavioral sickness response [21, 31–33].

In contrast, research has shown that anti-inflammatory cytokines in the CNS inhibit sickness behaviors and suppress peripheral immunity. As with proinflammatory cytokines, there are of course numerous cytokines in the CNS that serve an anti-inflammatory role such as (but not limited to) interleukin-10 (IL-10), interleukin-1 receptor antagonist (IL-1ra), interleukin-4 (IL-4), and interleukin-13 (IL-13) (for review, see [34]). Indeed, the central regulation of inflammatory processes—including sickness behaviors—is an intricate process involving a delicate balance between the expressions of both pro- and anti-inflammatory constituents in the CNS [35]. The anti-inflammatory cytokine, IL-10, has received a great deal of attention for its role in the CNS in counter-regulating the immune response (see [36] for a review). For instance, central administration of IL-10 blocks the reduction in social interaction produced by peripheral LPS administration [20] and inhibits the aphagia produced by central administration of the HIV envelope glycoprotein gp-120 [37]. IL-10 is also upregulated site-specifically within the CNS following immune activation, which is thought to prevent the development of certain neurodegenerative diseases with an inflammatory component [38]. The anti-inflammatory action of IL-10 is so powerful that researchers have recently employed adenoviral delivery of IL-10 as a method for inhibiting neuropathic pain [39–41]. Thus, endogenous expression of IL-10 appears to play a unique role in counter-regulating centrally mediated neuroinflammation, and central injection of IL-10 inhibits the expression of sickness behaviors during acute illness.

Though IL-10 is one prominent example, there are a wide variety of endogenous peptides that have anti-inflammatory properties. For instance, alpha-melanocyte-stimulating hormone (alpha-MSH) has been shown to blunt or reverse inflammatory-related changes evoked by stress [42] and in response to other inflammatory conditions (for review, see [43]). Similarly, the IL-1 superfamily includes two members with properties that can be considered anti-inflammatory. The first member, IL-1ra, is an endogenous peptide ligand that binds to the IL-1 type I receptor without any known signaling consequences, thereby blocking the proinflammatory actions of endogenous IL-1. The second, IL-1 receptor type II, is a soluble receptor that, when released from the cell, binds to IL-1 and consequently prevents signaling of IL-1. Similar mechanisms of counter-regulation exist within the TNF family via TNF-binding protein—a soluble peptide that binds TNF-α and prevents interaction with its functional receptors. These hardwired mechanisms of regulation and counter-regulation among cytokine signaling systems emphasize the importance of understanding the complete inflammatory milieu under the conditions of interest, which has led to greater utilization of measurement techniques that permit analysis of multiple inflammatory factors within the same samples (e.g., [44]).

5.2.2 Historical Overview of Stress-Dependent Changes in Central Cytokines

Since cytokines are expressed at exquisitely low levels in the uninjured brain, isolation and detection of cytokines under nonpathological conditions has been difficult, therefore rendering the discovery of stress-induced alterations in cytokine levels problematic. Nevertheless, there is ample evidence that cytokines such as IL-1 serve a normal physiological function in brain. For instance, the expression of IL-1 in the CNS follows a circadian rhythm that is independent of corticosterone (CORT) secretion and appears to play a key role in the initiation of sleep [45–49]. IL-1 is also increased in brain after consumption of novel food in rodent models [50]. Furthermore, IL-1 signaling is self-propagating and requires very low levels of expression to exert functional consequences on target cells [17]. Thus, small fluctuations in cytokine expression tend to exert a disproportionately large impact on target cells, at least in part through feed-forward actions of cytokine-receptor interactions and the ability of local injections of cytokine within the CNS to affect cytokine expression at distal targets (e.g., [51–53]).

The impact of stress on cytokine expression and other inflammatory processes in brain has been an important line of inquiry due to its implications for disease states of the CNS and beyond. When it comes to stress-dependent changes in inflammatory processes, much more is known about IL-1 in brain and its modulation by stress than any other cytokine. For this reason, we will use these studies as a case example of how stress exposure, and the hormonal systems activated by stress, interacts with inflammatory processes in brain. Although some initial papers in the mid-1990s reported that immobilization led to increased IL-1 expression using cellular techniques (in situ hybridization and immunohistochemistry), the bulk of what we know about IL-1 regulation by stress comes from studies utilizing gross brain dissections followed by IL-1 protein detection via ELISA. Using this strategy, for example, in one of our original studies, we demonstrated that adrenalectomized (ADX) subjects exhibited increased IL-1 protein in rat brain following inescapable tailshock relative to non-stressed ADX controls, with these stress effects not evident in sham-operated and stressed subjects [54]. At the time, these findings were important because they were the first to demonstrate that stressors were capable of significantly increasing IL-1 protein in brain. Within a few years, however, tissue extraction and assay sensitivity had improved to the point where it was no longer necessary to ADX rats in order to measure and detect tailshock-induced increases in IL-1 protein [45]. Furthermore, Nguyen et al. [45] also demonstrated the profound impact of endogenous CORT to inhibit brain cytokine responses evoked by stress, showing that both the magnitude and spatial distribution of IL-1 changes provoked by stress were augmented in ADX rats. Since then, our lab and others have gone on to successfully examine cytokine expression using gross dissections and ELISA detection (e.g., [22, 55–57]) or at the level of gene expression using RT-PCR [58–61], which together further support the view that stress challenges have the ability to drive cytokine changes in the uninjured brain. These changes, however, do not appear to be a universal response to all stress challenges [55, 61].

Indeed, through an exhaustive series of studies, we have examined IL-1 protein expression in the hypothalamus (a key stress-responsive site) following most commonly used laboratory stress challenges. This exploration was necessary in order to resolve discrepancies in the literature and address the highly appropriate critiques provided by experts in the field. Specifically, the primary issues we have worked to clarify can be summarized by three fundamental questions: first, does the increase in IL-1 produced by stress represent a universal response to all stress challenges? If this were the case, then we would expect systematic fluctuations in IL-1 (or other cytokines) in response to all stress challenges. Though early studies examining IL-1 protein failed to provide support for this viewpoint, our recent studies [61] may suggest otherwise. Secondly, is there a specific feature of the stress challenge (psychological distress, physical exertion, nociception) that is predictive of cytokine changes in brain? If this were true, then we would expect cytokine changes to vary across specific dimensions of the stress challenge in easily discernible ways. Such studies are ongoing in our laboratory. Finally, is there a threshold of stress that is necessary to increase cytokine expression in brain? This scenario is perhaps the most difficult to identify because, to our knowledge, there is no objective method for quantifying the intensity of a stress challenge, which would be essential for identification of a specific threshold.

Though we have no definitive answer to the conceptual issues raised above, we find them to be important theoretical considerations for understanding the relationship between stress and inflammatory processes in brain. Nevertheless, we have conducted a comprehensive series of experiments to try and tease these issues apart, as summarized below.

Based on studies focusing on IL-1 protein in whole hypothalamic blocks as the key measure of stress-related neuroinflammation, we know that stress challenges which would likely be characterized as psychological in nature such as social defeat [61], predator odor exposure [62], simple restraint in a Plexiglas tube [63], or maternal separation of guinea pig pups [64] have failed to alter hypothalamic IL-1 protein. Similar null effects have been observed following physiological stressors such as glucoprivic challenges evoked by injection of 2-deoxyglucose, or insulin-induced hypoglycemia [63]. However, when simple restraint in a Plexiglas tube was imposed on an orbital shaker as a method of "amplifying" the stress response (similar to [1, 63]), or combined with an insulin-induced hypoglycemia challenge [63], IL-1 was significantly increased in response to both "compound" stress challenges (restraint + oscillation; restraint + hypoglycemia), yet unaffected by the individual ones (restraint alone; hypoglycemia alone). These findings do not discriminate between a categorical model (point 2 above) and a threshold model (point 3 above), but they do provide evidence that substantive changes in IL-1 protein probably relate in meaningful ways to the severity of the stress challenge. Importantly, plasma CORT was examined in all of these studies and was not positively or negatively predictive of IL-1 changes [63]. This dissociation between IL-1 changes and CORT levels was true from simple observation of patterns in the data, as well as with more formalized statistical analyses performed later [65]. Additionally, one net outcome of these studies is that stress challenges involving exposure to an actively aversive stimulus (e.g., tailshock, footshock) appear to reliably produce telltale signs of inflammation such as elevated IL-1 [54, 61], which is in large part why many ongoing studies employ these types of stressors to pursue mechanisms of stress-dependent neuroinflammation (see [61] for further discussion).

As previously mentioned, the bulk of studies having examined IL-1 protein or mRNA changes following stressor exposure generally used gross tissue dissections of the hypothalamus, hippocampus, or cortex (e.g., [55, 56]), each of which are vastly heterogeneous structures. These early studies also indicated that IL-1 was probably the most stress-responsive of the classic proinflammatory cytokines in the CNS [56], an effect that we have validated more recently [61]. From these studies, IL-6 and TNF were either unaltered by stress exposure or were reduced as a result of stress, probably reflecting a CORT-mediated suppression (particularly for TNF-α). Given the need for greater spatial resolution (which would require enhanced sensitivity for measurement of cytokines in smaller tissue punches), we switched to RT-PCR for detection of inflammatory-related factors at the level of gene expression. This strategy additionally provided us with the advantage of being capable of examining numerous inflammatory factors within the same CNS structure. Accordingly, in a recent study [58], we examined a range of cytokines and other inflammatory markers in several brain structures after footshock, with a group of LPS-injected rats included as a control for assay sensitivity. Using this technique, it was again demonstrated that stress exposure resulted in significant IL-1 activation in the hypothalamus, as evidenced by increased IL-1 mRNA expression. Additional key outcomes of this experiment were that IL-1 changes in the hypothalamus were associated with increased CD14 expression (a cognate receptor for LPS), as well as decreased expression of CD200 receptor (which controls microglial activation state via interactions with CD200 on neurons). Combined, these receptor-related changes suggest that microglia may be primed/activated by stress, which could ultimately account for priming effects that we and others had observed previously [6, 7, 9]. Moreover, these changes were restricted to the hypothalamus (no effects in hippocampus, cortex, or pituitary gland) and were highly reproducible (replicated in four successive experiments [58]).

Taken together, these gene expression changes intimated that increased IL-1 provoked by stress may be part of a larger stress-induced inflammatory process involving microglial activation. In support of this hypothesis, administration of minocycline (a tetracycline antibiotic that appears to selectively target microglia as an inhibitor) was found to block the increase in IL-1 protein [57] as well as other gene expression changes observed after footshock [58]. These findings complement prior studies suggesting that repeated restraint in mice led to microglial proliferation, which is often a downstream consequence of early microglial activation events [66]. Indeed, work from our lab and others has shown that stress exposure produces signs of microglial activation [58, 67], priming [6], and proliferation [66]. While these studies imply that microglia may be the cellular source of IL-1 produced by stress, there is also some evidence to suggest otherwise. For example, Kwon et al. [67] showed that repeated exposure to immobilization led to morphological alterations in microglia indicative of activation, yet their studies seemed to support the notion that IL-1 was expressed in neurons (not microglia or astrocytes) based on single-label immunohistochemistry. Another recent study also reported a similar localization of IL-6 in neurons after acute restraint in rats [68]. Thus, it seems likely that multiple cell types produce cytokines in response to stress, but the precise location, cell type, and cytokine being expressed may vary as a function of species, strain, prior experience, or the nature of the stress challenge imposed.

Regardless of whether IL-1 changes are expressed by neurons, astrocytes, or microglia, it is clear that IL-1 changes in response to stress (1) are some of the most widely observed and reproducible changes across stress challenges; (2) are associated with other, more subtle cellular changes indicative of neuroinflammation; and (3) will have a functional impact upon all cells in the local cellular microenvironment since IL-1 receptors are expressed broadly across cell type [69]. Whether the cellular source of cytokine changes produced by stress varies as a function of the nature or duration of the stress challenge or the cytokine being examined remains to be determined. Furthermore, despite these strong advances in our understanding of stress-related neuroinflammation, sensitivity of measurement and approach remain key challenges for the field. In particular, due to the low intrinsic expression of cytokines under these conditions, future studies utilizing immunohistochemical approaches and microdialysis to detect picogram quantities that are expressed/released at the cellular level will be critical for clarifying the functional relationship between stress and/or alcohol exposure and the functional role of cytokines in the CNS. Such determinations will have to be made with the strictest of controls and the highest sensitivity possible, with full acknowledgement that species, strain, stress challenge, and timing are likely to be key variables that significantly impact the outcomes and conclusions that are drawn.

Given the widespread effects that increased central inflammation may have, it is therefore important to understand how exactly stress exposure relates to inflammatory processes in brain and the consequences of stress-related neuroinflammation for CNS dysfunction/function. With this in mind, we will turn our attention toward mechanisms involved in regulation and counter-regulation of brain inflammatory responses evoked by stress. Ultimately, a clear understanding of the acute mechanisms by which stress leads to alterations in brain inflammatory processes will hopefully reveal novel insights and therapeutic targets for preventing the progressive impact of stress-dependent inflammation on CNS function and, relevant to the present context, provide guidance for understanding the involvement of neuroinflammation in alcohol-related behavioral adaptations and brain damage.

5.2.3 The Interaction Between Stress Hormones and Stress-Dependent Inflammatory Processes

Based on a long history of studies examining regulation of cytokines and other immunological processes by norepinephrine (NE) and CORT (reviewed more extensively below), and the obvious impact of stress challenges on these transmitter/hormonal systems, early studies examined the role of adrenergic receptor activation as a key driver of IL-1 changes produced by stress. For instance, intracerebroventricular (icv) injection of the β-adrenergic agonist isoproterenol [70] or other agents that facilitate noradrenergic transmission [71] markedly increase the expression of mRNA for IL-1 in the hypothalamus, an effect which may occur via receptors on microglia [72]. In this latter study, icv injections of isoproterenol increased mRNA for IL-1 in a number of hypothalamic nuclei, including the paraventricular, ventromedial, dorsomedial, and medial mammillary nuclei. A series of pharmacological studies performed by [60] showed that increased IL-1 protein in brain (hypothalamus, hippocampus, etc.) occurs via activation of β-adrenergic receptor activation since isoproterenol reproduced and propranolol reversed the effects of stress on IL-1 protein. These effects were synonymous with findings from our lab [57], which replicated and extended the pharmacology by showing that desipramine (a NE reuptake inhibitor) increased both basal and stress-evoked IL-1 concentrations in brain. Furthermore, while administration of α1-adrenergic receptor antagonists had no effect on stress-induced IL-1 expression in the CNS, these compounds completely reversed the increase in *plasma* cytokines produced by tailshock (Johnson et al. [60]). These intriguing findings suggest that plasma and brain cytokine responses may be controlled by different adrenergic receptor mechanisms, with α1 receptor activation modulating plasma cytokine changes and β-adrenergic receptor activation influencing central cytokine changes. Perhaps even more importantly, these findings showed that plasma cytokine changes are dissociable from the brain cytokine changes and support the notion that neither effect appears to depend on the other. This is a crucial separation given the bidirectional interactions that occur between central and peripheral immune processes [29]. However, it should be noted that elevations in plasma concentrations of cytokines are highly variable across experiments and laboratories, making it difficult to establish clear functional relationships between plasma and brain cytokines. Whether this is due to low ambient expression, differential sample handling/processing, or the inherently unstable form of proteinaceous cytokines remains unclear.

Moving beyond pharmacological studies, the central mechanistic issues then turned toward neural circuits responsible for driving stress-dependent changes in IL-1 protein. With the NE system being the most logical target, Johnson et al. [60] injected the neurotoxin N-(2-chloroethyl)-N-ethyl-2-bromobenzylamine hydrochloride (DSP-4) into rats 3 weeks prior to tailshock exposure in order to selectively ablate central noradrenergic systems. This treatment was previously shown to target largely ascending adrenergic fibers through destruction of cells primarily in the locus coeruleus, which is a principle source of noradrenergic input to the hippocampus. Having used this procedure, NE levels in the hippocampus were reduced by approximately 87%, while the DSP-4 lesions only reduced NE levels in the hypothalamus by about 20%. The authors found that the lesion, as expected, blocked the increase in IL-1 protein in the hippocampus produced by tailshock, while stress-induced increases in IL-1 in the hypothalamus (where DSP-4 lesions had a lesser impact on NE levels) were largely unaffected by the lesion. It is probably important to note here that one key difference between studies conducted by Johnson's group and our lab is that their studies use Fischer rats, while our studies use Sprague Dawley rats. The net result of these subject differences is that Fisher rats show a much larger and broader spatial distribution of IL-1 changes in brain than Sprague Dawley rats. These strain differences are probably due to the well-documented, hyperadrenergic nature of Fisher rats relative to Sprague Dawley rats, as the hippocampus has not been a cytokine-responsive site in our studies (e.g., [55, 73]).

In order to more selectively target noradrenergic cells projecting largely from the nucleus tractus solitarius (NTS) to the hypothalamus and other forebrain structures, we recently performed a series of studies examining the impact of 6-hydroxydopamine lesions of the ventral noradrenergic bundle (VNAB) 10 days prior to stress exposure. As predicted based upon connectivity studies [74–77], this lesion led to approximately 70% depletion of NE in the paraventricular nucleus (PVN) and other structures, while other catecholamines were largely unaffected. Despite this NE depletion, rats bearing lesions of the VNAB unexpectedly exhibited an equivalent increase in IL-1 gene expression in the PVN after footshock relative to sham-lesioned controls (Blandino and Deak, unpublished observations). While these findings demonstrated that noradrenergic cells in the NTS likely do not modulate stress-induced increases in IL-1 in the hypothalamus, they do not conclusively determine the role of noradrenergic cells in the NTS as drivers of IL-1 changes in the PVN and suggest that the relationship between stress exposure and cytokine changes may be more complex than originally thought. Regardless, the evidence to date suggests that NE release is a key driver of inflammatory processes in at least some brain structures during times of stress.

In contrast to the actions of NE, which appear to stimulate the expression of IL-1 and other cytokines, CORT appears to constrain cytokine gene and protein expression during times of stress, at least in broad strokes. Indeed, there is a plethora of studies demonstrating that removal of endogenous glucocorticoids through ADX

[45, 54] or by injection of glucocorticoid synthesis inhibitors [58] enhances the expression of IL-1 and other cytokines in brain as a result of stress exposure. These findings parallel earlier studies showing that ADX enhanced the plasma IL-6 response evoked by exposure to a novel environment [78] and fit with the canonical viewpoint that glucocorticoids have powerful anti-inflammatory properties. While we do not wish to dispute this viewpoint, it is interesting to note that anti-inflammatory actions of CORT are most pronounced at high and supraphysiological concentrations, whereas lower concentrations of CORT appear to have some immune-potentiating effects (e.g., [6]). Whether these low-dose facilitation effects relate more directly to the timing of CORT injection relative to cytokine measurements, or represent differential tissue sensitivity to glucocorticoids, remains to be determined [79, 80]. Moreover, ambient levels of CORT taken during or immediately after stress exposure do not appear to be predictive of central IL-1 expression [63, 65]. Regardless, the point here is that the long-held assumption that glucocorticoids exert unilateral, anti-inflammatory actions does not broadly account for the literature writ large.

With that said, there are a variety of cellular mechanisms by which glucocorticoids impact gene expression for cytokines during times of stress, particularly in the case of IL-1 gene expression. Many of these effects are produced through interaction of the CORT-receptor complex with repressor sites in the promoter region of the IL-1 gene, including the nGRE repressor site and interference with nuclear factor κ-light chain-enhancer of activated B cells (NFkB) signaling. In addition, CORT has been shown to block cAMP response element-binding (CREB) phosphorylation, prevent NFkB binding to the DNA, and destabilize mRNA for IL-1 [81–83]. As such, CORT can interfere with IL-1 gene expression (and expression of other cytokines) through numerous pathways.

When taken together, a simple framework for understanding the interaction between the major stress-responsive systems (SNS and HPA axis) and inflammatory consequences of stress is that neuroinflammatory consequences of stress are mechanistically intertwined between the stimulatory actions of the SNS and the inhibitory actions of CORT (see [84] for a review), though much work remains to be done in this area. This concept is depicted in Fig. 5.1, which illustrates the basic mechanisms and pathways that appear to be involved in IL-1 regulation by stress. Although not a primary focus here, it is important to note that others have established prostaglandins as another highly stress-responsive, proinflammatory mediator within the CNS. The conversion of arachidonic acid to a particular prostanoid occurs via a COX-2 dependent pathway during times of inflammation. Upregulation of COX-2 expression has been observed as a result of stressor exposure in several brain regions [85], as has increased activity of other components of the pathway responsible for prostaglandin synthesis [86]. Furthermore, it seems that NE and CORT appear to exert opposing changes in prostanoid synthesis, with NE having been shown to stimulate prostaglandin E2 (PGE2) and CORT conversely inhibiting PGE2 (see [87] for review). Taken together, these data suggest that the overarching mechanisms governing neuroinflammatory consequences of stress may generalize across multiple inflammatory signaling families.

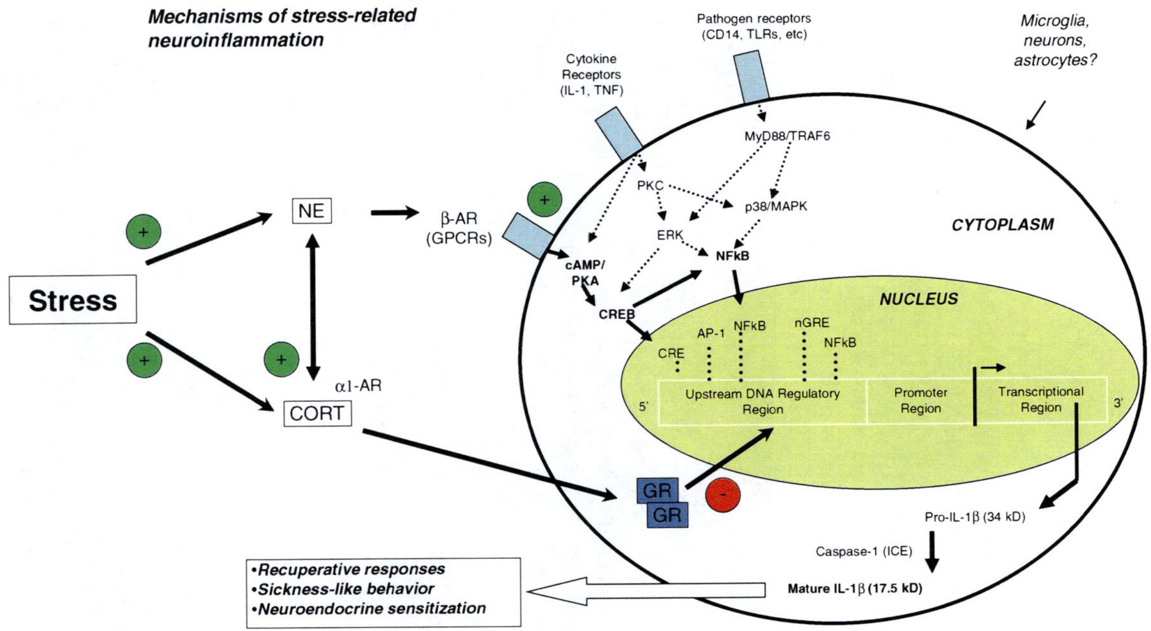

Fig. 5.1 Mechanisms of stress-related neuroinflammation. Exposure to stress challenges is associated with activation of both the sympathetic nervous system (SNS) and the hypothalamic–pituitary–adrenal (HPA) axis. These two principal stress-responsive systems involve the release of norepinephrine (NE) centrally and the release of corticosterone (CORT) from the adrenal glands. Results from mechanistic studies suggest that interaction of NE with β-adrenergic receptors leads to increased expression of the cytokine interleukin-1-beta (IL-1) in key regions of the CNS. Though the specific intracellular signaling pathways leading to IL-1 expression and the cell types expressing IL-1 have not yet been determined, evidence suggests that multiple signaling pathways and cell types may be involved in stress-dependent cytokine expression. CORT, on the other hand, appears to negatively influence the expression of IL-1 and other cytokines through interaction with glucocorticoid responsive elements (GREs) in the promoter region of cytokine genes. There are also reciprocal interactions between NE and CORT, whereby NE is a key driver in central regulation of the HPA axis (via α1-adrenergic receptors) and CORT tempers the release of NE. Ultimately, the release of mature IL-1 in the CNS has been linked to the expression of sickness-like behaviors, alteration of neuroendocrine processes including sensitization of the HPA axis, and other recuperative responses

Though the review above is largely centered on the role of NE as a key driver of inflammatory processes during times of stress, we would be remiss if we did not also point out alternative mechanisms that may be involved in regulation of neuroinflammation by stress. For instance, Nair and Bonneau [66] showed that the microglial proliferation associated with repeated stress was attenuated by administration of MK-801. Considering that glutamate plays an equally important role in the initiation of the stress response [88], and excitatory amino acids are directly coupled to inflammatory signaling pathways [86, 89, 90], it is not surprising that the general level of cellular excitation in the CNS microenvironment appears to be predictive of cytokine release [16, 91]. With that said, it is tempting to conclude that the relative state of neuronal excitation of a brain site under stressful conditions may show a strong association with IL-1 gene expression or the expression of other cytokines. However, our recent studies [61] (also Hueston et al. in prep) suggest that c-fos expression does not perfectly align with IL-1 expression but can serve as a useful comparator for understanding the relationship between cytokine expression and neuronal activation patterns more generally under certain conditions. In sum, the regulation of cytokine expression by stress is highly complex and probably involves a wide variety of ligands within the CNS, with a fair degree of convergence in signaling pathways known to be involved in cytokine gene expression.

In attempting to examine the stress-dependent changes in cytokines and the mechanisms responsible for these alterations, one of the major challenges associated with these studies has been the confounding impact of prior cranial surgery. This is largely to be expected due to the cellular trauma associated with implantation of guide cannula, microinjection of drugs, or insertion of injectors/electrodes to produce permanent lesions (e.g., [92]). However, designing studies without use of these conventional neuroscience techniques has proven to be quite difficult for several reasons. For instance, in microinjection studies, we have observed large and reliable increases in IL-1 protein expression after injection of sterile physiological saline into the third ventricle relative to noninjected (cannulated) controls despite agonizing attention to sterile process at all levels of the procedure. In fact, these changes are so large that they have masked our ability to detect increased IL-1 protein after injection of microgram quantities of LPS into the third ventricle (relative to vehicle-injected controls). The second, more subtle, problem associated with cranial surgery is how the surgical experience impacts reactivity to later stress challenges. We have other unpublished studies demonstrating marginal increases in basal IL-1 expression 10 days after cranial surgery in sites distal to the microinjection, which appeared to relegate subjects nonresponsive to the stress challenge. That is, rats receiving prior cranial surgery failed to show stress-dependent changes in cytokines, which have been highly reproducible in all studies except those involving cranial surgery. While intracisterna magna injections under brief anesthesia probably offer one viable alternative to cranial surgery, this procedure is also not without impact on central inflammatory processes and certainly impacts other more acute, labile stress measures [adrenocorticotropic hormone (ACTH), CORT, and behavioral indices of stress/anxiety].

With that said, these problems appear to be unique to situations where cranial surgery is performed and endogenous cytokine responses are being measured as dependent variables. The use of cranial surgery to implant guide cannula for microinjection of compounds with anti-inflammatory properties to try and reverse some behavioral or neurochemical process that is expected to be driven by stress-dependent cytokines has been much more successful. While this may seem ironic at first, it should be noted that central injection of anti-inflammatory compounds would be expected to reduce inflammatory consequences evoked by the stress challenge as well as any ongoing inflammation evoked by cannula-associated damage and/or delivery of the vehicle as well. We raise these issues as a cautionary note to others who seek to pursue stress-dependent changes in cytokines and/or their behavioral consequences. Aside from closely monitoring sterile procedure, we recommend incorporation of ultimate controls (i.e., unoperated subjects) where possible to clarify the influence of invasive procedures on the outcomes of interest.

5.3 Ethanol and Neuroinflammation

5.3.1 Ethanol Exposure Alters Both Peripheral and Central Cytokine Expression

Research has consistently demonstrated that alcohol exposure is capable of inducing alterations in the expression of many cytokines across a variety of locations, including the plasma, liver, and brain (e.g., see [93]). Historically, the effects of alcohol on peripheral cytokine expression have been well investigated in terms of the role that immune signaling plays in liver and organ damage following long-term alcohol consumption in humans. Taken together, these types of studies have generally shown that in humans, following chronic heavy alcohol consumption, levels of cytokines such as TNF-α, IL-6, and IL-1 are significantly elevated in the plasma (for review, see Table 2 of [94]). More specifically, it has been hypothesized that initial alcohol-induced increases in cytokines leads to the activation of T helper cytokines, with increased T helper 1 (Th1) cytokines being associated with early liver damage (hepatitis) and a switch from expression of Th1 to Th2 cytokines related to more pronounced liver disease (fibrosis/cirrhosis) (see [93] for review).

The influence of alcohol exposure on peripheral cytokine expression has also received considerable attention in animal models, with the effects of alcohol exposure on central cytokine levels beginning to receive increasing research interest as well. The results from animal studies investigating alterations in peripheral and central cytokines following alcohol exposure as well as antigen-stimulated production of cytokines after alcohol administration are presented in Table 5.1. These studies have been organized according to location of cytokine change (i.e., central versus peripheral), species under investigation (rats versus mice), and type of alcohol exposure (acute versus chronic).

We will first consider alterations in peripheral cytokines, which have primarily been examined as a means of understanding the effects of ethanol on peripheral organ damage and increased susceptibility to infection. In most instances, these experiments have explored the effects of alcohol exposure on antigen-stimulated

Table 5.1 Studies investigating alcohol–cytokine interactions

Species	Ethanol exposure	Results	Reference
Central			
Rat	Acute (cultured astrocytes treated with 50–300 mM)	Acute exposure of rat cortical astrocyte culture to EtOH lowered TNF-α expression at 300 mM, while stimulating IL-6 secretion at 100 mM and above	Sarc et al. [95]
	Acute (cultured microglia stimulated with 10, 50, and 100 mM)	EtOH-stimulated cells exhibited increased release of both TNF-α and IL-1β at all the concentrations used at 7, 24, and 48 h, relative to baseline	Fernandez-Lizarbe et al. [96]
	Chronic (50–300 mM × 7 days)	• Chronic exposure of rat cortical astrocytes to 50, 100, or 200 mM EtOH significantly elevated IL-6, with IL-6 decreased at concentrations greater than 200 mM EtOH • TNF-α secretion was dose-dependently reduced by EtOH	Sarc et al. [95]
	Chronic (5 g/kg i.g. on 3 consecutive days)	EtOH exposure on PND 7–9 resulted in a significant increase in levels of TNF-α and IL-1β in both cerebral cortex and hippocampus compared with controls	Tiwari and Chopra [97]
	Chronic (10 weeks of once daily10 g/kg EtOH i.g.)	Elevated levels of TNF-α and IL-1β in hippocampus and cerebral cortex of EtOH-consuming rats compared to controls	Tiwari et al. [98]
	Chronic (4–6 weeks liquid EtOH diet)	Chronic EtOH diet led to an increase in TNF-α and IL-6 expression in the hypothalamus, pituitary, and ovary of female rats	Emanuele et al. [99]
	Chronic (5 mos of liquid diet exposure OR cultured astrocytes exposed to 75 mM EtOH for 7 days)	• Chronic EtOH exposure upregulated expression of IL-1β and COX-2 in the cortex of EtOH-fed rats relative to controls • Cultured astrocytes exposed to long-term EtOH also exhibited increased production of IL-1β and COX-2	Valles et al. [100]
	EtOH withdrawal (following chronic intermittent EtOH exposures)	Preexposure to systemic LPS, central IL-1β, or central TNF-α prior to chronic intermittent EtOH exposure resulted in sensitized withdrawal-induced anxiety	Breese et al. [101]

(continued)

Table 5.1 (continued)

Species	Ethanol exposure	Results	Reference
Mouse	Acute (cultured microglia stimulated with 10, 50, and 100 mM)	In wild-type mice, EtOH-stimulated cells exhibited increased release of TNF-α, whereas TLR-deficient mice did not show this EtOH-related increase	Fernandez-Lizarbe et al. [96]
	Acute (1 day 5 g/kg i.g.)	• Increased brain expression of TNF-α and MCP-1 mRNA by an acute EtOH exposure, with cytokine gene expression either decreased or unchanged in the liver • LPS-induced cytokine production was increased in EtOH challenged versus vehicle mice	Qin et al. [102]
	Acute (BEC of 100 mg/dl, i.p.)	Acute dose of EtOH increased levels of TNF-α, IL-1β, and IL-6 protein in the hypothalamus 48 h after exposure	Emanuele et al. [103]
	Acute (100 mM)	Exposure of BV-2 microglial cells to EtOH significantly decreased LPS-induced IL-1β release	Lee et al. [104]
	Chronic (5 months ETOH-containing water, 10%)	Chronic EtOH consumption upregulated TNF-α, IL-1β, and IL-6 mRNA expression in cortex in WT mice compared to pair-fed WT controls, with EtOH-related increases not observed in TLR4-knockout mice	Alfonso-Loeches et al. [105]
	Chronic (daily 5 g/kg EtOH, i.g., for 10 days)	Increased expression of TNF-α and MCP-1 mRNA from EtOH exposure alone was reported in brain, with no changes in LPS-induced cytokine expression observed in relation to the EtOH exposure	Qin et al. [102]

(continued)

Table 5.1 (continued)

Species	Ethanol exposure	Results	Reference
Peripheral			
Rat	Acute (7 h i.v. EtOH infusion after priming dose)	In rats administered EtOH, TNF-α secretion was generally reduced in alveolar macrophages challenged in vitro with LPS relative to control rats	D'Souza et al. [106]
	Acute (2 g/kg i.g.)	Acute EtOH challenged resulted in significantly attenuated endotoxin-induced TNF levels in serum compared to vehicle rats	Honchel et al. [107]
	Acute (BEC levels of 75–175 mg/dl)	Endotoxin-induced serum levels of TNF-α decreased dose-dependently by prior EtOH administration	D'Souza et al. [108]
	Acute	EtOH exposure attenuated LPS-induced production of TNF in serum and lung	Nelson et al. [109]
	Chronic (daily10 g/kg EtOH i.g. for 10 weeks)	Increased TNF-α and IL-1β in serum/sciatic nerve of EtOH-exposed rats relative to controls	Tiwari et al. [110]
	Chronic (Kupffer cells, 4 weeks of daily 5 g/kg EtOH i.g.)	Chronic EtOH exposure increased TNF-α production in Kupffer cells after LPS stimulation compared to pair-fed controls	Enomoto et al. [111]
	Chronic (20 weeks of EtOH liquid diet)	• EtOH diet increased levels of TNF-α, IL-1β, and IL-10 in serum • EtOH diet also further augmented LPS-induced increases in IL-10 and IL-1β	Valles et al. [112]
	Chronic (4 weeks of EtOH diet)	Long-term EtOH consumption increased TNF-α, and decreased IL-1β, production after LPS stimulation in Kupffer cells	Kishore et al. [113]
	Chronic (Kupffer cells of rats on liquid diet of 17–35% EtOH for 2+ days)	Chronic EtOH exposure increased expression of TNF-α following LPS stimulation	Kishore et al. [114]
	Chronic (12–14 weeks liquid EtOH diet)	Both spontaneous and LPS-stimulated TNF-α secretion was generally reduced in alveolar macrophages of rats consuming EtOH, relative to pair-fed control rats	D'Souza et al. [106]
	Chronic (6 weeks liquid diet)	Long-term EtOH consumption resulted in exacerbated TNF serum levels in response to an endotoxin challenge, relative to pair-fed rats	Honchel et al. [107]

(continued)

Table 5.1 (continued)

Species	Ethanol exposure	Results	Reference
Mouse	Acute (6 g/kg i.g.)	In WT mice, levels of all cytokines measured (except for IL-10 and MIP-2) were decreased by previous EtOH exposure in response to *E. coli* with far fewer affected by EtOH in TLR4 mutant mice	Bhatty et al. [115]
	Acute (binge drinking model 4 or 6 g/kg i.g.)	Acute EtOH decreased immunogen-induced activation of proinflammatory cytokines for several hours after *E. coli* administration. In serum, at 21 h post-EtOH, IL-1β and IL-6 were increased in EtOH plus LPS mice	Pruett et al. [116]
	Acute (6 g/kg i.g.)	EtOH alone did not exact any changes on cytokines levels, while EtOH and poly I:C in combination suppressed serum TNF-α, but caused no changes in serum IL-1β, IL-6, or IL-10	Glover et al. [117]
	Acute (6 g/kg i.g.)	Acute EtOH (in vivo) delivered 30 min before LPS administration generally blocked production of proinflammatory cytokines by LPS	Pruett and Fan [118]
	Acute (cultured murine macrophages treated with either 10, 50, 100 mM EtOH)	• EtOH exposure alone generally increased release of TNF-α and IL-1β • EtOH in combination with LPS attenuated LPS-induced TNF-α production • EtOH exposure alone led to an increase release of IL-10	Fernandez-Lizarbe et al. [119]
	Acute (binge EtOH, i.g.)	Suppression of LPS-induced cytokine production in peritoneal fluid and serum	Dai and Pruett [120]
	Acute (3–6 g/kg i.g.)	Acute exposure suppressed LPS-induced serum IL-6 levels in a dose-dependent manner	Pruett and Pruett [121]
	Acute (treatment of macrophage cell culture line with 0.05–0.4% wt/v EtOH)	Acute EtOH exposure suppressed the LPS-induced increases in TNF-α levels	Dai et al. [122]
	Acute (2.9 g/kg i.p.)	Splenic macrophages isolated 3 h after EtOH exposure exhibited reduced IL-6 and TNF-α production in response to ligands on TLR4, TLR2, and TLR9	Goral and Kovacs [123]

(continued)

Table 5.1 (continued)

Species	Ethanol exposure	Results	Reference
	Acute (6 g/kg, i.g.)	In vivo, cytokines induced by either LPS or poly I:C were generally suppressed by EtOH (e.g., IL-6, IL-12), while IL-10 was increased, if changed at all. Results from the in vitro preparation where not the same as in the in vivo model	Pruett et al. [124]
	Acute (150 μl i.p.)	Splenic macrophages isolated for 3 or 24 h after EtOH exposure demonstrated decreased LPS-induced IL-6 production; effect disappeared after 48 h and no IL-6 changes in non-LPS-treated animals were observed	Goral et al. [125]
	Acute (6 g/kg i.g.)	Binge EtOH exposure suppressed poly I:C-induced inflammation in peritoneal macrophages (e.g., decreased IL-6 and IL-12 mRNA)	Pruett et al. [126]
	Acute (6 g/kg i.g.)	• Acute EtOH exposure generally suppressed the immune activating effects of several TLR ligands, although the exact cytokine/chemokine suppressed depended on serum versus peritoneal sample and ligand of interest • IL-10 production tended to be raised by EtOH exposure	Pruett et al. [127]
	Acute (4–7 g/kg i.g.)	Binge EtOH exposure decreased poly I:C-induced production of many cytokines in serum (TNF-α, IFN-β, IFN-γ, IL-6, IL-9, IL-12, and IL-15), yet increased poly I:C production of IL-10	Pruett et al. [128]
	Acute (4 g/kg i.p.)	EtOH exposure lowered LPS-stimulated IL-12 protein levels in blood and lung, but raised IL-10 production	Mason et al. [129]
	Acute (5–7 g/kg i.g.)	Binge EtOH exposure lowered *P. acnes*-induced TNF-α production in peritoneal samples	Vinson et al. [130]
	Acute (6 g/kg i.g.)	Acute EtOH exposure significantly decreased of IL-1β, IL-2, and IL-4 in response to SRBC challenge compared to non-EtOH-exposed controls	Han and Pruett [131]

(continued)

Table 5.1 (continued)

Species	Ethanol exposure	Results	Reference
	Chronic (daily 5 g/kg EtOH, i.g., for 10 days)	In liver, increased expression of TNF-α, MCP-1, IL-1β, and IL-10 mRNA was observed to the EtOH exposure alone, with LPS-induced increases in TNF-α exacerbated by repeated EtOH exposure	Qin et al. [102]
	Chronic (4 weeks EtOH-containing water, 20%)	Long-term exposure to EtOH resulted in general tolerance to EtOH-related suppression of LPS-induced cytokine production in serum and peritoneal fluid	Dai and Pruett [120]
	Chronic (7 weeks EtOH diet)	No changes were observed in liver cytokines from exposure to EtOH diet alone LPS-induced increases in cytokines were more pronounced in EtOH-fed mice compared to controls, particularly among IL-10 mutants	Hill et al. [132]
	Chronic (11 weeks of 5% EtOH liquid diet)	Slight but significant decrease in LPS- and concanavalin A-induced production of several cytokines in both splenocytes and thymocytes	Wang et al. [133]

Notes:
Numbers listed in brackets refer to reference number in bibliography
Abbreviations used: blood ethanol concentration (*BEC*); cycloxygenase-2 (*COX-2*); *Escherichia coli* (*E. coli*); ethanol (*EtOH*); interferon-beta (*IFN-β*); interferon-gamma (*IFN-γ*); interleukin-1 beta (*IL-1β*), interleukin-2 (*IL-2*); interleukin-4 (*IL-4*); interleukin-6 (*IL-6*); Interleukin-9 (*IL-9*); interleukin-10 (*IL-10*); interleukin-12 (*IL-12*); interleukin-15 (*IL-15*); intravenously (*i.v.*); intragastrically (*i.g.*); intraperitoneally (*i.p.*); lipopolysaccharide (*LPS*); monocyte chemoattractant protein-1 (*MCP-1*); macrophage inflammatory protein-2 (*MIP-2*); *Propionibacterium acnes* (*P. acnes*); polyinosinic:polycytidylic acid (*poly I:C*); postnatal days (*PND*); toll-like receptor(s) (*TLR*); toll-like receptor-4 (*TLR-4*); tumor necrosis factor (*TNF*); tumor necrosis factor-alpha (*TNF-α*); sheep red blood cells (*SRBC*)

immune responses and in general have reported that alcohol tends to have an overall dampening effect. It is noteworthy to mention that this trend parallels what was observed in the stress-immune literature when it was in its infancy as well. For instance, in mice, it has been found that acute binge-like doses (4–7 g/kg) of ethanol delivered intragastrically (i.g.) are capable of attenuating antigen-stimulated production of several proinflammatory cytokines in both serum and peritoneal samples (e.g., see [120, 128, 130]). Similarly, studies with rats have also shown that acute in vivo alcohol exposure suppresses later stimulation of proinflammatory cytokines by immunogens in serum (e.g., [107, 109]) and in lung (e.g., [109]). One way by which acute alcohol might decrease the proinflammatory response is through its interaction with toll-like receptors (TLRs), which are bound by immune challenges such as LPS and polyriboinosinic polyribocytidylic acid (poly I:C). This mechanism is supported by the observation that these attenuations in immunogen-precipitated proinflammatory cytokine production have been shown to depend upon the immunogen of interest. Specifically, ethanol has been shown to differentially impact the effects of ligands to the various TLR subtypes (i.e., LPS as a TLR-4 ligand; poly I:C as a TLR-3 ligand) (see [127]), with these effects also dependent upon the type of sample examined (i.e., serum versus peritoneal lavage versus alveolar sample). Additionally, though in vitro preparations also tend to show similar effects to the in vivo studies, differential effects of ethanol on antigen-stimulated cytokine production may also be observed in these two preparations (e.g., see [124]). In contrast to proinflammatory cytokines, in several instances, peripheral production of IL-10 (an anti-inflammatory cytokine) following immune challenge was shown to be *increased* after acute ethanol exposure in mice [124, 127, 128]. Together, a suppression of proinflammatory cytokines coupled with enhancement of anti-inflammatory cytokines, would lead to an immunosuppressive state, that is hypothesized to then act as a mechanism by which antigen exposure may lead to peripheral tissue damage.

While acute ethanol exposure has been shown to suppress peripheral production of proinflammatory cytokines after an immune challenge, in general, more chronic administration of ethanol has been reported to result in the opposite situation—an enhancement in production of these cytokines in response to an immunogen. In rats, long-term forced intake of ethanol via a liquid diet or ethanol-containing water source has been reported to increase levels of various proinflammatory cytokines in the periphery. For example, Valles et al. [112] reported increased IL-1β in the serum following LPS challenge in rats fed an ethanol-containing liquid diet for 20 weeks, compared to pair-fed rats. Similarly, only 6 weeks of exposure to an ethanol diet lead to enhanced endotoxin-induced increases in serum TNF-α levels in rats [107], while 4 weeks of daily ethanol (5 g/kg, i.g.) produced comparable results [111]. Parallel effects have also been observed using mice, with either 10 repeated days of ethanol gavage (5 g/kg) [102] or 7 weeks exposure to an ethanol-containing diet [132] exacerbating LPS-stimulated production of TNF-α in liver relative to vehicle-exposed controls, although these long-term effects of ethanol have not been ubiquitously observed (see [133]).

More recently, researchers have begun to investigate the impact of ethanol exposure on antigen-stimulated cytokine expression in the brain. In a recent study using mice, for instance, it was reported that acute ethanol exposure further enhanced the production of brain cytokines following a systemic LPS challenge [102]. In the same study, a more chronic regimen of ethanol exposure (5 g/kg i.g. for 10 days) was also found to potentiate levels of LPS-induced central cytokine production, with these increases lasting much longer in the brain than in the periphery. Additionally, it has been observed that microglial cells exposed to ethanol and then given LPS exhibited reduced levels of IL-1β relative to microglia not exposed to ethanol but challenged with LPS [104]. Clearly, more studies are needed in order to truly understand the effects that ethanol has on central production of cytokines in response to an immune challenge, with factors such as species, in vivo versus in vitro models, and dose influencing the results obtained.

Although an understanding of the involvement of cytokines in alcoholism-related tissue damage is of great importance, more recently researchers have begun to recognize the role that cytokines may play in the development of addictive processes

[93, 134, 135]. In particular, it is becoming apparent that alcohol exposure has the potential to alter peripheral or brain cytokines levels in the absence of an immune challenge, which may, in turn, impact behavioral responses to alcohol, stress-related processes [e.g., HPA axis; extrahypothalamic corticotropin-releasing hormone (CRH)], or physiological processes (e.g., SNS). Although the literature in this area is limited and has not always reported alcohol-induced changes in cytokines (e.g., see [117, 132]), there is mounting evidence to suggest that alcohol in and of itself is capable of significantly altering both peripheral and central cytokine expression. For example, Tiwari and colleagues have shown that chronic (10 weeks) exposure to ethanol induces increased levels of TNF-α and IL-1β in both the hippocampus and cortex in rats [98], as well as in the serum and sciatic nerve [110]. The same group also demonstrated that early postnatal exposure to ethanol on postnatal days (PND) seven to nine increased expression of TNF-α and IL-1β in the cortex and hippocampus [97]. Enhancement of cytokine expression in the serum [112] and cortex [100] following long-term ethanol intake has also reported by another group, and with mice as well [105]. Furthermore, when acute and chronic administration of ethanol was compared, Qin et al. [102] reported that both a single ethanol exposure, or ten repeated exposures to ethanol, resulted in increased levels of TNF-α and MCP-1 in brain, whereas liver expression of TNF-α, MCP-1, IL-1β, and IL-10 mRNA in liver was significantly decreased by ethanol administration.

In terms of stress-responsivity and immunological function, the hypothalamus is a brain region that is of great importance, particularly in the production of cytokines. When male mice were exposed to an acute systemic injection of ethanol, hypothalamic protein levels of TNF-α, IL-1β, and IL-6 were significantly increased 48 h later [103]. Similarly, when female rats were given long-term oral exposure to an ethanol diet, TNF-α and IL-6 levels were increased in both the hypothalamus and pituitary [99]. Indeed, recent data from our laboratory has shown that an acute binge dose of ethanol (4 g/kg) increased expression of IL-6 mRNA in several brain regions 3 h after exposure, including the PVN and hippocampus, whereas IL-1 mRNA levels were decreased, at this same time point (Deak et al. in prep). Recent in vitro work with both cultured astrocytes [95, 100] and microglia [119] has also demonstrated the capability of ethanol exposure alone to influence cytokine expression in brain, although the directionality of these effects seems to depend upon many factors, including cytokine of interest, dose of ethanol used, and duration of exposure.

Importantly, the timing of the assessment of cytokine changes in relation to ethanol exposure may be a key factor dictating the type of alterations observed in either peripheral or central inflammatory responses. Research that has focused specifically on the effects of withdrawal from ethanol administration on immune function has shown, for example, in response to a LPS challenge during alcohol withdrawal in humans monocyte derived cytokines, IL-1β, increased yet TNF-α and IL-12 decreased in controls. In another study [136], human alcoholics were shown to have decreased peripheral blood levels of CD4+ and CD8+ T cells during acute intoxication and withdrawal, as well as decreased levels of natural killer (NK) cells when compared to healthy controls, with, however, acute withdrawal increasing these subjects' blood monocyte levels. In contrast, Laso and colleagues [137] found increased peripheral blood levels of CD4+ T cells and NK cells during withdrawal from chronic alcoholism, but this difference may be due to the length of time after drinking cessation, which was much longer in the latter study (9 months versus 1 week). In other studies, Riikonen and colleagues [138] observed an increase in cerebellar microglia following intermittent ethanol exposure, which they interpreted as an involvement of microglia in brain atrophy in alcoholism. Kim et al. [139] found that certain blood cytokine (IL-10, IL-12, IFN-γ) levels were increased in humans during hangover, while others [140] have found that serum levels of IL-8 are increased 36 h after acute ethanol administration. Cytokine expression in the blood, however, does not necessarily reflect what is occurring in the brain. For instance, while LPS-induced increases in serum TNF-α in rats were found to last less than a day, these increases persisted for at least 10 months in the brain [141]. Interestingly, in an animal model of withdrawal sensitization, it was also recently found that immune challenges may impact behavioral expression of ethanol withdrawal. When an immune challenge (LPS exposure) or direct application of a cytokine (IL-1B or TNF-α) was administered prior to repeated intermittent ethanol exposure, withdrawal-related anxiogenesis was sensitized in the rats that had previously received an immune challenge relative to controls [101].

5.3.2 *Multiple Mechanisms May Contribute to the Production of Proinflammatory Cytokines in the CNS After Acute Ethanol Exposure*

Proinflammatory cytokines were initially characterized for their role in communication between immune cells in the body and only in more recent years have been recognized as critical signaling molecules within the CNS. Since cytokines are produced directly by cells in the CNS as well as by immune cells in the body, there are at least three specific mechanisms by which acute ethanol administration might lead to increased expression of cytokines in the CNS. Although described as discrete mechanisms, these should not be regarded as mutually exclusive mechanisms by which acute ethanol influences central cytokines.

The first possibility is that ethanol may increase cytokines via direct actions on neurons or glia in the CNS. As a lipophilic compound, ethanol passes readily across the blood–brain barrier where it exerts its effects through direct actions on parenchyma of the CNS. In this regard, ethanol may directly activate cells in the CNS (neurons, microglia, or astrocytes) to produce IL-1 or other cytokines since each of these cell types participates in central cytokine production under varying conditions [16, 30], and active crosstalk between neurons and glia seems to promote adaptive CNS functioning [142]. Of particular relevance in the present context is the recent finding that acute ethanol exposure has been shown to activate microglia as evidenced by changes in reactive oxygen species in vitro [143]. Such telltale signs of microglial activation are often accompanied by increased cytokine production by microglia, suggesting these cells may be a potential source of cytokines during acute ethanol withdrawal. Central expression of IL-1 is also known to increase during excessive neural activity such as that seen during seizure activity [62, 144]. Considering that ethanol withdrawal leads to increased neuronal activity that oftentimes approaches seizure levels (particularly when ethanol use/abuse is chronic [145]), it is quite possible that sustained high levels of neuronal activity produced during withdrawal from ethanol exposure might increase central cytokine expression.

Yet another possible mechanism for ethanol-associated alterations in cytokines may be through increased cytokine expression peripherally, thereby activating immune-to-brain communication pathways. For example, acute ethanol exposure has been shown to increase expression of proinflammatory cytokines (particularly IL-1 and IL-6) in the liver [112]. The liver is densely innervated by the vagus nerve, and the vagus is a well-established route of neural communication between the periphery and the CNS (e.g., [146]). In this scenario, proinflammatory cytokines produced and released in blood or organs bind to receptors on lymphoid tissue associated with vagal paraganglia, activate vagal transmission, and induce de novo synthesis of cytokines in projection regions of the vagus nerve, particularly in the NTS and hypothalamus [146]. At a functional level, electrical stimulation of the vagus nerve increases IL-1 expression in the hypothalamus [147], and subdiaphragmatic vagotomy blocks IL-1 responses in the hypothalamus produced by low doses of LPS or IL-1 injection [148, 149]. Similar effects have been observed with the sciatic [150] and glossopharyngeal [151, 152] nerves as well, suggesting that peripheral nerves may be common paths for immune-to-brain communication. A comparable mechanism might be proposed in the case of voluntary ethanol intake or during intragastric intubation where ethanol might act as an irritant to the GI tract (e.g., [153])—production of cytokines by intestinal epithelial cells may bind to receptors on peripheral nerves innervating the gut, thereby activating immune-to-brain signaling pathways and increasing central cytokine expression. In any case, these scenarios would indicate that central expression of cytokines during acute ethanol exposure and/or withdrawal could occur secondary to peripheral production of cytokines rather than due to a direct action of ethanol on the CNS parenchyma.

Finally, withdrawal from acute ethanol may increase cytokine production through an NE-dependent pathway. Acute ethanol withdrawal elicits robust activation of the HPA axis as well as the autonomic nervous system [154, 155]. Increased NE activity in the hypothalamus during withdrawal contributes to this HPA activation and may account for a variety of withdrawal-related changes in behavior [156]. Interestingly, increased NE release during withdrawal does not appear to be unique to alcohol, as withdrawal from chronic morphine has been shown to activate catecholamine-containing cells in brainstem autonomic nuclei, leading to NE release in the PVN and ultimately activation of the HPA axis [157]. Thus, there is considerable evidence to suggest that hypothalamic NE is increased during withdrawal from ethanol and other substances, ultimately contributing to withdrawal-related activation of HPA axis activity and expression of withdrawal-associated behaviors. As previously discussed (see above), NE has also been shown to be a powerful regulator of IL-1 expression, particularly in the hypothalamus. Therefore, it is a possibility that increased hypothalamic noradrenergic activity during ethanol withdrawal may induced expression of cytokines, providing a direct neurochemical-to-cytokine pathway by which acute ethanol withdrawal leads to increased cytokine expression. Indeed, this is an intriguing hypothesis that fits well with current theories of addiction/withdrawal [155].

5.3.3 Reciprocal Interactions Between Alcohol Effects and Stress Challenges: Timing of Events and Levels of Analysis Yield Many Interesting Questions

There is currently a rich literature investigating the interaction between stress and alcohol exposure. In particular, there has been a vast number of studies that have sought to elucidate the role that stressor exposure has on oral consumption of ethanol in both humans and in animal models (for review, see [158–161]). Although this is an especially important avenue of research, the influence of stressors on neural, hormonal, and behavioral responses to later ethanol access/exposure is also of importance. Of course, there is the potential for multiple (and not mutually exclusive) mechanisms to contribute to the effect of stressors on alcohol responsivity, with stress- and/or alcohol-induced alterations in central/peripheral cytokines just one of many possibilities.

Since studies examining alterations in cytokines following ethanol exposure alone (i.e., without immunological challenge) have, in fact, shown that ethanol is capable of altering central expression of some cytokines (see above), these alcohol-induced changes in cytokine levels may themselves contribute to changes in behavior during intoxication and/or withdrawal. In this sense, one might argue that cytokine changes observed during withdrawal could be the cause and/or consequence of distress produced by withdrawal, thereby raising interesting questions about whether anti-inflammatory agents might have "therapeutic" benefits for ameliorating the adverse consequences of ethanol withdrawal. If this is the case, then one might expect the pattern of cytokine expression to vary as a function of the phase of alcohol exposure (i.e., some cytokines might be elevated during intoxication, while others might relate more to alcohol withdrawal). Indeed, recent results from our laboratory have observed that both i.g. and intraperitoneal (i.p.) administration of ethanol changes cytokine expression in several brain regions. More specifically, a 4-g/kg challenge of ethanol resulted in increased expression of IL-6 mRNA in hypothalamus and hippocampus (as well as in the periphery) during peak intoxication, whereas IL-1 mRNA levels were decreased during intoxication and significantly increased during acute withdrawal from the ethanol exposure (Deak et al. in prep).

Based on these findings, one would predict that if cytokines are responsible for behavioral changes following ethanol exposure, then pharmacological studies should be able to uncover the mechanisms by which cytokines influence ethanol-induced behavioral alterations. More specifically, pharmacological blockade of cytokine signaling should be capable of attenuating behavioral consequences of ethanol exposure, either through more general suppression of cytokine actions or more specifically via antagonism of the action of particular cytokines. More recently, we have begun to explore this possibility by examining ethanol withdrawal-related behaviors. In animal models, acute ethanol withdrawal involves expression of a constellation of behavioral sequela, including hypoactivity, anxiogenesis, reduced social interactions, decreased food and water intake, and fever [162]. As described above, these behavioral alterations closely resemble sickness behaviors expressed following immune challenge. The possibility remains, therefore, that induction of cytokines following exposure to ethanol may play a key role in the initiation of these behaviors. Current work from our laboratory [171] has begun to explore this possibility. Briefly, rats were given a large acute ethanol challenge (4 g/kg) in order to induce an acute withdrawal (or hangover) state [163–165], which was indexed as reduced exploration and social investigation in a modified social investigation task previously validated in our laboratory [59]. In two different experiments, we explored the ability of indomethacin or IL-1ra to reverse these acute withdrawal-associated behaviors, with these pharmacological treatments administered following the alcohol challenge, but prior to behavioral testing. Although administration of these anti-inflammatory substances did not successfully block expression of acute withdrawal-associated behaviors under these particular circumstances, these results do not preclude the involvement of cytokines in the expression of ethanol-related behaviors since several variables (such as dose of drugs used, timing of drug application relative to ethanol exposure, and the ability of drugs to distribute broadly throughout the CNS) could have impacted the results obtained.

Together, the studies described above provide some context for understanding how ethanol-induced cytokines might play a role in orchestration of ethanol-related behavioral changes. Another very interesting line of inquiry is how ethanol-related cytokine changes might impact the response to later stress challenge, or vice versa. In this regard, stress exposure before, during, or after intoxication (during acute withdrawal or beyond) may be significantly impacted by the individual's recent and/ or lifelong history of alcohol exposure. This is relevant because cytokines are an important mechanism by which the HPA axis becomes sensitized [7]. Furthermore, as stressors themselves (such as footshock) have been shown to increase cytokine expression [57, 58, 63], exposure to stress during intoxication or withdrawal could impact cytokine levels and feedback onto alcohol-induced alterations. Breese and colleagues have explored this possibility, for example, using a stress-sensitization model of ethanol withdrawal-related behaviors. Whereas a single 5-day exposure to an ethanol-containing diet did not result in acute withdrawal-induced anxiogenesis in a social interaction paradigm, this group has shown that repeated intermittent exposures to an alcohol diet did sensitize the withdrawal response and result in exacerbated anxiety during withdrawal [166]. Furthermore, repeated exposure to stress substituted for previous alcohol exposure, ultimately leading to significant withdrawal-related anxiogenesis after just one bout of ethanol diet exposure, which in and of itself, did not produce withdrawal-induced anxiety [167]. Recently, work from their laboratory examined cytokines as a mechanism underlying this stress sensitization of ethanol withdrawal behaviors, with repeated administrations of LPS or several other cytokines substituting for repeated stress or ethanol diet exposures in the expression of sensitization of withdrawal-related anxiety [101].

In Fig. 5.2, a potential framework for viewing the interactions between alcohol exposure, stress-related processes and inflammatory systems has been presented. Within our own laboratory, we have begun to explore the possible interaction

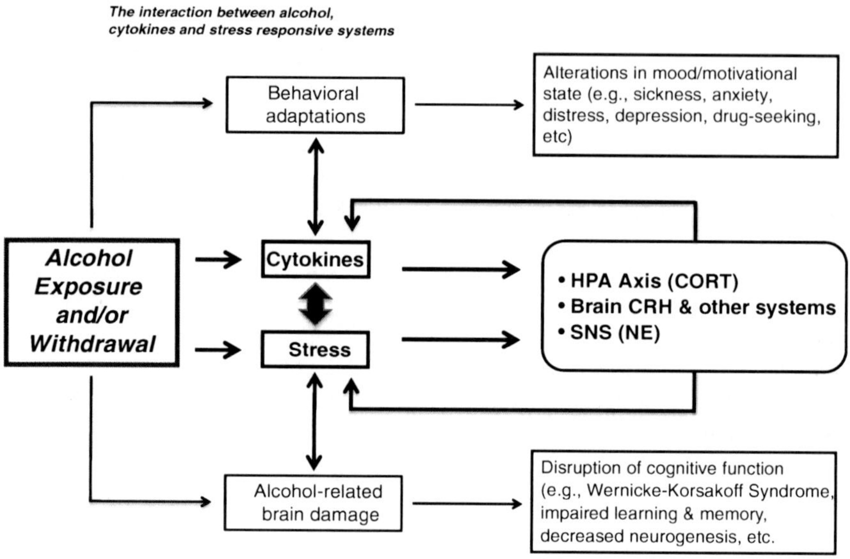

Fig. 5.2 The interaction between alcohol, cytokines, and stress-responsive systems. Exposure to and/or withdrawal from alcohol has the capacity to directly impact cytokine expression and is also associated with activation of stress-responsive systems. Given the powerful influence that cytokines exert over behavioral processes (e.g., sickness behaviors), it is likely that alcohol-induced changes in central cytokines play a key role in mediating alcohol-related behavioral adaptations. A growing body of research points toward long-term elevations in cytokines as a result of chronic alcohol use and abuse, which serve as a harbinger of alcohol-related brain damage. Given the cyclic interactions between alcohol exposure, cytokines, and stress-responsive systems, future studies delineating more precise mechanisms of central cytokine regulation by alcohol may hold promise for preventing the adverse consequences of long-term alcohol exposure

between stress, alcohol, and cytokines by investigating the HPA axis response to stress during withdrawal from an acute ethanol challenge since previous work has shown enhanced HPA activation to stressors during ethanol withdrawal [112, 168] (but see also [169]). Furthermore, we have been interested in whether any possible stress-related alterations in HPA activation are related to cytokine responses evoked by the stress and/or ethanol exposure itself. Briefly, we have recently reported that moderate acute stressors (e.g., restraint stress or exposure to a novel environment) imposed on adult male rats during peak withdrawal from an acute ethanol challenge were shown to significantly exacerbate stress-induced increases in plasma CORT [170]. While this initial work observed a stress hyperresponsive state during acute withdrawal, at this time, these changes have not been directly linked to ethanol-induced alterations in the expression of cytokines following stressor exposure [170]. Of course, the timing between ethanol exposure, stress exposure, and measurement of cytokines is certainly of importance and therefore does not necessarily rule out the involvement of cytokines in ethanol-withdrawal-related stress hyperresponsivity. Future studies in our lab will continue to explore these possibilities.

5.4 Conclusions and Future Directions

Our overarching goal here was to provide a broad overview of mechanisms and tribulations surrounding examination of stress-related inflammatory processes as well as the fledging literature surrounding alcohol–cytokine interactions. The central message we hoped to convey is that, to fully understand the interaction between alcohol and cytokines in brain and their potential implications for alcohol-related brain pathology, one must take into account the multitude of integrative, system-level interactions that are impacted by alcohol exposure, including, but not limited to, stress-responsive systems, liver–gut interactions, and the presence/absence of associated immunogenic stimuli (bacteria, viruses, etc.), as each of these will be determinants of cytokine expression in brain. Moreover, we assert that there is a wide body of literature now indicating a role for cytokines and other inflammatory processes in mediating the deleterious consequences of long-term alcohol exposure, whereas there is a relative paucity of studies seeking to identify early-term (acute), dynamic changes in cytokines in relation to alcohol, as well as how such effects might transform over the course of developmental history of the subject or in response to a growing number of alcohol exposures across the lifetime. Such studies will be instrumental in forging a near-certain link between initial alcohol exposure and the ultimate, adverse consequences of end-stage alcoholism.

Acknowledgements Supported by NIH grant number R21AA016305-01 to T.D., the Developmental Exposure Alcohol Research Center (DEARC; P50AA017823), National Science Foundation (NSF grant #0822129), and the Center for Development and Behavioral Neuroscience at Binghamton University. Any opinions, findings, and conclusions or recommendations expressed in this chapter are those of the author(s) and do not necessarily reflect the views of the above stated funding agencies. The authors have no conflicts of interest to declare.

References

1. Dhabhar FS, McEwen BS (1997) Acute stress enhances while chronic stress suppresses cell mediated immunity in vivo: a potential role for leukocyte trafficking. Brain Behav Immun 11:286–306
2. Deak T et al (1997) Evidence that brief stress may induce the acute phase response in rats. Am J Physiol 273:R1998–R2004
3. Deak T, Nguyen KT, Fleshner M, Watkins LR, Maier SF (1999) Acute stress may facilitate recovery from a subcutaneous bacterial challenge. Neuroimmunomodulation 6:344–354
4. Fleshner M, Nguyen KT, Cotter CS, Watkins LR, Maier SF (1998) Acute stressor exposure both suppresses acquired immunity and potentiates innate immunity. Am J Physiol 275:R870–R878
5. Frank MG, Baratta MV, Sprunger DB, Watkins LR, Maier SF (2007) Microglia serve as a neuroimmune substrate for stress-induced potentiation of CNS pro-inflammatory cytokine responses. Brain Behav Immun 21:47–59
6. Frank MG, Miguel ZD, Watkins LR, Maier SF (2010) Prior exposure to glucocorticoids sensitizes the neuroinflammatory and peripheral inflammatory responses to *E. coli* lipopolysaccharide. Brain Behav Immun 24:19–30

7. Johnson JD et al (2002) Prior stressor exposure primes the HPA axis. Psychoneuroendocrinology 27:353–365
8. Johnson JD, O'Connor KA, Hansen MK, Watkins LR, Maier SF (2003) Effects of prior stress on LPS-induced cytokine and sickness responses. Am J Physiol 284:R422–R432
9. Johnson JD, O'Connor KA, Watkins LR, Maier SF (2004) The role of IL-1b in stress-induced sensitization of proinflammatory cytokine and corticosterone responses. Neuroscience 127:569–577
10. Sawchenko PE, Li HY, Ericsson A (2000) Circuits and mechanisms governing hypothalamic responses to stress: a tale of two paradigms. Prog Brain Res 122:61–78
11. Turnbull AV, Rivier C (1995) Regulation of the HPA axis by cytokines. Brain Behav Immun 9:253–275
12. Turnbull AV, Rivier CL (1999) Regulation of the hypothalamic-pituitary-adrenal axis by cytokines: actions and mechanisms of action. Physiol Rev 79:1–71
13. Smagin GN, Swiergiel AH, Dunn AJ (1996) Peripheral administration of interleukin-1 increases extracellular concentrations of norepinephrine in rat hypothalamus: comparison with plasma corticosterone. Psychoneuroendocrinology 21:83–93
14. van Dam AM, Brouns M, Louisse S, Berkenbosch F (1992) Appearance of interleukin-1 in macrophages and in ramified microglia in the brain of endotoxin-treated rats: a pathway for the induction of non-specific symptoms of sickness? Brain Res 588:291–296
15. Martinez F, Abril ER, Earnest DL, Watson RR (1992) Ethanol and cytokine secretion. Alcohol 9:455–458
16. Tringali G et al (1996) Evidence for the neuronal origin of immunoreactive interleukin-1 beta released by rat hypothalamic explants. Neurosci Lett 219:143–146
17. Vitkovic L, Bockaert J, Jacque C (2000) "Inflammatory" cytokines: neuromodulators in normal brain? J Neurochem 74:457–471
18. Avitsur R, Cohen E, Yirmiya R (1997) Effects of interleukin-1 on sexual attractivity in a model of sickness behavior. Physiol Behav 63:25–30
19. Avitsur R, Yirmiya R (1999) The immunobiology of sexual behavior: gender differences in the suppression of sexual activity during illness. Pharmacol Biochem Behav 64:787–796
20. Bluthe RM et al (1999) Central injection of IL-10 antagonizes the behavioural effects of lipopolysaccharide in rats. Psychoneuroendocrinology 24:301–311
21. Dantzer R (2001) Cytokine-induced sickness behavior: mechanisms and implications. Ann N Y Acad Sci 933:222–234
22. Deak T, Bellamy C, Bordner KA (2005) Protracted increases in core body temperature and interleukin-1 following acute administration of lipopolysaccharide: implications for the stress response. Physiol Behav 85:296–307
23. Deak T et al (2005) Behavioral responses during the forced swim test are not affected by anti-inflammatory agents or acute illness induced by lipopolysaccharide. Behav Brain Res 160:125–134
24. Plata-Salaman CR (1994) Meal patterns in response to the intracerebroventricular administration of interleukin-1 beta in rats. Physiol Behav 55:727–733
25. Plata-Salaman CR, Ffrench-Mullen JM (1992) Intracerebroventricular administration of a specific IL-1 receptor antagonist blocks food and water intake suppression induced by interleukin-1 beta. Physiol Behav 51:1277–1279
26. Hart BL (1988) Biological basis of the behavior of sick animals. Neurosci Biobehav Rev 12:123–137
27. Johnson DR, O'Connor JC, Hartman ME, Tapping RI, Freund GG (2007) Acute hypoxia activates the neuroimmune system, which diabetes exacerbates. J Neurosci 27:1161–1166
28. Plata-Salaman CR, Oomura Y, Kai Y (1988) Tumor necrosis factor and interleukin-1 beta: suppression of food intake by direct action in the central nervous system. Brain Res 448:106–114
29. Maier SF, Watkins LR (1998) Cytokines for psychologists: implications of bidirectional immune-to-brain communication for understanding behavior, mood, and cognition. Psychol Rev 105:83–107

30. Breder CD, Dinarello CA, Saper CB (1988) Interleukin-1 immunoreactive innervation of the human hypothalamus. Science 240:321–324
31. Cadet P et al (2003) Cyclic exercise induces anti-inflammatory signal molecule increases in the plasma of Parkinson's patients. Int J Mol Med 12:485–492
32. Wandinger KP et al (1999) Effects of amantadine treatment on in vitro production of interleukin-2 in de-novo patients with idiopathic Parkinson's disease. J Neuroimmunol 98:214–220
33. Tringali G, Dello Russo C, Preziosi P, Navarra P (1998) Hypothalamic interleukin-1 in physiology and pathology. Toxicol Lett 102–103:295–299
34. Opal SM, DePalo VA (2000) Anti-inflammatory cytokines. Chest 117:1162–1172
35. Anisman H (2004) Considering cytokine panels. Brain Behav Immun 18:221–222
36. Dantzer R (2004) Cytokine-induced sickness behaviour: a neuroimmune response to activation of innate immunity. Eur J Pharmacol 500:399–411
37. Barak O et al (2002) Involvement of brain cytokines in the neurobehavioral disturbances induced by HIV-1 glycoprotein120. Brain Res 933:98–108
38. Arimoto T et al (2007) Interleukin-10 protects against inflammation-mediated degeneration of dopaminergic neurons in substantia nigra. Neurobiol Aging 28:894–906
39. Milligan ED et al (2005) Controlling pathological pain by adenovirally driven spinal production of the anti-inflammatory cytokine, interleukin-10. Eur J Neurosci 21:2136–2148
40. Milligan ED et al (2005) Controlling neuropathic pain by adeno-associated virus driven production of the anti-inflammatory cytokine, interleukin-10. Mol Pain 1:9
41. Soderquist RG et al (2010) Release of plasmid DNA-encoding IL-10 from PLGA microparticles facilitates long-term reversal of neuropathic pain following a single intrathecal administration. Pharm Res 27:841–854
42. Milligan ED et al (1998) The long term acute phase-like responses that follow acute stressor exposure are blocked by alpha-melanocyte stimulating hormone. Brain Res 810:48–58
43. Lasaga M, Debeljuk L, Durand D, Scimonelli TN, Caruso C (2008) Role of alpha-melanocyte stimulating hormone and melanocortin 4 receptor in brain inflammation. Peptides 29: 1825–1835
44. Maslanik T, Bernstein-Hanley I, Helwig B, Fleshner M (2012) The impact of acute-stressor exposure on splenic innate immunity: a gene expression analysis. Brain Behav Immun 26:142–149
45. Nguyen KT et al (2000) Timecourse and corticosterone sensitivity of the brain, pituitary, and serum interleukin-1b protein response to acute stress. Brain Res 859:193–201
46. Hansen M, Krueger J (1997) Subdiaphragmatic vagotomy blocks the sleep and fever-promoting effects of interleukin-1B. Am J Physiol 273:R1246–R1253
47. Krueger JM, Walter J, Dinarello CA, Wolff SM, Chedid L (1984) Sleep promoting effects of endogenous pyrogen (interleukin-1). Am J Physiol 246:994–999
48. Opp MR, Krueger JM (1991) Interleukin-1 receptor antagonist blocks interleukin-1-induced sleep and fever. Am J Physiol 260:453–457
49. Taishi P, Bredow S, Guha-Thakurta N, Obal F, Krueger J (1997) Diurnal variations of interleukin-1b mRNA and b-actin mRNA in rat brain. J Neuroimmunol 75:69–74
50. Hansen MK, Taishi P, Chen Z, Krueger JM (1998) Cafeteria feeding induces interleukin-1beta mRNA expression in rat liver and brain. Am J Physiol 274:793–798
51. Allan SM et al (2000) Cortical cell death induced by IL-1 is mediated via actions in the hypothalamus of the rat. Proc Natl Acad Sci USA 97:5580–5585
52. Grundy R, Rothwell N, Allan S (2002) Site-specific actions of interleukin-1 on excitotoxic cell death in the rat striatum. Brain Res 926:142–148
53. Proescholdt M et al (2002) Intracerebroventricular but not intravenous interleukin-1b induces widespread vascular-mediated leukocyte infiltration and immune signal mRNA expression followed by brain-wide glial activation. Neuroscience 112:731–749
54. Nguyen KT et al (1998) Exposure to acute stress induces brain interleukin-1b protein in the rat. J Neurosci 18:2239–2246
55. Deak T, Bellamy C, D'Agostine L (2003) Exposure to forced swim stress does not alter central production of IL-1. Brain Res 972:53–63

56. O'Connor K et al (2003) Peripheral and central proinflammatory cytokine response to a severe acute stressor. Brain Res 991:123–132
57. Blandino P Jr, Barnum CJ, Deak T (2006) The involvement of norepinephrine and microglia in hypothalamic and splenic IL-1beta responses to stress. J Neuroimmunol 173:87–95
58. Blandino P Jr et al (2009) Gene expression changes in the hypothalamus provide evidence for regionally-selective changes in IL-1 and microglial markers after acute stress. Brain Behav Immun 23:958–968
59. Arakawa H, Blandino P Jr, Deak T (2009) Central infusion of interleukin-1 receptor antagonist blocks the reduction in social behavior produced by prior stressor exposure. Physiol Behav 98:139–146
60. Johnson JD et al (2005) Catecholamines mediate stress-induced increases in peripheral and central inflammatory cytokines. Neuroscience 135:1295–1307
61. Hueston CM et al (2011) Stress-dependent changes in neuroinflammatory markers observed after common laboratory stressors are not seen following acute social defeat of the Sprague Dawley rat. Physiol Behav 104:187–198
62. Plata-Salaman CR et al (2000) Neither acute nor chronic exposure to a naturalistic (predator) stressor influences the interleukin-1beta system, tumor necrosis factor-alpha, transforming growth factor-beta1, and neuropeptide mRNAs in specific brain regions. Brain Res Bull 51:187–193
63. Deak T et al (2005) Stress-induced increases in hypothalamic IL-1: a systematic analysis of multiple stressor paradigms. Brain Res Bull 64:541–556
64. Hennessy MB et al (2004) Responses of guinea pig pups during isolation in a novel environment may represent stress-induced sickness behaviors. Physiol Behav 81:5–13
65. Barnum CJ, Blandino P Jr, Deak T (2008) Social status modulates basal IL-1 concentrations in the hypothalamus of pair-housed rats and influences certain features of stress reactivity. Brain Behav Immun 22:517–527
66. Nair A, Bonneau RH (2006) Stress-induced elevation of glucocorticoids increases microglia proliferation through NMDA receptor activation. J Neuroimmunol 171:72–85
67. Kwon M et al (2008) The repeated immobilization stress increases IL-1beta immunoreactivities in only neuron, but not astrocyte or microglia in hippocampal CA1 region, striatum and paraventricular nucleus. Neurosci Lett 430:258–263
68. Jankord R et al (2010) Stress activation of IL-6 neurons in the hypothalamus. Am J Physiol Regul Integr Comp Physiol 299:R343–R351
69. Licinio J, Frost P (2000) The neuroimmune-endocrine axis: pathophysiological implications for the central nervous system cytokines and hypothalamus-pituitary-adrenal hormone dynamics. Braz J Med Biol Res 33:1141–1148
70. Yamaguchi T, Kuraishi Y, Minami M, Yabuuchi K, Satoh M (1991) Involvement of central b-adrenoceptors in the induction of hypothalamic interleukin-1b mRNA by methamphetamine. Neurosci Res 12:432–439
71. Maruta E et al (1997) Beta2-adrenoceptors on the glial cells mediate the induction of interleukin-1beta mRNA in the rat brain. Brain Res Mol Brain Res 49:291–294
72. Yabuuchi K et al (1997) Intracerebroventricular injection of isoproterenol produces its analgesic effect through interleukin-1b production. Eur J Pharmacol 334:133–140
73. Porterfield VM et al (2011) Rat strain differences in restraint stress-induced brain cytokines. Neuroscience 188:48–54
74. Sawchenko PE, Swanson LW (1981) Central noradrenergic pathways for the integration of hypothalamic neuroendocrine and autonomic responses. Science 214:685–687
75. Sawchenko PE, Swanson LW (1982) The organization of noradrenergic pathways from the brainstem to the paraventricular and supraoptic nuclei in the rat. Brain Res 257:275–325
76. Swanson LW, Hartman BK (1975) The central adrenergic system. An immunofluorescence study of the location of cell bodies and their efferent connections in the rat utilizing dopamine-beta-hydroxylase as a marker. J Comp Neurol 163:467–505
77. Swanson LW, Sawchenko PE (1983) Hypothalamic integration: organization of the paraventricular and supraoptic nuclei. Annu Rev Neurosci 6:269–324
78. Morrow LE, McClellan JL, Conn CA, Kluger MJ (1993) Glucocorticoids alter fever and IL-6 responses to psychological stress and to lipopolysaccharide. Am J Physiol 264:R1010–R1016

79. Frank MG, Thompson BM, Watkins LR, Maier SF (2012) Glucocorticoids mediate stress-induced priming of microglial pro-inflammatory responses. Brain Behav Immun 26:337–345
80. Sorrells SF, Sapolsky RM (2007) An inflammatory review of glucocorticoid actions in the CNS. Brain Behav Immun 21:259–272
81. Necela BM, Cidlowski JA (2004) Mechanisms of glucocorticoid receptor action in noninflammatory and inflammatory cells. Proc Am Thorac Soc 1:239–246
82. Watkins LR, Hansen MK, Nguyen KT, Lee JE, Maier SF (1999) Dynamic regulation of the proinflammatory cytokine interleukin-1b: molecular biology for non-molecular biologists. Life Sci 65:449–481
83. Quan N, He L, Lai W, Shen T, Herkenham M (2000) Induction of IkappaBalpha mRNA expression in the brain by glucocorticoids: a negative feedback mechanism for immune-to-brain signaling. J Neurosci 20:6473–6477
84. Deak T (2007) From hippocampus to dorsal horn: the pervasive impact of IL-1 on learning and memory spans the length of the neuroaxis. Brain Behav Immun 21:746–747
85. Yamagata K, Andreasson KI, Kaufmann WE, Barnes CA, Worley PF (1993) Expression of a mitogen-inducible cyclooxygenase in brain neurons: regulation by synaptic activity and glucocorticoids. Neuron 11:371–386
86. Garcia-Bueno B, Madrigal JL, Perez-Nievas BG, Leza JC (2008) Stress mediators regulate brain prostaglandin synthesis and peroxisome proliferator-activated receptor-gamma activation after stress in rats. Endocrinology 149:1969–1978
87. Garcia-Bueno B, Caso JR, Leza JC (2008) Stress as a neuroinflammatory condition in brain: damaging and protective mechanisms. Neurosci Biobehav Rev 32:1136–1151
88. Ziegler DR, Cullinan WE, Herman JP (2005) Organization and regulation of paraventricular nucleus glutamate signaling systems: *N*-methyl-D-aspartate receptors. J Comp Neurol 484:43–56
89. Yirmiya R, Goshen I (2011) Immune modulation of learning, memory, neural plasticity and neurogenesis. Brain Behav Immun 25:181–213
90. Zhang XM, Zhu J (2011) Kainic acid-induced neurotoxicity: targeting glial responses and glia-derived cytokines. Curr Neuropharmacol 9:388–398
91. Tringali G et al (1997) The release of immunoreactive interleukin-1 beta from rat hypothalamic explants is modulated by neurotransmitters and corticotropin-releasing hormone. Pharmacol Res 36:269–273
92. Holguin A et al (2007) Characterization of the temporo-spatial effects of chronic bilateral intrahippocampal cannulae on interleukin-1beta. J Neurosci Methods 161:265–272
93. Crews FT et al (2006) Cytokines and alcohol. Alcohol Clin Exp Res 30:720–730
94. Achur RN, Freeman WM, Vrana KE (2010) Circulating cytokines as biomarkers of alcohol abuse and alcoholism. J Neuroimmune Pharmacol 5:83–91
95. Sarc L, Wraber B, Lipnik-Stangelj M (2011) Ethanol and acetaldehyde disturb TNF-alpha and IL-6 production in cultured astrocytes. Hum Exp Toxicol 30:1256–1265
96. Fernandez-Lizarbe S, Pascual M, Guerri C (2009) Critical role of TLR4 response in the activation of microglia induced by ethanol. J Immunol 183:4733–4744
97. Tiwari V, Chopra K (2011) Resveratrol prevents alcohol-induced cognitive deficits and brain damage by blocking inflammatory signaling and cell death cascade in neonatal rat brain. J Neurochem 117:678–690
98. Tiwari V, Kuhad A, Chopra K (2009) Suppression of neuro-inflammatory signaling cascade by tocotrienol can prevent chronic alcohol-induced cognitive dysfunction in rats. Behav Brain Res 203:296–303
99. Emanuele N, LaPaglia N, Kovacs EJ, Emanuele MA (2005) Effects of chronic ethanol (EtOH) administration on pro-inflammatory cytokines of the hypothalamic-pituitary-gonadal (HPG) axis in female rats. Endocr Res 31:9–16
100. Valles SL, Blanco AM, Pascual M, Guerri C (2004) Chronic ethanol treatment enhances inflammatory mediators and cell death in the brain and in astrocytes. Brain Pathol 14:365–371
101. Breese GR et al (2008) Repeated lipopolysaccharide (LPS) or cytokine treatments sensitize ethanol withdrawal-induced anxiety-like behavior. Neuropsychopharmacology 33:867–876
102. Qin L et al (2008) Increased systemic and brain cytokine production and neuroinflammation by endotoxin following ethanol treatment. J Neuroinflammation 5:10

103. Emanuele NV, LaPaglia N, Kovacs EJ, Emanuele MA (2005) The impact of burn injury and ethanol on the cytokine network of the mouse hypothalamus: reproductive implications. Cytokine 30:109–115
104. Lee H et al (2004) Ethanol selectively modulates inflammatory activation signaling of brain microglia. J Neuroimmunol 156:88–95
105. Alfonso-Loeches S, Pascual-Lucas M, Blanco AM, Sanchez-Vera I, Guerri C (2011) Pivotal role of TLR4 receptors in alcohol-induced neuroinflammation and brain damage. J Neurosci 30:8285–8295
106. D'Souza NB, Nelson S, Summer WR, Deaciuc IV (1996) Alcohol modulates alveolar macrophage tumor necrosis factor-alpha, superoxide anion, and nitric oxide secretion in the rat. Alcohol Clin Exp Res 20:156–163
107. Honchel R et al (1992) Tumor necrosis factor in alcohol enhanced endotoxin liver injury. Alcohol Clin Exp Res 16:665–669
108. D'Souza NB, Bagby GJ, Nelson S, Lang CH, Spitzer JJ (1989) Acute alcohol infusion suppresses endotoxin-induced serum tumor necrosis factor. Alcohol Clin Exp Res 13:295–298
109. Nelson S, Bagby G, Summer WR (1989) Alcohol suppresses lipopolysaccharide-induced tumor necrosis factor activity in serum and lung. Life Sci 44:673–676
110. Tiwari V, Kuhad A, Chopra K (2011) Amelioration of functional, biochemical and molecular deficits by epigallocatechin gallate in experimental model of alcoholic neuropathy. Eur J Pain 15:286–292
111. Enomoto N et al (2003) Prevention of ethanol-induced liver injury in rats by an agonist of peroxisome proliferator-activated receptor-gamma, pioglitazone. J Pharmacol Exp Ther 306:846–854
112. Valles SL et al (2003) Chronic ethanol consumption enhances interleukin-1-mediated signal transduction in rat liver and in cultured hepatocytes. Alcohol Clin Exp Res 27:1979–1986
113. Kishore R, Hill JR, McMullen MR, Frenkel J, Nagy LE (2002) ERK1/2 and Egr-1 contribute to increased TNF-alpha production in rat Kupffer cells after chronic ethanol feeding. Am J Physiol Gastrointest Liver Physiol 282:G6–G15
114. Kishore R, McMullen MR, Nagy LE (2001) Stabilization of tumor necrosis factor alpha mRNA by chronic ethanol: role of A+U-rich elements and p38 mitogen-activated protein kinase signaling pathway. J Biol Chem 276:41930–41937
115. Bhatty M, Jan BL, Tan W, Pruett SB, Nanduri B (2011) Role of acute ethanol exposure and TLR4 in early events of sepsis in a mouse model. Alcohol 45:795–803
116. Pruett SB et al (2010) Innate immunity and inflammation in sepsis: mechanisms of suppressed host resistance in mice treated with ethanol in a binge-drinking model. Toxicol Sci 117:314–324
117. Glover M, Cheng B, Fan R, Pruett S (2009) The role of stress mediators in modulation of cytokine production by ethanol. Toxicol Appl Pharmacol 239:98–105
118. Pruett SB, Fan R (2009) Ethanol inhibits LPS-induced signaling and modulates cytokine production in peritoneal macrophages in vivo in a model for binge drinking. BMC Immunol 10:49
119. Fernandez-Lizarbe S, Pascual M, Gascon MS, Blanco A, Guerri C (2008) Lipid rafts regulate ethanol-induced activation of TLR4 signaling in murine macrophages. Mol Immunol 45:2007–2016
120. Dai Q, Pruett SB (2006) Different effects of acute and chronic ethanol on LPS-induced cytokine production and TLR4 receptor behavior in mouse peritoneal macrophages. J Immunotoxicol 3:217–225
121. Pruett BS, Pruett SB (2006) An explanation for the paradoxical induction and suppression of an acute phase response by ethanol. Alcohol 39:105–110
122. Dai Q, Zhang J, Pruett SB (2005) Ethanol alters cellular activation and CD14 partitioning in lipid rafts. Biochem Biophys Res Commun 332:37–42
123. Goral J, Kovacs EJ (2005) In vivo ethanol exposure down-regulates TLR2-, TLR4-, and TLR9-mediated macrophage inflammatory response by limiting p38 and ERK1/2 activation. J Immunol 174:456–463

124. Pruett SB, Fan R, Zheng Q, Schwab C (2005) Differences in IL-10 and IL-12 production patterns and differences in the effects of acute ethanol treatment on macrophages in vivo and in vitro. Alcohol 37:1–8
125. Goral J, Choudhry MA, Kovacs EJ (2004) Acute ethanol exposure inhibits macrophage IL-6 production: role of p38 and ERK1/2 MAPK. J Leukoc Biol 75:553–559
126. Pruett SB, Schwab C, Zheng Q, Fan R (2004) Suppression of innate immunity by acute ethanol administration: a global perspective and a new mechanism beginning with inhibition of signaling through TLR3. J Immunol 173:2715–2724
127. Pruett SB, Zheng Q, Fan R, Matthews K, Schwab C (2004) Ethanol suppresses cytokine responses induced through Toll-like receptors as well as innate resistance to *Escherichia coli* in a mouse model for binge drinking. Alcohol 33:147–155
128. Pruett SB, Fan R, Zheng Q (2003) Acute ethanol administration profoundly alters poly I:C-induced cytokine expression in mice by a mechanism that is not dependent on corticosterone. Life Sci 72:1825–1839
129. Mason CM, Dobard E, Kolls JK, Nelson S (2000) Ethanol and murine interleukin (IL)-12 production. Alcohol Clin Exp Res 24:553–559
130. Vinson RB, Carroll JL, Pruett SB (1998) Mechanism of suppressed neutrophil mobilization in a mouse model for binge drinking: role of glucocorticoids. Am J Physiol 275: R1049–R1057
131. Han YC, Pruett SB (1995) Mechanisms of ethanol-induced suppression of a primary antibody response in a mouse model for binge drinking. J Pharmacol Exp Ther 275:950–957
132. Hill DB et al (2002) A role for interleukin-10 in alcohol-induced liver sensitization to bacterial lipopolysaccharide. Alcohol Clin Exp Res 26:74–82
133. Wang Y, Huang DS, Giger PT, Watson RR (1994) Influence of chronic dietary ethanol on cytokine production by murine splenocytes and thymocytes. Alcohol Clin Exp Res 18:64–70
134. Crews FT, Vetreno RP (2011) Addiction, adolescence, and innate immune gene induction. Front Psychiatry 2:19
135. Crews FT, Zou J, Qin L (2011) Induction of innate immune genes in brain create the neurobiology of addiction. Brain Behav Immun 25(Suppl 1):S4–S12
136. Kutscher S et al (2002) Concomitant endocrine and immune alterations during alcohol intoxication and acute withdrawal in alcohol-dependent subjects. Neuropsychobiology 45:144–149
137. Laso FJ et al (1996) Long lasting immunological effects of ethanol after withdrawal. Cytometry 26:275–280
138. Riikonen J et al (2002) Intermittent ethanol exposure increases the number of cerebellar microglia. Alcohol Alcohol 37:421–426
139. Kim DJ et al (2003) Effects of alcohol hangover on cytokine production in healthy subjects. Alcohol 31:167–170
140. Gonzalez-Quintela A et al (2000) Influence of acute ethanol intake and alcohol withdrawal on circulating levels of IL-6, IL-8, IL-10 and IL-12. Cytokine 12:1437–1440
141. Qin L et al (2007) Systemic LPS causes chronic neuroinflammation and progressive neurodegeneration. Glia 55:453–462
142. Polazzi E, Contestabile A (2002) Reciprocal interactions between microglia and neurons: from survival to neuropathology. Rev Neurosci 13:221–242
143. Colton CA, Snell-Callahan J, Chernyshev ON (1998) Ethanol induced changes in superoxide anion and nitric oxide in cultured microglia. Alcohol Clin Exp Res 22:710–716
144. Vezzani A, Moneta D, Richichi C, Perego C, De Simoni MG (2004) Functional role of proinflammatory and anti-inflammatory cytokines in seizures. Adv Exp Med Biol 548:123–133
145. Trevisan LA, Boutros N, Petrakis IL, Krystal JH (1998) Complications of alcohol withdrawal: pathophysiological insights. Alcohol Health Res World 22:61–66
146. Goehler LE et al (2000) Vagal immune-to-brain communication: a visceral chemosensory pathway. Auton Neurosci 85:49–59
147. Hosoi T, Okuma Y, Nomura Y (2000) Electrical stimulation of afferent vagus nerve induces IL-1b expression in the brain and activates HPA axis. Am J Physiol 279:R141–R147

148. Hansen MK et al (2000) Effects of vagotomy on serum endotoxin, cytokines, and corticosterone after intraperitoneal lipopolysaccharide. Am J Physiol 278:R331–R336
149. Hansen MK et al (2000) Effects of vagotomy on lipopolysaccharide-induced brain interleukin-1beta protein in rats. Auton Neurosci 85:119–126
150. Herzberg U, Sagen J (2001) Peripheral nerve exposure to HIV viral envelope protein gp120 induces neuropathic pain and spinal gliosis. J Neuroimmunol 116:29–39
151. Romeo H, Tio D, Rahman S, Chiappelli F, Taylor A (2001) The glossopharyngeal nerve as a novel pathway in immune-to-brain communication: relevance to neuroimmune surveillance of the oral cavity. J Neuroimmunol 115:91–100
152. Romeo H, Tio D, Taylor A (2003) Effects of glossopharyngeal nerve transection on central and peripheral cytokines and serum corticosterone induced by localized inflammation. J Neuroimmunol 136:104–111
153. Fleming S et al (2001) Pro- and anti-inflammatory gene expression in the murine small intestine and liver after chronic exposure to alcohol. Alcohol Clin Exp Res 25:579–589
154. Adinoff B, Iranmanesh A, Veldhuis J, Fisher L (1998) Disturbances of the stress response: the role of the HPA axis during alcohol withdrawal and abstinence. Alcohol Health Res World 22:67–72
155. Koob GF (2003) Alcoholism: allostasis and beyond. Alcohol Clin Exp Res 27:232–243
156. Linniola M, Mefford I, Nutt D, Adinoff B (1987) NIH conference: alcohol withdrawal and noradrenergic function. Ann Intern Med 107:875–889
157. Benavides M, Laorden ML, Garcia-Borron JC, Milanes MV (2003) Regulation of tyrosine hydroxylase levels and activity and Fos expression during opioid withdrawal in the hypothalamic PVN and medulla oblongata catecholaminergic cell groups innervating the PVN. Eur J Neurosci 17:103–112
158. Becker HC, Lopez MF, Doremus-Fitzwater TL (2011) Effects of stress on alcohol drinking: a review of animal studies. Psychopharmacology (Berl) 218:131–156
159. Pohorecky LA (1981) The interaction of alcohol and stress. A review. Neurosci Biobehav Rev 5:209–229
160. Pohorecky LA (1990) Interaction of ethanol and stress: research with experimental animals – an update. Alcohol Alcohol 25:263–276
161. Pohorecky LA (1991) Stress and alcohol interaction: an update of human research. Alcohol Clin Exp Res 15:438–459
162. Becker HC (2000) Animal models of alcohol withdrawal. Alcohol Res Health 24:105–113
163. Doremus TL, Brunell SC, Varlinskaya EI, Spear LP (2003) Anxiogenic effects during withdrawal from acute ethanol in adolescent and adult rats. Pharmacol Biochem Behav 75:411–418
164. Doremus-Fitzwater TL, Spear LP (2007) Developmental differences in acute ethanol withdrawal in adolescent and adult rats. Alcohol Clin Exp Res 31:1516–1527
165. Varlinskaya EI, Spear LP (2004) Acute ethanol withdrawal (hangover) and social behavior in adolescent and adult male and female Sprague–Dawley rats. Alcohol Clin Exp Res 28:40–50
166. Overstreet DH, Knapp DJ, Breese GR (2002) Accentuated decrease in social interaction in rats subjected to repeated ethanol withdrawals. Alcohol Clin Exp Res 26:1259–1268
167. Breese GR, Knapp DJ, Overstreet DH (2004) Stress sensitization of ethanol withdrawal-induced reduction in social interaction: inhibition by CRF-1 and benzodiazepine receptor antagonists and a 5-HT1A-receptor agonist. Neuropsychopharmacology 29:470–482
168. Rylkova D, Shah HP, Small E, Bruijnzeel AW (2009) Deficit in brain reward function and acute and protracted anxiety-like behavior after discontinuation of a chronic alcohol liquid diet in rats. Psychopharmacology (Berl) 203:629–640
169. Zhao Y, Weiss F, Zorrilla EP (2007) Remission and resurgence of anxiety-like behavior across protracted withdrawal stages in ethanol-dependent rats. Alcohol Clin Exp Res 31:1505–1515
170. Buck HM, Hueston CM, Bishop C, Deak T (2011) Enhancement of the hypothalamic–pituitary–adrenal axis but not cytokine responses to stress challenges imposed during withdrawal from acute alcohol exposure in Sprague–Dawley rats. Psychopharmacology (Berl) 218:203–215
171. Richey L, Doremus-Fitzwater TL, Bush HM, Deak T (2012) Acute illness-induced behavioral alterations are similar to those observed during withdrawal from acute alcohol exposure. Pharmacol Biochem Behav 103:281–294

Chapter 6
Mood Disorders and Immunity

Adam K. Walker, Robert Dantzer, and Keith W. Kelley

6.1 Introduction: The Link Between Substance Abuse and Depression

Depression is a common comorbid condition in substance abuse disorders. In particular, comorbidity between alcoholism and depression occurs with extremely high prevalence in the general population. Early studies investigating this dual diagnosis issue reported alcohol abuse to occur in almost a quarter of patients with affective disorders [1, 2]. Prevalence rises as high as 32% if other substance abuse disorders are included [2]. These findings made it clear that enormous challenges lie ahead in tracing the individual etiology of substance abuse and depression, let alone their interaction. Clearly, this dilemma was to produce further difficulties in providing adequate and appropriate treatment for sufferers.

Little has changed in 25 years since these early epidemiological findings were published. Indeed, a recent study estimated comorbidity rates of alcohol dependence and depression to be 17% [3]. This is unsurprising given the extremely heterogenous symptoms that characterize depression and substance abuse. For example, clinical depression can be manifested as either sleeping too much or not enough [4–6]. Similarly, alcoholism has been associated with both food restriction and

A.K. Walker
Integrative Immunology and Behavior Program, Department of Animal Sciences, University of Illinois at Urbana-Champaign, 227 Edward R. Madigan Laboratory, 1201 West Gregory Drive, Urbana, IL 61801, USA

R. Dantzer • K.W. Kelley, Ph.D. (✉)
Integrative Immunology and Behavior Program, Department of Animal Sciences, University of Illinois at Urbana-Champaign, 227 Edward R. Madigan Laboratory, 1201 West Gregory Drive, Urbana, IL 61801, USA

Department of Medical Pathology, University of Illinois at Urbana-Champaign, 227 Edward R. Madigan Laboratory, 1201 West Gregory Drive, Urbana, IL 61801, USA
e-mail: kwkelley@illinois.edu

C. Cui et al. (eds.), *Neural-Immune Interactions in Brain Function and Alcohol Related Disorders*, DOI 10.1007/978-1-4614-4729-0_6,

overeating [7]. In addition to these problems, a number of studies have demonstrated that remission rates of depressive symptoms vary dramatically depending on whether substance abuse precedes or follows the onset of depressive symptoms in clinical samples (reviewed in [8]).

Typical approaches to examining possible causal factors for the interrelationships between alcoholism and depression have largely focused around examining social risk factors, gene-specific susceptibilities, and neurotransmission mechanisms. Social explanations nominally argue that extended substance abuse often causes disruptions in interpersonal relationships, unemployment, reduced physical health, and legal issues, all of which can send individuals spiraling into depression (reviewed in [9]). Unfortunately, this approach fails to provide a holistic explanation for the comorbidity given that substance abuse is often initiated as a self-medicating device following the onset of depression [10]. Other studies have determined that the correlation between addiction and depression remains even after accounting for a range of these social and environmental factors [11, 12]. Similarly, examination of genetic overlaps between alcoholism and depression has yielded putative candidate mechanisms [13–15]. Once again, however, no specific gene or set of genes have solely been linked to either disorder. Based upon studies with monozygotic twins, many have concluded that there is only a moderate causal role for genetics in either depression or alcoholism (e.g., [16]). Finally, many studies have been carried out on possible commonalities of neurochemical dysfunction affecting serotoninergic, gamma-aminobutyric acid (GABA), and noradrenergic deficiencies. However, antidepressant medications aimed at targeting such neurotransmitters still produce nonresponse or only partial response rates of almost 50% within clinical settings [17]. While there is little doubt that social risk factors, gene-specific susceptibilities, and neurotransmission mechanisms contribute to substance abuse disorders, depression, and their comorbidity, they are clearly insufficient to explain both the pathogenesis of each disorder and their consistent overlap. Furthermore, such approaches have only limited heuristic application. Clearly, research needs to diversify into additional contributive mechanisms. One such area of research which is contributing both valuable mechanistic insights and greater heuristic methodologies to this comorbidity of depression and substance abuse is that of psychoneuroimmunology.

The social and physical environment of individuals plays an important role in the development of both alcoholism and clinical depression, particularly when considering the interaction of the environment with susceptible or resistant genes [13, 18]. One environmental factor that has recently been posited as a contributor to these disorders is the microbial environment and innate immune system. As such, it has been recently posited that neuroimmune mechanisms play a strong role in substance abuse and depression. Progress in inflammation research during the past two decades has led to a growing acceptance that inflammation plays an influential role in the development of depression. Even more recently, a burgeoning area of investigation has demonstrated a strong relationship between the innate immune system and substance abuse disorders [19–21]. For instance, ethanol (EtOH) and other addictive substances are known to activate monocytes and microglia [22]; intraperitoneal lipopolysaccharide (LPS) administration to mice promotes alcohol consumption and alters alcohol reward and aversion [23]; and chronic EtOH exposure elevates brain innate immune genes and increases depressive-like behaviors in mice [24].

Monoamines such as dopamine and serotonin have been strongly implicated in both depression and substance abuse. Depressed patients and alcohol abusers commonly display symptoms of anhedonia, psychomotor slowing, fatigue, and sleep disturbance. These symptoms are considered to largely involve dopaminergic transmission, which is depleted in both depression and substance abuse. Cytokines have been demonstrated to play an important role in the regulation of dopaminergic transmission. For instance, these symptoms are commonly observed in individuals undergoing cytokine therapy for somatic illnesses [25]. These subjects also show altered glucose metabolism in the basal ganglia, which is a region rich in dopamine and responsible for many of these cognitive, emotional, and reward states [26]. The basal ganglia have also been demonstrated to be able to express a substantially high level of cytokines [27]. The increase in neuroinflammation occurring in both patients suffering from depression and addiction represents the potential for an immune-mediated change in dopaminergic transmission. Similarly, serotonin deficiencies have been reported in depressed individuals as well as alcoholics, with cytokines again being a key factor. Interleukin (IL)-1 and tumor necrosis factor (TNF)-α substantially increase serotonin uptake and bias the serotonin precursor tryptophan away from pathways that result in serotonin production [28].

Glial cells, such as microglia and astrocytes, are increased in the brain of alcoholics. These cells are largely responsible for the signaling of proinflammatory cytokines known to be elevated in depressed individuals. Ethanol has been shown to activate toll-like receptor (TLR) 4 expressed on microglia, resulting in increased nuclear factor (NF)κβ transcript of immune genes [29]. Similar effects on TLR4 occur in response to infection or exposure to the bacterial mimetic LPS [30], which is known to result in increases in depressive-like behaviors. Thus, the characteristic elevation in cytokine production and increased activity of central glia that occurs in both clinical depression and addiction may be responsible for the regulatory changes in monoamine transmission observed in these disorders.

Finally, a great deal of research has been reported in relation to sleep disturbances in patients addicted to alcohol as well as those with clinical depression, and this change contributes substantially to comorbidity of these disorders. Sleep disturbances are common symptoms in both depression and alcoholism [31, 32]. Clark and colleagues [33] showed that individuals with comorbid alcoholism and depression display elevated density of rapid eye movement (REM) sleep, resulting in a total sleep time reduction. It is no surprise that the degree of sleep deprivation in these subjects is also associated with relapse rates. A recent avenue of investigation in regard to sleep regulation has been the involvement of cytokines with the IL-1 and TNF gene families being most often implicated [34]. Irwin et al. [31] investigated whether antagonism of TNF-α by etanercept could normalize sleep patterns in individuals with alcohol dependence currently abstaining from alcohol. Indeed, etanercept administration dramatically reduced REM sleep density to levels approaching those of the control groups, highlighting the importance of the immune system in the regulation of symptoms that occur in both alcohol addiction and clinical depression.

While these findings are promising, our understanding of the neuroimmune mechanisms driving depression and substance abuse comorbidity remains in its infancy. Regardless, understanding the differential roles of the immune system in addiction and depression holds the promise to explain how and why substance abuse occurs so often on a backdrop of depression and vice versa. It also allows for the testing of very clear hypotheses with implied causality, such as the increase in depression or alcoholism following medical illness or immune therapies, elevations in inflammatory markers in individuals with depression or substance abuse, and increases in behavioral and emotional symptomologies of depression following immune activation in healthy controls. While other chapters in this book will focus more directly on inflammatory markers considered to influence alcoholism and other drug addictions, it is important to provide a foundation of knowledge in regard to the neuroimmune factors demonstrated to regulate depression. Here we will attempt to provide a cohesive account of the known pathways believed to underscore the relationship between inflammation and depression.

6.2 Sickness and Depression

The dissociation between behaviors that signify sickness and those that connote depression lies central to the area of inflammation-induced depression. During illness, a number of behavioral symptoms are manifested. People feel tired and sleep more, they lie around and become less mobile, motivation decreases, appetite is reduced, grooming habits fall, social isolation increases, and concentration becomes difficult. These behavioral symptoms are easily identified as "sickness" in someone who is physically ill or who presents with a fever. These physiological changes represent adaptive responses aimed at conserving energy to fight off infection and minimize the spread of the pathogen that caused the disease. Importantly, each of these behaviors can describe symptoms of an individual with clinical depression as well (see Table 6.1). In addition to these behavioral symptoms and cognitive decline, there is an increase in negative affect and anxiety. It is this emotional and behavioral overlap between sickness and depressive-like behaviors that first directed researchers towards the possible role of the immune system in depression. As early as 1989, Dantzer and Kelley proposed that the emergence of sickness behaviors during infection is not a coincidental occurrence but reflects direct action of the immune system on the brain to regulate behavioral responses to infection [35]. A number of subsequent publications confirmed that centrally administered cytokines are capable of inducing sickness behaviors in animals, such as reduced locomotor activity, increased sleep, decreased social exploration, and less consumption of both food and water [36–38]. The suggestion that the same, or similar, neuroimmune processes which are responsible for the modulation of sickness behaviors may also be responsible for depressive-like behavior given the strength of their similarities soon

Table 6.1 Comparison of sickness and depressive-like behaviors

Sickness behaviors	Depressive-like behaviors
Lethargy	Lethargy
Anorexia	Overeating/anorexia
Increased sleep	Increased sleep/sleep disturbance
Reduced grooming	Reduced grooming
Concentration difficulties	Neurovegetation
Psychomotor retardation	Cognitive decline
Decreased libido	Decreased libido
Reduced locomotion	Anhedonia
Hyperalgesia	Motivation loss
Decreased social exploration	Decreased social exploration
Negative effect	Negative effect

Note the similarities between both behavioral sequelae; however, more persistent cognitive and mood disruption is usually associated with depressive-like behaviors. Similarly, perturbations in the normal functioning of some behaviors with depression can appear more complex with a greater range of outcomes compared to sickness behaviors. For instance, sickness behaviors are usually characterized by sleeping more and anorexia, whereas in depression, individuals tend to either exhibit anorexia or can overeat as well as sleep more or exhibit sleep difficulties

followed. Convincing preclinical data in this vein was presented by Raz Yirmiya's laboratory in 1996. Systemic injection of LPS to rats resulted in reduced motivation and anhedonia as measured by preference for saccharin solution and sexual behaviors, which were rescued by chronic treatment with the antidepressant imipramine [39]. Importantly, early studies such as this one conceptualized depression as another facet of sickness. Important work following this early conceptualization of depression and sickness allowed researchers and clinicians to dissociate between the two behavioral sequelae.

While the recognition that sickness and depressive behaviors appear quite identical has enhanced our insight into the etiology of depression, these similarities pose a predicament when investigating inflammation-induced depression. At the center of the debate lie two fundamental arguments. The first being that in order to advance our understanding of the etiology of depression, we must be able to discriminate between the adaptive behavioral responses that accompany sickness and the often maladaptive nature of depressive-like behaviors. While an argument that depressive-like behaviors have adaptive evolutionary roots certainly holds merit [40], there is no question that in the context of modern western medicine, clinical depression disrupts social networks, impairs functional activities, and places a burden on the community. It is for this reason that many believe that a distinction between the two is necessary in experimental settings when studying inflammation-induced depressive-like behavior. Others, however, believe it is within the blurring of sickness and depressive-like behaviors that most progress will occur. The various approaches to investigating inflammation-induced depression in humans exemplify this dilemma nicely. Studies that examine medically ill patients with depression capitalize on the overlap between sickness and depressive-like behaviors in this population. However, there is danger that the presence of sickness can inflate depressive-like effects in these studies. Conversely, approaches that examine individuals with clinical depression in the absence of medical illness make it easier to conclude that neuroimmune pathways contribute to the onset of depression alone but run the risk of underappreciating the role of inflammation.

Preclinical rodent models used to study inflammation-induced depressive-like behavior must be carefully controlled to be able to separate sickness from depressive-like behaviors. On the one hand, rodent studies allow for greater control, which can enhance dissociation between sickness and depressive-like behaviors. Importantly, rodent models permit an in-depth biochemical evaluation of the brain, which is not possible in humans. Indeed, animal studies routinely measure depressive-like behaviors after sickness behaviors have abated [41–43]. It is widely recognized that the depressive-like effects that occur as a result of inflammation can remain even after sickness behaviors have been terminated. This allows for animal models to investigate the more chronic outcomes of depression following acute inflammation in the absence of sickness. For instance, animals exposed intraperitoneally to the cytokine inducer LPS show increased depressive-like symptoms such as anhedonia, as measured by sucrose preference, and immobility during forced swim (FST) and tail suspension (TST) tests at 24 h following injection. This is a time point when sickness behaviors, such as reduced locomotor activity and social exploration, no longer differ from controls [41–45]. Similarly, infection in mice by Bacille Calmette–Guerin (BCG) produces chronic induction of sickness markers such as body weight loss and reduced food intake and locomotor activity for up to 3 days postinfection. At 7 days after the exposure, these sickness markers do not differ from controls, but depressive-like behaviors on the FST and TST are maintained [41, 42]. Importantly, the ability for animal studies to identify the biochemical and molecular influences driving this persistence of depressive-like behavior is represented in these studies as well. For instance, indoleamine 2,3-dioxygenase (IDO) is a rate-limiting enzyme involved in the metabolism of tryptophan and is considered to play an influential role in inflammation-induced depression. Administration of an IDO antagonist, 1-methyl tryptophan (1-MT), abrogates the onset of depressive-like behaviors, while sickness behaviors remain, as is the case in IDO-deficient mice [42].

The critical importance of time following introduction of a systemic inflammatory stimulus has indicated that inflammation-induced cytokine action on the brain produces complex changes in behavior beyond mere sickness behavior. We would argue, however, that caution is required when interpreting such data. While enhanced control in the animal laboratory is a necessity that aids the isolation of specific biochemical pathways that give rise to inflammation-induced depression, it is important to ensure this does not come at the cost of ecological validity. In humans, depression commonly emerges on the background of illness, and illness on the backdrop of depression. As such, some degree of overlap between sickness and depressive-like behaviors is likely to enhance representation of the human condition in rodent studies despite increased difficulty in identifying their respective neurobiological correlates. While there is no clear line which can be drawn to dissociate sickness and depressive-like behaviors, answers as to which behavioral subset is being measured in any given study lie somewhere in between the two

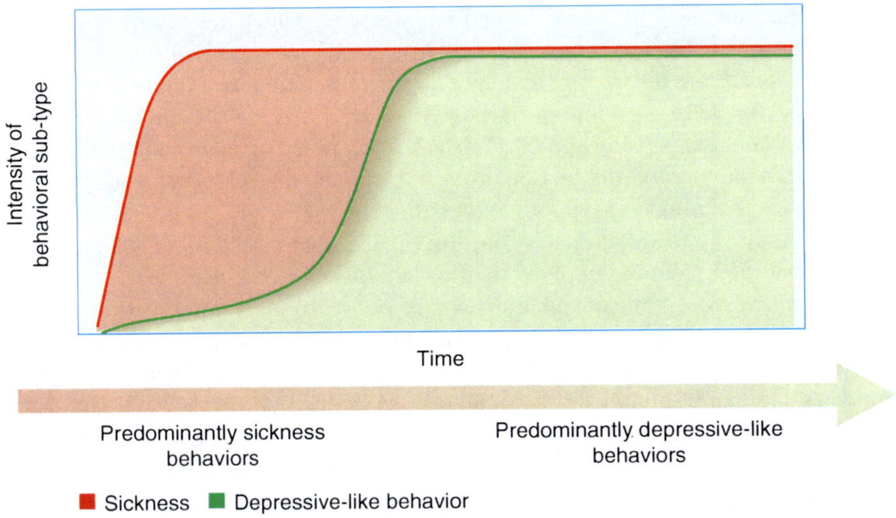

Fig. 6.1 The relative onset, duration, and intensity of sickness behaviors and depressive-like behaviors following inflammation. Typically, inflammation results in the early onset of predominantly sickness behaviors, which are maintained. A gradual increase in depressive-like behaviors and mood and cognitive disturbances then emerge with their peak phenotypic characteristics occurring some time after the onset of sickness behaviors. Importantly, the overlap between the two behavioral subsets often remains, highlighting the difficulty in distinguishing sickness behaviors from depressive-like behaviors following inflammation

approaches outlined here. As with the intricate co-expression of alcoholism with depression, sickness and depression need to be examined both separately and as a whole for progress in treatment of the disorder to occur. A graphical representation of the overlap in timing between sickness and depressive-like behaviors is provided in Fig. 6.1.

6.3 Peripheral Inflammation Versus Neuroinflammation

The old concept that the brain is relatively immune-privileged and that inflammation is largely confined to the periphery is far too simplistic. Importantly, inflammation itself has several facets including, but not limited to, localized inflammation, systemic inflammation, neuroinflammation, and sepsis [46]. It is now recognized that neuroinflammation is not simply the transposition of peripheral inflammation to the brain but has its own separate, distinct, and additional processes. Importantly, neuroinflammation must not be regarded as merely the presence of proinflammatory cytokines but should be considered to require changes in central nervous system (CNS) function [47]. In the absence of brain injury the molecular signature of inflammation emanates from resident immune cells, which results in functional changes such as behavioral consequences. It is in this way that neuroinflammation can be set apart from the normal regulatory roles of the immune system in the brain responsible for daily functioning. Behavioral alterations are a common outcome measure of such changes in the CNS. It is this capacity of neuroinflammation to produce measurable and distinct changes in behavior that has led to the field of research devoted to linking inflammation with depression.

As the majority of inflammatory insults occur in the periphery, a sophisticated communication system has evolved to allow exquisite cross talk between the brain and periphery via immune signaling. In response to an infection, phagocytic macrophages and liver Kupffer cells stimulate the production of proinflammatory cytokines, such as TNF-α, IL-6, and IL-1β. These cytokines can induce their own synthesis and that of other cytokines and can act on the brain via several pathways such as via the vagus, in the blood via circumventricular organs and specific transport mechanisms, chemokines, volume transmission, and central cytokine production.

Peripheral cytokines can take action on the brain via the targeting of neural afferents, including vagal nerves that innervate organs of the abdominal cavity in the case of intraperitoneal administration of immune stimulants [48–50]. IL-1β receptors are expressed within the sensory neurons of the hepatic vagus nerve, and their activation stimulates vagal sensory activity. The primary vagal projection area is the nucleus of the tractus solitarius (NTS). The earliest finding that the vagus can project immune information to the brain was reported by Wan and colleagues [51]. They determined immunohistochemically that peripheral LPS administration resulted in additional activation of the supraoptic nucleus, arcuate nucleus, and ventrolateral region of the brain stem compared to central LPS administration, suggestive of a vagal pathway for immune signaling. Norepinephrinergic activation leads to afferent catecholaminergic signaling from the NTS and ventrolateral medulla to the periventricular nucleus (PVN) [52], which in turn activates the hypothalamic–pituitary–adrenal (HPA) axis. Thus, activation of stress response systems is initiated, which assists in directing mood and behavior.

Blood-borne cytokines are also known to produce a central response by acting on cytokine receptors within the brain. However, given their size and hydrophilic nature, trafficking to the brain was once considered to be isolated to circumventricular organs. In 1983, Blatteis et al. demonstrated that lesioning of a circumventricular organ, the anteroventral third ventricle, abrogated the cytokine-induced febrile response [53]. Thus, it was established that circumventricular organs represent a mechanism for brain–immune communication, although some controversy remained as to whether lesioning simply blocked the neuroimmune signaling pathway or impaired the function of the surrounding brain regions [54]. Since then, it has been established that cytokines including IL-1α and TNF-α are able to be transported across the blood–brain barrier (BBB) beyond simply enhanced leakiness of the BBB. In fact, intact IL-1α and TNF-α can pass across the BBB via specific transport mechanisms which enhance saturability of the BBB [55, 56]. In addition to their intact transportation across the BBB, peripheral cytokines can also stimulate the release of secondary mediators such as nitric oxide and E2 series prostaglandins (PGE2s) [57], which more freely pass across the BBB. PGE2s are able to diffuse to nearby target sites involved in neuroendocrine and body temperature regulation, including the PVN, and catecholaminergic brainstem nuclei [50, 58, 59].

In addition to immune-driven stress response activation, neuroinflammation can also alter neuronal activity in regard to both the inhibition or activation of action potentials and suppression or expression of neurotransmitter release, reuptake, and metabolism. For instance, numerous studies have associated increases in inflammatory cytokines, particularly TNF-α, IL-1β, IL-1α, and IL-6, with inhibition of long-term potentiation (LTP) [60–63]. Such immune-mediated inhibition of LTP is associated with impaired hippocampal-dependent memory [64]. Convincing data regarding the importance of neuroinflammation for the regulation of LTP, and hence cognitive performance, has been recently published by Marina Lynch's laboratory in Dublin. They demonstrated that single-Ig-interleukin-1 related receptor (SIGIRR)-deficient mice exhibit impaired performance on cognitive tests such as novel object recognition and Y-maze exploration [65]. SIGIRR is a member of the TLR family. TLRs recognize microbial components and initiate signal transduction pathways resulting in anti-inflammatory outcomes [66]. Associated with the impaired performance on the novel object recognition and Y-maze exploration tasks was deficient LTP and elevated IL-1α in these SIGIRR-deficient mice [65]. Inflammation-induced stress and metabolic and neurotransmission pathways that may lead to depressive outcomes will be discussed later in this chapter.

Importantly, regions of the brain responsible for emotional regulation, executive functioning, and their consequential behavioral outcomes are sensitive to inflammation [22]. The extended limbic system, primitively responsible for fear and pleasure responses, stress, memory, and learning, has been shown to be modulated by immune signaling. Early work established that there is a high density of IL-1 receptors in the dentate gyrus and pyramidal cell layer of the hippocampus, the hypothalamus [67], and basolateral amygdala [68]. Additionally, the prefrontal cortex, involved in executive control of appropriate behavioral responses to stimuli, is vulnerable to inflammatory insults, especially during times of stress [69]. It is therefore no surprise that mood, memory, learning, executive function, and behavior are affected in inflammation-associated depression and together work to characterize what we refer to as depression. It must be emphasized, however, that the appearance of a particular sensitivity of these limbic processes (excluding the PFC) to an inflammatory insult may only exist as a reflection of these regions being among the first to respond to peripheral inflammation due to their circumventricular localization. Indeed, once cytokines and secondary mediators have diffused to the brain side of the BBB, a process known as volume transmission slowly diffuses cytokines throughout the brain, whereby any region with receptors for cytokines will respond in turn [70]. Figure 6.2 shows the major pathways through which the peripheral immune system reaches the brain.

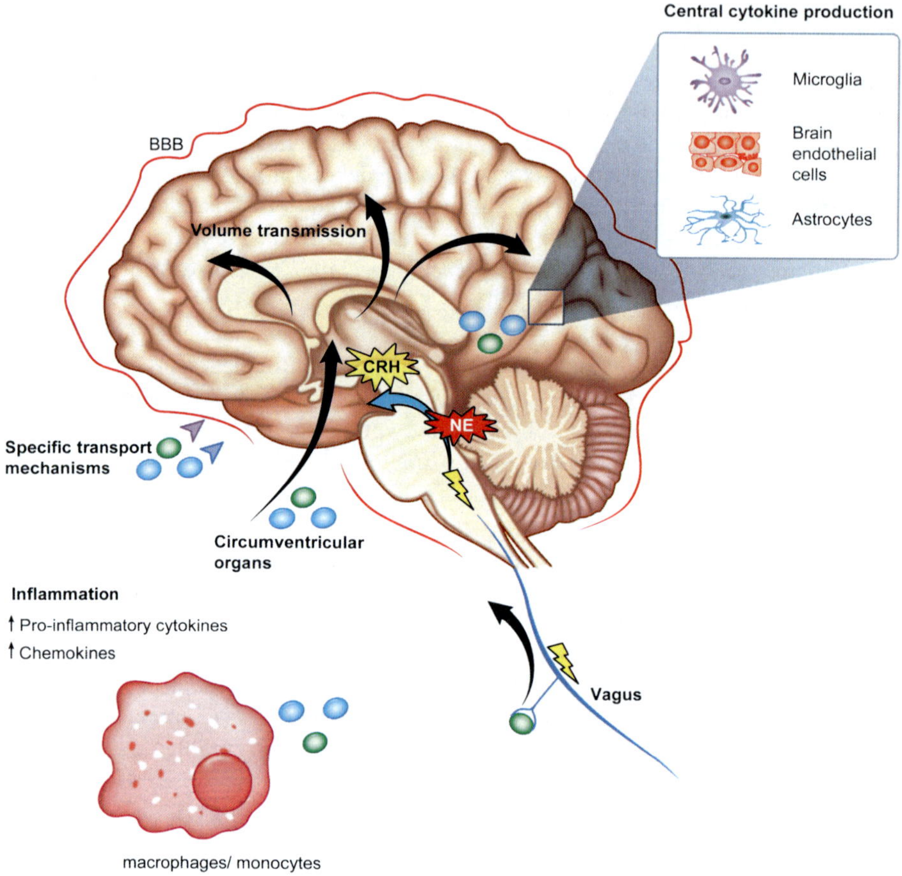

Fig. 6.2 Immune–brain communication pathways. Peripheral inflammation activates the release of proinflammatory cytokines and chemokines which can access the brain directly via circumventricular organs or pass across the blood–brain barrier (BBB) using specific transport mechanisms. Proinflammatory cytokines, especially IL-1β, can bind to receptors on afferent vagal nerves and transmit information to the brain ending in the NTS and ventrolateral medulla causing cytokine release and activation of norepinephrinergic pathways which direct towards the hypothalamus to activate the hypothalamic–pituitary–adrenal (HPA) axis. Microglia, astrocytes and brain endothelial cells can produce central cytokine release in response to inflammatory stimuli. Finally, once cytokines have entered the brain, a process referred to as volume transmission works to diffuse these cytokines throughout the brain, targeting cells with receptors viable for responding to immune signals

6.4 Clinical Models of Inflammation-Induced Depression

Physicians have long puzzled over the co-occurrence of depression with physical illness, particularly because it became apparent that such changes in patient mood went beyond the mere distress of being unwell. It was clear that the symptoms of depression observed in these individuals were manifested as a result of the physical changes that occurred during sickness. This empirical observation led to the conclusion that depression must not be restricted to being a disorder of the mind but is also a direct reflection of the health status of an individual. Unfortunately, the physiologic mechanisms that formed the basis for the association between mood and health were completely unknown. It was only in the 1980s that the first convincing evidence to link the immune system with behavioral regulation and emotionality emerged. These early findings demonstrated that administration of newly cloned and expressed recombinant cytokines for therapeutic purposes in fighting infection and suppressing tumor growth also induced activation of the HPA axis [71–73] as well as depressive-like symptoms which were considered as just another component of sickness in these early days. Nevertheless, a scientific basis for the link between the brain, immune system, and behavior was beginning to become established. There are now numerous publications that show increased depressive symptoms in clinically unwell populations [25, 74–76], and a growing number of methodological approaches are now commonly employed to examine and characterize the nature of this immune–brain relationship in humans. These approaches can be broadly categorized into (1) the study of patients with a physical illness who develop depression, (2) the study of individuals with depression in the absence of any major physical illness, or (3) the study of healthy subjects by correlating mood with their cytokine profiles at baseline or following a minor inflammatory stimulus. Some of the major findings using each of these approaches are summarized below.

Given the high prevalence of depression in patients suffering from physical illness, it is no surprise that many researchers have taken advantage of this subject population to understand and characterize immune regulation of mood and behavior—especially the etiology of inflammation-induced depression. However, the field began with early descriptions of the cell distributions in patients with depression in the absence of any physical ailment. Maes et al. reported depressive patient populations to be characterized by greater numbers of white blood cells (WBCs), monocytes, class II MHC HLA-DR, and T lymphocytes, as well as a negative correlation between depression severity and mitogen-induced lymphocyte responsivity [77]. Since then, significant progress has been made in emphasizing the causality of inflammation in depression with detailed studies focusing on the ability of cytokines to induce depression. For instance, a number of investigators have capitalized on hepatitis C or malignant melanoma patients undergoing innate cytokine treatment with interferon (IFN)-α, which has been demonstrated to increase depressive symptoms and act as a biomarker for depression severity. IFN-α is known to induce symptoms of depression in 30–50% of these patients [25, 75, 78] with negative and anxious affect and cognitive symptoms occurring later in treatment following the onset of psychomotor decline [79]. Furthermore, while nearly all patients show neurovegetative symptoms, in the form of fatigue, sleep disorders, and reduced appetite, only a portion of patients display cognitive and psychological symptoms of depression. The differences in timing and variability in responses between neurovegetative symptoms and depressed mood are suggestive of the need for enduring cytokine activation to yield changes in neuronal function and produce pronounced effects on cognition, mood, and behavior. Notably, in 2001, Andrew Miller's group reported a pronounced increase in the development of diagnosable depression in patients with malignant melanoma undergoing IFN-α treatment [75]. In accord with their hypothesis, pretreatment with paroxetine, a selective serotonin reuptake inhibitor (SSRI), reduced the incidence of major depression in patients undergoing IFN-α therapy [75]. Work of this kind has been pivotal in demonstrating therapeutic links between cytokine activation and monoamine pharmacology in clinical populations. Similar effects continue to be reported. For instance, Prather and colleagues [80] observed nondepressed hepatitis C patients undergoing IFN-α therapy over a period of 4 months. During this time, 22% of patients developed major depression, and IFN-α therapy was associated with time-related elevations in Beck Depression Inventory (BDI) scores. Furthermore, bidirectional relationships between circulating IL-6 concentrations and scores on depression scales were observed, reiterating the role of cytokines in mood and behavior. Functional magnetic resonance imaging (fMRI) studies have demonstrated changes in brain activity for hepatitis C patients receiving IFN-α treatment [81]. These patients demonstrate significant activation of the dorsal region of the anterior cingulate cortex (ACC) when conducting a visuospatial attention task. Furthermore, this degree of activation positively correlated with their errors on this task [81]. A similar fMRI examination of hepatitis C patients undergoing IFN-α therapy revealed that those patients who developed clinical depression exhibited elevated limbic activation and indeed an altered pattern of activation compared to patients with endogenous depression [82]. This may suggest underlying differences between inflammation-induced depression and other manifestations of the disorder.

In addition to the elevated prevalence of depression in individuals suffering from inflammatory illness described above, immune involvement in the comorbidity of depression with other illnesses has also been reported. For instance, diabetic populations exhibit at least a 7% increase in the prevalence of depression [83] compared to nondiabetic groups. A meta-analysis examining comorbid depression in adults with diabetes has even reported the odds of depression being double that of nondiabetic comparison groups in well-controlled studies [84]. Importantly, diabetes has been associated with an increase in circulating proinflammatory cytokines such as IL-6 and TNF-α [85], with the latter cytokine contributing substantially to insulin resistance. These and other inflammatory markers that occur in diabetics are known to contribute to the pathogenesis of depression. Similarly, congestive heart failure (CHF) is also associated with a significant increase in the prevalence of depression (15–36%) [86]. Cytokines play an important role in heart failure progression following myocardial injury (e.g., myocardial infarction), promoting prolonged activation of the cytokine cascade. Of the cytokines that have been implicated in the pathogenesis of CHF, IL-1β, IL-2, TNF-α, and IL-6 have also been implicated in the pathogenesis of depression. These findings are suggestive of a common immune-driven pathway and the exacerbation of both symptomologies [86]. Thus, comorbidity of depression among a range of somatic illnesses outside of those traditionally considered to be inflammatory is likely to derive, at least in part, from immune dysregulation.

Examination of individuals with depression who do not present with any apparent physical illness has bolstered support for the notion that the immune system is implicated in the etiology of depression. Such studies have identified circulating cytokines, particularly IL-1β and IL-6, to be elevated in depressives compared to healthy controls and usually correlate positively with mood ratings [87–90]. There is also evidence to indicate that not only is there an increase in proinflammatory cytokines in these patients but that the circadian rhythms of these cytokines are disrupted. This finding is indicative of a major disruption in the expression of these cytokines in sufferers of depression [87]. Confirmation of increased immune activity in patients with depression is a robust and replicable effect as demonstrated by Zorilla et al.'s [91] meta-analysis focusing on immune markers in studies of depressed individuals. They showed consistent reporting of increased IL-6, WBC counts, and elevated CD4/CD8 ratios in depressives. Note that the relationship between the immune system and depression is not simplistic as evidenced by the reduced T cell and natural killer (NK) cell function which also emerged from this meta-analysis. The dissociation between these findings remains rather unelucidated and likely represents the differential effects of long-term stress and depression on cognition, mood, and behavior, as well as that of immune susceptibility to disease. Nonetheless, these findings certainly implicate inflammation in clinical depression. However, they fail to provide any firm evidence that inflammation and neurotransmitter production may co-regulate each other to produce noticeable changes in mood and behavior. Such skepticism regarding the interrelationship between the immune system and neuronal function in depressed patients has been approached with studies examining the effect of antidepressants on immune parameters in these patients. An early study by Sluzewska et al. [92] showed that serum IL-6 concentrations and the acute phase protein alpha-1-acid glycoprotein (AGP) were normalized following treatment with fluoxetine. The ability of antidepressants to reduce proinflammatory cytokine profiles and inflammatory markers in patients with depression has since been replicated [93–95], suggestive of an intricate relationship between cytokines and neurotransmitter function. However, a number of studies have also reported normalization of depressive symptoms by antidepressants without normalization of cytokine profiles. A meta-analytic study investigating cytokine levels in response to antidepressant treatment found that across 22 publications, a significant reduction in IL-1β was consistently reported following antidepressant medication, but consistent reductions in TNF-α and IL-6 could not be convincingly argued [96]. Therefore, while there appears to be a link between cytokines and monoamine pharmacology, the exact nature of the relationship requires further elucidation and, as we will argue later in this chapter, may not represent the primary mediator in inflammation-induced depression.

Investigating healthy controls is a valuable exercise in strengthening our understanding of how immune regulation might result in changes in mood. It stands to reason that if the immune system is integral in regulating mood and behavior, then relationships between cytokines and behavior in healthy controls should also exist, albeit at lower concentrations than in clinical populations. One such study [97] examined well-functioning aged participants, capitalizing on the naturally occurring increase in inflammation that occurs with age. Penninx and colleagues observed that circulating levels of IL-6, TNF-α, and CRP significantly and positively correlated with increases in depressed mood ratings. Medically healthy subjects may also be challenged with a minor inflammatory stimulus to determine its consequence on mood and behavior. Low-dose LPS administration as well as typhoid vaccination in healthy participants has confirmed that low-grade inflammation increases transient symptoms of depressed mood [98–101]. For instance, Harrison and colleagues [98] administered typhoid vaccination to healthy male volunteers and found that depression-like scores on mood questionnaires significantly increased 3 h following vaccination. MRI results showed that these individuals also displayed reduced connectivity of the subgenual ACC to the amygdala and medial prefrontal cortex among other regions during an emotional face processing task, which was modulated by IL-6. A similar study by Wright et al. [101] demonstrated healthy volunteers to display increases in negative mood following Salmonella typhi vaccine, which significantly correlated with elevations in plasma IL-6. These studies further strengthen the functional biological importance of immune–brain–behavior relationships and assist in elucidating the basic underlying principles and mechanisms for inflammation-induced depression.

The two human-based approaches described above (i.e., examination of individuals suffering with depression with and without medical illness) indicate a strong role for systemic inflammation in the pathogenesis of symptoms of clinical depression. However, while the data are convincing, ethical and practical constraints limit the ability of most human studies to directly examine mechanisms, especially those in the brain. Too often, peripheral inflammatory markers are relied upon, which may not be entirely reflective of what is happening in the brain or at least do not elucidate the central processes that are involved. In order to identify specific central mechanisms, animal studies are required. In animal and ex vivo models, the biological pathways underlying neuroimmune regulation of mood and behavior in the brain can be thoroughly examined. These preclinical studies form the foundation from which future efficacious immune-targeted therapies for depression can be developed for the human population.

Finally, relationships between the immune system and depression cannot be considered to be unidirectional. We have thus far discussed a top-down approach, which is the way inflammation can contribute to symptoms of depression. However, there are numerous examples of mood- and behavior-altering immune status. Uncovering this phenomenon in clinical studies has largely occurred through the use of longitudinal prospective studies. For example, examination of healthy older men and women revealed that baseline BDI-II scores predicted a 6-year change in circulating IL-6 levels but baseline IL-6 did not predict future BDI-II scores [102]. Similarly, a recent longitudinal study in 1,420 children from 9 years old onward demonstrated that cumulative depressive episodes and depression variables on the child and adolescent psychiatric assessment and young adult psychiatric assessment predicted later C-reactive protein (CRP) levels but CRP levels did not predict later depression status [103]. Thus, the relationship between mood and immunity is intricate and bidirectional.

6.5 Animal Models of Inflammation-Induced Depression

Rodent models of inflammation-induced depression have become a valuable tool for investigating mechanisms involved in the pathogenesis of the human disorder. Proposed pathways believed to be involved in inflammation-induced depression, such as monoamine transmission, cytokine production, cell trafficking, microglial activation, tryptophan metabolism, and HPA axis activation, are largely conserved throughout evolution. Of note, these investigations have sometimes been sidetracked because of important differences that exist between species. For example, human IFN-α is not biologically active in rodents in the same way it is in humans. This was exemplified by Loftis and colleagues [104] who administered pegylated human IFN-α to Lewis rats and were unable to observe sickness or depressive-like behaviors in these rats. IFN-α in the brain and liver was examined via protein analysis with western blot and determined that signal transducer and activator of transcription (STAT1) protein failed to be phosphorylated. The major difference appears to be that IFN-α-induced sickness in rodents requires mediation via opioid receptors [93]. It should be noted that species differences among rodents also yield striking variability in immune responsivity, especially in regard to cytokine sensitivity with mice appearing less sensitive than rats [105–107].

Humans and animals have been getting sick since the beginning of time. It is therefore reasonable to propose that effective ways of responding to infection, both physiologically and behaviorally, were developed early in the evolutionary history of both man and animals. Existence of these common pathways allows for animal models to inform many human pathologies. Rodent models permit more intensive examination of such pathways, particularly in the brain, acute versus chronic infection, and transgenic models.

Human studies are typically restricted to investigating peripheral immune processes, making it difficult to differentiate the effects of peripheral inflammation versus neuroinflammation. Hence, the major limitation of human-based approaches is that the processes that direct emotionality and behavior reside within the brain, and as such, many human studies require speculation regarding how peripheral cytokines may gain access to the brain and act on extended limbic and prefrontal processes. Peripheral inflammation in rodents has proven to yield reliable changes in behavior, especially in regard to depressive-like symptomologies such as immobility, anhedonia measured by decreased preference for drinking a sweetened solution, and learned helplessness [108, 109]. In contrast to this important limitation in most human research, animal models of inflammation-induced depression can more effectively differentiate peripheral inflammation from neuroinflammation and the requirements of each to induce changes in behavior. A number of studies have demonstrated that administration of the cytokine inducer LPS, or cytokines such as TNF and IL-1, directly into the brain is sufficient to produce sickness and depressive-like behaviors [41–43, 110–112]. Furthermore, central administration of cytokine antagonists [36, 113–115] or anti-inflammatory agents such as minocycline [43] blocks the behavioral sequelae of sickness and depression-like behavior that follow peripheral cytokine or LPS injection. One such early study confirmed this important link between the peripheral immune system and the brain. In 1992, Kent and colleagues showed that IL-1ra administered to the periphery blocks the development of physical markers of sickness such as body temperature rise and oxygen consumption following peripheral administration of IL-1. Importantly, intracerebroventricular administration of IL-1ra was able to abrogate the depressive-like effects, such as food-motivated behavior and social exploration, in response to either centrally or peripherally administered IL-1 [115]. Thus, animal studies like this one have directly demonstrated that cytokines activated in the periphery are able to exert their effects in the brain. Please note that the high degree of overlap between sickness and depressive behaviors is no coincidence, and it requires careful consideration when trying to isolate depressive-like behaviors from sickness behaviors in the laboratory. This will be covered in detail in the section below.

By far the most sophisticated animal models for investigating the precise mechanisms responsible for inflammation-induced depression have emerged in only recent years. The advent of genetically modified mice has allowed researchers to ask more direct questions regarding the mechanisms responsible for inflammation-induced depression. A great deal of work has employed genetic knockout mice for specific cytokines in order to determine the consequences on behavior. For instance, Bluthé and colleagues [36] found that a number of cytokines can play common roles in inflammation-induced depression and sickness. For example, IL-1β-converting enzyme (ICE)-deficient mice are largely resistant to centrally administered LPS, showing only a mild reduction in food-motivated behavior. These data implicate IL-1β as a primary cytokine responsible for the regulation of sickness behaviors [116]. The fact that ICE deletion did not completely abrogate LPS-induced sickness behavior was subsequently explained by further use of knockout strains. IL-1R1 knockout mice administered either LPS or IL-1β either peripherally or centrally have been demonstrated to be resistant to the sickness-inducing effects of IL-1β but not LPS in regard to social exploration at 2, 4, and 6 h postinjection and immobility at 2 and 4 h postinjection. It was only when TNF-α action was blocked centrally that these sickness behaviors were attenuated in response to LPS. Therefore, TNF-α replaced the role of IL-1β in this murine knockout model, highlighting the plasticity of function in the immune system and illuminating the roles of these proinflammatory cytokines in inducing inflammation-induced sickness. IL-6 knockout mice have also been extensively employed with a number of studies showing anxiety-related and depressive-like behaviors to be attenuated [36, 117]. However, these data are less convincing in IL-6-deficient mice with at least one study reporting no change in inflammation-induced depressive-like behavior [118]. Additionally, cytokine receptor knockout models have been instrumental in answering complex questions regarding the interactive effects of cytokines in producing behavioral outcomes. Some cytokines, such as IL-1 and TNF-α, bind to multiple receptors to cause different effects. Murine knockout models have been used to determine which receptor is responsible for changes in behavior. For instance, examination of CD14-, TLR2-, and TLR4-deficient mice has shown TLR4 to be responsible for the anorexic effects of LPS, whereas TLR2 is involved in the response to gram-positive bacterial products, and CD14 is involved in both [119]. Palin et al. [120] recently reported that only TNFR1 knockout mice but not TNFR2 knockout mice fail to present with sickness behaviors following central administration of TNF-α. Thus, TNFR1 is responsible for the mediation of TNF-α-induced sickness behavior. In addition to cytokine knockout murine models, a great deal of progress has been made by using IDO knockout mice. IDO is an enzyme that metabolizes tryptophan. Inflammation-induced catabolism of tryptophan produces cytotoxic metabolites believed to play a leading role in the pathogenesis of depression. Indeed, investigation of IDO-deficient mice has yielded interesting findings as to the relative roles of IDO, cytokines, and serotonin in inflammation-induced depression [42]. Importantly, IFN-γ receptor-deficient mice also demonstrated this cytokine to be a crucial contributor to the activation of IDO following a live bacterial infection [41]. The use of transgenic models to determine the roles of cytokines and tryptophan metabolism will be discussed in greater detail later in this chapter.

Animal and human biomedical research often complements each other, and both are needed to advance and improve human health. It is the translation of these murine findings into efficacious clinical practices that is important and should be considered as the ultimate goal. Clearly, rodent models serve to advise and direct clinical avenues of investigation, confirm findings in humans, and more precisely define the mechanisms that drive inflammation-induced depression. This is why greater emphasis is being directed towards such translation from preclinical animal research into the clinical field and vice versa.

6.6 Mechanisms of Inflammation-Induced Depression

Several mechanisms have been proposed to account for the biological basis of depression. The most common of these include monoamine deficiencies, stress-induced neuroendocrine alterations, central cytokine action, and activation of the kynurenine pathway of tryptophan metabolism. The following sections briefly describe and evaluate these pathways with a particular focus on inflammation-induced depression.

6.6.1 The Monoamine Deficiency Hypothesis of Depression: Is Serotonin Involved in Everything but Responsible for Nothing?

No discussion of the proposed causal pathways of depression would be complete without beginning with the monoamine deficiency hypothesis. For many decades, this has been the dominant account of the biological components that underlie depression. Serotonin remains the primary target for psychopharmacological treatments of depression. The main pivot of this theory is the pharmacological property of selective serotonin reuptake inhibitors (SSRIs) that are supposed to exert their antidepressant properties by increasing the bioavailability of serotonin at the synapse. However, examination of the literature must result in the conclusion that the monoamine deficiency hypothesis simply cannot explain depression in its entirety nor that it may even be the most dominant pathway to depression. Therefore, while any account of depression would be incomplete without discussing monoamine deficiency, critical evaluation is required. It is our intention to use this predominant theory as a means to highlight the need for research to continue to expand into other promising pathways, which are to be discussed subsequently. We will briefly describe the monoamine deficiency hypothesis and the psychopharmacological approaches which have been inappropriately applied as support for the hypothesis and discuss the experimental findings in regard to inflammation-induced depression.

Simply put, the monoamine deficiency hypothesis proposes that major depressive disorders are caused by a reduction in the neurotransmission of monoamines such as noradrenaline, dopamine, and serotonin. This can result from a global reduction in monoamine levels or via ineffective functioning of the serotonin transporter, which removes serotonin from the synaptic cleft back into the synaptic boutons, diminishing free serotonin concentrations. The monoamine deficiency hypothesis arose following observations that treatments which target monoamine neurotransmission, such as tricyclic antidepressants and monoamine oxidase inhibitors (MAOIs), alleviated the symptoms of depression. The newer generation of antidepressants (SSRIs) approved in the 1980s and the newer norepinephrine and dopamine specific reuptake inhibitors act similarly on receptors and also increase central monoamine concentrations. Despite major symptoms being reduced in some patients with clinical depression following treatment with these monoamine-targeted antidepressants, such pharmacological approaches are ineffective in at least one-third of clinically depressed patients [121]. Although the claim that the ability of monoamine-targeted antidepressants to reduce depressive symptoms is primary evidence for the role of these monoamines in depression, the theory suffers from teleological problems. Lacasse and Leo [122] exemplify this problem nicely by asserting that aspirin's ability to alleviate headaches does not prove that headaches result from low concentrations of aspirin in the brain. Such flawed logic regarding aspirin would still provide greater support for this hypothesis than that used for antidepressants given that aspirin can produce fast-acting alleviation of headaches. In contrast, despite the acute action of monoamine-targeted antidepressants on increasing monoamine levels at neuronal synapses, 2–3 weeks of continued antidepressant treatment is required for noticeable changes in depressive symptoms [123]. To date, no explanation can fully account for this disconnect between the acute effect of antidepressants on monoamine levels and the delayed effects on depressive symptoms. One possible explanation is that antidepressants may increase the levels of growth factors such as brain-derived neurotrophic factor (BDNF) in patients with depression [124], thus promoting neuronal stem cell differentiation and the formation of new neuronal networks, which takes time to develop and take effect.

Pertinent to alcoholism, BDNF appears to play a role in alcohol consumption. Hensler and colleagues [125] reported that BDNF knockdown mice exhibit elevated consumption and preference for ethanol compared to wild-type controls but no difference for other nonalcoholic tastants. Thus, monoamine deficiency in depressives and alcoholics may stem from reduced BDNF in these individuals. Given that acute ethanol exposure increases both serotoninergic and dopaminergic projections, alcoholism may represent a self-medicating exercise to increase monoamine concentrations and carry over increases in BDNF to maintain elevated levels. This, however, proves unsuccessful given that serotonin and dopamine responses to ethanol habituate over time. Regardless, drugs targeting other neurotransmission systems such as ketamine and MK-801, which antagonize NMDA receptors, have been able to produce acute antidepressant effects in humans and rodents [126–128] with evidence that these too can elevate BDNF [126].

Finally, efforts to measure markers of monoamines in patients with clinical depression have been plagued with inconsistency [129]. Attempts to induce depression by reducing monoamine concentrations have also failed to produce consistent findings [129–131]. Despite many scientists remaining unconvinced that the primary cause of clinical depression is merely a central deficiency in monoamine concentrations [123, 132, 133], monoamine deficiencies are still likely to play a role in depression. This role, however, may be less robust than first believed. It is probable that simply modulating monoamine pathways is sufficient to modulate a range of diverse symptoms not limited to clinical depression but inclusive of schizophrenia, anxiety, and other disorders [123, 134].

A number of studies have shown monoamines to also be implicated in the pathophysiology of inflammation-induced depression. These data should not be ignored given the ability of cytokines to activate pathways leading to metabolism of tryptophan into serotonin. For instance, Raison et al. [135] found that IFN-α treatment of hepatitis C patients corresponded with elevated levels of IL-6 in the CSF. These increased concentrations of IL-6 were, in turn, negatively correlated with 5-hydroxyindoleacetic acid (5-HIAA), the major serotonin metabolite. The association between IL-6 and 5-HIAA was predictive of depression severity according to established psychological inventories. These findings reflect those of numerous studies that demonstrate the capacity of immunotherapies to increase monoamine turnover in humans [75, 136]. Evidence demonstrating significant changes in turnover rates of monoamine concentrations in models of inflammation-induced depression certainly supports the idea that the immune system is a contributor to the pathogenesis of depression. However, no direct evidence has surfaced that would implicate monoamine deficiency as the primary cause of depression, and inconsistencies also plague the area of inflammation-induced depression in regard to the specific roles of serotonin, noradrenaline, and dopamine [137]. In light of this lack of concrete evidence, we would argue that while monoamine deficiency is likely to contribute to depression, the available data clearly indicate that other key players are involved.

The ability of cytokines to regulate monoamine concentrations [138–140] and for anti-inflammatory antibiotics to produce antidepressant effects [141] suggests that monoamine deficiency may be a downstream consequence of other processes initiated by the synthesis of proinflammatory cytokines. An important study supportive of this proposition was published in 2009. O'Connor and colleagues [43] investigated the role of the tryptophan-degrading enzyme IDO in a murine model of inflammation-induced depression. IDO is activated by IFN-γ and TNF-α. It degrades tryptophan along the kynurenine pathway. The kynurenine pathway will be explored in greater depth below; however, it is important to note that IDO metabolizes tryptophan from which the resulting metabolites can proceed down a number of pathways. These include the production of serotonin metabolites as well as neurotoxic metabolites depending on which arm the cascade follows (see Fig. 6.3). This study showed that LPS-induced depressive-like behaviors in mice were associated with reduced plasma tryptophan concentrations independent of changes in brain serotonin turnover. Studies such as these are emerging that expand on potential biochemical targets other than monoamine deficiencies to treat clinical depression. These pathways are now being investigated more intensively and include the cytokine hypothesis of depression, neuroendocrine perturbations, and the kynurenine pathway. Importantly, the ability of these pathways to alter serotonin action in the brain is clear, indicative of the ubiquitous involvement of the neurotransmitters in many central functions without any required specificity to major depression alone.

6.6.2 Cytokines, Growth Factors, and Depression

The association between stress and depression with impaired immune function precipitated the suggestion that cytokine dysregulation may be responsible for symptoms of clinical depression [142–144]. Cytokines, such as IFN-γ, activate astrocytes and microglia resulting in the production of more cytokines from these cells [145]. Furthermore, IFN-γ reduces the synthesis of neuroprotective growth factors like insulin-like growth factor (IGF-I) [146]. It can be easily imagined that this feed-forward process that increases proinflammatory cytokines and reduces critical growth factors in the brain could produce profound changes in immunocompromised and genetically susceptible individuals. However, what are the precise lines of evidence in support of this, and how might central cytokines produce changes in mood and behavior? The lines of evidence in support of the cytokine hypothesis of depression are severalfold. Firstly, cytokines produce alterations in neurochemistry and growth factors, akin to those produced by traditional stressors, which reliably induce symptoms of depression. Secondly, depression in humans is associated with increases in circulating cytokines, suggestive of their elevated action in the brain. Thirdly, peripheral and central cytokine administration to rodents induces behavioral symptoms analogous to those observed in patients with depression. Lastly, anti-inflammatory agents administered to both humans and rodents can alleviate depressive-like behaviors. Each of these lines of evidence is discussed below with a particular focus on elucidating the actions that cytokines may be having on the brain to produce such changes in behavior.

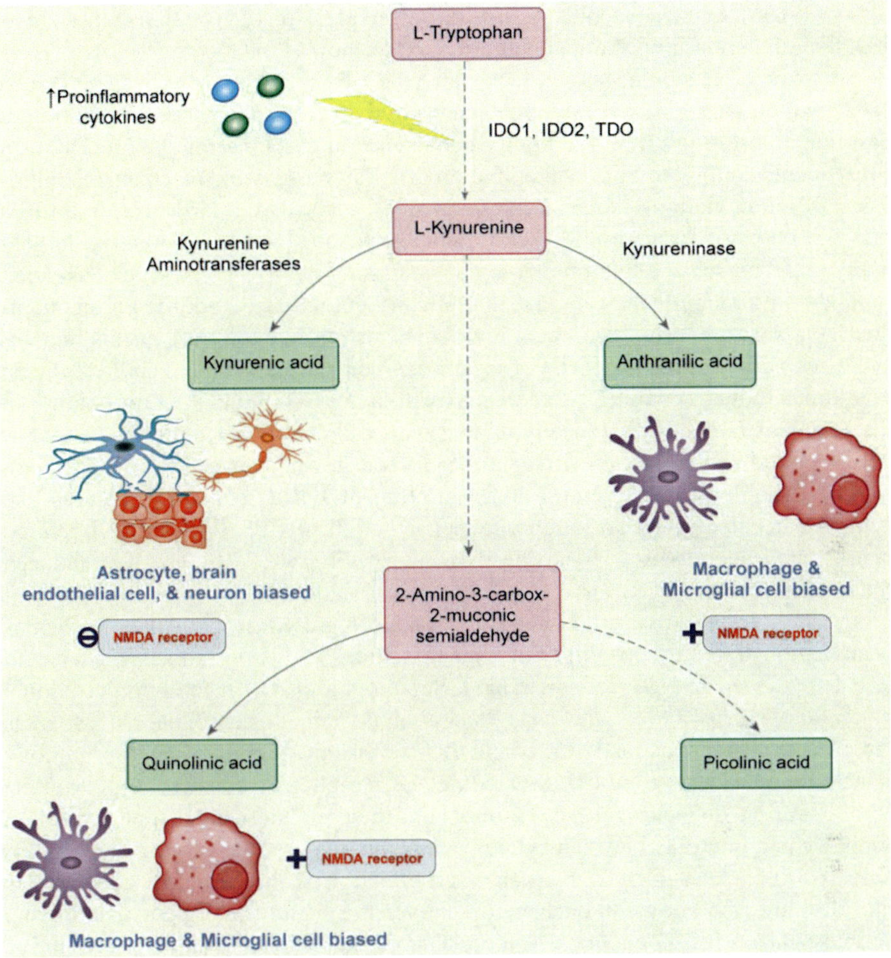

Fig. 6.3 Kynurenine pathway of tryptophan metabolism in response to inflammation. Tryptophan is metabolized intracellularly by IDO1, IDO2, and TDO, producing kynurenine. Astrocytes, brain endothelial cells, and neurons appear to be biased towards metabolizing kynurenine into kynurenic acid, which antagonizes NMDA receptors and can be neuroprotective. Macrophages and microglia appear biased towards metabolizing kynurenine into downstream neurotoxic metabolites anthranilic, quinolinic, and picolinic acid, which agonize NMDA receptors. Note that the relative roles of IDO1, IDO2, and TDO in producing downstream metabolites responsible for inflammation-induced depression have not yet been fully elucidated

Cytokines have been demonstrated to produce a vast array of changes to brain function beyond the previously described effects on monoamine availability. As with traditional stressors, cytokines stimulate activation of the HPA axis. Acute cytokine administration, both centrally and peripherally, results in elevated expression of CRH, adrenocorticotropin-releasing hormone (ACTH), and glucocorticoids

[71, 147–150]. Glucocorticoids, in turn, reduce expression of growth factors such as BDNF and disrupt hippocampal neurogenesis (reviewed in [151]).

Cytokine effects on negative feedback of the HPA axis have also been described [152], which suggests that the circadian pathways needed for stress regulation at baseline and during stress are highly dependent on cytokine regulation. Later, in subsequent sections of this chapter, the precise neuroendocrine mechanisms implicated in inflammation-induced depression will be discussed. However, it suffices that the known neural responses following exposure to typical stressors largely apply in the presence of an immunological stressor. These responses are considered to be driven predominantly by proinflammatory cytokines. In addition to neuroendocrine activation, proinflammatory cytokine activation in the CNS results in compromised neurotrophic support. For instance, Guan and Fang [153] reported almost ubiquitous reductions in BDNF, nerve growth factor (NGF), and neurotrophin-3 in the brains of rats intraperitoneally treated with LPS. Similarly, our laboratory has characterized the neuroprotective properties of IGF-I (reviewed in [154]). We recently established that central administration of IGF-I reduces depressive-like behavior in mice given an intraperitoneal injection of LPS [155, 156]. Impaired neurotrophic support also appears to contribute to IFN-α-, IL-1β-, and IL-6-induced reductions in neurogenesis [157–159]. This may occur by the ability of proinflammatory cytokines to induce resistance to growth factor receptors, as has been demonstrated by the ability of both IL-1β and TNF-α to induce resistance to IGF-I (reviewed in [154]). Interestingly, Seguin et al. [160] recently reported that systemic TNF-α administration reduces hippocampal neurogenesis, whereas repeated intra-hippocampal IL-1β and IL-6 infusion appears to have promitotic effects. While these contrasting outcomes of peripheral versus central cytokine administration on neurogenesis require confirmation, these data hint towards a sophisticated interplay between cytokines and the processes governing neural plasticity. Cytokines, such as TNF-α, and dysregulation of inflammatory chemokines like IL-8 are also known to enhance oxidative stress, glutamatergic activation and apoptosis [161–163]. Functionally, these changes in the brain manifest as cognitive decline and behavioral symptoms which characterize depression [28].

The second line of evidence in support of the cytokine hypothesis of depression revolves around the well-documented increases in circulating proinflammatory cytokines in patients with depression. Proinflammatory cytokines have been shown to be elevated in sufferers of depression with and without medical illness, as previously discussed. The most documented cytokines measured in these patients include IL-1β, TNF-α, and IL-6 [88, 144, 164]. A number of studies have also reported increases in the acute phase protein high-sensitivity C-reactive protein (hs-CRP) in patients with depression, whereby the severity of depressive episodes correlates positively with hs-CRP levels [165]. hs-CRP is a marker of inflammation due largely to the action of IL-6 in increasing its synthesis in the liver [166]. Reports of these increases in proinflammatory cytokines and inflammatory markers in patients with depression are being published with increased regularity.

The primary cytokine candidates as indicated by measures of depressive patients, IL-1β, TNF-α, and IL-6, have all been shown to be instrumental to the onset of depressive-like behaviors in rodent studies [36, 109, 117]. Inhibition of these cytokines is often shown to lead to attenuation of depressive-like behaviors or the co-opting of their roles in depressive-like behavior by other cytokines [36, 133]. Similarly, antibiotic treatment with minocycline convincingly blocks neuroinflammation and abrogates cytokine-induced sickness and depressive-like behaviors in mice following LPS challenge [43, 45].

Importantly, examination of both peripheral and central administration of cytokines has led to a better understanding of the communication networks that exist between the brain and periphery for the immune system. Cytokines are vital communicators in signaling infection to the brain and regulating the subsequent behavioral responses. The assumption that peripheral cytokines can act on the brain has been shown to occur via several pathways, including via HPA activation at the level of the hypothalamus, activation of afferent vagal fibers, passage into the CNS via circumventricular organs, and active transport across the BBB [167]. And, of course, it is now well accepted that cytokines in the periphery activate cytokines in the brain. Therefore, the local synthesis of cytokines in the brain is also important for inducing behavioral changes. The signaling cascade of each of these pathways has been described in Sect. 6.2 and is summarized in Fig. 6.2.

Finally, acute stress alone can induce the synthesis of cytokines. For instance, peripheral IL-6 and DNA binding of NFκβ increase following exposure to the Trier Social Stress Test and are further elevated in patients with depression compared to healthy controls [168]. Animal studies have also demonstrated that typical stress paradigms elevate levels of brain IL-1β [169, 170]. We will explore the relationship between the immune system and neuroendocrine stress responsivity in the following section, focusing primarily on the ability of cytokines to regulate HPA axis function. However, bear in mind that the HPA axis also plays a strong regulatory role for cytokines as well and their communication network is not unidirectional.

6.6.3 Stress/Neuroendocrine Pathways in the Context of Immune Activation and Depression

The neuroendocrine system has been implicated for decades in affective disorders including anxiety and depression. The impact of stress on the development of depression has been well established, with many studies demonstrating that stressful life events often precede the onset of depression (e.g., [171, 172]). Thus, neuroendocrine regulation of stress responses has been widely investigated and shown to be associated with the disorder at almost all of the fundamental levels of the HPA axis: centrally released CRH, peripherally released ACTH and cortisol, as well as glucocorticoid receptor (GR) involvement in negative feedback. Numerous studies have demonstrated depression and temperament changes to be positively correlated with cortisol concentrations in infants [171, 173], adolescents [174], and adults [175, 176]. Brain imaging studies are consistent with these findings and have demonstrated depressive patients to have significantly smaller hippocampal volumes, suggestive of the prolonged neurotoxic effect of HPA axis activation [177, 178]. Furthermore, dysfunction of CRH neuronal circuitries has been implicated in affective disorders in humans [179].

Neuroendocrine dysfunction has also been extensively studied in the context of inflammation-induced depression. This is no surprise given that cytokines and cytokine-inducing stimuli such as LPS activate the HPA axis and vice versa. This neuroimmune–neuroendocrine interaction has led to a body of work implicating immune-driven regulation of the HPA axis in the pathogenesis of depression. Studies in humans have reported an associative increase in both cytokines and glucocorticoids in depressive patients. For example, Humphreys et al. [180] reported a positive correlation between ACTH concentrations and LPS-stimulated IL-6 in depressive patients, but no such relationship existed for control subjects. These findings represent the associated chronic hyperactivity of the HPA axis with systemic inflammation commonly observed in depressive patients, such as the association between elevated hs-CRP and depressive symptomology reported by Raison et al. [135] in chronic fatigue patients. In vitro administration of TNF-α has been shown to disproportionately increase GRβ protein isoform over GRα in HeLaS3 and lymphoid cells. GRα binds hormones, translocates them to the nucleus, and regulates gene transcription, whereas GRβ does not bind known ligands and attenuates GRα action [181]. This change in expression between the two isoforms is predictive of glucocorticoid resistance development, which often characterizes HPA axis dysfunction in depression patients [181]. Concentrations of circulating TNF-α have been suggested to be particularly involved in depression severity scores and vasoconstriction responses (a marker of HPA axis function) in antidepressant-resistant depressives [182]. Examination of metastatic cancer patients has also determined associative increases in IL-6 and cortisol in those patients with depression [183].

Rodent studies have similarly demonstrated associative increases in HPA axis activity and depressive-like behavior following immune activation [170]. Similarly, paradigms involving neonatal immune activation have demonstrated increased propensities to develop stress-related behaviors alongside hypersecretion of corticosterone, elevated hippocampal cytokines, and altered GR abundance in adulthood [184–186]. Such animal studies have been useful in dissociating the relationship between the actions of cytokines and the HPA axis in the brain to provide a more direct account of how these factors may influence behavior. Raz Yirmiya's laboratory has been particularly influential in this regard, reporting brain IL-1β to be responsible for depressive-like behavior in relation to stress physiology. They have determined that psychological stress results in decreased hippocampal neurogenesis alongside significant increases in hippocampal IL-1β protein, depressive-like behavior, and ACTH and corticosterone release [169, 170]. The importance of IL-1 in producing these behavioral and HPA axis outcomes was confirmed both genetically using IL-1r knockout mice and pharmacologically using IL-1ra. In IL-1R1 knockout mice stress-induced brain, behavioral and endocrine perturbations observed in the wild-type mice were all attenuated [169, 170]. More importantly, these perturbations were abolished if wild-type mice were adrenalectomized [170]. These studies clearly established that a chronic stressor increases expression of

IL-1β protein in the brain. Genetic or pharmacological antagonism of IL-1 prevents development of depressive-like behaviors—all of which are linked through the HPA axis. One explanation posited for the association between HPA axis activation and depression onset is the neurodegenerative effects of glucocorticoids. Elevated corticosterone concentrations in rats have been shown to reduce neurogenesis in the hippocampus [187, 188], which coincides with increased depressive-like behavior in these animals. Antidepressant treatments increase hippocampal neurogenesis, and these progenitor cells develop and mature into neurons as opposed to glial cells [189]. Human studies have supported these effects in depression patients, whereby smaller hippocampal volumes were associated with poorer clinical outcomes [190]. Over a 3-year period, antidepressant treatment also increased hippocampal volume in these patients.

Some degree of inconsistency regarding the exact causal relationship between cytokine and neuroendocrine parameters in depression remains. For instance, Jehn and colleagues [191] found that comorbidity of cancer and depression was typified by positive correlations between IL-6 concentrations and depression severity, whereas negative correlations were observed between cortisol and severity of depression. The authors concluded that immune and endocrine involvements in depression are independent of one another. Further confusion arises in the face of the apparent paradox that exists in relation to inflammation-induced depression and HPA axis activity. Glucocorticoids have well-characterized anti-inflammatory properties. As such, an immunological stressor, such as infection or a mimetic such as LPS which works to activate the HPA axis, will result in downstream glucocorticoid release. This secretion of glucocorticoids should, in turn, attenuate the originating inflammation. However, this is often not the case when studying for patients of depression showing enhanced immune activation, and hence, additional processes must also be at play such as the proinflammatory nature of cytokines and glucocorticoid resistance (discussed below).

It is important to recognize that cytokine regulation of HPA axis function is complex and requires a multifaceted approach to studying such complexity. This is particularly so when considering the multitude of actions that cytokines can impart on GR function. Proinflammatory cytokines inhibit GR function via translocation, GR protein–protein interactions, induction of GR isoforms that reduce access to hormone, and GR binding to its DNA response element (see [192, 193]). Additionally, p38 mitogen-activated protein kinase (MAPK) modulation of GR transcriptional activity also appears to be a crucial intermediary for cytokine-induced inhibition of GR function [61]. The obvious potential for immune-focused therapies to reduce GR inhibition in depression patients has not gone unnoticed. As such, stimulation of nuclear extracts of cells with celecoxib (an anti-inflammatory agent that acts via COX-2 pathways) increases GR nuclear localization and binding to glucocorticoid response element via a reduction of phosphorylated p38 MAPK [194]. Given the number of ways through which immune parameters can alter GR activity, it is clear that simply measuring downstream outcomes such as cortisol is unlikely to entirely elucidate the relationship between the immune system and neuroendocrine axis in regard to inflammation-induced depression.

While a great deal of research has focused on cytokine mediation of the HPA axis in inflammation-induced depression, emerging evidence indicates that cytokines are also involved in regulation of the hypothalamic–pituitary–gonadal (HPG) axis. Gonadal hormones are potent behavioral mediators, with increased testosterone expression resulting in elevated aggression [195, 196]. A number of studies have reported associations between low testosterone and major depression [197], which has been demonstrated to improve following testosterone treatment [198]. The HPG axis interacts with both the neuroimmune system and the HPA axis. Many studies have demonstrated that LPS administration to rodents suppresses HPG function, with lower frequency rates of gonadotropin-releasing hormone (GnRH) pulses being described [199] as well as reduced luteinizing hormone (LH) concentrations [200]. The primary cytokines that mediated this suppression were IL-1β and TNF-α, which exerted their effects via cyclooxygenase formation of arachidonic acid and prostaglandins [199–201]. Additionally, female attractiveness and sexual behavior appear to be mediated by these cytokines during inflammation [202, 203], with particular involvement of IL-1 via the synthesis of prostaglandins [204]. Indeed, IL-1 and TNF-α appear to play the most potent role in HPG axis regulation of the cytokines, whereby they can exert effects on centrally released GnRH, as well as gonadal secretion of estradiol and Leydig steroidogenesis. Notably, gonadal production of IFNs also appears to act as autocrine and paracrine regulators (reviewed in [205]).

The HPA axis is also widely acknowledged to impact on HPG function and vice versa [206]. While HPA and cytokine pathways have been recognized to impact on HPG activity for some time [201], it is only recently that scientists have begun to question whether these pathways are as independent of one another as originally conceived. Matsuwaki et al. [207] recently determined that the production of glucocorticoids following TNF-α injection to rats counteracts the suppressive effects of cytokines on GnRH pulse generation. A number of studies have also demonstrated that neonatal immune activation (a critical period of HPA axis and HPG axis development) produces long-term alterations in reproductive fitness, suppresses gonadal hormones, and changes behavior in adulthood [208–210] with likely involvement of the HPA axis in regulating these outcomes. While elucidation of the precise relationships between neuroinflammation, the HPA axis, and the HPG axis remains in its infancy, such examination may become fruitful for investigating the pathogenesis of inflammation-induced depression. Sex differences are reliably observed in regard to the prevalence of depression [211], and one can speculate that within the cross talk of these systems may lie the answer to why depression is 1.68 times more prevalent among women than among men [212].

6.6.4 Kynurenine Pathway of Depression

A promising theory for the pathogenesis of inflammation-induced depression is the kynurenine pathway of tryptophan metabolism. Tryptophan, the least abundant of the essential amino acids, has three basic fates: (1) it is used for general protein synthesis, (2) it is used for serotonin synthesis, or (3) it is catabolized via the kynurenine pathway to produce biologically active metabolites. Tryptophan is the biochemical precursor and rate-limiting substrate for the synthesis of serotonin, which is why tryptophan metabolism has been an attractive avenue of investigation for subscribers to the monoamine deficiency hypothesis of depression. If tryptophan is excessively metabolized, the reservoir from which tryptophan hydroxylase is able to synthesize serotonin is reduced. The end result can be lower concentrations of brain serotonin. Indeed, low levels of circulating tryptophan have been consistently reported in patients suffering depression [213–217]. Three enzymes are responsible for the catabolism of tryptophan: tyrosine hydroxylase leads to the synthesis of serotonin, whereas IDO and tryptophan 2,3-dioxygenase (TDO) lead to the synthesis of kynurenine. Thus, low tryptophan availability for serotonin synthesis was originally explained by activation of TDO/IDO and was seen to fit comfortably within the monoamine deficiency hypothesis of depression.

In spite of the coincident increase in IDO activation and reduced serotonin concentrations in depression patients, there is evidence to indicate that the driving mechanism behind inflammation-induced depression in regard to tryptophan metabolism is unlikely to be caused by reductions in either tryptophan or serotonin. Instead, these behavioral changes may be caused by downstream neurotoxic metabolites produced from the kynurenine pathway. Kynurenine itself is biologically inactive but can be further metabolized into biologically active metabolites. These are produced from three arms of the pathway. Kynurenine aminotransferases can metabolize kynurenine into kynurenic acid (KA). This metabolite has been shown to be largely neuroprotective as it antagonizes *N*-methyl-D-aspartate (NMDA) and α-7 nicotinic acetylcholine receptors. Degradation of kynurenine into KA occurs in astrocytes, brain endothelial cells, and possibly neurons [218, 219]. Alternatively, kynurenine can be further metabolized into either anthranilic acid by kynurinase or further degraded into downstream quinolinic acid. Anthranilic acid and quinolinic acid, both of which generate oxidative radicals, are NMDA receptor agonists and can be neurotoxic. Microglia and macrophages are biased towards the arm of the kynurenine pathway that results in the production of anthranilic and quinolinic acids [218, 220]. Thus, the production of metabolites that serve either neuroprotective or neurotoxic roles is largely dependent on the cell type in which tryptophan is being metabolized. A recent study by Myint and colleagues [221] supports the notion that neurotoxic metabolites may be responsible for depressive symptoms. They observed plasma kynurenine to tryptophan ratios to be increased in depressed patients compared to controls. Depressed patients also demonstrated reduced KA to kynurenine ratio concentrations, suggestive of reduced neuroprotection in these individuals. A similar study has recently reported elevated plasma concentrations of kynurenine and lower concentrations of kynurenic acid in schizophrenic patients, which correspond to clinical symptomology [222]. Interestingly, after 6 weeks of antipsychotic treatment, these comparative concentrations began to normalize. A diagrammatic representation of the kynurenine pathway is presented in Fig. 6.3.

The fact that immune cells, such as microglia and macrophages, are biased towards producing the neurotoxic metabolites anthranilic and quinolinic acid represents promise for the area of inflammation-induced depression. Indeed, the relationship between the immune system and IDO is strong. IDO has both immunoregulatory and tolerogenic mechanisms. For instance, IDO activity produces tryptophan-depleted microenvironments that limit the proliferation of T cells and enhance their susceptibility to apoptosis [223], yet IDO is also important for limiting allogeneic rejection of the fetus during pregnancy [223–225]. Conversely, inflammation represents a potent activator of IDO primarily via the actions of the cytokines IFN-γ and TNF-α [41]. Hence, neuroinflammation in depressed individuals is likely to activate microglia and macrophages that, in turn, increase their intracellular IDO enzymatic activity and the release of these neurotoxic metabolites. A number of clinical studies have reported increased IDO activation and quinolinic acid to be associated with depressive symptoms and inflammatory markers. For instance, patients with malignant melanoma undergoing IFN-α treatment display increased plasma kynurenine to tryptophan ratios that are associated with elevated levels of neopterin (a marker of macrophage activation) [226]. Importantly, both duration and degree of elevation in these markers were significantly greater in antidepressant-free patients who developed major depression compared to those who failed to develop depression. Similarly, cancer and hepatitis C patients undergoing cytokine therapy have been shown to exhibit lower tryptophan to large neutral amino acid ratios and higher kynurenine concentrations which are predictive of depressive scores in these patients [227–229]. A recent study also determined that plasma kynurenine and neopterin were higher in patients with major depressive disorder who have attempted suicide as opposed to major depressive disorder patients in the absence of suicide attempts as well as healthy controls, indicative of the kynurenine pathway being implicated as a potential marker of severity of depression [230].

Animal studies have been instrumental in demonstrating that the onset of depressive-like behaviors following immune activation is reliant upon IDO activity. Intraperitoneal administration of BCG to mice results in prolonged increases in serum IFN-γ alongside elevated IDO activity in both the lung and brain [231]. These animals also show peripheral and central reductions in tryptophan. This model of BCG infection in mice has been effective in establishing that IDO activation following inflammation critically determines the behavioral outcome of these mice with depressive-like behaviors being measurable 7 days following the administration of BCG [41, 42]. BCG-treated mice also exhibit elevated levels of IDO mRNA and 3-hydroxyanthranilic acid (the precursor to quinolinic acid), indicative of the rise in neurotoxic metabolites during inflammation [42]. The causative effect of IDO in depressive-like behavior has been confirmed by genetically or pharmaceutically blocking IDO activity. Administration of the competitive antagonist of IDO 1-methyl tryptophan (1-MT) normalizes the kynurenine to tryptophan ratio peripherally and centrally and prevents the onset of depressive-like behavior [42]. Similarly, IDO-deficient mice develop no signs of depressive-like behavior following BCG infection despite obvious signs of sickness [42]. Cytokine signaling appears to be crucial for IDO activation for inflammation-induced depression, and most data suggest some cytokines are more influential than others in this regard. In particular, IFN-γ and TNF-α are critical activators of IDO. While IFN-γ, IL-1β, TNF-α, and IL-6 have all been shown to increase during inflammation in these models [41, 42, 231], it is IFN-γ receptor-deficient mice that fail to present with depressive-like behavior following BCG [41]. Similarly, the TNF-α antagonist, etanercept, has been shown to be capable of attenuating IDO activation and depressive-like behaviors [41]. BCG exposure to IDO-deficient mice has established that while these mice fail to present with depressive-like behavior, their cytokine profiles remain normal [42]. Thus, it would appear that the primary role of proinflammatory cytokines in inflammation-induced depression is to activate IDO metabolism of tryptophan.

Murine models have been used to confirm that inflammation-induced depressive-like behaviors are dependent upon IDO activation independently of changes in central serotonin turnover. These data stand in stark contrast to the notion that tryptophan metabolism is a means to which serotonin neurotransmission is depleted and thus is the cause for depressive symptomologies. Administration of LPS or BCG to mice increases the kynurenine to tryptophan ratio in circulation [43]. However, kynurenine and tryptophan behavior in the brain tells a different story. These mice display increased central kynurenine in the absence of any concurrent reduction in tryptophan. Moreover, brain tryptophan and serotonin turnover actually increase [43, 232]. Several human studies are supportive of these data. Wichers et al. [228] found that hepatitis C patients undergoing IFN-α therapy experience a concurrent rise in serum kynurenine to tryptophan ratios and depressive scores. However, the tryptophan to competing amino acids ratio did not differ, suggesting that tryptophan availability to the brain is not altered. This suggestion has since been confirmed by Raison et al. [229] who showed that hepatitis C patients undergoing IFN-α treatment demonstrate decreased circulating tryptophan concentrations but no change in cerebrospinal (CSF) tryptophan concentrations. Importantly, kynurenine and quinolinic acid were increased in the CSF of these patients that correlated with depressive inventory scores. Thus, it would appear that compensatory mechanisms in the brain work to ameliorate the fall in circulating tryptophan. The precise mechanisms responsible for this phenomenon remain to be elucidated, but one possible explanation is the co-induction of tryptophanyl-tRNA synthetase (WRS) with IDO following IFN-γ stimulation [233]. WRS catalyzes the attachment of tryptophan to its cognate transfer RNA molecule. This produces a tryptophan reservoir from which protein can be synthesized, protecting the cell against tryptophan starvation. For a review of the role of kynurenine pathway in inflammation-induced depression that can provide greater detail than is available to us here, see [234].

It is important to note that IDO is not the only enzyme responsible for the metabolism of tryptophan into kynurenine. As mentioned above, TDO which is responsible for the metabolism of dietary tryptophan was long thought to reside almost solely in the liver. Recently, it was determined that TDO is also present in other tissues, including the brain. For instance, an examination of TDO-deficient mice found that these mice exhibit elevated levels of brain tryptophan, serotonin, and hydroxyindoleacetic acid, which corresponded with reduced anxiety-related behavior [235]. TDO is strongly activated by cortisol, a hormone that is commonly dysregulated in affective disorders [175, 176]. Thus, the potential for TDO to play a role in inflammation-induced depression is likely, but this possibility requires confirmation.

In addition to TDO, another enzyme, indoleamine 2,3-dioxygenase 2 (IDO2), has been discovered [236]. The primary human IDO2 transcript is derived from 11 exons and encompasses a 74-kB region of chromosome 8p12.21 [236], which is encoded adjacent to IDO1 located at 8p12-p11. IDO2 is the more ancient of the two enzymes, and it catabolizes tryptophan with lower efficiency. The expression of IDO1 and IDO2 proteins differs in response to various stimuli, and they do not coexist in all cell types, which may indicate that they are not functionally redundant despite their 43% homology at the amino acid level. Clearly, much more research is required to determine the respective roles of IDO1, IDO2, and TDO in inflammation-induced depression before the precise cause for changes in kynurenine metabolites in inflammation-induced depression will be known.

Finally, it should be noted that the several glutamatergic pathways influenced by kynurenine and its metabolites are also affected by ethanol, and thus, yet another common pathway between alcoholism and depression emerges. Alcohol abuse increases neuroinflammation and proinflammatory cytokines in the brain and liver via TLR4 activation [29], increasing activity of IDO and TDO. For instance, chronic alcohol exposure to rats elevates hepatic TDO activity resulting in increased urinary kynurenine [237]. In humans, short-term alcohol exposure has been reported to decrease tryptophan concentration and lead to an elevation in kynurenine [238]. Hence, a person experiencing depression or a depressive episode and binge drinks may be likely to potentiate depressive symptoms due to the effects of short-term ethanol on kynurenine metabolites. One potential problem with this hypothesis resides in the opposing effects of ethanol and kynurenine metabolites on neurotransmitter function. For example, while kynurenine metabolites such as quinolinic acid serve as agonists of NMDA receptor activity and increase glutamatergic transmission, ethanol antagonizes NMDA receptors, decreases glutamatergic transmission, and elevates inhibitory GABA. While these downstream outcomes on neurotransmission appear contradictory to the commonality of comorbidity between alcoholism and depression, they may indeed hold the key for prolonged alcoholism in depressed individuals. Indeed, alcohol abuse may become a compensatory self-medicating mechanism for reducing the mood and cognitive effects of increased IDO activity in individuals with depression. Habituation to the effects of alcohol may lead to a spiraling effect of greater and greater alcohol abuse in an attempt to combat the effects of kynurenine metabolites on glutamatergic activity. While this possibility for the well-known comorbidity between alcoholism and depression appears reasonable, a great deal more research is required to confirm this proposition.

6.7 Summary and Conclusion

Understanding the relationship between substance abuse and depression is a complicated endeavor. The overlap in the manifestation of these disorders along with their behavioral and biological similarities is akin to trying to solve a Rubik's cube. Recent advancements in the area of brain, behavior, and immunity, however, have revealed one possible face of the puzzle. We now recognize that multiple neuroimmune mechanisms are dysregulated in the pathogenesis of both addiction and depression. Moreover, there is potential for these neuroimmune events to contribute to both disorders, leading to their comorbidity. Research into inflammation-induced sickness and depression is more developed than that of inflammation-induced substance abuse, and we have provided an overview of this more developed area in this chapter.

The impact of inflammation on emotionality, cognition, and behavior is clear. We can readily observe sickness behaviors such as anorexia, lethargy, and sexual decline in response to infection and now have the capacity with animal models to differentiate between sickness and the subsequent development of similar depression-like behaviors. Furthermore, neuroinflammation is believed to cause the onset of depressive-like behaviors via several putative causal mechanisms. Inflammation-induced neuroendocrine perturbations have been closely linked to depression most likely via the development of glucocorticoid resistance. Cytokines are now accepted as playing a strong role in the regulation of behavior, signaling the onset of sickness behavior, activating the HPA axis, and stimulating microglial IDO activity. Tryptophan metabolism down the kynurenine pathway is a promising avenue of research. Recent experiments have demonstrated immune-regulated production of neurotoxic metabolites and oxidative damage using both pharmacologic and genetic models that have implicated IDO as a necessary component in the pathogenesis of inflammation-induced depression. Finally, while we have argued against the dominance of the monoamine deficiency hypothesis in this chapter, this does not necessitate that this pathway has no contribution to the onset of inflammation-induced depression. Indeed, serotoninergic pathways either affect or are affected by each of the mechanisms discussed, as is the case for all of these pathways. It is likely that multiple mechanisms intersect with all the symptoms of depression and contribute to its multifactorial pathogenesis—although this argument becomes facile without any attempt to systematically understand the nature of their interaction. The challenge that remains is to differentiate between the relative influences of some mechanisms over other potentially spurious secondary mediators. Such a systems-based approach will yield greater insight into inflammation-induced depression and provide a firm foundation from which to explore its comorbidity with substance abuse in the broader context of inflammation.

Acknowledgments This work was supported by National Institutes of Health grant RO1 AG029573 to Keith W. Kelley.

References

1. Hasin DS, Endicott J, Lewis C (1985) Alcohol and drug abuse in patients with affective syndromes. Compr Psychiatry 26:283–295
2. Regier DA, Farmer ME, Rae DS, Locke BZ, Keith SJ, Judd LL, Goodwin FK (1990) Comorbidity of mental disorders with alcohol and other drug abuse: results from the Epidemiological Catchment Area (ECA) Study. JAMA 264:2511–2518

3. Boschloo L, Vogelzangs N, Smit JH, van den Brink W, Veltman DJ, Beekman ATF, Penninx BWJH (2011) Comorbidity and risk indicators for alcohol use disorders among persons with anxiety and/or depressive disorders: findings from the Netherlands Study of Depression and Anxiety (NESDA). J Affect Disord 131:233–242
4. Buysse DJ, Angst J, Gamma A, Ajdacic V, Eich D, Rössler W (2008) Prevalence, course, and comorbidity of insomnia and depression in young adults. Sleep 31:473–480
5. Jim HSL, Small B, Faul LA, Franzen J, Apte S, Jacobsen PB (2011) Fatigue, depression, sleep, and activity during chemotherapy: daily and intraday variation and relationships among symptom changes. Ann Behav Med 42:321–333. doi:10.1007/s12160-011-9294-9
6. Rodin J, McAvay G, Timko C (1988) A longitudinal study of depressed mood and sleep disturbances in elderly adults. J Gerontol 43:P45–P53
7. Sinha R, Robinson J, Merikangas K, Wilson GT, Rodin J, O'Malley S (1996) Eating pathology among women with alcoholism and/or anxiety disorders. Alcohol Clin Exp Res 20:1184–1191
8. Swendsen JD, Merikangas KR (2000) The comorbidity of depression and substance use disorders. Clin Psychol Rev 20:173–189
9. Boden JM, Fergusson DM (2011) Alcohol and depression. Addiction 106(5):906–914
10. Kessler RC, Price RH (1993) Primary prevention of secondary disorders: a proposal and agenda. Am J Comm Psychol 21:607–633
11. Fergusson DM, Boden JM, Horwood LJ (2009) Tests of causal links between alcohol abuse or dependence and major depression. Arch Gen Psychiatry 66:260–266
12. Paljarvi T, Koskenvuo M, Poikolainen K, Kauhanen J, Sillinmaki L, Makela P (2009) Binge drinking and depressive symptoms: a 5 year population-based cohort study. Addiction 104:1168–1178
13. Nurnberger JI, Foroud T, Flury L, Su J, Meyer ET, Hu K, Crowe R, Edenberg H, Goate A, Bierut L, Reich T, Schuckit M, Reich W (2001) Evidence for a locus on chromosome 1 that influences vulnerability to alcoholism and affective disorder. Am J Psychiatry 158:718–724
14. Fu Q, Heath AC, Bucholz KK, Nelson E, Goldberg J, Lyons MJ, True WR, Jacob T, Tsuang MT, Eisen SA (2002) Shared genetic risk of major depression, alcohol dependence, and marijuana dependence. Arch Gen Psychiatry 59:1125–1132
15. Kuo PH, Neale MC, Walsh D, Patterson DG, Riley B, Prescott CA, Kendler KS (2010) Genome-wide linkage scans for major depression in individuals with alcohol dependence. J Psychiatr Res 44:616–619
16. Kendler KS, Kessler RC, Walters EE, MacLean CJ, Sham PC, Neale MC, Heath AC, Eaves LJ (1995) Stressful life events, genetic liability and onset of an episode of major depression in women. Am J Psychiatry 152:833–842
17. Fava M, Davidson KG (1996) Definition and epidemiology of treatment-resistant depression. Psychiatr Clin North Am 19:179–200
18. Lesch KP (2004) Gene–environment interaction and the genetics of depression. J Psychiatry Neurosci 29:174–184
19. Agrawal A, Scherrer JF, Lynskey MT, Sartor CE, Grant JD, Autry AE, Adachi M, Nosyreva E, Na ES, Los MF, Cheng PF, Kavalali ET (2011) NMDA receptor blockade at rest triggers rapid behavioural antidepressant responses. Nature 475:91–97
20. Miguel-Hidalgo JJ, Nithuairisg S, Stockmeier C, Rajkowska G (2007) Distribution of ICAM-1 immunoreactivity during aging in the human orbitofrontal cortex. Brain Behav Immun 21:100–111
21. Haddad JJ (2004) Alcoholism and neuro-immune-endocrine interactions: physiochemical aspects. Biochem Biophys Res Commun 323:361–371
22. Crews FT, Zou J, Qin L (2011) Induction of innate immune genes in brain create the neurobiology of addiction. Brain Behav Immun 25:S4–S12
23. Blednov YA, Benavidez JM, Geil C, Perra S, Morikawa H, Harris RA (2011) Activation of inflammatory signaling by lipopolysaccharide produces a prolonged increase of voluntary alcohol intake in mice. Brain Behav Immun 25(Suppl 1):S92–S105

24. Stevenson JR, Schroeder JP, Nixon K, Besheer J, Crews FT, Hodge CW (2009) Abstinence following alcohol drinking produces depression-like behavior and reduced hippocampal neurogenesis in mice. Neuropsychopharmacology 34:1209–1222
25. Capuron L, Miller AH (2004) Cytokines and psychopathology: lessons from interferon-alpha. Biol Psychiatry 56:819–824
26. Capuron L, Pagnoni G, Lawson DH, Demetrashvili M, Woolwine BJ, Kilts CD, Bremner JD, Nemeroff CB, Miller AH (2002) Altered fronto-pallidal activity during high-dose interferon-alpha treatment as determined by positron emission tomography. Abstr Soc Neurosci 498:5
27. Haas HS, Schaenstein K (1997) Neuroimmunomodulation via limbic structures: the neuro-anatomy of psychoimmunology. Prog Neurobiol 51:195–222
28. Miller AH, Maletic V, Raison CL (2009) Inflammation and its discontents: the role of cytokines in the pathophysiology of major depression. Biol Psychiatry 65:732–741
29. Alfonso-Loeches S, Pascual-Lucas M, Blanco AM, Sanchez-Vera I, Guerri C (2010) Pivotal role of TLR4 receptors in alcohol-induced neuroinflammation and brain damage. J Neurosci 30:8285–8295
30. Lu YC, Yeh WC, Ohashi PS (2008) LPS/TLR4 signal transduction pathway. Cytokine 42:145–151
31. Irwin MR, Olmstead R, Valladares EM, Crabb Breen E, Ehlers CL (2009) Tumor necrosis factor antagonism normalizes rapid eye movement sleep in alcohol dependence. Biol Psychiatry 66:191–195
32. Ford DE, Kamerow DB (1989) Epidemiologic study of sleep disturbances and psychiatric disorders: an opportunity for prevention? JAMA 262:1479–1484
33. Clark CP, Gillin JC, Golshan S, Demodena A, Smith TL, Danowski S, Irwin M, Schuckit M (1998) Increased REM sleep density at admission predicts relapse by three months in primary alcoholics with a lifetime diagnosis of secondary depression. Biol Psychiatry 43:601–607
34. Opp MR, Toth LA (2003) Neural-immune interactions in the regulation of sleep. Front Biosci 8:d768–d779
35. Dantzer R, Kelley KW (1989) Stress and immunity: an integrated view of relationships between the brain and the immune system. Life Sci 44:1995–2008
36. Bluthé RM, Michaud B, Poli V, Dantzer R (2000) Role of IL-6 in cytokine-induced sickness behavior: a study with IL-6 deficient mice. Physiol Behav 70:367–373
37. Opp MR, Krueger JM (1991) Interleukin 1-receptor antagonist blocks interleukin 1-induced sleep and fever. Am J Physiol 260:R453–R457
38. Palin K, Bluthé RM, McCusker RH, Moos F, Dantzer R, Kelley KW (2007) TNF alpha-induced sickness behavior in mice with functional 55 kD TNF receptors is blocked by central IGF-I. J Neuroimmunol 187:55–60
39. Yirmiya R (1996) Endotoxin produces a depressive-like episode in rats. Brain Res 711:163–174
40. Dantzer R, Kelley KW (2007) Twenty years of research on cytokine-induced sickness behavior. Brain Behav Immun 21:153–160
41. O'Connor JC, André C, Wang Y, Lawson MA, Szegedi SS, Lestage J, Castanon N, Kelley KW, Dantzer R (2009) Interferon-γ and tumor necrosis factor-α mediate the upregulation of indoleamine 2,3-dioxygenase and the induction of depressive-like behavior in mice in response to Bacillus Calmette-Guerin. J Neurosci 29:4200–4209
42. O'Connor JC, Lawson MA, André C, Briley EM, Szegedi SS, Lestage J, Castanon N, Herkenham M, Dantzer R, Kelley KW (2009) Induction of IDO by Bacilli Calmette-Guerin is responsible for development of murine depressive-like behavior. J Immunol 182: 3202–3212
43. O'Connor JC, Lawson MA, André C, Moreau M, Lestage J, Castanon N, Kelley KW, Dantzer R (2009) Lipopolysaccharide-induced depressive-like behavior is mediated by indoleamine 2,3-dioxygenase activation in mice. Mol Psychiatry 14:511–522
44. Frenois F, Moreau M, O'Connor J, Lawson M, Micon C, Lestage J, Kelley KW, Dantzer R, Castanon N (2007) Lipopolysaccharide induces delayed Fos B/Delta Fos B immunostaining within the mouse extended amygdala, hippocampus and hypothalamus, that parallel the expression of depressive-like behavior. Psychoneuroendocrinology 32:516–531

45. Henry CJ, Huang Y, Wynne A, Hanke M, Himler J, Bailey MT, Sheridan JF, Godbout JP (2008) Minocycline attenuates lipopolysaccharide (LPS)-induced neuroinflammation, sickness behavior, and anhedonia. J Neuroinflammation 5:15
46. Kelley KW, Aubert A, Dantzer R (2011) Inflammation and behavior. In: Demas GE, Nelson RJ (eds) Eco-immunology and behavior. Oxford University Press, Oxford
47. Kelley KW, Dantzer R (2011) Alcoholism and inflammation: neuroimmunology of behavioral and mood disorders. Brain Behav Immun 25:S13–S20. doi:10.1016/j.bbi.2010.12.013
48. Bluthé RM, Walter V, Parnet P, Layé S, Lestage J, Verrier D, Poole S, Stenning BE, Kelley KW, Dantzer R (1994) Lipopolysaccharide induces sickness behaviour in rats by a vagal mediated mechanism. C R Acad Sci III 317:499–503
49. Watkins LR, Weirtelak EP, Goehler LE, Smith KP, Martin D, Maier SF (1994) Characterization of cytokine induced hyperalgesia. Brain Res 654:15–26
50. Watkins LR, Maier SF, Goehler LE (1995) Cytokine-to-brain communication: a review and analysis of alternative mechanisms. Life Sci 57:1011–1026
51. Wan W, Janz L, Vriend CY, Sorensen CM, Greenberg AH, Nance DM (1993) Differential induction of c-Fos immunoreactivity in hypothalamus and brain stem nuclei following central and peripheral administration of endotoxin. Brain Res Bull 32:581–587
52. Elmquist JK, Saper CB (1996) Activation of neurons projecting to the paraventricular hypothalamic nucleus by intravenous lipopolysaccharide. J Comp Neurol 374:315–331
53. Blatteis CM, Bealer SL, Hunter WS, Llanos-Q J, Ahokas RA, Mashburn TA Jr (1983) Suppression of fever after lesions of the anteroventral third ventricle in guinea pigs. Brain Res Bull 11:519–526
54. Quan N, Banks WA (2007) Brain-immune communications pathways. Brain Behav Immun 21:727–735
55. Banks WA, Kastin AK (1991) Blood to brain transport of interleukin links the immune and central nervous systems. Life Sci 48:PL117–PL121
56. Gutierrez EG, Banks WA, Kastin AJ (1993) Murine tumor necrosis factor alpha is transported from blood to brain in the mouse. J Neuroimmunol 47:169–176
57. Blatteis CM (2000) The afferent signalling of fever. J Physiol 526(Pt 3):470
58. Dantzer R (2004) Cytokine-induced sickness behaviour: a neuroimmune response to activation of innate immunity. Eur J Pharmacol 500:399–411
59. Konsman JP, Parnet P, Dantzer R (2002) Cytokine induced sickness behaviour: mechanisms and implications. Trends Neurosci 25:154–159
60. Tancredi V, D'Antuono M, Café C, Giovedì S, Bué MC, D'Arcangelo G, Onofri F, Benfenati F (2000) The inhibitory effects of interleukin-6 on synaptic plasticity in the rat hippocampus are associated with an inhibition of mitogen-activated protein kinase ERK. J Neurochem 75:634–643
61. Wang X, Wu H, Miller AH (2004) Interleukin 1a (IL-1a) induced activation of p38 mitogen-activated protein kinase inhibits glucocorticoid receptor function. Mol Psychiatry 9:65–75
62. Lynch MA (2004) Long-term potentiation and memory. Physiol Rev 84:87–136
63. Griffin R, Nally R, Nolan Y, McCartney Y, Linden J, Lynch MA (2006) The age-related attenuation in long-term potentiation is associated with microglial activation. J Neurochem 99(4):1263–1272
64. Gemma C, Bickford PC (2007) Interleukin-1beta and caspase-1: players in the regulation of age-related cognitive dysfunction. Rev Neurosci 18(2):137–148
65. Costello DA, Watson MB, Cowley TR, Murphy N, Murphy Royal C, Garlanda C, Lynch MA (2011) Interleukin-1alpha and HMGB1 mediate hippocampal dysfunction in SIGIRR-deficient mice. J Neurosci 31:3871–3879
66. Takeda K, Akira S (2005) Toll-like receptors in innate immunity. Int Immunol 17:1–14
67. Farrar WL, Kilian PL, Ruff MR, Hill JM, Pert CB (1987) Visualization and characterization of interleukin 1 receptors in brain. J Immunol 139:459–463
68. Ericsson A, Liu C, Hart RP, Sawchenko PE (1995) Type 1 interleukin-1 receptor in the rat brain: distribution, regulation, and relationship to sites of IL-1-induced cellular activation. J Comp Neurol 361:681–698

69. de Pablos RM, Villarán RF, Argüelles S, Herrera AJ, Venero JL, Ayala A, Cano J, Machado A (2006) Stress increases vulnerability to inflammation in the rat prefrontal cortex. J Neurosci 26:5709–5719
70. Dantzer R (2001) Cytokine-induced sickness behavior: where do we stand? Brain Behav Immun 15:7–24
71. Besedovsky H, del Rey A, Sorkin E, Dinarello CA (1986) Immunoregulatory feedback between interleukin-1 and glucocorticoid hormones. Science 233:652–654
72. Berkenbosch F, van Oers J, del Rey A, Tilders F, Besedovsky H (1987) Corticotropin-releasing factor-producing neurons in the rat activated by interleukin-1. Science 238:524–526
73. Sapolsky R, Rivier C, Yamamoto G, Plotsky P, Vale W (1987) Interleukin-1 stimulates the secretion of hypothalamic corticotropin-releasing factor. Science 238:522
74. Dalton SO, Laursen TM, Ross L, Mortensen PB, Johansen C (2009) Risk for hospitalization with depression after a cancer diagnosis: a nationwide, population-based study of cancer patients in Denmark from 1973 to 2003. J Clin Oncol 27:1440–1445
75. Musselman DL, Lawson DH, Gumnick JF, Manatunga AK, Penna S, Goodkin RS, Greiner K, Nemeroff CB, Miller AH (2001) Paroxetine for the prevention of depression induced by high-dose interferon alfa. N Engl J Med 344:961–966
76. Schäfer A, Scheurlen M, Seufert J, Keicher C, Weiss brich B, Rieger P, Kraus MR (2010) Platelet serotonin (5-HT) levels in interferon-treated patients with hepatitis C and its possible association with interferon-induced depression. J Hepatol 52:10–15
77. Maes M, Lambrechts J, Bosmans E, Jacobs J, Suy E, Vandervorst C (1992) Evidence for a systemic immune activation during depression: results of leukocyte enumeration by flow cytometry in conjunction with monoclonal antibody staining. Psychol Med 22:45–53
78. Capuron L (1999) Prediction of the depressive effects of interferon alfa therapy by the patient's initial affective state. N Engl J Med 340:1370
79. Capuron L, Gumnick JF, Musselman DL, Lawson DH, Reemsnyder A, Nemeroff CB, Miller AH (2002) Neurobehavioral effects of interferon-α in cancer patients: phenomenology and paroxetine responsiveness of symptom dimensions. Neuropsychopharmacology 26:643–652
80. Prather AA, Rabinovitz M, Pollock BG, Lotrich FE (2009) Cytokine-induced depression during IFN-alpha treatment: the role of IL-6 and sleep quality. Brain Behav Immun 23:1109–1116
81. Capuron L, Pagnoni G, Demetrashvili M, Woolwine BJ, Nemeroff CB, Berns GS, Miller AH (2005) Anterior cingulate activation and error processing during interferon alpha treatment. Biol Psychiatry 58:190–196
82. Marks DH, Adineh M, Wang B, Gupta S (2007) Use of fMRI to predict psychiatric adverse effects of interferon treatment for Hepatitis C – preliminary report. Neuropsychiatr Dis Treat 3:655–667
83. Goldney RD, Fischer LJ, Phillips PJ, Wilson DH (2004) Diabetes, depression, and quality of life: a population study. Diabetes Care 17:1066–1070
84. Anderson RJ, Clouse RE, Freedland KE, Lustman PJ (2001) The prevalence of comorbid depression in adults with diabetes. Diabetes Care 24:1069–1078
85. Pickup JC, Chusney GD, Thomas SM, Burt D (2000) Plasma interleukin-6, tumour necrosis factor α and blood cytokine production in type 2 diabetes. Life Sci 67:291–300
86. Pasic J, Levy WC, Sullivan MD (2003) Cytokines in depression and heart failure. Psychosom Med 65:181–193
87. Alesci S, Martinez PE, Kelkar S, Ilias I, Ronsaville DS, Listwak SJ, Ayala AR, Licinio J, Gold HK, Kling MA, Chrousos GP, Gold PW (2005) Major depression is associated with significant diurnal elevations in plasma IL-6 levels, a shift of its circadian rhythm, and loss of physiologic complexity in its secretion: clinical implications. J Clin Endocrinol Metab 90:2522–2530
88. Frommberger UH, Bauer J, Haselbauer P, Fraulin A, Riemann D, Berger M (1997) Interleukin-6-(IL-6) plasma levels in depression and schizophrenia: comparison between the acute state and after remission. Eur Arch Psychiatry Clin Neurosci 247:228–233

89. Maes M, Lin A, Delmeire L, Van Gastel A, Kenis G, De Jongh R, Bosmons E (1999) Elevated serum interleukin-6 (IL-6) and IL-6 receptor concentrations in posttraumatic stress disorder following accidental man-made traumatic events. Biol Psychiatry 45:833–839
90. Schlatter J, Ortuño F, Cervera-Enguix S (2004) Monocytic parameters in patients with dysthymia versus major depression. J Affect Disord 78:243–247
91. Zorrilla EP, Luborsky L, McKay JR, Rosenthal R, Houldin A, Tax A, McCorkle R, Seligman DA, Schmidt K (2001) The relationship of depression and stressors to immunological assays: a meta-analytic review. Brain Behav Immun 15(3):199–226
92. Sluzewska A, Rybakowski JK, Laciak M, Mackiewicz A, Sobieska M, Wiktorowicv K (1995) Interleukin-6 serum levels in depressed patients before and after treatment with fluoxetine. Ann N Y Acad Sci 762:474–476
93. Basterzi AD, Aydemir C, Kisa C, Aksaray C, Tuzer V, Yazici K, Goka E (2005) IL-6 levels decrease with SSRI treatment in patients with major depression. Hum Psychopharmacol 20:473–476
94. O'Brien SM, Scott LV, Dinan TG (2006) Antidepressant therapy and C-reactive protein levels. Br J Psychiatry 188:449–452
95. O'Brien SM, Scully P, Fitzgerald P, Scott LV, Dinan TG (2007) Plasma cytokine profiles in depressed patients who fail to respond to selective serotonin reuptake inhibitor therapy. J Psychiatr Res 41:326–331
96. Hannestad J, DellaGioia N, Bloch M (2011) The effect of antidepressant medication treatment on serum levels of inflammatory cytokines: a meta-analysis. Neuropsychopharmacology 36:2452–2459. doi:10.1038/npp.2011.132
97. Penninx BWJH, Kritchevsky SB, Yaffe K, Newman AB, Simonsick EM, Rubin S, Ferrucci L, Harris T, Pahor M (2003) Inflammatory markers and depressed mood in older persons: results from the Health, Aging and Body Composition Study. Biol Psychiatry 54:566–572
98. Harrison NA, Brydon L, Walker C, Gray MA, Steptoe A, Critchley HD (2009) Inflammation causes mood changes through alterations in subgenual cingulate activity and mesolimbic connectivity. Biol Psychiatry 66:407–414
99. Reichenberg A, Yirmiya R, Schuld A, Kraus T, Haack M, Morag A, Pollmächer T (2001) Cytokine-associated emotional and cognitive disturbances in humans. Arch Gen Psychiatry 58:445–452
100. Strike PC, Wardle J, Steptoe A (2004) Mild acute inflammatory stimulation induces transient negative mood. J Psychosom Res 57:189–194
101. Wright CE, Strike PC, Brydon L, Steptoe A (2005) Acute inflammation and negative mood: mediation by cytokine activation. Brain Behav Immun 19:345–350
102. Stewart JC, Rand KL, Muldoon MF, Kamarck TW (2009) A prospective evaluation of the directionality of the depression-inflammation relationship. Brain Behav Immun 23:936–944
103. Copeland WE, Shanahan L, Worthman C, Angold A, Costello EJ (2012) Cumulative depression episodes predict later C-reactive protein levels: a prospective analysis. Biol Psychiatry 71:15–21
104. Loftis JM, Wall JM, Pagel RL, Hauser P (2006) Administration of pegylated interferon-α-2a or -2b does not induce sickness behavior in Lewis rats. Psychoneuroendocrinology 31:1289–1294
105. Hori T, Nakashima T, Take S, Kaizuka Y, Mori T, Katafuchi T (1991) Immune cytokines and regulation of body temperature, food intake and cellular immunity. Brain Res Bull 27:309–313
106. Thompson CM, Proctor DM, Suh M, Haws LC, Hebert CD, Mann JF, Shertzer HG, Hixon JG, Harris MA (2012) Comparison of the effects of hexavalent chromium in the alimentary canal of F344 and B6C3F1 mice following exposure in drinking water: implications for carcinogenic models of action. Toxicol Sci 125:79–90
107. Suzuki M, Fujii E, Kato C, Yamazaki M, Adachi K, Sugimoto T, Doi K (2008) Differences in bone responses to recombinant human granulocyte colony-stimulating factor between mice and rats. J Toxicol Sci 33:245–249

108. Sammut S, Goodall G, Muscat R (2001) Acute interferon-α administration modulates sucrose consumption in the rat. Psychoneuroendocrinology 26:261–272
109. Dunn AJ, Swiergel AH (2005) Effects of interleukin-1 and endotoxin in the forced swim and tail suspension tests in mice. Pharmacol Biochem Behav 81:688–693
110. Connor TJ, Song C, Leonard BE, Merali Z, Anisman H (1998) An assessment of the effects of central interleukin-1β, -2, -6, and tumor necrosis factor-α administration on some behavioural, neurochemical, endocrine and immune parameters in the rat. Neuroscience 84:923–933
111. Huang Y, Henry CJ, Dantzer R, Johnson RW, Godbout JP (2008) Exaggerated sickness behavior and brain proinflammatory cytokine expression in aged mice in response to intracerebroventricular lipopolysaccharide. Neurobiol Aging 29:1744–1753
112. Palin K, McCusker RH, Strle K, Moos F, Dantzer R, Kelley KW (2008) Tumor necrosis factor-α-induced sickness behavior is impaired by central administration of an inhibitor of c-jun N-terminal kinase. Psychopharmacology 197:629–635
113. Kent S, Bluthé RM, Kellet KW, Dantzer R (1992) Sickness behavior as a new target for drug development. Trends Pharmacol Sci 13:24–28
114. Kent S, Bluthé RM, Dantzer R, Hardwick AJ, Kelley KW, Rothwell NJ, Vannice JL (1992) Different receptor mechanisms mediate the pyrogenic and behavioral effects of interleukin-1. Proc Natl Acad Sci USA 89:9117–9120
115. Kent S, Kelley KW, Dantzer R (1992) Effects of lipopolysaccharide on food-motivated behavior in the rat are not blocked by an interleukin-1 receptor antagonist. Neurosci Lett 145:83–86
116. Burgess W, Gheusi G, Yao J, Johnson RW, Dantzer R, Kelley KW (1998) Interleukin-1β-converting enzyme-deficient mice resist central but not systemic endotoxin-induced anorexia. Am J Physiol 274:R1829–R1833
117. Chourbaji S, Urani A, Inta I, Sanchis-Segura C, Brandwein C, Zink M, Schwaninger M, Gass P (2006) IL-6 knockout mice exhibit resistance to stress-induced development of depression-like behaviors. Neurobiol Dis 23:587–594
118. Swiergiel AH, Dunn AJ (2006) Feeding, exploratory, anxiety- and depression-related behaviors are not altered in interleukin-6-deficient male mice. Behav Brain Res 171(1):94–108
119. von Meyenburg B, Hrupka H, Arsenijevic D, Schwartz GJ, Landmann R, Langhans W (2004) Role for CD14, TLR2, and TLR4 in bacterial product-induced anorexia. Am J Physiol Regul Integr Comp Physiol 287:R298–R305
120. Palin K, Bluthe RM, McCusker RH, Levade T, Moos F, Dantzer R, Kelley KW (2009) The type 1 TNF receptor and its associated adapter protein, FAN, are required for TNF alpha-induced sickness behavior. Psychopharmacology 201:549–556
121. Mathew SJ, Charney DS (2009) Publication bias and the efficacy of antidepressants. Am J Psychiatry 166:140–145
122. Lacasse JR, Leo J (2005) Serotonin and depression: a disconnect between the advertisements and the scientific literature. PLoS Med 2:e392
123. Hindmarch I (2001) Expanding the horizons of depression: beyond the monoamine hypothesis. Hum Psychopharmacol 16:203–218
124. Shimizu E, Hashimoto K, Okamura N, Koike K, Komatsu N, Kumakiri C, Nakazato M, Watanabe H, Shinoda N, Okada S, Iyo M (2003) Alterations of serum levels of brain-derived neurotrophic factor (BDNF) in depressed patients with or without antidepressants. Biol Psychiatry 54:70–75
125. Hensler JG, Ladenheim EE, Lyons WE (2003) Ethanol consumption and serotonin-1A (5-HT1A) receptor function in heterozygous BDNF(+/−) mice. J Neurochem 85:1139–1147
126. Autry AE, Adachi M, Nosyreva E, Na ES, Los MF, Cheng PF, Kavalali ET, Monteggia LM (2011) NMDA receptor blockade at rest triggers rapid behavioural antidepressant responses. Nature 15:91–95
127. Berman RM, Cappiello A, Anand A, Oren DA, Heninger GR, Charney DS, Krystal JH (2000) Antidepressant effects of ketamine in depressed patients. Biol Psychiatry 47:351–354

128. Maj J, Rogoz Z, Skuza G, Sowinska H (1992) Effects of MK-801 and antidepressant drugs in the forced swimming test in rats. Eur Neuropsychopharmacol 2:37–41
129. Salomon RM, Miller HL, Krystal JH, Heninger GR, Charney DS (1997) Lack of behavioral effects of monoamine depletion in healthy subjects. Biol Psychiatry 41:58–64
130. Heninger G, Delgado P, Charney D (1996) The revised monoamine theory of depression: a modulatory role for monoamines, based on new findings from monoamine depletion experiments in humans. Pharmacopsychiatry 29:2–11
131. Delgado PL, Miller HL, Salomon RM, Licinio J, Krystal JH, Moreno FA, Heninger GR, Charney DS (1999) Tryptophan-depletion challenge in depressed patients treated with desipramine or fluoxetine: implications for the role of serotonin in the mechanism of antidepressant action. Biol Psychiatry 46:212–220
132. Berman RM, Sanacora G, Anand A, Roach LM, Fasula MK, Finkelstein CO, Wachen RM, Oren DA, Heninger GR, Charney DS (2002) Monoamine depletion in unmedicated depressed subjects. Biol Psychiatry 51:469–473
133. Nestler EJ (1998) Antidepressant treatments in the 21st Century. Biol Psychiatry 44:526–533
134. Den Boer JA, Bosker FJ, Slaap BR (2000) Serotonergic drugs in the treatment of depressive and anxiety disorders. Hum Psychopharmacol 15:315–336
135. Raison CL, Lin JMS, Reeves WC (2009) Association of peripheral inflammatory markers with chronic fatigue in a population-based sample. Brain Behav Immun 23:327–337
136. Bonaccorso S, Marino V, Biondi M, Grimaldi F, Ippoliti F, Maes M (2002) Depression induced by treatment with interferon-alpha in patients affected by hepatitis C virus. J Affect Disord 72:237–241
137. Blackburn-Munro G, Blackburn-Munro RE (2001) Chronic pain, chronic stress and depression: coincidence or consequence? J Neuroendocrinol 13:1009–1023
138. Jarskog FL, Xiao H, Wilkie MB, Lauder JM, Gilmore JH (1997) Cytokine regulation of embryonic rat dopamine and serotonin neuronal survival in vitro. Int J Dev Neurosci 15:711–716
139. Zhu CB, Blakely RD, Hewlett WA (2006) The proinflammatory cytokines interleukin-1beta and tumor necrosis factor-alpha activate serotonin transporters. Neuropsychopharmacology 31:2121–2131
140. Zalcman S, Green-Johnson JM, Murray L, Nance DM, Dyck D, Anismanc H, Greenberg AH (1994) Cytokine-specific central monoamine alterations induced by interleukin-1, -2 and -6. Brain Res 643:40–49
141. Pae CU, Marks DM, Han C, Patkar AA (2008) Does minocycline have antidepressant effect? Biomed Pharmacother 62:308–311
142. Fuchs D, Weiss G, Wachter H (1993) Neopterin, biochemistry and clinical use as a marker of cellular immune reaction. Int Arch Allergy Immunol 101:1–6
143. Maes M, Meltzer HY, Scharpe S, Bosmans E, Suy E, De Meester I, Calabrese J, Cosyns P (1993) Relationships between lower plasma L-tryptophan level and immune-inflammatory variables in depression. Psychiatry Res 49:151–165
144. Maes M, Meltzer HY, Bosmans E, Bergmans R, Vandoolaeghe E, Ranjan R, Desnyder R (1995) Increased plasma concentrations of interleukin-6, soluble interleukin-6, soluble interleukin-2 and transferrin receptor in major depression. J Affect Disord 34:301–309
145. Häusler KG, Prinz M, Nolte C, Weber JR, Schumann RR, Kettenmann H, Hanisch UK (2002) Interferon-γ differentially modulates the release of cytokines and chemokines in lipopolysaccharide- and pneumococcal cell wall-stimulated mouse microglia and macrophages. Eur J Neurosci 16:2113–2122
146. Arkins S, Rebeiz N, Brunke-Reese DL, Biragyn A, Kelley KW (1995) Interferon-gamma inhibits macrophage insulin-like growth factor-I synthesis at the transcriptional level. Mol Endocrinol 9:350–360
147. Dunn AJ (2000) Cytokine activation of the HPA axis. Ann N Y Acad Sci 917:608–617

148. Silverman MN, Miller AH, Biron CA, Pearce BD (2004) Characterization of an interleukin-6 and adrenocorticotropin-dependent, immune-to-adrenal pathway during viral infection. Endocrinology 145:3580–3589
149. Sullivan GM, Canfield SM, Lederman S, Xiao E, Ferin M, Wardlaw SL (1997) Intracerebroventricular injection of interleukin-1 suppresses peripheral lymphocyte function in the primate. Neuroimmunomodulation 4:12–18
150. Watanabe T, Morimoto A, Murakami N (1991) ACTH response in rats during biphasic fever induced by interleukin-1. Am J Physiol 261:R11014–R11108
151. Krishnan V, Nestler EJ (2010) Linking molecules to mood: new insight into the biology of depression. Am J Psychiatry 167:1305–1320
152. Chesnokova V, Melmed S (2002) Minireview: neuro-immuno-endocrine modulation of the hypothalamic-pituitary-adrenal (HPA) axis by gp130 signaling molecules. Endocrinology 143:1571–1574
153. Guan Z, Fang J (2006) Peripheral immune activation by lipopolysaccharide decreases neurotrophins in the cortex and hippocampus in rats. Brain Behav Immun 20:64–71
154. Kelley KW, Weigent DA, Kooijman R (2007) Protein hormones and immunity. Brain Behav Immun 21:384–392
155. Park SE, Lawson M, Dantzer R, Kelley KW, McCusker RH (2011) Insulin-like growth factor-I peptides act centrally to decrease depression-like behavior of mice treated intraperitoneally with lipopolysaccharide. J Neuroinflammation 21:179
156. Park SE, Dantzer R, Kelley KW, McCusker RH (2011) Central administration of insulin-like growth factor-I decreases depressive-like behavior and brain cytokine expression in mice. J Neuroinflammation 9:8–12
157. Kaneko N, Kudo K, Mabuchi T, Takemoto K, Fujimaki K, Wati H, Iguchi H, Tezuka H, Kanba S (2006) Suppression of cell proliferation by interferon-alpha through interleukin-1 production in adult rat dentate gyrus. Neuropsychopharmacology 31:2619–2626
158. Monje ML, Toda H, Palmer TD (2003) Inflammatory blockade restores adult hippocampal neurogenesis. Science 302:1760–1765
159. Vallières L, Campbell IL, Gage FH, Sawchenko PE (2002) Reduced hippocampal neurogenesis in adult transgenic mice with chronic astrocytic production of interleukin-6. J Neurosci 22:486–492
160. Seguin JA, Brennan J, Mangano E, Hayley S (2009) Proinflammatory cytokines differentially influence adult hippocampal cell proliferation depending upon the route and chronicity of administration. Neuropsychiatr Dis Treat 5:5–14
161. Tolosa L, Caraballo-Miralles V, Olmos G, Llado J (2011) TNF-α potentiates glutamate-induced spinal cord motoneuron death via NF-KB. Mol Cell Neurosci 46:176–186
162. Villani GR, Gargiulo N, Faraonio R, Castaldo S, Gonzalez Y, Reyero E, Di Natale P (2007) Cytokines, neurotrophins, and oxidative stress in brain disease from mucopolysaccharidosis IIIB. J Neurosci Res 85:612–622
163. Zhu W, Zheng H, Shao X, Wang W, Yao Q, Li Z (2010) Excitotoxicity of TNF alpha derived from KA activated microglia on hippocampal neurons in vitro and in vivo. J Neurochem 23:386–396
164. Howren MB, Lamkin DM, Suls J (2009) Associations of depression with C-reactive protein, IL-1, and IL-6: a meta-analysis. Psychosom Med 71:171–186
165. Liukkonen T, Silvennoinen-Kassinen S, Jokelainen J, Räsänend P, Leinoneng M, Meyer-Rochowe VB, Timonen M (2006) The association between C-reactive protein levels and depression: results from the Northern Finland 1966 birth cohort study. Biol Psychiatry 60:825–830
166. Heinrich PC, Castell JV, Andus T (1990) Interleukin-6 and the acute phase response. Biochem J 265:621–636
167. Anisman H, Merali Z, Poulter MO, Hayley S (2005) Cytokines as a precipitant of depressive illness: animal and human studies. Curr Pharm Des 11:963–972

168. Pace TWW, Mletzko TC, Alagbe O, Musselman DL, Nemeroff CB, Miller AH, Heim CM (2006) Increased stress-induced inflammatory responses in male patients with major depression and increased early life stress. Am J Psychiatry 163:1630–1633
169. Goshen I, Yirmiya R, Iverfeldt K, Weidenfeld J (2003) The role of endogenous interleukin-1 in stress-induced adrenal activation and adrenalectomy induced adrenocorticotropic hormone hypersecretion. Endocrinology 144:4453–4458
170. Goshen I, Yirmiya R (2007) The role of pro-inflammatory cytokines in memory processes and neural plasticity. In: Ader R (ed) Psychoneuroimmunology, 4th edn. Academic, Amsterdam, pp 337–377
171. Kendler KS, Karkowski LM, Prescott CA (1999) Causal relationships between stressful life events and the onset of major depression. Am J Psychiatry 156:837–841
172. Luby JL, Heffelfinger A, Mrakotsky C, Brown K, Hessler M, Spitznagel E (2003) Alterations in stress cortisol reactivity in depressed preschoolers relative to psychiatric and no-disorder comparison groups. Arch Gen Psychiatry 60:1248–1255
173. Tennes K, Downey K, Vernadakis A (1977) Urinary cortisol excretion rates and anxiety in normal one year old infants. Psychosom Med 39:178–187
174. Guerry JD, Hastings PD (2011) In search of HPA axis dysregulation in child and adolescent depression. Clin Child Fam Psychol Rev 14:135–160
175. Bhagwagar Z, Hafizi S, Cowen PJ (2003) Increase in concentration of waking salivary cortisol in recovered patients with depression. Am J Psychiatry 160:1890–1891
176. Pariante CM, Lightman SL (2008) The HPA axis in major depression: classical theories and new developments. Trends Neurosci 31:464–468
177. Koolschijn PCMP, van Haren NEM, Lensvelt-Mulders GJLM, Hulshoff Pol HE, Kahn RS (2009) Brain volume abnormalities in major depressive disorder: a meta-analysis of magnetic resonance imaging studies. Hum Brain Mapp 30:3719–3735
178. den Heijer T, Tiemeier H, Luijendijk HJ, van der Lijn F, Koudstaal PJ, Hofman A, Breteler MMB (2011) A study of the bidirectional association between hippocampal volume on magnetic resonance imaging and depression in the elderly. Biol Psychiatry 70:191–197
179. Heim C, Nemeroff CB (1999) The impact of early adverse experiences on brain systems involved in the pathophysiology of anxiety and affective disorders. Biol Psychiatry 46:1509–1522
180. Humphreys D, Schlesinger L, Lopez M, Araya AV (2006) Interleukin-6 production and deregulation of the hypothalamic-pituitary-adrenal axis in patients with major depressive disorders. Endocrine 30:371–376
181. Webster JC, Oakley RH, Jewell CM, Cidlowski JA (2001) Proinflammatory cytokines regulate human glucocorticoid receptor gene expression and lead to the accumulation of the dominant negative β isoform: a mechanism for the generation of glucocorticoid resistance. Proc Natl Acad Sci USA 98:6865–6870
182. Fitzgerald P, O'Brien SM, Scully P, Rijkers K, Scott LV, Dinan TG (2006) Cutaneous glucocorticoid receptor sensitivity and pro-inflammatory cytokine levels in antidepressant-resistant depression. Psychol Med 36:37–43
183. Jehn CF, Kuehnhardt D, Bartholomae A, Pfeiffer S, Krebs M, Regierer AC, Schmid P, Possinger K, Rath BC (2006) Biomarkers of depression in cancer patients. Cancer 107:2723–2729
184. Shanks N, Larocque S, Meaney MJ (1995) Neonatal endotoxin exposure alters the development of the hypothalamic-pituitary-adrenal axis: early illness and later responsivity to stress. J Neurosci 15:376–384
185. Walker AK, Nakamura T, Byrne R, Naicker S, Tynan RJ, Hodgson DM (2009) Neonatal lipopolysaccharide and adult stress exposure predisposes rats to anxiety-like behavior and blunted corticosterone responses: implications for the double-hit hypothesis. Psychoneuroendocrinology 34:1515–1525
186. Walker AK, Nakamura T, Hodgson DM (2010) Neonatal lipopolysaccharide exposure alters central cytokine responses to stress in adulthood in Wistar rats. Stress 13(6):506–515

187. Hajszan T, Dow A, Warner-Schmidt JL, Szigeti-Buck K, Sallam NL, Parducz A, Leranth A, Duman RS (2009) Remodeling of hippocampal spine synapses in the rat learned helplessness model of depression. Biol Psychiatry 65:392–400
188. Pittenger C, Duman RS (2008) Stress, depression, and neuroplasticity: a convergence of mechanisms. Neuropsychopharmacology 33:88–109
189. Malberg JE, Eisch AJ, Nestler EJ, Duman RS (2000) Chronic antidepressant treatment increases neurogenesis in adult rat hippocampus. J Neurosci 20:9104–9110
190. Frodl T, Jäger M, Smajstrlova I, Born C, Bottlender R, Palladino T, Reiser M, Möller HJ, Meisenzahl EM (2008) Effect of hippocampal and amygdala volumes on clinical outcomes in major depression: a 3-year prospective magnetic resonance imaging study. J Psychiatry Neurosci 33:423–430
191. Jehn CF, Kühnhardt D, Bartholomae A, Pfeiffer S, Schmid P, Possibger K, Flath BC, Lüftner D (2010) Association of IL-6, hypothalamus–pituitary–adrenal axis function, and depression in patients with cancer. Integr Cancer Ther 9:270–275
192. Pace T, Hu F, Miller AH (2007) Cytokine-effects on glucocorticoid receptor function: relevance to glucocorticoid resistance and the pathophysiology and treatment of major depression. Brain Behav Immun 21:9–19
193. Pace TW, Miller AH (2009) Cytokines and glucocorticoid receptor signaling. Relevance to major depression. Ann N Y Acad Sci 1179:86–105
194. Hu F, Wang X, Pace TWW, Wu H, Miller AH (2005) Inhibition of COX-2 by celecoxib enhances glucocorticoid receptor function. Mol Psychiatry 10:426–433
195. Klinesmith J, Kasser T, McAndrew FT (2006) Guns, testosterone, and aggression: an experimental test of a mediational hypothesis. Psychol Sci 17:568
196. Newman ML, Josephs RA (2009) Testosterone as a personality variable. J Res Pers 43:258–259
197. Schweiger U, Deuschle M, Weber B, Korner A, Lammers CH, Schmeider J, Gotthardt U, Heuser I (1999) Testosterone, gonadotropin, and cortisol secretion in male patients with major depression. Psychosom Med 61:292–296
198. Perry PJ, Yates WR, Williams RD, Anderson AE, MacIndoe JH, Lund BC, Holman TL (2002) Testosterone therapy in late-life major depression in males. J Clin Psychiatry 63:1096–1101
199. Yoo MJ, Nishihara M, Takahashi M (1997) Tumor necrosis factor a mediates endotoxin induced suppression of gonadotropin-releasing hormone pulse generator activity in the rat. Endocr J 44:141–148
200. Ebisui O, Fukata J, Tominaga T, Murakami N, Kobayashi H, Segawa H, Muro S, Naito Y, Nakai Y, Maui Y, Nishida T, Imura H (1992) Roles of interleukin-1a and 21b in endotoxin-induced suppression of plasma gonadotropin levels in rats. Endocrinology 130:3307–3313
201. Rivest S, Rivier C (1993) Interleukin-1b inhibits the endogenous expression of the early gene c-fos located within the nucleus of LH-RH neurons and interferes with hypothalamic LH-RH release during proestrus in the rat. Brain Res 613:132–142
202. Avitsur R, Cohen E, Yirmiya R (1998) Effects of interleukin-1 on sexual attractivity in a model of sickness behavior. Physiol Behav 63:25–30
203. Avitsur R, Yirmiya R (1999) Cytokines inhibit sexual behavior in female rats: I. Synergistic effects of tumor necrosis factor α and interleukin-1. Brain Behav Immun 13:14–32
204. Avitsur R, Wiedenfeld J, Yirmiya R (1999) Cytokines inhibit sexual behavior in female rats: II. Prostaglandins mediate the suppressive effects of interleukin-1β. Brain Behav Immun 13:33–45
205. Imura H, Fukata JI, Mori T (1991) Cytokines and endocrine function: an interaction between the immune and neuroendocrine systems. Clin Endocrinol 35:107–115
206. Li XF, Bowe JE, Lightman SL, O'Byrne KT (2005) Role of corticotropin-releasing factor receptor-2 in stress-induced suppression of pulsatile luteinizing hormone secretion in the rat. Endocrinology 146:318–322

207. Matsuwaki T, Suzuki M, Yamanouchi K, Nishihara M (2004) Glucocorticoid counteracts the suppressive effect of tumor necrosis factor-alpha on the surge of luteinizing hormone secretion in rats. J Endocrinol 181:509–513
208. Li XF, Kinsey-Jones JS, Knox AM, Wu XQ, Tahsinsoy D, Brain SD, Lightman SL, O'Byrne KT (2007) Neonatal lipopolysaccharide exposure exacerbates stress-induced suppression of luteinizing hormone pulse frequency in adulthood. Endocrinology 148:5984–5990
209. Knox AM, Li XF, Kinsey-Jones JS, Wilkinson ES, Wu XQ, Cheng YS, Milligan SR, Lightman SL, O'Byrne KT (2009) Neonatal lipopolysaccharide exposure delays puberty and alters hypothalamic Kiss1 and Kiss1r mRNA expression in the female rat. J Neuroendocrinol 21:683–689
210. Walker AK, Hiles SA, Sominsky L, Mclaughlin EA, Hodgson DM (2011) Neonatal lipopolysaccharide exposure impairs sexual development and reproductive success in the Wistar rat. Brain Behav Immun 25:674–684
211. Weissman MN, Bland R, Joyce PR, Newman S, Wells JE, Wittchen HU (1993) Sex differences in rates of depression: cross-national perspectives. J Affect Disord 29:77–84
212. Kessler RC, McGonagle KA, Zhao S, Nelson CB, Hughes M, Eshleman S, Wittchen HU, Kendler KS (1994) Lifetime and 12-month prevalence of DSM-III-R psychiatric disorders in the United States: results from the National Comorbidity Survey. Arch Gen Psychiatry 51:8–19
213. Cowen PJ, Parry-Billings M, Newsholme EA (1989) Decreased plasma tryptophan levels in major depression. J Affect Disord 16:27–31
214. Maes M, Minner B, Suy E (1991) The relationships between the availability of l-tryptophan to the brain, the spontaneous HPA-axis activity, and the HPA-axis responses to dexamethasone in depressed patients. J Amino Acids 1:57–65
215. Maes M, Schotte C, D'Hondt P, Claes M, Vandewoude M, Scharpe S, Cosyns P (1991) Biological heterogeneity of melancholia: results of pattern recognition methods. J Psychiatr Res 25:95–108
216. Quintana J (1992) Platelet serotonin and plasma tryptophan decreases in endogenous depression. Clinical, therapeutic, and biological correlations. J Affect Disord 24:55–62
217. Song C, Lin A, Bonaccorse S, Heide C, Verkerk R, Kenis G, Bosmans E, Scharpé S, Whelan A, Cosyns P, de Jongh R, Maes M (1998) The inflammatory response system and the availability of plasma tryptophan in patients with primary sleep disorders and major depression. J Affect Disord 49:211–219
218. Guilleman GJ, Smythe G, Takikawa O, Brew BJ (2005) Expression of indoleamine 2,3-dioxygenase and production of quinolinic acid by human microglia, astrocytes and neurons. Glia 49:15–23
219. Kwidzinski E, Bechmann I (2007) IDO expression in the brain: a double-edged sword. J Mol Med 85:1351–1359
220. Heyes MP, Achim CL, Wiley CA, Major EO, Saito K, Markey SP (1996) Human microglia convert L-tryptophan into the neurotoxin quinolinic acid. Biochem J 320:595–597
221. Myint AM, Kim YK, Verkerk R, Scharpé S, Syeinbusch H, Leonard B (2007) Kynurenine pathway in major depression: evidence of impaired neuroprotection. J Affect Disord 98:143–151
222. Myint AM, Schwarz MJ, Verkerk R, Mueller HH, Zach J, Scharpé S, Steinbusch HWM, Leonard BE, Kim YK (2011) Reversal of imbalance between kynurenic acid and 3-hydroxykynurenine by antipsychotics in medication-naïve and medication-free schizophrenic patients. Brain Behav Immun 25:1576–1581. doi:10.1016/j.bbi.2011.05.005
223. Munn DH, Zhou M, Attwood JT, Bondarev I, Conway SJ, Marshall B, Brown C, Mellor AL (1998) Prevention of allogeneic fetal rejection by tryptophan catabolism. Science 281:1191–1193
224. Mellor AL, Munn DH (2000) Immunology at the maternal–fetal interface: lessons for T cell tolerance and suppression. Annu Rev Immunol 18:367–391

225. Mellor AL, Sivakumar J, Chandler P, Smith K, Molina H, Mao D, Munn DH (2001) Prevention of T cell-driven complement activation and inflammation by tryptophan catabolism during pregnancy. Nat Immunol 2:64–68
226. Capuron L, Neurauter G, Musselman DL, Lawson DH, Nemeroff CB, Fuchs D, Miller AH (2003) Interferon-alpha-induced changes in tryptophan metabolism: relationship to depression and paroxetine treatment. Biol Psychiatry 54:906–914
227. Capuron L, Ravaud A, Neveu PJ, Miller AH, Maes M, Dantzer R (2002) Association between decreased serum tryptophan concentrations and depressive symptoms in cancer patients undergoing cytokine therapy. Mol Psychiatry 7:468–473
228. Wichers MC, Koek GH, Robays G, Verkerk R, Scharpé S, Maes M (2005) IDO and interferon-α-induced depressive symptoms: a shift in hypothesis from tryptophan depletion to neurotoxicity. Mol Psychiatry 10:538–544
229. Raison CL, Dantzer R, Kelley KW, Lawson MA, Wool Wine BJ, Vogt G, Spivey JR, Saito K, Miller AH (2010) CSF concentrations of brain tryptophan and kynurenines during immune stimulation with IFN-α: relationship to CNS immune responses and depression. Mol Psychiatry 15:393–403
230. Sublette ME, Galfalvy HC, Fuchs D, Lapidus M, Grunebaum MF, Oquendo MA, Mann JJ, Postolache TT (2011) Plasma kynurenine levels are elevated in suicide attempters with major depressive disorder. Brain Behav Immun 25:1272–1278
231. Moreau M, Lestage J, Verrier D, Mormede C, Kelley KW, Dantzer R, Castanon N (2005) Bacille calmette-guerin inoculation induces chronic activation of peripheral and brain indoleamine 2,3-dioxygenase in mice. J Infect Dis 192:537–544
232. Godbout JP, Moreau M, Lestage J, Chen J, Sparkman NL, O'Connor J, Castanon N, Kelley KW, Dantzer R, Johnson RW (2008) Aging exacerbates depressive-like behavior in mice in response to activation of the peripheral innate immune system. Neuropsychopharmacology 33:2341–2351
233. Yadav MC, Burudi EM, Alirezaei M, Flynn CC, Watry DD, Lanigan CM, Fox HS (2007) IFN-gamma-induced IDO and WRS expression in microglia is differentially regulated by IL-4. Glia 55:1385–1396
234. Müller N, Schwarz MJ (2007) The immune-mediated alteration of serotonin and glutamate: towards an integrated view of depression. Mol Psychiatry 12:988–1000
235. Funakoshi H, Kanai M, Nakamura T (2011) Modulation of tryptophan metabolism, promotion of neurogenesis and alteration of anxiety-related behavior in tryptophan 2,3-dioxygenase-deficient mice. Int J Tryptophan Res 4:7–18
236. Metz R, Du Hadaway JB, Kamasani U, Laury-Kleintop L, Muller AJ, Prendergast JC (2007) Novel tryptophan catabolic enzyme IDO2 is the preferred biochemical target of the antitumor indoleamine 2,3-dioxygenase inhibitory compound D-1-methyl-tryptophan. Cancer Res 67:7082–7087
237. Branchey L, Lieber CS (1982) Activation of tryptophan pyrrolase after chronic alcohol administration. Subst Alcohol Actions Misuse 3:225–229
238. Badawy AA-B, Dougherty DM, Marsh-Richard DM, Steptoe A (2009) Activation of liver tryptophan pyrrolase mediates the decrease in tryptophan availability to the brain after acute alcohol consumption in normal subjects. Alcohol Alcohol 44:267–271

Chapter 7
Neuroimmune Cross Talk and HIV-Associated Neurocognitive Disorders

JoEllyn McMillan and Howard E. Gendelman

7.1 Introduction

7.1.1 HIV Infection and Neurologic Complications

Human immunodeficiency virus (HIV) infection is a global and devastating epidemic affecting the health, well-being, and social fabric of many of the world's populations. UNAIDS estimates that 34 million people are infected with HIV with 2.7 million new infections yearly. The majority of infections are in sub-Saharan Africa (http://www.who.int/hiv/data/en/). Viral infection is persistent despite vigorous host immune responses [1–6]. HIV enters its CD4+ T lymphocyte, and mononuclear phagocyte (MP; dendritic cells, monocytes, macrophages, and microglial cells) targets through cell surface interaction of its envelope protein, gp120, and with its CD4 and chemokine receptor/co-receptors, CCR5 and CXCR4 [7–11]. Following cell entry, HIV RNA is reverse transcribed and integrated into the host cell's genome. Transcription of viral genes is controlled by interactions between HIV-1 regulatory proteins and host cell transcription factors such as nuclear factor kappa beta (NF-κB) [12–15]. Viral assembly proceeds at the cell surface or in subcellular organelles and the viral RNA encapsulated by interactions with Gag, GagPol, and envelope proteins, with subsequent viral budding and release of mature virions [16–23]. Ongoing HIV infection results in profound CD4+ T cell losses with immune impairments resulting in a range of opportunistic infections, metabolic disorders, and malignancies [3, 5, 24–26]. Less appreciated are primary manifestations of viral replication that include its effects on the central nervous system (CNS) [27–29].

J. McMillan • H.E. Gendelman, M.D. (✉)
Center for Neurodegenerative Disorders and Department of Pharmacology and Experimental Neuroscience, University of Nebraska Medical Center, 985880 Nebraska Medical Center, Omaha, NE 68198-5880, USA
e-mail: hegendel@unmc.edu

C. Cui et al. (eds.), *Neural-Immune Interactions in Brain Function and Alcohol Related Disorders*, DOI 10.1007/978-1-4614-4729-0_7,

Indeed, although HIV primarily affects immune function and integrity [3, 24–26], virus-associated effects on the nervous system are a significant cause of comorbidity during the course of disease [27, 29, 30]. Prior to the introduction of combination antiretroviral therapy (cART), opportunistic infections (OIs) and advanced cognitive, motor, and behavioral abnormalities commonly occurred as HIV disease advanced and was associated with virus-induced progressive immunosuppression. While cART has reduced the prevalence of OIs along with severity of virus-induced nervous system disorders, both remain active albeit less severe [27].

7.1.2 Neurological Manifestations of Disease

Continuing in the cART era, HIV infection is tightly associated with a spectrum of cognitive impairments termed HIV-associated neurocognitive disorders (HAND). Symptoms range from more common asymptomatic and mild cognitive-motor impairments to the most severe but least common HIV-associated dementia (HAD) [31–34]. HAD interferes with the person's ability to perform day-to-day activities, and as such, patients are unable to work or care for themselves; such patients exhibit severe neurocognitive impairment, emotional disturbances, and motor abnormalities [35]. Since the advent of cART and with the increasing longevity of infected people, the incidence of HAD has dropped to less than 7 %. In milder forms of HAND, difficulties in learning new information and in prospective memory, attention deficit, reduced cognition, and disruptions in executive functions, such as evaluating, planning, and abstract reasoning, are common, and verbal fluency is impaired. The initial patchy nature of brain involvement is reflected in variation of the severity and spectrum of performance decline among individuals. Such less profound cognitive impairments adversely impact employment, medication adherence, and other aspects of daily life [36, 37].

The exact mechanisms for HAND and what underlies its persistence is not completely understood. Likely, continued glial activation and peripheral effects on innate immune function are disease pathobiologic components [28]. The duration of HIV-1 infection, nadir CD4+ counts of <200, and repeated viral load peaks affect neuroinflammatory responses and as such may herald the development of HAND [38–40]. Genotypic and phenotypic variations in the virus may affect the neuropathogenicity of HIV. For example, HIV clades (clades A and C) are perhaps less neuropathogenic than clades B and D l [41–44]. Recent works suggest specific clade differences in Tat-induced apoptosis in neurons. C-Tat was shown to be less neurotoxic than B-Tat likely as a result of changes in the dicysteine motif within Tat's neurotoxic region [45]. In addition, ART resistance is associated with reduced neurocognitive performance [46]. The variation in susceptibility to development of HAND strongly suggests involvement of host genetic factors. Polymorphisms in the CCR2 receptor, the monocyte chemotactic protein-1 (MCP-1) promoter, apolipoprotein E, and tumor necrosis factor alpha (TNFα) have all been associated with progression of neurocognitive impairment [47–53]. Other factors such as aging, concurrent substance abuse, and hepatitis C virus infection can contribute to HAND onset and severity [54–64].

The relationship between cognitive impairments and neuropathology is not absolute by any measure. Although the most severe disorder, HAD is associated, in measure, with an encephalitis; this is not true for milder disorders. Common in the pre-cART era, HIV encephalitis is characterized by robust viral CNS replication; formation of multinucleated giant cells; widespread reactive astrogliosis, microgliosis, and neuronal degeneration; loss of synaptic connections; and myelin pallor [29, 65, 66] (Fig. 7.1). With current widespread use of cART, a subtle neuropathology has emerged that is linked only to synaptodendritic damage [67]. Activation of macrophages and microglial cells is observed; however, the inflammation appears to be shifted from the basal ganglia to the hippocampus and adjacent entorhinal and temporal cortex areas [68, 69]. Neuropathology associated with HIV-1 infection is certainly metabolic, resulting from the production of viro- and cellular toxins released from infected and immune activated MP [70]. Control of viral growth and of disease is maintained by effector and regulatory T cell surveillance [71] (Fig. 7.1).

7.1.3 *Drugs of Abuse and HIV*

Viral transmission and comorbid conditions can also be complicated by coexisting conditions such as hepatitis C virus infection and drug abuse [61, 72–74]. Substance abuse can increase the risk of HIV-1 infection by promotion of risky behaviors such as needle sharing and sexual promiscuity. In addition, substance abuse may accelerate HAND [75–77]. HIV-1 infected patients who also abuse drugs commonly show neuropathologic features reflective of an encephalitis (leukocyte infiltration and glial activation) [78, 79], increased microglial activation [80] and giant cell formation [81], and disruption of the blood–brain barrier (BBB) [78]. Thus, substance abuse may act with HIV-1 to promote development of neurological disorders [82].

The drugs most commonly abused in HIV-1 infected individuals include alcohol, opioids, cocaine, and methamphetamine (Meth) [68, 83], often in combination. Alcohol use among infected people is the most common with 55 % of HIV seropositive patients in one study fitting the criteria for current or past alcohol abuse [84]. Of special concern, alcohol addiction can adversely affect the treatment of hepatitis C infection in HIV-1 seropositive patients and can interact pharmacologically with the nucleoside reverse transcriptase inhibitor abacavir, resulting in increased plasma drug concentration, which may lead to increased drug-related toxicity [85].

7.1.3.1 Opioids

Opioid drugs of abuse are known to have a broad array of functional effects on the cells of the immune system. Opioids such as morphine can enhance production of chemokines and induce expression of the CC-chemokine receptor type 5 (CCR5)

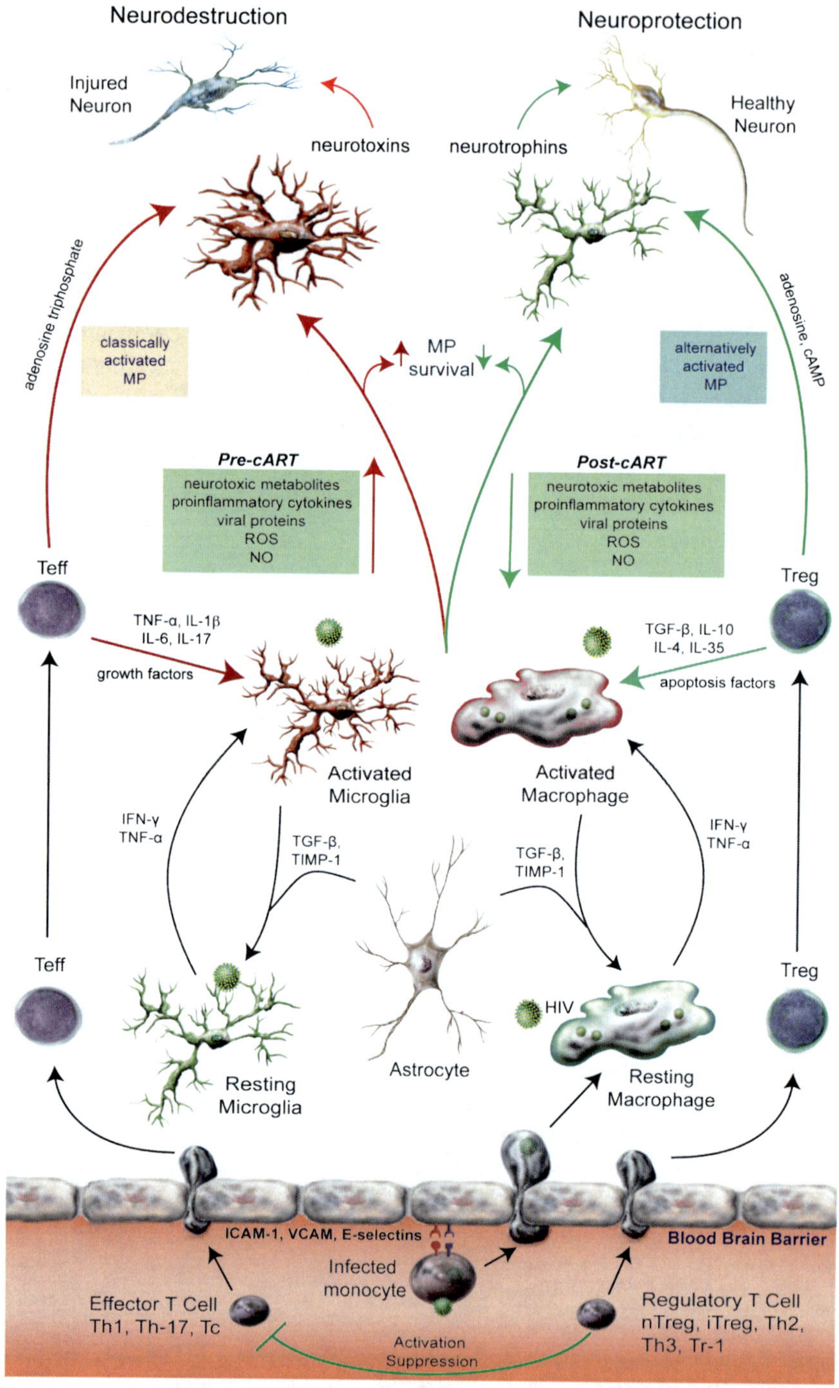
Neurodestruction
Neuroprotection
Injured Neuron
Healthy Neuron
neurotoxins
neurotrophins
adenosine triphosphate
adenosine, cAMP
classically activated MP
alternatively activated MP
MP survival
Pre-cART
neurotoxic metabolites
proinflammatory cytokines
viral proteins
ROS
NO
Post-cART
neurotoxic metabolites
proinflammatory cytokines
viral proteins
ROS
NO
Teff
Treg
TNF-α, IL-1β
IL-6, IL-17
growth factors
TGF-β, IL-10
IL-4, IL-35
apoptosis factors
Activated Microglia
Activated Macrophage
IFN-γ
TNF-α
TGF-β,
TIMP-1
TGF-β,
TIMP-1
IFN-γ
TNF-α
Teff
Treg
HIV
Astrocyte
Resting Microglia
Resting Macrophage
ICAM-1, VCAM, E-selectins
Blood Brain Barrier
Infected monocyte
Effector T Cell
Th1, Th-17, Tc
Regulatory T Cell
nTreg, iTreg, Th2,
Th3, Tr-1
Activation
Suppression

and the CXC chemokine receptor type 4 (CXCR4) on monocytes and T cells [86–90]. Furthermore, increased HIV replication was reported in opioid-exposed human lymphoid, monocyte, and microglial cells. Whether opioid modulation of chemokine production and chemokine receptor expression can affect the progression of AIDS and development of neurodegeneration in vivo is a subject of interest and a focus of animal studies including the simian immune virus (SIV)-macaque model of the disease with data sets sometimes conflicting [91, 92].

7.1.3.2 Cocaine

Epidemiological studies in HIV-1 seropositive cocaine abusers link cocaine abuse to disease progression [93]. Cocaine mediates its effects on HIV-1 infection through a number of pathways [94]. Viral replication is enhanced by cocaine in MP and astrocytes with modulation of cytokine expression to T-helper 2 (Th2) responses by interleukin 10 (IL-10). Cocaine increases the transmigration of HIV-1-infected monocytes and dendritic cells across the BBB through enhanced endothelial cell adhesion molecules and production of matrix metalloproteinase (MMP) expression [95–97].

7.1.3.3 Meth

HIV-1-infected individuals with concurrent Meth abuse demonstrate a higher degree of neuropsychological impairment [61]. These patients also show higher plasma viral loads, possibly due to lack of adherence to ART or changes in drug metabolism [98–102]. More direct interactions between Meth and HIV-1 infection for the CNS are suggested by studies demonstrating long-lasting dopamine depletion in the cortex and striatum and decreased dopamine and homovanillic acid in the basal ganglia in Meth-HIV-infected people [103–107]. Because of the dual effects of HIV-1 and Meth on the striatal dopaminergic system, an increased risk for developing other neurodegenerative disorders, such as Parkinson's disease, may be seen in future years [107–109]. Other interactive effects of Meth and HIV on neuronal damage are suggested by increased neuronal apoptosis and levels of autophagosome and multilamellar bodies seen during cell treatments of combinations of Meth and Tat [110].

Fig. 7.1 Innate and adaptive immunity in HIV-1 neuropathogenesis in the era of cART. A schematic of the cascade of neuropathologic and immunologic events during progressive HIV-1 infection of the CNS. Activated monocytes are attracted to the CNS by chemokines released from activated astrocytes and resident microglia. Virus is spread to neighboring glial cells with concomitant neuroinflammation. Within the infected brain both proinflammatory (Teff, CTL, M1 microglia) and anti-inflammatory (Treg, M2 microglia) aspects of the immune response are evident. Disease progresses when the proinflammatory environment becomes dominant. Combination antiretroviral therapy (cART) decreases host viral load leading to reduced levels of proinflammatory cytokines and neurotoxic factors in the CNS, which favors neuronal survival (reproduced with permission from Kraft-Terry et al. [71])

7.1.4 Changing Epidemiologic Patterns of HAND in the Era of cART

With effective and widespread use of cART, HIV-1 infection has become a chronic disorder. The increasing longevity of people with HIV infection has meant that age is now identified as a risk factor for development of HAD [51] and development of motor and other neurological signs [111]. An association between chronic HIV-1 infection and increased risk of development of Alzheimer's disease (AD) and other neurodegenerative disorders was recently suggested by epidemiological studies [112]. Increased cerebral deposition of Aβ and tau hyperphosphorylation and neurofibrillary pathology are seen earlier than expected in brain tissue from HIV-positive patients [113–115]. Magnetic resonance spectroscopy (MRS) studies have shown greater age-related increases in glial activation in frontal white matter in virus-infected, ART-naïve patients compared to virus-seronegative controls [116]. The mechanism for enhanced age-related neurodegeneration in infected patients may be related to the actions of HIV-1 proteins such as Tat to inhibit Aβ degrading enzymes such as neprilysin seen in AD [115, 117] and to interfere with microglial phagocytosis of Aβ [118]. Long-term treatment with nelfinavir and saquinavir may promote cerebral accumulation of Aβ through inhibition of insulin-degrading enzyme or proteasome peptidase [119, 120]. Nucleoside reverse transcriptase inhibitors may promote neuronal damage through induction of mitochondrial dysfunction and oxidative stress [121, 122]. Downregulation of leptin gene transcription in adipose tissue by cART could affect neurocognitive function through inhibition of leptin-mediated development of learning and memory [123, 124]. In addition, long-term cART effects such as insulin resistance, hyperlipidemia, and fat redistribution may promote development of atherosclerotic disease in HIV-positive patients [125, 126].

7.2 The Neuropathogenesis of HIV Infection

A metabolic encephalopathy associated with HIV-1 infection is initiated by infection of brain MPs including perivascular and parenchymal brain macrophages and microglia. The encephalopathy is sustained by a cascade of inflammatory and neurotoxic responses induced by the secretion of proinflammatory cytokines, chemokines, platelet-activating factor, arachidonic acid and metabolites, nitric oxide, quinolinic acid, progeny virions, and viral proteins. Neuronal damage can be elicited by exposure to excitatory neurotransmitters such as glutamate or to cytokines and neurotoxins produced by infected and activated brain MP.

7.2.1 *Viral and Cellular Toxins*

HIV-1 entry into the CNS occurs early in disease and after systemic infection and only later, after chronic viral infection in the periphery, initiates events leading to neurological damage. Although HIV-1 is carried to the CNS in infected macrophages responding to and inducing neuroinflammatory responses and resulting from initial host responses to infection, the interplay between what is ongoing in the periphery and CNS impairments are not completely understood. The likely scenario is that peripheral immune activation drives brain disease through diffusible proinflammatory factors that readily enter the brain and cause neuronal impairments [28]. This is perpetuated, as once in the CNS, the virus can infect MP and astrocytes in a restricted manner. Infected cells can release viral proteins that have known neurotoxic effects, including envelope proteins and accessory proteins such as Vpr, Tat, and Nef.

These proteins can elicit neurotoxic effects through direct interaction with neurons and through interactions with nonneuronal cells such as astrocytes, macrophages and microglial cells, and brain microvascular endothelial and neural progenitor cells. Induction of apoptosis, damage due to oxidative stress, Ca^{2+} influx, and membrane compositional changes can lead to neuronal injury and cell death [127]. These events can be initiated through direct interaction of HIV proteins with chemokine receptors and activation of excitatory neurotransmitter receptors such as the N-methyl-D-aspartate-type glutamate (NMDA) receptor [128–132]. HIV-1gp120 can bind to chemokine receptors such as CXCR4 and CCR5 on neurons, increasing intracellular Ca^{2+} and leading to cell death. Both gp120 and the HIV regulatory protein Tat (a viral transactivator of transcription) can stimulate NMDA receptors on neurons, resulting in excitotoxicity with subsequent Ca^{2+} dysregulation and cell death [131, 133]. HIV accessory proteins such as Vpr and Nef can induce neuronal apoptosis subsequent to the formation of ion channels in the cell membrane (Vpr) or by altering the conductance of potassium channels (Nef) [127, 134–139].

Viral proteins can indirectly affect neuronal integrity by their interactions with cells that serve support functions. Thus, HIV-1gp120 can interact with the chemokine receptors CXCR4 and CCR5 on macrophages, microglia, and astrocytes, leading to activation and release of cellular neurotoxic factors [138, 140–151]. TNFα and interleukin-1 beta (IL-1β) released from gp120-stimulated macrophages and microglia can further stimulate macrophages and astrocytes to release L-cysteine and glutamate, which overstimulate NMDA receptors on neurons. Induction of neuronal apoptosis may occur by interaction of TNFα with the TNFR1 receptor and activation of caspase 8. IL-1β can stimulate astrocytes to release glutamate and stromal cell-derived factor-1 alpha (SDF-1α), which can induce neuronal apoptosis. Production of nitric oxide through gp120-mediated activation of inducible nitric oxide synthase in astrocytes can lead to peroxynitrite-dependent neuronal damage [152, 153]. Furthermore, activation of the transcription factor p53 in neurons, microglia, and astrocytes has been reported in HAD patients [154, 155]; and studies in p53-deficient mice suggest that such activation may play a role in HIV-1gp120-mediated neurotoxicity [155–157].

Tat can indirectly cause neuronal damage through stimulation of the release of a range of proinflammatory mediators from activated glial cells and MP. Tat stimulates increased secretion of cytokines such as TNFα, interleukins, and chemokines, such as CC-chemokine ligand 2 (CCL2)/MCP-1 and CXC chemokine 10 (CXCL10)/interferon inducible protein-10 (IP-10) from glial cells and macrophages. In addition, infiltration of monocytes into the brain is induced by upregulation of astrocyte production of chemoattractant proteins, such as MCP-1.

Vpr and Nef can induce indirect neurotoxicity by adverse interactions with astrocytes and by affecting the function of the BBB. Vpr is cytotoxic to both astrocytes and brain microvascular endothelial cells. Nef may increase permeability of the BBB by activation of MMP and may induce leukocyte transmigration by stimulating astrocyte production of CCL2/MCP-1 [158–160].

HIV proteins can also affect the properties of neural progenitor cells. HIV-1gp120 has been observed to induce quiescence in neural progenitor cells [161, 162]. Furthermore, HIV-infected progenitor cells exhibit increased expression of glial cell markers and morphological changes that indicate glial rather than neuronal progression [163]. Thus, the already limited capacity for de novo neuronal production in the adult brain may be further compromised by HIV infection.

7.2.2 HIV Target Cells in the Nervous System

The target cells for HIV-1 are CD4+ T lymphocytes and MP. In addition to their inability to eradicate the virus, MP act as the viral storage sites and the means by which the virus is disseminated from the periphery to the CNS [164–166]. HIV-1 infection of the CNS is initiated by diapedesis of infected macrophages that occur at the BBB interface. Release of proinflammatory mediators and chemokines by activated microglial cells and perivascular macrophages results in continuous recruitment of circulating monocytes to the brain and underlies the cascade of immunologic events that lead to the development of an encephalitis. Production of the chemoattractant MCP-1 by activated glial cells affects the integrity of the BBB and enhances MP infiltration [167, 168]. The proinflammatory environment induced by continued viral infection leads to further MP activation and secretion of proinflammatory cytokines, chemokines, and arachidonic acid derivatives that compromise the function of the BBB and enhance neurodegeneration [169]. The proinflammatory environment can also induce the production of the chemokine fractalkine by neurons and endothelial cells, which recruit CD16+ monocyte subsets to sites of inflammation and induces these cells to produce IL-6, MCP-1, and MMP-9, opening the way for additional MP infiltration.

The mechanisms whereby MPs effect neurodegeneration are complex and overlapping. Viral proteins such as gp120, Tat, Vpr, and Nef can induce the production of cytokines and chemokines by astrocytes and microglia and can promote the release of a broad range of proinflammatory and cytotoxic mediators from macrophages. In addition, viral infection itself can elicit secretion of a broad array of neurotoxic factors from activated MP. TNFα is a proinflammatory cytokine released from macrophages and microglia that can cause further macrophage/microglia activation and recruitment, activate astrocytes, potentiate glutamate toxicity, and increase BBB permeability by enhancing endothelial cell expression of adhesion molecules and MCP-1 [170, 171]. The inflammatory cascade induced by TNFα can result in cytoskeletal remodeling in macrophages and increased cytokine secretion [172]. IL-1β is a neurotoxic cytokine produced by activated macrophages and, like TNFα, can induce expression of adhesion molecules on brain endothelial cells and astrocytes to increase monocyte infiltration. TNFα and IL-1β also increase astrocyte secretion of the neutrophil chemoattractant IL-8. Other proinflammatory cytokines produced by activated macrophages/microglia such as interferon gamma (IFN-γ) and macrophage colony-stimulating factor increase the release of the chemokines nitric oxide, RANTES (CCL5), macrophage inhibitory protein-1 alpha (MIP-1α), MIP-1β, and eicosanoids.

Polarization of MP into M1 (classically activated) or M2 (alternatively activated) phenotype determines the response to infectious disease (Fig. 7.1) [173]. A well-accepted hypothesis is that MP polarization is central in the development of HAND. M1 responses include secretion of proinflammatory cytokines such as TNFα, IL-1β, and IFN-γ and are driven by viral infection and immune activation. These responses, while of relatively short duration, affect HIV-1 replication and neurotoxicity. Antiviral M2 responses are driven through MP activation by IL-4 and IL-13 and include secretion of anti-inflammatory cytokines such as IL-10 and IFN-β. These responses are of more long-term duration and result in viral suppression and maintenance of neural homeostasis without changes in proviral DNA or protein production [174, 175]. Whether MP polarization is reversible and phenotypes are interchangeable as a cause of or a consequence of disease is unclear. In vitro studies [175, 176] suggest that MP can switch activation states based upon expression of CD163 and CD206 and secretion of CCL3 and CCL18, which suggests that MP polarization may regulate HIV-1 infection and viral replication in vivo.

7.2.3 *Neurotoxicity*

Ultimately, it is the interaction among virus, monocytes/macrophages, neurons, glial cells, the BBB, and cells of the adaptive immune system that determines the severity of neurotoxicity associated with HAND.

7.2.3.1 Neurons

HIV-1 related viral proteins can cause direct damage to neuronal function and structural integrity. Viral proteins such as gp120, Tat, Vpr, and Nef can elicit direct neurotoxic effects through interaction with chemokine receptors and excitatory receptors such as the NMDA receptor. In addition, activation of neuroinflammatory cells by

HIV-infected monocytes/macrophages results in the release of chemokines, cytokines and neurotoxins that can, in association with HIV proteins, damage synaptodendritic associations of pyramidal neurons and lead to neuronal dysfunction and apoptosis [177]. Loss of specific populations of neurons in the neocortex, putamen, globus pallidus, and hippocampus has been associated with memory loss and corticostriatal damage with resultant motor damage in patients with HIV encephalopathy (HIVE) [178]. Recent studies have also demonstrated that HIV proteins and neuroinflammation may interfere with neurogenesis in the dentate gyrus, thus contributing to overall neurodegeneration [179–183]. Interaction of the viral envelope protein gp120 with the CXCR4 receptor has been shown to inhibit migration of fetal neuroprogenitor cells in vitro and to decrease proliferation of adult neuronal stem cells in vitro and in vivo [161, 162]. Other HIV coat proteins can promote quiescence of human neuroprogenitor cells through activation of p21 and p27 and downregulation of ERK [179]. The release of proinflammatory cytokines such as TNFα and IL-1β by infected macrophages and glial cells can also affect neurogenesis. Recent studies have shown that TNFα can promote differentiation of neuroprogenitor cells into astrocytes, thus decreasing neurogenesis [163]. In addition, IL-1β has been shown to decrease proliferation of neuronal stem cells in the dentate gyrus of rats [184]. Thus, HIV-1 infection in the CNS may not only result in neuronal injury but may also impair normal regenerative processes, further exacerbating neurodegeneration.

7.2.3.2 Glial Cells

The important role that macrophages and microglia play in HIV-1 infection of the CNS and development of neurological disorders in HIV-1 infection is well-known and appreciated [185]. However, other glial cell types such as astrocytes may play a crucial role in the development of HIV-induced neurotoxicity. Recent studies have demonstrated that restricted viral infection of astrocytes in HIV dementia may occur in most patients with HAD [186]. Latent HIV infection in astrocytes increases production of neurotoxic and excitatory factors such as CCL2 and glutamate that can lead to glial cell and neuronal apoptosis. Toxic factors produced by latently infected astrocytes can also be passed to other astrocytes through gap junctions [187]. Recent studies also suggest that toxic mediators transferred from a few infected astrocytes to noninfected astrocytes by gap junctions can cause apoptosis of brain endothelial cells and affect the formation and signaling of astrocyte end feet, thus compromising BBB integrity [188]. Production of immunomodulatory cytokines and chemokines by HIV-infected astrocytes also occurs. Secretion of chemokine ligand-10 and MMP-9 by astrocytes can serve to recruit immune cells and contribute to BBB degradation [189, 190]. Furthermore, moderate increases in expression of TNFα and IL-1β by astrocytes [191] are detected in the CNS of patients with HIVE, indicating that secretion of proinflammatory cytokines by astrocytes can contribute to the neuroinflammatory response.

7.2.3.3 Adaptive Immune Responses

In addition to the well-studied innate immune response, there is strong evidence for an adaptive immune response to HIV infection in the brain. Infected macrophages and microglial cells in the CNS secrete proinflammatory and neurotoxic cytokines and chemokines that attract activated lymphocytes [primarily CD4+ T-helper 1 (Th-1) and CD8+ cytotoxic (CTL) lymphocytes and natural killer (NK) cells] to the site of injury [192–196]. Activated lymphocytes in turn further activate microglial cells via IFN-γ and induction of major histocompatibility (MHC) class II proteins to elicit an adaptive immune response [197]. The adaptive immune response (primarily CTL) in the majority of cases acts to control HIV (and SIV) replication in the CNS; however, it is insufficient to eradicate the virus [198]. Unfortunately, the recruitment of lymphocytes susceptible to HIV-1 infection into the brain (CD4+ T cells) as well as the continued activity of CTL can accelerate HIV-1 replication and lead to development of HIV-1 encephalitis and secondary OIs.

While CTL are the most effective in controlling HIV infection in the CNS, CD4+CD25+ regulatory T cells (Treg) and effector T cells (Teff) may also have an important role in the control of HAND (Fig. 7.1) [199–201]. Treg are well recognized for their immunomodulatory functions and are active in suppressing autoimmune reactions and attenuating T cell-mediated immunity [202, 203]. The role of Treg in control of HAND is not entirely clear. Some studies have suggested that Treg are detrimental in HIV infection because of their immunosuppressive role, allowing HIV-1 progression [204, 205]. However, a beneficial function has also been ascribed to Treg because of their ability to suppress chronic immune activation and neuroinflammation [199, 201, 204]. Thus divergent roles for Treg early in infection and late in disease may apply. Early in infection, Treg suppress the immune response to HIV-1 infection, while later in infection Treg suppress chronic immune activation that may contribute to HAND [204, 206–209].

In contrast to Treg, Teff promote inflammatory responses and enhance immune recognition. Using an animal model of HIVE, it was demonstrated that adoptive transfer of Teff to mice injected intracerebrally with bone marrow macrophages infected with HIV-1/VSV-pseudotyped virus could promote development of HIV-induced focal encephalitis [201]. Furthermore, adoptive transfer of Treg to these mice suppressed the development of HIV-1 induced lesions and reduced HIV-1 levels in the brain.

Initiation of cART in HIV-infected patients has greatly improved CD4+ T cell counts and led to partial restoration of immune function. However, in approximately 25 % of patients, initiation of ART can induce a variety of inflammatory symptoms, termed immune reconstitution inflammatory syndrome (IRIS) [210]. IRIS is manifested in patients with underlying OIs or tumors that are responsive to ART and is usually self-limiting [210, 211]. While the immunopathogenesis of IRIS is unclear, studies have suggested an imbalance in reconstitution of Teff and Treg contributes to its development [211, 212]. Thus, independent of the original OI or tumor, rapid expansion of antigen-specific CD4+ T cells upon initiation of ART may initiate a broad inflammatory response to residual antigen, manifesting as IRIS.

7.2.4 Blood–Brain Barrier

Under normal physiologic conditions, the BBB acts to restrict and regulate the passage of substances from the peripheral blood compartment to the brain parenchyma (reviewed by [213, 214]). The function of the barrier is determined by endothelial cell tight junctions [215–220]. Its functional properties are governed by its few fenestrations, low pinocytotic activity, luminal efflux transporters, integrity of the capillary basement membrane, and activities of closely associated pericytes and astrocytes [215, 219, 221–223]. Under conditions of brain infection and tissue damage, immune cells can enter the brain to remove infecting microbes or necrotic tissue. Leukocyte transmigration across the endothelial layer occurs by integrin b1-mediated processes followed by penetration through the basement membrane by the action of MMPs [224]. The interaction of leukocytes with endothelial cells and their ability to pass through the basement membrane is affected by secretion of cytokines by activated glial cells that can alter basement membrane permeability and by enhanced expression of adhesion molecules on endothelial cells induced by HIV-1 viral proteins such as Tat [225–227].

Leukocyte transmigration occurs by interactions between the leukocyte and the cells of the BBB. Interactions of leukocytes with selectins on the surface of BBB endothelial cells are followed by responses to chemokines that promote leukocyte binding to endothelial cell adhesion molecules such as ICAM-1 and VCAM-1 [228–233]. Diapedesis across the endothelial layer is mediated by binding molecules such as junctional adhesion molecules, platelet endothelial cell adhesion molecule 1 and CD99 [234–238]. Evidence suggests that increased expression of these molecules is associated with development of CNS pathology in HIV infection [167, 239, 240]. In addition, CCL2-dependent enhanced migration of HIV-infected monocytes occurs as well as enhanced expression of MMP-2 and MMP-9, which disrupts the integrity of the BBB [167].

Neuroimmune substances such as cytokines, chemokines, excitatory amino acids, and neuropeptides can disrupt the BBB and cause endothelial cell and neuronal toxicity. Elevated levels of cytokines such as TNFα, IFNγ, IL-1β, and IL-6 and the chemokine CCL2 have been detected in sera, brain, and cerebrospinal fluid (CSF) of HIVE patients. These substances are produced by HIV-1-infected and HIV protein-activated macrophages, microglial cells, and astrocytes and are known modulators of BBB function [239, 241, 242].

7.3 Models for HIV-1 Neuropathogenesis

7.3.1 Laboratory Models

Laboratory models have been key in elucidating the interaction of HIV-1 with the cells of the immune system. Cell culture model systems have been used extensively for determining mechanisms of action of HIV-1 virus and viral proteins as well as the potential antiviral activities of ART. Human or macaque peripheral blood lymphocytes and leukocytes can be isolated from donors and cultured in the presence of cytokines and activators to induce differentiation and select for specific cell subtypes [200, 201, 243–250]. The responses of these cells to HIV-1 proteins and HIV-1 infection in regard to cytokine production and chemokine response can be determined. The cellular response of HIV-1 replication requires appropriate cellular growth factors that are linked to viral protein production and reverse transcriptase activity [251]. BBB function and integrity can be modeled using a monolayer culture of human brain microvascular endothelial cells. By culturing the endothelial cells on a microporous membrane and astrocytes on the opposite side of the membrane, the effect of activated monocytes on BBB integrity and the transmigration of the monocytes across the BBB construct was examined [252]. Cultures of human, simian, rodent, or feline astrocytes and microglia are used to determine glial cell responses to HIV-1 viral proteins [253–257]. Rodent, simian, and feline neuronal cultures and cell lines have been used to determine neuroexcitotoxicity and neuronal apoptosis in response to HIV-1 proteins [254, 258–261]. By coculturing glial cells and neuronal cells, the interactive response to HIV-1 infection of glial cells and the effect of glial cell production of cytokines and cytotoxic mediators on neuronal integrity can be determined.

7.3.2 Animal Models

The HIV-1 encephalitis mouse model relies on the use of isolated and cultured human lymphocytes that have been infected with HIV-1 virus and administered intracranially to immune deficient mice. This mouse model has been used to determine effective ART therapy and the interaction between the innate and adaptive immune responses in CNS responses [262, 263]. Use of ex vivo systems has also been explored for expansion of CD8+ T cells from patients on cART to select for those cells with the highest cytolytic activity and reinfusion of these cells into patients to restore immune responsiveness [264].

A particular challenge for the study of mechanisms of HIV neurodegeneration and development of new therapeutics is the specificity of HIV-1 infection for humans. The most widely used animal model has been the SIV-macaque model that accurately mimics many of the key clinical and pathological features of HIV infection including development of encephalitis and neurological impairment [177, 265–267]. In addition, insertion of portions of the HIV gene into SIV, creating hybrid simian-human immunodeficiency virus (SHIV), allows for studying the actions of HIV genes in vivo in monkeys. The SIV/macaque model has been valuable for studying innate and adaptive immune functions and their involvement in the development of late-stage CNS disease [267–269]. This model has also proven valuable for the discovery of therapeutic options and identification of mechanisms of therapeutic insufficiency and viral latency [270].

A second model that recapitulates most aspects of HIV infection in humans is feline immunodeficiency virus (FIV) infection of domestic and free-ranging large cats [271]. FIV is a naturally occurring lentivirus similar to HIV that expresses conserved tropism for the chemokine CXCR4 receptor and rapidly penetrates the CNS, infecting macrophages and microglial cells [272]. The use of both in vivo and feline cell culture systems for FIV has led to the identification of interactions among neuronal, astrocytic, T cell, and BBB components in development of neurological disease and provides a system for investigating potential therapeutic interventions [273–276].

Because of the expense of these models and limitations in their use, model systems using "humanized" and transgenic rodents have been developed for studying HIV-1 pathogenesis and neuroAIDS [277]. Development of humanized mice became possible with the discovery of severe combined immunodeficiency (SCID) mice [278] that contained an autosomal recessive mutation in the *prkdc* (DNA-dependent protein kinase, DNA-PK) gene resulting in deficiencies of mature T and B lymphocytes. This mutation allowed for the transplantation of foreign tissues into these mice and the development of animal models using transplanted human thymus, fetal liver, and peripheral blood lymphocytes [279, 280]. The immune cells from the immunocompetent human tissues are able to reconstitute the immune system of the SCID mouse. A SCID murine model of encephalitis, wherein HIV-1-infected human monocyte-derived macrophages (MDM) are injected into the brains of CB-17/scid immunodeficient mice, reproduces the key features of human neuropathology [281]. A modification of this model was developed to study the dynamics between HIV-1-infected macrophages and the adaptive immune response. In this modified model, nonobese diabetic (NOD)-SCID mice are repopulated with human peripheral blood lymphocytes (huPBL) and then injected intracranially with HIV-1-infected human MDM [262, 263]. The huPBL-HIVE model has demonstrated the role of the innate and adaptive immune systems in modulating the degree of CNS damage and repair. Long-term reconstitution of immune deficient mice with human immune cells has recently been demonstrated [282–284]. These "humanized" mice support chronic infection with HIV-1 and demonstrate hallmarks of HIV-1 infection, namely, loss of CD4+ cells, development of HIV-1-specific CTL, and humoral antiviral immune responses. In addition, CNS pathologies associated with human cell infiltration, microglial activation, and CD8+ cell depletion are observed in these mice following HIV-1 infection [283]. Although there are questions regarding the interaction between the murine system and human cell grafts that might limit mechanistic interpretations, the ability of these mice to survive long-term can help address questions of the interactions of aging and chronic HIV infection in the development of neurodegeneration and neuronal functional deficits [277].

Transgenic rodent models have also been developed as a tool for studying the effects of HIV-1 infection on CD4 expressing cells and the systemic effects of HIV-1 gene products. Rats engineered to express human CD4, CCR5, and cyclin T1 were observed to support limited HIV-1 expression which could be suppressed by treatment with the antiretroviral drugs enfuvirtide (peptidic fusion inhibitor) and efavirenz (nonnucleoside reverse transcriptase inhibitor) [285, 286]. In other transgenic rodent models, HIV-1 genes have been inserted into the mouse genome, and these mice have proven useful in defining tissue-specific pathology of specific HIV-1 gene products (reviewed by [287]). These transgenic mouse models have been useful in elucidating the roles of Tat, Vpr, and envelope proteins in development of neurotoxicity and neurodegeneration [277].

7.4 Translational Human Studies

7.4.1 Bioimaging

Neuroimaging studies have been used extensively to evaluate changes in the structure and function of the brain in patients with HAND. Because it is noninvasive, it presents the opportunity for both comparative measurements and monitoring progressive changes over time [288, 289]. Structural imaging using computed tomography (CT) and magnetic resonance imaging (MRI) has been useful in detecting OIs and neurodegeneration in patients with HAD but is not useful for detecting early cognitive impairments. CT in particular has been used to study the effects of HIV in the brains of infected children due to its ready availability, lower cost, and decreased likelihood for the use of anesthesia. In adults, MRI has proven to be more sensitive than CT in detection of progressive multifocal leukoencephalopathy lesions [290]. Structural MRI has revealed common findings of atrophy of gray matter in the basal ganglia, white matter lesions, ventricular enlargement, and cerebral atrophy in HIV-1-infected patients that correlate with disease progression (reviewed by [289]). Other scanning techniques that use ionizing radiation such as single photon emission computed tomography (SPECT) and positron emission tomography (PET) can be used to identify focal perfusion defects but are not suitable for diagnosing HIVE [291–293]. Imaging techniques that are useful during the early stages of neuroAIDS are those that can provide both functional and biochemical information. These techniques include MRS, PET, and functional, perfusion and diffusion MRI. PET techniques using radiolabeled tracers that target glial cell receptors, dopamine transporters, and dopamine receptors have been used to determine neuroinflammation and dopaminergic deficits in patients with HAND [294–296]. The PET substrate [^{18}F]flourodeoxyglucose is used to measure glucose uptake in brain regions as an indication of local metabolism [289]. Using this substrate, a progression of early stage hypermetabolism in the basal ganglia to hypometabolism in the medial frontal lobes during onset of motor dysfunction in HIV-infected patients was deduced [297–299]. Perfusion and functional MRI provide the ability to measure cerebral perfusion changes during the performance of specific cognitive tasks. Functional MRI (fMRI) studies, using blood-oxygen-level-dependent (BOLD) contrast, have detected increased activation in the frontal lobes of HIV patients with mild cognitive impairments [300–302] and suggest that increased utilization of brain reserve is needed to maintain normal cognitive function in these patients. Further studies have reported a correlation between disease severities as determined by nadir CD4 cell counts and increased viral load and enhanced activation during attention-related tasks in the right prefrontal and parietal regions and the left cerebellar region [303].

Proton and phosphorous MRS have been used clinically and experimentally to study brain metabolism in patients with HIV-associated brain injury and OIs and neoplasms in AIDS patients. MRS studies have observed subtle changes in choline-containing compounds and the medial glial marker M1 in white matter in early brain injury. In more advanced brain injury, a decrease in N-acetyl compounds and increases in choline-containing compounds and M1 are observed [304–306]. Recently, MRS tests demonstrated persistence of cognitive impairment and brain injury in patients on chronic cART [307]. This study demonstrated that brain inflammatory changes are common among infected people; whereas, neuronal injury is closely associated with individuals in whom cognitive impairment is seen. The observations uncovered in this study suggest that despite the widespread use of cART, HAND and brain injury persist. With continued development of bioimaging techniques to assess regional brain differences, specific metabolite profiles linked to neuronal injury can be identified. Imaging methods such as functional and structural MRI and MRS may also provide longitudinal measurements for disease progression.

7.4.2 *Biomarkers*

The persistence of HAND has led to the search for viral and host cellular biomarkers that may be indicators of effective cART and disease progression. Letendre and Ellis [308] have reviewed viral and host biomarkers in CSF and their response to ART. While a definitive diagnosis of HAND is not yet available, a panel of host CSF biomarkers has been proposed. This includes a viral and immune activation profiling and indicators of neuronal injury [67, 309]. CSF levels of viral RNA are reduced in most patients by cART and remain suppressed as long as plasma viral levels remain low and immune activation is suppressed. Other studies that monitored HIV RNA in the CSF and plasma and genotyped HIV-1 proteins such as pol or env suggested that drug resistant mutations are present in the CSF but not plasma in patients taking ART [310]. The presence of indicators of macrophage/microglial activation in the CSF is well known. Early studies identified elevated CSF levels of β-2 microglobulin and neopterin in patients with encephalitis. Elevated protein levels were reduced in patients on monotherapy or cART; however, in only 55 % of patients were neopterin levels reduced to normal, suggesting residual macrophage activation that is not suppressed by ART. The presence of soluble CD14 (sCD14), a marker of trafficking lymphocytes, in CSF has also been monitored in response to ART. In patients with severe HAND, cART has been shown to decrease sCD14 levels in the CSF [311]. Elevated levels of TNFα and the chemokines MIP-1α, MIP-1β, and RANTES in the CSF are seen during HAND and not affected by cART [312, 313]. Presence of the neuronal biomarker neurofilament light (NFL) has also been measured in CSF of HAND patients. Higher NFL levels have been reported in patients with HAND, and these levels can be reduced by initiation of cART [314, 315]. Interruption of cART resulted in a rebound in NFL levels in CSF [316].

Other protein biomarkers that can indicate changes in BBB integrity, neuronal excitotoxicity, astrocyte activation, and oxidative stress are seen in CSF. CSF biomarkers may serve as an indicator of the need for regimen changes in cART or inclusion of anti-inflammatory or neuroprotective adjunct therapies.

7.5 Antiretroviral Drugs and the Nervous System

7.5.1 Treatment

cART has achieved remarkable progress in reducing HIV-1 morbidity and mortality. The trends toward initiating ART earlier in disease and in certain comorbid conditions such as hepatitis B and C infection have served to greatly reduce the occurrence of HAD, and ART initiation can dramatically improve cognitive dysfunction in ART-naïve HAD patients [311, 317]. ART initiation in these patients decreases viral load and levels of neurotoxic compounds in plasma and CSF.

However, widespread use of ART for more than a decade has still not reduced the prevalence of HAND. Low penetration of ART into the CNS, complicated by lack of medication adherence by patients, may partially explain the persistence of HAND. To enter the CNS, drugs must pass the BBB or the blood CSF barrier for entry into the CSF. Physicochemical factors such as high protein binding and low lipophilicity can serve to reduce penetration across the BBB for some ART [312]. In addition, many ART are substrates for P-glycoprotein and other ATP-dependent efflux pumps on the luminal surface of brain endothelial cells that serve to prevent entry of many drugs into the CNS [318–320]. In an animal model of HIVE, inhibition of P-glycoprotein was effective at enhancing penetration of ART into the CNS and reducing CNS HIV-1 viral load [321]. Thus, inhibition of drug efflux transporters may enhance penetration of ART into the CNS, improving viral inhibition; however, increased CNS toxicity related to increased drug levels needs to be explored.

7.5.2 Long-Acting Nanoformulated ART (nanoART)

A broad range of nanoformulated medicines are currently under development for treatment of CNS diseases [322, 323]. These formulations have been designed for improving drug penetration across the BBB and enhancing therapeutic efficacy. Drug penetrance across the BBB can be improved by using nanoformulations to interfere with endothelial efflux pumps and metabolizing enzymes or by optimizing nanoART with respect to size, shape, and protein and lipid coatings to facilitate ART uptake, release, and transport across the BBB [324]. Conjugation of ritonavir nanoparticles with Tat peptide increased the transport of drug nanoparticles across the brain endothelial cells and resulted in CNS drug levels 800-fold higher than when free drug was administered [325]. Additional studies are exploring the use of cell-based nanoparticle delivery for ART. In this context, nanoformulated ART are taken up by MP and stored in nondegrading recycling endosomal compartments for release at sites of injury in the CNS [326–328]. Such formulations administered parenterally have proven effective in reducing viral infectivity in humanized mouse models of HIV-1 infection (our unpublished results).

7.5.3 Adjunctive Therapies

The pathophysiology of HAND shares many characteristics with that of other neurodegenerative disorders. While HAND is initiated by virus, a cascade of continued inflammation mediated by peripheral and central immune effector cells, neurotoxicity, proinflammatory cytokines and chemokines, and reduced levels of trophic factors serves to promote disease progression. Thus, therapies that have been designed for treating other neurodegenerative diseases are being increasingly utilized for treatment of HAND. A small phase I–II clinical trial using the L-type calcium channel antagonist nimodipine showed a trend toward patient improvement on quantitative neuropsychological testing and less painful peripheral neuropathy compared to placebo [329]. A second recently completed clinical trial using the NMDA receptor antagonist memantine showed promising positive changes by MRS, but minimal long-term improvement in patient performance using neuropsychological tests [330, 331]. Clinical trials have also been carried out using antioxidant compounds such as the vitamin E analog OPC-14117, lipoic acid, selegiline, thioctic acid, and CP-1189; however, no improvements in performance on cognitive tests compared to treatment with placebo were observed [332–337]. Trends toward improved cognitive performance were seen in clinical trials using inhibitors of platelet-activating factor (PAF) receptor. PAF, when released from NMDA-stimulated postsynaptic neurons, can increase presynaptic glutamate release and effect neuronal damage and is capable of inducing production of inflammatory cytokines such as TNF [338, 339]. Other clinical trials have shown trends toward improvement in neuropsychological performance and brain metabolism using valproate and lithium as inhibitors of the glycogen synthase kinase 3β (GSK-3β), an enzyme complex that can induce neuronal apoptosis when overactivated by Tat and PAF [336, 340]. The use of growth factors such as platelet-derived growth factor (PDGF) for reducing neurotoxicity has also been suggested by in vitro studies. In these studies human neuroblastoma cells and rat neuronal cells were protected from gp120-induced apoptosis by PDGF through downregulation of the proapoptotic protein bax and inhibition of cytochrome c release from mitochondria [168, 341]. Use of therapies such as these to modulate the immune response indirectly in conjunction with ART intervention to reduce viral load will likely provide the best approach to controlling HAND.

7.6 HIV and cART in Children and Adolescents

7.6.1 HIV-1 Encephalopathy in Children and Adolescents

The incidence of mother to child (vertical) transmission of HIV-1 has been greatly diminished by the advent of cART [342–347]. However in resource-limited parts of the world, where access to ART and compliance with protocols to limit perinatal exposure is inadequate, HIV-1 infection in children is still a major public health crisis [348, 349]. HIV-1-associated progressive encephalopathy (PE) is a defining clinical feature of HIV-1-associated neurological disease in children and is analogous to HAND in adults. As with HAND, there are differing levels of neurological impairment ranging from asymptomatic to minor impairment to HAD [350–356]. PE can cause impairment of brain growth and development, leading to motor dysfunction and a decline or plateau in neurodevelopment and neurocognition [350, 352–355, 357–359]. The direct and indirect effects of HIV-1 infection, as well as ART toxicities, are associated with increased incidences of seizures, stroke, myelopathies and neuropathies, OIs, CNS lymphomas, and neuropsychiatric disorders.

HIVE in children has many of the same pathological features seen in adults with AIDS and correlates roughly with PE. Activation of monocytes/macrophages is associated with CNS inflammatory cell infiltration consisting of infected microglia and macrophages and activated astrocytes, lymphocytes, and plasma cells. As in adults, the presence of multinucleated giant cells is common, and myelin pallor of the cerebral hemispheres associated with reactive astrocytosis is seen. HIV-1 infection of monocytes/macrophages and latent and restricted infection of astrocytes occur. However, unlike HIV-1-infected adults, over 90 % of children who die of HIV-1 infection exhibit basophilic mineralization in the basal ganglia and deep cerebral white matter. These deposits consist of calcium salts and other minerals such as iron and have been hypothesized to result from transient local disruptions in the BBB [360, 361]. In addition, infiltration of the parenchymal blood vessel walls in the hemispheric white matter and basal ganglia by lymphocytes, monocytes, and multinucleated giant cells can be observed in 25–30 % of children with AIDS.

7.6.2 Combination ART and Neuropathological Disease in Children

HIV-1 viral load and the presence of activated monocytes with accompanying inflammation in the CNS have been greatly reduced by the use of antiretroviral drugs. The early use of a single antiretroviral agent has been supplanted by initial therapeutic regimens that consist of combinations of antiretroviral drugs in order to limit spread and progression of the disease and to reduce development of drug resistance. Indeed the use of cART has resulted in a dramatic decline in severe forms of HIV-1 encephalopathy in children. Improvements in neurodevelopment upon initiation of cART in children with encephalopathy and the preservation of neurocognitive functioning in neurologically healthy children have been described [362–364]. These improvements, however, appear to be limited to the use of non-nucleoside reverse transcriptase inhibitors rather than protease inhibitors [365, 366]. This effect may be due to the lower CNS penetration of many protease inhibitors and the use of CNS penetrating ART is recommended. Timing of ART initiation has been shown to be important in improving clinical outcomes in HIV-1 infected infants. In addition, treatment of infants before the age of 6 months may prevent early-onset of severe HIV encephalopathy. However, while development of severe PE can be prevented with the use of cART, the occurrence of milder forms of neurodevelopmental delay and neurobehavioral problems may be increasing, especially in older children [367, 368]. Critical comprehensive assessment of neurodevelopment in HIV-1-infected children is difficult, especially in developing countries. Studies on the interactive roles of environment and viral infection in mild developmental delay and cognitive impairment are urgently needed.

7.7 Future Perspectives

There are several key questions that remain in better understanding HIV neurobiology, in improving treatment, in positively effecting drug delivery, and in improving disease diagnostics and monitoring. In regard to disease, viral diversity and molecular signatures for CI are paramount. For treatment, improving cART regimens and brain penetrance will prove pivotal if HAND can be prevented. An important parallel query that remains is a thorough examination of brain region-specific ART neurotoxicity and the effects that long-term peripheral immune activation may play on the nervous system. Key to addressing this query is finding which drugs alone or in combination affect disease as well as when, how, and where disease occurs. Yet another future need revolves around the development of adjunctive therapies; those that can work as additives to existing cART regiments or perhaps can improve cognitive function in synergy through targeting parallel disease-related pathways. This perhaps can be facilitated through improved drug delivery paradigms including works developed from our own laboratories using nanomedicines. The idea for both improved drug delivery paradigms and adjunctives (used alone or in combination) is that they are targeted to attenuate neuronal damage and its subsequent links to CI. It is fully understood that to date, neuroprotective and immune modulatory strategies have all but failed and none are used as part of any neuroAIDS therapeutic plan. This perhaps would be improved when and if specific biomarkers are found for disease and can be used to differentiate what disease components are caused by HIV itself or are secondary effects on aging or on other neurodegenerative disorders such as Alzheimer's and Parkinson's diseases. Last, there is no gold standard for determining adherence to cART and other medications. This is yet another important issue, as adherence to treatment regimens may also influence the development of HAND and affects its severity.

7.8 Conclusions

The ultimate eradication of HIV disease can potentially be realized through effective immunization. So far, a vaccine to prevent infection has not been realized. Until then, research efforts need to focus on better understanding virus-host interactions, viral diversity and persistence, and tropisms. To this end, viral reservoirs including those maintained in the CNS are of vital importance to both comorbid disease manifestations and to viral eradication. To better understand disease mechanisms and antiretroviral biodistribution and to find drugs or drug combinations that specifically effect nervous system function will prove key in improved outcomes for disease in any infected human viral host.

References

1. Borrow P, Tishon A, Oldstone MB (1991) Infection of lymphocytes by a virus that aborts cytotoxic T lymphocyte activity and establishes persistent infection. J Exp Med 174:203–212
2. Fauci AS (1993) Multifactorial nature of human immunodeficiency virus disease: implications for therapy. Science 262:1011–1018
3. Haase AT (1999) Population biology of HIV-1 infection: viral and CD4+ T cell demographics and dynamics in lymphatic tissues. Annu Rev Immunol 17:625–656. doi:10.1146/annurev.immunol.17.1.625
4. Pantaleo G et al (1994) Major expansion of CD8+ T cells with a predominant V beta usage during the primary immune response to HIV. Nature 370:463–467. doi:10.1038/370463a0
5. Roy S, Devadas K, Dhawan S, Buch S (2011) In: Gendelman HE et al (eds) The neurology of AIDS. Oxford University Press, New York, pp 22–35
6. Schmitz JE et al (1999) Control of viremia in simian immunodeficiency virus infection by CD8+ lymphocytes. Science 283:857–860
7. Alkhatib G, Broder CC, Berger EA (1996) Cell type-specific fusion cofactors determine human immunodeficiency virus type 1 tropism for T-cell lines versus primary macrophages. J Virol 70:5487–5494
8. Berger EA (1997) HIV entry and tropism: the chemokine receptor connection. AIDS 11(Suppl A):S3–S16
9. Berger EA (1998) HIV entry and tropism. When one receptor is not enough. Adv Exp Med Biol 452:151–157
10. Deng H et al (1996) Identification of a major co-receptor for primary isolates of HIV-1. Nature 381:661–666. doi:10.1038/381661a0
11. Dragic T et al (1996) HIV-1 entry into CD4+ cells is mediated by the chemokine receptor CC-CKR-5. Nature 381:667–673. doi:10.1038/381667a0
12. Cullen BR (1991) Regulation of HIV-1 gene expression. FASEB J 5:2361–2368
13. Cullen BR, Greene WC (1989) Regulatory pathways governing HIV-1 replication. Cell 58:423–426
14. Nabel G, Baltimore D (1987) An inducible transcription factor activates expression of human immunodeficiency virus in T cells. Nature 326:711–713. doi:10.1038/326711a0
15. Naghavi MH, Estable MC, Schwartz S, Roeder RG, Vahlne A (2001) Upstream stimulating factor affects human immunodeficiency virus type 1 (HIV-1) long terminal repeat-directed transcription in a cell-specific manner, independently of the HIV-1 subtype and the core-negative regulatory element. J Gen Virol 82:547–559

16. Bieniasz PD (2006) Late budding domains and host proteins in enveloped virus release. Virology 344:55–63. doi:10.1016/j.virol.2005.09.044
17. Gomez CY, Hope TJ (2006) Mobility of human immunodeficiency virus type 1 Pr55Gag in living cells. J Virol 80:8796–8806. doi:10.1128/JVI.02159-05
18. Gross I et al (2000) A conformational switch controlling HIV-1 morphogenesis. EMBO J 19:103–113. doi:10.1093/emboj/19.1.103
19. Hermida-Matsumoto L, Resh MD (2000) Localization of human immunodeficiency virus type 1 Gag and Env at the plasma membrane by confocal imaging. J Virol 74:8670–8679
20. Jouvenet N et al (2006) Plasma membrane is the site of productive HIV-1 particle assembly. PLoS Biol 4:e435
21. Klein KC, Reed JC, Lingappa JR (2007) Intracellular destinies: degradation, targeting, assembly, and endocytosis of HIV Gag. AIDS Rev 9:150–161
22. Morita E, Sundquist WI (2004) Retrovirus budding. Annu Rev Cell Dev Biol 20:395–425. doi:10.1146/annurev.cellbio.20.010403.102350
23. Yuan X, Yu X, Lee TH, Essex M (1993) Mutations in the N-terminal region of human immunodeficiency virus type 1 matrix protein block intracellular transport of the Gag precursor. J Virol 67:6387–6394
24. Gottlieb GS et al (2002) Equal plasma viral loads predict a similar rate of CD4+ T cell decline in human immunodeficiency virus (HIV) type 1- and HIV-2-infected individuals from Senegal, West Africa. J Infect Dis 185:905–914. doi:10.1086/339295
25. Kahn JO, Walker BD (1998) Acute human immunodeficiency virus type 1 infection. N Engl J Med 339:33–39. doi:10.1056/NEJM199807023390107
26. Laurence J (1993) T-cell subsets in health, infectious disease, and idiopathic CD4+ T lymphocytopenia. Ann Intern Med 119:55–62
27. Heaton RK et al (2011) HIV-associated neurocognitive disorders before and during the era of combination antiretroviral therapy: differences in rates, nature, and predictors. J Neurovirol 17:3–16. doi:10.1007/s13365-010-0006-1
28. Kraft-Terry SD, Stothert AR, Buch S, Gendelman HE (2010) HIV-1 neuroimmunity in the era of antiretroviral therapy. Neurobiol Dis 37:542–548. doi:10.1016/j.nbd.2009.12.015, pii: S0969-9961(09)00370-2
29. Yadav A, Collman RG (2009) CNS inflammation and macrophage/microglial biology associated with HIV-1 infection. J Neuroimmune Pharmacol 4:430–447. doi:10.1007/s11481-009-9174-2
30. Kraft-Terry S et al (2010) Proteomic analyses of monocytes obtained from Hispanic women with HIV-associated dementia show depressed antioxidants. Proteomics Clin Appl 4:706–714. doi:10.1002/prca.201000010
31. Antinori A et al (2007) Updated research nosology for HIV-associated neurocognitive disorders. Neurology 69:1789–1799. doi:10.1212/01.WNL.0000287431.88658.8b
32. Dore G, McDonald A, Li Y, Kaldor J, Brew B (2003) Marked improvement in survival following AIDS dementia complex in the era of highly active antiretroviral therapy. AIDS (Lond, Engl) 17:1539–1545
33. Sacktor N (2002) The epidemiology of human immunodeficiency virus-associated neurological disease in the era of highly active antiretroviral therapy. J Neurovirol 8(Suppl 2):115–121. doi:10.1080/13550280290101094
34. Woods SP, Moore DJ, Weber E, Grant I (2009) Cognitive neuropsychology of HIV-associated neurocognitive disorders. Neuropsychol Rev 19:152–168. doi:10.1007/s11065-009-9102-5
35. Boisse L, Gill MJ, Power C (2008) HIV infection of the central nervous system: clinical features and neuropathogenesis. Neurol Clin 26:799–819, x. doi:10.1016/j.ncl.2008.04.002
36. Grant I (2008) Neurocognitive disturbances in HIV. Int Rev Psychiatry 20:33–47. doi:10.1080/09540260701877894
37. Grant I, Sacktor N (2012) In: Gendelman HE et al (eds) The neurology of AIDS. Oxford University Press, New York, pp 488–503
38. Cole MA et al (2007) Longitudinally preserved psychomotor performance in long-term asymptomatic HIV-infected individuals. Neurology 69:2213–2220. doi:10.1212/01.WNL.0000277520.94788.82

39. Ellis RJ et al (2011) CD4 nadir is a predictor of HIV neurocognitive impairment in the era of combination antiretroviral therapy. AIDS 25:1747–1751. doi:10.1097/QAD.0b013e32834a40cd
40. Tozzi V et al (2007) Persistence of neuropsychologic deficits despite long-term highly active antiretroviral therapy in patients with HIV-related neurocognitive impairment: prevalence and risk factors. J Acquir Immune Defic Syndr 45:174–182. doi:10.1097/QAI.0b013e318042e1ee
41. Gupta J et al (1985) Specific identification of human papillomavirus type in cervical smears and paraffin sections by in situ hybridization with radioactive probes: a preliminary communication. Int J Gynecol Pathol 4:211–218
42. Liner KJ 2nd, Hall CD, Robertson KR (2007) Impact of human immunodeficiency virus (HIV) subtypes on HIV-associated neurological disease. J Neurovirol 13:291–304. doi:10.1080/13550280701422383
43. Rao VR et al (2008) HIV-1 clade-specific differences in the induction of neuropathogenesis. J Neurosci 28:10010–10016. doi:10.1523/JNEUROSCI.2955-08.2008, pii: 28/40/10010
44. Sacktor N et al (2009) HIV subtype D is associated with dementia, compared with subtype A, in immunosuppressed individuals at risk of cognitive impairment in Kampala, Uganda. Clin Infect Dis 49:780–786. doi:10.1086/605284
45. Mishra M, Vetrivel S, Siddappa NB, Ranga U, Seth P (2008) Clade-specific differences in neurotoxicity of human immunodeficiency virus-1 B and C Tat of human neurons: significance of dicysteine C30C31 motif. Ann Neurol 63:366–376. doi:10.1002/ana.21292
46. Hightower GK et al (2009) Select resistance-associated mutations in blood are associated with lower CSF viral loads and better neuropsychological performance. Virology 394:243–248. doi:10.1016/j.virol.2009.08.007, pii: S0042-6822(09)00487-5
47. Corder EH et al (1998) HIV-infected subjects with the E4 allele for APOE have excess dementia and peripheral neuropathy. Nat Med 4:1182–1184. doi:10.1038/2677
48. Cutler RG et al (2004) Involvement of oxidative stress-induced abnormalities in ceramide and cholesterol metabolism in brain aging and Alzheimer's disease. Proc Natl Acad Sci USA 101:2070–2075. doi:10.1073/pnas.0305799101
49. Letendre S et al (2004) The monocyte chemotactic protein-1–2578G allele is associated with elevated MCP-1 concentrations in cerebrospinal fluid. J Neuroimmunol 157:193–196. doi:10.1016/j.jneuroim.2004.08.028
50. Quasney MW et al (2001) Increased frequency of the tumor necrosis factor-alpha-308 A allele in adults with human immunodeficiency virus dementia. Ann Neurol 50:157–162
51. Valcour V et al (2004) Higher frequency of dementia in older HIV-1 individuals: the Hawaii Aging with HIV-1 Cohort. Neurology 63:822–827
52. Llorente A et al (2006) Effects of polymorphisms of chemokine receptors on neurodevelopment and the onset of encephalopathy in children with perinatal HIV-1 infection. Appl Neuropsychol 13:180–189. doi:10.1207/s15324826an1303_6
53. Singh KK et al (2003) Genetic influence of CCR5, CCR2, and SDF1 variants on human immunodeficiency virus 1 (HIV-1)-related disease progression and neurological impairment, in children with symptomatic HIV-1 infection. J Infect Dis 188:1461–1472. doi:10.1086/379038
54. Applebaum AJ et al (2009) The impact of neuropsychological functioning on adherence to HAART in HIV-infected substance abuse patients. AIDS Patient Care STDS 23:455–462. doi:10.1089/apc.2008.0181
55. Baum MK et al (2009) Crack-cocaine use accelerates HIV disease progression in a cohort of HIV-positive drug users. J Acquir Immune Defic Syndr 50:93–99. doi:10.1097/QAI.0b013e3181900129
56. Chang L, Ernst T, Speck O, Grob CS (2005) Additive effects of HIV and chronic methamphetamine use on brain metabolite abnormalities. Am J Psychiatry 162:361–369. doi:10.1176/appi.ajp. 162.2.361
57. Cherner M et al (2004) Effects of HIV-1 infection and aging on neurobehavioral functioning: preliminary findings. AIDS 18(Suppl 1):S27–S34
58. Fama R, Rosenbloom MJ, Nichols BN, Pfefferbaum A, Sullivan EV (2009) Working and episodic memory in HIV infection, alcoholism, and their comorbidity: baseline and 1-year follow-up examinations. Alcohol Clin Exp Res 33:1815–1824. doi:10.1111/j.1530-0277.2009.01020.x

59. Fein G, Fletcher DJ, Di Sclafani V (1998) Effect of chronic alcohol abuse on the CNS morbidity of HIV disease. Alcohol Clin Exp Res 22:196S–200S
60. Hinkin CH et al (2007) Drug use and medication adherence among HIV-1 infected individuals. AIDS Behav 11:185–194. doi:10.1007/s10461-006-9152-0
61. Rippeth JD et al (2004) Methamphetamine dependence increases risk of neuropsychological impairment in HIV infected persons. J Int Neuropsychol Soc 10:1–14. doi:10.1017/S1355617704101021
62. Rosenbloom MJ et al (2007) Alcoholism, HIV infection, and their comorbidity: factors affecting self-rated health-related quality of life. J Stud Alcohol Drugs 68:115–125
63. Toussi SS et al (2009) Short communication: methamphetamine treatment increases in vitro and in vivo HIV replication. AIDS Res Hum Retroviruses 25:1117–1121. doi:10.1089/aid.2008.0282
64. Valcour VG, Shikuma CM, Watters MR, Sacktor NC (2004) Cognitive impairment in older HIV-1-seropositive individuals: prevalence and potential mechanisms. AIDS 18(Suppl 1):S79–S86, pii: 00002030-200418001-00012
65. Gendelman H, Lipton S, Tardieu M, Bukrinsky M, Nottet H (1994) The neuropathogenesis of HIV-1 Infection. J Leukoc Biol 56:389–398
66. Lawrence DM, Major EO (2002) HIV-1 and the brain: connections between HIV-1-associated dementia, neuropathology and neuroimmunology. Microbes Infect 4:301–308. doi:10.1016/S1286-4579(02)01542-3
67. Ellis R, Langford D, Masliah E (2007) HIV and antiretroviral therapy in the brain: neuronal injury and repair. Nat Rev Neurosci 8:33–44. doi:10.1038/nrn2040
68. Anthony IC, Bell JE (2008) The neuropathology of HIV/AIDS. Int Rev Psychiatry 20:15–24. doi:10.1080/09540260701862037
69. Anthony IC, Ramage SN, Carnie FW, Simmonds P, Bell JE (2005) Influence of HAART on HIV-related CNS disease and neuroinflammation. J Neuropathol Exp Neurol 64:529–536
70. Zheng J, Gendelman HE (1997) The HIV-1 associated dementia complex: a metabolic encephalopathy fueled by viral replication in mononuclear phagocytes. Curr Opin Neurol 10:319–325
71. Kraft-Terry SD, Buch SJ, Fox HS, Gendelman HE (2009) A coat of many colors: neuroimmune crosstalk in human immunodeficiency virus infection. Neuron 64:133–145. doi:10.1016/j.neuron.2009.09.042, pii: S0896-6273(09)00753-3
72. Bell JE, Brettle RP, Chiswick A, Simmonds P (1998) HIV encephalitis, proviral load and dementia in drug users and homosexuals with AIDS. Effect of neocortical involvement. Brain 121:2043–2052
73. Clifford DB et al (2009) Effects of active HCV replication on neurologic status in HIV RNA virally suppressed patients. Neurology 73:309–314. doi:10.1212/WNL.0b013e3181af7a10
74. Letendre S et al (2007) Pathogenesis of hepatitis C virus coinfection in the brains of patients infected with HIV. J Infect Dis 196:361–370. doi:10.1086/519285
75. Hauser KF et al (2006) Impact of opiate-HIV-1 interactions on neurotoxic signaling. J Neuroimmune Pharmacol 1:98–105. doi:10.1007/s11481-005-9000-4
76. Hu S, Sheng WS, Lokensgard JR, Peterson PK (2005) Morphine potentiates HIV-1 gp120-induced neuronal apoptosis. J Infect Dis 191:886–889. doi:10.1086/427830, pii: JID33502
77. Khurdayan VK et al (2004) Preferential vulnerability of astroglia and glial precursors to combined opioid and HIV-1 Tat exposure in vitro. Eur J Neurosci 19:3171–3182. doi:10.1111/j.0953-816X.2004.03461.x
78. Bell JE, Arango JC, Anthony IC (2006) Neurobiology of multiple insults: HIV-1-associated brain disorders in those who use illicit drugs. J Neuroimmune Pharmacol 1:182–191. doi:10.1007/s11481-006-9018-2
79. Davies J et al (1998) HIV-associated brain pathology: a comparative international study. Neuropathol Appl Neurobiol 24:118–124
80. Arango JC, Simmonds P, Brettle RP, Bell JE (2004) Does drug abuse influence the microglial response in AIDS and HIV encephalitis? AIDS 18(Suppl 1):S69–S74

81. Martinez AJ et al (1995) The neuropathology and epidemiology of AIDS. A Berlin experience. A review of 200 cases. Pathol Res Pract 191:427–443
82. Pitcher J, Shimizu S, Burbassi S, Meucci O (2010) Disruption of neuronal CXCR4 function by opioids: preliminary evidence of ferritin heavy chain as a potential etiological agent in neuroAIDS. J Neuroimmunol 224:66–71. doi:10.1016/j.jneuroim.2010.05.006
83. Nath A et al (2008) Evolution of HIV dementia with HIV infection. Int Rev Psychiatry 20:25–31. doi:10.1080/09540260701861930
84. Hegyi Z, Kis G, Hollo K, Ledent C, Antal M (2009) Neuronal and glial localization of the cannabinoid-1 receptor in the superficial spinal dorsal horn of the rodent spinal cord. Eur J Neurosci 30:251–262. doi:10.1111/j.1460-9568.2009.06816.x, pii: EJN6816
85. McDowell JA, Chittick GE, Stevens CP, Edwards KD, Stein DS (2000) Pharmacokinetic interaction of abacavir (1592U89) and ethanol in human immunodeficiency virus-infected adults. Antimicrob Agents Chemother 44:1686–1690
86. Bidlack JM et al (2006) Opioid receptors and signaling on cells from the immune system. J Neuroimmune Pharmacol 1:260–269. doi:10.1007/s11481-006-9026-2
87. Eisenstein TK, Rahim RT, Feng P, Thingalaya NK, Meissler JJ (2006) Effects of opioid tolerance and withdrawal on the immune system. J Neuroimmune Pharmacol 1:237–249. doi:10.1007/s11481-006-9019-1
88. Finley MJ et al (2008) Bi-directional heterologous desensitization between the major HIV-1 co-receptor CXCR4 and the kappa-opioid receptor. J Neuroimmunol 197:114–123. doi:10.1016/j.jneuroim.2008.04.021
89. Machelska H, Stein C (2006) Leukocyte-derived opioid peptides and inhibition of pain. J Neuroimmune Pharmacol 1:90–97. doi:10.1007/s11481-005-9002-2
90. Molina PE (2006) Opioids and opiates: analgesia with cardiovascular, haemodynamic and immune implications in critical illness. J Intern Med 259:138–154. doi:10.1111/j.1365-2796.2005.01569.x
91. Donahoe RM et al (2009) Probable deceleration of progression of Simian AIDS affected by opiate dependency: studies with a rhesus macaque/SIVsmm9 model. J Acquir Immune Defic Syndr 50:241–249. doi:10.1097/QAI.0b013e3181967354
92. Kumar R et al (2006) Chronic morphine exposure causes pronounced virus replication in cerebral compartment and accelerated onset of AIDS in SIV/SHIV-infected Indian rhesus macaques. Virology 354:192–206. doi:10.1016/j.virol.2006.06.020
93. Baldwin GC et al (1997) Marijuana and cocaine impair alveolar macrophage function and cytokine production. Am J Respir Crit Care Med 156:1606–1613
94. Buch SJ, Yai H, Roy S (2012) In: Gendelman HE et al (eds) The neurology of AIDS. Oxford University Press, New York, pp 354–361
95. Dhillon N et al (2008) Molecular mechanism(s) involved in the synergistic induction of CXCL10 by human immunodeficiency virus type 1 Tat and interferon-gamma in macrophages. J Neurovirol 14:196–204. doi:10.1080/13550280801993648, pii: 794315853
96. Gan X et al (1999) Cocaine enhances brain endothelial adhesion molecules and leukocyte migration. Clin Immunol 91:68–76. doi:10.1006/clim.1998.4683
97. Zhang L et al (1998) Cocaine opens the blood–brain barrier to HIV-1 invasion. J Neurovirol 4:619–626
98. Ellis RJ et al (2003) Increased human immunodeficiency virus loads in active methamphetamine users are explained by reduced effectiveness of antiretroviral therapy. J Infect Dis 188:1820–1826. doi:10.1086/379894
99. Halkitis PN, Kutnick AH, Slater S (2005) The social realities of adherence to protease inhibitor regimens: substance use, health care and psychological states. J Health Psychol 10:545–558. doi:10.1177/1359105305053422
100. King MA et al (2009) Human peripheral blood leucocyte non-obese diabetic-severe combined immunodeficiency interleukin-2 receptor gamma chain gene mouse model of xenogeneic graft-versus-host-like disease and the role of host major histocompatibility complex. Clin Exp Immunol 157:104–118. doi:10.1111/j.1365-2249.2009.03933.x, pii: CEI3933
101. Moss AR et al (2004) Adherence to highly active antiretroviral therapy in the homeless population in San Francisco: a prospective study. Clin Infect Dis 39:1190–1198. doi:10.1086/424008

102. Pal D et al (2011) Efflux transporters- and cytochrome P-450-mediated interactions between drugs of abuse and antiretrovirals. Life Sci 88:959–971. doi:10.1016/J.Lfs.2010.09.012
103. Kumar AM et al (2009) Human immunodeficiency virus type 1 in the central nervous system leads to decreased dopamine in different regions of postmortem human brains. J Neurovirol 15:257–274. doi:10.1080/13550280902973952
104. Sekine Y et al (2003) Association of dopamine transporter loss in the orbitofrontal and dorsolateral prefrontal cortices with methamphetamine-related psychiatric symptoms. Am J Psychiatry 160:1699–1701
105. Sekine Y et al (2006) Brain serotonin transporter density and aggression in abstinent methamphetamine abusers. Arch Gen Psychiatry 63:90–100. doi:10.1001/archpsyc.63.1.90
106. Volkow ND et al (2001) Loss of dopamine transporters in methamphetamine abusers recovers with protracted abstinence. J Neurosci 21:9414–9418
107. Volkow ND et al (2001) Association of dopamine transporter reduction with psychomotor impairment in methamphetamine abusers. Am J Psychiatry 158:377–382
108. Garwood ER, Bekele W, McCulloch CE, Christine CW (2006) Amphetamine exposure is elevated in Parkinson's disease. Neurotoxicology 27:1003–1006. doi:10.1016/j.neuro.2006.03.015
109. Tse W et al (2004) Movement disorders and AIDS: a review. Parkinsonism Relat Disord 10:323–334. doi:10.1016/j.parkreldis.2004.03.001
110. Qi L et al (2011) Programmed neuronal cell death induced by HIV-1 tat and methamphetamine. Microsc Res Tech 74:1139–1144. doi:10.1002/jemt.21006
111. Valcour V et al (2008) Aging exacerbates extrapyramidal motor signs in the era of highly active antiretroviral therapy. J Neurovirol 14:362–367. doi:10.1080/13550280802216494
112. Ikezu T (2009) The aging of human-immunodeficiency-virus-associated neurocognitive disorders. J Neuroimmune Pharmacol 4:161–162. doi:10.1007/s11481-009-9155-5
113. Anthony IC, Ramage SN, Carnie FW, Simmonds P, Bell JE (2006) Accelerated Tau deposition in the brains of individuals infected with human immunodeficiency virus-1 before and after the advent of highly active anti-retroviral therapy. Acta Neuropathol 111:529–538. doi:10.1007/s00401-006-0037-0
114. Esiri MM, Biddolph SC, Morris CS (1998) Prevalence of Alzheimer plaques in AIDS. J Neurol Neurosurg Psychiatry 65:29–33
115. Rempel HC, Pulliam L (2005) HIV-1 Tat inhibits neprilysin and elevates amyloid beta. AIDS 19:127–135
116. Ernst T, Chang L (2004) Effect of aging on brain metabolism in antiretroviral-naive HIV patients. AIDS 18(Suppl 1):S61–S67
117. Daily A, Nath A, Hersh LB (2006) Tat peptides inhibit neprilysin. J Neurovirol 12:153–160. doi:10.1080/13550280600760677
118. Giunta B et al (2008) HIV-1 TAT inhibits microglial phagocytosis of Abeta peptide. Int J Clin Exp Pathol 1:260–275
119. Hamel FG, Fawcett J, Tsui BT, Bennett RG, Duckworth WC (2006) Effect of nelfinavir on insulin metabolism, proteasome activity and protein degradation in HepG2 cells. Diabetes Obes Metab 8:661–668. doi:10.1111/j.1463-1326.2005.00546.x
120. Piccinini M et al (2005) The HIV protease inhibitors nelfinavir and saquinavir, but not a variety of HIV reverse transcriptase inhibitors, adversely affect human proteasome function. Antivir Ther 10:215–223
121. Schweinsburg BC et al (2005) Brain mitochondrial injury in human immunodeficiency virus-seropositive (HIV+) individuals taking nucleoside reverse transcriptase inhibitors. J Neurovirol 11:356–364. doi:10.1080/13550280591002342
122. Valcour V, Shiramizu B (2004) HIV-associated dementia, mitochondrial dysfunction, and oxidative stress. Mitochondrion 4:119–129. doi:10.1016/j.mito.2004.05.009
123. Giralt M et al (2006) HIV-1 infection alters gene expression in adipose tissue, which contributes to HIV-1/HAART-associated lipodystrophy. Antivir Ther 11:729–740
124. Shanley LJ, Irving AJ, Harvey J (2001) Leptin enhances NMDA receptor function and modulates hippocampal synaptic plasticity. J Neurosci 21:RC186

125. Grunfeld C (2008) Insulin resistance in HIV infection: drugs, host responses, or restoration to health? Top HIV Med 16:89–93
126. Mallewa JE et al (2008) HIV-associated lipodystrophy: a review of underlying mechanisms and therapeutic options. J Antimicrob Chemother 62:648–660. doi:10.1093/jac/dkn251
127. Mattson MP, Haughey NJ, Nath A (2005) Cell death in HIV dementia. Cell Death Differ 12(Suppl 1):893–904. doi:10.1038/sj.cdd.4401577
128. Dreyer EB, Kaiser PK, Offermann JT, Lipton SA (1990) HIV-1 coat protein neurotoxicity prevented by calcium channel antagonists. Science 248:364–367
129. Meucci O et al (1998) Chemokines regulate hippocampal neuronal signaling and gp120 neurotoxicity. Proc Natl Acad Sci USA 95:14500–14505
130. Zheng J et al (1999) Intracellular CXCR4 signaling, neuronal apoptosis and neuropathogenic mechanisms of HIV-1-associated dementia. J Neuroimmunol 98:185–200, pii: S0165572899000491
131. Chandra T et al (2005) Molecular interactions of the type 1 human immunodeficiency virus transregulatory protein Tat with N-methyl-D-aspartate receptor subunits. Neuroscience 134:145–153. doi:10.1016/j.neuroscience.2005.02.049
132. Gonzalez-Scarano F, Martin-Garcia J (2005) The neuropathogenesis of AIDS. Nat Rev Immunol 5:69–81. doi:10.1038/nri1527
133. Erdmann NB, Whitney NP, Zheng J (2006) Potentiation of excitotoxicity in HIV-1 associated dementia and the significance of glutaminase. Clin Neurosci Res 6:315–328. doi:10.1016/j.cnr.2006.09.009
134. Jones GJ et al (2007) HIV-1 Vpr causes neuronal apoptosis and in vivo neurodegeneration. J Neurosci 27:3703–3711. doi:10.1523/JNEUROSCI.5522-06.2007, pii: 27/14/3703
135. Kort JJ, Jalonen TO (1998) The nef protein of the human immunodeficiency virus type 1 (HIV-1) inhibits a large-conductance potassium channel in human glial cells. Neurosci Lett 251:1–4
136. Piller SC, Ewart GD, Premkumar A, Cox GB, Gage PW (1996) Vpr protein of human immunodeficiency virus type 1 forms cation-selective channels in planar lipid bilayers. Proc Natl Acad Sci USA 93:111–115
137. Piller SC, Jans P, Gage PW, Jans DA (1998) Extracellular HIV-1 virus protein R causes a large inward current and cell death in cultured hippocampal neurons: implications for AIDS pathology. Proc Natl Acad Sci USA 95:4595–4600
138. Sabbah EN, Roques BP (2005) Critical implication of the (70–96) domain of human immunodeficiency virus type 1 Vpr protein in apoptosis of primary rat cortical and striatal neurons. J Neurovirol 11:489–502. doi:10.1080/13550280500384941
139. Trillo-Pazos G, McFarlane-Abdulla E, Campbell IC, Pilkington GJ, Everall IP (2000) Recombinant nef HIV-IIIB protein is toxic to human neurons in culture. Brain Res 864:315–326
140. Adamson DC et al (1996) Immunologic NO synthase: elevation in severe AIDS dementia and induction by HIV-1 gp41. Science 274:1917–1921
141. Bukrinsky MI et al (1995) Regulation of nitric oxide synthase activity in human immunodeficiency virus type 1 (HIV-1)-infected monocytes: implications for HIV-associated neurological disease. J Exp Med 181:735–745
142. Garden GA et al (2002) Caspase cascades in human immunodeficiency virus-associated neurodegeneration. J Neurosci 22:4015–4024. doi:20026351, pii: 22/10/4015
143. Giulian D (1990) Microglia, cytokines, and cytotoxins: modulators of cellular responses after injury to the central nervous system. J Immunol Immunopharmacol 10:15–21
144. Giulian D, Corpuz M, Chapman S, Mansouri M, Robertson C (1993) Reactive mononuclear phagocytes release neurotoxins after ischemic and traumatic injury to the central nervous system. J Neurosci Res 36:681–693
145. Giulian D, Vaca K, Noonan CA (1990) Secretion of neurotoxins by mononuclear phagocytes infected with HIV-1. Science 250:1593–1596
146. Giulian D et al (1996) Study of receptor-mediated neurotoxins released by HIV-1-infected mononuclear phagocytes found in human brain. J Neurosci 16:3139–3153
147. Kaul M, Garden GA, Lipton SA (2001) Pathways to neuronal injury and apoptosis in HIV-associated dementia. Nature 410:988–994

148. Kaul M, Lipton SA (1999) Chemokines and activated macrophages in HIV gp120-induced neuronal apoptosis [in process citation]. Proc Natl Acad Sci USA 96:8212–8216
149. Lipton SA, Gendelman HE (1995) Seminars in medicine of the Beth Israel Hospital, Boston. Dementia associated with the acquired immunodeficiency syndrome. N Engl J Med 332:934–940. doi:10.1056/NEJM199504063321407
150. Yeh MW et al (2000) Cytokine-stimulated, but not HIV-infected, human monocyte-derived macrophages produce neurotoxic levels of l-cysteine. J Immunol 164:4265–4270
151. Lipton SA (1992) Models of neuronal injury in AIDS: another role for the NMDA receptor? Trends Neurosci 15:75–79
152. Li W, Galey D, Mattson MP, Nath A (2005) Molecular and cellular mechanisms of neuronal cell death in HIV dementia. Neurotox Res 8:119–134
153. Walsh KA et al (2004) Antioxidant protection from HIV-1 gp120-induced neuroglial toxicity. J Neuroinflammation 1:8. doi:10.1186/1742-2094-1-8
154. Mukerjee R et al (2008) Involvement of the p53 and p73 transcription factors in neuroAIDS. Cell Cycle 7:2682–2690
155. Jayadev S et al (2007) The glial response to CNS HIV infection includes p53 activation and increased expression of p53 target genes. J Neuroimmune Pharmacol 2:359–370. doi:10.1007/s11481-007-9095-x
156. Garden GA et al (2004) HIV associated neurodegeneration requires p53 in neurons and microglia. FASEB J 18:1141–1143. doi:10.1096/fj.04-1676fje
157. Jayadev S et al (2011) Transcription factor p53 influences microglial activation phenotype. Glia 59:1402–1413. doi:10.1002/glia.21178
158. Acheampong EA et al (2005) Human immunodeficiency virus type 1 Nef potently induces apoptosis in primary human brain microvascular endothelial cells via the activation of caspases. J Virol 79:4257–4269. doi:10.1128/JVI.79.7.4257-4269.2005
159. Bergonzini V et al (2009) Nef and cell signaling transduction: a possible involvement in the pathogenesis of human immunodeficiency virus-associated dementia. J Neurovirol 15:238–248. doi:10.1080/13550280902939748
160. Kohleisen B et al (1999) Stable expression of HIV-1 Nef induces changes in growth properties and activation state of human astrocytes. AIDS 13:2331–2341
161. Ni HT et al (2004) High-level expression of functional chemokine receptor CXCR4 on human neural precursor cells. Brain Res Dev Brain Res 152:159–169. doi:10.1016/j.devbrainres.2004.06.015
162. Okamoto S et al (2007) HIV/gp120 decreases adult neural progenitor cell proliferation via checkpoint kinase-mediated cell-cycle withdrawal and G1 arrest. Cell Stem Cell 1:230–236. doi:10.1016/j.stem.2007.07.010
163. Keohane A, Ryan S, Maloney E, Sullivan AM, Nolan YM (2010) Tumour necrosis factor-alpha impairs neuronal differentiation but not proliferation of hippocampal neural precursor cells: role of Hes1. Mol Cell Neurosci 43:127–135. doi:10.1016/j.mcn.2009.10.003
164. Kanmogne G (2012) In: Gendelman HE et al (eds) The neurology of AIDS. Oxford University Press, New York, pp 246–254
165. Martinez-Skinner A, Nowacek AS, McMillan J, Gendelman HE (2012) In: Gendelman HE et al (eds) The neurology of AIDS. Oxford University Press, New York, pp 257–274
166. Tardieu M, Boutet A (2002) HIV-1 and the central nervous system. Curr Top Microbiol Immunol 265:183–195
167. Eugenin EA et al (2006) CCL2/monocyte chemoattractant protein-1 mediates enhanced transmigration of human immunodeficiency virus (HIV)-infected leukocytes across the blood–brain barrier: a potential mechanism of HIV-CNS invasion and NeuroAIDS. J Neurosci 26:1098–1106. doi:10.1523/JNEUROSCI.3863-05.2006, pii: 26/4/1098
168. Peng F et al (2008) Mechanisms of platelet-derived growth factor-mediated neuroprotection – implications in HIV dementia. Eur J Neurosci 28:1255–1264. doi:10.1111/j.1460-9568.2008.06444.x
169. Ricardo-Dukelow M et al (2007) HIV-1 infected monocyte-derived macrophages affect the human brain microvascular endothelial cell proteome: new insights into blood–brain barrier dysfunction for HIV-1-associated dementia. J Neuroimmunol 185:37–46. doi:10.1016/j.jneuroim.2007.01.004, pii: S0165-5728(07)00005-7

170. Brabers NA, Nottet HS (2006) Role of the pro-inflammatory cytokines TNF-alpha and IL-1beta inHIV-associateddementia.EurJClinInvest36:447–458.doi:10.1111/j.1365-2362.2006.01657.x, pii: ECI1657
171. Persidsky Y, Ramirez SH, Haorah J, Kanmogne GD (2006) Blood–brain barrier: structural components and function under physiologic and pathologic conditions. J Neuroimmune Pharmacol 1:223–236. doi:10.1007/s11481-006-9025-3
172. Kadiu I, Ricardo-Dukelow M, Ciborowski P, Gendelman HE (2007) Cytoskeletal protein transformation in HIV-1-infected macrophage giant cells. J Immunol 178:6404–6415, pii: 178/10/6404
173. Benoit M, Desnues B, Mege JL (2008) Macrophage polarization in bacterial infections. J Immunol 181:3733–3739
174. Cassol E, Cassetta L, Alfano M, Poli G (2010) Macrophage polarization and HIV-1 infection. J Leukoc Biol 87:599–608. doi:10.1189/jlb.1009673
175. Cassol E, Cassetta L, Rizzi C, Alfano M, Poli G (2009) M1 and M2a polarization of human monocyte-derived macrophages inhibits HIV-1 replication by distinct mechanisms. J Immunol 182:6237–6246. doi:10.4049/jimmunol.0803447
176. Stout RD, Suttles J (2004) Functional plasticity of macrophages: reversible adaptation to changing microenvironments. J Leukoc Biol 76:509–513. doi:10.1189/jlb.0504272
177. Crews L, Lentz MR, Gonzalez RG, Fox HS, Masliah E (2008) Neuronal injury in simian immunodeficiency virus and other animal models of neuroAIDS. J Neurovirol 14:327–339. doi:10.1080/13550280802132840
178. Moore DJ et al (2006) Cortical and subcortical neurodegeneration is associated with HIV neurocognitive impairment. AIDS 20:879–887. doi:10.1097/01.aids.0000218552.69834.00
179. Krathwohl MD, Kaiser JL (2004) HIV-1 promotes quiescence in human neural progenitor cells. J Infect Dis 190:216–226. doi:10.1086/422008
180. Lawrence DM et al (2004) Human immunodeficiency virus type 1 infection of human brain-derived progenitor cells. J Virol 78:7319–7328. doi:10.1128/JVI.78.14.7319-7328.2004
181. Reinecke JB, Peng H, Huang Y, Chen Q, Zheng JC (2012) In: Gendelman HE et al (eds) The neurology of AIDS. Oxford University Press, New York, pp 173–186
182. Tran PB, Miller RJ (2005) HIV-1, chemokines and neurogenesis. Neurotox Res 8:149–158
183. van Marle G, Power C (2005) Human immunodeficiency virus type 1 genetic diversity in the nervous system: evolutionary epiphenomenon or disease determinant? J Neurovirol 11:107–128. doi:10.1080/13550280590922838, pii: K306R212512Q0Q45
184. Mathieu P et al (2010) The more you have, the less you get: the functional role of inflammation on neuronal differentiation of endogenous and transplanted neural stem cells in the adult brain. J Neurochem 112:1368–1385. doi:10.1111/j.1471-4159.2009.06548.x
185. Koenig S et al (1986) Detection of AIDS virus in macrophages in brain tissue from AIDS patients with encephalopathy. Science 233:1089–1093
186. Churchill MJ et al (2009) Extensive astrocyte infection is prominent in human immunodeficiency virus-associated dementia. Ann Neurol 66:253–258. doi:10.1002/ana.21697
187. Eugenin EA, Berman JW (2007) Gap junctions mediate human immunodeficiency virus-bystander killing in astrocytes. J Neurosci 27:12844–12850. doi:10.1523/JNEUROSCI.4154-07.2007
188. Eugenin EA, Clements JE, Zink MC, Berman JW (2011) Human immunodeficiency virus infection of human astrocytes disrupts blood–brain barrier integrity by a gap junction-dependent mechanism. J Neurosci 31:9456–9465. doi:10.1523/JNEUROSCI.1460-11.2011
189. Ju SM et al (2009) Extracellular HIV-1 Tat up-regulates expression of matrix metalloproteinase-9 via a MAPK-NF-kappaB dependent pathway in human astrocytes. Exp Mol Med 41:86–93, pii: 20092284
190. Williams R et al (2009) Proinflammatory cytokines and HIV-1 synergistically enhance CXCL10 expression in human astrocytes. Glia 57:734–743. doi:10.1002/glia.20801
191. Xing HQ et al (2009) In vivo expression of proinflammatory cytokines in HIV encephalitis: an analysis of 11 autopsy cases. Neuropathology 29:433–442
192. Bissel SJ et al (2008) Systemic and brain macrophage infections in relation to the development of simian immunodeficiency virus encephalitis. J Virol 82:5031–5042. doi:10.1128/JVI.02069-07

193. Hutter-Saunders JAL, Poluektova L, Dhawan S, Gendelman HE (2012) In: Gendelman HE et al (eds) The neurology of AIDS. Oxford University Press, New York, pp 131–145
194. Mankowski JL, Clements JE, Zink MC (2002) Searching for clues: tracking the pathogenesis of human immunodeficiency virus central nervous system disease by use of an accelerated, consistent simian immunodeficiency virus macaque model. J Infect Dis 186(Suppl 2):S199–S208. doi:10.1086/344938
195. Sadagopal S et al (2008) Enhancement of human immunodeficiency virus (HIV)-specific CD8+ T cells in cerebrospinal fluid compared to those in blood among antiretroviral therapy-naive HIV-positive subjects. J Virol 82:10418–10428. doi:10.1128/JVI.01190-08
196. Shieh TM et al (2001) Functional analyses of natural killer cells in macaques infected with neurovirulent simian immunodeficiency virus. J Neurovirol 7:11–24. doi:10.1080/135502801300069593
197. Aloisi F et al (1999) Relative efficiency of microglia, astrocytes, dendritic cells and B cells in naive CD4+ T cell priming and Th1/Th2 cell restimulation. Eur J Immunol 29:2705–2714
198. Pantaleo G, Koup RA (2004) Correlates of immune protection in HIV-1 infection: what we know, what we don't know, what we should know. Nat Med 10:806–810. doi:10.1038/nm0804-806
199. Gong N et al (2011) Brain ingress of regulatory T cells in a murine model of HIV-1 encephalitis. J Neuroimmunol 230:33–41. doi:10.1016/j.jneuroim.2010.08.014, pii: S0165-5728(10)00378-4
200. Huang X, Reynolds AD, Mosley RL, Gendelman HE (2009) CD 4+ T cells in the pathobiology of neurodegenerative disorders. J Neuroimmunol 211:3–15. doi:10.1016/j.jneuroim.2009.04.006, pii: S0165-5728(09)00140-4
201. Liu J et al (2009) Neuromodulatory activities of CD4+ CD25+ regulatory T cells in a murine model of HIV-1-associated neurodegeneration. J Immunol 182:3855–3865. doi:10.4049/jimmunol.0803330, pii: 182/6/3855
202. Sakaguchi S (2000) Regulatory T cells: key controllers of immunologic self-tolerance. Cell 101:455–458
203. Takahashi T et al (2000) Immunologic self-tolerance maintained by CD25(+)CD4(+) regulatory T cells constitutively expressing cytotoxic T lymphocyte-associated antigen 4. J Exp Med 192:303–310
204. Seddiki N, Kelleher AD (2008) Regulatory T cells in HIV infection: who's suppressing what? Curr Infect Dis Rep 10:252–258
205. Tenorio AR et al (2009) The effect of aging on T-regulatory cell frequency in HIV infection. Clin Immunol 130:298–303. doi:10.1016/j.clim.2008.10.001
206. Ansari AA, Pattanapanyasat K, Pereira LE (2007) Autoimmunity and HIV/simian immunodeficiency virus infection: a two edged sword. Hepatol Res 37(Suppl 3):S389–S395. doi:10.1111/j.1872-034X.2007.00225.x
207. Klatt NR et al (2008) Availability of activated CD4+ T cells dictates the level of viremia in naturally SIV-infected sooty mangabeys. J Clin Invest 118:2039–2049. doi:10.1172/JCI33814
208. O'Connell K, Siliciano RF (2008) Immune alteration fends off AIDS. Nat Med 14:1016–1018. doi:10.1038/nm1008-1016
209. Silvestri G, Paiardini M, Pandrea I, Lederman MM, Sodora DL (2007) Understanding the benign nature of SIV infection in natural hosts. J Clin Invest 117:3148–3154. doi:10.1172/JCI33034
210. French MA et al (2000) Immune restoration disease after the treatment of immunodeficient HIV-infected patients with highly active antiretroviral therapy. HIV Med 1:107–115
211. Sharma SK, Soneja M (2011) HIV & immune reconstitution inflammatory syndrome (IRIS). Indian J Med Res 134:866–877. doi:10.4103/0971-5916.92632
212. Mahnke YD et al (2012) Selective expansion of polyfunctional pathogen-specific CD4+ T cells in HIV-1-infected patients with immune reconstitution inflammatory syndrome. Blood. doi:10.1182/blood-2011-09-380840
213. Ballabh P, Braun A, Nedergaard M (2004) The blood–brain barrier: an overview: structure, regulation, and clinical implications. Neurobiol Dis 16:1–13. doi:10.1016/j.nbd.2003.12.016

214. Banks WA (2012) In: Gendelman HE et al (eds) The neurology of AIDS. Oxford University Press, New York, pp 189–206
215. Gloor SM et al (2001) Molecular and cellular permeability control at the blood–brain barrier. Brain Res Brain Res Rev 36:258–264
216. Pardridge WM (1986) Blood–brain barrier transport of nutrients. Nutr Rev 44(Suppl):15–25
217. Risau W (1991) Induction of blood–brain barrier endothelial cell differentiation. Ann N Y Acad Sci 633:405–419
218. Risau W, Wolburg H (1990) Development of the blood–brain barrier. Trends Neurosci 13:174–178
219. Rubin LL, Staddon JM (1999) The cell biology of the blood–brain barrier. Annu Rev Neurosci 22:11–28. doi:10.1146/annurev.neuro.22.1.11
220. Strelow LI, Janigro D, Nelson JA (2001) The blood–brain barrier and AIDS. Adv Virus Res 56:355–388
221. Goldstein GW (1988) Endothelial cell-astrocyte interactions. A cellular model of the blood–brain barrier. Ann N Y Acad Sci 529:31–39
222. Pardridge WM, Boado RJ, Farrell CR (1990) Brain-type glucose transporter (GLUT-1) is selectively localized to the blood–brain barrier. Studies with quantitative western blotting and in situ hybridization. J Biol Chem 265:18035–18040
223. Shen SS, Zhang WD (2010) ABC transporters and drug efflux at the blood–brain barrier. Rev Neurosci 21:29–53
224. Agrawal S et al (2006) Dystroglycan is selectively cleaved at the parenchymal basement membrane at sites of leukocyte extravasation in experimental autoimmune encephalomyelitis. J Exp Med 203:1007–1019
225. Nottet HS et al (1996) Mechanisms for the transendothelial migration of HIV-1-infected monocytes into brain. J Immunol 156:1284–1295
226. Persidsky Y, Buttini M, Limoges J, Bock P, Gendelman HE (1997) An analysis of HIV-1-associated inflammatory products in brain tissue of humans and SCID mice with HIV-1 encephalitis. J Neurovirol 3:401–416
227. Pu H et al (2003) HIV-1 Tat protein upregulates inflammatory mediators and induces monocyte invasion into the brain. Mol Cell Neurosci 24:224–237
228. Liu Y et al (2004) Regulation of leukocyte transmigration: cell surface interactions and signaling events. J Immunol 172:7–13
229. Muller WA (2003) Leukocyte-endothelial-cell interactions in leukocyte transmigration and the inflammatory response. Trends Immunol 24:327–334
230. van Buul JD, Hordijk PL (2004) Signaling in leukocyte transendothelial migration. Arterioscler Thromb Vasc Biol 24:824–833. doi:10.1161/01.ATV.0000122854.76267.5c
231. Engelhardt B (2008) Immune cell entry into the central nervous system: involvement of adhesion molecules and chemokines. J Neurol Sci 274:23–26. doi:10.1016/J.Jns.2008.05.019
232. Engelhardt B (2008) The blood-central nervous system barriers actively control immune cell entry into the central nervous system. Curr Pharm Des 14:1555–1565
233. Nottet HSLM (1999) Interactions between macrophages and brain microvascular endothelial cells: role in pathogenesis of HIV-1 infection and blood–brain barrier function. J Neurovirol 5:659–669
234. Del Maschio A et al (1999) Leukocyte recruitment in the cerebrospinal fluid of mice with experimental meningitis is inhibited by an antibody to junctional adhesion molecule (JAM). J Exp Med 190:1351–1356
235. Ebnet K, Suzuki A, Ohno S, Vestweber D (2004) Junctional adhesion molecules (JAMs): more molecules with dual functions? J Cell Sci 117:19–29. doi:10.1242/jcs.00930
236. Martin-Padura I et al (1998) Junctional adhesion molecule, a novel member of the immunoglobulin superfamily that distributes at intercellular junctions and modulates monocyte transmigration. J Cell Biol 142:117–127
237. Ostermann G, Weber KS, Zernecke A, Schroder A, Weber C (2002) JAM-1 is a ligand of the beta(2) integrin LFA-1 involved in transendothelial migration of leukocytes. Nat Immunol 3:151–158. doi:10.1038/ni755

238. Schenkel AR, Mamdouh Z, Chen X, Liebman RM, Muller WA (2002) CD99 plays a major role in the migration of monocytes through endothelial junctions. Nat Immunol 3:143–150. doi:10.1038/ni749
239. Buckner CM, Luers AJ, Calderon TM, Eugenin EA, Berman JW (2006) Neuroimmunity and the blood–brain barrier: molecular regulation of leukocyte transmigration and viral entry into the nervous system with a focus on neuroAIDS. J Neuroimmune Pharmacol 1:160–181. doi:10.1007/s11481-006-9017-3
240. Seilhean D et al (1997) Astrocytic adhesion molecules are increased in HIV-1-associated cognitive/motor complex. Neuropathol Appl Neurobiol 23:83–92
241. Banks WA, Ercal N, Price TO (2006) The blood–brain barrier in neuroAIDS. Curr HIV Res 4:259–266
242. Cartier L, Hartley O, Dubois-Dauphin M, Krause KH (2005) Chemokine receptors in the central nervous system: role in brain inflammation and neurodegenerative diseases. Brain Res Brain Res Rev 48:16–42. doi:10.1016/j.brainresrev.2004.07.021
243. Gendelman HE et al (1990) Regulation of HIV replication in infected monocytes by IFN-alpha. Mechanisms for viral restriction. J Immunol 145:2669–2676
244. Gendelman HE, Genis P, Jett M, Zhai QH, Nottet HS (1994) An experimental model system for HIV-1-induced brain injury. Adv Neuroimmunol 4:189–193
245. Huang ZB et al (1993) Infection of macrophages with lymphotropic human immunodeficiency virus type 1 can be arrested after viral DNA synthesis. J Virol 67:6893–6896
246. Colleton BA et al (2009) Primary human immunodeficiency virus type 1-specific CD8+ T-cell responses induced by myeloid dendritic cells. J Virol 83:6288–6299. doi:10.1128/JVI.02611-08
247. Lubong Sabado R et al (2009) In vitro priming recapitulates in vivo HIV-1 specific T cell responses, revealing rapid loss of virus reactive CD4 T cells in acute HIV-1 infection. PLoS One 4:e4256. doi:10.1371/journal.pone.0004256
248. Sabado RL et al (2007) Pathways utilized by dendritic cells for binding, uptake, processing and presentation of antigens derived from HIV-1. Eur J Immunol 37:1752–1763. doi:10.1002/eji.200636981
249. Malaspina A, Rinaldo CR, Sekaly RP, Flores J, D'Souza PM (2011) "In vitro systems to characterize the immune response to HIV-1 and HIV-1 vaccine candidates", NIAID Workshop Report, Bethesda, August 4, 2010. Vaccine 29:4647–4653. http://dx.doi.org/10.1016/j.vaccine.2011.04.035
250. Co JG, Witwer KW, Gama L, Zink MC, Clements JE (2011) Induction of innate immune responses by SIV in vivo and in vitro: differential expression and function of RIG-I and MDA5. J Infect Dis 204:1104–1114. doi:10.1093/infdis/jir469
251. Kalter DC et al (1991) Enhanced HIV replication in macrophage colony-stimulating factor-treated monocytes. J Immunol 146:298–306
252. Persidsky Y, Gendelman HE (1997) Development of laboratory and animal model systems for HIV-1 encephalitis and its associated dementia. J Leukoc Biol 62:100–106
253. Shah A, Singh DP, Buch S, Kumar A (2011) HIV-1 envelope protein gp120 up regulates CCL5 production in astrocytes which can be circumvented by inhibitors of NF-kappaB pathway. Biochem Biophys Res Commun 414:112–117. doi:10.1016/j.bbrc.2011.09.033
254. Meeker RB, Poulton W, Feng WH, Hudson L, Longo FM (2011) Suppression of immunodeficiency virus-associated neural damage by the p75 neurotrophin receptor ligand, LM11A-31, in an in vitro feline model. J Neuroimmune Pharmacol 7(2):388–400. doi:10.1007/s11481-011-9325-0
255. Bethel-Brown C et al (2011) HIV-1 Tat-mediated induction of platelet-derived growth factor in astrocytes: role of early growth response gene 1. J Immunol 186:4119–4129. doi:10.4049/jimmunol.1002235
256. Overholser ED, Babas T, Zink MC, Barber SA, Clements JE (2005) CD4-independent entry and replication of simian immunodeficiency virus in primary rhesus macaque astrocytes are regulated by the transmembrane protein. J Virol 79:4944–4951. doi:10.1128/JVI.79.8.4944-4951.2005

257. Peng H et al (2011) HIV-1-infected and immune-activated macrophages induce astrocytic differentiation of human cortical neural progenitor cells via the STAT3 pathway. PLoS One 6:e19439. doi:10.1371/journal.pone.0019439
258. Chen L et al (2011) HIV-1gp120 induces neuronal apoptosis through enhancement of 4-aminopyridine-senstive outward K+ currents. PLoS One 6:e25994. doi:10.1371/journal.pone.0025994
259. Podhaizer EM et al (2011) Morphine and gp120 toxic interactions in striatal neurons are dependent on HIV-1 strain. J Neuroimmune Pharmacol. doi:10.1007/s11481-011-9326-z
260. White MG et al (2011) Parallel high throughput neuronal toxicity assays demonstrate uncoupling between loss of mitochondrial membrane potential and neuronal damage in a model of HIV-induced neurodegeneration. Neurosci Res 70:220–229. doi:10.1016/j.neures.2011.01.013
261. Zhang Y et al (2012) Sigma-1 receptor agonists provide neuroprotection against gp120 via a change in bcl-2 expression in mouse neuronal cultures. Brain Res 1431:13–22. doi:10.1016/j.brainres.2011.10.053
262. Gorantla S, Che M, Gendelman HE (2005) Isolation, propagation, and HIV-1 infection of monocyte-derived macrophages and recovery of virus from brain and cerebrospinal fluid. Methods Mol Biol 304:35–48. doi:10.1385/1-59259-907-9:035
263. Poluektova L et al (2004) Neuroregulatory events follow adaptive immune-mediated elimination of HIV-1-infected macrophages: studies in a murine model of viral encephalitis. J Immunol 172:7610–7617, pii: 172/12/7610
264. Chapuis AG et al (2011) HIV-specific CD8+ T cells from HIV+ individuals receiving HAART can be expanded ex vivo to augment systemic and mucosal immunity in vivo. Blood 117:5391–5402. doi:10.1182/blood-2010-11-320226
265. Clements JE, Mankowski JL, Gama L, Zink MC (2008) The accelerated simian immunodeficiency virus macaque model of human immunodeficiency virus-associated neurological disease: from mechanism to treatment. J Neurovirol 14:309–317. doi:10.1080/13550280802132832
266. Fox HS (2008) Virus-host interaction in the simian immunodeficiency virus-infected brain. J Neurovirol 14:286–291. doi:10.1080/13550280802132824
267. Zink MC, Mankowski JL, Graham DR, Gama L, Clements JE (2012) In: Gendelman HE et al (eds) The neurology of AIDS. Oxford University Press, New York, pp 365–376
268. Roberts ES et al (2006) Host response and dysfunction in the CNS during chronic simian immunodeficiency virus infection. J Neurosci 26:4577–4585. doi:10.1523/JNEUROSCI.4504-05.2006
269. Roberts ES et al (2003) Induction of pathogenic sets of genes in macrophages and neurons in NeuroAIDS. Am J Pathol 162:2041–2057. doi:10.1016/S0002-9440(10)64336-2
270. Zink MC et al (2010) Simian immunodeficiency virus-infected macaques treated with highly active antiretroviral therapy have reduced central nervous system viral replication and inflammation but persistence of viral DNA. J Infect Dis 202:161–170. doi:10.1086/653213
271. Meeker RB (2012) In: Gendelman HE et al (eds) The neurology of AIDS. Oxford University Press, New York, pp 377–390
272. Fletcher NF, Meeker RB, Hudson LC, Callanan JJ (2011) The neuropathogenesis of feline immunodeficiency virus infection: barriers to overcome. Vet J 188:260–269. doi:10.1016/j.tvjl.2010.03.022
273. Elder JH, Lin YC, Fink E, Grant CK (2010) Feline immunodeficiency virus (FIV) as a model for study of lentivirus infections: parallels with HIV. Curr HIV Res 8:73–80
274. Elder JH et al (2008) Molecular mechanisms of FIV infection. Vet Immunol Immunopathol 123:3–13. doi:10.1016/j.vetimm.2008.01.007
275. Van Rompay KK (2010) Evaluation of antiretrovirals in animal models of HIV infection. Antiviral Res 85:159–175. doi:10.1016/j.antiviral.2009.07.008
276. Yamamoto JK, Sanou MP, Abbott JR, Coleman JK (2010) Feline immunodeficiency virus model for designing HIV/AIDS vaccines. Curr HIV Res 8:14–25
277. Poluektova L (2012) In: Gendelman HE et al (eds) The neurology of AIDS. Oxford University Press, New York, pp 414–430

278. Bosma GC, Custer RP, Bosma MJ (1983) A severe combined immunodeficiency mutation in the mouse. Nature 301:527–530
279. Borkow G (2005) Mouse models for HIV-1 infection. IUBMB Life 57:819–823. doi:10.1080/15216540500459642
280. Stoddart CA et al (2007) Validation of the SCID-hu Thy/Liv mouse model with four classes of licensed antiretrovirals. PLoS One 2:e655. doi:10.1371/journal.pone.0000655
281. Tyor WR, Power C, Gendelman HE, Markham RB (1993) A model of human immunodeficiency virus encephalitis in scid mice. Proc Natl Acad Sci USA 90:8658–8662
282. Brainard DM et al (2009) Induction of robust cellular and humoral virus-specific adaptive immune responses in human immunodeficiency virus-infected humanized BLT mice. J Virol 83:7305–7321. doi:10.1128/JVI.02207-08
283. Gorantla S et al (2010) CD8+ cell depletion accelerates HIV-1 immunopathology in humanized mice. J Immunol 184:7082–7091. doi:10.4049/jimmunol.1000438
284. Watanabe S et al (2007) Humanized NOD/SCID/IL2R{gamma}null mice transplanted with hematopoietic stem cells under nonmyeloablative conditions show prolonged life spans and allow detailed analysis of human immunodeficiency virus type 1 pathogenesis. J Virol 81:13259–13264. doi:10.1128/jvi.01353-07
285. Goffinet C, Allespach I, Keppler OT (2007) HIV-susceptible transgenic rats allow rapid preclinical testing of antiviral compounds targeting virus entry or reverse transcription. Proc Natl Acad Sci USA 104:1015–1020. doi:10.1073/pnas.0607414104
286. Michel N et al (2009) Human cyclin T1 expression ameliorates a T-cell-specific transcriptional limitation for HIV in transgenic rats, but is not sufficient for a spreading infection of prototypic R5 HIV-1 strains ex vivo. Retrovirology 6:2. doi:10.1186/1742-4690-6-2
287. Anderson ER, Xiong H, Gendelman HE (2006) In: Friedman H, Specter S, Bendinelli M (eds) In vivo models of HIV disease and control. Springer, New York, pp 19–43
288. Chang L, Feger U, Ernst TM (2012) In: Gendelman HE et al (eds) The neurology of AIDS. Oxford University Press, New York, pp 763–797
289. Tate DF, Khedraki R, McCaffrey D, Branson D, Dewey J (2011) The role of medical imaging in defining CNS abnormalities associated with HIV-infection and opportunistic infections. Neurotherapeutics 8:103–116. doi:10.1007/s13311-010-0010-4
290. Whiteman ML et al (1993) Progressive multifocal leukoencephalopathy in 47 HIV-seropositive patients: neuroimaging with clinical and pathologic correlation. Radiology 187:233–240
291. Brass LM et al (1994) The role of single photon emission computed tomography brain imaging with 99mTc-bicisate in the localization and definition of mechanism of ischemic stroke. J Cereb Blood Flow Metab 14(Suppl 1):S91–S98
292. Rosci MA et al (1992) Methods for detecting early signs of AIDS dementia complex in asymptomatic HIV-1-infected subjects. AIDS 6:1309–1316
293. Schwartz RB et al (1994) SPECT imaging of the brain: comparison of findings in patients with chronic fatigue syndrome, AIDS dementia complex, and major unipolar depression. AJR Am J Roentgenol 162:943–951
294. Chang L et al (2008) Greater than age-related changes in brain diffusion of HIV patients after 1 year. J Neuroimmune Pharmacol 3:265–274. doi:10.1007/s11481-008-9120-8
295. Hammoud DA et al (2005) Imaging glial cell activation with [11C]-R-PK11195 in patients with AIDS. J Neurovirol 11:346–355. doi:10.1080/13550280500187351
296. Wang GJ et al (2004) Decreased brain dopaminergic transporters in HIV-associated dementia patients. Brain 127:2452–2458. doi:10.1093/brain/awh269
297. Georgiou MF et al (2008) Analysis of the effects of injecting drug use and HIV-1 infection on 18F-FDG PET brain metabolism. J Nucl Med 49:1999–2005. doi:10.2967/jnumed.108.052688
298. von Giesen HJ et al (2000) Potential time course of human immunodeficiency virus type 1-associated minor motor deficits: electrophysiologic and positron emission tomography findings. Arch Neurol 57:1601–1607

299. Hinkin CH et al (1995) Cerebral metabolic change in patients with AIDS: report of a six-month follow-up using positron-emission tomography. J Neuropsychiatry Clin Neurosci 7:180–187
300. Castelo JM, Sherman SJ, Courtney MG, Melrose RJ, Stern CE (2006) Altered hippocampal-prefrontal activation in HIV patients during episodic memory encoding. Neurology 66:1688–1695. doi:10.1212/01.wnl.0000218305.09183.70
301. Chang L et al (2001) Neural correlates of attention and working memory deficits in HIV patients. Neurology 57:1001–1007
302. Melrose RJ, Tinaz S, Castelo JM, Courtney MG, Stern CE (2008) Compromised fronto-striatal functioning in HIV: an fMRI investigation of semantic event sequencing. Behav Brain Res 188:337–347. doi:10.1016/j.bbr.2007.11.021
303. Chang L et al (2004) Adaptation of the attention network in human immunodeficiency virus brain injury. Ann Neurol 56:259–272. doi:10.1002/ana.20190
304. Chang L et al (1999) Highly active antiretroviral therapy reverses brain metabolite abnormalities in mild HIV dementia. Neurology 53:782–789
305. Chang L, Ernst T, Leonido-Yee M, Walot I, Singer E (1999) Cerebral metabolite abnormalities correlate with clinical severity of HIV-1 cognitive motor complex. Neurology 52:100–108
306. Chang L et al (2004) A multicenter in vivo proton-MRS study of HIV-associated dementia and its relationship to age. Neuroimage 23:1336–1347. doi:10.1016/j.neuroimage.2004.07.067
307. Harezlak J et al (2011) Persistence of HIV-associated cognitive impairment, inflammation, and neuronal injury in era of highly active antiretroviral treatment. AIDS 25:625–633. doi:10.1097/QAD.0b013e3283427da7
308. Letendre S, McCutchan JA, Ellis RJ, Best BM, Capparelli EV (2012) In: Gendelman HE et al (eds) The neurology of AIDS. Oxford University Press, New York, pp 962–977
309. Price RW et al (2007) Biomarkers of HIV-1 CNS infection and injury. Neurology 69:1781–1788. doi:10.1212/01.wnl.0000278457.55877.eb
310. Canestri A et al (2010) Discordance between cerebral spinal fluid and plasma HIV replication in patients with neurological symptoms who are receiving suppressive antiretroviral therapy. Clin Infect Dis 50:773–778. doi:10.1086/650538
311. Gendelman HE et al (1998) Suppression of inflammatory neurotoxins by highly active antiretroviral therapy in human immunodeficiency virus-associated dementia. J Infect Dis 178:1000–1007
312. Letendre S, Ellis RJ (2012) In: Gendelman HE et al (eds) The neurology of AIDS. Oxford University Press, New York, pp 825–846
313. Letendre SL, Lanier ER, McCutchan JA (1999) Cerebrospinal fluid beta chemokine concentrations in neurocognitively impaired individuals infected with human immunodeficiency virus type 1. J Infect Dis 180:310–319. doi:10.1086/314866
314. Abdulle S et al (2007) CSF neurofilament protein (NFL) – a marker of active HIV-related neurodegeneration. J Neurol 254:1026–1032. doi:10.1007/s00415-006-0481-8
315. Mellgren A et al (2007) Antiretroviral treatment reduces increased CSF neurofilament protein (NFL) in HIV-1 infection. Neurology 69:1536–1541. doi:10.1212/01.wnl.0000277635.05973.55
316. Gisslen M, Rosengren L, Hagberg L, Deeks SG, Price RW (2005) Cerebrospinal fluid signs of neuronal damage after antiretroviral treatment interruption in HIV-1 infection. AIDS Res Ther 2:6. doi:10.1186/1742-6405-2-6
317. Swindells S, Sandkovsky U (2012) In: Gendelman HE et al (eds) The neurology of AIDS. Oxford University Press, New York, pp 945–961
318. Enting RH et al (1998) Antiretroviral drugs and the central nervous system. AIDS 12:1941–1955
319. Lee CG et al (1998) HIV-1 protease inhibitors are substrates for the MDR1 multidrug transporter. Biochemistry 37:3594–3601. doi:10.1021/bi972709x
320. Rao VV et al (1999) Choroid plexus epithelial expression of MDR1 P glycoprotein and multidrug resistance-associated protein contribute to the blood-cerebrospinal-fluid drug-permeability barrier. Proc Natl Acad Sci USA 96:3900–3905

321. Spitzenberger TJ et al (2007) Novel delivery system enhances efficacy of antiretroviral therapy in animal model for HIV-1 encephalitis. J Cereb Blood Flow Metab 27:1033–1042. doi:10.1038/sj.jcbfm.9600414, pii: 9600414
322. Roy U, Balkundi S, McMillan J, Gendelman HE (2012) In: Gendelman HE et al (eds) The neurology of AIDS. Oxford University Press, New York, pp 987–998
323. Wong HL, Wu XY, Bendayan R (2011) Nanotechnological advances for the delivery of CNS therapeutics. Adv Drug Deliv Rev 64(7):686–700. doi:10.1016/j.addr.2011.10.007
324. Nowacek A, Gendelman HE (2009) NanoART, neuroAIDS and CNS drug delivery. Nanomedicine (Lond) 4:557–574. doi:10.2217/nnm.09.38
325. Rao KS, Reddy MK, Horning JL, Labhasetwar V (2008) TAT-conjugated nanoparticles for the CNS delivery of anti-HIV drugs. Biomaterials 29:4429–4438. doi:10.1016/j.biomaterials.2008.08.004, pii: S0142-9612(08)00567-X
326. Dou H et al (2006) Development of a macrophage-based nanoparticle platform for antiretroviral drug delivery. Blood 108:2827–2835. doi:10.1182/blood-2006-03-012534
327. Dou H et al (2009) Macrophage delivery of nanoformulated antiretroviral drug to the brain in a murine model of neuroAIDS. J Immunol 183:661–669. doi:10.4049/jimmunol.0900274
328. Kadiu I, Gendelman HE (2011) Human immunodeficiency virus type 1 endocytic trafficking through macrophage bridging conduits facilitates spread of infection. J Neuroimmune Pharmacol 6(4):658–675. doi:10.1007/s11481-011-9298-z
329. Lipton SA (1993) Prospects for clinically tolerated NMDA antagonists: open-channel blockers and alternative redox states of nitric oxide. Trends Neurosci 16:527–532
330. Schifitto G et al (2007) Memantine and HIV-associated cognitive impairment: a neuropsychological and proton magnetic resonance spectroscopy study. AIDS 21:1877–1886. doi:10.1097/QAD.0b013e32813384e8
331. Zhao Y et al (2010) Memantine for AIDS dementia complex: open-label report of ACTG 301. HIV Clin Trials 11:59–67. doi:10.1310/hct1101-59
332. Clifford DB et al (2002) A randomized clinical trial of CPI-1189 for HIV-associated cognitive-motor impairment. Neurology 59:1568–1573
333. Kagan VE et al (1992) Dihydrolipoic acid – a universal antioxidant both in the membrane and in the aqueous phase. Reduction of peroxyl, ascorbyl and chromanoxyl radicals. Biochem Pharmacol 44:1637–1649
334. Sacktor N et al (2000) Transdermal selegiline in HIV-associated cognitive impairment: pilot, placebo-controlled study. Neurology 54:233–235
335. Schifitto G et al (1999) Randomized trial of the platelet-activating factor antagonist lexipafant in HIV-associated cognitive impairment. Neurological AIDS Research Consortium. Neurology 53:391–396
336. Schifitto G et al (2009) Lithium therapy for human immunodeficiency virus type 1-associated neurocognitive impairment. J Neurovirol 15:176–186. doi:10.1080/13550280902758973, pii: 909764548
337. Tang LH, Aizenman E (1993) Allosteric modulation of the NMDA receptor by dihydrolipoic and lipoic acid in rat cortical neurons in vitro. Neuron 11:857–863
338. Gelbard HA et al (1994) Platelet-activating factor: a candidate human immunodeficiency virus type 1-induced neurotoxin. J Virol 68:4628–4635
339. Perry SW et al (1998) Platelet-activating factor receptor activation. An initiator step in HIV-1 neuropathogenesis. J Biol Chem 273:17660–17664
340. Maggirwar SB, Tong N, Ramirez S, Gelbard HA, Dewhurst S (1999) HIV-1 Tat-mediated activation of glycogen synthase kinase-3beta contributes to Tat-mediated neurotoxicity. J Neurochem 73:578–586
341. Peng F et al (2008) Platelet-derived growth factor protects neurons against gp120-mediated toxicity. J Neurovirol 14:62–72. doi:10.1080/13550280701809084
342. Mofenson LM (2000) Technical report: perinatal human immunodeficiency virus testing and prevention of transmission. Committee on Pediatric Aids. Pediatrics 106:E88
343. Mofenson LM, Fowler MG (1999) Interruption of materno-fetal transmission. AIDS 13(Suppl A):S205–S214

344. Gibb DM et al (2003) Decline in mortality, AIDS, and hospital admissions in perinatally HIV-1 infected children in the United Kingdom and Ireland. BMJ 327:1019. doi:10.1136/bmj.327.7422.1019
345. Judd A et al (2007) Morbidity, mortality, and response to treatment by children in the United Kingdom and Ireland with perinatally acquired HIV infection during 1996–2006: planning for teenage and adult care. Clin Infect Dis 45:918–924. doi:10.1086/521167
346. McConnell MS et al (2005) Trends in antiretroviral therapy use and survival rates for a large cohort of HIV-infected children and adolescents in the United States, 1989–2001. J Acquir Immune Defic Syndr 38:488–494
347. Selik RM, Lindegren ML (2003) Changes in deaths reported with human immunodeficiency virus infection among United States children less than thirteen years old, 1987 through 1999. Pediatr Infect Dis J 22:635–641. doi:10.1097/01.inf.0000073241.01043.9c
348. Mintz M et al (1995) Pediatric HIV infection in Elista, Russia: interventional strategies. Am J Public Health 85:586–588
349. Mintz M, Sharer LR, Civitello LA (2012) In: Gendelman HE et al (eds) The neurology of AIDS. Oxford University Press, New York, pp 875–906
350. Belman AL et al (1988) Pediatric acquired immunodeficiency syndrome. Neurologic syndromes. Am J Dis Child 142:29–35
351. Blanche S et al (1990) Longitudinal study of 94 symptomatic infants with perinatally acquired human immunodeficiency virus infection. Evidence for a bimodal expression of clinical and biological symptoms. Am J Dis Child 144:1210–1215
352. Epstein LG et al (1986) Neurologic manifestations of human immunodeficiency virus infection in children. Pediatrics 78:678–687
353. Mintz M (1994) Clinical comparison of adult and pediatric NeuroAIDS. Adv Neuroimmunol 4:207–221
354. Mintz M (1999) Clinical features and treatment interventions for human immunodeficiency virus-associated neurologic disease in children. Semin Neurol 19:165–176. doi:10.1055/s-2008-1040834
355. Mintz M, Epstein LG (1992) Neurologic manifestations of pediatric acquired immunodeficiency syndrome: clinical features and therapeutic approaches. Semin Neurol 12:51–56. doi:10.1055/s-2008-1041157
356. Tardieu M et al (2000) HIV-1-related encephalopathy in infants compared with children and adults. French Pediatric HIV Infection Study and the SEROCO Group. Neurology 54:1089–1095
357. Brouwers P et al (1994) Interrelations among patterns of change in neurocognitive, CT brain imaging and CD4 measures associated with anti-retroviral therapy in children with symptomatic HIV infection. Adv Neuroimmunol 4:223–231
358. Pollack H et al (1997) Impaired early growth of infants perinatally infected with human immunodeficiency virus: correlation with viral load. J Pediatr 130:915–922
359. Schneider E et al (2008) Revised surveillance case definitions for HIV infection among adults, adolescents, and children aged <18 months and for HIV infection and AIDS among children aged 18 months to <13 years – United States, 2008. MMWR Recomm Rep 57:1–12
360. Epstein LG et al (1987) Expression of human immunodeficiency virus in cerebrospinal fluid of children with progressive encephalopathy. Ann Neurol 21:397–401. doi:10.1002/ana.410210413
361. Sharer LR et al (1986) Pathologic features of AIDS encephalopathy in children: evidence for LAV/HTLV-III infection of brain. Hum Pathol 17:271–284
362. Chiriboga CA, Fleishman S, Champion S, Gaye-Robinson L, Abrams EJ (2005) Incidence and prevalence of HIV encephalopathy in children with HIV infection receiving highly active anti-retroviral therapy (HAART). J Pediatr 146:402–407. doi:10.1016/j.jpeds.2004.10.021
363. McCoig C et al (2002) Effect of combination antiretroviral therapy on cerebrospinal fluid HIV RNA, HIV resistance, and clinical manifestations of encephalopathy. J Pediatr 141:36–44. doi:10.1067/mpd.2002.125007
364. Shanbhag MC et al (2005) Neurocognitive functioning in pediatric human immunodeficiency virus infection: effects of combined therapy. Arch Pediatr Adolesc Med 159:651–656. doi:10.1001/archpedi.159.7.651

365. Lindsey JC, Malee KM, Brouwers P, Hughes MD (2007) Neurodevelopmental functioning in HIV-infected infants and young children before and after the introduction of protease inhibitor-based highly active antiretroviral therapy. Pediatrics 119:e681–e693. doi:10.1542/peds.2006-1145
366. Tamula MA, Wolters PL, Walsek C, Zeichner S, Civitello L (2003) Cognitive decline with immunologic and virologic stability in four children with human immunodeficiency virus disease. Pediatrics 112:679–684
367. Koekkoek S et al (2006) Effects of highly active antiretroviral therapy (HAART) on psychomotor performance in children with HIV disease. J Neurol 253:1615–1624. doi:10.1007/s00415-006-0277-x
368. Nozyce ML et al (2006) A behavioral and cognitive profile of clinically stable HIV-infected children. Pediatrics 117:763–770. doi:10.1542/peds.2005-0451

Section II
Alcohol and Neuroimmune Interactions

Chapter 8
Innate Immune Signaling and Alcoholism

Ryan P. Vetreno and Fulton T. Crews

8.1 Introduction

Chronic use of alcohol and other drugs of abuse have been demonstrated to activate the innate immune system of the CNS. This activation, in turn, may lead to progressive changes to neurobiology that underlie loss of behavioral control and increased limbic negative affect common in addiction. Ethanol exerts effects on the systemic adaptive and innate immune systems [1]. Adaptive immunity comprises a class of highly specialized lymphocytic cells that recognize, remember, and target specific pathogens through the production of antibodies. This system is protective in that it maintains an immunological memory of the pathogen in the lymphocytes in preparation for future insults. In contrast, the innate immune system mounts a nonspecific immune response to a pathogen via secretion of cytokines and other chemical messages that activate cells. This chapter will focus on the effects of alcohol on innate immunity because the brain possesses few lymphocytes and, consequently, minimal adaptive immunity. Although beyond the scope of this chapter, it is important to note that ethanol causes the gut to become "leaky" resulting in the release of the endotoxin lipopolysaccharide (LPS). The increased gut permeability to LPS contributes to liver inflammation and secretion of proinflammatory cytokines [e.g., tumor necrosis factor-alpha (TNFα)] into the blood. The blood is then transported into the brain where TNFα and other cytokines are induced causing neuroinflammation that persists long after the resolution of the peripheral inflammatory response [2] (see Fig. 8.1).

The continued and often escalating negative events associated with persistently increased alcohol and drug consumption that result in addiction are well known [3, 4].

R.P. Vetreno • F.T. Crews, Ph.D. (✉)
Department of Pharmacology and Psychiatry, The Bowles Center for Alcohol Studies,
The University of North Carolina at Chapel Hill, The School of Medicine,
CB #7178, 1021 Thurston-Bowles Building, Chapel Hill, NC 27599-71780, USA
e-mail: fulton_crews@med.unc.edu

C. Cui et al. (eds.), *Neural-Immune Interactions in Brain Function and Alcohol Related Disorders*, DOI 10.1007/978-1-4614-4729-0_8,

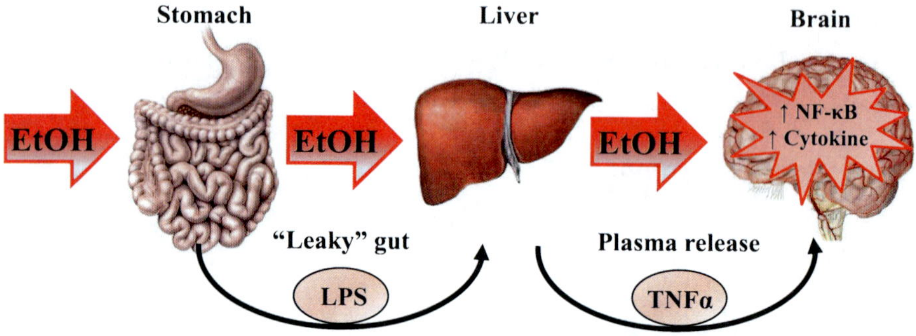

Fig. 8.1 Systemic ethanol-induced activation of cytokine signaling in the brain. Consumed ethanol (EtOH) enters the stomach and makes it "leaky" allowing lipopolysaccharide (LPS) to enter the blood. The circulating EtOH and LPS lead to liver inflammation, which results in the production and secretion of tumor necrosis factor-alpha (TNFα) and other proinflammatory cytokines. These proinflammatory cytokines enter the brain through the blood and increase neuroinflammation through the synthesis of nuclear factor kappa-light-chain-enhancer of activated B cells (NF-κB) and other cytokines

Addiction involves excessive, compulsive drug use despite negative consequences (e.g., a loss of behavioral control over drug use). Initially, alcohol dependence was characterized by both neurological and physiological symptoms of withdrawal that included hyperexcitability, anxiety, pyrexia, and sickness behavior. However, more contemporary definitions of alcohol dependence focus on the loss of behavioral control, preoccupation with alcohol, and increased impulsivity, anxiety, drug craving, and depression-like negative affect. Studies also find increased immediate reward-seeking behavior and preoccupation/perseverative sensitization to alcohol stimuli that promote compulsive habitual behavior [5, 6]. Thus, contemporary views of addiction define it as a chronic neurobiological disorder involving genetic, psychosocial, and environmental factors that contribute to the complex dysfunction of alcoholism and alcohol dependence. These emerging views have led the diagnoses of substance abuse and substance dependence, which are currently used in the diagnostic and statistical manual of mental disorders (DSM)-IV text revision, to be combined into a single substance use disorder in the DSM-V.

Although it is well accepted that alcohol and drug intoxication change neurochemistry and behavior, the persistent effects of drugs of abuse on long-term immune and behavioral alterations are only beginning to be understood. Drug dependence and addiction involve a disruption of the normal balance between self-control mechanisms and emotional drive. Brain regions vulnerable to the harmful effects of alcohol and drugs of abuse are the frontal cortex and the limbic system. The frontal cortex regulates decision-making and other executive functions, such as motivation, planning and goal setting, and impulse inhibition. Limbic structures, such as the amygdala and hippocampus, contribute to memory, emotion, and mood. Continued use of alcohol and other drugs of abuse alters cortical and limbic neurobiology, culminating in a loss of attention, impaired decision-making, and increased impulsivity and anxious urgency that escalates a progressive loss of behavioral control over drug use. Thus, a key element of the behavioral pathology of addiction and substance dependence centers on the loss of frontal cortical executive behavioral control and flexibility and mounting limbic anxiety and urgency.

8.2 Cornerstones of Addiction: Loss of Frontal Cortical Behavioral Control and Increased Limbic System Negative Affect

The loss of behavioral control that characterizes addiction involves the reduction of frontal cortical regulation of attention and cognitive flexibility and increased limbic system anxiety and negative affect. The frontal lobes modulate decision-making and other executive functions, such as motivation, planning and goal setting, impulse inhibition, and regulation of limbic drive. Behavioral control mechanisms mediated by the frontal lobes involve blockade of impulsive errors as well as the prediction of future rewards [7]. In addition, the frontal lobes are involved in the retention of long-term emotional memories and adjustment of emotion to conform to societal norms allowing for successful integration into society. In agreement, alcohol dependence is often associated with progressive social isolation and/or increased time in heavy drinking social environments. Thus, impaired frontal cortical function is a key cognitive component involved in the development of drug dependence and the neurobiology of addiction.

The limbic system consists of the amygdala and hippocampus, as well as the septal nuclei, limbic cortex, and anterior thalamic nuclei. Together, these interconnected structures support a variety of functions, including emotion, anxiety, need and urgency, and long-term memory. Drug-induced negative affect, altered mood, and increased anxiety are the emotive components of addiction. Limbic system urgency propagates impulsive behaviors [8]. Impaired frontal cortical and limbic system function is a key element of drug dependence and the neurobiology of addiction. In the sections to follow, we will provide compelling evidence supporting a role for activation of the innate immune system in the development of addiction symptomology, including reduced behavioral flexibility, increased negative affect and anxiety, and drug-seeking behavior.

Dependence on alcohol and other drugs of abuse leads to a loss of attention, poor decision-making, and increased impulsivity and anxious urgency that promote the progressive loss of behavioral control over drug use. Individuals addicted to alcohol and other drugs of abuse demonstrate impaired decision-making on tasks that involve delayed reward and cognitive flexibility, which is indicative of frontal cortex dysfunction [9–13]. The loss of behavioral control associated with drug addiction likely involves relearning deficits and increased perseveration. Using animal models of drug abuse, our laboratory and others have found that ethanol [14, 15] and cocaine [16, 17] exposure results in persistent reversal learning deficits, which parallels the frontal cortical dysfunction and a loss of behavioral flexibility observed in human alcoholics. The progression from abuse to addiction involves increased

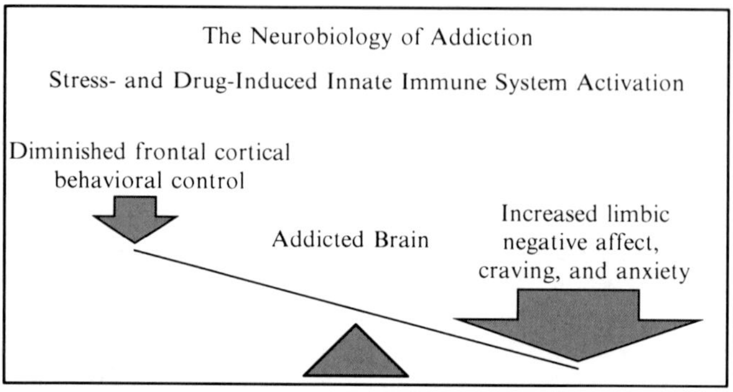

Fig. 8.2 The neurobiology of addiction. A simplified schematic distinguishing the frontal cortical and limbic system behavioral changes that characterize drug-induced activation of the innate immune system. Both drug addiction and stress activate the innate immune system in the brain, which reduces glutamate transporters leading to prefrontal cortex hyperexcitability [19, 20] and diminished cortical behavioral control and behavioral inflexibility [21, 22]. Concurrently, induction of innate immune genes in limbic brain regions increases negative affect, craving, and anxiety-like behaviors prompting further drug abuse and self-medication [23]. The long-term deleterious effects of prolonged alcohol abuse are diminished activation of frontal cortical behavioral control circuits that lead to a progressive loss of attention and poor decision-making that combine with increased negative affect and anxiety that motivate further drug-taking behaviors. Taken together, these innate immune gene-induced behavioral changes characterize the drug-addicted brain

drug wanting, negative emotional urgency, and decreased behavioral control [11, 18]. Although it has been established that drug intoxication leads to altered behavior, the importance of persistent drug-induced changes to neurobiology that result in behavioral change has only recently begun to be appreciated. This chapter will present evidence that innate immune gene induction diminishes frontal cortical behavioral control while simultaneously increasing limbic system anxiety and negative affect to create the neurobiology of addiction (see Fig. 8.2).

8.3 The Innate Immune System of the CNS

Although the brain has historically been considered an immune-privileged organ, innate immune responses are increasingly linked to CNS disease [24, 25]. While astrocytes, oligodendrocytes, and neurons are all formed from neuroectoderm, microglia are brain innate immune cells that originate from the mesoderm, migrating to the brain early in fetal development [26]. Microglia are part of the innate immune system that express a series of pattern recognition receptors to bind highly conserved pathogen-associated molecular patterns (PAMPs) present on invading microorganisms [27, 28]. In addition to sensing and responding to PAMPs, microglia were recently discovered to recognize endogenous danger signals. Although the innate immune system consists of several pattern recognition receptor families, the Toll-like receptor (TLR) family has received the most attention with regard to ethanol- and drug-induced neuroinflammation. In general, receptors of this family consist of an extracellular domain of leucine-rich repeat motifs, a transmembrane domain that determines receptor location, and an intracellular domain containing the Toll/interleukin-1 receptor. To date, 13 TLRs have been identified (i.e., TLRs 1–13) in mammals [28, 29]. Recent interest has also begun to highlight a role for another pattern recognition receptor, receptor for advanced glycation end products (RAGE), in neuroinflammation associated with drug addiction because of its involvement in Alzheimer's disease and cancer [30–32]. This multiligand transmembrane receptor is ubiquitously expressed at relatively low levels throughout the CNS, but its expression is increased as inflammatory signals accumulate in tissue [32] (see Fig. 8.3).

More recently, TLRs and RAGE have been found to detect and respond to neuroinflammation associated with danger-associated molecular patterns (DAMPs), which are also endogenous innate immune activators. Although research is only beginning to identity potential DAMPs, likely examples include high-mobility group box 1 (HMGB1 [33]), heat shock proteins [34], and extracellular matrix breakdown proteins [35]. The danger pattern that has received the most attention to date is HMGB1, which is a ubiquitous evolutionarily conserved protein present in the nucleus of most eukaryotic cells. Structurally, HMGB1 consists of three distinct domains: a tandem of high-mobility group box domains (Box A and Box B) and a C-terminal tail. As a nuclear protein, it is involved in DNA bending and transcription stability [36, 37]. However, upon its secretion from cells, by either active release from neuroimmune cells or passive release by leaky necrotic cells, HMGB1 exerts cytokine activity [32]. It is likely that alcohol, drugs of abuse, and stress also lead to the release of DAMPs that activate TLRs and RAGE, which contribute to innate immune gene induction [38] in addiction (see Fig. 8.4). Although defense against invading pathogens is key to the survival of all mammals, timely resolution of the immune response is critical to prevent inflammation-induced tissue damage. However, persistent activation of the innate immune system can lead to chronic inflammation, which is associated with the development and progression of multiple diseases, including Parkinson's disease, Alzheimer's disease, and multiple sclerosis [43].

8.3.1 Microglia and Astrocytes: The Vanguard of the Innate Immune System

Microglia, the resident innate immune macrophage-like cells of the CNS, and astrocytes contribute important metabolic, trophic, synaptic, and other brain functions, including responding to and amplifying innate immune responses in the CNS [44, 45]. Microglia communicate to sense the environment, remove unnecessary extracellular matrix and damaged tissue, regulate healing responses to insult, and produce proinflammatory cascades that defend against infection. In the healthy brain, ramified or "resting" microglia contribute to the integration of sensory systems and

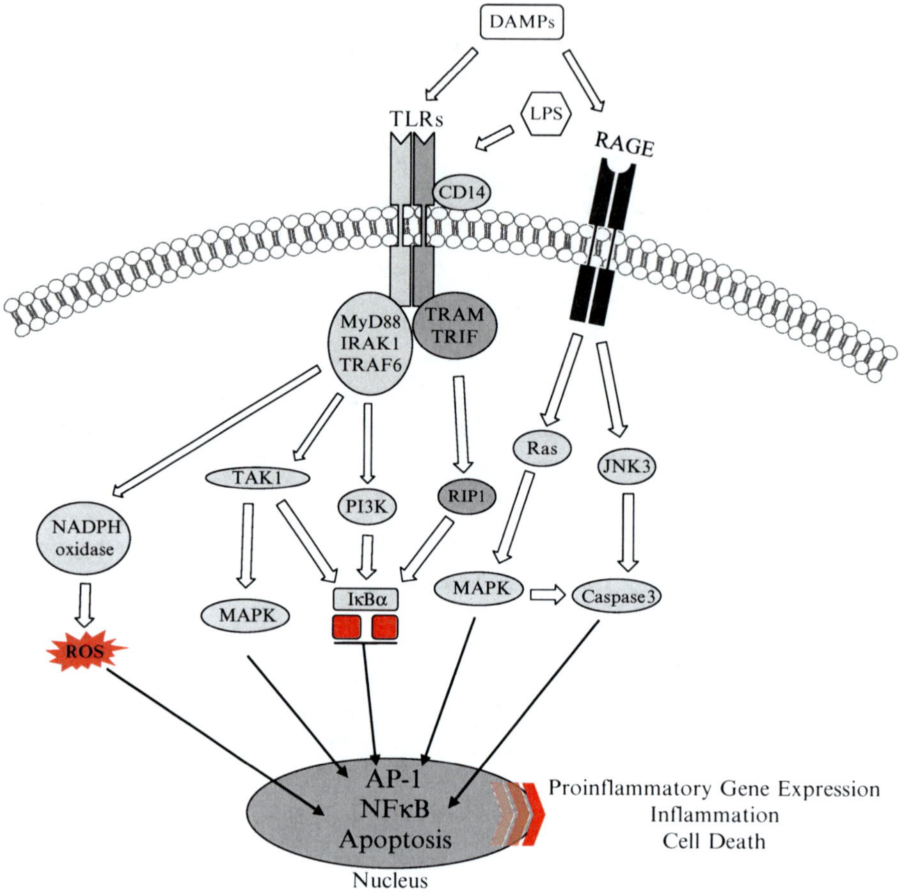

Fig. 8.3 Innate immune signaling cascade and evidence for involvement in alcohol-induced neurodegeneration. A simplified schematic of the Toll-like receptor (TLR) and RAGE receptor signaling cascades. Stimulation of TLRs leads to the generation of ROS and downstream activation of NF-κB. Similarly, activation of the RAGE receptor leads to caspase-3 induction and downstream activation of NF-κB. The production of NF-κB leads to the secretion of proinflammatory gene expression, neuroinflammation, and cell death. *AP-1* activator protein-1, *HMGB1* high-mobility group box-1, *IκBα* nuclear factor of kappa light polypeptide gene enhancer in B-cells inhibitor-alpha, *IRAK* interleukin-1 receptor-associated kinase, *JNK3* c-Jun N-terminal kinases, *MAPK* mitogen-activated protein kinase, *MyD88* myeloid differentiation primary response gene 88, *NADPH oxidase* nicotinamide adenine dinucleotide phosphate-oxidase, *NF-κB* nuclear factor kappa-light-chain-enhancer of activated B cells, *PI3K* phosphatidylinositol 3-kinase, *RAGE* receptor for advanced glycation end products, *RIP1* receptor interacting protein, *ROS* reactive oxygen species, *TAK1* transforming growth factor β-activated kinase 1, *TRAF* TNF receptor-associated factor, *TRAM* TRIF-related adaptor molecule, *TRIF* TIR domain-containing adaptor-inducing interferon-β

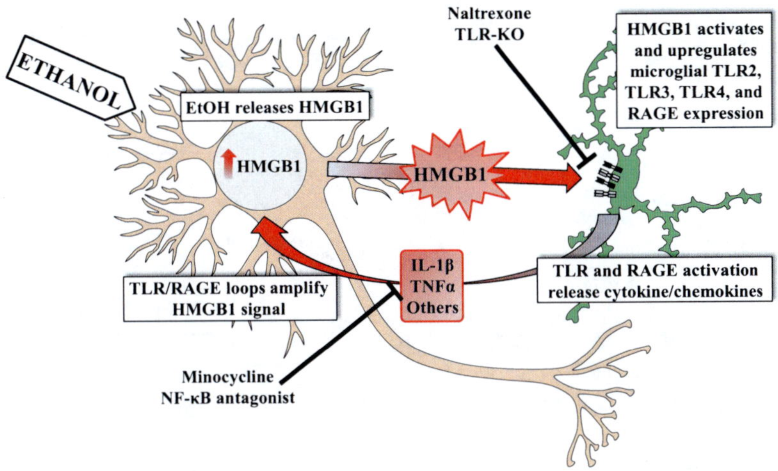

Fig. 8.4 Involvement of HMGB1 in ethanol-induced neuroinflammation. Simplified schematic demonstrating the cyclic nature of ethanol-induced activation of the endogenous TLR/RAGE agonist and high-mobility group box-1 (HMGB1). Ethanol exposure stimulates nuclear release of HMGB1 into the extracellular environment where it activates and upregulates the expression of Toll-like receptors (TLRs) and receptors for advance glycation end products (RAGE) on microglia. Activation of TLRs and RAGE leads to secretion of proinflammatory molecules [e.g., interleukin-1β (IL-1β), tumor necrosis factor-alpha (TNFα)] that further increase and amplify HMGB1 release in a cyclic fashion. Administration of naltrexone or genetic knockout of TLR4 blocks microglial activation and neuronal cell death [39–41]. Similarly, administration of minocycline, an antibiotic that blocks microglial activation [40, 42], and BHT, an NF-κB antagonist [19], inhibits microglial activation and secretion of proinflammatory cytokines and chemokines

overall survey of the brain milieu. In response to an endogenous or exogenous insult, microglia undergo distinct morphological and gene expression changes associated with cellular activation [46]. During the initial stages of activation, microglia become enlarged and increase the expression of cell matrix and cell adhesion proteins. They also begin to secrete signaling molecules that upregulate the expression of major histocompatibility complex (MHC) proteins and TLRs. In addition, activated microglia secrete cytokines [e.g., TNFα, interleukin (IL)-1β, and IL-6], which are small cell-signaling molecules that regulate cell movement toward sites of tissue damage and initiate proinflammatory-oxidative activation. These signaling molecules lead to further activation in an autocrine and paracrine fashion that amplifies microglial activation (see Fig. 8.5a). As activation continues to progress, microglia progress toward mitosis, proliferation, and phagocytic oxidative surges that dissolve and engulf waste [46]. Highly activated microglia are termed "amoeboid" and are characterized by a phagocytic, rounded macrophage-like morphology (see Fig. 8.5b). These phagocytic microglia are further characterized as either proinflammatory (M1) or anti-inflammatory (M2), depending on whether they are functioning to "kill" or "repair," respectively [51, 52]. Astrocytes are also pivotal components of the neuroimmune system as they also secrete cytokines and inflammatory modulators

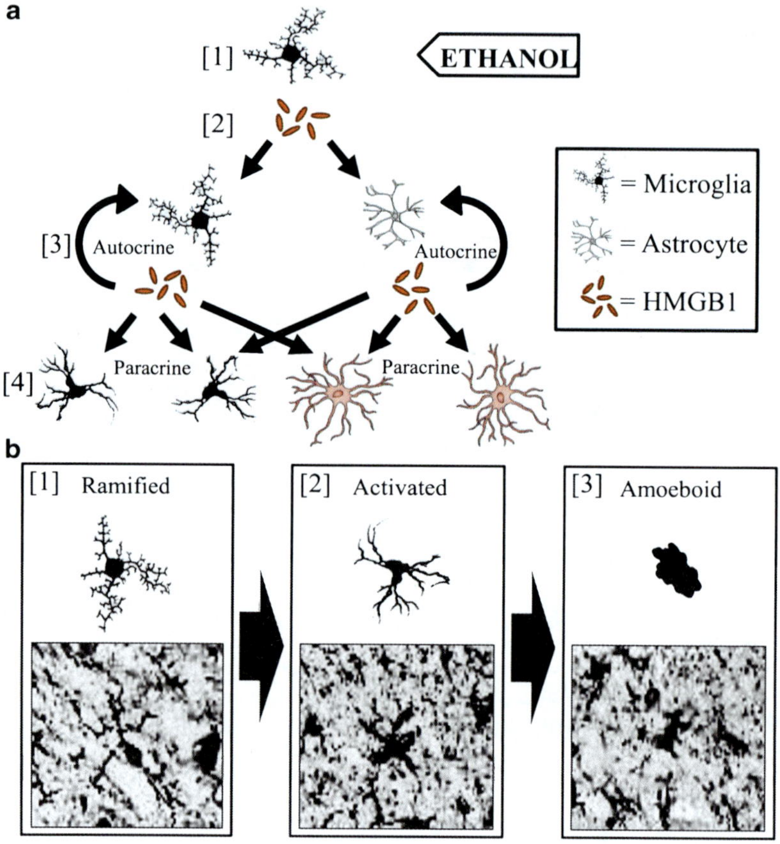

Fig. 8.5 Glial response to ethanol in the brain. (**a**) Simplified schematic of ethanol-induced HMGB1 activation and proliferation of glial cells. Ethanol exposure activates microglia and astrocytes [*1*], leading to secretion of HMGB1 to extracellular environment [*2*]. The secreted HMGB1 further activates microglia and astrocytes in an autocrine and paracrine fashion that in turn leads to further HMGB1 secretion [*3*]. The cyclic release of HMGB1 results in glial cell proliferation and further activation [*4*]. (**b**) Representative figures depicting characteristic stages of microglial activation. [*1*] Ramified or "resting" microglia are characterized by long, highly ramified processes with comparatively small cell bodies. Activated microglia are characterized by swollen, truncated processes and enlarged cell bodies. Amoeboid or "phagocytic" microglia are characterized by large, amoeba-like cell body with no or few small processes [47, 48]. Photomicrographs depict stages of microglial activation in postmortem human brain tissue [49]. Ethanol- and stress-activated microglia secrete NF-κB target genes but do not become the phagocytic oxidative burst macrophage phenotype associated with severe neurodegeneration diseases [50]

following insult-induced activation [53, 54]. In contrast to microglia, activation of astrocytes causes them to become transiently "reactive," which involves cellular enlargement, fibrillary process elongation and swelling, and increased expression of glial fibrillary acidic protein. In addition to secretion of inflammatory mediators, reactive astrocytes also modify the extracellular matrix, which provides cellular structure and segregation [55]. Together, microglia and astrocytes are the major constituents of the brain innate immune system.

Exposure to alcohol, drugs of abuse, and stress activates microglia and astrocytes [2, 56, 57], which might lead to glial adaptive conditioning [58]. Glial adaptive conditioning occurs when the brain is exposed to a persistent low level of noxious insult resulting in neuroprotection by conditioning the brain (i.e., increased protection) for future insults [1]. However, repeated bouts of exposure to these stimuli also lead to persistent activation of microglia and astrocytes [59] that secrete proinflammatory cytokines that promote neuronal damage and cell death [60]. Interestingly, microglial dysfunction has been found to contribute to trichotillomania, an obsessive–compulsive disorder associated with hair pulling [61], supporting a role for altered glial function in the neurobiology of addiction. In another study, administration of prednisone, a synthetic glucocorticoid, increased anxiety-like behavior on the elevated plus-maze as well as microglial populations in the frontal cortex and hippocampus [62]. Similarly, in human postmortem frontal cortical tissue of depressed individuals that committed suicide, astrocytes were found to exhibit enlarged cell bodies and more ramified processes, which is indicative of activated astrocytes [63]. Thus, microglia and astrocytes likely contribute to the neurobiology of addiction by inducing behavioral change associated with altered adaptive conditioning and/or persistent activation following repeated exposures to drugs of abuse and stress.

8.4 Stress, Alcohol, and Other Drugs of Abuse Foster Addiction and Stimulate NF-κB Transcription of Innate Immune Genes

Nuclear factor kappa-light-chain-enhancer of activated B cells (NF-κB) is a key transcription factor involved in the induction of innate immune genes in microglia and astrocytes. Although neuronal NF-κB has been implicated in some studies [64], others suggest that NF-κB does not activate neuronal gene transcription [65]. Regardless, stimuli such as stress, cytokines, oxidative free radicals, bacterial and viral antigens, and other molecules increase NF-κB expression that lead to downstream transcription of chemokines, cytokines, oxidases, and proteases. For instance, restraint stress increases the expression of NF-κB, cytokines, prostaglandin E2, and cyclooxygenase-2 (COX-2) in the CNS [66, 67]. In addition, LPS exposure leads to the reversal of acute glucocorticoid anti-inflammatory responses to proinflammatory NF-κB activation in the cortex [68]. In a similar fashion, our laboratory has determined that ethanol increases NF-κB-DNA binding in the brain in vivo [19] and in vitro in hippocampal–entorhinal cortex slice culture [69]. Furthermore, our laboratory and others have found that ethanol increases the transcription of NF-κB target genes, including chemokine monocyte chemoattractant protein-1 (MCP-1, CCL2 [49]), proinflammatory cytokines (e.g., TNFα, IL-1β, and IL-6 [2]), proinflammatory oxidases (inducible nitric oxide synthase [57, 70], COX [57, 71], and NOX [50]), and proteases (TNFα-converting enzyme [TACE] and tissue plasminogen activator [tPA] [70]). Finally, all addictive drugs cause chronic basal elevations of glucocorticoids [62] that could contribute to activation of the brain NF-κB cascade. Thus, activation of NF-κB by alcohol, stress, and other drugs of abuse is a common molecular mechanism involving innate immune gene induction that is consistent with a stress-drug synergy culminating in progressive increases in loss of behavioral control and addiction.

8.5 Positive Loops Amplify NF-κB Innate Immune Gene Induction Through Autocrine and Paracrine Processes

Alcohol- and stress-induced activation of the NF-κB pathway spreads through positive loops (see Fig. 8.6) that increase activation in an autocrine and paracrine fashion [70]. In autocrine signaling, secreted molecules activate receptors on the secreting cell that results in further activation of that cell. In contrast, paracrine signaling involves the secreted molecule activating cells that are nearby resulting in the spread of the signal across cells. Alcohol-induced upregulation of reactive oxygen species (ROS) results in NF-κB-induced transcription of TNFα and TACE, the TNFα activating protease, as well as increased expression of the TNF receptor (TNFR1). The release of TNFα stimulates further NF-κB transcription through activation of TNFR1 and is amplified in an autocrine and paracrine fashion.

Alcohol and stress signal through the NF-κB signaling pathway. Rats exposed to restraint stress evidence increased expression of the NF-κB target genes, TACE and TNFα mRNA, within 30 min and free mature brain TNFα levels within 1 h [66]. Similarly, an acute dose of ethanol activates NF-κB within 30 min in rat brain [73] and cultured human astrocytes [74]. Although NF-κB is expressed in most cells, NF-κB in the CNS is transcriptionally active primarily in glia [65]. Other proinflammatory interactions include NOX-ROS activation of proteases such as TACE that release mature TNFα [75], which contributes to ethanol-induced increased transcription of the NF-κB target gene, TNFα [76]. These positive loops of activation lead to persistently increased oxidative free radicals from oxidases, increased TLR expression and formation of endogenous TLR agonists, and chemokine and cytokine receptor upregulation [50, 70]. In the frontal cortex, this perpetual signaling leads to hyperexcitability [19]. Ethanol and stress-activated microglia secrete NF-κB target genes but do not become the phagocytic oxidative burst macrophage phenotype associated with severe neurodegeneration diseases [50]. Taken together, these studies indicate that persistent exposure to stress and drugs of abuse leads to sustained activation of NF-κB loops that lead to transcription of innate immune genes that appear to contribute to the development and sustainment of the neurobiology of addiction.

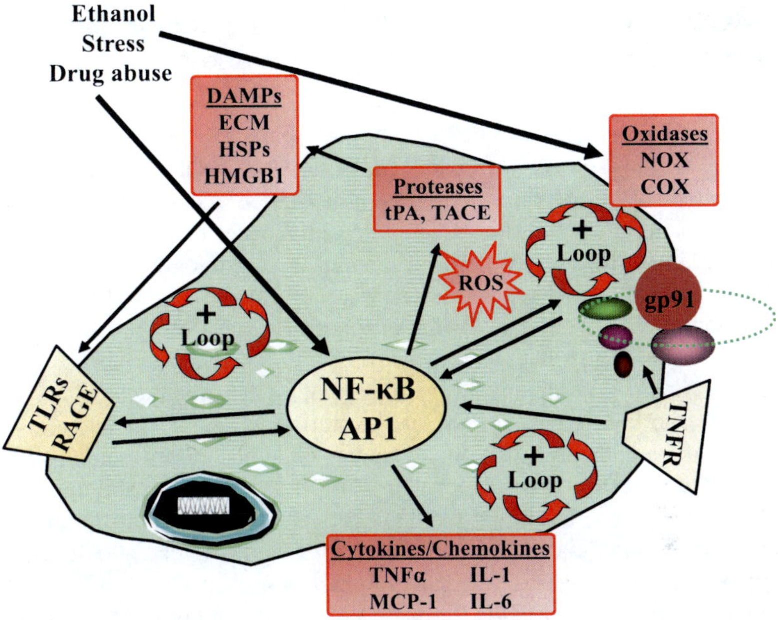

Fig. 8.6 Transcription of NF-κB increases the expression of chemokines, cytokines, oxidases, and proteases. The transcription factor NF-κB is involved in the induction of innate immune genes [49, 50, 57, 70, 71]. Stimuli such as stress, drugs of abuse, bacteria, viruses, trauma, and other factors all increase NF-κB–DNA binding and transcription. Reactive oxygen species resulting from oxidases such as NADPH oxidase or ethanol metabolism increase NF-κB transcription of $NOX2^{phox}$ (gp91), a key NOX catalytic subunit [72] that produces ROS [50]. Loops of activation occur through induction of genes that stimulate further NF-κB activation leading to autocrine and paracrine amplification and persistent signals. Cytokines and chemokines, such as TNFα, IL1β, IL6, and MCP-1 as well as their receptors (TNFR in figure), are also induced resulting in amplification loops. Toll-like receptors are increased by ethanol [40, 57] as are other damage-associated molecular pattern receptors and their agonists resulting in the formation of positive activation loops [40]. Toll-like receptors and HMGB1 interact to create another activation-amplification loop. Persistent and repeated activation occurs through positive cycles of activation. These loops spread innate immune signaling across the brain causing altered neurocircuitry and neurobiology. Abbreviations: *AP-1* activator protein-1, *COX* cyclooxygenase, *DAMPs* damage (or danger)-associated molecular patterns, *ECM* extracellular matrix, *gp91* NADPH oxidase flavocytochrome b components, *HMGB1* high-mobility group box 1, *HSPs* heat shock proteins, *IL-1* interleukin-1, *MCP-1* monocyte chemoattractant protein-1, *NF-κB* nuclear factor kappa-light-chain-enhancer of activated B cells, *NOX* nicotinamide adenine dinucleotide phosphate (NADPH) oxidases, *TACE* TNFα-converting enzyme, *TLR* Toll-like receptor, *TNFα* tumor necrosis factor-alpha, *tPA* tissue plasminogen activator, *RAGE* receptor for advanced glycation end products, *ROS* reactive oxygen species

8.6 Alcohol and Other Drugs of Abuse Activate the Innate Immune System Through Astrocytes and Microglia

There is an increasing body of literature supporting the hypothesis that alcohol and other drugs of abuse activate astrocytes and microglia. Through a series of elegant studies, Guerri's laboratory has demonstrated both in vivo and in vitro that chronic ethanol treatment induces astrocytic and microglial activation [39, 57, 60, 77, 78]. For instance, in vitro administration of ethanol at physiologically relevant levels activates TLR4/type I IL-1 receptor signaling pathways in cultured astrocytes in a time-dependent fashion similar to what occurs following LPS and IL-1β administration. Activation of this pathway leads to induction of NF-κB and activator protein-1 (AP-1) that is accompanied by an upregulation of inducible NOS and COX-2 expression in astrocytes that, in turn, leads to further downstream activation of the NF-κB signaling pathway [39]. Interestingly, ethanol-induced activation of the NF-κB pathway in astrocytes is blocked in TLR4-knockout mice and mice exposed to small interfering (si) RNA that target TLR4 [57], supporting a role of TLR4 in ethanol-induced inflammation. Further support for the involvement of TLR4 receptor activation in ethanol-induced inflammation was demonstrated in astrocyte cultures wherein siRNA against MD-2 and CD14, which are both critical component of the TLR4 signaling cascade, blocked downstream NF-κB induction [57].

Toll-like and IL-1 receptors are also expressed on microglia, and the expression of these receptors is induced during innate immune activation. Ethanol exposure causes microglia to adopt an activated state characterized by multiple radiating processes and an amoeboid morphology in human studies [49] and animal models [59, 78]. Indeed, Guerri's laboratory [78] found that ethanol activates microglia, leading to TLR4 induction and downstream secretion of cytokines and inflammatory mediators that culminate in neuronal cell death, an effect that was blocked in TLR4-knockout mice. These studies are consistent with ethanol-induced induction of innate immunity leading to altered behavior and causing neurodegeneration related to addiction. In support of the role of innate immunity in addiction, treatment with indomethacin, an anti-inflammatory drug, reduces chronic intermittent ethanol induction of brain innate immune genes (iNOS and COX-2) in astrocytes and reduces markers of cell death and behavioral dysfunction [79]. In addition, administration of minocycline, an antibiotic that blocks microglial activation [40, 42], reduces ethanol-induced inflammation and neural cell death [40].

In addition to alcohol, opiates are known addictive drugs, and endogenous opioid receptors and agonists clearly contribute to the neurobiology of addiction [80]. Opiate antagonists are used in the treatment of both alcohol and opiate addiction. Interestingly, a potential therapeutic mechanism of opiate antagonists appears to involve blockade of innate immune gene activation. Studies have found that opiate antagonists blunt LPS-induced immune responses [81] and protect dopaminergic neurons from microglial-mediated toxicity via inhibition of microglial activation and reduced NOX formation of ROS [82, 83]. Other studies have demonstrated that opiate antagonists block TLR4 activation of innate immune transcription, which is a site of action in innate immune loops [41, 84]. Thus, opiate antagonist therapy might exert some of its beneficial effects through blockade of innate immune gene induction, and anti-inflammatory drugs might serve as anti-addiction agents. Given the progressive nature of innate immune gene induction and addictive behaviors, therapeutic treatments aimed at reducing the induction of these genes might be advantageous for the alleviation of addiction. These studies support a role for innate immune induction in glia as important factors that contribute to the neurobiology of addiction.

8.7 Innate Immune Molecules Mimic Addiction-Like Behavior

A key element of the behavioral pathology of addiction and substance dependence is the increased anxiety and negative affect that accompanies mounting drug craving [80, 85]. A significant body of literature exists demonstrating that induction of innate immune genes alters behavior. Perhaps the most serious consequence of cancer treatment with proinflammatory type 1 interferon-α and IL-2 is the development of severe depression requiring treatment with antidepressant medication [86–88]. In these cases, the severity of depressive symptoms led approximately 45% of patients treated with interferon-α to terminate treatment due to increased suicidal ideation. Although these human cases are extreme examples, several other studies have linked negative affect and anxiety to innate immune activation. Indeed, human patients with chronic hepatitis C that received treatment with interferon-α evidenced serious side effects, including increased depression, anxiety, and cognitive impairment [89]. Similarly, bacterial endotoxin induces sickness behavior and negative affect across multiple species. Eisenberger and colleagues recently demonstrated that infusion of LPS into healthy humans reduced reward responses and increased depressed affect [90]. Breese and colleagues [87] demonstrated that the effects of ethanol and stress on anxiety-like behavior can be mimicked by brain injections of the chemokine MCP-1, the cytokine TNFα, or the TLR4 agonist LPS. These findings are consistent with the notion that stress and ethanol act through induction of innate immune genes to progressively increase negative affect and anxiety. In addition, systemic administration of LPS-induced sickness behavior as evidenced by reduced locomotor activity, diminished social interaction, and impaired cognitive function, which provides a model of depression [23, 91]. The discovery that direct injection of innate immune molecules into the brain facilitates stress and addiction-like anxiety further supports a role for the involvement of the innate immune system in the neurobiology of addiction.

8.8 Alcoholism Increases Markers of Innate Immune Activation in Human Brain

Postmortem human brain studies provide insight into the neurobiology of alcoholism and addiction. Our laboratory found upregulated levels of MCP-1 (CCL2) in the ventral tegmental area, substantia nigra, hippocampus, and amygdala of postmortem

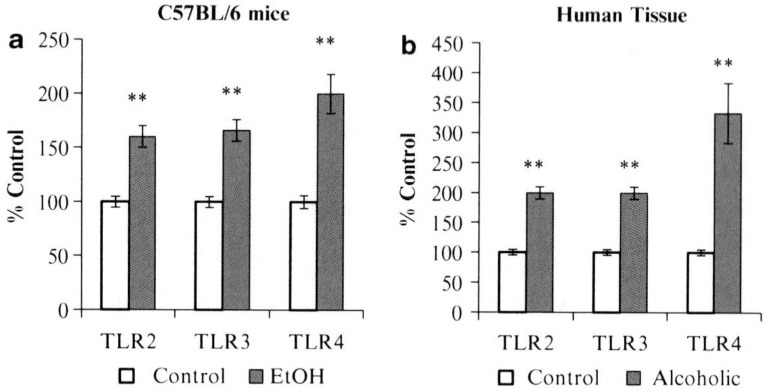

Fig. 8.7 Increased expression of Toll-like receptors (TLRs) in the mouse brain following chronic ethanol treatment and postmortem human alcoholic brain [40]. (**a**) Male C57BL/6 mice were treated with ethanol (5 g/kg, i.g.) for 10 days. Ethanol treatment significantly increased the number of TLR2-, TLR3-, and TLR4-immunoreactive cells in the orbitofrontal cortex, relative to matched controls. (**b**) Postmortem human alcohol tissue from the orbitofrontal cortex has significantly more TLR2-, TLR3-, and TLR4-immunoreactive cells than age-matched controls. $^{**}p<0.01$, relative to corresponding control group

human alcoholic brain tissue [49]. A similar increase in MCP-1 expression was found in mice exposed to chronic ethanol treatment [71]. Similarly, our laboratory recently found increased TLR2, TLR3, and TLR4 expression in the orbitofrontal cortex of human alcoholic and mice exposed to chronic ethanol [40] (see Fig. 8.7). These MCP-1 and TLR studies indicate that rodent paradigms model the innate immune-induced neurobiological changes that occur in the addicted human brain. In another postmortem human alcoholic brain study, Okvist and colleagues [92] reported increased NF-κB nuclear binding of p50 subunits with 479 NF-κB-driven genes upregulated in the frontal cortex but not motor cortex. In addition, a human gene expression analysis conducted on postmortem tissue [93] revealed altered expression of a group of cell adhesion genes that is consistent with altered extracellular membrane components and innate immune activation. Studies of human methamphetamine addicts and animal models of methamphetamine addiction also demonstrate persistently elevated levels of innate immune proteins, such as MCP-1 [94]. Thus, postmortem human alcoholic brain and animal studies are consistent with the hypothesis that increased innate immune gene expression occurs in the addicted brain.

Substantial evidence has accumulated over the last decade implicating the innate immune system in ethanol-induced neurodegeneration. Recent findings implicate TLRs and RAGE in the neuroinflammation associated with ethanol, which are likely activated through danger signals. A potential signaling molecule that likely plays a role is HMGB1, which belongs to the class of molecules known as DAMPs. Upon upregulation of HMGB1, it is released from the nucleus to the cytoplasm and is then secreted into the extracellular space where it exerts its cytokine-like effects [95–97].

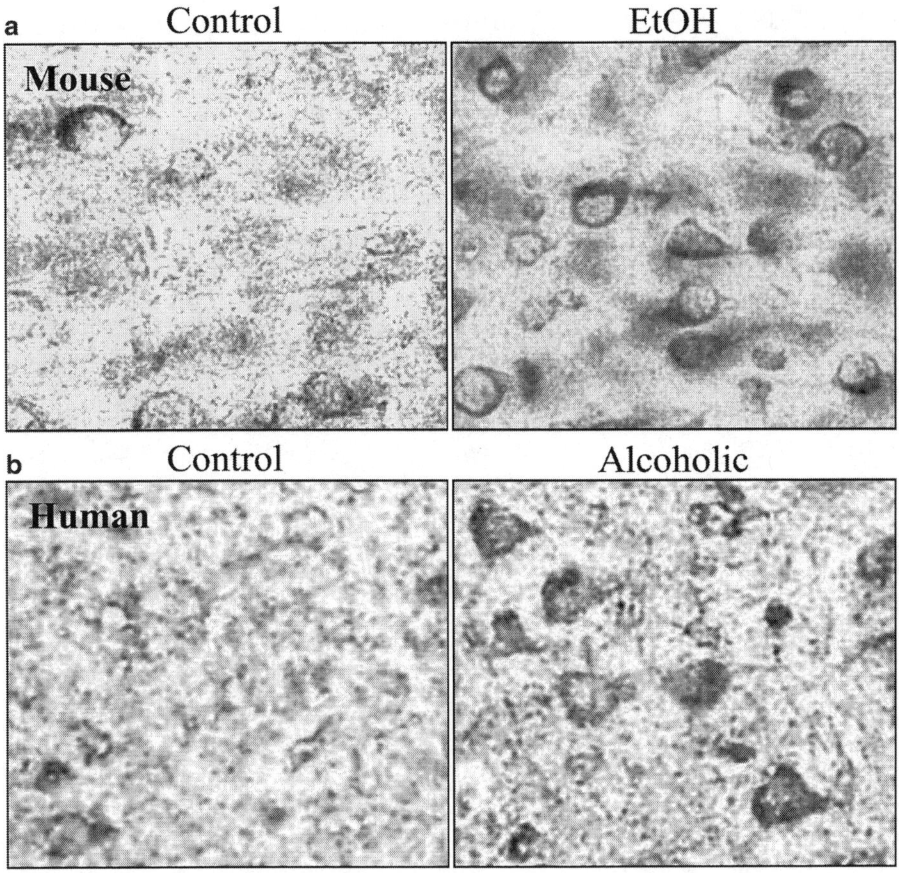

Fig. 8.8 High-mobility group box-1 (HMGB1) expression in the frontal cortex is increased in mice exposed to chronic ethanol and postmortem human alcoholic brain [40]. (**a**) Male C57BL/6 mice were treated with ethanol (5 g/kg, i.g.) for 10 days. Ethanol treatment significantly increased HMGB1-immunoreactive cells in the orbitofrontal cortex than matched controls. (**b**) Postmortem human alcohol tissue from the orbitofrontal cortex has significantly more HMGB1-immunoreactive cells than age-matched controls

Following its release, HMGB1 activates TLRs, including TLR2 and TLR4, and RAGE, which facilitate inflammatory responses in a cyclic fashion [98–101] (see Fig. 8.4). Recently, our laboratory demonstrated that HMGB1 expression is elevated in postmortem human alcoholic brain tissue and mice exposed to chronic ethanol [40] (see Fig. 8.8). These data implicate HMGB1 in the propagation of the alcohol-induced innate immune responses, but it remains to be determined whether HMGB1 is involved in overall addiction-induced innate immune activation or whether it is specific to alcoholism. Taken together, these data demonstrate that alcohol activates the innate immune system providing further evidence that this pathway is involved in the neurobiology of addiction.

8.9 Addiction-Induced Expression of Innate Immune Genes Creates the Neurobiology of Addiction

Although innate immunity plays a major role in neuroinflammation, recent studies have demonstrated that it also modulates cognitive and emotive functioning [91, 102–104]. Given that ethanol promotes innate immune gene induction [49, 50], it is plausible that ethanol-induced activation of innate immunity will alter behavior. Indeed, innate immunity activation via ethanol appears to contribute to addiction-like behaviors. For instance, Pascual and colleagues [77] demonstrated that ethanol-induced innate immune activation in mice impaired short- and long-term memory for object recognition. This behavioral impairment was accompanied by a reduction of H3 and H4 histone acetylation as well as histone acetyltransferase activity in the frontal cortex, striatum, and hippocampus. Interestingly, ethanol neither impaired behavioral performance nor altered histone activity in TLR4 knockout mice. Another inflammatory pathway, mediated by the RAGE receptor, might be involved in the memory impairments associated with chronic alcohol exposure since neuroinflammation associated with increased expression of this receptor is implicated in the memory impairments that accompany Alzheimer's disease [105–108]. Thus, it appears as though the innate immune system, particularly the TLRs and RAGE, is involved in the neurocognitive effects of ethanol.

The innate immune system also appears to modulate alcohol consumptive behavior. Gene expression studies of genetically paired rats and mice that differ primarily in their preference for ethanol consumption reveal that NF-κB, its regulatory proteins, and many innate immune genes are central to high-ethanol drinking behaviors [109]. Beta-2 microglobulin (B2M), an NF-κB target gene involved in MHC immune signaling [110], and other transcriptional regulation transcriptomes (i.e., RNA molecules) were significantly overexpressed in high-ethanol-preferring rodents compared to low-ethanol-preferring strains, suggesting that B2M might induce a predisposition for excessive ethanol consumption [109]. In addition, work from Blednov and colleagues provided interesting and novel data supporting the hypothesis that innate immune genes regulate ethanol drinking behavior [111, 112]. Across multiple strains of transgenic mice with innate immune gene deletion, these animals universally drink significantly less ethanol than matched controls across multiple ethanol drinking paradigms. Recently, Blednov and colleagues [113] also demonstrated that innate immune activation through LPS can induce long-lasting increases in ethanol drinking. Indeed, strains of mice show varied innate immune responses to LPS that correspond to increases in the consumption of ethanol. Furthermore, a single injection of LPS is capable of producing a delayed but long-lasting increase in ethanol consumption even in strains of high drinking mice. Recent data has revealed that the TLR family, especially TLR4, modulates ethanol intake. Administration of a $GABA_A\alpha2$ siRNA vector into the central nucleus of the amygdala of alcohol-preferring rats diminished binge drinking, which was associated with reduced expression of TLR4 [114]. Interestingly, the neuronal location of $GABA_A\alpha2$ receptors suggests that the influence of TLR4 on binge drinking is at least partially mediated by neurons. Taken together, these findings are consistent with a modulatory role of innate immune genes in alcohol preference and consumption.

8.9.1 Innate Immune Genes Increase Negative Affect and Reduce Hippocampal Neurogenesis

Human patients with major depressive disorder evidence increased blood inflammatory markers, and treatment with antidepressant medication is associated with a reduction of these markers. In addition to increased innate immune gene expression, patients with depression evidence diminished hippocampal volumes relative to age-matched controls [115]. In animal studies, endotoxin-induced increases in innate immune genes increase depression-like behavior [23]. Furthermore, immune activation leads to induction of microglial tryptophan metabolism that could reduce serotonin thereby contributing to depression [23]. Recent findings suggest that reductions of adult hippocampal neurogenesis may underlie depression and provide an index of mood and negative affect that allow for molecular studies. Both alcoholism and depression may be mediated by changes in adult hippocampal neurogenesis [116, 117]. Indeed, stress, multiple addictive drugs, and other factors that precipitate depression also reduce neurogenesis [118–120].

Although neurogenesis is primarily an early developmental process with most neurons generated during the prenatal and early postnatal periods, it continues to occur throughout adulthood in discrete brain regions, including the forebrain subventricular zone and subgranular zone of the hippocampal dentate gyrus. The generation and functional integration of nascent neurons into preexisting adult neural circuits are believed to enable the hippocampus to adapt to novel and more complex situations [121]. Indeed, the contribution of hippocampal neurogenesis to learning and memory [122] as well as mood and affective state [123] is supported by many studies. These findings are consistent with depression-associated diminution of hippocampal neurogenesis and antidepressant-induced increases in neurogenesis, hippocampal volume in humans, and reversal of depressive symptomology [124].

Many of the factors that increase depression-like behaviors also reduce neurogenesis [125]. Recent research has revealed that activation of NF-κB is necessary for stress-induced inhibition of neurogenesis and induction of depression-like behaviors [117], such as the social defeat model of depression [126]. Interestingly, administration of the antioxidant butylated hydroxytoluene (BHT) to rats blocks the NF-κB cascade preventing the reduction of hippocampal neurogenesis associated with binge ethanol exposure [19]. In addition, antidepressant efficacy in rodent behavioral models is dependent upon hippocampal neurogenesis [127]. In animal studies, endotoxin-induced increases in innate immune genes reduce hippocampal neurogenesis [23]. TLRs are necessary components of both ethanol neurotoxicity [57] and innate immune-induced depressive behavior and reduction of neurogenesis [23]. We have found that chronic ethanol increases brain innate immune genes,

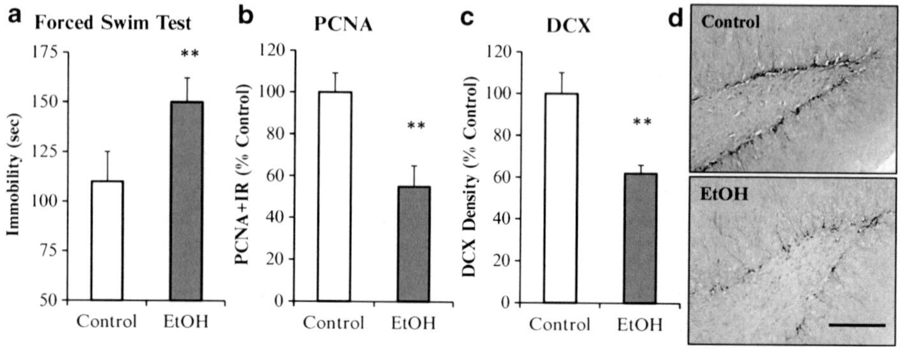

Fig. 8.9 Chronic ethanol self-administration induces depression-like behavior and inhibits hippocampal neurogenesis [128]. C57BL/6J mice self-administered either ethanol (10 %, v/v) or water for 28 days. (**a**) Increased immobility (seconds) on the forced swim test provides an index of depression-like behavior. Following a period of abstinence from chronic ethanol consumption, mice evidenced increased immobility time relative to controls. This demonstrates that ethanol exposure induces a depression-like state. (**b**) Ethanol self-administration decreased PCNA, a marker of cell proliferation, in the neurogenic region of the hippocampal dentate gyrus. (**c**) Doublecortin (DCX) expression, a marker of neurogenesis, was also reduced in the dentate gyrus of mice exposed to chronic ethanol self-administration. (**d**) Representative photomicrographs depicting the reduced DCX expression. Reduced progenitor cell proliferation and neurogenesis is associated with increased depression-like behavior. Scale bar = 200 μm. $^{**}p<0.01$, relative to corresponding control group

reduces brain neurogenesis, and increases depression-like behavior. Indeed, mice that self-administer ethanol in a chronic drinking model evidence depression-like behavior during abstinence that was associated with reduced neurogenesis [128] (see Fig. 8.9). Ethanol-induced loss of neurogenesis parallels the onset of depression-like behavior and is reversed via treatment with antidepressants. Similarly, stress-induced IL-1β reduces neurogenesis and increases depression-like behaviors [117]. Inhibition of neurogenesis is associated with increased negative affect and depression, which are key elements in the neurobiology of addiction. Thus, neurogenesis affects mood, and reduced neurogenesis is associated with innate immune gene induction, drug-induced negative affect, and depression-like behavior.

8.9.2 Frontal Cortical Hyperexcitability and Behavioral Inflexibility

The frontal cortex regulates decision-making and other executive functions, such as motivation, planning and goal setting, and behavioral flexibility. Studies of social drinkers have found that the heaviest binge drinkers reported more evidence of negative mood and performed poorly on executive functioning tasks [129, 130]. The frontal cortex regulates mood and cognition through reciprocal glutamatergic connections with multiple brain regions. In astrocytes, ethanol exposure induces NF-κB transcription leading to increased expression of proinflammatory innate immune genes [69, 70, 79] and reduced astrocyte glutamate transport [131]. The increased extracellular glutamate levels lead to increased neuronal excitation, microglial activation, and excitotoxicity [69, 132]. Indeed, the increased glutamate excitotoxicity likely contributes to ethanol-induced increases of caspase-3 and COX-2 in the frontal cortex [57, 71]. Glutamatergic hyperexcitability has also been demonstrated to occur in the cocaine- and stimulant-addicted brain [20]. The involvement of innate immune genes in the reduction of glutamate transporters and subsequent hyperexcitability inactivates the frontal cortex and contributes to the neurobiology of addiction [19] as evidenced by increased behavioral inflexibility [21].

Frontal cortical dysfunction is common in the alcoholic brain [8] and is manifest in impulsivity and behavioral inflexibility. In animal models, reversal learning tasks are used to assess frontal cortical dysfunction. During reversal learning, flexible behavior in response to outcomes that do not match those predicted by the preceding cues is necessary to successfully complete the task [22]. Indeed, proper frontal cortex function is necessary to weigh the value of decisions and is important when new learning and/or behavior is necessary. This behavioral paradigm models the perseveration and inability of human drug addicts to learn new behaviors. Using animal models of binge drinking, our laboratory found that ethanol exposure induces persistent deficits in reversal learning in rats [15] and in adult mice following binge drinking [14]. Interestingly, these reversal learning deficits are common to drug addiction as others have demonstrated that rats exposed to either self-administration of cocaine or passive cocaine injections are abnormally slow to learn the reversal shift even though they learn initial contingencies at a normal rate [16, 17].

The involvement of the frontal cortex in reversal learning is critical given that lesion produces reversal learning deficits similar in nature to chronic drug abuse-induced deficits [133]. In addition, frontal cortical dysfunction results in perseveration and repetition of previously learned behaviors due to failure to associate new information (e.g., negative consequences) into decision-making. Although there is no direct evidence of innate immunity on reversal learning, the persistence of addiction clearly overlaps with the persistence of innate immune gene induction [2, 50]. Together, the increased hyperexcitability of the frontal cortex, coupled with the loss of behavioral control due to innate immune upregulation, is consistent with innate immune gene induction culminating in the neurobiology of addiction.

8.10 The Adolescent Brain Is Vulnerable to Drug-Induced Frontal Cortical Damage

The adolescent developmental period is a critical time of neural maturation that encompasses the transition from childhood to adulthood. The frontal cortex and the limbic system undergo prominent pruning and reorganization during adolescence. For instance, absolute prefrontal cortex gray matter volumes decline in humans [134, 135] as well as in rats [136] during adolescence. Similarly, a substantial loss of synapses, especially excitatory glutamatergic inputs to the prefrontal cortex, occurs during the adolescent period in humans and nonhuman primates [137, 138]. In contrast to such adolescent-associated pruning, dopaminergic and serotonergic inputs to the prefrontal cortex increase to peak levels well above those observed earlier or later in life [139, 140]. In a similar fashion, cholinergic innervation of the prefrontal cortex increases at this time point, ultimately reaching mature levels in rats [141] and humans [142].

Maturation of the frontal cortex during adolescence is paralleled by the development of behavioral control [143]. Adolescence is characterized by increased risk-taking, environmental exploration, novelty and sensation seeking, and social interaction as well as higher levels of activity and play behavior that promote the acquisition of skills necessary for maturation and independence [143, 144]. Unfortunately, adolescence is also a period for the initiation of drug experimentation and addiction due to the vulnerability of the developing frontal cortex [145]. The effects of alcohol on the adolescent brain are different from those observed in adulthood. Adolescents are less sensitive to the sedative effects of alcohol [146], which allows them to binge drink. In contrast, they are more vulnerable to alcohol-induced neurotoxicity [145, 147]. The increased sensitivity of the adolescent brain to alcohol-induced toxicity, coupled with the dynamic synaptic remodeling that characterizes this stage, might strengthen the learning components of heavy drinking behaviors and perpetuate the loss of important self-control and goal setting components of the maturing brain's executive centers.

Innate immune gene activation in the brain persists for long periods [2, 50] consistent with the persistence of addiction. This persistent nature is likely amplified in the adolescent brain [144] because of their increased ethanol consumption [146] and greater vulnerability to the neurotoxic effects of alcohol [145, 147]. Animal models of adolescent binge drinking have found both impaired reversal learning [14] and innate immune gene induction [79] that persists into adulthood (see Fig. 8.10). Initiation of alcohol and drug use during adolescent maturation of the frontal cortical behavioral control circuitry likely sets the stage for the development of lifelong addiction. These data suggest that the adolescent brain might be more vulnerable to the neurocognitive effects of ethanol-induced innate immunity activation.

8.11 Summary

Over time, continued drug use and stress lead to the development of addiction-promoting behaviors. The neurobiology of addiction involves inhibition of frontal cortical behavioral control and increased limbic system negative affect. Stress and repeated drug use increase NF-κB transcription of chemokines, cytokines, oxidases, proteases, and other innate immune genes in a cyclic fashion that spreads across the CNS. Evidence supports the hypothesis that the spread and persistence of innate immune gene expression in the CNS creates the neurobiology of addiction

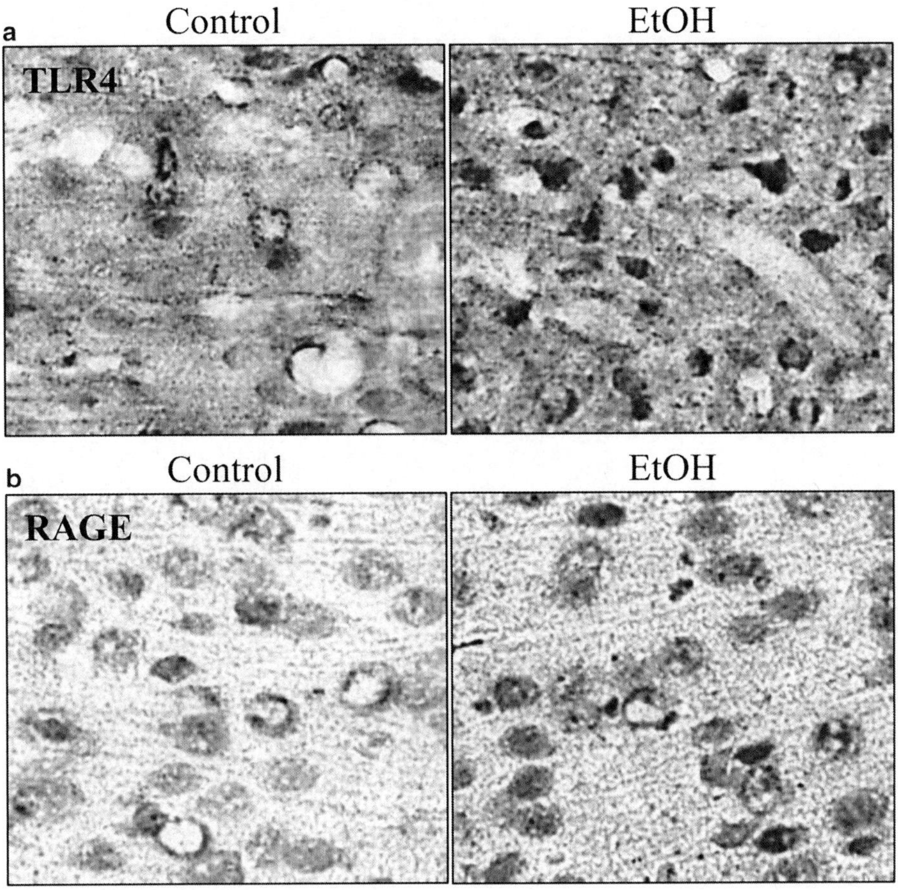

Fig. 8.10 TLR4 and RAGE expressions are increased in the frontal cortex of adult rats exposed to adolescent binge ethanol exposure. Male Wistar rats were treated with ethanol (5 g/kg, i.g.) on a 2-day on/2-day off binge ethanol exposure paradigm for the totality of adolescence [i.e., postnatal day (P) 25–55]. Following a period of abstinence, brain tissue was collected in adulthood (P80). (**a**) Binge ethanol exposure during adolescence persistently upregulated TLR4 expression in the adult prefrontal cortex, relative to controls. (**b**) Adolescent binge ethanol exposure led to persistently upregulated RAGE expression in the adult prefrontal cortex, relative to controls

(see Fig. 8.11). Animal studies demonstrate that ethanol-induced activation of innate immunity leads to behavioral changes that involve increased impulsivity, anxiety, and negative affect that increase immediate reward-seeking behavior. Furthermore, hyperexcitability and behavioral inflexibility are associated with innate immune gene expression in the frontal cortex. The emerging view that innate immunity underlies addiction has important ramifications for the early onset of drug and alcohol abuse. Furthermore, it suggests that therapies aimed at suppressing innate immune activation in the CNS might represent a novel therapeutic approach to the treatment of addiction and/or anxiety disorders.

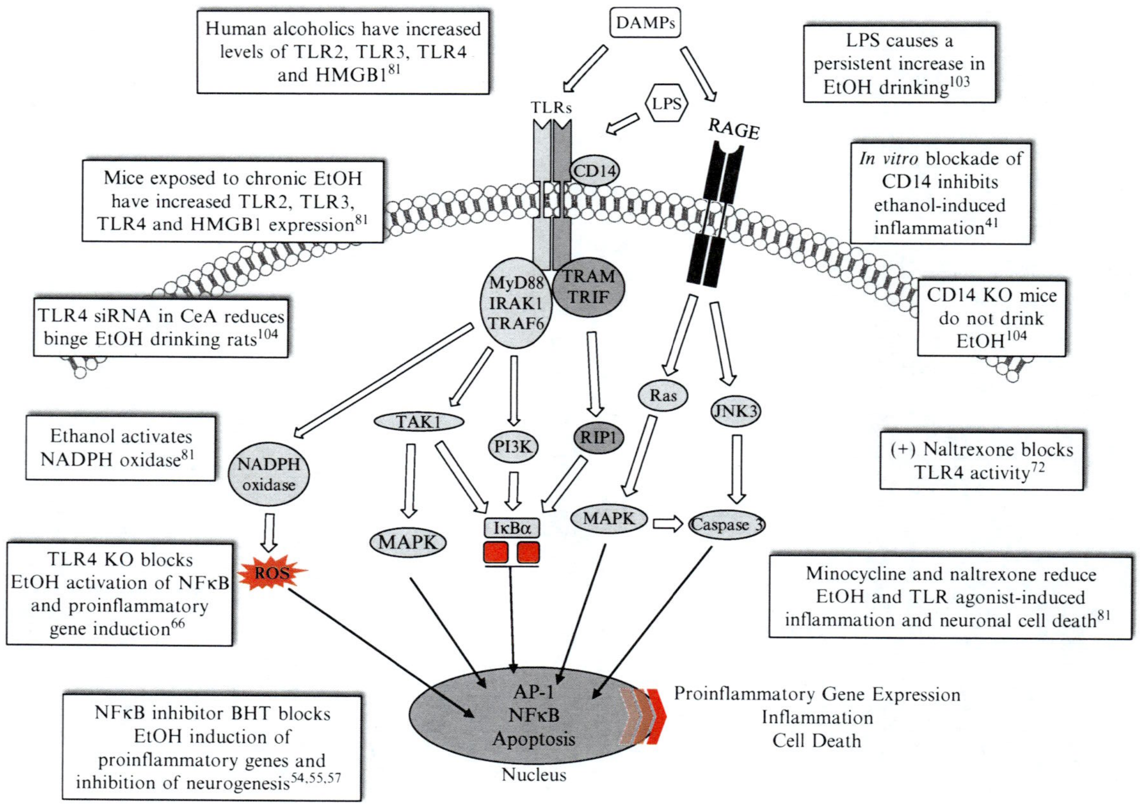

Fig. 8.11 Summary of supportive data indicating a role for the innate immune signaling cascade in alcohol-induced neurodegeneration. The *white boxes* provide summaries of the converging lines of evidence supporting a role for alcohol-induced activation of the innate immune system

Acknowledgments The authors wish to acknowledge support from the Bowles Center of Alcohol Studies, School of Medicine, The University of North Carolina at Chapel Hill, the National Institute of Health, and the National Institute on Alcoholism and Alcohol Abuse through AA020023, AA020022, AA019767, AA11605, and AA007573.

References

1. Crews FT et al (2006) Cytokines and alcohol. Alcohol Clin Exp Res 30:720–730
2. Qin L et al (2007) Systemic LPS causes chronic neuroinflammation and progressive neurodegeneration. Glia 55:453–462
3. Li TK (2008) Quantifying the risk for alcohol-use and alcohol-attributable health disorders: present findings and future research needs. J Gastroenterol Hepatol 23(Suppl 1):S2–S8
4. Gilson AM (2010) The concept of addiction in law and regulatory policy related to pain management: a critical review. Clin J Pain 26:70–77
5. Kalivas PW, Volkow ND (2005) The neural basis of addiction: a pathology of motivation and choice. Am J Psychiatry 162:1403–1413
6. Koob GF, Le Moal M (2005) Plasticity of reward neurocircuitry and the 'dark side' of drug addiction. Nat Neurosci 8:1442–1444
7. Schoenbaum G, Shaham Y (2008) The role of orbitofrontal cortex in drug addiction: a review of preclinical studies. Biol Psychiatry 63:256–262
8. Crews FT, Boettiger CA (2009) Impulsivity, frontal lobes and risk for addiction. Pharmacol Biochem Behav 93:237–247
9. Bechara A, Dolan S, Hindes A (2002) Decision-making and addiction (part II): myopia for the future or hypersensitivity to reward? Neuropsychologia 40:1690–1705
10. Ersche KD, Roiser JP, Robbins TW, Sahakian BJ (2008) Chronic cocaine but not chronic amphetamine use is associated with perseverative responding in humans. Psychopharmacology (Berl) 197:421–431
11. Jentsch JD, Taylor JR (1999) Impulsivity resulting from frontostriatal dysfunction in drug abuse: implications for the control of behavior by reward-related stimuli. Psychopharmacology (Berl) 146:373–390
12. Rogers RD et al (1999) Dissociable deficits in the decision-making cognition of chronic amphetamine abusers, opiate abusers, patients with focal damage to prefrontal cortex, and tryptophan-depleted normal volunteers: evidence for monoaminergic mechanisms. Neuropsychopharmacology 20:322–339
13. Rogers RD, Robbins TW (2001) Investigating the neurocognitive deficits associated with chronic drug misuse. Curr Opin Neurobiol 11:250–257
14. Coleman LG Jr, He J, Lee J, Styner M, Crews FT (2011) Adolescent binge drinking alters adult brain neurotransmitter gene expression, behavior, brain regional volumes, and neurochemistry in mice. Alcohol Clin Exp Res 35:671–688
15. Obernier JA, White AM, Swartzwelder HS, Crews FT (2002) Cognitive deficits and CNS damage after a 4-day binge ethanol exposure in rats. Pharmacol Biochem Behav 72:521–532
16. Calu DJ, Roesch MR, Stalnaker TA, Schoenbaum G (2007) Associative encoding in posterior piriform cortex during odor discrimination and reversal learning. Cereb Cortex 17:1342–1349
17. Schoenbaum G, Saddoris MP, Ramus SJ, Shaham Y, Setlow B (2004) Cocaine-experienced rats exhibit learning deficits in a task sensitive to orbitofrontal cortex lesions. Eur J Neurosci 19:1997–2002
18. Robinson TE, Berridge KC (2003) Addiction. Annu Rev Psychol 54:25–53
19. Crews F et al (2006) BHT blocks NF-kappaB activation and ethanol-induced brain damage. Alcohol Clin Exp Res 30:1938–1949
20. Reissner KJ, Kalivas PW (2010) Using glutamate homeostasis as a target for treating addictive disorders. Behav Pharmacol 21:514–522

21. Gruber AJ et al (2010) More is less: a disinhibited prefrontal cortex impairs cognitive flexibility. J Neurosci 30:17102–17110
22. Stalnaker TA, Takahashi Y, Roesch MR, Schoenbaum G (2009) Neural substrates of cognitive inflexibility after chronic cocaine exposure. Neuropharmacology 56(Suppl 1):63–72
23. Kelley KW, Dantzer R (2011) Alcoholism and inflammation: neuroimmunology of behavioral and mood disorders. Brain Behav Immun 25(Suppl 1):S13–S20
24. Hanisch UK, Johnson TV, Kipnis J (2008) Toll-like receptors: roles in neuroprotection? Trends Neurosci 31:176–182
25. Lehnardt S (2010) Innate immunity and neuroinflammation in the CNS: the role of microglia in Toll-like receptor-mediated neuronal injury. Glia 58:253–263
26. Ginhoux F et al (2010) Fate mapping analysis reveals that adult microglia derive from primitive macrophages. Science 330:841–845
27. Iwasaki A, Medzhitov R (2004) Toll-like receptor control of the adaptive immune responses. Nat Immunol 5:987–995
28. Takeda K, Kaisho T, Akira S (2003) Toll-like receptors. Annu Rev Immunol 21:335–376
29. Medzhitov R (2001) Toll-like receptors and innate immunity. Nat Rev Immunol 1:135–145
30. Sparvero LJ et al (2009) RAGE (receptor for advanced glycation endproducts), RAGE ligands, and their role in cancer and inflammation. J Transl Med 7:17
31. Han SH, Kim YH, Mook-Jung I (2011) RAGE: the beneficial and deleterious effects by diverse mechanisms of actions. Mol Cells 31:91–97
32. Sims GP, Rowe DC, Rietdijk ST, Herbst R, Coyle AJ (2010) HMGB1 and RAGE in inflammation and cancer. Annu Rev Immunol 28:367–388
33. Garg AD et al (2010) Immunogenic cell death, DAMPs and anticancer therapeutics: an emerging amalgamation. Biochim Biophys Acta 1805:53–71
34. Vabulas RM et al (2002) The endoplasmic reticulum-resident heat shock protein Gp96 activates dendritic cells via the Toll-like receptor 2/4 pathway. J Biol Chem 277:20847–20853
35. Scheibner KA et al (2006) Hyaluronan fragments act as an endogenous danger signal by engaging TLR2. J Immunol 177:1272–1281
36. Thomas JO (2001) HMG1 and 2: architectural DNA-binding proteins. Biochem Soc Trans 29:395–401
37. Paull TT, Haykinson MJ, Johnson RC (1993) The nonspecific DNA-binding and -bending proteins HMG1 and HMG2 promote the assembly of complex nucleoprotein structures. Genes Dev 7:1521–1534
38. Huang W, Tang Y, Li L (2010) HMGB1, a potent proinflammatory cytokine in sepsis. Cytokine 51:119–126
39. Blanco AM, Valles SL, Pascual M, Guerri C (2005) Involvement of TLR4/type I IL-1 receptor signaling in the induction of inflammatory mediators and cell death induced by ethanol in cultured astrocytes. J Immunol 175:6893–6899
40. Crews FT, Qin L, Sheedy D, Vetreno RP, Zou J (In Press) HMGB1/TLR receptor danger signaling increases brain neuroimmune activation in alcohol dependence. Biological Psychiatry
41. Hutchinson MR et al (2008) Non-stereoselective reversal of neuropathic pain by naloxone and naltrexone: involvement of toll-like receptor 4 (TLR4). Eur J Neurosci 28:20–29
42. Plane JM, Shen Y, Pleasure DE, Deng W (2010) Prospects for minocycline neuroprotection. Arch Neurol 67:1442–1448
43. Glass CK, Saijo K, Winner B, Marchetto MC, Gage FH (2010) Mechanisms underlying inflammation in neurodegeneration. Cell 140:918–934
44. Streit WJ, Mrak RE, Griffin WS (2004) Microglia and neuroinflammation: a pathological perspective. J Neuroinflammation 1:14
45. Farina C, Aloisi F, Meinl E (2007) Astrocytes are active players in cerebral innate immunity. Trends Immunol 28:138–145
46. Graeber MB (2010) Changing face of microglia. Science 330:783–788
47. Kreutzberg GW (1996) Microglia: a sensor for pathological events in the CNS. Trends Neurosci 19:312–318
48. Raivich G et al (1999) Neuroglial activation repertoire in the injured brain: graded response, molecular mechanisms and cues to physiological function. Brain Res Brain Res Rev 30:77–105

49. He J, Crews FT (2008) Increased MCP-1 and microglia in various regions of the human alcoholic brain. Exp Neurol 210:349–358
50. Qin L et al (2008) Increased systemic and brain cytokine production and neuroinflammation by endotoxin following ethanol treatment. J Neuroinflammation 5:10
51. Colton CA (2009) Heterogeneity of microglial activation in the innate immune response in the brain. J Neuroimmune Pharmacol 4:399–418
52. Michelucci A, Heurtaux T, Grandbarbe L, Morga E, Heuschling P (2009) Characterization of the microglial phenotype under specific pro-inflammatory and anti-inflammatory conditions: effects of oligomeric and fibrillar amyloid-beta. J Neuroimmunol 210:3–12
53. Aschner M (1998) Astrocytes as mediators of immune and inflammatory responses in the CNS. Neurotoxicology 19:269–281
54. Aschner M, Sonnewald U, Tan KH (2002) Astrocyte modulation of neurotoxic injury. Brain Pathol 12:475–481
55. Ridet JL, Malhotra SK, Privat A, Gage FH (1997) Reactive astrocytes: cellular and molecular cues to biological function. Trends Neurosci 20:570–577
56. Lee H et al (2004) Ethanol selectively modulates inflammatory activation signaling of brain microglia. J Neuroimmunol 156:88–95
57. Alfonso-Loeches S, Pascual-Lucas M, Blanco AM, Sanchez-Vera I, Guerri C (2010) Pivotal role of TLR4 receptors in alcohol-induced neuroinflammation and brain damage. J Neurosci 30:8285–8295
58. Shpargel KB et al (2008) Preconditioning paradigms and pathways in the brain. Cleve Clin J Med 75(Suppl 2):S77–S82
59. McClain JA et al (2011) Adolescent binge alcohol exposure induces long-lasting partial activation of microglia. Brain Behav Immun 25(Suppl 1):S120–S128
60. Valles SL, Blanco AM, Pascual M, Guerri C (2004) Chronic ethanol treatment enhances inflammatory mediators and cell death in the brain and in astrocytes. Brain Pathol 14:365–371
61. Greer JM, Capecchi MR (2002) Hoxb8 is required for normal grooming behavior in mice. Neuron 33:23–34
62. Gonzalez-Perez O, Ramos-Remus C, Garcia-Estrada J, Luquin S (2001) Prednisone induces anxiety and glial cerebral changes in rats. J Rheumatol 28:2529–2534
63. Torres-Platas SG et al (2011) Astrocytic hypertrophy in anterior cingulate white matter of depressed suicides. Neuropsychopharmacology 36:2650–2658
64. Ang E et al (2001) Induction of nuclear factor-kappaB in nucleus accumbens by chronic cocaine administration. J Neurochem 79:221–224
65. Mao XR, Moerman-Herzog AM, Chen Y, Barger SW (2009) Unique aspects of transcriptional regulation in neurons–nuances in NFkappaB and Sp1-related factors. J Neuroinflammation 6:16
66. Madrigal JL, Moro MA, Lizasoain I, Lorenzo P, Leza JC (2002) Stress-induced increase in extracellular sucrose space in rats is mediated by nitric oxide. Brain Res 938:87–91
67. Madrigal JL et al (2003) Relationship between cyclooxygenase-2 and nitric oxide synthase-2 in rat cortex after stress. Eur J Neurosci 18:1701–1705
68. Munhoz CD, Sorrells SF, Caso JR, Scavone C, Sapolsky RM (2010) Glucocorticoids exacerbate lipopolysaccharide-induced signaling in the frontal cortex and hippocampus in a dose-dependent manner. J Neurosci 30:13690–13698
69. Zou J, Crews F (2006) CREB and NF-kappaB transcription factors regulate sensitivity to excitotoxic and oxidative stress induced neuronal cell death. Cell Mol Neurobiol 26:385–405
70. Zou J, Crews F (2010) Induction of innate immune gene expression cascades in brain slice cultures by ethanol: key role of NF-kappaB and proinflammatory cytokines. Alcohol Clin Exp Res 34:777–789
71. Knapp DJ, Crews FT (1999) Induction of cyclooxygenase-2 in brain during acute and chronic ethanol treatment and ethanol withdrawal. Alcohol Clin Exp Res 23:633–643
72. Cao Q, Mak KM, Lieber CS (2005) Cytochrome P4502E1 primes macrophages to increase TNF-alpha production in response to lipopolysaccharide. Am J Physiol Gastrointest Liver Physiol 289:G95–G107
73. Ward RJ et al (1996) Identification of the nuclear transcription factor NFkappaB in rat after in vivo ethanol administration. FEBS Lett 389:119–122

74. Davis RL, Syapin PJ (2004) Ethanol increases nuclear factor-kappa B activity in human astroglial cells. Neurosci Lett 371:128–132
75. Pietri M et al (2005) Reactive oxygen species-dependent TNF-alpha converting enzyme activation through stimulation of 5-HT2B and alpha1D autoreceptors in neuronal cells. FASEB J 19:1078–1087
76. Thakur V, McMullen MR, Pritchard MT, Nagy LE (2007) Regulation of macrophage activation in alcoholic liver disease. J Gastroenterol Hepatol 22(Suppl 1):S53–S56
77. Pascual M, Balino P, Alfonso-Loeches S, Aragon CM, Guerri C (2011) Impact of TLR4 on behavioral and cognitive dysfunctions associated with alcohol-induced neuroinflammatory damage. Brain Behav Immun 25(Suppl 1):S80–S91
78. Fernandez-Lizarbe S, Pascual M, Guerri C (2009) Critical role of TLR4 response in the activation of microglia induced by ethanol. J Immunol 183:4733–4744
79. Pascual M, Blanco AM, Cauli O, Minarro J, Guerri C (2007) Intermittent ethanol exposure induces inflammatory brain damage and causes long-term behavioural alterations in adolescent rats. Eur J Neurosci 25:541–550
80. Koob GF, Volkow ND (2010) Neurocircuitry of addiction. Neuropsychopharmacology 35:217–238
81. Liu B et al (2000) Systemic infusion of naloxone reduces degeneration of rat substantia nigral dopaminergic neurons induced by intranigral injection of lipopolysaccharide. J Pharmacol Exp Ther 295:125–132
82. Liu B, Du L, Hong JS (2000) Naloxone protects rat dopaminergic neurons against inflammatory damage through inhibition of microglia activation and superoxide generation. J Pharmacol Exp Ther 293:607–617
83. Qin L, Liu Y, Qian X, Hong JS, Block ML (2005) Microglial NADPH oxidase mediates leucine enkephalin dopaminergic neuroprotection. Ann N Y Acad Sci 1053:107–120
84. Hutchinson MR et al (2010) Evidence that opioids may have toll-like receptor 4 and MD-2 effects. Brain Behav Immun 24:83–95
85. Breese GR, Overstreet DH, Knapp DJ (2005) Conceptual framework for the etiology of alcoholism: a "kindling"/stress hypothesis. Psychopharmacology (Berl) 178:367–380
86. O'Connor MF, Irwin MR, Seldon J, Kwan L, Ganz PA (2007) Pro-inflammatory cytokines and depression in a familial cancer registry. Psychooncology 16:499–501
87. Breese GR et al (2008) Repeated lipopolysaccharide (LPS) or cytokine treatments sensitize ethanol withdrawal-induced anxiety-like behavior. Neuropsychopharmacology 33:867–876
88. Capuron L, Ravaud A (1999) Prediction of the depressive effects of interferon alfa therapy by the patient's initial affective state. N Engl J Med 340:1370
89. McNutt MD et al (2012) Neurobehavioral effects of interferon-alpha in patients with hepatitis-C: symptom dimensions and responsiveness to paroxetine. Neuropsychopharmacol 37:1444–1454
90. Eisenberger NI et al (2010) Inflammation-induced anhedonia: endotoxin reduces ventral striatum responses to reward. Biol Psychiatry 68:748–754
91. Dantzer R, O'Connor JC, Freund GG, Johnson RW, Kelley KW (2008) From inflammation to sickness and depression: when the immune system subjugates the brain. Nat Rev Neurosci 9:46–56
92. Okvist A et al (2007) Neuroadaptations in human chronic alcoholics: dysregulation of the NF-kappaB system. PLoS One 2:e930
93. Liu J et al (2006) Patterns of gene expression in the frontal cortex discriminate alcoholic from nonalcoholic individuals. Neuropsychopharmacology 31:1574–1582
94. Loftis JM, Choi D, Hoffman W, Huckans MS (2011) Methamphetamine causes persistent immune dysregulation: a cross-species, translational report. Neurotox Res 20:59–68
95. Hock R, Furusawa T, Ueda T, Bustin M (2007) HMG chromosomal proteins in development and disease. Trends Cell Biol 17:72–79
96. Muller S, Ronfani L, Bianchi ME (2004) Regulated expression and subcellular localization of HMGB1, a chromatin protein with a cytokine function. J Intern Med 255:332–343
97. Maroso M et al (2010) Toll-like receptor 4 and high-mobility group box-1 are involved in ictogenesis and can be targeted to reduce seizures. Nat Med 16:413–419

98. Park JS et al (2004) Involvement of toll-like receptors 2 and 4 in cellular activation by high mobility group box 1 protein. J Biol Chem 279:7370–7377
99. Yang L et al (2010) Role of phospholipase D2/phosphatidic acid signal transduction in micro- and delta-opioid receptor endocytosis. Mol Pharmacol 78:105–113
100. Rauvala H, Rouhiainen A (2010) Physiological and pathophysiological outcomes of the interactions of HMGB1 with cell surface receptors. Biochim Biophys Acta 1799:164–170
101. Volz HC, Kaya Z, Katus HA, Andrassy M (2010) The role of HMGB1/RAGE in inflammatory cardiomyopathy. Semin Thromb Hemost 36:185–194
102. Yirmiya R, Goshen I (2011) Immune modulation of learning, memory, neural plasticity and neurogenesis. Brain Behav Immun 25:181–213
103. Okun E et al (2010) Toll-like receptor 3 inhibits memory retention and constrains adult hippocampal neurogenesis. Proc Natl Acad Sci USA 107:15625–15630
104. Hanke ML, Kielian T (2011) Toll-like receptors in health and disease in the brain: mechanisms and therapeutic potential. Clin Sci 121:367–387
105. Arancio O et al (2004) RAGE potentiates Abeta-induced perturbation of neuronal function in transgenic mice. EMBO J 23:4096–4105
106. Fang F et al (2010) RAGE-dependent signaling in microglia contributes to neuroinflammation, Abeta accumulation, and impaired learning/memory in a mouse model of Alzheimer's disease. FASEB J 24:1043–1055
107. Maczurek A, Shanmugam K, Munch G (2008) Inflammation and the redox-sensitive AGE-RAGE pathway as a therapeutic target in Alzheimer's disease. Ann N Y Acad Sci 1126:147–151
108. Wilson JS et al (2009) Anti-RAGE and Abeta immunoglobulin levels are related to dementia level and cognitive performance. J Gerontol A Biol Sci Med Sci 64:264–271
109. Mulligan MK et al (2006) Toward understanding the genetics of alcohol drinking through transcriptome meta-analysis. Proc Natl Acad Sci USA 103:6368–6373
110. Pahl HL (1999) Activators and target genes of Rel/NF-kappaB transcription factors. Oncogene 18:6853–6866
111. Blednov YA et al (2005) Perturbation of chemokine networks by gene deletion alters the reinforcing actions of ethanol. Behav Brain Res 165:110–125
112. Blednov YA et al (2012) Neuroimmune regulation of alcohol consumption: behavioral validation of genes obtained from genomic studies. Addict Biol 17:108–120
113. Blednov YA et al (2011) Activation of inflammatory signaling by lipopolysaccharide produces a prolonged increase of voluntary alcohol intake in mice. Brain Behav Immun 25(Suppl 1):S92–S105
114. Liu J et al (2011) Binge alcohol drinking is associated with GABAA alpha2-regulated Toll-like receptor 4 (TLR4) expression in the central amygdala. Proc Natl Acad Sci USA 108:4465–4470
115. Videbech P, Ravnkilde B (2004) Hippocampal volume and depression: a meta-analysis of MRI studies. Am J Psychiatry 161:1957–1966
116. Crews FT, Nixon K (2003) Alcohol, neural stem cells, and adult neurogenesis. Alcohol Res Health 27:197–204
117. Koo JW, Duman RS (2008) IL-1beta is an essential mediator of the antineurogenic and anhedonic effects of stress. Proc Natl Acad Sci USA 105:751–756
118. Tanapat P, Hastings NB, Rydel TA, Galea LA, Gould E (2001) Exposure to fox odor inhibits cell proliferation in the hippocampus of adult rats via an adrenal hormone-dependent mechanism. J Comp Neurol 437:496–504
119. Malberg JE, Duman RS (2003) Cell proliferation in adult hippocampus is decreased by inescapable stress: reversal by fluoxetine treatment. Neuropsychopharmacology 28:1562–1571
120. Gregus A, Wintink AJ, Davis AC, Kalynchuk LE (2005) Effect of repeated corticosterone injections and restraint stress on anxiety and depression-like behavior in male rats. Behav Brain Res 156:105–114
121. Kempermann G (2002) Why new neurons? Possible functions for adult hippocampal neurogenesis. J Neurosci 22:635–638
122. Shors TJ et al (2001) Neurogenesis in the adult is involved in the formation of trace memories. Nature 410:372–376

123. Malberg JE, Eisch AJ, Nestler EJ, Duman RS (2000) Chronic antidepressant treatment increases neurogenesis in adult rat hippocampus. J Neurosci 20:9104–9110
124. Dranovsky A, Hen R (2006) Hippocampal neurogenesis: regulation by stress and antidepressants. Biol Psychiatry 59:1136–1143
125. Johnson SA, Fournier NM, Kalynchuk LE (2006) Effect of different doses of corticosterone on depression-like behavior and HPA axis responses to a novel stressor. Behav Brain Res 168:280–288
126. Christoffel DJ et al (2011) IkappaB kinase regulates social defeat stress-induced synaptic and behavioral plasticity. J Neurosci 31:314–321
127. Santarelli L et al (2003) Requirement of hippocampal neurogenesis for the behavioral effects of antidepressants. Science 301:805–809
128. Stevenson JR et al (2009) Abstinence following alcohol drinking produces depression-like behavior and reduced hippocampal neurogenesis in mice. Neuropsychopharmacology 34:1209–1222
129. Townshend JM, Duka T (2003) Mixed emotions: alcoholics' impairments in the recognition of specific emotional facial expressions. Neuropsychologia 41:773–782
130. Weissenborn R, Duka T (2003) Acute alcohol effects on cognitive function in social drinkers: their relationship to drinking habits. Psychopharmacology (Berl) 165:306–312
131. Zou JY, Crews FT (2005) TNF alpha potentiates glutamate neurotoxicity by inhibiting glutamate uptake in organotypic brain slice cultures: neuroprotection by NF kappa B inhibition. Brain Res 1034:11–24
132. Ward RJ et al (2009) Neuro-inflammation induced in the hippocampus of 'binge drinking' rats may be mediated by elevated extracellular glutamate content. J Neurochem 111:1119–1128
133. Schoenbaum G, Roesch MR, Stalnaker TA (2006) Orbitofrontal cortex, decision-making and drug addiction. Trends Neurosci 29:116–124
134. Sowell ER, Thompson PM, Holmes CJ, Jernigan TL, Toga AW (1999) In vivo evidence for post-adolescent brain maturation in frontal and striatal regions. Nat Neurosci 2:859–861
135. Sowell ER, Thompson PM, Tessner KD, Toga AW (2001) Mapping continued brain growth and gray matter density reduction in dorsal frontal cortex: Inverse relationships during post-adolescent brain maturation. J Neurosci 21:8819–8829
136. van Eden CG, Kros JM, Uylings HB (1990) The development of the rat prefrontal cortex. Its size and development of connections with thalamus, spinal cord and other cortical areas. Prog Brain Res 85:169–183
137. Huttenlocher PR (1984) Synapse elimination and plasticity in developing human cerebral cortex. Am J Ment Defic 88:488–496
138. Zecevic N, Bourgeois JP, Rakic P (1989) Changes in synaptic density in motor cortex of rhesus monkey during fetal and postnatal life. Brain Res Dev Brain Res 50:11–32
139. Kalsbeek A, Voorn P, Buijs RM, Pool CW, Uylings HB (1988) Development of the dopaminergic innervation in the prefrontal cortex of the rat. J Comp Neurol 269:58–72
140. Rosenberg DR, Lewis DA (1994) Changes in the dopaminergic innervation of monkey prefrontal cortex during late postnatal development: a tyrosine hydroxylase immunohistochemical study. Biol Psychiatry 36:272–277
141. Gould E, Woolf NJ, Butcher LL (1991) Postnatal development of cholinergic neurons in the rat: I. Forebrain. Brain Res Bull 27:767–789
142. Kostovic I (1990) Structural and histochemical reorganization of the human prefrontal cortex during perinatal and postnatal life. Prog Brain Res 85:223–239, discussion 239–240
143. Ernst M, Romeo RD, Andersen SL (2009) Neurobiology of the development of motivated behaviors in adolescence: a window into a neural systems model. Pharmacol Biochem Behav 93:199–211
144. Spear LP (2000) The adolescent brain and age-related behavioral manifestations. Neurosci Biobehav Rev 24:417–463
145. Crews F, He J, Hodge C (2007) Adolescent cortical development: a critical period of vulnerability for addiction. Pharmacol Biochem Behav 86:189–199
146. Silveri MM, Spear LP (1998) Decreased sensitivity to the hypnotic effects of ethanol early in ontogeny. Alcohol Clin Exp Res 22:670–676
147. Monti PM et al (2005) Adolescence: booze, brains, and behavior. Alcohol Clin Exp Res 29:207–220

Chapter 9
Role of Toll-Like Receptor 4 in Alcohol-Induced Neuroinflammation and Behavioral Dysfunctions

Consuelo Guerri and María Pascual

9.1 Introduction

Until quite recently, the central nervous system (CNS) has been considered an autonomous unit nourished by blood and shielded not only from circulating immune cells but also from pathogens and toxins originating from circulation. However, it is now well established that immune surveillance occurs in the normal CNS and that inflammatory responses can take place in the neural plasticity and disease context. The CNS presents a well-organized series of innate immune reactions in response to systematic bacterial and viral infections and to cerebral injury [1, 2]. Activation of the innate immune system is an important component of inflammatory response.

The neuroimmune system plays essential roles in both maintenance of tissue homeostasis and response to infection and injury. Neuroinflammatory stimuli induce beneficial effects, and inflammation is linked to tissue repair processes. Emerging evidence also indicates that the immune system positively regulates learning, memory, neural plasticity, and neurogenesis [3, 4]. However, sustained inflammation resulting in tissue implies persistence of an inflammatory stimulus or failure in normal resolution mechanisms. The cytokine production which characterizes this condition can impair the beneficial role under quiescent conditions, resulting in impaired memory, neural plasticity, and neurogenesis [4]. A persistent stimulus may result from either environmental factors or the formation of endogenous factors (e.g., pathogens, protein aggregates, toxins, psychological stress) that are perceived by the immune system as "danger" signals.

C. Guerri (✉) • M. Pascual
Department of Cell Pathology, Príncipe Felipe Research Center,
Avda. Autopista del Saler 16-3 (junto Oceanográfico), 46012 Valencia, Spain
e-mail: guerri@cipf.es

C. Cui et al. (eds.), *Neural-Immune Interactions in Brain Function and Alcohol Related Disorders*, DOI 10.1007/978-1-4614-4729-0_9,

Inflammatory responses to infectious agents are initiated by pattern recognition receptors that bind to *pathogen-associated molecular patterns* (*PAMPs*, stranger signals), such as Toll-like receptors (TLRs) which recognize a diverse set of pathogen-associated molecules not present in the host [5]. Recent evidence indicates that TLRs are vital players in this orchestrated form of immune response in the CNS and that deregulation of this immunological response against CNS-associated antigens may play important roles in brain damage and neurodegeneration [6, 7].

TLRs were originally discovered for the role they play in dorsal–ventral axis development in *Drosophila* embryos [8], and plants, insects, and vertebrates all use homologous mechanisms that rely on Toll recognition to coordinate an immune response [9]. Recent studies indicate that TLRs respond not only to microbial infection but also to the host tissue injury [10] participating in infectious diseases, CNS autoimmunity, neurodegeneration, and neural damage [7, 10, 11]. In the CNS, glial cells express numerous TLRs [12, 13] and allow the recognition of different pathogens and potentially endogenous TLR agonists. The immediate activation of resident glia via TLRs likely serves as an amplification pathway to maximize the proinflammatory responses within the CNS compartment. Indeed, studies on TLR-deficient mice have clearly shown the critical role of TLR activation in several neuropathological disorders [14, 15]. New evidence has further established the actions of TLR signaling in modulating multiple processes in the developing and adult CNS, including axonal growth and structural plasticity [16, 17], as well as regulation of behaviors such as learning and anxiety [17, 18].

Alcohol abuse can cause behavioral dysfunction and brain damage. Current evidence demonstrates that the participation of TLR4 in these effects is an important target of alcohol action in the brain and provides a new perspective of the neuroimmune mechanisms in alcohol-related disorders. In this chapter, we review the mechanisms by which ethanol, acting as a TLR4 agonist, activates TLR4 signaling in glial cells which leads to neuroinflammation, brain injury, behavioral alteration, and addictive-like behavior.

9.2 Glial Cells as a Hallmark of Brain Inflammation: Effects of Ethanol

It is increasingly recognized that the most important CNS disorders are not merely defined by the enigmatic emergence of dysfunctional neurons but are in fact orchestrated largely by glial-cell-controlled inflammatory processes. In multiple archetypical inflammatory CNS diseases, and in Alzheimer's disease (AD), stroke, and Parkinson's diseases (PD), they have been implicated in inflammatory disease processes, and the involvement of glial cells has been clearly shown [11, 19, 20].

It was only recently that glial cells were considered to contribute to merely feeding and supporting neurons. However, rapidly accumulating evidence suggests glial cells are, in fact, very important players in CNS development, repair [21], and neurotransmission [22]. They are also functional elements in CNS vascularization [23], inflammation, neuroprotection [24], and also innate immune responses [25]. Astroglial cells respond to various injuries and insults [26–28] and are capable of proliferating throughout life, acting as neural stem cells with the potential of generating neurons, even in the adult brain [29].

In the human brain, glial cells outnumber neurons by a factor of approximately 10, and this number includes both the microglia that mediates inflammatory responses in the CNS and macroglia. The latter type belongs to the class of oligodendrocytes, which form the insulating myelin sheaths within the CNS, or to the class of astrocytes, these being most paradigmatic glial cells which are most abundant, star-shaped, and perform important physiological functions. Both microglia and astrocytes are responsible for providing immune functions within the brain and are essential elements in the immune response in the nervous system. Resident microglia constitute 12 % of the cells in the CNS and are considered to be the resident immune system. Microglia regulates neuronal development, differentiation, and survival using immune mechanisms to elicit apoptosis or proliferation. Microglia enforce the programmed elimination of neurons through the properties related to CNS homeostasis [30, 31], and they are necessary to elicit the differentiation and migration of neural precursor cells [32]. Under physiological conditions, microglia exhibit a deactivated phenotype that is associated with the production of anti-inflammatory and neurotrophic factors [33, 34]. Microglia switch to an activated phenotype in response to pathogen invasion or tissue damage, promoting an inflammatory response that serves to further engage the immune system and to initiate tissue damage. Considerable evidence indicates that microglial activation contributes to neuronal damage in neurodegenerative diseases [11]. In response to certain environmental toxins and endogenous proteins, microglia can enter an over-activated state and release reactive oxygen species (ROS) that cause neurotoxicity [35]. Pattern recognition receptors, expressed on the microglial surface, seem to be one of the primary, common pathways by which diverse toxin signals are transduced into ROS production [36].

Astrocytes constitute one of the major populations of glial cells in the CNS and are proposed to be actively engaged in diverse CNS functions, including control of water and ion homeostasis, uptake of neurotransmitters and excitatory amino acids, and growth factor production allowing for the induction and maintenance of neurite outgrowth, synapse formation, and neuronal survival [37]. Astrocytes also contribute to the establishment and maintenance of the blood–brain barrier (BBB) [38] by modulating the migration of monocytes and lymphocytes across the BBB [39]. In addition, these cells respond vigorously to brain injury and are postulated to play an important role in the fine-tuning of brain inflammation [40]. Indeed, injury to CNS damage is inevitably accompanied by astrocytic hypertrophy, proliferation, and altered gene expression, a process commonly referred to as reactive astrogliosis [41, 42]. This process is associated with inflammation [43]. However, depending on the disease context, astrogliosis can be seen as a positive event that promotes neuronal and glial survival via the production of neurotrophins and growth factors or as a negative influence on regeneration via the inhibition of neuronal and glial growth and migration. The diffuse nature of reactive astrogliosis in the CNS suggests a role for either soluble mediators, such as cytokines, or the presence of an integrated astrocyte-to-astrocyte syncytium that permits information to be transferred across extended distances.

The cytokines for which evidence is most compelling in the initiation and modulation of reactive astrogliosis include IL-1β, TNF-α, IFN-γ, and TGF-β. Activation of astrocytes, in response to a neuropathologic process, triggers the production of inflammatory cytokines, increases the expression of major express receptors for all these cytokines, and each appears to affect a different histocompatibility complex II and augments the production of free radicals [25]. This functional reprogramming may prove essential for maintaining homeostasis and the local regulation of inflammatory and immune reaction. Among cytokines, IL-1β has been considered an important mediator of inflammatory responses in the CNS, is involved in AD [44, 45], and is a produced under conditions of brain damage, disease, or stress [46, 47]. This cytokine is initially released by glial cells, acting on astrocytes and microglia to induce production of additional cytokines and growth factors, thereby promoting inflammatory activity in the brain [48]. Indeed when directly injected into the CNS, IL-1β promotes glial scarring or astrogliosis, which suggests a potential role for IL-1β in mediating astrocytic hypertrophy upon neuronal damage [49].

Although the primary stimuli for cytokine expression in the CNS are largely unknown, stress or injury condition might activate the innate immune response and TLRs by triggering the activation of transcription factors, such as nuclear factor-kappa B (NF-κB), and the induction of cytokines (see below), which initiates a series of inflammatory responses in the brain.

Finally, as regards the relevance of CNS injury and recovery, the dual astrogliogenic response has been clearly illustrated with mice in which reactive astrocytes were selectively targeted for ablation in the injured brain [50, 51]. The results of these experiments show that astrocyte loss is associated with a significant loss of neurons adjacent to the injury site, which can be prevented by chronic glutamate receptor blockade, thus demonstrating an essential role for both astrocytes and neuronal survival. Loss of astrocytes was also associated not only with the prolonged infiltration of $CD45^+$ leukocytes but also with the BBB's failure to reestablish itself following injury. However, astrocyte loss was also associated with a pronounced increase in local neurite outgrowth, indicating that reactive astrogliosis restricts neuronal outgrowth at these same sites. Taken together, these data illustrate the complex properties of reactive astrogliosis and also suggest that understanding the mechanisms underlying this response may help identify potential therapies to selectively encourage repair in the injured CNS.

Substantial evidence demonstrates that the impact of alcohol on glial cells causes astrogliosis [52] and loss of astrocytes in some brain areas, such as the prefrontal cortex and the hippocampus of human alcoholics' brains [53–56]. Changes in myelination can also occur during chronic alcoholism since the gene expression of several myelin-associated genes is downregulated in alcoholics' brains [57, 58]. In addition, ethanol is capable of activating microglia in vitro in cells in culture [59] and in vivo in both alcoholics' brains [60] and experimental animals with chronic or acute alcohol intake [52, 61].

The results provided above suggest that glial cells are important mediators of ethanol actions in the brain and that they might participate in neuroinflammation, neurodegeneration [52, 62, 63], and alcohol dependence [56].

9.3 TLR4 Expression in the CNS and Signaling Response

9.3.1 TLR4 Signaling Response

TLRs, type I integral membrane glycoproteins, are members of a large superfamily that includes interleukin-1 receptors (IL-1Rs) [64, 65]. TLRs and IL-1Rs have a conserved region of ~200 amino acids in their cytoplasmic tails known as the Toll/IL-1R (TIR) domain [65]. In contrast, the extracellular regions of TLRs and IL-1Rs differ markedly: the extracellular region of TLRs contains leucine-rich repeats motifs; however, the extracellular region of IL-1Rs comprises three immunoglobulin-like domains [65]. Activation of both IL-1R and TLR shares downstream signaling molecules, which culminate in the activation of the NF-κB and the AP-1 nuclear factors, which are transcription factors that regulate the expression of a wide array of genes involved in immune responses [66]. A diagram depicting the major steps involved in the TLR4 and IL-1RI signaling pathways is presented in Fig. 9.1.

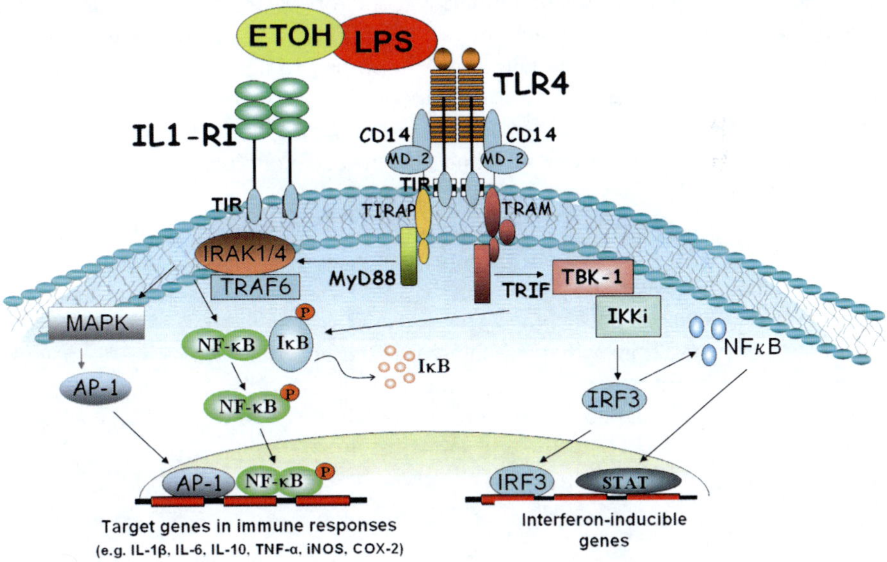

Fig. 9.1 Schematic representation of the TLR4 and IL-1RI signaling pathways. LPS or ethanol activates the MyD88-dependent pathway, which triggers a fast downstream stimulation of NF-κB, and the MAPK and AP-1 nuclear factors, leading to the induction of inflammatory mediators (iNOS, COX-2) and cytokine (IL-1β, TNF-α, IL-6, IL-8) genes. The second pathway involves the MyD88-independent pathway, in which adaptors TRAM and TRIF participate, and it activates the IRF-3 and a number of IFN-β-dependent genes (adapted from Blanco and Guerri [63])

TLRs and other PAMP receptors recognize molecular patterns. TLR4 is well known for its response to lipopolysaccharide (LPS), an outer cell wall component of gram-negative bacteria [67]. To confer a signal, TLR4 also requires its extracellular binding partner MD-2, or myeloid differentiation factor 2, which associates before ligand-induced signaling takes place [67, 68]. Mice lacking MD-2 do not respond to LPS [68]. CD14 is a protein that acts as a co-receptor and its function is to enhance the sensitivity of the MD-2/TLR4 signaling complex [69]. In addition to LPS and its variants, a number of exogenous and endogenous molecules have caught our attention because of their TLR4-binding properties (Table 9.1). Although the exact way that TLR4 recognizes its many ligands is a long-standing question, the structural analyses of the TLR4-MD-2-ligand complexes may help us to assess the TLR4 activation mechanism by different molecules [70–72].

TLR response involves the recruitment and activation of complex intracellular signaling cascades which culminate with the induction of cytokines and other inflammatory mediators (Fig. 9.1). Upon ligand binding, TLR4 undergoes dimerization and recruitment with different adaptor proteins into the membrane microdomain lipid rafts [73]. These adaptors couple two distinct signaling pathways. The first pathway utilizes adaptors MAL and TIRAP (MyD88-adaptor-like and TIR-associated proteins) and activates serine/threonine IL-1R-associated kinases (IRAK) and triggers a fast downstream stimulation of NF-κB, along with the activation of mitogen-activated protein kinases (MAPK) and the AP-1 nuclear factor. The AP-1 and RelA/P50 factors of NF-κB have been shown to regulate the transcription of proinflammatory cytokine and inflammatory mediators [74]. The MyD88-dependent signaling pathway is an important activator of NF-κB and the subsequent neuroregulatory effects of NF-κB signaling. It is important to note that NF-κB not only contributes to inflammatory reactions and apoptotic cell death after brain injury and stroke [75] but is also essential for neuron survival, and its activation may protect neurons against oxidative stress or ischemic neurodegeneration [76].

The second pathway, which is activated by TLR4, involves adaptors TRAM (Toll-receptor-associated molecule) and TRIF (TIR domain-containing adaptor-inducing IFN-β). This pathway triggers the expression of interferon response transcription factor IFN regulatory factor-3 (IRF-3) and a number of IFN-β-dependent genes [65] (Fig. 9.1). Under some conditions, TRIF-mediated IFN-β release can counteract inflammation, whereas the MyD88-dependent pathway is usually proinflammatory. But when and why the IFN-β anti-inflammatory pathway is induced is not well understood. The criteria for TRIF versus MyD88 signaling are unknown, but common TLR4 ligands appear to utilize the same pathway, or even both pathways, consistently [10].

9.3.2 *TLR4 in the CNS*

TLRs are expressed mainly in the immune system-related cell types including B cells, mast cells, natural killer cells, regulatory T cells, macrophages, monocytes,

Table 9.1 Ligands that activate or inhibit the TLR4 pathway

TLR4 activation		TLR4 inhibition	
LPS	Major component of Gram-negative bacterial cell wall. LPS binding leads to a rapid recruitment of different proteins, which triggers activation of two signaling pathways (NF-κB and mitogen-activated protein kinase pathways), which induce inflammatory cytokines [55]	Resveratrol	A polyphenolic phytoalexin with antioxidant and anti-inflammatory properties either by a Myd88-dependent or by a TRIF-dependent pathway [184]
Opioid agonist (e.g., morphine)	Morphine activates TLR4 signaling, which involves the recruitment of MyD88-dependent and MyD88-independent signaling cascades potentiated the morphine analgesia [170]	Opioid antagonist (e.g., naloxone)	Naloxone causes a significant change in the MD-2 residues, which are critical to initiating TLR4/MD-2 signaling [170]
Amyloid-β 42 peptide	Learning and memory impairment in Alzheimer disease are associated with inflammation (e.g., increased the expression of TLR4 and NF-κB mRNA and protein in brain) [185]	Curcumin	A phytochemical (polyphenol) with anti-inflammatory effects by inhibiting both MyD88- and TRIF-dependent pathways [186]
Saturated fatty acids	Using mechanism by which saturated fatty acids differentially modulate TLR4-induced inflammatory responses through the regulation of ROS-dependent dimerization and recruitment of TLR4 to lipid rafts [187]	Cinnamaldehyde	An essential oil isolated from *Cinnamomum* trees with anti-inflammatory and antibacterial properties suppresses the activation of NF-κB and IRF3 [188]
Heat shock proteins (e.g., HSP70)	TLR4 is the major receptor for HSP70, its activation induces proinflammatory cytokine production in autocrine manner by activating both MyD88- and TRIF-dependent signaling pathways [189]	Vitamin D3	It was shown to decrease TLR4 mRNA and protein expression in human monocytes [190] and led to a decreased production of cytokines and NF-κB activity

dendritic cells, neutrophils, and basophils, in addition to some nonimmune cells such as epithelial and endothelial cells. However, the presence of TLRs, including TLR4, has also been demonstrated in the CNS, while recent data have evidenced the impact of TLR response in the CNS on infectious and noninfectious CNS diseases/ injuries, influencing inflammatory responses during tissue injury and autoimmunity. The expression of TLR4 has been found in glial cells, particularly microglia [77], astrocytes [78, 79], and oligodendrocytes [15], and some studies have shown the expression of TLR4 in neurons [6, 15]. Nevertheless, the neuronal TLR4 response is mediated by a noncanonical pathway since no activation of NF-κB occurs [15], suggesting that none of the known TLR4-induced signaling, such as MyD88 and TRIF, is activated in these cells as a result of TLR4 activation by LPS.

As mentioned earlier, microglia in the CNS is mainly responsible for the early control of infections and recruitment of the adaptive immune system cells required for pathogen clearance. Engagement of TLRs provides an important mechanism by which microglia are able to sense both pathogen- and host-derived ligands in the CNS. Activation of microglia can normally occur through the TLR4 signaling pathway, which leads to the induction of cytokines and other inflammatory mediators [59]. These neuromodulatory inflammatory mediators can act on both astrocytes to induce secondary inflammatory signals [80] and neurons, which may have deleterious effects, as demonstrated in cells and animal models [81, 82]. Indeed, some studies have demonstrated that a conditioned medium of ethanol-treated activated microglia induces apoptosis in cultured neurons, suggesting that TLR4 activation in microglia by acute ethanol intake can induce an inflammatory environment in the brain which may result in neuronal damage [59]. These results further support the role of TLR4 in neurotoxicity induced by activated microglia since a microglia-deficient TLR4 function protects neurons in culture from ethanol-induced damage [59].

TLRs signaling can also influence multiple dynamic processes in the developing and adult CNS, including neurogenesis, axonal growth, and structural plasticity in the absence of any underlying infectious etiology. In addition, TLRs are implicated in the regulation of certain behaviors including learning, memory, and anxiety. However, little is known about if and how TLR signaling interacts with other signaling pathways involved in developmental and adult neuroplasticity. Glutamate is the major excitatory neurotransmitter in the CNS, and activation of both AMPA and NMDA subtypes of glutamate receptors is essential for synaptic plasticity/learning and memory. Neurons lacking either TLR2 or TLR4 exhibit increased resistance to the adverse effects of energy deprivation in an experimental model in which activation of glutamate receptors played a role in the cell death process [15]. Interestingly, TLR3 deficiency affects the levels of activated cAMP response element-binding (CREB), a transcription factor that performs important roles in learning and memory which is known to be activated in response to glutamate receptor stimulation [83]. Another signaling pathway which one or more TLRs may interact with is the phosphoinositide 3-kinase (PI3 kinase)/Akt pathway that is activated by neurotrophic factors, such as the brain-derived neurotrophic factor (BDNF) and insulin-like growth factors (IGFs). Because several TLRs (TLR2 and TLR4 in particular) increase the vulnerability of neurons to oxidative and metabolic stresses, it may well be expected that BDNF and IGF-1 signaling counteracts the adverse effects of TLR2 and 4 signaling [17].

9.4 Implications of TLR4 in Neurodegenerative Diseases

As previously commented, the inflammatory response represents a highly regulated biological program that enables the innate and adaptive immune systems to effectively deal with rapidly dividing microbial pathogens but also involves the production of factors that are capable of inducing significant tissue pathology. Inflammatory responses to infectious agents are typically initiated by TLRs, which recognize a diverse set of pathogen-associated molecules that are not present in the host. These pattern recognition receptors have more recently been found to be also capable of responding to endogenously derived molecules, such as the components released from necrotic cells (danger signals) and by molecules that may be formed as a result of pathogenic mechanisms. Growing evidence has demonstrated the role of neuroinflammation as a pathogenic factor in many neurodegenerative diseases, including AD, PD, amyotrophic lateral sclerosis, and multiple sclerosis (see review [11]). In all these disorders, activation not only of the innate immune cells in the CNS, such as microglia and astrocytes, but also of TLRs in glial cells significantly contributes to the initiation and progression of inflammatory responses in the brain.

The significance of TLR4 in neurodegenerative diseases has been clearly demonstrated in experimental animals in which absence of TLR4 protects or ameliorates the effects observed in animal models of several chronic inflammatory diseases such as AD, multiple sclerosis, and ischemic brain stroke [84–87]. Furthermore, TLR4 polymorphisms are associated with several human age-related diseases, including atherosclerosis [88], type-2 diabetes [89], and rheumatoid arthritis [90].

Evidence of the role of TLR4 in AD stems from findings demonstrating that both microglia and astrocytes are apparently capable of activating microglia and astrocytes through TLR4 (together with CD14 and MD2 in microglia), leading to the activation of signal-dependent transcription factors that triggers the expression of downstream inflammatory response genes [91]. Consistently with this, the mice carrying a nonfunctional TLR4 crossed with a mouse AD model (APP/PS1 double transgenic mice) showed production of fewer inflammatory cytokines [92], while the mice carrying mutant TLR4 crossed with AD transgenic mice exhibited more Aβ plaques [93]. It is noteworthy that a polymorphism in the TLR4 extracellular domain has been reported to be associated with protection against late-onset AD in an Italian population [94], suggesting that sterile inflammation may influence AD pathology through TLR4 signaling. Activation of TLRs, such as TLR2, also seems to participate since the mice lacking TLR2 crossed with APP/PS1 transgenic AD mice have been reported to show a delay in Aβ deposition and improved behavior in memory tests [95].

As regards other neurodegenerative diseases such as PD, several lines of evidence suggest that inflammatory mediators such as NO, TNF-α, IL-1β, activation of NADPH oxidase, and production of oxygen radical species derived from nonneuronal cells modulate the progression of neuronal cell death in PD [96]. For instance, direct application of LPS (specific ligand of TLR4) to neurons has little effect on gene expression or survival. In contrast, the conditioned media from LPS-treated microglia are neurotoxic, and this effect is enhanced when conditioned media are added to mixed cultures of astrocytes and neurons [97]. These results indicate that activation of microglia TLR4 is the primary initial responder to produce mediators such as TNF-α and IL-1β that activate astrocytes [80] which, in turn, can cause dopaminergic neurotoxicity. Accordingly, TLR4 and CD14 expressions were found to increase in the CNS of the MPTP mouse model of Parkinson's-like symptoms [98].

In summary, the above findings suggest the role of glial cells and TLR4 signaling in brain damage and in neurodegeneration [11]. Consistently, elimination of TLR4 protects focal cerebral ischemia by abolishing TLR4 signaling [99] and diminishing the neuronal apoptotic levels in AD [6]. It is intriguing to note that current data suggest a link between the innate immune system/TLR response and cytokine induction with behavioral alterations [100], as indicated by a new dimension of the innate immune system in the brain.

9.5 Participation of TLR4 in Ethanol-Induced Innate Immune Dysfunction, Neuroinflammation, and Brain Damage

9.5.1 Role of TLRs in the Ethanol-Induced Immune System Dysfunctions and in Inflammatory Damage

Human and experimental animal studies have revealed that ethanol intake affects immune and inflammatory systems [101], resulting in specific defects in the cellular components of the innate immune responses to bacterial and viral pathogens [101]. Acute ethanol impairs immune functions by interfering with inflammatory response and increasing the risk of infectious complications in trauma and burn patients [102, 103]. A higher incidence of infectious diseases, bacterial pneumonia in particular, is observed in alcoholics [104].

Several studies have also demonstrated the action of ethanol on TLRs. Thus, acute ethanol treatment has been shown to suppress TLR3 [105] and TLR4 [106] responses in monocytes or macrophages in vitro by decreasing the synthesis and secretion of numerous cytokines [107]. Conversely, prolonged alcohol treatment significantly augments both TLR8 and TLR4 signaling by increasing TNF-α production in human monocytes [108]. Similarly, chronic ethanol treatment increases the innate immune response in the liver by participating in the development of alcoholic liver disease [109]. The gastrointestinal tract seems to be the initial target of ethanol since alcohol consumption impairs the intestinal mucosa's barrier function [110] by increasing the plasma levels of LPS (a component of the gram-negative bacterial wall) which, in turn, activate TLR4 in Kupffer cells (resident macrophages in the liver). TLR4 signaling leads to NF-κB stimulation and to inflammatory and fibrogenic cytokine production, triggering hepatocyte injury [109]. The role of TLR4 in the development of alcohol liver diseases is exemplified by the demonstration that suppressing the TLR4 function attenuates the ethanol-induced hepatocellular damage in mice [111]. However, while TLR4 deficiency is protective, MyD88 deficiency fails to prevent alcohol-induced liver damage and inflammation, suggesting that protection occurs via the TLR4-MyD88-independent pathway [112] (see Fig. 9.1). Furthermore, TLR4 also appears to participate in poor ethanol-induced sepsis prognosis [113] and in pulmonary inflammation following ethanol and burn injury [114].

9.5.2 TLR4 Is a Molecular Target for Ethanol-Induced Neuroinflammation and Brain Damage

As commented above, although the CNS has been viewed for many years as an immune-privileged organ, it is now well accepted that immune surveillance occurs in the normal CNS and that inflammatory responses can and do take place in the disease context. The CNS launches an organized series of innate immune responses during both localized and systemic infections [115], and recent evidence suggests a role of innate immunity in response to injury [1, 2]. It is important to emphasize that although some inflammatory stimuli and the release of cytokines cause beneficial effects (e.g., phagocytosis of debris and apoptotic cells) and that inflammation is linked to tissue repair processes, uncontrolled inflammation would result in neuroinflammation, reduced neuroprotection and neuronal repair, and increased neurodegeneration.

In general, neuroinflammation or inflammation of the CNS is characterized by increased glial activation, proinflammatory cytokine concentration, and, in some cases, BBB permeability, allowing leukocyte invasion [116]. Disruption of the BBB allows the influx of peripheral immune cells into the CNS at the injury site, thus contributing to the neuroinflammation associated in ischemia, edema, and AD [117–119]. Immune cells respond to injuries by eliminating debris and by synthesizing and releasing a host of powerful regulatory substances, like complements, cytokines, chemokines, nitric oxide, and reactive oxygen. [116]. These substances have both beneficial and harmful effects on the cellular environment by creating further damage.

Several studies support that CNS resident cells, as glial cells, are capable of mounting a quick and effective response to control infection until peripheral adaptive immune system cells can be recruited. Both microglia and astroglia express TLR2, TLR4, TLR5, and TLR9 [120–122]; respond functionally with cytokine and chemokine production [120, 122]; and are capable of contributing to an inflammatory environment in the CNS [25, 123] following a variety of infectious or inflammatory insults [123].

Data emerging in recent years demonstrate that microglia and astrocytes in culture respond to ethanol by secreting cytokines and other inflammatory mediators [59, 79] and that they contribute to an inflammatory environment in the brains of alcohol-fed animals [52, 124]. The mechanisms involved in these effects take place through the activation of TLR4/IL-1RI in glial cells. Indeed, it has been shown that ethanol, at physiological relevant concentrations (10 or 50 mM), is capable of rapidly inducing the activation of the TLR4/IL-1RI signal-transduction pathways [79] in astrocytes similar to when cells are stimulated with LPS and IL-1β [125–128]. Thus, upon ethanol stimulation, a rapid phosphorylation (within 10 min) of IRAK, p44/p42 ERK, SAPK/JNK, and p-38 MAPK occurs, along with the subsequent downstream activation of transcription factors NF-κB and AP-1 and the upregulation of iNOS and COX-2 noted after a 30-min ethanol treatment. In line with the hypothesis that ethanol mediates inflammatory events by activating TLR4/IL-1RI, blocking the activation of these receptors with neutralizing Abs in astrocytes [79], or downregulated TLR4 by siRNA, abolishes the ethanol-induced inflammatory mediators in microglia and astroglial in primary culture [59, 79]. These results suggest that, by acting as a TLR4 agonist, ethanol triggers a similar fast response as LPS [59, 79], although cytokine release is lower in cells treated with ethanol than in cells stimulated with LPS. The in vitro results were further confirmed in animals treated with ethanol. The results of Alfonso-Loeches et al. [52] demonstrate that chronic ethanol intake triggers neuroinflammatory damage by activating microglia and astroglial cells and by also increasing both caspase-3 activity and proinflammatory mediators (iNOS, COX-2, IL-1β, TNF-α, IL-6) in the cerebral cortex of female wild-type (WT) mice. However, chronic ethanol-treated mice lacking TLR4 are protected against ethanol-induced glial activation, induction of inflammatory mediators, and apoptosis. In agreement with these in vivo findings, the conditioned medium of ethanol-treated microglia induces apoptosis in cultured neurons, while the microglia-deficient TLR4 function protects neurons in culture from ethanol-induced damage [59]. These findings support the critical role of glial activation and the TLR4 response in neuroinflammation, brain injury, and possibly in neurodegeneration induced by chronic ethanol intake.

However, considering that TLR4 is activated by endogenous molecules (see Table 9.1), it is plausible that despite ethanol possibly initiating inflammatory events by activating TLR4 in glial cells, amplification of inflammatory events may occur after cell injury via a mechanism in which endogenous ligands activate TLRs, thus contributing to brain damage (see Fig. 9.2). Indeed, TLRs are considered important inducers and contributors in several neurodegenerative diseases [11], and elimination of TLR4 protects against several brain insults and neurodegeneration [6, 15, 99, 129].

The above finding suggests that unlike some neuroinflammatory disorders characterized by cytokine production, which are associated with disruption of the BBB and with peripheral immune cells influx into the CNS, ethanol can directly activate glial TLR4, as demonstrated in microglia and astrocytes in primary culture, triggering inflammatory mediators and causing neuroinflammation and brain damage in chronically ethanol-fed animals. Conversely, a transient exposure to low/moderate levels of ethanol in rats can induce not only microglia activation [59] but also the release of proinflammatory and anti-inflammatory cytokines (e.g., IL-10) in both glial cells [59]

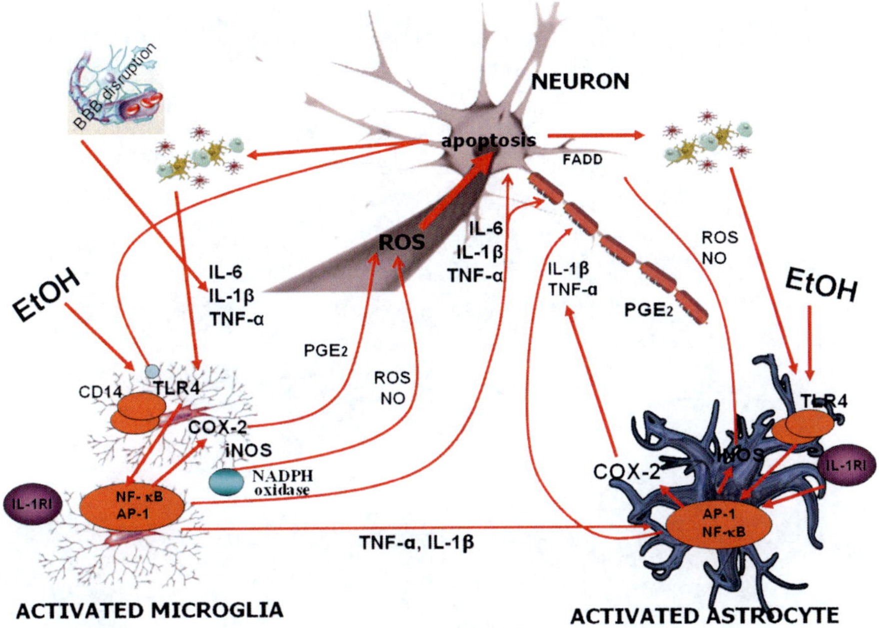

Fig. 9.2 Potential mechanism of ethanol-induced brain damage. Ethanol activates inflammatory signaling cascades (e.g., TLR4, IL-1RI) in astroglia and microglia, which trigger the release of inflammatory mediators and cytokines (iNOS, COX-2, IL-1β, TNF-α), leading to brain damage and apoptotic neuronal death, as well as myelin disarrangements (adapted from Alfonso-Loeches and Guerri [62])

and peritoneal macrophages [130], suggesting that one ethanol dose may offer some beneficial actions for the immune system response. Indeed, a number of studies have demonstrated that cytokine production elicits physiological and behavioral changes that are adaptive and beneficial to the host organism (e.g., [131]).

Demyelination is also a hallmark of diverse neurodegenerative autoimmune diseases, including multiple sclerosis [132] and acute disseminated encephalomyelitis [133]. In addition, alcohol causes demyelination and myelin disruptions in alcoholics, as demonstrated by cerebral white matter atrophy, reduction in myelin-related genes, and spreading demyelination in postmortem [134–137]. Neuroimaging studies have also revealed that chronic alcohol abuse disrupts the white matter microstructure as manifested by abnormally low fractional anisotropy in the corpus callosum and throughout the cortex [138]. The mechanisms by which ethanol induces myelin disruptions are presently unknown. Recent findings suggest the participation of TLR4 in the effect that ethanol has on myelin disruptions. Studies from our group reveal that chronic ethanol treatment downregulates the proteins involved in myelination, alters myelin morphology, reduces the number of myelin basic protein positive fibers, and causes oligodendrocyte death in the cerebral cortex and corpus callosum of WT [139]. In contrast, most myelin alterations were not observed in brains from ethanol-treated TLR4-deficient mice. Electron microscopy studies

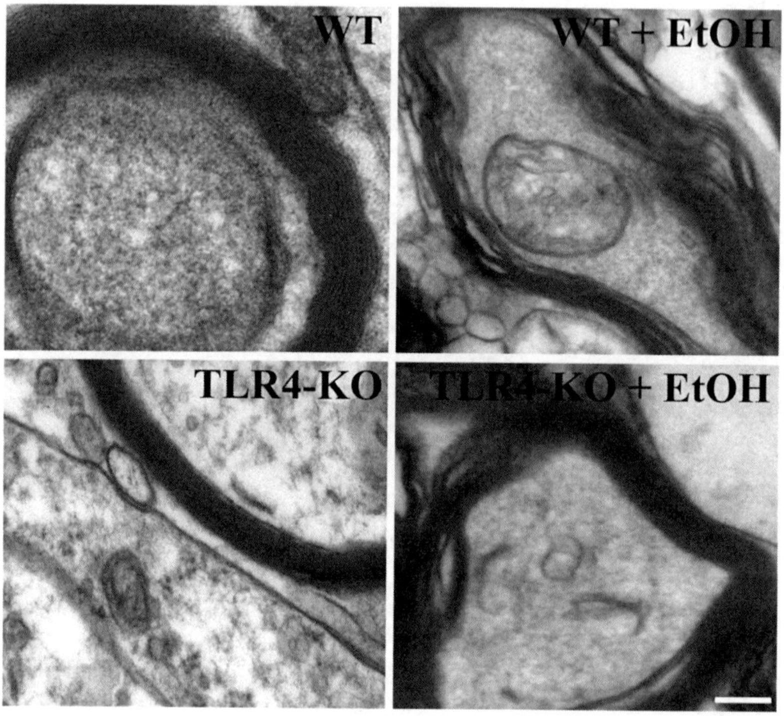

Fig. 9.3 Electron microscopic examination of the myelinated axons of the corpus callosum of WT and TLR4$^{-/-}$ KO mice treated with or without ethanol for 5 months. The WT mice treated with chronic ethanol show myelin sheath disarrangements when compared with their control counterparts, while minimal changes in myelin disruptions were noted in the ethanol-treated TLR4$^{-/-}$ KO mice. The scale bar is 100 nm. These images have been provided by Dr. Jaime Renau-Piqueras at the La Fe Research Hospital Center (Valencia, Spain)

have further revealed myelin sheath disarrangements in the cerebral cortex and corpus callosum of ethanol-treated mice, but only small focal fiber disruptions were noted in the same brain areas of ethanol-treated TLR4$^{-/-}$ mice (see Fig. 9.3). These results highlight the importance of TLR4 and the inflammatory environment in the myelin alterations induced by alcohol abuse. However, other mechanisms might also participate in ethanol-induced myelin disruptions since some myelin alterations have been noted in ethanol-treated TLR4$^{-/-}$ mice. Among the mechanisms which may participate, we find participation of the inflammasome, which is critical for caspase-1 activation and the proteolytic processing of pro-IL-1β [140], and the action of other TLRs [141]. Another possible alternative is that the TLR4-induced activation of macrophages or microglia could sequester the iron required for proper development and myelination, leading to demyelination in the brains of chronically ethanol-fed animals, as demonstrated in spinal cord injury [142]. Finally, further research in this area is required to improve our knowledge on the relationship between TLR4 and demyelination.

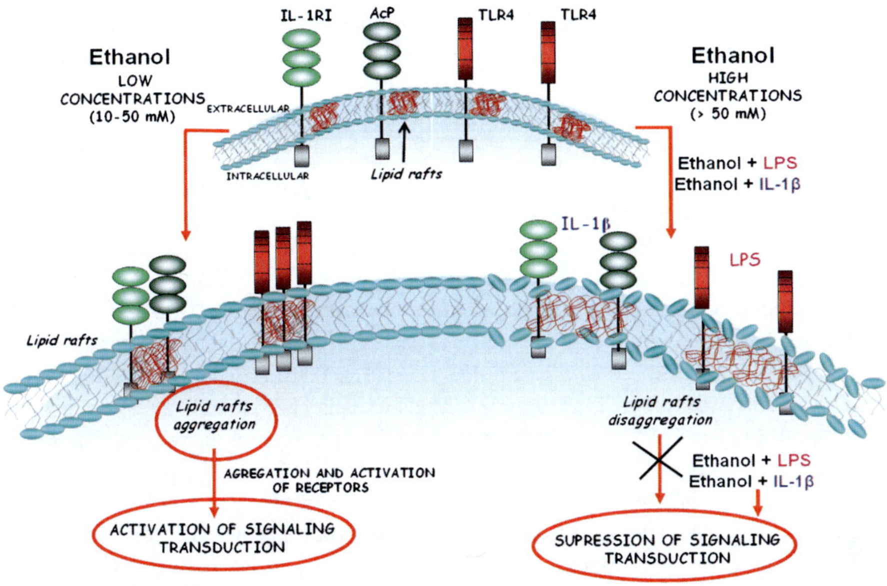

Fig. 9.4 Ethanol mediates TLR4/IL-1RI activation through its interactions with lipid rafts. Ethanol at low concentrations (10–50 mM) might facilitate TLR4 and IL-1RI aggregation and recruitment into lipid rafts, leading to the clustering of these receptors and the activation of their signal transduction. Conversely, ethanol at high concentrations (>50 mM) can perturb membrane lipids, including lipid rafts, resulting in a disruption of receptor clustering which suppresses TLR4 and IL-1RI activation by its ligands binding and signaling (adapted from Blanco and Guerri [63])

9.5.3 Molecular Mechanisms of Ethanol-Induced TLR4 Activation: Role of Lipid Rafts

At present, it is not clear how ethanol interacts with TLR4 to either activate or inhibit its signaling response depending on the cell type and ethanol concentration. It is well established that ethanol interacts with membrane lipids and influences the function of membrane proteins [143]. It has been proposed [52, 62, 63] that, depending on the ethanol concentration and the physicochemical characteristics of the cell membrane, ethanol can facilitate or disrupt the recruitment of these receptors. A high ethanol concentration can perturb membrane lipid microdomains, such as lipid rafts, and result in a disruption of the receptor function. However, low ethanol concentrations can facilitate the aggregation and interaction of proteins within the membrane, thus allowing their activation and signaling through lipid rafts. Furthermore, ethanol in the presence of a TLR4-ligand, such as LPS, is capable of inhibiting a TLR4 signaling response. A schematic model of this hypothesis is presented in Fig. 9.4.

Lipid rafts are cholesterol/sphingomyelin-enriched membrane microdomains, which are involved in the recruitment of molecules that are implicated in signal transduction [144], and are now recognized as important sites for initial immune cell activation. The presence of TLR2 and TLR4 within lipid rafts and their subsequent clustering in response to LPS [73, 145] support their role in the innate immune response [73, 145]. Disruption of lipid rafts leads to the inhibition of TLR4 internalization and signaling [73]. High ethanol concentrations act as lipid rafts disrupting agents, interfere with lipid raft clustering, and lead to the suppression of TLR4 signaling in murine macrophages [146, 147]. However, low ethanol concentrations promote receptor recruitment into lipid rafts, leading to dimerization and signaling. Furthermore, ethanol in the presence of other stimuli (e.g., LPS, cytokines) interferes with the recruitment of TLRs into lipid rafts, leading to an inhibitory, rather than an additive, effect on TLR signaling transduction [148] (see Fig. 9.4). This may explain not only those findings demonstrating the suppressing effects of ethanol on the cytokine-induced iNOS expression [149] and LPS-induced NO production [150] in glial cells but also an increasing risk of infections in alcoholics [151].

In summary, in vitro studies support the idea that ethanol indirectly activates TLR4 receptors. Experimental evidence demonstrates that, by interacting with membrane microdomains *lipid rafts*, ethanol can recruit TLR4 receptors into these microdomains by activating their signaling response. Experimental evidence also shows that the TLR4 co-receptors CD14 and MD-2 participate in ethanol-induced TLR4 activation since knocking down these proteins, by using siRNAs, reduces ethanol-induced pERK and the nuclear activation of the NFκB-p65 subunit [52]. These results suggest that ethanol-induced in vivo TLR4 activation could be mediated by the same mechanism of the in vitro findings. Indeed, glial activation and cytokine production are observed after a transient acute ethanol dose and in the absence of brain damage [59]. However, under chronic ethanol treatment, although ethanol could initiate the inflammatory events by activating TLR4 in glial cells, amplification of the inflammatory events could occur after cell injury via a mechanism by which endogenous ligands activate these receptors, thus contributing to brain damage.

9.6 TLR4 Response Is Associated with Cognitive and Behavior Alterations as well as with Addictive-Like Behavior

9.6.1 Immune Response and Cognitive Deficits

A number of studies have demonstrated a link between activation of innate immune genes with behavioral alterations. For instance, cognitive and behavior alterations have been observed in patients with chronic neuroinflammation/neurodegeneration (e.g., AD) [11]. Depression has also been reported in people with cancer treated with proinflammatory IFN [152], or severe depressive behavioral symptoms have been associated with high IL-1 plasma levels [153]. Animal studies have also shown that administration of LPS into the fourth rat ventricle causes glial activation and degeneration of hippocampal pyramidal neurons and that these effects are associated with spatial memory [154–156]. Similarly, the intracerebroventricular administration of LPS enhances the proinflammatory cytokine response (IL-1β, IL-6, and TNF-α) and the TLR4 expression in the cerebellum; these events are associated with an alteration in locomotor activity and social behavior [157]. These findings suggest that brain inflammatory mediators may lead to behavior alterations. Accordingly, attenuation of microglia and IL-1 signaling protects mice from acute ethanol-induced sedation and/or motor impairment [158]. Likewise, administration of indomethacin, an anti-inflammatory compound, reduces the expression of inflammatory mediators and protects both neural death and short- and long-term behavioral dysfunction [159].

Other studies have also demonstrated an association between ethanol-induced activation of the glial cells in the frontal cortex with cognitive dysfunction and anxiety-like behavior during withdrawal. However, mice lacking TLR4 receptors are protected against ethanol-induced inflammatory damage, cognitive dysfunctions, or behavioral-associated effects [18]. Consistently with these results, neuroinflammation induced by LPS administration is also associated with long-term cognitive deficits and locomotor activity impairment. Deletion of COX-2 [160] or NOS-2 [161], two important downstream effectors of TLR4 signaling [63, 64], abolishes or attenuates cognitive and motor activity dysfunctions in mice. Furthermore, TLR4-deficient mice subjected to subacute stress show better behavior conditions and a minor brain inflammatory response than WT mice [75] and are protected from cognitive impairment upon immunization with the amyloid-β peptide [162]. These results suggest activation of the innate immune system or TLRs in the brain might cause some cognitive and behavior dysfunctions.

9.6.2 Addictive-Like Behavior

Addiction can be defined as a primary, chronic, neurobiological disease with genetic, psychosocial, and environmental factors influencing its development and manifestation. Although genetics may contribute to the high rate of addictive-like behavior, studies into postmortem human brains support the notion that the expression of some genes of the immune system are upregulated in alcohol-addictive individuals. Thus, monocyte chemoattractant protein-1 (MCP-1), a member of the β-chemokine subfamily, has been seen to increase in the ventral tegmental area, the substantia nigra, the hippocampus, and the amygdala of alcoholics [60]. Similarly, variations in gene NF-κB1 [163], which encodes a subunit of transcription factor NF-κB [164], as well as changes in the DNA-binding levels of NF-κB [164], appear to not only be associated with alcohol dependence but especially contribute to an earlier onset of the disease. Similarly, human methamphetamine addicts and methamphetamine-treated mice show persistent increases in innate immune proteins [165].

Animal models indicate that the glial TLR4 response is an important contributor to the neurobiology of addiction. For example, the $GABA_A$ receptor, a well-recognized contributor in vulnerability to excessive alcohol drinking [166], has been recently implicated in the stimulation of the TLR4 pathway either directly (as a ligand) or indirectly (through chemokine/cytokine networks) [167]. Recent data also demonstrate that the $GABA_A$ α2-regulated TLR4 expression in the central nucleus of the amygdala contributes to binge drinking and to early neuroadaptation in excessive drinking [168].

The association found between innate immune gene induction and drug addiction has also been demonstrated using different transgenic knockout (KO) mice. Mutant mice lacking the genes related to the immune function show diminished alcohol consumption [169]. For instance, systemic administration of LPS enhances alcohol-induced motor impairment [170] and alcohol consumption in both C57BL/6J(B6) inbred mice and B6xNZBF1 hybrid mice [171]. However, no increase of alcohol intake has been observed in those mice lacking CD14, a key component of TLR4 signaling. Blednov et al. [172] further showed that LPS treatment decreases ethanol-conditioned taste aversion but does not alter ethanol-conditioned place preference (B6xNZBF1 mice), although LPS is able to lower the neuronal firing rate in the ventral tegmental area. These results suggest that activation of immune signaling promotes alcohol consumption and alters certain aspects of alcohol reward/aversion.

Alterations in the immune function and cytokine production within the brain have also been associated with anxiety-like behavior and addiction. Breese et al. [173] demonstrated that repeated LPS or TNF-α treatment sensitizes ethanol withdrawal-induced anxiety-like behavior. These results suggest that cytokines may contribute to a cumulative adaptation, which sensitizes anxiety-like behavior upon withdrawal from alcohol exposure. Recent data also provide evidence of the role of glial TLR4 in the anxiety-like behavior. Thus, Pascual et al. [18] demonstrated that chronic ethanol intake induces anxiety-like behavior within a 15-day withdrawal period in WT mice, which still maintain microglia and astroglia activation. Nevertheless, mice lacking TLR4 receptors are protected against ethanol-induced reactive gliosis and anxiety-like behavior [18]. Anxiety induced by repeated social defeat is also associated with microglia reactivity, along with upregulation of glial surface receptor, CD14/TLR4, and cytokine production in the brain (e.g., IL-1β) [174]. These findings suggest that TLR4-dependent glial activation and proinflammatory cytokine release may contribute and maintain ethanol-induced anxiety-like behavior during the withdrawal period.

In short, the results provided above suggest that immune signaling activation might participate in alcohol consumption and in certain behavioral aspects of alcohol addiction.

9.7 Impact of Opioids and Alcohol on TLR4 Signaling: Role of Naltrexone

Experimental evidence has demonstrated the action of opiates in inducing proinflammatory glial activation by modulating wide-ranging opioid pharmacology aspects [175, 176]. Indeed, several in vitro and in vivo findings have shown that morphine can trigger microglial and astroglial activation [175] which, in turn, promotes a proinflammatory response. This inflammatory response seems to contribute to several pharmacological and behavioral aspects of opioids, such as reduced acute opioid analgesia, enhanced development of opioid tolerance, hyperalgesia, and allodynia following chronic opioid administration [175, 177]. For example, morphine is able to enhance analgesia by blocking proinflammatory cytokine actions [178], inhibitors for glial activation, such as fluorocitrate, minocycline, or ibudilast, modulate morphine effects [179–181].

Recent evidence indicates that opioid-induced glial activation occurs via TLR4 activation, and not by the stereoselective μ-opioid receptors. Administration of naloxone, an opioid antagonist, ameliorates LPS-induced glial activation and behavior, suggesting that opioids promote TLR4 signaling [181, 182]. Accordingly, TLR4 inhibition has been associated with the concomitant potentiation of opioid analgesia and attenuates the development of opioid tolerance, hyperalgesia, withdrawal, or addictive-like behavior [181]. Blocking the TLR4 signaling pathway with genetic $TLR4^{-/-}$ KO potentiates the magnitude and duration of morphine analgesia [176]. The competitive opioid antagonist, naltrexone, suppresses microglial signaling, thereby decreasing the production of proinflammatory cytokines and ROS [182, 183]. Interestingly, naltrexone is used to treat both alcohol and opiate addictions [191], suggesting that both addictive drugs share some common mechanisms of action. For instance, both drugs trigger TLR4 signaling, while naltrexone attenuates glial activation, along with inflammatory cytokine production and some behavioral effects. A recent study demonstrated that pharmacologically inhibiting TLR4 signaling with naloxone or genetic deficiency of TLR4 or MyD88 significantly reduces acute alcohol-induced sedation and motor impairment in mice [192]. These findings provide new evidence that TLR4-MyD88 signaling is involved in the acute behavioral actions of alcohol in mice and may also participate in alcohol dependence. Indeed, administration of naltrexone and disulfiram to alcohol-dependent patients ameliorates the ethanol-induced activation of the immune response [193]. These results suggest that the beneficial actions of naltrexone and disulfiram on alcohol dependence may be mediated in part via inhibition of the TLR4 signaling, preventing the glial activation, the immune response, and the behavioral dysfunctions.

9.8 Conclusion

In the last few years, new data have emerged that link the neuroimmune system response and ethanol actions in the brain. Previous results have shown that the genes encoding the proteins involved in "immune/stress responses" exhibit a differential gene expression in the frontal cortex when comparing human alcoholics and nonalcoholics [57]. The expression of these genes in the brain has also been related to a genetic predisposition for alcohol consumption in mice, indicating a role for proinflammatory mediators in regulating alcohol intake [194]. Recent findings confirm that ethanol treatment increases cytokine levels in rodents [52, 60, 124] and in the brains of human alcoholics [60]. These findings also demonstrate that ethanol is capable of activating glial TLR4, a receptor of the innate immune system, by triggering a signaling inflammatory response and glial activation and by causing neuroinflammation, brain damage, and behavioral alterations. Blocking these receptors in vitro in glial cultures and in vivo protects against ethanol-induced glial activation, induction of inflammatory mediators, and apoptosis. Upregulation of the genes in the Toll and Imd innate immune signaling pathways in *Drosophila* exposed to ethanol [195] further supports a link between alcohol and immune mediators to invertebrates. Novel insights into the interactions of ethanol with TLR4 indicate that the innate immune system might also participate in the cognitive and behavioral dysfunction associated with alcohol abuse. These novel findings in the role of the innate immune system on the effects of ethanol highlight new avenues for future research into the prevention and treatment of alcohol-related disorders.

Acknowledgments We thank M. March for excellent professional assistance in preparing the figures in this manuscript. The authors' experimental work has been supported by grants from the Spanish Ministry of Science and Innovation (SAF 2009-07503), the Spanish Ministry of Health, the Carlos III Institute (RTA Network, RD06/0001/1119), PNSD (2010I037), PROMETEO/2009/072, and General Direct Drug Dependence, GV.

References

1. Owens T, Babcock AA, Millward JM, Toft-Hansen H (2005) Cytokine and chemokine interregulation in the inflamed or injured CNS. Brain Res Brain Res Rev 48:178–184
2. Tanga FY, Nutile-McMenemy N, DeLeo JA (2005) The CNS role of Toll-like receptor 4 in innate neuroimmunity and painful neuropathy. Proc Natl Acad Sci USA 102:5856–5861
3. Ben Menachem-Zidon O et al (2011) Astrocytes support hippocampal-dependent memory and long-term potentiation via interleukin-1 signaling. Brain Behav Immun 25:1008–1016
4. Yirmiya R, Goshen I (2011) Immune modulation of learning, memory, neural plasticity and neurogenesis. Brain Behav Immun 25:181–213
5. Takeuchi O, Akira S (2010) Pattern recognition receptors and inflammation. Cell 140: 805–820
6. Tang SC et al (2008) Toll-like receptor-4 mediates neuronal apoptosis induced by amyloid beta-peptide and the membrane lipid peroxidation product 4-hydroxynonenal. Exp Neurol 213:114–121
7. Campbell A (2004) Inflammation, neurodegenerative diseases, and environmental exposures. Ann N Y Acad Sci 1035:117–132
8. Anderson KV, Jurgens G, Nusslein-Volhard C (1985) Establishment of dorsal-ventral polarity in the Drosophila embryo: genetic studies on the role of the Toll gene product. Cell 42:779–789
9. Medzhitov R, Preston-Hurlburt P, Janeway CA Jr (1997) A human homologue of the Drosophila Toll protein signals activation of adaptive immunity. Nature 388:394–397

10. Buchanan MM, Hutchinson M, Watkins LR, Yin H (2010) Toll-like receptor 4 in CNS pathologies. J Neurochem 114:13–27
11. Glass CK, Saijo K, Winner B, Marchetto MC, Gage FH (2010) Mechanisms underlying inflammation in neurodegeneration. Cell 140:918–934
12. Olson JK, Miller SD (2004) Microglia initiate central nervous system innate and adaptive immune responses through multiple TLRs. J Immunol 173:3916–3924
13. Jack CS et al (2005) TLR signaling tailors innate immune responses in human microglia and astrocytes. J Immunol 175:4320–4330
14. Caso JR et al (2007) Toll-like receptor 4 is involved in brain damage and inflammation after experimental stroke. Circulation 115:1599–1608
15. Tang SC et al (2007) Pivotal role for neuronal Toll-like receptors in ischemic brain injury and functional deficits. Proc Natl Acad Sci USA 104:13798–13803
16. Okun E et al (2010) Toll-like receptor 3 inhibits memory retention and constrains adult hippocampal neurogenesis. Proc Natl Acad Sci USA 107:15625–15630
17. Okun E, Griffioen KJ, Mattson MP (2011) Toll-like receptor signaling in neural plasticity and disease. Trends Neurosci 34:269–281
18. Pascual M, Balino P, Alfonso-Loeches S, Aragon CM, Guerri C (2011) Impact of TLR4 on behavioral and cognitive dysfunctions associated with alcohol-induced neuroinflammatory damage. Brain Behav Immun 25:S80–S91
19. Nagele RG, Wegiel J, Venkataraman V, Imaki H, Wang KC (2004) Contribution of glial cells to the development of amyloid plaques in Alzheimer's disease. Neurobiol Aging 25:663–674
20. Wu DC et al (2002) Glial cell response: a pathogenic factor in Parkinson's disease. J Neurovirol 8:551–558
21. Wurmser AE, Palmer TD, Gage FH (2004) Neuroscience. Cellular interactions in the stem cell niche. Science 304:1253–1255
22. Araque A, Perea G (2004) Glial modulation of synaptic transmission in culture. Glia 47:241–248
23. Downie LE, Pianta MJ, Vingrys AJ, Wilkinson-Berka JL, Fletcher EL (2007) Neuronal and glial cell changes are determined by retinal vascularization in retinopathy of prematurity. J Comp Neurol 504:404–417
24. Borlongan CV et al (2000) Glial cell survival is enhanced during melatonin-induced neuroprotection against cerebral ischemia. FASEB J 14:1307–1317
25. Dong Y, Benveniste EN (2001) Immune function of astrocytes. Glia 36:180–190
26. Kornyei Z, Czirok A, Vicsek T, Madarasz E (2000) Proliferative and migratory responses of astrocytes to in vitro injury. J Neurosci Res 61:421–429
27. Herrmann JE et al (2008) STAT3 is a critical regulator of astrogliosis and scar formation after spinal cord injury. J Neurosci 28:7231–7243
28. Chen Y, Swanson RA (2003) Astrocytes and brain injury. J Cereb Blood Flow Metab 23:137–149
29. Sanai N et al (2004) Unique astrocyte ribbon in adult human brain contains neural stem cells but lacks chain migration. Nature 427:740–744
30. Marin-Teva JL et al (2004) Microglia promote the death of developing Purkinje cells. Neuron 41:535–547
31. Wakselman S et al (2008) Developmental neuronal death in hippocampus requires the microglial CD11b integrin and DAP12 immunoreceptor. J Neurosci 28:8138–8143
32. Aarum J, Sandberg K, Haeberlein SL, Persson MA (2003) Migration and differentiation of neural precursor cells can be directed by microglia. Proc Natl Acad Sci USA 100: 15983–15988
33. Nakajima K et al (2001) Neurotrophin secretion from cultured microglia. J Neurosci Res 65:322–331
34. Streit WJ, Mrak RE, Griffin WS (2004) Microglia and neuroinflammation: a pathological perspective. J Neuroinflammation 1:14

35. Dimayuga FO et al (2007) SOD1 overexpression alters ROS production and reduces neurotoxic inflammatory signaling in microglial cells. J Neuroimmunol 182:89–99
36. Qin L et al (2005) Interactive role of the toll-like receptor 4 and reactive oxygen species in LPS-induced microglia activation. Glia 52:78–84
37. Haberg A et al (2001) Differences in neurotransmitter synthesis and intermediary metabolism between glutamatergic and GABAergic neurons during 4 hours of middle cerebral artery occlusion in the rat: the role of astrocytes in neuronal survival. J Cereb Blood Flow Metab 21:1451–1463
38. Prat A, Biernacki K, Wosik K, Antel JP (2001) Glial cell influence on the human blood–brain barrier. Glia 36:145–155
39. Weiss JM, Berman JW (1998) Astrocyte expression of monocyte chemoattractant protein-1 is differentially regulated by transforming growth factor beta. J Neuroimmunol 91:190–197
40. Dietrich PY, Walker PR, Saas P (2003) Death receptors on reactive astrocytes: a key role in the fine tuning of brain inflammation? Neurology 60:548–554
41. Vargas MR, Johnson JA (2010) Astrogliosis in amyotrophic lateral sclerosis: role and therapeutic potential of astrocytes. Neurotherapeutics 7:471–481
42. Maragakis NJ, Rothstein JD (2006) Mechanisms of disease: astrocytes in neurodegenerative disease. Nat Clin Pract Neurol 2:679–689
43. Little AR, O'Callagha JP (2001) Astrogliosis in the adult and developing CNS: is there a role for proinflammatory cytokines? Neurotoxicology 22:607–618
44. Griffin WS, Liu L, Li Y, Mrak RE, Barger SW (2006) Interleukin-1 mediates Alzheimer and Lewy body pathologies. J Neuroinflammation 3:5
45. Mrak RE, Griffin WS (2005) Potential inflammatory biomarkers in Alzheimer's disease. J Alzheimers Dis 8:369–375
46. Fan L et al (1995) Experimental brain injury induces expression of interleukin-1 beta mRNA in the rat brain. Brain Res Mol Brain Res 30:125–130
47. Patel HC, Boutin H, Allan SM (2003) Interleukin-1 in the brain: mechanisms of action in acute neurodegeneration. Ann N Y Acad Sci 992:39–47
48. Merrill JE, Benveniste EN (1996) Cytokines in inflammatory brain lesions: helpful and harmful. Trends Neurosci 19:331–338
49. John GR et al (2004) Interleukin-1beta induces a reactive astroglial phenotype via deactivation of the Rho GTPase-Rock axis. J Neurosci 24:2837–2845
50. Faulkner JR et al (2004) Reactive astrocytes protect tissue and preserve function after spinal cord injury. J Neurosci 24:2143–2155
51. Sofroniew MV et al (1999) Genetically-targeted and conditionally-regulated ablation of astroglial cells in the central, enteric and peripheral nervous systems in adult transgenic mice. Brain Res 835:91–95
52. Alfonso-Loeches S, Pascual-Lucas M, Blanco AM, Sanchez-Vera I, Guerri C (2010) Pivotal role of TLR4 receptors in alcohol-induced neuroinflammation and brain damage. J Neurosci 30:8285–8295
53. Ikegami Y et al (2003) Increased TUNEL positive cells in human alcoholic brains. Neurosci Lett 349:201–205
54. Korbo L (1999) Glial cell loss in the hippocampus of alcoholics. Alcohol Clin Exp Res 23:164–168
55. Kril JJ, Halliday GM, Svoboda MD, Cartwright H (1997) The cerebral cortex is damaged in chronic alcoholics. Neuroscience 79:983–998
56. Miguel-Hidalgo JJ et al (2002) Glia pathology in the prefrontal cortex in alcohol dependence with and without depressive symptoms. Biol Psychiatry 52:1121–1133
57. Liu J et al (2006) Patterns of gene expression in the frontal cortex discriminate alcoholic from nonalcoholic individuals. Neuropsychopharmacology 31:1574–1582
58. Lewohl JM, Wixey J, Harper CG, Dodd PR (2005) Expression of MBP, PLP, MAG, CNP, and GFAP in the human alcoholic brain. Alcohol Clin Exp Res 29:1698–1705
59. Fernandez-Lizarbe S, Pascual M, Guerri C (2009) Critical role of TLR4 response in the activation of microglia induced by ethanol. J Immunol 183:4733–4744

60. He J, Crews FT (2008) Increased MCP-1 and microglia in various regions of the human alcoholic brain. Exp Neurol 210:349–358
61. Qin L et al (2008) Increased systemic and brain cytokine production and neuroinflammation by endotoxin following ethanol treatment. J Neuroinflammation 5:10
62. Alfonso-Loeches S, Guerri C (2011) Molecular and behavioral aspects of the actions of alcohol on the adult and developing brain. Crit Rev Clin Lab Sci 48:19–47
63. Blanco AM, Guerri C (2007) Ethanol intake enhances inflammatory mediators in brain: role of glial cells and TLR4/IL-1RI receptors. Front Biosci 12:2616–2630
64. Akira S, Takeda K (2004) Functions of toll-like receptors: lessons from KO mice. C R Biol 327:581–589
65. O'Neill LA, Bowie AG (2007) The family of five: TIR-domain-containing adaptors in Toll-like receptor signalling. Nat Rev Immunol 7:353–364
66. Yang Y et al (2007) Lipopolysaccharide (LPS) regulates TLR4 signal transduction in nasopharynx epithelial cell line 5-8F via NFkappaB and MAPKs signaling pathways. Mol Immunol 44:984–992
67. Shimazu R et al (1999) MD-2, a molecule that confers lipopolysaccharide responsiveness on Toll-like receptor 4. J Exp Med 189:1777–1782
68. Nagai Y et al (2002) Essential role of MD-2 in LPS responsiveness and TLR4 distribution. Nat Immunol 3:667–672
69. Gioannini TL et al (2004) Isolation of an endotoxin-MD-2 complex that produces Toll-like receptor 4-dependent cell activation at picomolar concentrations. Proc Natl Acad Sci USA 101:4186–4191
70. Kim HM et al (2007) Crystal structure of the TLR4-MD-2 complex with bound endotoxin antagonist Eritoran. Cell 130:906–917
71. Ohto U, Fukase K, Miyake K, Satow Y (2007) Crystal structures of human MD-2 and its complex with antiendotoxic lipid IVa. Science 316:1632–1634
72. Park BS et al (2009) The structural basis of lipopolysaccharide recognition by the TLR4-MD-2 complex. Nature 458:1191–1195
73. Triantafilou M, Miyake K, Golenbock DT, Triantafilou K (2002) Mediators of innate immune recognition of bacteria concentrate in lipid rafts and facilitate lipopolysaccharide-induced cell activation. J Cell Sci 115:2603–2611
74. Kawai T, Akira S (2011) Toll-like receptors and their crosstalk with other innate receptors in infection and immunity. Immunity 34:637–650
75. Caso JR et al (2008) Toll-like receptor 4 is involved in subacute stress-induced neuroinflammation and in the worsening of experimental stroke. Stroke 39:1314–1320
76. Sarnico I et al (2009) NF-kappaB p50/RelA and c-Rel-containing dimers: opposite regulators of neuron vulnerability to ischaemia. J Neurochem 108:475–485
77. Lehnardt S et al (2003) Activation of innate immunity in the CNS triggers neurodegeneration through a Toll-like receptor 4-dependent pathway. Proc Natl Acad Sci USA 100:8514–8519
78. Blanco AM, Perez-Arago A, Fernandez-Lizarbe S, Guerri C (2008) Ethanol mimics ligand-mediated activation and endocytosis of IL-1RI/TLR4 receptors via lipid rafts caveolae in astroglial cells. J Neurochem 106:625–639
79. Blanco AM, Valles SL, Pascual M, Guerri C (2005) Involvement of TLR4/type I IL-1 receptor signaling in the induction of inflammatory mediators and cell death induced by ethanol in cultured astrocytes. J Immunol 175:6893–6899
80. Saijo K et al (2009) A Nurr1/CoREST pathway in microglia and astrocytes protects dopaminergic neurons from inflammation-induced death. Cell 137:47–59
81. Boje KM, Arora PK (1992) Microglial-produced nitric oxide and reactive nitrogen oxides mediate neuronal cell death. Brain Res 587:250–256
82. Chao CC, Hu S, Molitor TW, Shaskan EG, Peterson PK (1992) Activated microglia mediate neuronal cell injury via a nitric oxide mechanism. J Immunol 149:2736–2741
83. Okun E, Griffioen KJ, Mattson MP (2010) Toll-like receptor signaling in neural plasticity and disease. Trends Neurosci 34:269–281

84. Drexler SK, Foxwell BM (2010) The role of toll-like receptors in chronic inflammation. Int J Biochem Cell Biol 42:506–518
85. Landreth GE, Reed-Geaghan EG (2009) Toll-like receptors in Alzheimer's disease. Curr Top Microbiol Immunol 336:137–153
86. Marsh BJ, Williams-Karnesky RL, Stenzel-Poore MP (2009) Toll-like receptor signaling in endogenous neuroprotection and stroke. Neuroscience 158:1007–1020
87. Okun E et al (2009) Toll-like receptors in neurodegeneration. Brain Res Rev 59:278–292
88. Kiechl S et al (2002) Toll-like receptor 4 polymorphisms and atherogenesis. N Engl J Med 347:185–192
89. Manolakis AC et al (2011) TLR4 gene polymorphisms: evidence for protection against type 2 diabetes but not for diabetes-associated ischaemic heart disease. Eur J Endocrinol 165:261–267
90. Emonts M et al (2011) Polymorphisms in genes controlling inflammation and tissue repair in rheumatoid arthritis: a case control study. BMC Med Genet 12:36
91. Reed-Geaghan EG, Savage JC, Hise AG, Landreth GE (2009) CD14 and toll-like receptors 2 and 4 are required for fibrillar A{beta}-stimulated microglial activation. J Neurosci 29:11982–11992
92. Jin JJ, Kim HD, Maxwell JA, Li L, Fukuchi K (2008) Toll-like receptor 4-dependent upregulation of cytokines in a transgenic mouse model of Alzheimer's disease. J Neuroinflammation 5:23
93. Tahara K et al (2006) Role of toll-like receptor signalling in Abeta uptake and clearance. Brain 129:3006–3019
94. Minoretti P et al (2006) Effect of the functional toll-like receptor 4 Asp299Gly polymorphism on susceptibility to late-onset Alzheimer's disease. Neurosci Lett 391:147–149
95. Richard KL, Filali M, Prefontaine P, Rivest S (2008) Toll-like receptor 2 acts as a natural innate immune receptor to clear amyloid beta 1–42 and delay the cognitive decline in a mouse model of Alzheimer's disease. J Neurosci 28:5784–5793
96. Sawada M, Imamura K, Nagatsu T (2006) Role of cytokines in inflammatory process in Parkinson's disease. J Neural Transm Suppl 373–381
97. Castano A, Herrera AJ, Cano J, Machado A (1998) Lipopolysaccharide intranigral injection induces inflammatory reaction and damage in nigrostriatal dopaminergic system. J Neurochem 70:1584–1592
98. Panaro MA et al (2008) Expression of TLR4 and CD14 in the central nervous system (CNS) in a MPTP mouse model of Parkinson's-like disease. Immunopharmacol Immunotoxicol 30:729–740
99. Kilic U, Kilic E, Matter CM, Bassetti CL, Hermann DM (2008) TLR-4 deficiency protects against focal cerebral ischemia and axotomy-induced neurodegeneration. Neurobiol Dis 31: 33–40
100. Kelley KW, Dantzer R (2011) Alcoholism and inflammation: neuroimmunology of behavioral and mood disorders. Brain Behav Immun 25(Suppl 1):S13–S20
101. Szabo G, Mandrekar P (2009) A recent perspective on alcohol, immunity, and host defense. Alcohol Clin Exp Res 33:220–232
102. Germann G, Barthold U, Lefering R, Raff T, Hartmann B (1997) The impact of risk factors and pre-existing conditions on the mortality of burn patients and the precision of predictive admission-scoring systems. Burns 23:195–203
103. Messingham KA, Faunce DE, Kovacs EJ (2002) Alcohol, injury, and cellular immunity. Alcohol 28:137–149
104. Ruiz M et al (1999) Severe community-acquired pneumonia. Risk factors and follow-up epidemiology. Am J Respir Crit Care Med 160:923–929
105. Pruett SB, Schwab C, Zheng Q, Fan R (2004) Suppression of innate immunity by acute ethanol administration: a global perspective and a new mechanism beginning with inhibition of signaling through TLR3. J Immunol 173:2715–2724
106. Yamashina S et al (2000) Tolerance and sensitization to endotoxin in Kupffer cells caused by acute ethanol involve interleukin-1 receptor-associated kinase. Biochem Biophys Res Commun 277:686–690

107. Goral J, Choudhry MA, Kovacs EJ (2004) Acute ethanol exposure inhibits macrophage IL-6 production: role of p38 and ERK1/2 MAPK. J Leukoc Biol 75:553–559
108. Pang M, Bala S, Kodys K, Catalano D, Szabo G (2011) Inhibition of TLR8- and TLR4-induced Type I IFN induction by alcohol is different from its effects on inflammatory cytokine production in monocytes. BMC Immunol 12:55
109. Miller AM, Horiguchi N, Jeong WI, Radaeva S, Gao B (2011) Molecular mechanisms of alcoholic liver disease: innate immunity and cytokines. Alcohol Clin Exp Res 35:787–793
110. Parlesak A, Schafer C, Schutz T, Bode JC, Bode C (2000) Increased intestinal permeability to macromolecules and endotoxemia in patients with chronic alcohol abuse in different stages of alcohol-induced liver disease. J Hepatol 32:742–747
111. Uesugi T et al (2001) Delivery of IkappaB superrepressor gene with adenovirus reduces early alcohol-induced liver injury in rats. Hepatology 34:1149–1157
112. Hritz I et al (2008) The critical role of toll-like receptor (TLR) 4 in alcoholic liver disease is independent of the common TLR adapter MyD88. Hepatology 48:1224–1231
113. Bhatty M, Jan BL, Tan W, Pruett SB, Nanduri B (2011) Role of acute ethanol exposure and TLR4 in early events of sepsis in a mouse model. Alcohol 45:795–803
114. Bird MD et al (2010) Decreased pulmonary inflammation following ethanol and burn injury in mice deficient in TLR4 but not TLR2 signaling. Alcohol Clin Exp Res 34:1733–1741
115. Marcolini J, Nguyen M, Ericsson C (2002) Klebsiella pneumoniae brain abscess in a Taiwanese adult. J Infect 44:205–210
116. de Vries HE, Kuiper J, de Boer AG, Van Berkel TJ, Breimer DD (1997) The blood–brain barrier in neuroinflammatory diseases. Pharmacol Rev 49:143–155
117. Kalaria RN (1992) The blood–brain barrier and cerebral microcirculation in Alzheimer disease. Cerebrovasc Brain Metab Rev 4:226–260
118. Latour LL, Kang DW, Ezzeddine MA, Chalela JA, Warach S (2004) Early blood–brain barrier disruption in human focal brain ischemia. Ann Neurol 56:468–477
119. van Horssen J, Brink BP, de Vries HE, van der Valk P, Bo L (2007) The blood–brain barrier in cortical multiple sclerosis lesions. J Neuropathol Exp Neurol 66:321–328
120. Bowman CC, Rasley A, Tranguch SL, Marriott I (2003) Cultured astrocytes express toll-like receptors for bacterial products. Glia 43:281–291
121. Bsibsi M, Ravid R, Gveric D, van Noort JM (2002) Broad expression of Toll-like receptors in the human central nervous system. J Neuropathol Exp Neurol 61:1013–1021
122. Esen N, Tanga FY, DeLeo JA, Kielian T (2004) Toll-like receptor 2 (TLR2) mediates astrocyte activation in response to the Gram-positive bacterium Staphylococcus aureus. J Neurochem 88:746–758
123. Minagar A et al (2002) The role of macrophage/microglia and astrocytes in the pathogenesis of three neurologic disorders: HIV-associated dementia, Alzheimer disease, and multiple sclerosis. J Neurol Sci 202:13–23
124. Valles SL, Blanco AM, Pascual M, Guerri C (2004) Chronic ethanol treatment enhances inflammatory mediators and cell death in the brain and in astrocytes. Brain Pathol 14:365–371
125. Wilflingseder D et al (2004) HIV-1-induced migration of monocyte-derived dendritic cells is associated with differential activation of MAPK pathways. J Immunol 173:7497–7505
126. Wu D, Marko M, Claycombe K, Paulson KE, Meydani SN (2003) Ceramide-induced and age-associated increase in macrophage COX-2 expression is mediated through up-regulation of NF-kappa B activity. J Biol Chem 278:10983–10992
127. Buss H et al (2004) Constitutive and interleukin-1-inducible phosphorylation of p65 NF-{kappa}B at serine 536 is mediated by multiple protein kinases including I{kappa}B kinase (IKK)-{alpha}, IKK{beta}, IKK{epsilon}, TRAF family member-associated (TANK)-binding kinase 1 (TBK1), and an unknown kinase and couples p65 to TATA-binding protein-associated factor II31-mediated interleukin-8 transcription. J Biol Chem 279:55633–55643
128. Jiang Z et al (2003) Pellino 1 is required for interleukin-1 (IL-1)-mediated signaling through its interaction with the IL-1 receptor-associated kinase 4 (IRAK4)-IRAK-tumor necrosis factor receptor-associated factor 6 (TRAF6) complex. J Biol Chem 278:10952–10956

129. Salaria S et al (2007) Toll-like receptor pathway gene expression is associated with human immunodeficiency virus-associated neurodegeneration. J Neurovirol 13:496–503
130. Pascual M, Fernandez-Lizarbe S, Guerri C (2011) Role of TLR4 in ethanol effects on innate and adaptive immune responses in peritoneal macrophages. Immunol Cell Biol 89:716–727
131. O'Connor JC et al (2009) Interferon-gamma and tumor necrosis factor-alpha mediate the upregulation of indoleamine 2,3-dioxygenase and the induction of depressive-like behavior in mice in response to bacillus Calmette-Guerin. J Neurosci 29:4200–4209
132. Back SA et al (2005) Hyaluronan accumulates in demyelinated lesions and inhibits oligodendrocyte progenitor maturation. Nat Med 11:966–972
133. Kadhim H, De Prez C, Gazagnes MD, Sebire G (2003) In situ cytokine immune responses in acute disseminated encephalomyelitis: insights into pathophysiologic mechanisms. Hum Pathol 34:293–297
134. Paula-Barbosa MM, Tavares MA (1985) Long term alcohol consumption induces microtubular changes in the adult rat cerebellar cortex. Brain Res 339:195–199
135. Mayfield RD et al (2002) Patterns of gene expression are altered in the frontal and motor cortices of human alcoholics. J Neurochem 81:802–813
136. Lewohl JM et al (2000) Gene expression in human alcoholism: microarray analysis of frontal cortex. Alcohol Clin Exp Res 24:1873–1882
137. de la Monte SM (1988) Disproportionate atrophy of cerebral white matter in chronic alcoholics. Arch Neurol 45:990–992
138. Pfefferbaum A, Adalsteinsson E, Sullivan EV (2006) Dysmorphology and microstructural degradation of the corpus callosum: interaction of age and alcoholism. Neurobiol Aging 27:994–1009
139. Alfonso-Loeches S et al (2012) Toll-like receptor 4 participates in the myelin disruptions associated with chronic alcohol abuse. Glia 60:948–964
140. Schroder K, Tschopp J (2010) The inflammasomes. Cell 140:821–832
141. Marta M (2009) Toll-like receptors in multiple sclerosis mouse experimental models. Ann N Y Acad Sci 1173:458–462
142. Schonberg DL, McTigue DM (2009) Iron is essential for oligodendrocyte genesis following intraspinal macrophage activation. Exp Neurol 218:64–74
143. Peoples RW, Li C, Weight FF (1996) Lipid vs protein theories of alcohol action in the nervous system. Annu Rev Pharmacol Toxicol 36:185–201
144. Pike LJ (2003) Lipid rafts: bringing order to chaos. J Lipid Res 44:655–667
145. Triantafilou M, Morath S, Mackie A, Hartung T, Triantafilou K (2004) Lateral diffusion of Toll-like receptors reveals that they are transiently confined within lipid rafts on the plasma membrane. J Cell Sci 117:4007–4014
146. Dolganiuc A, Bakis G, Kodys K, Mandrekar P, Szabo G (2006) Acute ethanol treatment modulates Toll-like receptor-4 association with lipid rafts. Alcohol Clin Exp Res 30:76–85
147. Dai Q, Zhang J, Pruett SB (2005) Ethanol alters cellular activation and CD14 partitioning in lipid rafts. Biochem Biophys Res Commun 332:37–42
148. Fernandez-Lizarbe S, Pascual M, Gascon MS, Blanco A, Guerri C (2008) Lipid rafts regulate ethanol-induced activation of TLR4 signaling in murine macrophages. Mol Immunol 45:2007–2016
149. Wang JH, Sun GY (2001) Ethanol inhibits cytokine-induced iNOS and sPLA2 in immortalized astrocytes: evidence for posttranscriptional site of ethanol action. J Biomed Sci 8:126–133
150. Militante JD, Feinstein DL, Syapin PJ (1997) Suppression by ethanol of inducible nitric oxide synthase expression in C6 glioma cells. J Pharmacol Exp Ther 281:558–565
151. Boe DM, Nelson S, Zhang P, Quinton L, Bagby GJ (2003) Alcohol-induced suppression of lung chemokine production and the host defense response to Streptococcus pneumoniae. Alcohol Clin Exp Res 27:1838–1845
152. Friebe A et al (2010) Dose-dependent development of depressive symptoms during adjuvant interferon-{alpha} treatment of patients with malignant melanoma. Psychosomatics 51: 466–473

153. Khairova RA, Machado-Vieira R, Du J, Manji HK (2009) A potential role for pro-inflammatory cytokines in regulating synaptic plasticity in major depressive disorder. Int J Neuropsychopharmacol 12:561–578
154. Hauss-Wegrzyniak B, Lukovic L, Bigaud M, Stoeckel ME (1998) Brain inflammatory response induced by intracerebroventricular infusion of lipopolysaccharide: an immunohistochemical study. Brain Res 794:211–224
155. Hauss-Wegrzyniak B, Vannucchi MG, Wenk GL (2000) Behavioral and ultrastructural changes induced by chronic neuroinflammation in young rats. Brain Res 859:157–166
156. Hauss-Wegrzyniak B, Wenk GL (2002) Beta-amyloid deposition in the brains of rats chronically infused with thiorphan or lipopolysaccharide: the role of ascorbic acid in the vehicle. Neurosci Lett 322:75–78
157. Huang Y, Henry CJ, Dantzer R, Johnson RW, Godbout JP (2008) Exaggerated sickness behavior and brain proinflammatory cytokine expression in aged mice in response to intracerebroventricular lipopolysaccharide. Neurobiol Aging 29:1744–1753
158. Wu Y et al (2011) Attenuation of microglial and IL-1 signaling protects mice from acute alcohol-induced sedation and/or motor impairment. Brain Behav Immun 25(Suppl 1): S155–S164
159. Pascual M, Blanco AM, Cauli O, Minarro J, Guerri C (2007) Intermittent ethanol exposure induces inflammatory brain damage and causes long-term behavioural alterations in adolescent rats. Eur J Neurosci 25:541–550
160. Hayley S, Mangano E, Strickland M, Anisman H (2008) Lipopolysaccharide and a social stressor influence behaviour, corticosterone and cytokine levels: divergent actions in cyclooxygenase-2 deficient mice and wild type controls. J Neuroimmunol 197:29–36
161. Weberpals M et al (2009) NOS2 gene deficiency protects from sepsis-induced long-term cognitive deficits. J Neurosci 29:14177–14184
162. Vollmar P et al (2010) Active immunization with amyloid-beta 1–42 impairs memory performance through TLR2/4-dependent activation of the innate immune system. J Immunol 185:6338–6347
163. Edenberg HJ et al (2008) Association of NFKB1, which encodes a subunit of the transcription factor NF-kappaB, with alcohol dependence. Hum Mol Genet 17:963–970
164. Okvist A et al (2007) Neuroadaptations in human chronic alcoholics: dysregulation of the NF-kappaB system. PLoS One 2:e930
165. Loftis JM, Choi D, Hoffman W, Huckans MS (2010) Methamphetamine causes persistent immune dysregulation: a cross-species, translational report. Neurotox Res 20:59–68
166. Edenberg HJ et al (2004) Variations in GABRA2, encoding the alpha 2 subunit of the GABA(A) receptor, are associated with alcohol dependence and with brain oscillations. Am J Hum Genet 74:705–714
167. Zou J, Crews F (2010) Induction of innate immune gene expression cascades in brain slice cultures by ethanol: key role of NF-kappaB and proinflammatory cytokines. Alcohol Clin Exp Res 34:777–789
168. Liu J et al (2011) Binge alcohol drinking is associated with GABAA alpha2-regulated Toll-like receptor 4 (TLR4) expression in the central amygdala. Proc Natl Acad Sci USA 108:4465–4470
169. Blednov YA et al (2011) Neuroimmune regulation of alcohol consumption: behavioral validation of genes obtained from genomic studies. Addict Biol 17:108–120
170. Drugan RC, Wiedholz LM, Holt A, Kent S, Christianson JP (2007) Environmental and immune stressors enhance alcohol-induced motor ataxia in rat. Pharmacol Biochem Behav 86:125–131
171. Blednov YA et al (2011) Activation of inflammatory signaling by lipopolysaccharide produces a prolonged increase of voluntary alcohol intake in mice. Brain Behav Immun 25(Suppl 1):S92–S105
172. Blednov YA et al (2011) Loss of ethanol conditioned taste aversion and motor stimulation in knockin mice with ethanol-insensitive alpha2-containing GABA(A) receptors. J Pharmacol Exp Ther 336:145–154

173. Breese GR et al (2008) Repeated lipopolysaccharide (LPS) or cytokine treatments sensitize ethanol withdrawal-induced anxiety-like behavior. Neuropsychopharmacology 33:867–876
174. Wohleb ES et al (2011) beta-Adrenergic receptor antagonism prevents anxiety-like behavior and microglial reactivity induced by repeated social defeat. J Neurosci 31:6277–6288
175. Watkins LR, Hutchinson MR, Rice KC, Maier SF (2009) The "toll" of opioid-induced glial activation: improving the clinical efficacy of opioids by targeting glia. Trends Pharmacol Sci 30:581–591
176. Hutchinson MR et al (2010) Evidence that opioids may have toll-like receptor 4 and MD-2 effects. Brain Behav Immun 24:83–95
177. Hutchinson MR et al (2010) Evidence that tricyclic small molecules may possess toll-like receptor and myeloid differentiation protein 2 activity. Neuroscience 168:551–563
178. Hutchinson MR et al (2008) Proinflammatory cytokines oppose opioid-induced acute and chronic analgesia. Brain Behav Immun 22:1178–1189
179. Song P, Zhao ZQ (2001) The involvement of glial cells in the development of morphine tolerance. Neurosci Res 39:281–286
180. Hutchinson MR et al (2008) Minocycline suppresses morphine-induced respiratory depression, suppresses morphine-induced reward, and enhances systemic morphine-induced analgesia. Brain Behav Immun 22:1248–1256
181. Hutchinson MR et al (2009) Reduction of opioid withdrawal and potentiation of acute opioid analgesia by systemic AV411 (ibudilast). Brain Behav Immun 23:240–250
182. Liu B, Du L, Hong JS (2000) Naloxone protects rat dopaminergic neurons against inflammatory damage through inhibition of microglia activation and superoxide generation. J Pharmacol Exp Ther 293:607–617
183. Liu JG, Prather PL (2002) Chronic agonist treatment converts antagonists into inverse agonists at delta-opioid receptors. J Pharmacol Exp Ther 302:1070–1079
184. Sebai H et al (2010) Protective effect of resveratrol against LPS-induced extracellular lipoperoxidation in AR42J cells partly via a Myd88-dependent signaling pathway. Arch Biochem Biophys 495:56–61
185. Ding BJ et al (2011) Soybean isoflavone alleviates beta-amyloid 1–42 induced inflammatory response to improve learning and memory ability by down regulation of Toll-like receptor 4 expression and nuclear factor-kappaB activity in rats. Int J Dev Neurosci 29:537–542
186. Youn HS, Saitoh SI, Miyake K, Hwang DH (2006) Inhibition of homodimerization of Toll-like receptor 4 by curcumin. Biochem Pharmacol 72:62–69
187. Wong SW et al (2009) Fatty acids modulate Toll-like receptor 4 activation through regulation of receptor dimerization and recruitment into lipid rafts in a reactive oxygen species-dependent manner. J Biol Chem 284:27384–27392
188. Youn HS et al (2008) Cinnamaldehyde suppresses toll-like receptor 4 activation mediated through the inhibition of receptor oligomerization. Biochem Pharmacol 75:494–502
189. Chen T, Guo J, Han C, Yang M, Cao X (2009) Heat shock protein 70, released from heat-stressed tumor cells, initiates antitumor immunity by inducing tumor cell chemokine production and activating dendritic cells via TLR4 pathway. J Immunol 182:1449–1459
190. Equils O et al (2006) 1,25-Dihydroxyvitamin D inhibits lipopolysaccharide-induced immune activation in human endothelial cells. Clin Exp Immunol 143:58–64
191. Roozen HG et al (2006) A systematic review of the effectiveness of naltrexone in the maintenance treatment of opioid and alcohol dependence. Eur Neuropsychopharmacol 16:311–323
192. Wu Y et al (2011) Inhibiting the TLR4-MyD88 signalling cascade by genetic or pharmacologic strategies reduces acute alcohol dose-induced sedation and motor impairment in mice. Br J Pharmacol 165:1319–1329
193. Franchi S et al (2010) The effects of alcoholism pharmacotherapy on immune responses in alcohol-dependent patients. Int J Immunopathol Pharmacol 23:847–855
194. Mulligan MK et al (2006) Toward understanding the genetics of alcohol drinking through transcriptome meta-analysis. Proc Natl Acad Sci USA 103:6368–6373
195. Kong EC et al (2010) Ethanol-regulated genes that contribute to ethanol sensitivity and rapid tolerance in Drosophila. Alcohol Clin Exp Res 34:302–316

Chapter 10
Prenatal Alcohol Exposure: Impact on Neuroendocrine–Neuroimmune Networks

Tamara Bodnar and Joanne Weinberg

10.1 Introduction

Alcohol exposure *in utero* can have numerous adverse effects on a developing fetus. The term fetal alcohol spectrum disorder (FASD) refers to the broad spectrum of structural, neurocognitive, and behavioral abnormalities or deficits that can occur following prenatal alcohol (ethanol) exposure (PAE). At the most severe end of the spectrum is fetal alcohol syndrome (FAS), which involves the complete phenotype of characteristic facial anomalies, growth retardation, and central nervous system (CNS) abnormalities. Alcohol exposure at levels that result in some but not all components of the facial, growth, and CNS deficits, and with evidence of neurobehavioral abnormalities, is termed partial fetal alcohol syndrome (PFAS). In the absence of any facial anomalies or growth deficits, a range of effects can occur that may be primarily physical, termed alcohol-related birth defects (ARBD), or primarily neurological and/or neurobehavioral, termed alcohol-related neurodevelopmental disorder (ARND) [1].

Following the description of FAS by Jones and Smith in 1973, there has been extensive interest in investigating the effects of PAE on the function of the immune system. In general, studies on children exposed to alcohol prenatally have demonstrated impairments in immune competence in both innate and adaptive immunity. Adaptive immunity is MHC (major histocompatibility complex) restricted and can be classified as either cellular immunity, mediated by T lymphocytes, or humoral immunity, mediated by B lymphocytes. Innate immunity is not MHC restricted. Through phagocytes such as monocytes, macrophages, and polymorphonuclear

T. Bodnar • J. Weinberg, Ph.D. (✉)
Department of Cellular and Physiological Sciences, University of British Columbia, 2350 Health Sciences Mall, Vancouver, BC, Canada V6T 1Z3
e-mail: jweinberg@exchange.ubc.ca

C. Cui et al. (eds.), *Neural-Immune Interactions in Brain Function and Alcohol Related Disorders*, DOI 10.1007/978-1-4614-4729-0_10,

leukocytes (PMNL), as well as natural killer cells and mediators such as complement and C-reactive protein, innate immunity provides a first line of defense against many common pathogens. Importantly, interactions between the innate and adaptive components of the immune system are necessary to launch effective immune responses.

In this chapter, we will review the normal development of the immune system, a complicated and multisystem process, to provide the context for understanding how and when PAE can impact immune function. We will then review data from both clinical studies and animal models that have demonstrated adverse effects of PAE on neuroendocrine activity and immune competence. We will describe the sexual dimorphism of the normal immune response and discuss sex differences in alcohol's adverse effects on both neuroendocrine and immune functions. Examination of mechanisms underlying alcohol's adverse effects will focus primarily on neuroendocrine–neuroimmune interactions and will probe how PAE alters the interaction of hormonal and immune systems. Included in this section is a discussion of recent work from our laboratory utilizing an animal model of human rheumatoid arthritis, which reveals an increased susceptibility to inflammation in PAE offspring. Finally, we will examine fetal programming of stress and immune function by PAE and consider possible mechanisms mediating alcohol-induced fetal programming, including epigenetic mechanisms.

10.2 Normal Ontogeny of the Immune System

10.2.1 Timeline of Development

Although development of the immune system takes place throughout life, the basic framework of the immune system and immune competence develops before birth (see Fig. 10.1). The normal newborn is not immunologically naïve. Indeed, data have shown that neonatal T and B cells are matured to a stage that they can mount an antigen-specific response to antigens encountered *in utero*, such as the parasite *Ascaris* and tetanus toxoid vaccination [4, 5] (Table 10.1).

The first stage of fetal hematopoiesis involves the formation and development of hematopoietic stem cells in the yolk sac over the first month of fetal life. These stem cells migrate to the liver 3–4 weeks later and subsequently migrate to the thymus and spleen. Stem cells can be found in the bone marrow at 11–12 weeks of gestation [6, 7], and differentiated T and B cells can be identified in fetal blood [8] and spleen [9] as early as 14 weeks of gestation. Environmental influences can affect lymphopoiesis and cell migration as early as 7–10 weeks postconception [10]. The spleen is considered completely immunocompetent by 18 weeks of gestation. At this time, splenic T cells have adult levels of expression of CD3, CD4, and CD8 and are able to respond to mitogens [11], and splenic accessory cells are fully functional in delivering co-stimulatory signals [7]. However, fetuses still have fewer memory T cells than adults at this time [7]. Yolk-sac-derived pro-B cells develop in the liver and acquire surface immunoglobulin (Ig) M, IgD, and CD20 expression by 10–13 weeks of gestation [12]. B cells are detectable in lymph nodes from 16 to

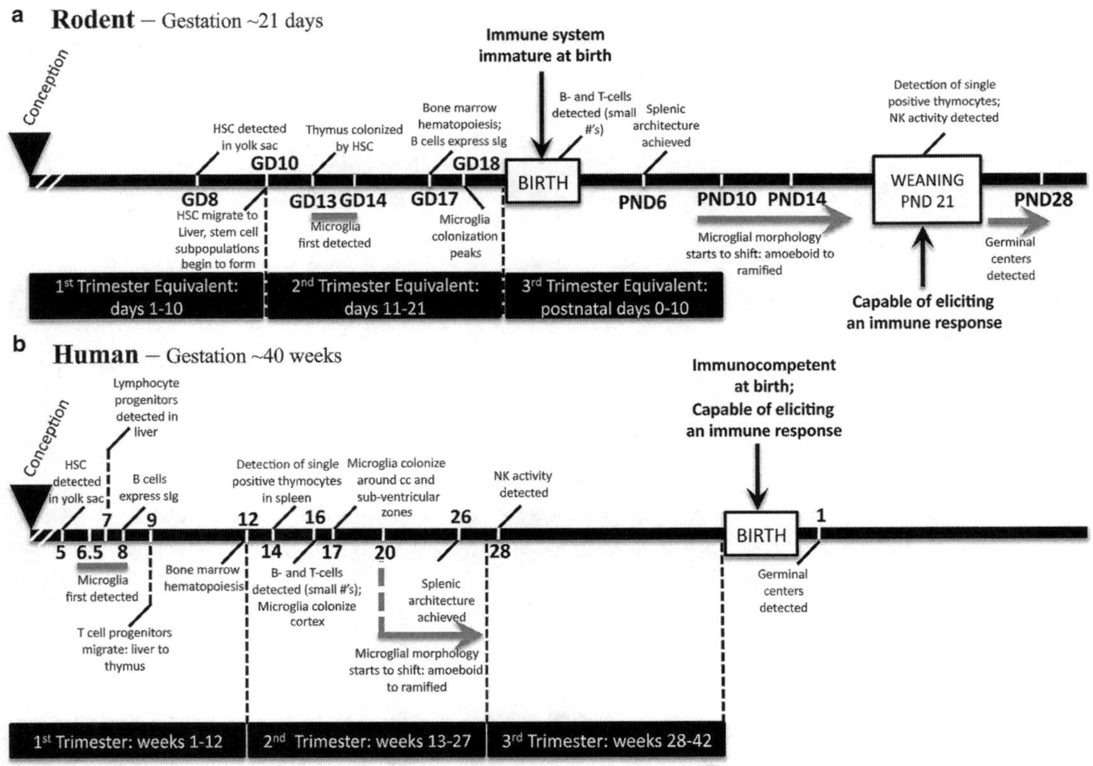

Fig. 10.1 Comparison timelines of key developmental immunological milestones in rodents and humans. Timelines showing many of the important immunological milestones in the rodent (**a**) and the human (**b**) fetus and newborn are illustrated [figure modeled on (Holsapple et al. [2])]. *HSC* hematopoietic stem cells, *sIg* surface immunoglobulin, *NK* natural killer cells, *CC* corpus callosum, *splenic architecture* delineation of *red* and *white* pulp areas (Information in this figure is drawn from a number of sources including Holladay and Smialowicz [250]); Holsapple et al. [2]; Bilbo and Schwarz [3]; Schwarz [213])

Table 10.1 Common immune terminology.

Immune term	Definition
Adaptive immune response	The response of specialized immune cells to foreign antigen. Adaptive immunity is activated approximately 3–7 days postinfection, allowing for the targeted elimination of pathogens as well as immunological memory. The adaptive immune system can be subdivided based on how immunity was achieved: (1) *naturally acquired immunity* develops following a natural occurring infection; (2) *artificially acquired immunity* develops as a result of an external or artificial challenge, as in the case of vaccination; (3) *passive immunity* develops through the transfer of antibodies and/or activated immune cells from a host, for example, from mother to infant through breast milk
Antibody	Protein produced by plasma B cells in response to infection or vaccination. Every antibody has a unique structure, allowing it to bind to a specific target substance, known as the *antigen*. Antibodies bind to antigens, neutralizing them and targeting them for elimination by phagocytic cells
B lymphocyte (B cell)	A subset of white blood cell that mediates humoral immunity. B cells are generated in the bone marrow, and each B cell has a unique B cell receptor, allowing it to bind to a particular antigen. Prior to activation, B cells play an immune surveillance role and do not start producing antibodies until they become fully activated. There are many types of B cells, each having a different function; these include memory B cells, plasma B cells, B-1 cells, and B-2 cells
Cell-mediated immunity	Includes the aspects of the adaptive immune response in which antigen-specific T cells have the primary role
Cluster of differentiation	Groups of antibodies that identify cell-surface markers. This method of identifying cell-surface markers is important for immunophenotyping of cells. The marker to which the antibody binds is termed CD, followed by a number (e.g., CD4, CD8). The surface markers expressed by a cell are informative as to the lineage and function of the cell. Thus, cell populations can be categorized based on the CD molecules they express. The symbols "+" and "–" are used to indicate whether the cell type expresses a given surface marker or not CD45+: all hematopoietic cells CD45+, CD3+: thymocytes, T cells CD45+, CD3+, CD4+: T helper cells CD45+, CD3+, CD8+: cytotoxic T cells CD45+, CD19+ or CD45+, CD20+, CD24+: B cells (CD20 is not expressed during the first and last stage of B cell development) Other important CD antigens: CD7: expressed by pluripotential hematopoietic cells, thymocytes, and T cells CD25: expressed by activated T cells, B cells, and monocytes CD43: expressed by leukocytes (except resting B cells)
Cytokines	Signaling molecules secreted by immune system cells as a means of cellular communication. Cytokines include proteins, peptides, and glycoproteins, and their defining feature is that they have a wide array of immunomodulating effects, including pro- and anti-inflammatory effects

(continued)

Table 10.1 (continued)

Immune term	Definition
Humoral immunity	Includes the antibody-mediated aspects of the adaptive immune response. The main cell types involved in humoral immunity are cells of B lymphocyte lineage. Humoral immunity, unlike cell-mediated immunity, can be transferred to a recipient via serum (cell-free)
IgE	Class of immunoglobulin involved in defending the body against parasites; also implicated in allergic reactions
IgG	Class of immunoglobulin produced by plasma B cells; most abundant serum immunoglobulin in humans
IgM	Class of immunoglobulin produced by B cells; first antibody to be secreted by B cells following exposure to an antigen
Immunoglobulin (Ig)	Antibody molecules (plasma proteins) belonging to the immunoglobulin superfamily, which includes an array of proteins with immunoglobulin domains and includes T and B cell receptors, antibodies, and MHC molecules. Immunoglobulins can exist in both membrane-bound and soluble forms. The specific antigen receptor on B lymphocytes is an example of a membrane-bound immunoglobulin. Soluble immunoglobulins are generally composed using variations of the same building blocks, known as heavy chains and light chains, with two heavy and two light chains per antibody Key examples of secreted immunoglobulins: *IgG*, *IgM*, *IgE*
Inflammation	Mechanism of the innate immune response resulting in the accumulation of fluid and white blood cells in response to physical trauma, infection, or irritation. Inflammation is a protective response designed to remove the noxious agent and initiate healing. Inflammation can be categorized as (1) *acute*, the early and transient response to harmful stimuli resulting in a cascade of responses including movement of plasma and white blood cells to the site of trauma/infection, and (2) *chronic*, persistent inflammation following persistent challenge or with autoimmune diseases and is associated with a shift in the immune cells present within the tissue as well as tissue destruction in addition to healing
Innate immune response	The nonspecific immune mechanisms that act as a first line of defense against pathogens. Innate immunity is a critical component of the immune response; however, it does not increase with repeated exposures to a pathogen, and it is not long lasting or protective against future encounters with the same pathogen
Leukocytes	White blood cells; critical cells of the immune system involved in defending the body against pathogens. Leukocytes include lymphocytes, monocytes, and polymorphonuclear leukocytes
Lymphocytes	A class of white blood cells involved in mediating adaptive immune responses. Lymphocytes can be divided into two main categories—*B lymphocytes* (B cells) and *T lymphocytes* (T cells)
Macrophage	Phagocytic cells that are of key importance in the innate immune response. Macrophages engulf foreign antigens and necrotic tissue, removing them from the environment. Macrophages also participate in the development of inflammation

(continued)

Table 10.1 (continued)

Immune term	Definition
Major histocompatibility complex (MHC)	Cluster of genes encoding cell-surface molecules (membrane glycoproteins) that mediate the interactions between leukocytes and other cells (includes other leukocytes)
Microglia	Nonneuronal cells within the brain. These cells are the resident macrophages of the central nervous system, scavenging the brain and spinal cord for infectious agents and damaged cells. Microglia also play many important roles during brain development including participating in neuronal pruning, phagocytosis of dying neurons, and producing cytokines
Monocyte	A type of leukocyte; the precursor to macrophages. Monocytes are attracted to damaged tissue from the circulation, and as they enter the tissue, through a process called leukocyte extravasation, they undergo changes, developing into macrophages
Pathogens	Infectious agents, including viruses, bacteria, or fungi that can cause disease in a host (human, animal, or plant)
T lymphocyte (T cell)	A subset of white blood cell that mediates cell-mediated immunity. These cells mature in the thymus and express unique cell-surface receptors (T cell receptors), allowing them to bind and recognize antigens. There are many different subsets of T lymphocytes, each undertaking specific functions in the cell-mediated immune response; these include T helper cells, cytotoxic T cells, and memory T cells
T_H1 cells	A subset of CD4+ T cells that are of key importance in adaptive immune functions including activating macrophages. These cells produce T_H1 cytokines, which are involved primarily in cell-mediated immunity and tend to produce proinflammatory responses. T_H1 cytokines include IL-2 and IFN-γ
T_H2 cells	A subset of CD4+ T cells that are of key importance in adaptive immune functions including stimulating B cells to produce antibody. These cells produce T_H2 cytokines, which are involved primarily in humoral immunity, and tend to be more anti-inflammatory. T_H2 cytokines include IL-3, IL-4, IL-5, and IL-13

17 weeks, in spleen at 16–21 weeks, and are abundant in bone marrow at 16–20 weeks of gestation [10].

10.2.2 Development of the Thymus

The structure of the human thymus develops as a result of epithelial–mesenchymal interactions. The endoderm of the third pharyngeal pouch differentiates into thymic epithelium. Neural crest cells then migrate and become the mesenchyme that will form the layers around the epithelial primordium of the thymus and the connective tissue framework [13]. T cell progenitors arrive in the thymus from the liver during the ninth week of gestation, and at about 10–12 weeks, a cortex and a medulla become differentiated [10]. Prothymocytes express CD7 and CD45 in the fetal liver at 7 weeks and CD3 at 8–9 weeks of gestation, when these cells migrate to the thymus [7]. Expression of the T cell receptors, CD4 and CD8, occurs around 12 weeks of gestation, and mature CD4+ and CD8+ (single positive) T cells leave the thymus after week 13. At 13–14 weeks, thymocytes acquire the ability to proliferate to most mitogens, and the T cell pool expands rapidly during 14–16 weeks of gestation [11, 14]. During the second trimester, susceptibility to environmental factors can disrupt the thymic development and T cell proliferation and lead to a defective T cell repertoire.

10.2.3 Comparison of Rodent and Human Timelines

Rodent gestation (~3 weeks) is substantially shorter than human gestation (~40 weeks) and as such, newborn rat and mouse pups are born developmentally and immunologically immature, at a time equivalent to the end of the second trimester in humans [15]. Therefore, studies in rodent models of PAE have focused primarily on the early events in ontogeny of the immune system, that is, during the late gestational period and early postnatal life (the first 10 days of postnatal life are roughly equivalent to the third trimester of human gestation). Typically, these models involve alcohol exposure for the full period of gestation. As the formation of a functional immune system requires sequential, yet synchronized, developmental events, including the coordinated development of immune cells and organs, which begins during early embryonic development, administration of alcohol to the rat or mouse throughout gestation encompasses all of the key early developmental events. Figure 10.1 presents a comparative timeline of immune system development in humans and rodents.

As shown in this figure, human gestation is approximately 40 weeks in duration with many of the key steps involved in maturation of the immune system being initiated in the first and second trimesters.

- At birth, the human neonate is immunocompetent and capable of initiating an immune response. However, the neonate is immunologically naïve, having yet to encounter antigens. Thus, while an immune response can be elicited, the strength of the response has not yet achieved adult levels at birth.
- In contrast, gestation in the rodent is approximately 21 days in duration, and in general, shorter gestation periods are associated with relatively immature immune systems at parturition [250]. At birth, the rodent immune system is very immature, and the pre-weanling rodent is incapable of initiating an immune response.
- Many key developmental milestones in the rodent are achieved during the equivalent of the third trimester in the human, which is approximately the first ten postnatal days in the rodent. By weaning (approximately postnatal day 21), an immune response to pathogenic challenge can be initiated; however, maturation of the immune system is still occurring, and an adult level response cannot yet be achieved.

- Of note, both the rodent and human fetus are released from intrauterine maternal influences at birth, notably the restraint on Th1 responses. However, maternal influences persist through lactation. As such, the maternal immunological milieu and response system can impact development of the fetal as well as the neonatal immune system.

10.3 Adverse Effects of PAE on the Immune System

10.3.1 Deficits in Immune Competence in Children with FASD

Clinical data on deficits in immune competence in children with FASD remain somewhat limited. An early review of 13 documented cases of FAS in infants and children by Johnson and colleagues [16] showed a higher incidence of minor infections, such as recurrent otitis media and respiratory infections, as well as major life-threatening bacterial infections, in alcohol-exposed compared to nonexposed children.

Inborn errors of immunocompetent cells in children with FAS result in immunodeficiency disorders or increased susceptibility to infections. Recurrent opportunistic infection and infection caused by ubiquitous microorganisms, such as bacteria, viruses, and fungi, typically occur with deficits in cell-mediated immunity, whereas deficits in B cells, immunoglobulins, complement, and phagocytes usually lead to infection by encapsulated and pyrogenic bacteria, such as *Haemophilus influenzae*, *Streptococcus pneumoniae*, and *Staphylococcus aureus*. In children with FASD, both types of recurrent infections have been reported, suggesting that both T and B cell-mediated immunities are compromised [16].

Subsequent studies confirmed an increased vulnerability to recurrent serious otitis media and recurrent upper respiratory infections, both of which have significant implications for hearing loss and, consequently, speech and language problems, as well as learning disabilities in children with FAS [17]. *In utero* alcohol exposure has also been shown to have an adverse impact on numerous parameters of immune function, including decreased eosinophil and neutrophil cell counts and leukocyte responses to mitogens, resulting in an increased incidence of hypogammaglobulinemia [16]. Interestingly, a study by Ammann and colleagues [18] described a number of characteristics of patients with the DiGeorge syndrome (a congenital immune deficiency syndrome) that were paralleled in children with FAS. Children with both syndromes show deficits in immune function, including altered T cell function. Immune status was also assessed in subjects who were part of a longitudinal cohort of 320 mother–child pairs [19]. Alcohol-exposed teens did not appear to have an increased susceptibility to infectious diseases compared to controls. However, adolescents showing dysmorphic features appeared to have increased rates of asthma (threefold), allergic rhinitis, and persistent skin rashes and/or eczema. As well, absolute lymphocyte counts, including counts of T cells (both T helper and T cytotoxic/suppressor) and NK cells, but not B cells, were increased in alcohol-exposed adolescents, regardless of dysmorphia or gender. It was suggested that this increase might have resulted from the acute stress of neuropsychological testing that had occurred during the morning, just prior to the blood draw. We will come back to this issue of stress, neuroendocrine activity, and immune function in FASD in Sects. 10.12 and 10.13 below.

A more recent study by Gauthier and colleagues [20] lends additional support to the early findings of alterations in immune competence following *in utero* alcohol exposure in the human population. This study addressed whether maternal alcohol use would increase the risk of sepsis in very low-birth-weight, alcohol-exposed newborns. It has been well established that premature newborns are at a higher risk of infection in early life, and despite improvements in neonatal intensive care, the rate of infant death due to sepsis remains high [21]. In addition, research in animal models suggests that lung development may be impaired following *in utero* alcohol exposure (see Sect. 10.7), and as a result, alcohol-exposed neonates may be at a higher risk for the development of infection during early life. This was substantiated by the findings of Gauthier and colleagues [20], which indicated that very low-birth-weight newborns exposed to alcohol *in utero* have a 15-fold higher incidence of early-onset sepsis (defined as presence of microorganisms in blood cultures collected within the first 72 h post-birth) as compared to the matched control group [20]. It was hypothesized that this increased risk in the alcohol-exposed neonates is due not only to immune alterations in the neonate but also to alcohol-induced impairments in immune function in the mother [20]. Adult alcohol exposure has been shown to impair immune function at many levels including negatively impacting the activation and response of macrophages and impairing lymphocyte proliferation and antibody production [22]. As a result, the immune milieu may also be significantly altered in the alcohol-consuming pregnant female, which may impact development, including development of the immune system in the offspring (see Sect. 10.2).

Gauthier and colleagues [23] have also examined the risk of infection in newborns of alcohol-consuming and control mothers in one of the largest clinical studies of this type ($n = 872$). Due to the large sample size, timing and degree of alcohol consumption before and during pregnancy could be assessed and linked to neonatal infection risk. The level of alcohol intake was found to be an important factor in predicting neonatal infection risk, with increasing levels of alcohol consumption by the mother 3 months prior to conception or during the first, second, or third trimesters of pregnancy resulting in a significantly increased risk. In addition, binge drinking, defined as consumption of seven or more drinks per week during the 3 months prior to conception or the second or third trimesters of pregnancy, was found to increase the risk of neonatal infection by approximately fourfold. When controlling for many possible confounds including maternal smoking, low maternal income, and being small for gestational age, high levels of maternal drinking (binge drinking), specifically during the second trimester, were found to increase the risk of infection by approximately fourfold, compared to that in unexposed newborns. Thus, while maternal drinking is often associated with factors such as concomitant smoking, drug use, or low socioeconomic status, which could themselves affect health outcomes, this study points toward a direct link between maternal alcohol consumption and decreased immune competence in the neonate. Interestingly, the second trimester appears to be particularly sensitive to the immunoteratogenic effects of alcohol, perhaps as during this period many immunological milestones are achieved in the fetus. As such, alcohol exposure during the second trimester may permanently impact the fetal immune system, resulting in an increased risk of infection at birth and potentially extending to an increased infection risk in later life.

While studies to date on children with FASD have provided clear evidence of alterations in immune function, detailed information regarding the effects of alcohol on the developmental time course of the naïve immune system, including effects on T and B cell development, as well as susceptibility to infections and autoimmune disorders can be difficult to obtain in human studies. Human studies are also limited by the lack of information regarding the level, timing, and duration of alcohol consumption, as well as confounding factors such as concomitant exposure to maternal smoking, drugs of abuse, and stress, all of which have been shown to impact development of the immune system [24, 25]. Animal models, which allow for the precise control of genetic and environmental factors, are as important for research on the immune system as they are for research on other aspects of alcohol's adverse effects.

10.3.2 Studies in Rodent and Primate Models Demonstrate Impaired Development and Function of the Immune System

To date, a large proportion of the research examining the effects of maternal alcohol consumption on immune competence has been conducted in rodent models, in which there is extensive knowledge of physiology, pharmacokinetics, and behavior. Rodent models have been invaluable in addressing fundamental immunological questions and have served as an important tool in shaping our understanding of autoimmune disorders, transplant rejection, and immune responses to infectious agents [26]. Additionally, many inbred, mutant, and congenic rodent strains have been established to address immune system function, and microsurgical techniques have been tailored for immunological probing [26].

Compared to rodent models of FASD, nonhuman primate models have distinct advantages due to the high degree of physiological, developmental, reproductive, and phylogenetic similarities to the human, making them especially important and relevant in assessing immunomodulatory agents [27, 28]. Nonhuman primates are also very similar to humans with regard to development and composition of the immune system. For example, important lymphoid tissues such as the spleen, tonsils, and thymus follow very similar developmental timelines in humans and nonhuman primates (reviewed in [2]). Additionally, humans and nonhuman primates are highly comparable in early postnatal development of the immune system, in that newborns of both species display lower levels of memory T cells and higher levels of naïve T cells, compared to adults [29]. This indicates that the immune systems of newborn humans and nonhuman primates are developmentally mature at birth but that antigenic contact has yet to occur in the protective environment of the womb in both species [29]. In addition, both susceptibility to infectious agents and the immune response are relatively conserved between humans and nonhuman primates. Species-specific functional differences in immune responses do exist, due to the fact that these responses are quite malleable and may evolve relatively rapidly alongside the rapidly evolving microbe environment, an environment that differs between species [30]. Nonetheless, due to the strong phylogenic similarities between humans and nonhuman primates, functional overlap of the immune systems is expected to be relatively high.

10.3.2.1 Rodent Models

Research using rodent models has confirmed the clinical findings of alterations in immune function following *in utero* alcohol exposure (reviewed in [31, 32]) and has increased our understanding of both the spectrum of effects in different organ systems and the mechanisms mediating the immunoteratogenic effects of alcohol. Interestingly, deficits in innate immunity have typically not been observed in animal studies. In contrast, marked deficits in adaptive immunity have been reported consistently in PAE offspring. PAE alters development of the thymus in both rats and mice. In a mouse model, maternal consumption of an ethanol-containing diet (25% ethanol-derived calories (EDC), BALs ~90 mg/dl) delayed development of the thymus [34], decreased thymus cell numbers, and diminished mitogen-induced cell proliferation in late-term fetuses [35]. Decreased thymus weight, size, and cell counts have also been observed at birth in a rat model (liquid diet exposure, 35% EDC, BALs ~79–127 mg/dl) [36]. These changes have been shown to persist throughout the preweaning period and into adolescence [35, 37–39], although data on mice suggested that recovery or catch-up in total thymocyte numbers may occur [37]. Decreased mitogen-induced proliferation of thymic cells was reported to persist until weaning in PAE male rats [36], but paradoxically, proliferation was shown to be greater than control levels in adolescence [40, 41]. The mechanisms underlying this increased thymocyte proliferation remain to be elucidated.

The adverse effects of alcohol on development of the thymus were confirmed by in vitro studies using organotypic cultures. Alcohol-treated organ cultures (alcohol concentrations of 0.2–0.8%) showed a concentration-dependent decrease in total cell numbers and percentages of immature fetal thymocytes (CD4+/CD8+) [42] likely due to increased apoptosis [34, 42–44]. Furthermore, thymic cell counts, total numbers of CD4+ and CD8+ cells, and numbers of immature CD8+/TCR+ and CD8+/CD45RC+ thymocytes were decreased postweaning and into early adolescence [39], suggesting that prenatal exposure to alcohol alters the later stages of thymocyte maturation.

The adverse effects of PAE last well into adulthood and appear to involve primarily alterations in cell-mediated rather than humoral immunity. PAE animals, typically exposed via maternal consumption of ethanol-containing liquid diets (~35–36% EDC), have decreased numbers of Thy1.2+, CD4+, CD8+, and IgG+ splenocytes [38, 45, 46]. In addition, data from both rodent and primate studies indicate that splenic lymphocytes taken from PAE males from adolescence through young adulthood show decreased proliferative responses to mitogens [36, 47–53], although in some studies, it was shown that the proliferative response may normalize by 8 months of age [51, 54]. Furthermore, deficits not only in the response of freshly isolated splenic T cells but also in the response of T lymphoblasts (obtained following treatment with concanavalin A [Con A]) to IL-2 or further Con A stimulation have been observed in PAE rats [49, 52, 54, 55]. In contrast, the changes in mitogen-induced proliferation of thymocytes reported in PAE animals appear to normalize in young adulthood [40].

Increased susceptibility to infections following PAE has also been reported in rodent models, in parallel with the human data, and provides further evidence of alterations in immune competence. For example, work by McGill and colleagues [56] found that mice exposed to alcohol during gestation and lactation (10–12% ethanol in drinking water) showed enhanced disease severity as well as increased and sustained pulmonary viral titers following influenza virus infection. Similarly, in a study of *Macaca nemestrina*, 4 of 18 (22%) animals exposed to once weekly oral doses of alcohol died or were euthanized after infectious disease or failure to thrive during the first year of life, whereas none of the controls died [33].

As noted, humoral immunity appears to be less affected by fetal alcohol exposure than cellular immunity. For example, in both rats [51] and nonhuman primates [33], the serum immunoglobulin response to immunization was shown to be unaffected by prenatal exposure to alcohol. On the other hand, abnormal development of B cell lineages in mouse bone marrow, spleen, and liver has been reported. Decreased numbers of splenic B cells and decreased B cell proliferative response to LPS (liquid diet, 25% EDC) [57], as well as delayed B cell maturation [58], have been observed. In addition, numbers of both immature and mature B cells in spleen and bone marrow were found to be decreased at birth, although most recovered to normal levels by 3–5 weeks after birth [59, 60].

10.3.2.2 Primate Models

Studies using nonhuman primates, including pigtailed macaques [61–63], vervet monkeys [64], and *M. nemestrina* [65], have demonstrated significant teratogenic effects of prenatal exposure to alcohol on numerous systems and behaviors (ethanol doses ranging from 0.3 to 4.1 g/kg maternal weight; BALs ~24–550 mg/dl). In general, alcohol-exposed offspring showed neuroanatomical, neurological, developmental, and facial anomalies similar to those seen in human FAS, with severity increasing as dose of alcohol increased [61–65]. Numerous cognitive and behavioral abnormalities were also observed, even in the absence of physical or facial anomalies, which increased with increasing dose of exposure, and were greater with earlier than with later gestational exposure [61]. At higher doses, facial features characteristic of FAS (1.8 g/kg) and microcephaly (4.1 g/kg) were observed [66].

In contrast to these important studies, few researchers have utilized nonhuman primate models to assess immune status following *in utero* alcohol exposure. A study by Grossmann and colleagues [33] was the first to examine immune function in the nonhuman primate *M. nemestrina*. Interestingly, the initial focus of the study was not immunological, but the focus shifted following death of a large proportion (4 of 18) of their alcohol-exposed animals from infectious disease. This high mortality rate from an otherwise treatable infection suggested profound alterations in immune competence in this population and prompted a review of immune status [33]. Exposure to alcohol (mean maternal BALs ~220 mg/dl) throughout gestation was shown to decrease T cell proliferative responses to tetanus toxin and result in lower tetanus toxoid titers after both initial vaccination and a booster, suggesting a deficit in the development of immunological memory. This finding clearly has important implications for the long-term health of the animals. In contrast, however, there were no significant effects of *in utero* alcohol exposure on total numbers of white blood cells, leukocyte subsets or monocyte phagocytic activity compared to that in control subjects.

10.4 Immunity at the Fetal–Placental Interface

Alterations in immunity at the fetal–placental interface could contribute to the increased incidence of spontaneous abortions and premature births that occur with maternal alcohol consumption. Under normal condition, immunity at the fetal–placental interface is biased toward humoral immunity, and cellular immunity is suppressed to prevent fetal rejection. In addition, cell-mediated immunity is skewed toward Th2 responses and production of cytokines such as IL-3, IL-4, IL-5, and IL-13, rather than toward Th1 responses and production of cytokines such as IL-2 and IFN-γ [67]. Th1/Th2 responses refer to the functionally polarized responses by CD4+ T helper (Th)- and CD8+ T cytotoxic (Tc)-cell subsets that depend on the cytokines they produce. Th1 cytokines are involved primarily in cell-mediated immunity and tend to produce proinflammatory responses, whereas Th2 cytokines are involved primarily in humoral immunity and tend to be more anti-inflammatory and to counteract Th1-mediated responses. The Th1/Th2 concept suggests that modulation of the relative contribution of Th1- or Th2-type cytokines regulates the balance between protection and immunopathology, as well as the development and/or the severity of some immunologic disorders.

It has been suggested that progesterone and CD4+ regulatory T cells are key factors enabling the fetus to evade immune rejection by the pregnant female and thus allowing the female to maintain pregnancy. Interestingly, progesterone appears to suppress cytotoxic activity of lymphocytes from pregnant women via interaction with progesterone receptors [68]. As well, CD4+/CD25+ regulatory T cells, which are suppressive in nature, may act at the interface between the fetus and the pregnant female to suppress allogeneic responses directed against the fetus [68–70]. *In utero* exposure to alcohol could alter the balance between regulatory T cells and T effector cells and, thus, contribute to the increased incidence of spontaneous abortions and premature births. This possibility is supported by observations of elevated cord-blood IgE concentrations in alcohol-exposed infants, indicating increased activity of Th2-type responses [71].

10.5 Lactational Transfer of Immunity

Human and animal milk contains cytokines and immunoglobulins that can pass to the newborn and likely play a role in augmenting immune defenses of neonates who are born with relatively immature immune systems [72, 73]. Alcohol consumption during pregnancy and/or lactation is known to alter maternal immune function. Of particular relevance, maternal alcohol consumption can result in alterations in the structure of the mammary gland and/or immune cells and soluble immune factors in the milk. A series of studies, using liquid diets with 36% EDC, have reported numerous alterations in the maternal mammary gland, including lower mammary gland weight and length, and histological changes such as a decreased percentage of alveolar epithelium, an increased percentage of connective tissue and alterations in the density of the mammary epithelium [74, 75]. Similarly, Vilaro and colleagues [76] reported that alcohol consumption decreased both absolute and relative mammary gland weight and mammary tissue protein content and reduced daily milk production. Immune components of the milk are also altered by alcohol consumption. Changes include altered numbers and distribution of T and B lymphocytes, primarily within the connective tissue compartment of the lactating mammary gland [77], increased IL-2 production by milk cells stimulated with Con A [78], decreased milk NK cell levels, and reduced specific IgG antibody levels following challenge with the intestinal parasite *Trichinella spiralis* [79].

Alterations in immune function of offspring nursed by alcohol-consuming dams have been reported. In a mouse model, with alcohol exposure (20% EDC, BALs ~250–290 mg/dl) during pregnancy and lactation or lactation only, offspring showed reduced numbers of splenic lymphocytes overall and, particularly, Thy 1.2+, CD4+, CD8+, and IgG+ lymphocyte subsets [38]. Deficits in lactational transfer of immunity have also been observed in the rat model utilizing *T. spiralis* challenge. Under normal conditions, T cell-mediated immunity to *T. spiralis* can be transferred from dam to neonate during lactation [251]. However, PAE neonates exhibited increased intestinal worm counts in response to *T. spiralis*, suggesting a decreased capacity to mount an immune response to the pathogen [79–81]. Abnormalities in the offspring involved depressed T and B cell-mediated responses such as lower serum IL-2 and tumor necrosis factor (TNF) contents and lower IgM and IgG antibody titers. Alterations in immune cells and soluble immune components in milk, as described above, could have direct effects on postnatal development and function of the offspring immune system and play a role in the depressed lactational immune transfer observed [79, 82]. Moreover, the detrimental effects of alcohol appear to increase across generations. That is, second-generation PAE pups mothered by first-generation PAE adults who themselves consumed alcohol during pregnancy showed significantly reduced proliferative responses to both *T. spiralis* antigen and stimulation with Con A, lower titers of serum antibodies, and lower percentages of T cells and cytotoxic T cells, compared to the first-generation PAE and pair-fed groups [83]. Of note, treating ethanol-consuming dams with an immunostimulatory drug, levamisole, during pregnancy and lactation was shown to reverse some of the deleterious effects of lactational transfer of immunity to suckling rat pups [84].

Rodent models such as these, in which alcohol exposure is extended to include the lactation period, are important in that they address one of the key limitations of rodent models in general, the fact that rodents are born developmentally immature, at a stage of development equivalent approximately to the end of the second trimester, compared to humans. The third trimester in the human is a sensitive period characterized by key development events including the development of immunocompetence [2]. The third trimester equivalent in the rodent is approximately the first 10 days postnatal. The finding that alcohol exposure during lactation (through maternal milk) results in long-term deficits in cellular immunity, in parallel with the human situation, supports the finding of high sensitivity to alcohol's immunoteratogenic effects during the third trimester of human gestation. These data have important implications for understanding possible deficits in immune function that might occur in children of women who consume alcohol in the late gestation period.

10.6 PAE and Increased Susceptibility to Tumors

One index of immune function in children following *in utero* alcohol exposure is the incidence of various forms of cancer. Studies have reported, for example, an increased incidence of malignancies, particularly those of embryonic origin, including neuroblastoma, ganglioneuroblastoma, medulloblastoma, and embryonal rhabdomyosarcoma in children ranging from birth to 12 years of age [53, 85], although a large case–control study found no evidence of an association between maternal smoking or drinking alcohol and risk of childhood germ cell tumors [86]. One case report found an association between the development of an adrenal cortical carcinoma and a history of FAS [87], and alcohol consumption has been identified as a prenatal risk factor for testicular cancer [88]. With only a few exceptions [89, 90], studies have reported an association between maternal alcohol consumption and an increased risk for or incidence of infant or childhood leukemia. A comprehensive review and meta-analysis of case–control studies found a link between alcohol consumption during pregnancy and increased risk of childhood leukemia, particularly acute myeloid leukemia (AML) [91]. Associations between maternal drinking and both acute lymphoblastic leukemia (ALL) [92] and acute nonlymphocytic leukemia [93] have also been reported. As well, positive dose–response associations between maternal alcohol consumption and both AML [92, 94] and ALL [94] have been observed. Interestingly, data suggest that a possible interaction between PAE and polymorphisms of carcinogen-metabolizing genes may influence the risk for ALL [95].

Increased susceptibility to malignancies has been demonstrated in animal models of FASD as well [53]. For example, susceptibility to an estrogen-induced prolactin-secreting tumor was increased in adult males prenatally exposed to alcohol (liquid diet, 35% EDC), as assessed by greater anterior pituitary weights and serum prolactin levels compared to those in control males [96]. Interestingly, whereas there were no effects of PAE on sensitivity to nitrosomethylbenzylamine (NMBA)-induced esophageal tumors, PAE males showed a marked decrease in thymus to body weight ratio as well as adrenal gland hyperplasia compared to controls, suggesting altered immune function [96]. Research by Hilakivi-Clarke and colleagues [75] investigated the effects of maternal alcohol consumption [liquid diet, low (7% EDC) and moderate (15% EDC) alcohol concentrations] on pregnancy estradiol levels and risk of 7,12-dimethylbenz(a)anthracene (DMBA)-induced mammary tumors in female offspring. Somewhat surprisingly, they found that estradiol levels were significantly higher in dams in the 7% but not the 15% EDC diet condition compared to control dams. Female offspring from both alcohol groups had a significantly greater incidence of tumors than controls, and females from the 15% EDC diet condition had greater tumor multiplicity (number of palpable tumors/animal) and higher mammary gland estrogen receptor-α (ER-α) expression than control offspring. While the role of estradiol in the increased risk for mammary tumors in PAE females requires further investigation, these data suggest that PAE may increase breast cancer risk. Recent findings by Polanco et al. [97] examining susceptibility to *N*-nitroso-*N*-methylurea (NMU)-induced tumors in PAE offspring support and extend these findings. Consistent with Hilakivi-Clarke et al. [75], these investigators found higher tumor multiplicity and decreased tumor latency in PAE females (dams consumed liquid ethanol diet, 35% EDC). In addition, PAE female offspring had more ER-α negative tumors and higher circulating estradiol and insulin-like growth factor (IGF)-1 levels compared to control females. As well, IGF-binding protein-5 (IGFBP5) mRNA and protein were decreased in tumors of PAE compared to control females. These findings suggest a more malignant tumor phenotype in PAE compared to control females.

Together, these data demonstrating a relationship between PAE and carcinogenesis have important implication for understanding health risks of individuals prenatally exposed to alcohol. The mechanisms underlying the increased cancer risk remain to be determined. Estradiol and components of the IGF system may be involved. In addition, compromised immune status of fetal alcohol-exposed individuals may play a role [98]. A depressed immune system may impair the body's ability to destroy cancer cells. Alternatively, defective neuroimmune feedback mechanisms may fail to control the immune response to challenge, leading to uncontrolled lymphocyte proliferation and increased risk for lymphomas and other hematopoietic cancers. Fetal alcohol-induced alterations in hormone levels or in neuroendocrine–neuroimmune interactions could also play a role in tumorigenesis. Furthermore, it may be that alcohol itself does not act as a carcinogen but rather acts as a cocarcinogen with a variety of chemical or hormonal agents and that susceptibility to cancer is increased in the context of a depressed immune system. Future work in this area is important for elucidating mechanisms underlying the increased risk for cancers in individuals prenatally exposed to alcohol.

10.7 Effects of PAE on the Pulmonary Immune System

Recent data suggest that maternal alcohol use increases the risk of extreme premature delivery by 35-fold, with many of these infants born with underdeveloped lungs [99]. Extensive investigation into the functioning of the vulnerable pulmonary system following *in utero* alcohol, particularly in the context of prematurity, has been conducted. The respiratory tract, being the conduit for vital oxygen to enter the lungs, is constantly challenged with foreign antigens, which are drawn into the system along with oxygen. As a result, proper functioning of the respiratory system requires both pro- and anti-inflammatory responses, both of which must be tightly controlled in order for the primary function of the respiratory system to be maintained.

Given the high rate of premature births in alcohol-exposed infants, it is not surprising that the incidence of sepsis and infection is higher. However, even when one controls for gestational age, alcohol-exposed newborns remain at a higher risk for infection compared to unexposed newborns [20, 23]. Recently there has been extensive investigation into the immune function of the developing lung, with the hypothesis that *in utero* alcohol exposure may alter the immune status of the lung, potentially resulting in an increased risk of early-life infection. Late gestational alcohol exposure has been shown to impact the immune status of the fetal ovine lung including decreasing surfactant levels and altering extracellular matrix composition, which may confer an increased risk of infection after birth [100]. Of note, at 9 weeks of age, many of the deficits associated with *in utero* alcohol exposure in the ovine lung appear to have normalized, suggesting that the lung may be capable of a certain degree of repair [101]. Yet the finding of increased infection risk in early life still warrants further investigation, as early life has been shown to be a critical period during which infection may permanently impact adult immune responses (see Sect. 10.14).

Investigation into the mechanism(s) underlying the altered pulmonary immune competence with *in utero* alcohol exposure has focused on the alveolar macrophage. The alveolar macrophage is the resident inflammatory cell of the lung. Given its connection to the outside world, the lung acts as a portal of entry for infection. Alveolar macrophages provide the first line of defense against foreign and infectious particles in the lung, by initiating and regulating the inflammatory cascade and participating in phagocytosis of infectious particles. Timely clearance of infectious agents by alveolar macrophages is important for decreasing the risk of serious infection. Alcohol-induced disruption of macrophage activity or function could underlie at least some of the immune deficits observed following *in utero* alcohol exposure. Of note, newborns are particularly vulnerable to infection, as neonatal alveolar macrophage numbers are depressed at birth and the existing alveolar macrophages show decreased phagocytic abilities and deficient chemotaxis [102]. As a result, any alcohol-associated impairments in alveolar macrophage function have the potential to significantly impact health during the vulnerable early postnatal period. Indeed, alcohol exposure is known to have a number of adverse effects on macrophage function, which include decreases in both the percentage of cells that phagocytose bacteria and total phagocytosis, increases in expression of early apoptotic markers, and increases in the levels of the fatty acid oxidation product, malonyldialdehyde (MDA), which is indicative of chronic oxidative stress [100, 103, 104]. Overall, these data suggest that *in utero* alcohol exposure results in alveolar macrophage dysfunction, limiting their ability to survive and adequately clear an infection, which would pose a significant health risk to the newborn.

Ping and colleagues [104] propose that decreased availability of the antioxidant glutathione may underlie, at least in part, the increase in oxidative stress in the newborn lung and that this may contribute to alveolar macrophage dysfunction. Glutathione is an essential antioxidant in the lung and is required for optimal functioning of alveolar macrophages [105]. *In utero* alcohol exposure is known to alter glutathione levels in other tissues, including the fetal brain and liver, resulting in increased oxidative stress [106, 107]. It is possible that glutathione levels may also be low in the lung of alcohol-exposed neonates, at a time when glutathione levels are expected to be high [108], and that this may underlie some of the alcohol-associated immune deficits in the lung. This was explored by Ping and colleagues [104] through administration of *S*-adenosyl-methionine (SAM), a key methyl donor in the methionine–homocysteine cycle, to the drinking water of alcohol-exposed and control animals, as SAM has been shown to restore glutathione levels following alcohol exposure [109]. SAM administration was found to significantly ameliorate many of the alcohol-induced deficits including increasing the level of alveolar macrophage phagocytosis per cell, decreasing expression of the early marker of apoptosis, and decreasing MDA, indicative of decreased oxidative stress, all to control levels [104]. The mechanism(s) underlying these SAM-mediated improvements remains to be determined. However, evidence suggests that the methionine–homocysteine cycle is disrupted with PAE and supplementation may be on approach for reversing some of the alcohol-induced deficits (see Sect. 10.16.1 for further discussion). In addition, antioxidant administration (e.g., glutathione) has been shown to restore growth, decrease neuronal apoptosis, and improve hepatic function [109–111] in animal models of PAE.

In a unique parallel to the clinical studies, which found a higher incidence of infection and sepsis in newborns exposed to alcohol *in utero* [20, 23], Gauthier and colleagues [112] examined the response to immune challenge using an *in vivo* model of group B *Streptococcus* pneumonia in a guinea pig model. *In utero* alcohol exposure was found to increase the amount of group B *Streptococcus* detected in the lung, increase systemic sepsis (increased *group B Streptococcus* levels in the blood), and impair phagocytosis of group B *Streptococcus*. Overall, and in support of the clinical findings, this suggests that *in utero* alcohol exposure significantly impairs the lung's ability to defend against bacterial infection in the newborn. Moreover, when SAM is administered to the alcohol-consuming dam, the newborn's response to immune challenge is greatly improved, lending further support to the hypothesis that alcohol exposure disrupts key components of the methionine–homocysteine cycle (see Sect. 10.16.1). Interestingly, Gauthier and colleagues [112] also showed that inhaled glutathione therapy, administered 6 h prior to group B *Streptococcus* challenge, significantly improved health outcome in the pup. Glutathione levels have been supplemented in a similar manner in premature human infants, as low glutathione levels are associated with later chronic lung disease [113]. These results point toward important potential clinical implications. Maternal supplementation with glutathione may be one eventual application, although this would require early recognition of alcohol consumption during pregnancy, which is not generally feasible. More hope, however, stems from the potential for administration of inhaled glutathione in preterm infants as well as in infants with confirmed alcohol exposure to potentially reduce the incidence of lung infection and adverse health outcomes in these newborns.

10.8 Paternal Alcohol Exposure and Offspring Immune Function

A relatively understudied area is the effects of paternal alcohol exposure on offspring immune function. Reduced birth weight and length, reduced litter size, and increased malformations have been reported in offspring sired by alcohol-consuming males [3, 114, 115], although fetal vulnerability to depressed body weight was shown to have a maternal genetic contribution as well [116]. In addition, data suggest that paternal alcohol exposure, either in combination with maternal alcohol exposure [117] or alone [118], may have adverse effects on offspring function, including immune function. It was shown, for example, that offspring sired by alcohol-consuming males (17.5–35% EDC) exhibit decreased locomotor activity, an effect opposite to that seen with maternal alcohol exposure; show poorer adaptation to stress as measured in the forced swim task; and show increased susceptibility to infection [119]. Utilizing *Pseudomonas aeruginosa*, a Gram negative opportunistic bacterium, and applying it topically to corneas that had been scarified, an increased severity of ocular infection and an increased incidence of perforated corneas were observed in both mice and rats in adulthood [117, 118]. Interestingly, in mice (liquid ethanol diets, 5–25% EDC), the increased susceptibility to infection occurred despite the presence of high titers of antibody specific to pseudomonas [118]. Further research is warranted on the role of paternal alcohol consumption in adverse developmental outcomes of their offspring.

10.9 Sexual Dimorphism in the Immune Response

Sex differences in responsiveness to immune challenges provide an additional level of complexity in understanding the data regarding the adverse effects of PAE. Sexual dimorphism of the immune system is well known [120, 121]. Both humoral and cell-mediated immune responses are more active in adult females than in males [120, 122]. For example, the thymus is larger in female than in male mice, and castration of young males leads to feminization (i.e., increased weights) of the immune organs. In response to immunization, women develop higher antibody titers than males, and they show a higher rate of transplant rejection. Females also represent the majority of patients affected with autoimmune disorders, ranging from 65 to 75% of patients with rheumatoid arthritis to 85% of patients with Hashimoto's thyroiditis and Grave's disease [121]. In humans, circulating levels of IgM are significantly elevated in girls compared to boys as early as 6 years of age, and juvenile rheumatoid arthritis can appear before 5 years of age. Gonadal hormones clearly play an important role in immune sexual dimorphism both in adulthood and during development. It appears that both the type and concentration of sex hormones within the microenvironment play a key role in lymphocyte maturation [120]. Both the immune organs and lymphocytes have receptors for the sex steroid hormones, linking the endocrine and immune systems [120, 121]. In addition, data suggest that other hormone systems and complex hormonal interactions can affect developing lymphocytes and regulate adult effector cells. Estrogen appears to be particularly important in the development of sexual dimorphism in the immune system, both through direct effects on immune cells and through modulation of other hormone systems, including adrenal glucocorticoids, thyroid hormones, growth hormone, and prolactin. These hormonal systems interact with each other and with the immune system to influence sexually dimorphic immune responses. Thymic hormones and cytokines generated by activated lymphocytes may also play a role in sexually dimorphic immune responses [120].

Sexual dimorphism related to fetal alcohol-associated alterations in behavior [123, 124], HPA function [124–126], neuroimmune interactions [127, 128], and susceptibility to inflammatory disorders [129] have been characterized (see further discussion below). Despite recognition that these sex differences exist, a large proportion of studies using rodent and nonhuman primate models of FASD are conducted only in males or with both sexes but too small a sample size to achieve the statistical power required to detect sexual dimorphism. Lending further support to the importance of considering sex differences are findings of sexual dimorphisms in the immune responses following PAE [49, 127, 128, 130, 131], many of which may also be impacted by fetal alcohol-induced, sex-specific HPA differences.

10.10 Mechanisms Underlying Alcohol Effects on the Developing Immune System

10.10.1 Direct Effects of Alcohol

Alcohol has a number of direct effects on the fetal immune system. Alcohol can directly disrupt the development of the fetal thymus. The thymus has a highly specified microenvironment provided by epithelial and mesenchymal cells. This environment is critical in attracting immature lymphoid precursors and enabling them to be selected and differentiated into mature T cells. Alcohol exposure during the period of thymus development inhibits the ontogeny of the thymic epithelium and disrupts the microenvironment necessary for T cell maturation, which in turn leads to impaired T cell-mediated immunity [13]. These data are interesting in light of the shared clinical characteristics of FAS and DiGeorge syndrome [18]. The latter is a congenital immune deficiency syndrome involving mainly T cells and caused by an abnormality in the development of several components of the thymus [132].

10.10.2 Indirect Effects of Alcohol

Alcohol can also alter immune function through numerous indirect effects.

10.10.2.1 Alcohol Effects on HPA Mediators of Thymic Function

Selective effects of alcohol on thymic corticotropin-releasing hormone (CRH) and pro-opiomelanocortin (POMC) gene expression in male fetuses have been reported. Significant increases in thymic CRH and decreases in thymic POMC gene expression were observed on G19 [133]. These changes appear to be unrelated to corticosterone (CORT) concentrations in the fetus or pregnant female but are possibly induced by the fetal testosterone surge. Similarly, the development of glucocorticoid receptor (GR) sites on thymocytes of PAE animals during the first 2 months of life was shown to differ from that in control animals [40]. This indicates a role for the glucocorticoid hormones and thus a possible role for prenatal or early-life stress in the differential thymic development described in PAE compared to control animals.

10.10.2.2 Altered IL-2/IL-2 Receptor Interactions

Altered IL-2/IL-2 receptor interactions may play a role in the development of altered immune function in PAE animals [49, 134, 135] (these studies utilized liquid diets, 35–36% EDC). For example, although PAE animals have normal levels of IL-2 production, IL-2 receptor expression and distribution, calcium influx into T cells, and binding of IL-2 to its receptor, they show a markedly reduced proliferative response to mitogens. In addition, the internalization and/or utilization of IL-2 by lymphoblasts is reduced, and the half-time for dissociation of IL-2 from its receptor is increased in T cells from PAE animals [135]. Therefore, impaired intracellular signaling events, mediated by altered IL-2/IL-2R interactions, may underlie, at least partly, the immune deficits observed in PAE animals.

10.10.2.3 Altered Neurotransmitter Function

Altered neurotransmitter regulation of immune function is another factor that could play a role in altered immune function of PAE animals. Prenatal exposure to alcohol (liquid diets, 36% EDC) was found to decrease concentrations of norepinephrine (NE) and β-adrenoreceptors in the lymphoid organs and diminish synaptic transmission in the spleen and thymus, but not the heart [55]. Altered noradrenergic synaptic transmission, including a higher rate of NE turnover leading to reduced NE concentration and reduced β-adrenoceptor density, could affect immune capacity, in terms of NE-mediated IL-2 secretion and cytotoxic T cell responses.

10.10.2.4 Altered Neuroendocrine–Neuroimmune Networks

Prenatal exposure to alcohol has been shown to have adverse effects on neuroendocrine function, which could in turn alter the bidirectional communication between neuroendocrine and neuroimmune networks. The nervous system, endocrine system, and immune system exist within a complex regulatory network of connections through nerve pathways, hormonal cascades, and cellular interactions [136–138] (Fig. 10.2). The CNS can regulate both the endocrine and immune systems directly, through autonomic innervation of lymphoid organs and tissues and endocrine glands, and indirectly, through release of neurotransmitters. In turn, the nervous system can "read" information from both the endocrine and immune systems. Cytokines produced by immune cells and hormones produced by endocrine glands may be similar or even identical in structure and function. Cytokines not only play an important role in regulating immune responses but also feed back to the CNS to affect neuronal function. Cytokines can also influence or stimulate hormone release from the hypothalamus and pituitary and endocrine glands and in themselves may have neuroendocrine effects. Similarly, hormones produced by the endocrine glands and pituitary not only feed back to the CNS to influence neural and endocrine function but also have immunoregulatory functions. Lymphocytes express receptors for hormones and neurotransmitters, including CRH, ACTH, glucocorticoids, norepinephrine, and epinephrine. The glucocorticoid hormones can exert profound influences on T cell function through their interaction with GRs on T cells, which modulate trafficking, homing, proliferation, activation, and apoptosis [139]. Of note, other endocrine systems besides the HPA axis are involved in mutual regulation of immune function. Thyroid hormones and estradiol are known to have activational effects on the immune response, whereas progesterone, testosterone, and dehydroepiandrosterone (DHEA) reciprocally inhibit immune responses. In addition, the autonomic nervous system can exert regulatory actions on the immune system. Noradrenergic neurotransmitters can target most immune cells including thymocytes, T lymphocytes, macrophages, and plasma cells, resulting in selective suppression of Th1-mediated inflammation and cellular immunity, which, in turn, will favor humoral immunity as well as protecting the organism from the adverse effects of proinflammatory cytokines. Synergistic effects of catecholamines and glucocorticoids have also been described [136]. Indeed, the HPA response is

Fig. 10.2 Bidirectional communication between the nervous, endocrine, and immune systems. (**a**) Neuroendocrine–neuroimmune circuitry under normal conditions. (**b**) Neuroendocrine–neuroimmune circuitry following prenatal alcohol exposure. *RH* releasing hormones act on cells of the anterior pituitary to stimulate synthesis and secretion of appropriate tropic hormones; corticotropin-releasing hormone (*CRH*) stimulates release of adrenocorticotropic hormone (*ACTH*); thyrotropin-releasing hormone (*TRH*) stimulates release of thyroid-stimulating hormone (*TSH*); gonadotropin-releasing hormones (*GnRH*) stimulate release of luteinizing hormone (*LH*) and follicle-stimulating hormone (*FSH*). The pituitary tropic hormones then stimulate peripheral endocrine glands to release their hormones: *CORT* (corticosterone/cortisol), thyroid hormones, triiodothyronine/tetraiodothyronine (*T3*, *T4*), and gonadal hormones (estradiol, E_2, testosterone, T)

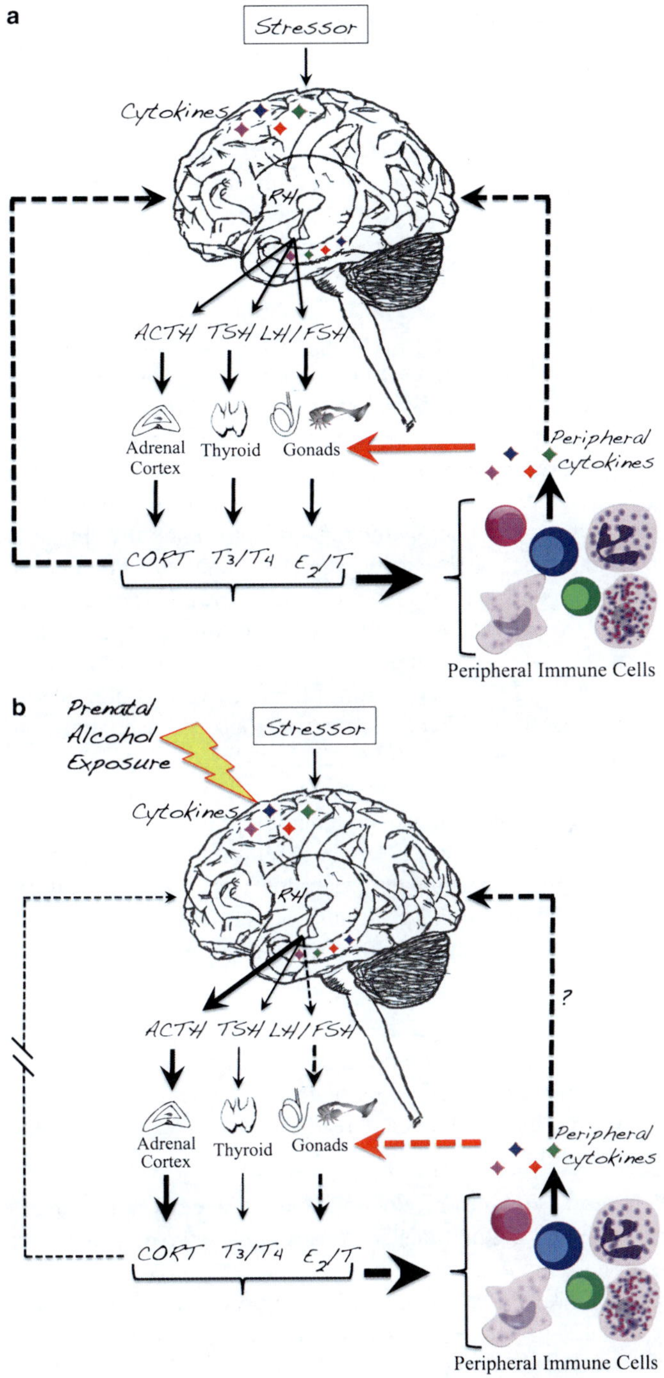
a
Stressor
Cytokines
RH
ACTH TSH LH/FSH
Adrenal Cortex
Thyroid
Gonads
Peripheral cytokines
CORT T3/T4 E2/T
Peripheral Immune Cells
b
Prenatal Alcohol Exposure
Stressor
Cytokines
RH
ACTH TSH LH/FSH
Adrenal Cortex
Thyroid
Gonads
?
Peripheral cytokines
CORT T3/T4 E2/T
Peripheral Immune Cells

coordinated with that of the autonomic nervous system to enable the organism to respond to stress. The autonomic response is extremely rapid and provides an immediate response to a stressor through initiation of the sympathetic "fight or flight" response (increased heart rate and blood pressure, pupil dilation, bronchodilation, glucose release). The HPA response is slower, and the secretion of the glucocorticoid hormones initiates numerous metabolic and behavioral effects that mediate effective coping with a stressor in the longer term. The HPA and autonomic systems appear to be regulated by similar neurotransmitters (e.g., acetylcholine, serotonin, norepinephrine, GABA). In addition, there is reciprocal stimulation of HPA and autonomic activity by CRH and norepinephrine and reciprocal actions of the glucocorticoids and catecholamines. The glucocorticoids are thought to feed back to restrain activity of both systems. Further, the activity and sensitivity of both systems are modulated by stress and circadian influences [140].

Reciprocal expression of receptors for hormones, neurotransmitters, and neuropeptides and shared hormonal/peptide ligands or products by immune, endocrine, and neural cells underlie the bidirectional communication that allows these systems to "speak a common biochemical language" [137] and thus influence each other to maintain homeostasis. It has been suggested that the immune system acts as a "peripheral receptor organ" able to transmit information to the brain about responses to external or internal antigenic stimuli [136]. Similarly, Blalock and Smith [137] suggest that through the sharing of ligands and receptors, the immune system can serve as a "sixth sense" to detect things the body cannot otherwise hear, see, smell, taste, or touch and to signal and mobilize the body to respond to pathogens, tumors, allergens, and other immune challenges.

Fig 10.2 illustrates the complex bidirectional communication between the nervous system, the endocrine system and the immune system. As shown, reciprocal expression of receptors for hormones, neurotransmitters, and neuropeptides and shared hormonal/peptide ligands or products by immune, endocrine, and neural cells underlie the bidirectional communication that occurs.

Figure 2a shows neuroendocrine–neuroimmune circuitry under normal conditions.

- The CNS can regulate the endocrine and immune systems directly, through autonomic innervation of lymphoid organs and tissues as well as endocrine glands, and indirectly, through release of neurotransmitters (not shown).
- In turn, the nervous system can "read" information from both the endocrine and immune systems.
  - Peripheral cytokines produced by immune cells not only play an important role in regulating immune responses but also provide feedback to the CNS to affect neuronal function.
  - Furthermore, cytokines can influence or stimulate hormone release from the hypothalamus and pituitary and endocrine glands and in themselves may have neuroendocrine effects.
  - Similarly, hormones produced by the endocrine glands (glucocorticoids, thyroid hormones, and the sex hormones) and pituitary (ACTH, TSH, LH, FSH)

not only provide feedback to the CNS to influence neural and endocrine function but also have immunoregulatory functions. Indeed, hormones produced by endocrine glands and cytokines produced by immune cells may be similar or even identical in structure and function.

- The sex hormones as well as many other hormones play a role in the sexually dimorphic nature of immune system function.
- There is evidence for local production of cytokines by microglia in the CNS in response to challenge, including stress.

Together, cytokine expression and activation of receptors in key brain areas such as the cortex, hippocampus, and hypothalamus have the potential to modulate the secretion of releasing hormones (RH) of the endocrine system and thus affect endocrine and immune responses to stress. Finally, mutual regulatory interactions between the autonomic nervous system and the HPA axis (not shown) play an important role in the organism's ability to respond to stressors, and at the same time, the autonomic nervous system can exert important regulatory actions on the immune system (not shown).

Figure 2b shows neuroendocrine–neuroimmune circuitry following prenatal alcohol exposure.

PAE impacts neuroendocrine and neuroimmune function and the interactions between these networks.

- Alterations in hormonal activity under basal conditions and in response to stressors:
  - Increased HPA activity, decreased thyroid function, and decreased levels of gonadal hormones. Altered hormone levels will differentially alter immune function.
  - However, PAE may increase facilitatory effects of estradiol (E_2) and decrease inhibitory effects of testosterone (T) on HPA activity (dotted arrow), further altering immune function.
  - Impaired CORT negative feedback (broken arrow) will further increase HPA activity and responsiveness to stressors.
- Alterations in peripheral cytokine production (in response to challenge) may alter how cytokine signals are transmitted (e.g., elaborated) to the brain (? in figure).
- Alterations in cytokine levels will alter feedback to HPA, thyroid, and gonadal systems (Red arrow).

PAE can link into this neuroimmune–neuro endocrine circuit to exert its teratogenic effects in numerous ways. Alcohol may (1) have direct effects on the ontogeny of the immune and endocrine systems and on CNS development and function; (2) act indirectly through effects on neuroendocrine function; and (3) disrupt the intimate bidirectional link between the neuroendocrine and neuro immune systems.

While several endocrine systems, including the adrenal, thyroid, and gonadal systems, interact with and modulate immune system function, the immune–HPA axis circuit is of particular interest for our work. The HPA axis mediates the neuroendocrine response to stressors, both systemic stressors that threaten homeostasis and/or survival and perceived threats or psychogenic stressors [141]. Inputs to the HPA axis provided by stressors and the endogenous circadian rhythm [142] act through central neural pathways to the paraventricular nuclei (PVN) of the hypothalamus, where CRH is synthesized. CRH (potentiated by arginine vasopressin [AVP]) [143, 144] stimulates the release of adrenocorticotropin (ACTH) from the anterior pituitary, which in turn stimulates synthesis and secretion of glucocorticoids from the adrenal cortex. The glucocorticoids, cortisol in humans and corticosterone in most rodents, have numerous metabolic and physiological effects and provide negative feedback to inhibit HPA activity at the level of the pituitary, PVN, hippocampus, prefrontal cortex, and other brain areas [142–144]. Glucocorticoids, acting on their receptors (mineralocorticoid receptors, MR; glucocorticoid receptors, GR), initiate metabolic and physiological responses that facilitate response to and coping with the stressor and, ultimately, dampen stress-activated defense reactions, including immune responses, to prevent them from overshooting and themselves causing harm [145, 146]. In the short term, the metabolic and physiological changes induced by the glucocorticoids promote survival (increased gluconeogenesis and blood pressure, suppressed immune and reproductive function). However, prolonged exposure to glucocorticoids can result in metabolic, cognitive, and immune dysfunction [147]. Thus, it is important that the HPA axis be tightly controlled through efficient feedback and efficient termination of the stress response; the ability to turn off the stress response is as important as the ability to respond initially [145]. The HPA axis acts in conjunction with the locus coeruleus–noradrenergic sympathetic adrenal medullary system (hereafter referred to as the LC-NE system) to maintain homeostasis. The LC-NE system is involved in the "fight or flight" response and enables the organism to react rapidly through the secretion of NE from sympathetic nerve terminals and epinephrine from the adrenal medulla. In contrast, through its various physiological and metabolic effects, the HPA axis acts over a longer time frame and helps orchestrate the response and adaptation of the body to the stressor.

10.11 Alcohol Consumption Alters HPA Activity of the Pregnant Female

Data from rodent models indicate that gestational alcohol consumption increases adrenal weights, basal corticosterone concentrations, and the corticosterone stress response in the pregnant dam [148]. This occurs as early as gestation day 11, persists throughout gestation, may increase as gestation progresses, and occurs even with low–moderate concentrations of alcohol (BALs 42–150 mg/dl) in the diet [148, 149]. The stimulatory effects of alcohol on basal hormone concentrations and the corticoid response to stressors can also extend through parturition, even when alcohol administration is discontinued prior to parturition [150]. Thus, regular consumption of high doses of alcohol during pregnancy not only raises the set point of HPA function in the pregnant female by increasing both basal and stress concentrations of CORT, but it also results in HPA hyperresponsiveness to stressors.

Because the pregnant female and fetus constitute an interrelated functional unit, alcohol-induced alterations in HPA activity in the pregnant female have implications for fetal HPA development. Corticosterone crosses the placenta, at least to some extent [151], resulting in suppression of endogenous fetal HPA activity. At the same time, alcohol can cross the placenta and directly activate the fetal HPA axis.

10.12 Effects of Maternal Alcohol Consumption on Offspring HPA Activity

The complex interaction of the opposing direct and indirect effects of alcohol is apparent in studies using rat models (liquid diets, ~36% EDC), with offspring showing differences from controls in HPA activity over the course of development. Fetuses exposed to alcohol have lower corticosterone concentrations than control fetuses on gestation day 19 [133] but elevated plasma corticosterone and β-endorphin levels at birth [150, 152–154]. Interestingly, throughout the preweaning period, PAE offspring exhibit blunted HPA and β-endorphin responses to a wide range of stressors including ether, novelty, saline injection, and cold stress [150, 154, 155]. In addition, PAE appears to alter the development of expression of CRH and POMC mRNAs in a sexually dimorphic manner, delaying and exaggerating the rise in CRH expression in female (but not male) pups and suppressing POMC mRNA concentrations in male (but not female) pups throughout the preweaning period [156]. Sexually dimorphic effects of PAE on these two important glucocorticoid-regulated genes may contribute to both the immediate and the long-term effects of prenatal alcohol on stress responsiveness of the offspring.

Interestingly, the blunted hormone responsiveness of PAE pups during preweaning life is a transient phenomenon. It appears that maternal alcohol consumption, in the long term, programs the fetal HPA axis such that HPA tone is increased throughout life. From weaning age on, although basal hormone levels are typically normal, PAE offspring are hyperresponsive to stressors and to drugs such as alcohol and morphine, showing increased HPA activation and/or delayed or deficient recovery following stress [157–160]. In addition, central HPA regulation appears to be altered under both basal and stress conditions [31, 124, 161]. Findings in both nonhuman primates and human infants are consistent with those in the rodent model. In rhesus monkeys, prenatal exposure to moderate amounts of alcohol induced higher plasma ACTH and marginally higher plasma cortisol responses to the stress of maternal separation [162]. Similarly, Jacobson and colleagues [163] found that heavy drinking during pregnancy and at conception is associated with higher post-stress (blood draw) cortisol concentrations in 13-month-old infants. Basal cortisol was also shown to be higher in 2-month-old infants exposed *in utero* to alcohol or cigarettes [164]. Moreover, Haley and colleagues [165] found that alcohol consumption during the early postconception period was related to increases in cortisol reactivity, elevated heart rate, and negative affect in infants in response to a modified "still-face" procedure, with differential effects in boys and girls. It was suggested that the effects of PAE on infant stress systems could underlie problems in cognitive and social–emotional functioning that are common among individuals exposed prenatally to alcohol.

10.12.1 Sexual Dimorphism of the HPA Response in PAE Offspring

Sexually dimorphic effects of PAE on HPA activity may extend into adulthood. Beyond the normal sexual dimorphism of HPA activity, differences in the response of PAE male and female offspring compared to their control counterparts are often observed and may vary depending on the nature of the stressor, and the time course and hormonal endpoint measured [130, 166–168]. For example, PAE males and females (liquid diets, 35–36% EDC) both exhibit increased hormone responses to stressors ranging from repeated restraint to footshock to immune challenges [130, 157, 159, 160, 169], stress-induced increases in expression of mRNAs for immediate early genes and CRH [160], and deficits in habituation to repeated restraint [170]. In contrast, in response to prolonged restraint or cold stress, HPA hyperactivity may be seen primarily in PAE males [167, 169], whereas in response to acute restraint or acute alcohol or morphine challenge, greater corticosterone and ACTH increases occur primarily in PAE females [157, 166, 171].

10.12.2 Role of Maternal Corticosterone in Offspring HPA Responsiveness

Whether the increased plasma corticosterone levels induced in the maternal female by alcohol consumption during pregnancy play a role in mediating HPA hyperresponsiveness in PAE offspring is as yet unresolved. Interestingly, maternal adrenalectomy (ADX) was found to dramatically increase the corticosterone response to restraint stress in offspring of pair-fed mothers but to have no effect on the corticosterone response to stress in PAE offspring [172]. Similarly, corticosterone treatment of ADX dams did not mimic the effect of PAE on HPA activity of offspring [173]. Based on these data, one could conclude that the increased corticosterone concentrations in pregnant females are not the primary mediator of offspring hyperresponsiveness. In contrast, ADX of the pregnant female was shown to reverse the effects of PAE on pituitary POMC mRNA concentrations in PAE offspring [174]. In addition, a compensatory increase in fetal CORT that was seen following maternal ADX of control females was attenuated in fetuses of ethanol-consuming dams, suggesting a direct effect of alcohol on fetal HPA activity [175]. Tritt and colleagues reported that removal of the adrenal cortex, but not the medulla, in the pregnant female prevents the growth-retarding effects of PAE and may reverse the delay in postpartum weight gain in PAE offspring [176]. Moreover, whereas the febrile response to IL-1β is normally attenuated in PAE offspring, maternal ADX completely eliminated the effect of PAE on this response [177]. These findings suggest that at least some of the adverse effects of alcohol may be mediated fully or partially through the adrenal gland and, possibly, through maternal–fetal HPA interactions. Further studies are needed to resolve the role of corticosteroids derived from the pregnant female on the altered stress responsiveness of their offspring. It is likely, however, that the adverse consequences of alcohol consumption during pregnancy on fetal development and programming of fetal HPA activity arise from a complex interaction between direct and indirect effects of alcohol on both the mother and the fetus.

10.13 PAE Alters the Effects of Stress on Immune Function

Chronic or repeated stress in adulthood provides a challenge to the immune system that differentially affects PAE and control animals. Data suggest that exposure to stressors may unmask specific deficits in the immune system that are not observed under nonstressed conditions. Furthermore, consistent with the findings that PAE may alter HPA responses to stress in a sexually dimorphic manner, sexually dimorphic effects of PAE on immune function may also be observed following stress.

In a rat model (liquid diet, 36% EDC), exposure to chronic intermittent stressors selectively downregulated the numbers of thymic and peripheral blood CD43+ and CD4+ T cells and marginally decreased the number of peripheral blood antigen-presenting cells, whereas CD43 antigen expression on peripheral blood T cells was selectively upregulated in PAE males but not females [178]. The possible role for estrogen as an immunoprotective hormone and testosterone as an immunosuppressive hormone in these sexually dimorphic immune responses remains to be determined.

One of the first demonstrations of fetal alcohol effects on immune function of females is the data from Halasz and colleagues [168] who found that the challenge of chronic alcohol exposure in adulthood selectively increased Con A-induced lymphocyte proliferation in PAE females, but not males. Consistent with these data, work from our laboratory [46] reported an interaction between prenatal alcohol and chronic cold stress. After 1 day of cold stress, PAE females but not males exhibited increased lymphocyte proliferation in response to pokeweed mitogen (a T cell-dependent B cell mitogen) and Con A challenge. In contrast, PAE males exposed to 1 or 3 days of cold stress had increased basal CORT concentrations compared to PAE males not exposed to cold. Altered interactions between T and B cells may underlie the alcohol-induced alterations in immune responsiveness observed in response to stressors.

10.14 PAE, Cytokines, and the HPA Axis

Cytokines secreted by immune cells, including IL-1, IL-2, IL-4, and IL-6, influence the function of hypothalamic neurosecretory and thermoregulatory neurons and pituitary cells [179–182], resulting in activation of the HPA axis and inducing "sickness behavior" [183, 184]. IL-1, IL-6, and TNF-α are also produced in the hypothalamus by microglia and macrophages [185, 186] and, thus, can directly influence neuroendocrine function. For example, IL-1 stimulates the release of CRH and AVP from the hypothalamus [187, 188], and IL-1, IL-6, and TNF-α stimulate ACTH secretion from the anterior pituitary [189–191]. During prolonged stress, cytokines exert effects at the level of the pituitary and adrenal glands [136]. In turn, the glucocorticoids play a major role in the stress-induced suppression of immune/inflammatory reactions.

PAE has been shown to alter neuroendocrine and behavioral responses to cytokines. For example, the LPS-induced febrile response was shown to be blunted in PAE males [192, 193]. It was suggested that a decreased hypothalamic IL-1β response to LPS administration, possibly due to an impaired release of endogenous pyrogens, could underlie the differential responsiveness observed [194–196]. Interestingly, both ADX and sham surgery in the pregnant female abrogated the effect of alcohol on the febrile response of female offspring to IL-1β, but only ADX had an effect on male offspring. This would suggest that maternal adrenal mediators play an important role in the blunted febrile response of PAE males and that non-adrenal mediators participate in modulation of thermoregulatory systems in PAE females [177, 193].

In parallel with the blunted hormonal responses to stressors observed in PAE animals during the preweaning period, preweaning PAE offspring exhibit reduced ACTH, β-EP, and TNF-α responses to the immune challenge of IL-1β and LPS. This reduction was found to persist into adolescence in PAE males, but not in females [197, 198]. An altered ability of IL-1 to stimulate secretion of ACTH and other peptides may underlie this reduced responsiveness [197]. Interestingly, ovariectomy prior to puberty eliminated the difference in the ACTH response between PAE and control females [130], suggesting that alcohol and female sex hormones may regulate HPA activity through a common pathway.

Following weaning and into adulthood, in parallel with the observed HPA hyper-responsiveness to stressors, hormonal responses to immune signals are also increased in PAE animals. Alcohol exposure *in utero* appears to induce a proinflammatory profile bias, suggested by the finding of greater ACTH and/or corticosterone responses to IL-1β and LPS in PAE compared to control offspring [130, 169, 199]. As well, embryos exposed *in vitro* to alcohol had greater tissue levels of TNF-α and IL-6 than control embryos [200]. PAE males also exhibited increased plasma concentrations of proinflammatory cytokines (IL-1β, TNF-α, IL-6) to LPS challenge following repeated restraint stress [31]. Importantly, however, corticosterone responses to LPS were comparable among groups. These data support and extend previous studies suggesting that although PAE animals may not differ in cytokine responses under basal or nonstressed conditions, they may be more vulnerable to the adverse effects of stress on immune function [31]. Furthermore, these data have important implications for understanding differential vulnerability to inflammatory disorders such as arthritis in PAE offspring, as discussed below (Sect. 10.15).

10.14.1 Effects of Alcohol Intake During Pregnancy on the Maternal Immune System

In order to understand fully how maternal alcohol intake can alter maternal and fetal cytokine profiles and immune status, it is helpful to understand the effects of adult alcohol intake in general. Alcohol use has been shown to impact the immune defense system, altering how the body responds to an infection (reviewed in [201]). These alterations can affect innate and adaptive immune responses, which in turn gives rise to an increased incidence of infections as well as alterations in antigen presentation postinfection [202]. Of note, the effects of alcohol on immune function appear to depend on the frequency of alcohol consumption. Acute alcohol administration has been shown to inhibit proinflammatory cell activation, an essential process in innate immune system activation, whereas chronic alcohol consumption is associated with increased proinflammatory cytokine activation (reviewed in [201]).

Chronic alcohol exposure also affects the maternal immune system, increasing the levels of proinflammatory cytokines including TNF-α, IL-1α, IL-1β, IL-6, IL-8, IL-2, and MCP-1 and giving rise to alcohol-induced physiological alterations and tissue damage (reviewed in [201]). To date, there is an absence of studies examining how an increased proinflammatory cytokine profile in the alcohol-consuming mother will affect the offspring. However, there is a growing body of research examining the effects of prenatal and/or early postnatal infection and the resulting increase in proinflammatory cytokine levels on the developing fetus [203–205]. As both PAE and early immune challenge are associated with increased proinflammatory cytokines, which can have programming effects on the developing fetus, the early inflammation literature has significant implications for understanding the immune effects of early alcohol exposure.

10.14.2 Early-Life Exposure to Maternally Derived Cytokines

It has been proposed that early-life exposure to maternally derived cytokines programs the developing fetus, affecting brain and immune system development and giving rise to altered cognitive and immune function in adulthood (reviewed in [3]). Increased levels of proinflammatory cytokines occur following naturally occurring maternal infections and have also been simulated in experimental models through direct injection of the mother with cytokines such as IL-6 [114] or maternal immune stimulation with the endotoxin LPS [203] or *Escherichia coli* [115]. An important question, however, remains to be answered—how the maternal cytokine signal is passed on to the fetus. The current literature is contradictory. Cytokines, being small proteins generally less than 50 kDa in size, have the potential to enter the amniotic fluid and affect the fetus [204]. However, while Gayle and colleagues [206] found that maternal inflammation results in elevated levels of IL-1β, TNF-α, and IL-6 in the amniotic fluid and the fetal brain, other studies suggest that there would be minimal transfer of IL-1β and TNF-α and only the potential for modest transfer of IL-6 between mother and fetus [207, 208]. However, a recent study involving radiolabeled IL-6 found that in rodents, this proinflammatory cytokine crosses to the fetus during mid-gestation (GD 11–13) but not late gestation (GD 17–19) [209]. Thus, it appears that cytokine passage to the fetus may depend on the cytokine in question as well as the gestational time point. Importantly, there may be additional, yet currently unknown, mechanisms in place that allow for the maternal cytokine signal to pass to the fetus. Late gestational (GD 18) maternal increases in cytokines, such as TNF-α, IL-1β, and IL-6, that are unable to cross the placenta stimulate a concomitant cytokine increase in the fetal brain [204]. As a result, it is hypothesized that there is a different mechanism in place in the fetus, resulting in local elaboration of the maternal cytokine signal [204]. While it may seem detrimental to increase cytokine levels in the fetal brain in response to maternal infection, this may be a required adaptive strategy, critical to fetal survival in the context of life-threatening infections but perhaps developmentally disadvantageous in the context of less severe infections.

10.14.2.1 Early-Life Exposure to Cytokines May Increase Vulnerability in Later Life

Regardless of the mode by which cytokines gain entry into the fetal system, many lines of evidence link early-life increases in cytokine levels with increased vulnerability to later-life infections and cytokine overproduction [205, 210, 211]. It is hypothesized that immune stimulation during sensitive periods of development may affect cytokine production, glial morphology, brain development, and neuroimmune responsiveness into adulthood (reviewed in [3]). In the case of *in utero* alcohol exposure, maternal increases in cytokines may be transmitted or elaborated on by the fetus, and alcohol may directly activate the fetal immune system resulting in increased proinflammatory cytokines during development. It is important to note that cytokines, in addition to their role in immune system activation, are also involved in normal brain development, including a role in neuronal plasticity, morphogenesis, growth, and differentiation [212, 213]. Cytokines play a role in mediating neuronal migration, synaptogenesis, synaptic pruning, and stem-cell fate during development [214, 215]. Abnormally high cytokine levels, however, may alter many of the important neuronal processes listed above, resulting in abnormal brain connectivity and ultimately increased immune system impairments and vulnerability to neurodevelopmental disorders [3, 205, 210].

10.14.2.2 Possible Role of Microglia in Neuroimmune Changes in Later Life

It is hypothesized that cytokine-induced alterations in microglia may underlie many of the neuroimmune system alterations and increased susceptibilities to cognitive disorders in adulthood (reviewed in [3, 204]). Microglia, resident macrophages of the CNS, are uniquely poised to retain immunological memory of early-life insults. Microglia arise by embryological day 14 in the rodent [216] and exist in an activated, amoeboid state, producing cytokines and contributing to neuronal development [217]. The precise role of microglia in the normal developing brain has yet to be determined; however, they likely play a role in processes such as neuronal pruning and phagocytosis of dying neurons and act as a source of nerve growth factor (NGF) and brain-derived neurotrophic factor (BDNF) [218–220]. Importantly, microglia are very long-lived cells [221]. Thus, early-life exposure to alcohol or immune insult may alter microglial morphology and activation, changes that may persist for long periods of time due to the long-lived nature of these immune cells amid a more ephemeral immune cell background. Importantly, while microglia arise early in development and exist in an increasingly activated state until postnatal day 6, by postnatal day 15, they transition into a quiescent, ramified state and remain this way throughout adulthood and only become activated with immune challenge [216]. The current hypothesis in the field is that early-life immune stimulation, during the developmental window in which microglia are activated, may prime these cells such that they are more easily stimulated in adulthood (reviewed in [3]). As a result, later-life immune challenge may unveil the neuroimmune system deficits, which were engrained in the system during early development. This "two-hit" hypothesis is particularly relevant in the context of PAE, where exposure to alcohol and maternally derived proinflammatory cytokines may have sensitized the fetal system, making it easier for it to be pushed out of the optimal range when further challenged.

10.15 PAE Alters the Development and Course of Adjuvant-Induced Arthritis

Recent studies in our laboratory have utilized an adjuvant-induced arthritis (AA) model to extend our understanding of the adverse effects of PAE on the functional status of the immune system [222]. AA is widely used as a model of human rheumatoid arthritis. Like rheumatoid arthritis, AA in the rodent is an inflammatory disease of the joints, especially the hind paws, shown to be mediated by CD4+ T cells. AA has been used to study disease pathogenesis, chronic pain, and stress and in the context of the involvement of the HPA axis in arthritis, neuroendocrine imbalance, and altered interactions between the neuroendocrine and neuro immune systems [223–226].

In view of the known effects of PAE on HPA and immune function and the interaction between the HPA axis and the immune system, we undertook a study to test the hypothesis that the autoimmune response and the severity of arthritis would be increased in PAE females compared to their control counterparts [222]. At approximately 55 days of age, females from prenatal alcohol, pair-fed, and *ad libitum*-fed control conditions were injected at the base of the tail with complete Freund's adjuvant. The incidence and clinical signs of arthritis, as well as immune markers and HPA hormone levels, were assessed over the course of inflammation. Our data demonstrate, for the first time, an adverse effect of PAE on the course and severity of adjuvant-induced arthritis in adulthood, indicating an important and clinically relevant long-term alteration in functional immune status. PAE females showed a delay in arthritis onset during the induction phase. However, during the resolution phase, PAE females showed a more prolonged course of disease and greater severity of inflammation compared to controls. In addition, PAE females showed a greater increase in basal ACTH levels than controls during the induction phase of arthritis. This increased HPA activation during induction could have played a role in the delayed onset of disease, either directly, by suppressing the inflammatory response, or indirectly, by suppressing the lymphocyte proliferative response to adjuvant. Consistent with this latter possibility and with previous work from our laboratory [49] and others [35, 48], we found an overall reduction in lymph node lymphocyte proliferative responses to Con A in PAE females. Interestingly, the differential elevation in basal HPA activity observed during the induction phase was not sustained over the course of testing. We saw graded corticosterone and ACTH responses at the peak of inflammation, with the greatest activation in adjuvant-injected animals that developed arthritis, lesser activation in adjuvant-injected animals that did not develop arthritis, and low levels of activity in saline-injected controls. During the resolution phase of disease, hormone levels did not differ among the three prenatal groups, and if anything, corticosterone and ACTH levels were somewhat lower in PAE compared to control animals with clinical signs of arthritis. It has been suggested [227] that low levels of glucocorticoid hormones in the context of high levels of inflammation may reflect a disconnect between the immune and endocrine systems. As noted above, we have evidence for such a disconnect in our previous work. That is, following exposure to repeated restraint stress, PAE animals had significantly increased and sustained elevations in plasma levels of IL-1β and TNF-α to LPS injection compared to controls but showed similar corticosterone responses to LPS [31]. These data suggest a possible disruption of the cytokine–HPA axis feedback circuit. It is possible that reduced HPA activity in PAE animals during the resolution phase, in the face of persisting inflammation, may be related to the prolonged course and severity of disease seen in PAE compared to control animals.

As noted above, later-life immune challenges may unveil neuroimmune system deficits, which were engrained in the system during early development. Our PAE model is a prime example of the "two-hit" hypothesis, where prenatal exposure to alcohol sensitizes the offspring neuroendocrine and neuroimmune systems, possibly through increased exposure to maternally derived proinflammatory cytokines, resulting in increased responsiveness to the later-life challenge of adjuvant injection.

10.16 PAE Reprograms Fetal HPA and Immune Function

One conceptual framework for understanding the long-term adverse effects of alcohol on neuroendocrine and neuroimmune function is that of fetal programming. Programming refers to the concept that environmental factors acting during sensitive periods of development can exert organizational effects on physiological/neurobiological systems, resulting in changes that can persist throughout life [228]. Data suggest that the HPA axis is particularly vulnerable to programming by prenatal and early-life events [228, 229]. The overall effect of early HPA programming is altered exposure to endogenous glucocorticoids throughout life [230, 231] which, in turn, can modify behavior, cognition, learning, memory, and emotion and predispose the individual to cardiovascular-, metabolic-, mental health-, and stress-related disorders and disorders of immune function, including the risk for rheumatoid arthritis [227, 232]. Although environmental factors play an important role in fetal programming, it is likely that perinatal environmental and genetic factors mutually influence each other in determining HPA activity and behavior later in life. Moreover, though the effects of programming are often long lasting, postnatal and later environmental events can modulate the effects of prenatal programming.

Fetal programming is generally thought to facilitate the organism's adaptation to the postnatal environment. However, programming can be detrimental if stimuli program systems to function outside their normal physiological range, leading to high allostatic load [233]. In this regard, because HPA dysregulation has been implicated in the pathogenesis of rheumatoid arthritis in humans and adjuvant-induced arthritis in animal models [226, 227], it has been suggested that HPA programming may be one potential mechanism through which early-life factors can predispose the individual to autoimmune diseases [200–203]. As noted, we and others have shown that alcohol exposure *in utero* programs the fetal HPA axis such that HPA tone is increased throughout life, resulting in increased HPA activation, delayed or deficient recovery [157, 160], and altered central HPA regulation [124, 161]. We postulate that imposition of the chronic inflammatory stress of adjuvant-induced arthritis on a system already sensitized by prenatal exposure to alcohol could underlie the increased autoimmune responses and the altered course and severity of disease that we observed in our model. Alcohol-induced disruptions of normal neuroendocrine–neuro immune interactions may provide an indirect route through which early-life experiences can have long-term effects on the immune system.

10.16.1 Possible Role of Epigenetic Mechanisms in Fetal Programming by Alcohol

Over the past decade, it has become apparent that epigenetic processes provide one mechanism for fetal programming of neurobiological systems [234]. Indeed, data suggest that investigation of possible epigenetic mechanisms as mediators of alcohol's adverse effects on the fetus provides a promising approach for understanding the complex phenotypes associated with FASD and the persistence of these characteristics into adulthood [235, 236].

Epigenetics refers to stable, but potentially reversible, alterations in a cell's genetic information that result from environmental influences. Epigenetic processes cause changes in gene expression without mutations or changes in the underlying DNA structure. Such changes persist through cell divisions for the remainder of the cell's life and are potentially heritable. Epigenetic mechanisms appear to function as mediators connecting the genome to environmental stimuli and exposures, such as *in utero* alcohol exposure [236].

In the cell nucleus, DNA is found spooled around histone proteins. Local changes in this packaging system can influence gene expression. Two major types of epigenetic modifications are methylation—addition of methyl groups to DNA building blocks (cytosine nucleotides) in a regulatory region of the gene called the promoter region—and the addition of various molecules or chemical groups to the histone proteins. Both DNA methylation and histone modifications are dynamic processes. Numerous enzymes can remove and/or replace these chemical modifications, which, in turn, results in increased or decreased expression of the gene. Activity and availability of these enzymes and chemical groups are influenced by environmental factors. Recent data support the possibility that adverse effects of alcohol exposure *in utero* may be mediated, at least partly, by epigenetic mechanisms [204, 205].

10.16.1.1 Alcohol Exposure During Preconception, Preimplantation, and Gastrulation Periods of Development

Some of the strongest evidence for epigenetic programming comes from data suggesting that both *preconception* and *preimplantation* alcohol exposure, when the embryo is not yet implanted in the uterus and thus not yet connected to the maternal system, can cause adverse effects.

Paternal alcohol consumption may be one route through which preconception effects of alcohol can occur. As noted (Sect. 10.8), paternal alcohol exposure has numerous adverse effects of offspring developmental outcome, including immune function. Of relevance to the present discussion, alterations in the expression of DNA methyltransferase 1 (DNMT1), a key enzyme catalyzing DNA methylation, in the sperm of alcohol-consuming paternal rats suggest a potential mechanism mediating paternal alcohol effects [207]. Interestingly, a correlation between chronic alcohol use and altered DNA methylation in sperm DNA from human volunteers has been reported [208].

Preconception effects of maternal alcohol consumption on birth weight and growth have been observed [237]. Moreover, Kaminen-Ahola and colleagues [210] found effects of preconception maternal alcohol consumption (peak BALs ~120 mg/dl) on DNA methylation of a promoter element for a gene called *Agouti viable yellow* (A^{vy}). DNA hypomethylation of a promoter element for A^{vy} is associated with constitutive expression of the *Agouti* gene, which is reflected phenotypically in a yellow coat color. Conversely, hypermethylation correlates with silencing of the *Agouti* gene and a brownish pseudo-agouti coat color, and intermediate methylation results in mottled coat color. Maternal alcohol consumption increased the probability of transcriptional silencing of A^{vy}, and the percentage of pseudo-agouti offspring more than doubled compared with that in the control group, consistent with a higher probability of hypermethylation and decreased expression of A^{vy}.

Several studies have examined the teratogenic effects of alcohol exposure during the preimplantation period (after the egg cell has been fertilized and before it implants in the uterus), which encompasses the first 4–6 days of mouse and rat development, corresponding roughly to the first 2 weeks of human pregnancy. Interestingly, a recent study by Haycock and Ramsay [238] found that the alleles inherited from the fathers were significantly less methylated in the placentas of alcohol-exposed animals compared with the placentas of control animals, suggesting effects on imprinting (expression of specific genes primarily from the alleles inherited from one rather than from both parents). Similarly, a recent study [212] reported strain-specific vulnerability to PAE via hippocampal parent-of-origin expression of deiodinase-III (*Dio3*), an imprinted thyroid hormone-inactivating gene. Moreover, alcohol effects on paternal and total expression of *Dio3* in the fetus and the adult male were paralleled by hippocampal-dependent behavioral alterations in male but not female offspring. Imprinting may be a novel mechanism underlying certain adverse effects of alcohol.

Global hypomethylation of fetal DNA, likely resulting from a direct inhibitory effect of alcohol on DNMT activity, was reported by Garro and colleagues [239] following alcohol exposure during *gastrulation* (onset of organ development, days 7–14 of gestation in the mouse and gestation weeks 3–8 in humans). Consistent with these findings, Liu and colleagues [214], utilizing whole mouse embryos in culture, observed not only delayed and reduced growth but also a more than tenfold increase in the number of genes with increased promoter DNA methylation on chromosomes 10 and X in alcohol-exposed (dose of ~400 mg/dl for 44 h beginning on gestation day 8.5) embryos with a neural tube defect compared with embryos without a neural tube defect. Importantly, alterations in DNA methylation were associated with significant changes in the expression of 84 genes. Consistent with these data, Zhou and colleagues [215] examined cultured neural stem cells (NSCs) exposed to alcohol in a "binge-like pattern" (6 h, 88 mM alcohol) and found retarded migration, neuronal formation, and growth processes, as well as altered methylation patterns of these cells. Alcohol exposure appeared to prevent the normal DNA methylation programming of key NSC genes, including genes related to neural development, neuronal receptors, and olfaction.

Consistent with the studies above, Guo and colleagues [240], using a rat model of FASD, demonstrated that moderate alcohol exposure (serum ethanol approximately 40–60 mM) during early neonatal life (third trimester equivalent) leads to an almost 50% reduction of CREB-binding protein expression in developing cerebellar neurons, as well as a reduction in histone acetylation. These findings suggest one possible mechanism underlying deficits in motor coordination found in children exposed to alcohol *in utero* and in animal models of FASD. Together, the studies discussed above suggest that the effects of alcohol exposure during critical developmental periods may be mediated by epigenetic mechanisms.

10.16.1.2 A Possible Role for Epigenetic Mechanisms in Intervention for FASD

Important research led by Dr. Jennifer Thomas and colleagues [217–219] has focused on the effects of *in utero* alcohol exposure on choline, a nutrient important in the methionine–homocysteine cycle. This cycle involves conversion of methionine to SAM with the subsequent generation of homocysteine and the regeneration of methionine through folate-dependent or folate-independent methylation of homocysteine. Of note, the folate-independent pathway involves the contribution of methyl groups from choline. Using well-established animal models involving either PAE or early postnatal exposure, these investigators demonstrated that choline supplementation during the period of alcohol exposure may reduce the severity of a number of fetal alcohol effects, including reductions in birth and brain weights, delays in eye opening and incisor eruption, and impairments in performance on learning and memory tasks. Given the importance of choline in the methionine–homocysteine cycle and the role of this cycle in generating methyl groups for reactions central to the creation of epigenetic marks, it is tempting to speculate that the protective effects of choline are at least in part mediated by epigenetic mechanisms.

10.16.2 Possible Role of Epigenetic Mechanisms in Immune Function and Inflammation

Accumulating evidence suggests that epigenetic mechanisms are likely involved in immune development and some immune system disorders [241]. T cell differentiation, activation, and memory have been shown to be under epigenetic control [242], and epigenetic mechanisms appear to play a role in both Th1 and Th2 differentiation and the development of regulatory T cells (Treg) and proinflammatory Th17 cells [243]. Recent data suggest, for example, that differences in immune responses of allergic and nonallergic children are evident at birth, suggesting that *in utero* environmental exposures can influence epigenetic programming of immune function. It has been suggested that a failure of epigenetic control may underlie an increased risk for allergic disease in infants, with greater immaturity of both Th1 and Treg function, and persistence of Th2 immune responses beyond gestation [243, 244]. Interestingly, dietary folate intake in pregnancy has been implicated in increased allergic risk of offspring through epigenetic mechanisms [245].

Epigenetic processes may play an important role in rheumatoid arthritis (RA). It has become clear that while genetic factors may predispose an individual to RA, the overall contribution of genetic factors is 50% or less, and thus, nongenetic factors, particularly epigenetic mechanisms, are likely important. The presence of hypomethylated cells in the synovial tissue of RA patients suggests that genomic hypomethylation has a role in the pathogenesis of RA. Support for this comes from the finding that mimicking hypomethylation by treating synovial fibroblasts with a DNMT1 inhibitor led to irreversible phenotypic changes such that normal synovial fibroblasts resembled activated rheumatoid arthritis synovial fibroblasts (RASFs) [225, 226]. Hyperacetylation may also be an underlying mechanism in RA. Histone deacetylase activity was found to be about twofold lower in tissue extracts from patients with RA compared to those with osteoarthritis [246, 247]. In addition, studies have suggested that altered expression and function of microRNAs (miRNA), small noncoding RNAs that act as posttranslational regulators, might also be involved in the pathogenesis of RA. Specifically, miR-155 and miR-146 were shown to be constitutively more highly expressed in RASFs than in synovial fibroblasts from patients with osteoarthritis and were upregulated by treating RASFs with TNF-α [247]. These miRNAs might play a role in the modulation of the destructive behavior of RASFs.

In summary, epigenetic processes appear to play a major role in the development of inflammatory and allergic disorders or diseases by directly influencing the expression of genes involved in immune function. Indirect effects on inflammatory diseases may occur through early-life programming of glucocorticoid sensitivity, either through environmental exposure or through adverse early-life events [248]. These findings have important implications for the development of drugs to target histone modifications as well as DNA methylating and demethylating enzymes [249].

10.16.3 Epigenetic Mechanisms May Influence Function Throughout Life

It is important to note that epigenetic mechanisms can act to influence the epigenome and gene expression not only during early developmental periods but throughout life. Although the organism is particularly vulnerable to environmental influences on the genome and on gene expression during periods of rapid growth and development, environmental events may influence the epigenome later in life as well. Thus, positive environmental factors, such as a healthy diet and lifestyle, can positively influence physiological and behavioral function over the life course and may possibly even rescue adverse effects of PAE or other early-life insults. Conversely, diet and other lifestyle factors may adversely influence the epigenome and possibly alter gene expression over time, as indicated by the discussion above on the effects of adult alcohol intake on the epigenome.

10.17 Summary and Conclusions

This chapter discusses the adverse effects of prenatal exposure to alcohol on offspring neuroendocrine and neuro immune function, with particular emphasis on the HPA axis, fetal programming, and interacting neuroendocrine–neuroimmune networks. Data are presented from both clinical studies and animal models to demonstrate the adverse effects of PAE on immune development of offspring. The adverse impact of alcohol at the fetal–placental interface and on lactational transfer of immunity has important clinical implications for understanding alcohol's adverse effects. The discussion of the effects of alcohol exposure *in utero* on the pulmonary immune system and on susceptibility to tumors provides unique examples that highlight specific deficits in immune function in alcohol-exposed offspring. A review of possible mechanisms underlying alcohol's effects on the developing immune system suggests that adverse effects of PAE on the interaction between neuroendocrine and neuroimmune systems play a major role. Maternal alcohol consumption alters HPA activity and regulation in the pregnant female and has long-term effects on her offspring. The fetal HPA axis is reprogrammed by alcohol exposure such that HPA tone is increased throughout life, with increased HPA activation and/or delayed or deficient recovery following stress, as well as altered HPA regulation under both basal and stress conditions, which may manifest in a sexually dimorphic manner. Altered neuroendocrine activity can impact immune regulation as well as the bidirectional communication among the nervous, endocrine, and immune systems. PAE can alter the effects of stress on the immune response as well as neuroendocrine and behavioral responses to cytokines. In addition, maternal alcohol consumption alters the cytokine profile during pregnancy. Increased levels of proinflammatory cytokines in the mother cause physiological changes, including changes in neuroendocrine function. Moreover, exposure to maternally derived cytokines plays a role in programming the developing fetus. While it is not clear how maternal cytokines gain entry into fetal systems, evidence links early-life exposure to increased cytokine levels with increased vulnerability to later-life infections and cytokine overproduction. Thus, fetal programming of HPA function is paralleled by fetal programming of immune function. It is likely that the complex interactions between direct and indirect effects of alcohol on both the mother and the fetus underlie the programming of neuroendocrine and neuroimmune systems and mediate the dysregulation of these systems in alcohol-exposed offspring. Epigenetic mechanisms may, at least in part, mediate the programming effects of alcohol on the fetus. Increasing evidence implicates epigenetic processes in mediating the adverse effects of alcohol exposure on numerous aspects of offspring development, including immune function and inflammation, and promising approaches to intervention have come from consideration of epigenetic mechanisms. Importantly, environmental events influence the epigenome not only during vulnerable periods of development but throughout the life course. This provides hope that the adverse effects of early-life insults such as PAE may be attenuated in the long term. An increased understanding of how PAE impacts the "common biochemical language" of the nervous, endocrine, and immune systems will guide us in the development of appropriate and targeted interventions and bring us closer to this goal.

Acknowledgements The research from our laboratory reported in this review is supported by grants NIH/NIAAA R37 AA007789 to JW and R21 AA AA016683 to JW (CoIs: S. Innis, A. Devlin); NIH/NIMH R24 MH081797 (Project 2) to JW (CoIs: M. Kobor, G.G. Meadows). TB is supported by a fellowship from the National Sciences and Engineering Research Council (NSERC CGS-D) and a University Graduate Fellowship. Versions of some sections of this chapter have appeared previously in published work, as cited.

References

1. Stratton K, Howe C, Battaglia F (1996) Fetal alcohol syndrome; diagnosis, epidemiology, prevention and treatment. National Academy Press, Washington
2. Holsapple MP, West LJ, Landreth KS (2003) Species comparison of anatomical and functional immune system development. Birth Defects Res B Dev Reprod Toxicol 68:321–334
3. Bilbo SD, Schwarz JM (2009) Early-life programming of later-life brain and behavior: a critical role for the immune system. Front Behav Neurosci 3:14
4. Gill TJ 3rd et al (1983) Transplacental immunization of the human fetus to tetanus by immunization of the mother. J Clin Invest 72:987–996
5. King CL et al (1998) B cell sensitization to helminthic infection develops in utero in humans. J Immunol 160:3578–3584
6. Migliaccio G et al (1986) Human embryonic hemopoiesis. Kinetics of progenitors and precursors underlying the yolk sac—liver transition. J Clin Invest 78:51–60
7. Holt PG, Jones CA (2000) The development of the immune system during pregnancy and early life. Allergy 55:688–697
8. Goldblatt D (1998) Immunisation and the maturation of infant immune responses. Dev Biol Stand 95:125–132
9. Peakman M, Buggins AG, Nicolaides KH, Layton DM, Vergani D (1992) Analysis of lymphocyte phenotypes in cord blood from early gestation fetuses. Clin Exp Immunol 90:345–350
10. West LJ (2002) Defining critical windows in the development of the human immune system. Hum Exp Toxicol 21:499–505
11. Kay HE, Doe J, Hockley A (1970) Response of human foetal thymocytes to phytohaemagglutinin (PHA). Immunology 18:393–396
12. Hofman FM, Danilovs J, Husmann L, Taylor CR (1984) Ontogeny of B cell markers in the human fetal liver. J Immunol 133:1197–1201
13. Bockman DE (1997) Development of the thymus. Microsc Res Tech 38:209–215
14. Berry SM et al (1992) Circulating lymphocyte subsets in second- and third-trimester fetuses: comparison with newborns and adults. Am J Obstet Gynecol 167:895–900
15. Zajac CS, Abel EL (1992) Animal models of prenatal alcohol exposure. Int J Epidemiol 21(Suppl 1):S24–S32
16. Johnson S, Knight R, Marmer DJ, Steele RW (1981) Immune deficiency in fetal alcohol syndrome. Pediatr Res 15:908–911
17. Church MW, Kaltenbach JA (1997) Hearing, speech, language, and vestibular disorders in the fetal alcohol syndrome: a literature review. Alcohol Clin Exp Res 21:495–512
18. Ammann AJ, Wara DW, Cowan MJ, Barrett DJ, Stiehm ER (1982) The DiGeorge syndrome and the fetal alcohol syndrome. Am J Dis Child 136:906–908
19. Oleson DR, Magee RM, Donahoe RM, Falek A, Coles CD (1998) Immunity and prenatal alcohol exposure. A pilot study in human adolescents. Adv Exp Med Biol 437:255–264

20. Gauthier TW, Manar MH, Brown LA (2004) Is maternal alcohol use a risk factor for early-onset sepsis in premature newborns? Alcohol 33:139–145
21. Lemons JA et al (2001) Very low birth weight outcomes of the National Institute of Child Health and Human Development Neonatal Research Network, January 1995 through December 1996. NICHD Neonatal Research Network. Pediatrics 107:E1
22. Jerrells TR, Eckardt MJ, Weinberg J (1990) Mechanisms of ethanol-induced immunosuppression. Prog Clin Biol Res 325:173–180
23. Gauthier TW, Drews-Botsch C, Falek A, Coles C, Brown LA (2005) Maternal alcohol abuse and neonatal infection. Alcohol Clin Exp Res 29:1035–1043
24. Abel EL, Hannigan JH (1995) Maternal risk factors in fetal alcohol syndrome: provocative and permissive influences. Neurotoxicol Teratol 17:445–462
25. Dietert RR, Piepenbrink MS (2008) The managed immune system: protecting the womb to delay the tomb. Hum Exp Toxicol 27:129–134
26. Suckow MA, Weisbroth SH, Franklin CL, NetLibrary Inc. (2006) The laboratory rat. In: American College of Laboratory Animal Medicine series xvi. Elsevier, Amsterdam, 912 p
27. Hendrickx AG, Makori N, Peterson P (2000) Nonhuman primates: their role in assessing developmental effects of immunomodulatory agents. Hum Exp Toxicol 19:219–225
28. Hendrickx AG, Makori N, Peterson P (2002) The nonhuman primate as a model of developmental immunotoxicity. Hum Exp Toxicol 21:537–542
29. Neubert RT, Webb JR, Neubert D (2002) Feasibility of human trials to assess developmental immunotoxicity, and some comparison with data on New World monkeys. Hum Exp Toxicol 21:543–567
30. Barreiro LB, Marioni JC, Blekhman R, Stephens M, Gilad Y (2010) Functional comparison of innate immune signaling pathways in primates. PLoS Genet 6:e1001249
31. Zhang X, Sliwowska JH, Weinberg J (2005) Prenatal alcohol exposure and fetal programming: effects on neuroendocrine and immune function. Exp Biol Med (Maywood) 230:376–388
32. Weinberg J (1994) Recent studies on the effects of fetal alcohol exposure on the endocrine and immune systems. Alcohol Alcohol Suppl 2:401–409
33. Grossmann A et al (1993) Immune function in offspring of nonhuman primates (*Macaca nemestrina*) exposed weekly to 1.8 g/kg ethanol during pregnancy: preliminary observations. Alcohol Clin Exp Res 17:822–827
34. Ewald SJ, Walden SM (1988) Flow cytometric and histological analysis of mouse thymus in fetal alcohol syndrome. J Leukoc Biol 44:434–440
35. Ewald SJ, Frost WW (1987) Effect of prenatal exposure to ethanol on development of the thymus. Thymus 9:211–215
36. Redei E, Clark WR, McGivern RF (1989) Alcohol exposure in utero results in diminished T-cell function and alterations in brain corticotropin-releasing factor and ACTH content. Alcohol Clin Exp Res 13:439–443
37. Ewald SJ (1989) T lymphocyte populations in fetal alcohol syndrome. Alcohol Clin Exp Res 13:485–489
38. Giberson PK, Blakley BR (1994) Effect of postnatal ethanol exposure on expression of differentiation antigens of murine splenic lymphocytes. Alcohol Clin Exp Res 18:21–28
39. Taylor AN, Tio DL, Chiappelli F (1999) Thymocyte development in male fetal alcohol-exposed rats. Alcohol Clin Exp Res 23:465–470
40. Chiappelli F, Tio D, Tritt SH, Pilati ML, Taylor AN (1992) Selective effects of fetal alcohol exposure on rat thymocyte development. Alcohol 9:481–487
41. Wong CM et al (1992) Prenatal exposure to alcohol enhances thymocyte mitogenic responses postnatally. Int J Immunopharmacol 14:303–309
42. Bray LA, Shao H, Ewald SJ (1993) Effect of ethanol on development of fetal mouse thymocytes in organ culture. Cell Immunol 151:12–23
43. Ewald SJ, Huang C, Bray L (1991) Effect of prenatal alcohol exposure on lymphocyte populations in mice. Adv Exp Med Biol 288:237–244
44. Ewald SJ, Shao H (1993) Ethanol increases apoptotic cell death of thymocytes in vitro. Alcohol Clin Exp Res 17:359–365

45. Ewald SJ, Huang C (1990) Lymphocyte populations and immune responses in mice prenatally exposed to ethanol. Prog Clin Biol Res 325:191–200
46. Giberson PK et al (1997) The effect of cold stress on lymphocyte proliferation in fetal ethanol-exposed rats. Alcohol Clin Exp Res 21:1440–1447
47. Monjan AA, Mandell W (1980) Fetal alcohol and immunity: depression of mitogen-induced lymphocyte blastogenesis. Neurobehav Toxicol 2:213–215
48. Norman DC, Chang MP, Castle SC, Van Zuylen JE, Taylor AN (1989) Diminished proliferative response of con A-blast cells to interleukin 2 in adult rats exposed to ethanol in utero. Alcohol Clin Exp Res 13:69–72
49. Weinberg J, Jerrells TR (1991) Suppression of immune responsiveness: sex differences in prenatal ethanol effects. Alcohol Clin Exp Res 15:525–531
50. Grossman CJ et al (1993) Sex differences and the effects of alcohol on immune response in male and female rats. Alcohol Clin Exp Res 17:832–840
51. Gottesfeld Z, Ullrich SE (1995) Prenatal alcohol exposure selectively suppresses cell-mediated but not humoral immune responsiveness. Int J Immunopharmacol 17:247–254
52. Jerrells TR, Weinberg J (1998) Influence of ethanol consumption on immune competence of adult animals exposed to ethanol in utero. Alcohol Clin Exp Res 22:391–400
53. Gottesfeld Z, Abel EL (1991) Maternal and paternal alcohol use: effects on the immune system of the offspring. Life Sci 48:1–8
54. Norman DC et al (1991) Changes with age in the proliferative response of splenic T cells from rats exposed to ethanol in utero. Alcohol Clin Exp Res 15:428–432
55. Gottesfeld Z, Christie R, Felten DL, LeGrue SJ (1990) Prenatal ethanol exposure alters immune capacity and noradrenergic synaptic transmission in lymphoid organs of the adult mouse. Neuroscience 35:185–194
56. McGill J et al (2009) Fetal exposure to ethanol has long-term effects on the severity of influenza virus infections. J Immunol 182:7803–7808
57. Wolcott RM, Jennings SR, Chervenak R (1995) In utero exposure to ethanol affects postnatal development of T- and B-lymphocytes, but not natural killer cells. Alcohol Clin Exp Res 19:170–176
58. Biber KL, Moscatello KM, Dempsey DC, Chervenak R, Wolcott RM (1998) Effects of in utero alcohol exposure on B-cell development in the murine fetal liver. Alcohol Clin Exp Res 22:1706–1712
59. Moscatello KM, Biber KL, Jennings SR, Chervenak R, Wolcott RM (1999) Effects of in utero alcohol exposure on B cell development in neonatal spleen and bone marrow. Cell Immunol 191:124–130
60. Robinson RS, Seelig LL Jr (2002) Effects of maternal ethanol consumption on hematopoietic cells in the rat fetal liver. Alcohol 28:151–156
61. Clarren SK, Bowden DM (1982) Fetal alcohol syndrome: a new primate model for binge drinking and its relevance to human ethanol teratogenesis. J Pediatr 101:819–824
62. Clarren SK, Astley SJ, Bowden DM (1988) Physical anomalies and developmental delays in nonhuman primate infants exposed to weekly doses of ethanol during gestation. Teratology 37:561–569
63. Clarren SK et al (1990) Neuroanatomic and neurochemical abnormalities in nonhuman primate infants exposed to weekly doses of ethanol during gestation. Alcohol Clin Exp Res 14:674–683
64. Burke MW, Palmour RM, Ervin FR, Ptito M (2009) Neuronal reduction in frontal cortex of primates after prenatal alcohol exposure. Neuroreport 20:13–17
65. Miller MW, Astley SJ, Clarren SK (1999) Number of axons in the corpus callosum of the mature *Macaca nemestrina*: increases caused by prenatal exposure to ethanol. J Comp Neurol 412:123–131
66. Sheller B, Clarren SK, Astley SJ, Sampson PD (1988) Morphometric analysis of *Macaca nemestrina* exposed to ethanol during gestation. Teratology 38:411–417
67. Lin H, Mosmann TR, Guilbert L, Tuntipopipat S, Wegmann TG (1993) Synthesis of T helper 2-type cytokines at the maternal-fetal interface. J Immunol 151:4562–4573

68. Szekeres-Bartho J, Philibert D, Chaouat G (1990) Progesterone suppression of pregnancy lymphocytes is not mediated by glucocorticoid effect. Am J Reprod Immunol 23: 42–43
69. Somerset DA, Zheng Y, Kilby MD, Sansom DM, Drayson MT (2004) Normal human pregnancy is associated with an elevation in the immune suppressive CD25+ CD4+ regulatory T-cell subset. Immunology 112:38–43
70. Aluvihare VR, Kallikourdis M, Betz AG (2004) Regulatory T cells mediate maternal tolerance to the fetus. Nat Immunol 5:266–271
71. Gonzalez-Quintela A, Vidal C, Gude F (2002) Alcohol-induced alterations in serum immunoglobulin e (IgE) levels in human subjects. Front Biosci 7:e234–e244
72. Juto P (1985) Human milk stimulates B cell function. Arch Dis Child 60:610–613
73. Zimecki M, Pierce-Cretel A, Spik G, Wieczorek Z (1987) Immunoregulatory properties of the proteins present in human milk. Arch Immunol Ther Exp (Warsz) 35:351–360
74. Steven WM, Bulloch B, Seelig LL Jr (1989) A morphometric study of the effects of ethanol consumption on lactating mammary glands of rats. Alcohol Clin Exp Res 13:209–212
75. Hilakivi-Clarke L et al (2004) In utero alcohol exposure increases mammary tumorigenesis in rats. Br J Cancer 90:2225–2231
76. Vilaro S, Vinas O, Remesar X, Herrera E (1987) Effects of chronic ethanol consumption on lactational performance in rat: mammary gland and milk composition and pups' growth and metabolism. Pharmacol Biochem Behav 27:333–339
77. Steven WM, Barron RA, Stewart GL, Seelig LL Jr (1991) The effects of maternal ethanol consumption on the distribution of leukocyte subsets in the lactating mammary gland of rats. Alcohol Alcohol 26:615–625
78. Na HR, Seelig LL Jr (1994) Effect of maternal ethanol consumption on in vitro tumor necrosis factor, interleukin-6 and interleukin-2 production by rat milk and blood leukocytes. Alcohol Clin Exp Res 18:398–402
79. Na HR, Chervenak R, Seelig LL Jr (1993) Effects of maternal alcohol consumption on milk and blood lymphocytes and IgG antibody levels from *Trichinella spiralis*-infected rats. Alcohol Clin Exp Res 17:532–538
80. Steven WM, Stewart GL, Seelig LL (1992) The effects of maternal ethanol consumption on lactational transfer of immunity to *Trichinella spiralis* in rats. Alcohol Clin Exp Res 16:884–890
81. Seelig LL Jr, Steven WM, Stewart GL (1996) Effects of maternal ethanol consumption on the subsequent development of immunity to *Trichinella spiralis* in rat neonates. Alcohol Clin Exp Res 20:514–522
82. Steven WM, Kumar SN, Stewart GL, Seelig LL Jr (1990) The effects of ethanol consumption on the expression of immunity to *Trichinella spiralis* in rats. Alcohol Clin Exp Res 14:87–91
83. Seelig LL Jr, Steven WM, Stewart GL (1999) Second generation effects of maternal ethanol consumption on immunity to *Trichinella spiralis* in female rats. Alcohol Alcohol 34:520–528
84. Steven WM, Stewart GL, Seelig LL Jr (1993) Effects of levamisole on ethanol-induced suppression of lactational immune transfer in rats. Alcohol Clin Exp Res 17:958–962
85. Becker H, Zaunschirm A, Muntean W, Domej W (1982) Fetal alcohol syndrome and malignant tumors. Wien Klin Wochenschr 94:364–365
86. Chen Z et al (2005) Risk of childhood germ cell tumors in association with parental smoking and drinking. Cancer 103:1064–1071
87. Hornstein L, Crowe C, Gruppo R (1977) Adrenal carcinoma in child with history of fetal alcohol syndrome. Lancet 2:1292–1293
88. Brown LM, Pottern LM, Hoover RN (1986) Prenatal and perinatal risk factors for testicular cancer. Cancer Res 46:4812–4816
89. Slater ME et al (2011) Maternal prenatal cigarette, alcohol and illicit drug use and risk of infant leukaemia: a report from the Children's Oncology Group. Paediatr Perinat Epidemiol 25:559–565
90. Rudant J et al (2008) Childhood hematopoietic malignancies and parental use of tobacco and alcohol: the ESCALE study (SFCE). Cancer Causes Control 19:1277–1290

91. Latino-Martel P et al (2010) Maternal alcohol consumption during pregnancy and risk of childhood leukemia: systematic review and meta-analysis. Cancer Epidemiol Biomarkers Prev 19:1238–1260
92. Shu XO et al (1996) Parental alcohol consumption, cigarette smoking, and risk of infant leukemia: a Childrens Cancer Group study. J Natl Cancer Inst 88:24–31
93. van Duijn CM, van Steensel-Moll HA, Coebergh JW, van Zanen GE (1994) Risk factors for childhood acute non-lymphocytic leukemia: an association with maternal alcohol consumption during pregnancy? Cancer Epidemiol Biomarkers Prev 3:457–460
94. MacArthur AC et al (2008) Risk of childhood leukemia associated with parental smoking and alcohol consumption prior to conception and during pregnancy: the cross-Canada childhood leukemia study. Cancer Causes Control 19:283–295
95. Infante-Rivard C, Krajinovic M, Labuda D, Sinnett D (2002) Childhood acute lymphoblastic leukemia associated with parental alcohol consumption and polymorphisms of carcinogen-metabolizing genes. Epidemiology 13:277–281
96. Gottesfeld Z, Trippe K, Wargovich MJ, Berkowitz AS (1992) Fetal alcohol exposure and adult tumorigenesis. Alcohol 9:465–471
97. Polanco TA, Crismale-Gann C, Reuhl KR, Sarkar DK, Cohick WS (2010) Fetal alcohol exposure increases mammary tumor susceptibility and alters tumor phenotype in rats. Alcohol Clin Exp Res 34:1879–1887
98. Penn I (1988) Tumors of the immunocompromised patient. Annu Rev Med 39:63–73
99. Sokol RJ et al (2007) Extreme prematurity: an alcohol-related birth effect. Alcohol Clin Exp Res 31:1031–1037
100. Sozo F et al (2009) Repeated ethanol exposure during late gestation alters the maturation and innate immune status of the ovine fetal lung. Am J Physiol Lung Cell Mol Physiol 296:L510–L518
101. Sozo F et al (2011) Effects of prenatal ethanol exposure on the lungs of postnatal lambs. Am J Physiol Lung Cell Mol Physiol 300:L139–L147
102. Bellanti JA, Zeligs BJ (1995) Developmental aspects of pulmonary defenses in children. Pediatr Pulmonol Suppl 11:79–80
103. Gauthier TW et al (2005) Fetal alcohol exposure impairs alveolar macrophage function via decreased glutathione availability. Pediatr Res 57:76–81
104. Ping XD, Harris FL, Brown LA, Gauthier TW (2007) In vivo dysfunction of the term alveolar macrophage after in utero ethanol exposure. Alcohol Clin Exp Res 31:308–316
105. Brown LA, Harris FL, Bechara R, Guidot DM (2001) Effect of chronic ethanol ingestion on alveolar type II cell: glutathione and inflammatory mediator-induced apoptosis. Alcohol Clin Exp Res 25:1078–1085
106. Amini SA, Dunstan RH, Dunkley PR, Murdoch RN (1996) Oxidative stress and the fetotoxicity of alcohol consumption during pregnancy. Free Radic Biol Med 21:357–365
107. Addolorato G et al (1997) Prenatal exposure to ethanol in rats: effects on liver energy level and antioxidant status in mothers, fetuses, and newborns. Alcohol 14:569–573
108. Jain A et al (1995) Glutathione metabolism in newborns: evidence for glutathione deficiency in plasma, bronchoalveolar lavage fluid, and lymphocytes in prematures. Pediatr Pulmonol 20:160–166
109. Seyoum G, Persaud TV (1994) In vitro effect of S-adenosyl methionine on ethanol embryopathy in the rat. Exp Toxicol Pathol 46:177–181
110. Devi BG, Henderson GI, Frosto TA, Schenker S (1993) Effect of ethanol on rat fetal hepatocytes: studies on cell replication, lipid peroxidation and glutathione. Hepatology 18:648–659
111. Heaton MB, Mitchell JJ, Paiva M (2000) Amelioration of ethanol-induced neurotoxicity in the neonatal rat central nervous system by antioxidant therapy. Alcohol Clin Exp Res 24: 512–518
112. Gauthier TW et al (2009) In utero ethanol exposure impairs defenses against experimental group B streptococcus in the term Guinea pig lung. Alcohol Clin Exp Res 33:300–306

113. Cooke RW, Drury JA (2005) Reduction of oxidative stress marker in lung fluid of preterm infants after administration of intra-tracheal liposomal glutathione. Biol Neonate 87:178–180
114. Samuelsson AM et al (2004) Prenatal exposure to interleukin-6 results in hypertension and increased hypothalamic-pituitary-adrenal axis activity in adult rats. Endocrinology 145: 4897–4911
115. Bilbo SD et al (2005) Neonatal infection-induced memory impairment after lipopolysaccharide in adulthood is prevented via caspase-1 inhibition. J Neurosci 25:8000–8009
116. Sittig LJ, Redei EE (2010) Paternal genetic contribution influences fetal vulnerability to maternal alcohol consumption in a rat model of fetal alcohol spectrum disorder. PLoS One 5:e10058
117. Hazlett LD, Barrett RP, Berk RS, Abel EL (1989) Maternal and paternal alcohol consumption increase offspring susceptibility to *Pseudomonas aeruginosa* ocular infection. Ophthalmic Res 21:381–387
118. Berk RS, Montgomery IN, Hazlett LD, Abel EL (1989) Paternal alcohol consumption: effects on ocular response and serum antibody response to *Pseudomonas aeruginosa* infection in offspring. Alcohol Clin Exp Res 13:795–798
119. Abel EL, Hazlett LS, Berk RS, Mutchnick MG (1990) Neuroimmunotoxic effects in offspring of paternal alcohol consumption. Prog Clin Biol Res 325:47–55
120. Grossman CJ (1990) Are there underlying immune-neuroendocrine interactions responsible for immunological sexual dimorphism? Prog Neuroendocrinol Immunol 3:75–82
121. Da Silva JA (1999) Sex hormones and glucocorticoids: interactions with the immune system. Ann N Y Acad Sci 876:102–117, discussion 117–108
122. Martin JT (2000) Sexual dimorphism in immune function: the role of prenatal exposure to androgens and estrogens. Eur J Pharmacol 405:251–261
123. McGivern RF, Clancy AN, Hill MA, Noble EP (1984) Prenatal alcohol exposure alters adult expression of sexually dimorphic behavior in the rat. Science 224:896–898
124. Weinberg J, Sliwowska JH, Lan N, Hellemans KG (2008) Prenatal alcohol exposure: foetal programming, the hypothalamic-pituitary-adrenal axis and sex differences in outcome. J Neuroendocrinol 20:470–488
125. Fonzo D, Mims RB, Nelson DH (1967) Estrogen influence on pituitary and adrenal function in the rat. Endocrinology 81:29–33
126. Williams TD, Carter DA, Lightman SL (1985) Sexual dimorphism in the posterior pituitary response to stress in the rat. Endocrinology 116:738–740
127. Spinedi E, Suescun MO, Hadid R, Daneva T, Gaillard RC (1992) Effects of gonadectomy and sex hormone therapy on the endotoxin-stimulated hypothalamo-pituitary-adrenal axis: evidence for a neuroendocrine-immunological sexual dimorphism. Endocrinology 131: 2430–2436
128. Spinedi E, Gaillard RC, Chisari A (2002) Sexual dimorphism of neuroendocrine-immune interactions. Front Horm Res 29:91–107
129. Van Eden W et al (2002) Balancing the immune system: Th1 and Th2. Ann Rheum Dis 61(Suppl 2):ii25–ii28
130. Lee S, Rivier C (1996) Gender differences in the effect of prenatal alcohol exposure on the hypothalamic-pituitary-adrenal axis response to immune signals. Psychoneuroendocrinology 21:145–155
131. Daneva T et al (1993) Transient sex-related changes in the mice hypothalamo-pituitary-adrenal (HPA) axis during the acute phase of the inflammatory process. Mediators Inflamm 2:123–127
132. Kirby ML, Bockman DE (1984) Neural crest and normal development: a new perspective. Anat Rec 209:1–6
133. Revskoy S, Halasz I, Redei E (1997) Corticotropin-releasing hormone and proopiomelanocortin gene expression is altered selectively in the male rat fetal thymus by maternal alcohol consumption. Endocrinology 138:389–396

134. Taylor AN et al (1993) Actions of alcohol on immunity and neoplasia in fetal alcohol exposed and adult rats. Alcohol Alcohol Suppl 2:69–74
135. Chang MP, Yamaguchi DT, Yeh M, Taylor AN, Norman DC (1994) Mechanism of the impaired T-cell proliferation in adult rats exposed to alcohol in utero. Int J Immunopharmacol 16:345–357
136. Besedovsky HO, Rey AD (2007) Physiology of psychoneuroimmunology: a personal view. Brain Behav Immun 21:34–44
137. Blalock JE, Smith EM (2007) Conceptual development of the immune system as a sixth sense. Brain Behav Immun 21:23–33
138. Besedovsky HO, del Rey A (1996) Immune-neuro-endocrine interactions: facts and hypotheses. Endocr Rev 17:64–102
139. Dhabhar FS, Miller AH, McEwen BS, Spencer RL (1996) Stress-induced changes in blood leukocyte distribution. Role of adrenal steroid hormones. J Immunol 157:1638–1644
140. Ulrich-Lai YM, Herman JP (2009) Neural regulation of endocrine and autonomic stress responses. Nat Rev Neurosci 10:397–409
141. Jankord R, Herman JP (2008) Limbic regulation of hypothalamo-pituitary-adrenocortical function during acute and chronic stress. Ann N Y Acad Sci 1148:64–73
142. Dallman MF et al (1987) Regulation of ACTH secretion: variations on a theme of B. Recent Prog Horm Res 43:113–173
143. Rivier CL, Plotsky PM (1986) Mediation by corticotropin releasing factor (CRF) of adenohypophysial hormone secretion. Annu Rev Physiol 48:475–494
144. Jones MT, Gillham B (1988) Factors involved in the regulation of adrenocorticotropic hormone/beta-lipotropic hormone. Physiol Rev 68:743–818
145. Sapolsky RM, Romero LM, Munck AU (2000) How do glucocorticoids influence stress responses? Integrating permissive, suppressive, stimulatory, and preparative actions. Endocr Rev 21:55–89
146. De Kloet ER, Vreugdenhil E, Oitzl MS, Joels M (1998) Brain corticosteroid receptor balance in health and disease. Endocr Rev 19:269–301
147. McEwen BS, Stellar E (1993) Stress and the individual. Mechanisms leading to disease. Arch Intern Med 153:2093–2101
148. Weinberg J, Bezio S (1987) Alcohol-induced changes in pituitary-adrenal activity during pregnancy. Alcohol Clin Exp Res 11:274–280
149. Weinberg J, Gallo PV (1982) Prenatal ethanol exposure: pituitary-adrenal activity in pregnant dams and offspring. Neurobehav Toxicol Teratol 4:515–520
150. Weinberg J (1989) Prenatal ethanol exposure alters adrenocortical development of offspring. Alcohol Clin Exp Res 13:73–83
151. Eguchi Y (1969) Interrelationships between the foetal and maternal hypophysial-adrenal axes in rats and mice. In: Bajusz E (ed) Physiology and pathology of adaptation mechanisms. Pergamon Press, New York, pp 3–27
152. Kakihana R, Butte JC, Moore JA (1980) Endocrine effects of maternal alcoholization: plasma and brain testosterone, dihydrotestosterone, estradiol, and corticosterone. Alcohol Clin Exp Res 4:57–61
153. Taylor AN, Branch BJ, Kokka N, Poland RE (1983) Neonatal and long-term neuroendocrine effects of fetal alcohol exposure. Monogr Neural Sci 9:140–152
154. Angelogianni P, Gianoulakis C (1989) Prenatal exposure to ethanol alters the ontogeny of the beta-endorphin response to stress. Alcohol Clin Exp Res 13:564–571
155. Taylor AN, Branch BJ, Nelson LR, Lane LA, Poland RE (1986) Prenatal ethanol and ontogeny of pituitary-adrenal responses to ethanol and morphine. Alcohol 3:255–259
156. Aird F, Halasz I, Redei E (1997) Ontogeny of hypothalamic corticotropin-releasing factor and anterior pituitary pro-opiomelanocortin expression in male and female offspring of alcohol-exposed and adrenalectomized dams. Alcohol Clin Exp Res 21:1560–1566
157. Taylor AN, Branch BJ, Van Zuylen JE, Redei E (1988) Maternal alcohol consumption and stress responsiveness in offspring. Adv Exp Med Biol 245:311–317

158. Lee S, Imaki T, Vale W, Rivier C (1990) Effect of prenatal exposure to ethanol on the activity of the hypothalamic-pituitary-adrenal axis of the offspring: importance of the time of exposure to ethanol and possible modulating mechanisms. Mol Cell Neurosci 1:168–177
159. Weinberg J (1993) Neuroendocrine effects of prenatal alcohol exposure. Ann N Y Acad Sci 697:86–96
160. Lee S, Schmidt D, Tilders F, Rivier C (2000) Increased activity of the hypothalamic-pituitary-adrenal axis of rats exposed to alcohol in utero: role of altered pituitary and hypothalamic function. Mol Cell Neurosci 16:515–528
161. Glavas MM, Ellis L, Yu WK, Weinberg J (2007) Effects of prenatal ethanol exposure on basal limbic-hypothalamic-pituitary-adrenal regulation: role of corticosterone. Alcohol Clin Exp Res 31:1598–1610
162. Schneider ML, Moore CF, Kraemer GW (2004) Moderate level alcohol during pregnancy, prenatal stress, or both and limbic-hypothalamic-pituitary-adrenocortical axis response to stress in rhesus monkeys. Child Dev 75:96–109
163. Jacobson SW, Bihun JT, Chiodo LM (1999) Effects of prenatal alcohol and cocaine exposure on infant cortisol levels. Dev Psychopathol 11:195–208
164. Ramsay DS, Bendersky MI, Lewis M (1996) Effect of prenatal alcohol and cigarette exposure on two- and six-month-old infants' adrenocortical reactivity to stress. J Pediatr Psychol 21:833–840
165. Haley DW, Handmaker NS, Lowe J (2006) Infant stress reactivity and prenatal alcohol exposure. Alcohol Clin Exp Res 30:2055–2064
166. Weinberg J (1988) Hyperresponsiveness to stress: differential effects of prenatal ethanol on males and females. Alcohol Clin Exp Res 12:647–652
167. Weinberg J (1992) Prenatal ethanol effects: sex differences in offspring stress responsiveness. Alcohol 9:219–223
168. Halasz I, Aird F, Li L, Prystowsky MB, Redei E (1993) Sexually dimorphic effects of alcohol exposure in utero on neuroendocrine and immune functions in chronic alcohol-exposed adult rats. Mol Cell Neurosci 4:343–353
169. Kim CK, Turnbull AV, Lee SY, Rivier CL (1999) Effects of prenatal exposure to alcohol on the release of adenocorticotropic hormone, corticosterone, and proinflammatory cytokines. Alcohol Clin Exp Res 23:52–59
170. Weinberg J, Taylor AN, Gianoulakis C (1996) Fetal ethanol exposure: hypothalamic-pituitary-adrenal and beta-endorphin responses to repeated stress. Alcohol Clin Exp Res 20:122–131
171. Taylor AN, Branch BJ, Liu SH, Kokka N (1982) Long-term effects of fetal ethanol exposure on pituitary-adrenal response to stress. Pharmacol Biochem Behav 16:585–589
172. Slone JL, Redei EE (2002) Maternal alcohol and adrenalectomy: asynchrony of stress response and forced swim behavior. Neurotoxicol Teratol 24:173–178
173. Lee S, Rivier C (1992) Administration of corticosterone to pregnant adrenalectomized dams does not alter the hypothalamic-pituitary-adrenal axis' activity of the offspring. Mol Cell Neurosci 3:118–123
174. Redei E, Halasz I, Li LF, Prystowsky MB, Aird F (1993) Maternal adrenalectomy alters the immune and endocrine functions of fetal alcohol-exposed male offspring. Endocrinology 133:452–460
175. Sinha P, Halasz I, Choi JF, McGivern RF, Redei E (1997) Maternal adrenalectomy eliminates a surge of plasma dehydroepiandrosterone in the mother and attenuates the prenatal testosterone surge in the male fetus. Endocrinology 138:4792–4797
176. Tritt SH, Tio DL, Brammer GL, Taylor AN (1993) Adrenalectomy but not adrenal demedullation during pregnancy prevents the growth-retarding effects of fetal alcohol exposure. Alcohol Clin Exp Res 17:1281–1289
177. Taylor AN, Tritt SH, Tio DL, Romeo HE, Yirmiya R (2002) Maternal adrenalectomy abrogates the effect of fetal alcohol exposure on the interleukin-1beta-induced febrile response: gender differences. Neuroendocrinology 76:185–192
178. Giberson PK, Weinberg J (1995) Effects of prenatal ethanol exposure and stress in adulthood on lymphocyte populations in rats. Alcohol Clin Exp Res 19:1286–1294

179. Cunningham ET Jr, De Souza EB (1993) Interleukin 1 receptors in the brain and endocrine tissues. Immunol Today 14:171–176
180. Zalcman S et al (1994) Cytokine-specific central monoamine alterations induced by interleukin-1, -2 and -6. Brain Res 643:40–49
181. Dunn AJ, Wang J (1995) Cytokine effects on CNS biogenic amines. Neuroimmunomodulation 2:319–328
182. Rivier C (1993) Effect of peripheral and central cytokines on the hypothalamic-pituitary-adrenal axis of the rat. Ann N Y Acad Sci 697:97–105
183. Watkins LR, Maier SF (2000) The pain of being sick: implications of immune-to-brain communication for understanding pain. Annu Rev Psychol 51:29–57
184. Dantzer R (2001) Cytokine-induced sickness behavior: mechanisms and implications. Ann N Y Acad Sci 933:222–234
185. Hetier E et al (1988) Brain macrophages synthesize interleukin-1 and interleukin-1 mRNAs in vitro. J Neurosci Res 21:391–397
186. Sebire G et al (1993) In vitro production of IL-6, IL-1 beta, and tumor necrosis factor-alpha by human embryonic microglial and neural cells. J Immunol 150:1517–1523
187. Suda T et al (1990) Interleukin-1 stimulates corticotropin-releasing factor gene expression in rat hypothalamus. Endocrinology 126:1223–1228
188. Chover-Gonzalez AJ, Lightman SL, Harbuz MS (1994) An investigation of the effects of interleukin-1 beta on plasma arginine vasopressin in the rat: role of adrenal steroids. J Endocrinol 142:361–366
189. Kehrer P, Turnill D, Dayer JM, Muller AF, Gaillard RC (1988) Human recombinant interleukin-1 beta and -alpha, but not recombinant tumor necrosis factor alpha stimulate ACTH release from rat anterior pituitary cells in vitro in a prostaglandin E2 and cAMP independent manner. Neuroendocrinology 48:160–166
190. Sharp BM et al (1989) Tumor necrosis factor-alpha is a potent ACTH secretagogue: comparison to interleukin-1 beta. Endocrinology 124:3131–3133
191. Lyson K, McCann SM (1991) The effect of interleukin-6 on pituitary hormone release in vivo and in vitro. Neuroendocrinology 54:262–266
192. Taylor AN, Tio DL, Yirmiya R (1999) Fetal alcohol exposure attenuates interleukin-1beta-induced fever: neuroimmune mechanisms. J Neuroimmunol 99:44–52
193. Taylor AN, Tio DL, Heng NS, Yirmiya R (2002) Alcohol consumption attenuates febrile responses to lipopolysaccharide and interleukin-1 beta in male rats. Alcohol Clin Exp Res 26:44–52
194. Ylikorkala O, Stenman UH, Halmesmaki E (1988) Testosterone, androstenedione, dehydroepiandrosterone sulfate, and sex-hormone-binding globulin in pregnant alcohol abusers. Obstet Gynecol 71:731–735
195. Yirmiya R, Pilati ML, Chiappelli F, Taylor AN (1993) Fetal alcohol exposure attenuates lipopolysaccharide-induced fever in rats. Alcohol Clin Exp Res 17:906–910
196. Yirmiya R, Tio DL, Taylor AN (1996) Effects of fetal alcohol exposure on fever, sickness behavior, and pituitary-adrenal activation induced by interleukin-1 beta in young adult rats. Brain Behav Immun 10:205–220
197. Lee S, Rivier C (1993) Prenatal alcohol exposure blunts interleukin-1-induced ACTH and beta-endorphin secretion by immature rats. Alcohol Clin Exp Res 17:940–945
198. Chiappelli F et al (1997) Fetal alcohol exposure augments the blunting of tumor necrosis factor production in vitro resulting from in vivo priming with lipopolysaccharide in young adult male but not female rats. Alcohol Clin Exp Res 21:1542–1546
199. Yirmiya R, Chiappelli F, Tio DL, Tritt SH, Taylor AN (1998) Effects of prenatal alcohol and pair feeding on lipopolysaccharide-induced secretion of TNF-alpha and corticosterone. Alcohol 15:327–335
200. Vink J, Auth J, Abebe DT, Brenneman DE, Spong CY (2005) Novel peptides prevent alcohol-induced spatial learning deficits and proinflammatory cytokine release in a mouse model of fetal alcohol syndrome. Am J Obstet Gynecol 193:825–829
201. Crews FT et al (2006) Cytokines and alcohol. Alcohol Clin Exp Res 30:720–730

202. Szabo G, Mandrekar P (2009) A recent perspective on alcohol, immunity, and host defense. Alcohol Clin Exp Res 33:220–232
203. Liverman CS et al (2006) Altered expression of pro-inflammatory and developmental genes in the fetal brain in a mouse model of maternal infection. Neurosci Lett 399:220–225
204. Jonakait GM (2007) The effects of maternal inflammation on neuronal development: possible mechanisms. Int J Dev Neurosci 25:415–425
205. Meyer U, Feldon J, Yee BK (2009) A review of the fetal brain cytokine imbalance hypothesis of schizophrenia. Schizophr Bull 35:959–972
206. Gayle DA et al (2004) Maternal LPS induces cytokines in the amniotic fluid and corticotropin releasing hormone in the fetal rat brain. Am J Physiol Regul Integr Comp Physiol 286:R1024–R1029
207. Zaretsky MV, Alexander JM, Byrd W, Bawdon RE (2004) Transfer of inflammatory cytokines across the placenta. Obstet Gynecol 103:546–550
208. Aaltonen R, Heikkinen T, Hakala K, Laine K, Alanen A (2005) Transfer of proinflammatory cytokines across term placenta. Obstet Gynecol 106:802–807
209. Dahlgren J, Samuelsson AM, Jansson T, Holmang A (2006) Interleukin-6 in the maternal circulation reaches the rat fetus in mid-gestation. Pediatr Res 60:147–151
210. Kaminen-Ahola N, Ahola A, Maga M, Mallitt KA, Fahey P, Cox TC, Whitelaw E, Chong S. Maternal ethanol consumption alters the epigenotype and the phenotype of offspring in a mouse model. PLoS Genet. 6:e1000811, 2010
211. Pang Y, Cai Z, Rhodes PG (2003) Disturbance of oligodendrocyte development, hypomyelination and white matter injury in the neonatal rat brain after intracerebral injection of lipopolysaccharide. Brain Res Dev Brain Res 140:205–214
212. Ader R (1998) Psychoneuroimmunology. ILAR J 39:27–29
213. Schwarz JM, Bilbo SD (2012) The immune system and the developing brain. In: McCarthy MM (ed) The developing brain. Morgan & Claypool Life Sciences, San Rafael, pp 1–118
214. Bajetto A, Bonavia R, Barbero S, Florio T, Schettini G (2001) Chemokines and their receptors in the central nervous system. Front Neuroendocrinol 22:147–184
215. Nakashima K, Taga T (2002) Mechanisms underlying cytokine-mediated cell-fate regulation in the nervous system. Mol Neurobiol 25:233–244
216. Ling EA, Wong WC (1993) The origin and nature of ramified and amoeboid microglia: a historical review and current concepts. Glia 7:9–18
217. Fujita S, Tsuchihashi Y, Kitamura T (1981) Origin, morphology and function of the microglia. Prog Clin Biol Res 59A:141–169
218. Ashwell K (1990) Microglia and cell death in the developing mouse cerebellum. Brain Res Dev Brain Res 55:219–230
219. Elkabes S, DiCicco-Bloom EM, Black IB (1996) Brain microglia/macrophages express neurotrophins that selectively regulate microglial proliferation and function. J Neurosci 16:2508–2521
220. Marin-Teva JL et al (2004) Microglia promote the death of developing Purkinje cells. Neuron 41:535–547
221. Town T, Nikolic V, Tan J (2005) The microglial "activation" continuum: from innate to adaptive responses. J Neuroinflammation 2:24
222. Zhang X et al (2012) Prenatal alcohol exposure alters the course and severity of adjuvant-induced arthritis in female rats. Brain Behav Immun 26:439–450
223. Harbuz MS, Rees RG, Lightman SL (1993) HPA axis responses to acute stress and adrenalectomy during adjuvant-induced arthritis in the rat. Am J Physiol 264:R179–R185
224. Chover-Gonzalez AJ et al (1999) Effects of stress on susceptibility and severity of inflammation in adjuvant-induced arthritis. Ann N Y Acad Sci 876:276–286
225. Chover-Gonzalez AJ, Jessop DS, Tejedor-Real P, Gibert-Rahola J, Harbuz MS (2000) Onset and severity of inflammation in rats exposed to the learned helplessness paradigm. Rheumatology (Oxford) 39:764–771
226. Bomholt SF, Harbuz MS, Blackburn-Munro G, Blackburn-Munro RE (2004) Involvement and role of the hypothalamo-pituitary-adrenal (HPA) stress axis in animal models of chronic pain and inflammation. Stress 7:1–14

227. Del Rey A et al (2010) When immune-neuro-endocrine interactions are disrupted: experimentally induced arthritis as an example. Neuroimmunomodulation 17:165–168
228. Owen D, Andrews MH, Matthews SG (2005) Maternal adversity, glucocorticoids and programming of neuroendocrine function and behaviour. Neurosci Biobehav Rev 29:209–226
229. Welberg LA, Seckl JR (2001) Prenatal stress, glucocorticoids and the programming of the brain. J Neuroendocrinol 13:113–128
230. Liu L, Li A, Matthews SG (2001) Maternal glucocorticoid treatment programs HPA regulation in adult offspring: sex-specific effects. Am J Physiol Endocrinol Metab 280:E729–E739
231. Matthews SG (2002) Early programming of the hypothalamo-pituitary-adrenal axis. Trends Endocrinol Metab 13:373–380
232. Colebatch AN, Edwards CJ (2011) The influence of early life factors on the risk of developing rheumatoid arthritis. Clin Exp Immunol 163:11–16
233. McEwen BS (1998) Stress, adaptation, and disease. Allostasis and allostatic load. Ann N Y Acad Sci 840:33–44
234. Weaver IC et al (2004) Epigenetic programming by maternal behavior. Nat Neurosci 7:847–854
235. Haycock PC (2009) Fetal alcohol spectrum disorders: the epigenetic perspective. Biol Reprod 81:607–617
236. Kobor M, Weinberg J (2011) Focus on: epigenetics and fetal alcohol spectrum disorders. Alcohol Res Health 34:29–37
237. Livy DJ, Maier SE, West JR (2004) Long-term alcohol exposure prior to conception results in lower fetal body weights. Birth Defects Res B Dev Reprod Toxicol 71:135–141
238. Haycock PC, Ramsay M (2009) Exposure of mouse embryos to ethanol during preimplantation development: effect on DNA methylation in the h19 imprinting control region. Biol Reprod 81:618–627
239. Garro AJ, McBeth DL, Lima V, Lieber CS (1991) Ethanol consumption inhibits fetal DNA methylation in mice: implications for the fetal alcohol syndrome. Alcohol Clin Exp Res 15:395–398
240. Guo W et al (2011) Alcohol exposure decreases CREB binding protein expression and histone acetylation in the developing cerebellum. PLoS One 6:e19351
241. Mathews HL, Janusek LW (2011) Epigenetics and psychoneuroimmunology: mechanisms and models. Brain Behav Immun 25:25–39
242. Cuddapah S, Barski A, Zhao K (2010) Epigenomics of T cell activation, differentiation, and memory. Curr Opin Immunol 22:341–347
243. Martino DJ, Prescott SL (2010) Silent mysteries: epigenetic paradigms could hold the key to conquering the epidemic of allergy and immune disease. Allergy 65:7–15
244. Martino D, Prescott S (2011) Epigenetics and prenatal influences on asthma and allergic airways disease. Chest 139:640–647
245. Miller RL (2008) Prenatal maternal diet affects asthma risk in offspring. J Clin Invest 118:3265–3268
246. Trenkmann M, Brock M, Ospelt C, Gay S (2010) Epigenetics in rheumatoid arthritis. Clin Rev Allergy Immunol 39:10–19
247. Strietholt S, Maurer B, Peters MA, Pap T, Gay S (2008) Epigenetic modifications in rheumatoid arthritis. Arthritis Res Ther 10:219
248. Chen P, Jiang T, Ouyang J, Cui Y, Chen Y (2009) Epigenetic programming of diverse glucocorticoid response and inflammatory/immune-mediated disease. Med Hypotheses 73:657–658
249. Szyf M (2010) Epigenetic therapeutics in autoimmune disease. Clin Rev Allergy Immunol 39:62
250. Holladay SD, Smialowicz RJ, (2000). Development of the murine and human immune system: differential effects of immunotoxicants depend on time of exposure. Environ Health Perspect 108 Suppl 3:463–473
251. Kumar SN, Stewart GL, Steven WM, Seelig LL (1989) Maternal to neonatal transmission of T-cell mediated immunity to Trichinella spiralis during lactation. Immunology 68(1):87–92

Chapter 11
Neuroimmune Mechanisms of Glia and Their Interplay with Alcohol Exposure Across the Lifespan

Paul D. Drew and Cynthia J.M. Kane

11.1 Introduction

A plethora of studies demonstrate a link between alcohol consumption and altered immune activity in the central nervous system (CNS). Alcohol-induced immune activity has been observed in animal models of fetal alcohol syndrome, in binge models associated with adolescence and adulthood, and in chronic models of alcohol consumption in adults and aged animals. The purpose of this chapter is to provide an overview of studies analyzing alcohol effects on immune activity in the CNS throughout the lifespan.

11.1.1 Alcohol Effects on Immune Activity in the CNS: An Overview

Alcohol abuse at all life stages can result in significant disability with tremendous personal and societal consequences. For example, fetal alcohol spectrum disorders (FASD) result from fetal exposure to alcohol and occur in approximately 2–5% of births in the United States [1–3]. Binge drinking at all ages, beginning in adolescence and continuing with chronic alcohol abuse in adults and the elderly, occurs at alarming rates. Alcohol abuse at each of these life stages is associated with immune responses in the CNS. This may contribute to neurodegeneration and impaired neurological function associated with excessive alcohol consumption. In addition, intriguing recent studies demonstrate that neuroimmune molecules including cytokines

P.D. Drew • C.J.M. Kane, Ph.D. (✉)
Department of Neurobiology and Developmental Sciences, University of Arkansas for Medical Sciences, 4301 W. Markham, #846, Little Rock, AR 72205, USA
e-mail: kanecynthiaj@uams.edu

C. Cui et al. (eds.), *Neural-Immune Interactions in Brain Function and Alcohol Related Disorders*, DOI 10.1007/978-1-4614-4729-0_11,

and chemokines modulate addiction to alcohol. This has resulted in an important paradigm shift linking immune response to alcohol-related disorders [4, 5].

Alcohol modulates the peripheral innate immune response in a very complex manner which depends on factors including the amount of alcohol consumed, the pattern of consumption (acute, binge, or chronic), and the tissue or organ under investigation [6]. One example of this complexity is that chronic alcoholics are commonly immunosuppressed and prone to infections and, yet, often exhibit inflammation in the liver. Although a variety of studies have addressed the effect of alcohol in peripheral immune cells, the effects of alcohol on CNS immune responses by glial cells have been less extensively investigated [7–9]. Initial studies indicated that alcohol altered microglial phagocytosis as well as glial production of proinflammatory molecules including nitric oxide (NO), COX-2, and cytokines. The transcription factors NF-κB and CREB were demonstrated to mediate the effects of alcohol on CNS inflammation and oxidative stress, and associated neuron cell death [10–16]. In addition to direct effects of alcohol on CNS glia, alcohol effects on peripheral immune cells can also alter CNS function. Peripheral immune cells are capable of entering the CNS, where they interact with CNS glia and collectively regulate both innate and adaptive immunity in the CNS. Little is known concerning the interactions of peripheral and CNS cells in response to alcohol, and future studies will undoubtedly address this important area of research.

11.1.2 Microglia

11.1.2.1 Overview

Microglia, together with astrocytes and oligodendrocytes, comprise the glial cell populations of the CNS. In the healthy CNS, microglia perform a variety of functions which contribute to CNS homeostasis including protection of neurons. However, in response to CNS infection, trauma, or disease, microglia are stimulated to undergo morphological and functional changes. The stimulated microglia produce inflammatory molecules important in the clearance of pathogens and debris. However, chronic production of these inflammatory molecules by microglia can also damage the CNS and contribute to the pathology associated with various neuroinflammatory and neurodegenerative disorders.

11.1.2.2 Ontogeny of Microglia

The origin of microglia has been empirically evaluated and debated for many years. Today, it is generally believed that, distinct from other glia and neurons, microglia are of hematopoietic origin [17]. Specifically, microglia are myeloid lineage cells. In the mouse, microglia are generated from primitive hematopoiesis in the fetal yolk sac and migrate to the CNS during early stages of fetal development. Among myeloid lineage cells, microglia are most closely related to macrophages and are commonly classified as specialized resident macrophages of the CNS [17]. Microglia and macrophages express many common markers including CD11b, CD14, and EGF-like module containing mucin-like hormone receptor-like 1 (EMR1) [18]. Microglial differentiation is similar to that of macrophages and requires colony stimulating factor 1 (CSF1), CSF1 receptor, and the transcription factor PU.1. The observation that PU.1-deficient mice, which lack development of myeloid cells [19], also lack microglia further supports the myeloid origin of microglia.

Controversy has also existed regarding whether there is turnover of microglia in the parenchyma and whether microglia can be replaced following microglial cell loss. It is unknown if monocytic cells from the periphery are capable of replenishing the microglial population. Bone marrow transfer studies in irradiated animals suggest that bone marrow-derived monocytes can migrate into the CNS and form microglia [20–24]. However, more recent parabiosis studies in which the circulatory systems of donor and recipient animals are connected indicated that few donor cells enter the CNS of the recipient in the absence of CNS irradiation [25, 26]. Circulating monocytes are known to infiltrate the CNS in animal models of disease associated with disruption of the blood–brain barrier including experimental autoimmune encephalomyelitis (EAE), a model of the human disease multiple sclerosis. However, although these infiltrating monocytes contribute to the development of disease, they do not appear to differentiate into microglia [27]. Collectively, these studies indicate peripheral monocytes do not significantly replenish microglia in the adult CNS. Furthermore, microglia are believed to be long-lived cells which are capable of proliferation in response to changes in the CNS. Under pathological conditions, for example, microglia undergo proliferation or microgliosis [28–31]. It is possible that microglia could be replenished from microglial cell progenitors present in the CNS parenchyma. However, little is known concerning this potential mechanism of generating microglia.

11.1.2.3 Function of Microglia in the Healthy CNS

Microglia constitute approximately 5–20% of all cells in the CNS. They are present throughout the CNS, but the relative density of these cells, as well as their morphology, varies significantly in different regions [32]. In addition, white matter contains fewer microglia than gray matter [33]. It is interesting to speculate that variations in microglial cell density and morphology may reflect functional differences which could be associated with regional differences in CNS function. In the adult CNS, microglia generally exhibit a ramified appearance characterized by a small soma with limited perinuclear cytoplasm and a number of finely branched processes [34]. Microglia appear to occupy defined territories that do not overlap with adjacent microglia. Although microglia in the healthy CNS have typically been referred to as quiescent, these cells in fact perform a variety of functions central to the maintenance and homeostasis of the CNS. Microglia perform a critical surveillance function in the healthy CNS. There, they also exhibit phagocytic behavior. Microglial phagocytosis in the healthy CNS includes phagocytosis of apoptotic cells which can occur at all life stages from early development through senescence [35]. In this manner, microglial surveillance removes cells and debris without activation of microglia and without associated inflammation. Phagocytic activity is further enhanced following microglial activation in response to neuroinflammatory or neurodegenerative conditions. Multiphoton and high-resolution electron microscopy studies indicate that microglia are highly dynamic in vivo in the healthy CNS, extending and retracting processes and sampling the local environment especially at the synapse [36, 37]. In this manner, microglia are responsive to changes in the macroenvironment and microenvironment of the CNS. Further, microglia respond to a variety of neurotransmitters, neurohormones, and neuromodulators with changes in metabolism and gene expression [18]. Microglia secrete a variety of growth factors critical to survival of CNS cells, including neurons, and thus maintain the health of these cells.

Microglia play important roles in synaptic establishment, structure, function, connectivity, and activity throughout life, from earliest CNS development through adolescence, adulthood, and aging. CNS development is associated with extensive plasticity involving formation and remodeling of synapses during activity-dependent synaptic pruning [38, 39]. Support for a role of microglia in these processes includes the physical association of microglia with developing as well as mature synapses [40–42]. Association of microglial processes with dendritic spines further supports a role of microglia in synaptic structure and function [38, 43]. Furthermore, microglia play a role in complement-mediated elimination of CNS synapses in the developing visual system [44, 45]. Microglial engulfment of dendritic spines and phagocytosis of synaptic structures occurs in other regions too, suggesting that synaptic pruning by microglia may be a common mechanism resulting in synapse elimination in the developing CNS [38, 46]. Neuroimmune molecules expressed by microglia, and in some cases astrocytes, are fundamental to synaptic development, structure, and function [47]. These include tumor necrosis factor (TNF)-α, interleukin (IL)-6, complement proteins, pentraxins, Dscam, major histocompatibility complex (MHC)-I proteins, and MHC-I binding receptors. In addition, the chemokine receptor CX3CR1 which is specific to microglia binds to the chemokine CX3CL1 present on neurons. CX3CR1 knockout mice exhibit fewer microglia and have deficient synapse formation and plasticity in the developing hippocampus [46]. Collectively, these studies indicate that microglia interact with neurons to modulate synapse development and plasticity. In summary, microglia are believed to be critical to neuronal integrity and to be capable of altering neuronal circuitry throughout the lifetime of an organism.

11.1.2.4 Response of Microglia to CNS Insult

Traditionally, the CNS was thought to be an immunoprivileged site where little or no immune activity occurs. This was based on observations that tissue grafted into the CNS demonstrated prolonged survival, that the blood–brain barrier limits movement of peripheral immune cells into the CNS, and that the CNS lacks lymphatic drainage. However, it is clear that peripheral immune cells are capable of entering the CNS and, in addition, that resident CNS cells including neurons and glia are capable of modulating immune responses of infiltrating macrophages and lymphocytes [48]. As described above, in the healthy CNS, microglia perform a variety of homeostatic functions. However, following CNS infection, injury, or inflammation, microglia become activated and these cells exhibit increased proliferation, motility, and phagocytosis, which collectively facilitates removal of pathogens and debris.

Activated microglia also undergo morphological changes converting from a ramified to an ameboid appearance. Activated microglia also can serve as antigen-presenting cells and produce a variety of cytokines, including TNF-α, IL-1β, and IL-6, and chemokines including MCP-1, MIP-1α, and MIP-1β, as well as NO and reactive oxygen species. In this manner, microglia play a role in innate immunity within the CNS parenchyma. In addition, microglial products alter T cell function and modulate adaptive immunity in the CNS. Although molecules including NO, TNF-α, and IL-1β may be toxic to pathogens, these agents can also be toxic to resident CNS cells including neurons. Thus, although microglia are normally protective in the CNS, microglia can contribute to CNS pathology under conditions of CNS inflammation [17, 34].

11.1.2.5 Microglia Morphology and Activation

Microglia that migrate into the parenchyma during CNS development initially exhibit an ameboid morphology [49, 50] (Fig. 11.1). After moving into the parenchyma, they differentiate and develop a ramified appearance common to microglia in the mature, healthy CNS. Relatively little is known regarding the factors that convert microglia from an ameboid to a ramified morphology in the developing CNS. However, astrocytic products found in astrocyte-conditioned medium as well as ATP and adenosine appear to contribute to the conversion of microglia to a ramified phenotype [52, 53]. Antibody neutralization studies suggest that TGF-β, M-CSF, and GM-CSF are astrocyte products involved in this morphological conversion of microglia [54].

Cells of the innate immune system including microglia respond to pathogens principally through Toll-like receptors (TLRs) which recognize conserved motifs expressed by pathogens. TLRs are capable of responding to endogenous molecules termed danger signals such as high-mobility group box 1 protein and heat-shock protein 70, which are normally sequestered from the immune response but are released under conditions including trauma or inflammation. Following activation of TLR signaling pathways, microglia exhibit an ameboid morphology (Fig. 11.1) and exhibit characteristics of activated microglia as described above including increased production of proinflammatory molecules, antigen presentation, and phagocytosis. However, it is clear that microglia, activation is not an all-or-none phenomenon and microglia can exhibit distinct levels of activation which are associated with distinct functional states [55–60]. In addition, microglia may respond to

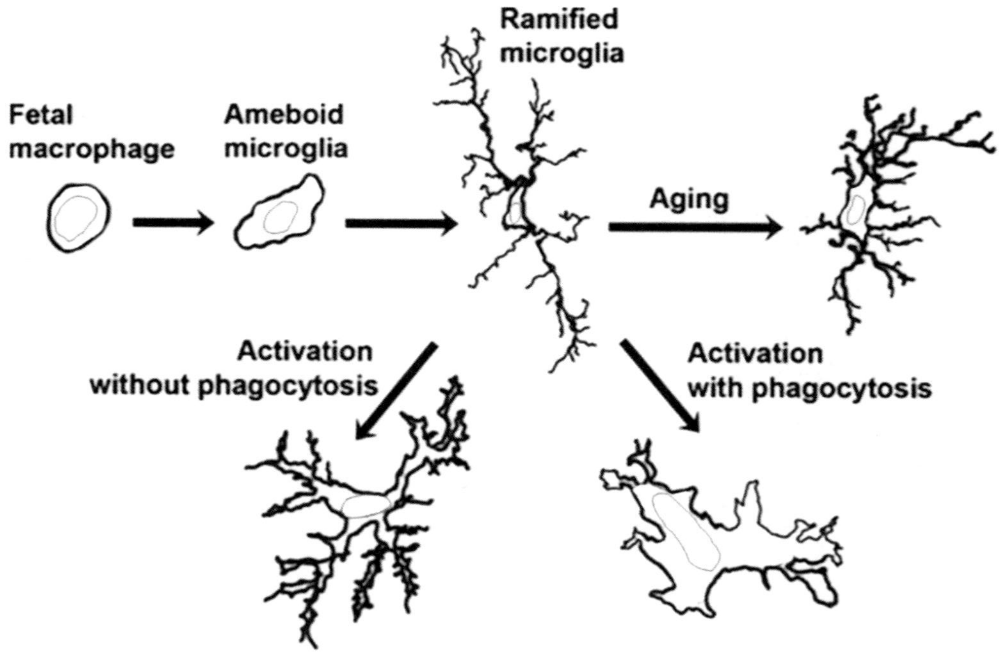

Fig. 11.1 Microglia undergo morphological changes across the lifespan from CNS development through aging. In addition, microglia respond to insult with activation-induced morphological changes. The morphological changes are paralleled by changes in function. Functional changes as well as the activation process continue to be elucidated. Figure modified from Streit et al. [51] with permission

distinct stimuli to undergo classic or alternative activation in a manner similar to M1 or M2 activation originally described in macrophages [34].

Following removal of inflammatory stimuli, microglia revert back toward a ramified phenotype. However, there are suggestions that microglia may not completely revert to an inactive state and may be more prone to subsequent activation [57]. The mechanisms which regulate the phenotype and response of post-activated microglia are unknown. Interestingly, microglia from aged animals exhibit significant regional morphological heterogeneity [61–63] (Fig. 11.1), as well as morphological markers characteristic of deterioration and senescence [51, 64–66]. Although there is a wealth of information concerning the mechanisms that regulate microglial cell activation, little is known concerning deactivation and suppression of microglial activation. In this regard, molecules such as TREM2 are believed to suppress microglial activation. Cytokines such as TGF-β and IL-10 suppress activation of these cells as do steroid hormones and other ligands that interact with nuclear receptors [17].

11.1.3 Astrocytes

Astrocytes comprise a highly heterogeneous cell population [67]. Distinct subclasses of astrocytes with distinct morphologies exist including fibrous and protoplasmic forms. The astrocyte composition varies regionally with distinctions between gray and white matter astrocytes. Astrocytes play an important role in CNS functions including energy metabolism, regulation of blood flow, ion metabolism, sequestration of neurotoxicants, neurotransmission, neurogenesis, and synapse formation, modulation, and plasticity [68–71]. Like microglia, astrocytes are capable of responding to infectious and traumatic stimuli, during which they undergo reactive astrogliosis. Reactive astrocytes are capable of producing many of the same proinflammatory molecules as microglia including cytokines and chemokines and reactive oxygen and nitrogen species. However, astrocytes are not phagocytic and exhibit limited or no antigen presentation function in vivo.

11.2 Effects of Alcohol on the Neuroimmune Response

The following sections will provide an overview of alcohol effects on immune response in the human CNS as well as fetal/neonatal, adolescent, adult, and aged rodents. The role of glia in mediating alcohol-induced toxicity to neurons as well as drinking behavior and addiction will be discussed. Molecular and cellular mechanisms mediating glial neuroimmune activity will be described. Finally, potential treatments for alcohol-induced neuropathologies will be considered.

11.2.1 Human Studies

Several studies have highlighted links between alcohol consumption, glial activity and phenotype, and immune molecules in the CNS. Circulating TNF-α, IL-1β, IL-6, and IL-12 are often observed in chronic alcoholics [72]. Peripheral blood monocytes release higher levels of these cytokines in individuals who chronically consume alcohol [73]. Alcoholic liver disease associated with chronic alcohol consumption is often characterized by high levels of circulating TNF-α and other cytokines. In addition, alcohol craving has been linked to levels of proinflammatory cytokines [74]. Proinflammatory cytokine levels are also increased under the conditions of hangover in humans [75].

Gene expression has been analyzed in human postmortem brain tissue in order to identify changes associated with chronic alcohol consumption. Microarray analysis of the frontal cortex revealed that a major difference in gene expression in alcoholic and non-alcoholic brains was a change in neuroimmune genes [76]. Human studies indicate that NF-κB expression is altered in the brains of alcoholics [77, 78]. These studies suggest that alterations of NF-κB-mediated inflammation and plasticity may play a role in alcohol dependence and cognitive deficits in alcoholics. In another study, aged alcoholics demonstrated a twofold to threefold increase in MCP-1 protein in the ventral tegmental area, substantia nigra, hippocampus, and amygdala compared to non-alcoholic individuals [79]. MCP-1 is a chemokine that attracts monocytic cells including microglia. Previous studies indicated that MCP-1-deficient mice exhibited altered CNS inflammation and drinking behavior, suggesting that MCP-1 may play a similar role in humans [80]. Notably, in the above study, there was no difference in the morphology of microglia in alcoholic individuals [79]. However, quantification of the microglial-specific markers Iba-1 or GluT5 in tissue sections revealed upregulation in alcoholic brains, suggesting that microglia reacted to alcohol exposure. A caveat raised by the investigators in this study [79] was a difference in the history of smoking in the alcohol versus the control group, which could not be further analyzed with the limited sample size. Nonetheless, it is clear that there was a difference in MCP-1 protein and microglial phenotype in individuals who chronically consumed alcohol. This may be particularly important because increased MCP-1 expression in the CNS of rodents has been demonstrated to alter hippocampal synaptic transmission [81]. Collectively, these studies suggest that alcohol-induced MCP-1 may result in decreased synaptic plasticity in alcoholics.

Alcohol induction of cytokine expression during human pregnancy has also been investigated [82] using fetal cord blood and maternal blood collected at parturition. Chronic drinking, compared to moderate or no drinking, led to high levels of the cytokines TNF-α, IL-1β, and IL-6 in both the maternal and fetal blood. This observation revealed that chronic alcohol use during pregnancy leads to synthesis of proinflammatory cytokines that freely cross the placenta. The maternal versus fetal source of the cytokines was not investigated, and either or both sources may be stimulated by chronic alcohol exposure.

11.2.2 Animal Studies

11.2.2.1 Fetal and Neonatal

FASD, the most common cause of mental retardation, is associated with persistent cognitive dysfunction as well as structural and behavioral anomalies [3, 83]. In addition, individuals exposed to alcohol in utero have an increased risk of alcohol abuse and addiction later in life. Alcohol-induced pathological damage to the immature neuroimmune system and alcohol effects on neuroimmune activity appear to contribute to this spectrum of alcohol teratology.

Fetal and neonatal rodent models have contributed the majority of our understanding of alcohol effects on the developing neuroimmune system. Neonatal studies in mice or rats mimic the third trimester of human brain development which is highly vulnerable to fetal alcohol damage [84–89]. Recent studies from our laboratory demonstrated that alcohol exposure in the neonatal mouse causes loss of microglial cells and impairs development of the normal microglial population in the immature cerebellum [90] (Fig. 11.2). In addition, the change in microglia was coincident with loss of Purkinje neurons. In this study, male and female mice were treated by gavage with 3.5 g/kg/day of alcohol (15%, w/v) on PD 3–5 and BAC was measured to be 250–325 mg/dl. The microglia that survived alcohol cytotoxicity exhibited altered morphology, with shorter and broader processes, reduced branching, and enlarged cell bodies compared to control animals. The morphologic changes suggest partial activation of the cells implying alteration of their normal function in the developing CNS. Loss of microglia is likely to be significant given the repertoire of microglial activity in the healthy CNS. Further, loss of cells is even more meaningful under circumstances in which the CNS is challenged by an outside insult. We hypothesize that direct alcohol impact on immature microglia may alter their developmental profile to cause persistent changes in microglial function that lasts into adolescence and adulthood.

Alcohol exposure in neonatal rat brain produces changes in cytokines for extended periods following treatment [91]. Neonatal rats treated with alcohol exhibit changes in the cytokines TNF-α, IL-1 β, and TGF-β1 in the cerebral cortex and hippocampus 19 days posttreatment. In this study, male rats were treated with 5.0 g/kg/day alcohol (12%, v/v) on PD7-9 with a BAC of 325 mg/dl. All three cytokines were increased twofold to sevenfold in both regions in alcohol-treated animals compared to control animals. Quantification of the expression of the transcription factor NF-κB p65 in the same tissues indicated a fourfold to sixfold elevation. Together, these findings strongly suggest that exposure to alcohol during a vulnerable period of brain development produces persistent changes in synthesis of neuroimmune molecules. Because alcohol treatment during this same window of development causes neuron and glial loss [85–90] and deficits in cognition and behavior [92], it is hypothesized that changes in the neuroimmune system contribute to the persistent neuropathologic and functional damage.

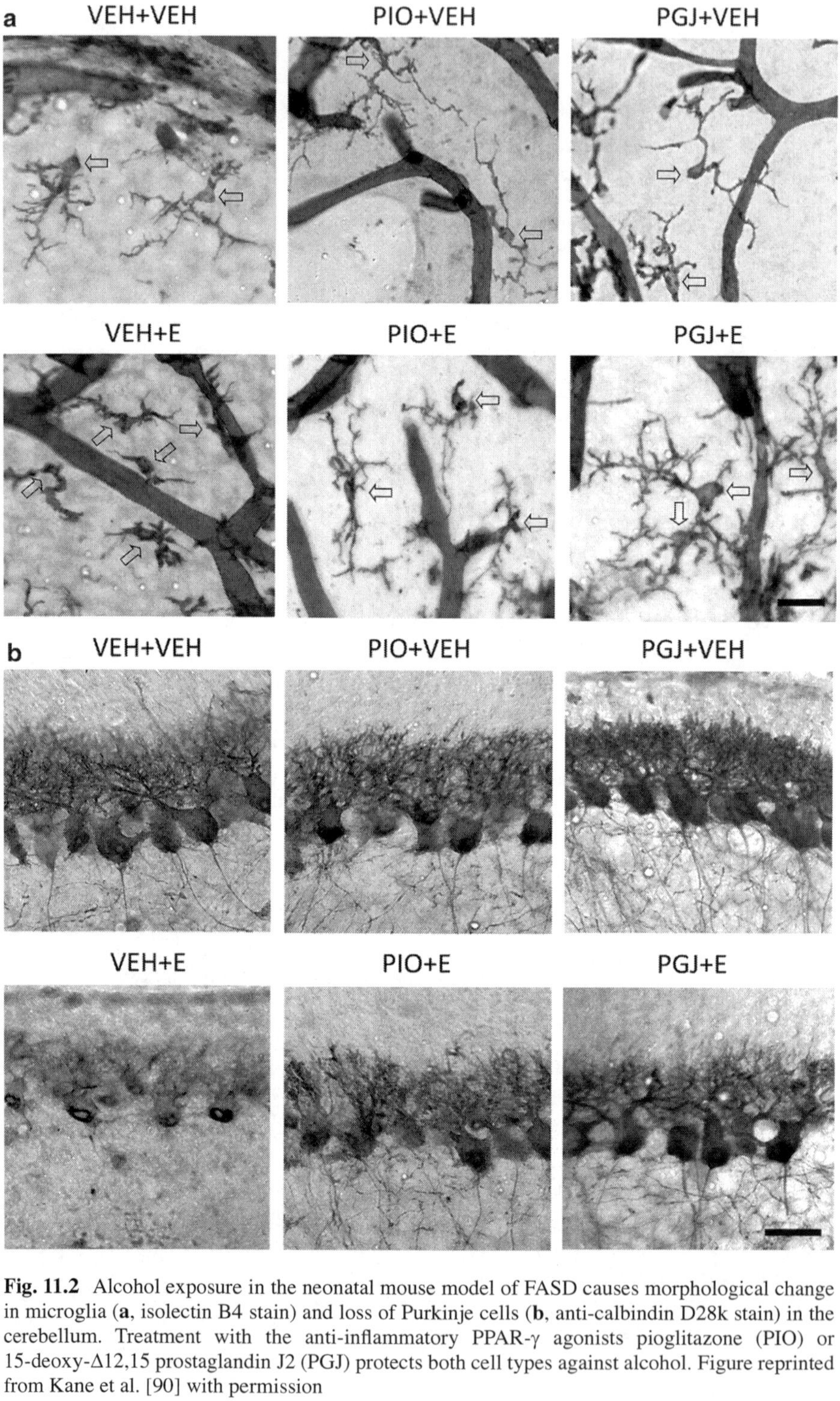

Fig. 11.2 Alcohol exposure in the neonatal mouse model of FASD causes morphological change in microglia (**a**, isolectin B4 stain) and loss of Purkinje cells (**b**, anti-calbindin D28k stain) in the cerebellum. Treatment with the anti-inflammatory PPAR-γ agonists pioglitazone (PIO) or 15-deoxy-Δ12,15 prostaglandin J2 (PGJ) protects both cell types against alcohol. Figure reprinted from Kane et al. [90] with permission

Oxidative stress is one facet of the innate immune response that often leads to neurodegeneration. It is well documented that alcohol treatment during gestation or in the neonatal rodent produces oxidative stress and neuronal loss. For example, in the study by Tiwari et al. [91], increased cytokine expression co-occurred with expression of the apoptosis marker activated caspase-3 in the cerebral cortex and hippocampus of alcohol-treated animals. In addition, several markers of oxidative stress were elevated in these regions including acetylcholinesterase, catalase, and superoxide dismutase activity, levels of reduced glutathione, NO production, and lipid peroxidation. A separate study indicates that even a single exposure to alcohol leads to activation of caspase-3, DNA fragmentation, and lipid peroxidation in the cerebellum [93] in PD7 rats given alcohol by inhalation at a BAC of 368 mg/dl. Thus, oxidative stress is an important component of the innate immune response to alcohol exposure in the developing CNS.

Intriguing studies have revealed that fetal exposure to alcohol compromises the ability of the neuroimmune system to respond to insult in adulthood [94]. Pregnant rats were fed alcohol (unspecified dose as 2.2–4.5%, v/v, in liquid diet with BAC reported as 37–96 mg/dl) from gestational day 7 to birth, and the male offspring were raised to adulthood. The innate immune response was then challenged with a stab wound into the brain parenchyma. Expression of TNF-α, glial fibrillary acidic protein (GFAP), ICAM-1, VCAM-1, and the microglial activation marker ED-1 were quantified 4 days post-injury. Animals that were exposed to alcohol during fetal life exhibited suppression of the normal neuroinflammatory increase in TNF-α, GFAP, and ICAM-1 in response to injury. Suppression of these molecules may reflect alcohol-induced long-term damage to astrocytes. In contrast, expression of VCAM-1 and ED1 following injury was potentiated by fetal alcohol exposure suggesting that alcohol-primed endothelial and microglial cell responses to a subsequent activating insult. Clearly, fetal exposure to alcohol has long-lasting consequences for the adult neuroimmune system. A series of interesting studies indicate that alcohol exposure during development results in reprogramming of the hypothalamic–pituitary–adrenal axis and associated alterations in immune function [95]. This may represent an important general mechanism by which alcohol alters immune activity. Suppression of the neuroimmune response following fetal exposure to alcohol may be mediated by teratogenic damage to the immature glial populations in the developing brain. These results directly suggest that the lasting effects may increase the vulnerability of the adult brain to insult, injury, infection, neurodegeneration, or disease. However, given the recently discovered interplay between neuroimmune molecules and alcohol drinking behavior, it also raises the possibility that the adolescent or adult brain may be more vulnerable to alcohol abuse and addiction.

11.2.2.2 Adolescence

The adolescent brain is characterized by an extensive repertoire of developmental processes including synaptic remodeling. From a behavioral standpoint, adolescence is characterized by less developed control of inhibition and decision-making proficiency relative to adults. Alcohol is also less sedating in adolescents than adults [96], and adolescents suffer less severe motor impairment than adults at equivalent blood alcohol concentrations [97]. These factors likely contribute to binge drinking behavior common to adolescents in which high levels of alcohol are consumed over short periods of time. The adolescent CNS is particularly vulnerable to the adverse effects of alcohol [98–100]. As a result, alcohol consumption during adolescence has powerful long-term consequences for the brain and behavior [101–103]. Given our growing understanding of the impact of alcohol in the adolescent brain, it is of high importance to fully understand the events, processes, and mechanisms involved.

Studies of adolescent rodents reinforce the importance of understanding the consequences of alcohol exposure during this stage of life. Ongoing neurogenesis in the dentate gyrus of the hippocampus and the subventricular zone is important to normal function in the adolescent as well as the adult brain [104]. The adolescent stage of brain development is particularly characterized by elevated levels of neurogenesis in the hippocampus. Alcohol exposure during adolescence has significant implications for the integrity of the neurogenic process in this region [105].

In a recent study [106], alcohol was administered to adolescent male rats for 4 days in a binge exposure model. The animals were PD35-39 when they were treated with 12.0 g/kg/day as a 25% (w/v) solution by gavage, achieving BAC of 297 mg/dl. The number of proliferating neurons decreased 2 days posttreatment, and survival of newly generated neurons was reduced 28 days posttreatment. Treatment with alcohol did not change the size of the proliferating cell population and did not alter the fraction of the population in the G1 or G2/M phase. However, the S phase was highly vulnerable to alcohol with reductions in the fraction of the cell population in the S phase, the duration of the S phase, and the duration of the cell cycle. Shortening of the cell cycle was suggested to induce a compensatory acceleration in the rate and incidence of neural progenitor cell proliferation, and accordingly, there was a measurable increase in the number of progenitor cells in the hippocampus. The investigators suggest that accelerated cell cycle progression may be consistent with decreased neurogenesis because neurogenesis is a composite of several processes including cell proliferation, differentiation, migration, and survival [106]. Evidence of increased proliferation in the hippocampus of the adolescent rat in this study is strikingly different than observations in the adult rat discussed below [13, 107].

Glial activation promotes inflammatory processes through secretion of cytokines and chemokines and enhancement of oxidative stress. Recruitment of additional glia into the inflammatory cascade and recruitment of peripheral immune cells into the parenchyma propel an initial beneficial response of glia into an excessive, detrimental inflammatory milieu. Neurogenesis and proliferation of neural progenitor cells are particularly vulnerable to disruption by an amplified innate immune response or neuroinflammation [108]. Cytokines including TNF-α, IL-1β, IL-6, and IFN-γ, as well as reactive oxygen and nitrogen species, produced by both microglia and astrocytes, inhibit renewal of the stem cell population and inhibit proliferation of progenitors [109–114]. TLR4 is expressed by neural progenitor cells and its activation with LPS inhibits proliferation through a NF-κB-dependent pathway [115].

Activated microglia have been specifically associated with suppression of neurogenesis [112, 116]. However, it is becoming increasingly clear that normal innate immune activity of glia can also be protective and enhance neurogenesis [57, 117, 118] depending on the level and duration of glial activation.

It is meaningful, then, that studies in the adolescent also reveal that alcohol treatment induces a partially activated microglial cell phenotype in the hippocampus [119] as reflected by a significant change in the morphology of microglia in alcohol-treated animals compared to controls. In this study, male mice PD35-39 were treated with 11–13 g/kg/day alcohol as a 25% (v/v) solution via gavage with a BAC range of 304–410 mg/dl. Activation was suggested to be less than maximum because there was not an increase in cells positive for MHC-II or the phagocytic marker ED-1. Additionally, there was no change in expression of TNF-α protein in hippocampal tissue quantified 2 days posttreatment. However, alcohol treatment stimulated proliferation of microglia, consistent with at least partial activation. The newly generated microglia survived at 30 days posttreatment. The consequences of an increase in the number of resident microglia are unknown, but particularly given their partially activated morphology, it will be important to expand this knowledge to fully understand interactions between microglial-derived factors, alcohol-induced neuropathology, and alcohol drinking behavior.

11.2.2.3 Adulthood

Chronic alcohol abuse can lead to neurodegeneration and associated cognitive impairment. Alcohol abuse and alcohol addiction can result in increased limbic negative emotion and depression, which can further exacerbate the addictive behavior. In addition, chronic alcohol abuse in adults can lead to impairment of other organ systems including damage to the gut and liver. Alcohol-induced damage to the gut can increase its permeability and allow the bacterial endotoxin LPS to enter the circulation [120]. Circulating LPS can stimulate the peripheral immune system, liver, and vasculature to release cytokines, chemokines, and other inflammatory molecules. Chronic exposure also leads to liver disease and elevation of circulating cytokines [72].

Adult rodent models have been used to investigate the interaction between the gut, the liver, and the CNS in disease across the lifespan [121, 122]. Peripheral injection of LPS in 56-day old adult mice increased brain, serum, and liver levels of TNF-α and levels of MCP-1 in the brain [123]. Peripheral injection of TNF-α mimicked the effects of LPS. Within the brain, LPS-induced TNF-α levels remained elevated as long as 10 months in contrast to peripheral serum and liver levels which rapidly returned to baseline. Mice with a null deletion mutation of the TNF receptor demonstrated that the effect of LPS or TNF-α in the brain was mediated through TNF-α and TNF receptors, respectively. Peripheral LPS injection also generated a significant change in microglial morphology in the substantia nigra, hippocampus, and cortex characteristic of the activated microglial phenotype. Evidence of concomitant LPS activation of microglia and production of parenchymal cytokines opens the possibility that LPS induction of TNF-α in the periphery may lead to activation of microglia and their long-term production of neuroimmune molecules.

In related experiments, alcohol treatment for 10 days increased TNF-α levels threefold to fourfold and MCP-1 levels almost 1.5-fold in the brain when analyzed 24 h posttreatment [124]. The study was performed in adult male mice given 5.0 g/kg/day alcohol (25%, w/v) by gavage with BAC measured at 304–316 mg/dl. Alcohol treatment did not alter the levels of IL-1β in the brain. One week posttreatment, the MCP-1 level remained elevated but the TNF-α level returned to baseline. Because not all of these immune molecules were increased in the brain in association with alcohol treatment, it is unlikely that their inflammatory activity is the only target of the selective alcohol-induced changes. The results support the emerging idea that these glial-derived molecules also have noninflammatory roles in the brain and may be regulated independent of each other. It should be noted that some studies show no effect of alcohol on expression of cytokines and chemokines in the CNS [125]. This result was observed in adult male rats gavaged with 10.0 g/kg/day alcohol (25%, w/v) with a BAC of 421 mg/dl. Thus, alcohol induction of neuroinflammation may depend on a variety of factors including the age and strain of animal and the experimental paradigm under which alcohol is administered.

Alcohol-induced neuropathology has recently been expanded to include changes in glia. In particular, evidence is accumulating for pathological changes in microglia and astrocytes. In early studies in adult rats exposed to alcohol, microglial changes were apparent. Exposure of adult male rats to chronic, intermittent alcohol from approximately 60 to 225 days of age increased the size of the microglial population in the molecular layer, but not the granular layer, of the cerebellar vermis [126]. Alcohol was administered in the drinking water as a 10% (v/v) solution (5.8 g/kg/day). The investigators did not analyze the activation phenotype of the cells with respect to either morphology or antigen expression. Proliferation of microglial cells and astrocytes throughout the hippocampus and in the motor and somatosensory cortex has been observed in adult male rats treated with alcohol for 4 days [127]. The animals were treated with 8.0–10.4 g/kg/day (25%, w/v) by gavage with a BAC range of 230–372 mg/dl. The majority of proliferative cells differentiated into microglia. Expression of the phagocytic marker ED-1 and the morphology of the microglial cells were not different in alcohol-treated and control animals, indicating that full microglial activation was not produced in this alcohol exposure paradigm. In these two studies, however, an increase in the number of microglia and astrocytes is consistent with transition to at least a partially activated glial state in response to alcohol insult.

In the adult mouse, evidence is being uncovered that alcohol-induced neuroimmune activity is linked to the neurobiology of addiction. For example, neuroimmune gene expression data in alcohol-naïve adult mice can be correlated with an alcohol drinking behavioral phenotype [128]. Meta-analysis of microarray results from 13 strains of adult mice with well-characterized high or low preference for alcohol consumption revealed 3,800 unique genes that were significantly, reproducibly changed in the brain in correlation with preference for alcohol. Among the molecules overexpressed in alcohol-preferring lines, several genes encoding neuroimmune and neuroinflammatory proteins were found including IL-1, IL-6, and β2-microglobulin. Transcription factors that are central to inflammatory responses were also overexpressed including NF-κB, CREBBP, PKC-ε, and MAP kinases. Thus, expression of cytokines and transcription factors that regulate neuroimmune function within the CNS is associated not only with classical neuroimmune and neuroinflammatory functions but also with the complex phenotype of preference for alcohol consumption. Furthermore, cytokines were demonstrated to contribute to alcohol withdrawal-induced anxiety behavior [129]. Thus, alcohol-induced inflammation may contribute to alcohol addiction by increasing alcohol consumption and by increasing anxiety associated with alcohol withdrawal, resulting in relapse toward alcohol abuse behavior.

Focused investigation using transgenic null mutant mice has linked specific neuroimmune molecules to drinking behavior [130]. The genetic targets for the study were identified by meta-analysis of existing microarray databases associated with predisposition to high alcohol consumption in mice [128] and rats [131] and compared with gene expression changes identified in human alcoholics [76, 132]. Based on selection criteria and given that the mechanisms of interest were immune and inflammatory processes, molecules in the IL-1, IL-6, MHC-I and II, TLR2, and TLR4 signaling pathways were selected for investigation. Drinking was assessed in adult mice with a null deletion mutation of β2-microglobulin, cathepsin S, cathepsin F, IL-1 receptor antagonist, CD14, or IL-6. Both preference and voluntary consumption were quantified using multiple drinking models. Deletion of any of the genes led to reduced preference and consumption in the 24-h two-bottle choice test, but only deletion of IL-1 receptor antagonist or cathepsin S led to reduced consumption in the limited access one-bottle test. It is somewhat surprising that knockout of any of these six molecules, selected from the broad category of neuroimmune factors, produced a significant diminution of alcohol intake. This result appears to be specific to these genes and does not suggest that null mutation of any neuroimmune gene will lead to reduced alcohol drinking. This is evidenced by other studies from the same investigators in which mice with a null mutation of MCP-1, CCR2, or CCL3/MIP-1α, but not those with a null mutation of CCR5, exhibited reduced alcohol preference and consumption [80]. A functional link between the nine genes whose null mutation reduced alcohol consumption is that each is either expressed by microglia or acts on microglia. These results provide key evidence that several neuroimmune molecules and multiple neuroimmune pathways are instrumental in determining alcohol drinking behavior in adult animals. Additionally, perhaps microglia in their normal role in the naïve, healthy, and adult CNS shape the framework of the neuroimmune system response to alcohol exposure.

11.2.2.4 Aging

Few studies have addressed the effects of alcohol in the aging brain. However, drinking is a major health problem in the elderly population. This population is more vulnerable to the detrimental effects of alcohol including increased sensitivity to acute motor impairment and more severe withdrawal symptoms. Chronic activation of microglia and heightened astrocyte reactivity are believed to contribute to pathologies associated with aging and age-related neurodegenerative diseases [108, 133–135]. The morphology of microglia changes with aging, including an increase in the prevalence of dystrophic microglia. The dystrophic cells exhibit enlargement of the cell body, deramification of branched processes, formation of spheroids, and fragmentation of the cell bodies. Additional changes to microglia with aging may include increased production of inflammatory molecules such as cytokines, chemokines, and reactive oxygen and nitrogen species, as well as altered mitochondrial function and increases in iron storage. Neurons in the aging CNS are provided with less trophic support in the form of reduced growth factor production and reduced synaptic function. Furthermore, neurogenesis dramatically declines with aging, thus restricting neuron regeneration. Alterations of microglial function may significantly contribute to these changes. Thus, it is intriguing to speculate that natural age-associated changes in microglia and astrocytes may enhance alcohol neurotoxicity and the negative effects of alcohol on the neuroimmune system. Given this background and the knowledge that a high incidence of binge drinking occurs in the elderly population, it is plausible that age-associated changes in microglial activation and production of neuroimmune molecules contribute to this alcohol drinking behavior.

Studies of alcohol effects on glia in aged animals are limited [136, 137] and employ different ages, strains, and alcohol exposure models. Two very long-term chronic treatment studies provide particular insights into alcohol impact on glia in the aging brain. In one study, male rats were given alcohol in liquid diet daily from 12 to 22 months of age [136]. Analysis in the molecular layer of the cerebellum revealed that the numerical density of GFAP-positive Bergmann glial fibers and OX-42-positive microglia was not different between alcohol exposed and control animals. There was no glial loss or activation observed. However, there are numerous reports in both humans and rodent models that astrocytes and microglia exhibit increased reactivity in the aged brain as described above. In this study, perhaps the glial activation response was already at a maximum in the control animals such that alcohol exposure could not elicit further activation. In another study, both sexes of alcohol-preferring AA rats were given alcohol daily (3.2–6.6 g/kg/day as a 10–12%, v/v, solution in drinking water) from 3 to 24 months of age [137]. Similar to the above study, there were no morphological differences in GFAP-positive cells between alcohol consuming and control animals, but less GFAP immunoreactivity was present in the Bergmann glia in animals that received alcohol. Chronic alcohol consumption may have reduced the number of Bergmann glial cells, caused retraction of fibers, or suppressed expression of GFAP protein. Major differences in experimental design between these two studies could account for the distinct observations.

More recent studies indicate that alcohol induces molecular markers of neuroinflammation in the aged brain. In humans, the chemokine MCP-1 and the microglial glial markers Iba-1 and GluT5 were increased in postmortem brain from alcoholic individuals with a mean age of 61–64 years [79]. In addition, unpublished data from our laboratory demonstrates induction of inflammatory molecules in specific brain regions in aged mice treated with alcohol compared to control animals. Further investigation is needed to understand the interaction of alcohol with the aging CNS and aging neuroimmune system.

11.2.3 Role of TLR4 in Alcohol Induction of Immune Activity in the CNS

Microglia respond to pathogens and endogenous danger signals through activation of TLRs. Recent studies have begun to evaluate the role of specific TLRs in modulating alcohol effects on CNS inflammation. For example, alcohol was demonstrated to induce the expression of the inflammatory molecules iNOS and COX-2 in primary astrocyte cultures. In these studies, alcohol rapidly increased the phosphorylation of IL-1R-associated kinase (IRAK), ERK1/2, stress-associated protein kinase/JNK, and p38 MAP kinase. Alcohol was further demonstrated to increase the activation of the transcription factors NF-κB and AP1. Activation of these signaling pathways and production of inflammatory molecules by alcohol were inhibited by antibodies specific for TLR4 and IL-1R, supporting a role of the TLR-IL-1R pathway in alcohol-induced production of inflammatory molecules by astrocytes [138]. Additional studies indicated that alcohol-induced phosphorylation of Src and TLR4 played a significant role in the production of inflammatory molecules like COX-2 in astrocytes [139]. Alcohol was also demonstrated to induce the translocation of TLR4 and/or IL-1R into lipid rafts in astrocytes [140]. These studies suggest that alcohol triggers association of TLR and IL-1R with lipid rafts which are subsequently endocytosed and stimulate signaling pathways which result in production of inflammatory molecules.

Studies utilizing TLR4-deficient mice were particularly informative in evaluating the role of TLR4 in alcohol-induced modulation of inflammation in the CNS. Alcohol induced the activation of ERK, JNK, and p38 MAPK, as well as NF-κB, iNOS, COX-2, and cytokine expression by primary microglia derived from wild-type mice, but not from TLR4-deficient animals. These studies further demonstrated that conditioned medium from alcohol-treated wild-type microglia increased the apoptosis of primary cortical neurons while conditioned medium from TLR4-deficient microglia induced less apoptosis of these neurons [141]. Additional studies indicated that chronic administration of alcohol increased the expression of the astrocytic marker GFAP and the microglial marker CD11b in the medial frontal cortex of wild-type mice. Alcohol induction of CD11b was completely abrogated in TLR4-deficient mice, and alcohol induction of GFAP was partially suppressed in these animals. The study was performed in female mice of unspecified age exposed to 4.0 g/kg/day alcohol as a 25% (v/v) solution via intraperitoneal injection. Homogenates of brain cortex also exhibited alcohol-induced increases in TNF-α, IL-1β, and IL-6 mRNA, as well as iNOS and COX-2 protein in wild type, but not TLR4 knockout mice. Altered NF-κB signaling in TLR4 knockout mice likely mediated the suppressed production of inflammatory molecules in these mice. This was demonstrated in studies indicating that IκB-α expression was suppressed and NF-κB expression was increased in the cerebral cortex of alcohol-treated wild-type animals but not TLR4-deficient animals. In addition, alcohol induced the expression of caspase-3, which is associated with apoptosis, in the frontal cortex of wild type but not TLR4-deficient mice [142]. In this study, female mice 7 weeks of age were treated with 12.8 g/kg/day alcohol (10%, v/v, in drinking water) for 5 months with a BAC range of 87–140 mg/dl. Collectively, these studies suggest that TLR4 plays a critical role in alcohol-induced glial activation in vitro and alcohol-induced neuroinflammation and neurotoxicity in vivo. Thus, alcohol-induced inflammation likely plays a significant role in alcohol-mediated neurotoxicity.

Recently, deletion of TLR4 was demonstrated to negate chronic alcohol consumption-mediated cognitive and anxiety-associated behavioral impairment [143]. This study was performed in 6- to 7-week-old male mice treated with 10.3 g/kg/day (10%, v/v, in drinking water) for 5 months with a BAC of 110 mg/dl. Since TLR4-deficient mice did not exhibit glial activation and cytokine production like wild-type mice, these studies further support a TLR4-mediated link between alcohol-induced neuroinflammation and behavioral and cognitive deficits. Interestingly, these studies also demonstrated that chronic alcohol administration decreased histone acetylation and histone acetyltransferase activity in the hippocampus, striatum, and cortex of wild-type but not TLR4 knockout mice. These studies suggest that TLR4-mediated alterations in chromatin configuration and epigenetic modifications may play an important role in modulating alcohol effects in the brain including cognitive and behavioral dysfunction. In other studies, TLR4-deficient mice demonstrated a reduced period of sedation and impaired motor activity compared to wild-type animals. These altered behavioral responses in TLR4 knockout mice were suggested not to be due to altered alcohol pharmacokinetics [144] based on a study in 10- to 14-week-old male mice given 2.0–4.5 g/kg alcohol as a single intraperitoneal injection, achieving BAC of 368–574 mg/dl. Interestingly, infusion of TLR4 siRNA into the central nucleus of the amygdala inhibited binge drinking behavior in alcohol-preferring rats. Other studies indicated that $GABA_A$ $\alpha 2$ modulates TLR4-mediated alterations in drinking behavior in rats [145]. Collectively, the studies outlined support the hypothesis that alcohol-induced activation of the innate immune response in the CNS is mediated, at least in part, by TLR4. In addition, TLR4 plays a significant role in modulating the behavioral and cognitive effects of alcohol.

Ligand binding to the TLR4 receptor activates two distinct signaling pathways, the MyD88-dependent and the MyD88-independent pathways. In the MyD88-dependent pathway, TLR4 interacts with the adaptor protein MyD88 which results in the activation of IRAK1/4 and TRAF6. Subsequently, the transcription factor NF-κB is activated followed by the induction of a variety of proinflammatory molecules such as cytokines and chemokines. In the MyD88-independent, or TRIF, pathway, the TLR4 complex interacts with the adaptor TRIF leading to activation of IKKTAK1 kinase and phosphorylation of the transcription factor IRF-3 as well as late activation of NF-κB. Activated IRF-3 induces the expression of type I interferons and interferon-responsive genes. Although it is clear that many of the effects of alcohol on the CNS are mediated by TLR4, little is known concerning whether these effects occur through MyD88-dependent or MyD88-independent pathways. Recent studies demonstrated that the sedative and motor impairment effects of alcohol were muted in both TLR4 and MyD88 knockout mice suggesting that the MyD88-dependent pathway is critical in controlling these alcohol-mediated effects. In other studies, alcohol was demonstrated to not only increase NF-κB but also IFR-3 activation in microglia

[141], which may suggest that the MyD88-independent and possibly the MyD88-dependent pathways are involved in these processes. Future studies utilizing MyD88 and TRIF knockout mice are required to fully assess the role of MyD88-dependent and MyD88-independent signaling in alcohol responses in the CNS.

11.2.4 Alcohol Effects on Neurons: Role of Immune Activation

Alcohol is known to have toxic effects on neurons. For example, prenatal treatment with alcohol is toxic to hippocampal neurons, cortical neurons, Purkinje and granule cell neurons of the cerebellum, and hypothalamic neurons [89, 146–151]. More recently, a primary cell culture model was utilized to evaluate the effects of alcohol on microglial activation and neuron cell viability [152]. Alcohol increased the expression of the cytokines TNF-α and IL-6 and the chemokines MIP-1 and MIP-2 in primary microglial cultures. Conditioned medium from alcohol-treated microglia increased the presence of nucleosomes in cultured mediobasal hypothalamic neurons indicating that these neurons were undergoing apoptosis. Immunoneutralization studies indicated that microglial-produced TNF-α played a critical role in the apoptosis of these neurons and, further, that cAMP suppressed TNF-α production by microglia to protect these hypothalamic neurons.

11.2.5 Treatment of Alcohol-Induced Neuropathologies with Immunomodulatory Drugs

As discussed above, there is now a wealth of literature indicating that alcohol induces the expression of inflammatory molecules which can contribute to the pathologies associated with alcohol abuse as well as alcohol-mediated behaviors including alcohol addiction. This suggests that agents that suppress inflammation may be effective in obviating these pathologies. Recently, studies have begun to investigate the effects of immunosuppressive agents on CNS pathologies associated with alcohol abuse. Using a neonatal mouse model of FASD, we found that PPAR-γ agonists, which were previously shown to suppress inflammation [153], protected cerebellar Purkinje neurons and microglia from the toxic effects of alcohol [90] (Fig. 11.2). Interestingly, PPAR-γ agonists were also recently demonstrated to suppress alcohol consumption and relapse to alcohol-seeking behavior in alcohol-preferring rats [154]. Resveratrol, a polyphenolic compound possessing antioxidant and neuroprotective qualities, was shown to protect cerebellar neurons in a rat model of FASD [93]. In another study utilizing a rat model of FASD, resveratrol was shown to suppress alcohol induction of cytokines, NF-κB, oxidative stress, and caspase-3 expression in the cortex and hippocampus, as well as alcohol-induced impairment of memory [91]. The nonsteroidal anti-inflammatory drug indomethacin was demonstrated to suppress the production of the molecules COX-2 and iNOS as well as suppress alcohol-induced behavioral deficits. These studies were performed using a binge model in adolescent rats which exhibited cognitive and motor deficits as adults [155]. The tetracycline antibiotic minocycline has previously been demonstrated to suppress immune responses and was recently demonstrated to reduce alcohol consumption in mice [156]. Minocycline was also shown to decrease alcohol-induced sedation but interestingly increased motor impairment in mice. Collectively, these studies suggest that immunosuppressive compounds have potential as therapeutics in alcohol-induced pathologies.

11.3 Conclusion

Alcohol abuse throughout the lifetime can result in neuropathology ranging from FASD to long-term deficits resulting from binge drinking at all ages beginning in adolescence, and the effects of chronic alcohol ingestion in adults and the elderly. Alcohol abuse at each of these life stages is associated with CNS inflammation as determined by glial morphology and expression of inflammatory molecules. The resulting neuroinflammation is postulated to contribute to alcohol-induced neuropathology. Further studies are needed to systematically compare the relative effects of alcohol-induced inflammation on CNS pathology over the lifespan.

Animal studies indicate that alcohol-stimulated inflammatory molecules play a critical role not only in neurodegeneration but also in modulation of alcohol consumption, addiction, and the behavioral effects of alcohol. These animal studies are supported by human studies demonstrating CNS inflammation in postmortem brain from alcoholics. Human studies also indicate a correlation between genes encoding inflammatory molecules and the genetic risk for alcoholism. Questions remain regarding how alcohol alters the complex interactions that occur between glia and neurons and how this contributes to glial activation, glial loss, neurodegeneration, reduced neurogenesis, and changes in synaptic plasticity and physiology that may underlie drinking behavior and addiction. Current studies suggest that cytokines, chemokines, oxidative stress, and TLR4 signaling can initiate or maintain alcohol-induced inflammation in the CNS. The efficacy of immunomodulatory therapeutics in preclinical animal models provides support for the hypothesized role of neuroinflammation in alcohol-induced neurodegeneration as well as alcohol drinking behavior and addiction. A more complete understanding of the mechanisms by which alcohol induces inflammation in the CNS is required for directed therapies for alcohol-mediated neuropathologies.

The studies outlined in this review have been seminal in defining our current understanding of the effects of alcohol on immune activity in the CNS throughout the lifespan. However, significant gaps in our knowledge continue to exist including the molecular mechanisms by which alcohol modulates CNS inflammation, the mechanisms by which immune molecules affect alcohol consumption and addiction, and the mechanisms by which alcohol alters the interplay between peripheral and CNS immune cells. Future studies are needed to address these and other important issues.

References

1. May PA, Gossage JP (2001) Estimating the prevalence of fetal alcohol syndrome. A summary. Alcohol Res Health 25(3):159–167
2. Sokol RJ, Delaney-Black V, Nordstrom B (2003) Fetal alcohol spectrum disorder. JAMA 290(22):2996–2999
3. May PA, Gossage JP, Kalberg WO, Robinson LK, Buckley D, Manning M, Hoyme HE (2009) Prevalence and epidemiologic characteristics of FASD from various research methods with an emphasis on recent in-school studies. Dev Disabil Res Rev 15(3):176–192
4. Cui C, Grandison L, Noronha A (2011) Neuroimmune mechanisms of brain function and alcohol related disorders. Brain Behav Immun 25(Suppl 1):S1–S3
5. Crews FT, Zou J, Qin L (2011) Induction of innate immune genes in brain create the neurobiology of addiction. Brain Behav Immun 25(Suppl 1):S4–S12
6. Goral J, Karavitis J, Kovacs EJ (2008) Exposure-dependent effects of ethanol on the innate immune system. Alcohol 42(4):237–247
7. Crews FT, Bechara R, Brown LA, Guidot DM, Mandrekar P, Oak S, Qin L, Szabo G, Wheeler M, Zou J (2006) Cytokines and alcohol. Alcohol Clin Exp Res 30(4):720–730
8. Crews FT, Nixon K (2009) Mechanisms of neurodegeneration and regeneration in alcoholism. Alcohol Alcohol 44(2):115–127
9. Blanco AM, Guerri C (2007) Ethanol intake enhances inflammatory mediators in brain: role of glial cells and TLR4/IL-1RI receptors. Front Biosci 12:2616–2630
10. Davis RL, Syapin PJ (2004) Ethanol increases nuclear factor-kappa B activity in human astroglial cells. Neurosci Lett 371(2–3):128–132
11. Zou J, Crews F (2006) CREB and NF-kappaB transcription factors regulate sensitivity to excitotoxic and oxidative stress induced neuronal cell death. Cell Mol Neurobiol 26(4–6):385–405
12. Knapp DJ, Crews FT (1999) Induction of cyclooxygenase-2 in brain during acute and chronic ethanol treatment and ethanol withdrawal. Alcohol Clin Exp Res 23(4):633–643
13. Crews F, Nixon K, Kim D, Joseph J, Shukitt-Hale B, Qin L, Zou J (2006) BHT blocks NF-kappaB activation and ethanol-induced brain damage. Alcohol Clin Exp Res 30(11):1938–1949
14. Zou J, Crews F (2010) Induction of innate immune gene expression cascades in brain slice cultures by ethanol: key role of NF-kappaB and proinflammatory cytokines. Alcohol Clin Exp Res 34(5):777–789
15. Blanco AM, Pascual M, Valles SL, Guerri C (2004) Ethanol-induced iNOS and COX-2 expression in cultured astrocytes via NF-kappa B. Neuroreport 15(4):681–685
16. Ward RJ, Zhang Y, Crichton RR, Piret B, Piette J, de Witte P (1996) Identification of the nuclear transcription factor NFkappaB in rat after in vivo ethanol administration. FEBS Lett 389(2):119–122
17. Saijo K, Glass CK (2011) Microglial cell origin and phenotypes in health and disease. Nat Rev Immunol 11(11):775–787
18. Kettenmann H, Hanisch UK, Noda M, Verkhratsky A (2011) Physiology of microglia. Physiol Rev 91(2):461–553
19. McKercher SR, Torbett BE, Anderson KL, Henkel GW, Vestal DJ, Baribault H, Klemsz M, Feeney AJ, Wu GE, Paige CJ, Maki RA (1996) Targeted disruption of the PU.1 gene results in multiple hematopoietic abnormalities. EMBO J 15(20):5647–5658
20. Eglitis MA, Mezey E (1997) Hematopoietic cells differentiate into both microglia and macroglia in the brains of adult mice. Proc Natl Acad Sci USA 94(8):4080–4085
21. Priller J, Flugel A, Wehner T, Boentert M, Haas CA, Prinz M, Fernandez-Klett F, Prass K, Bechmann I, de Boer BA, Frotscher M, Kreutzberg GW, Persons DA, Dirnagl U (2001) Targeting gene-modified hematopoietic cells to the central nervous system: use of green fluorescent protein uncovers microglial engraftment. Nat Med 7(12):1356–1361

22. Hickey WF, Kimura H (1988) Perivascular microglial cells of the CNS are bone marrow-derived and present antigen in vivo. Science 239(4837):290–292
23. Ransohoff RM (2007) Microgliosis: the questions shape the answers. Nat Neurosci 10(12):1507–1509
24. Hickey WF (1991) Migration of hematogenous cells through the blood–brain barrier and the initiation of CNS inflammation. Brain Pathol 1(2):97–105
25. Ajami B, Bennett JL, Krieger C, Tetzlaff W, Rossi FM (2007) Local self-renewal can sustain CNS microglia maintenance and function throughout adult life. Nat Neurosci 10(12):1538–1543
26. Mildner A, Schmidt H, Nitsche M, Merkler D, Hanisch UK, Mack M, Heikenwalder M, Bruck W, Priller J, Prinz M (2007) Microglia in the adult brain arise from Ly-6ChiCCR2+ monocytes only under defined host conditions. Nat Neurosci 10(12):1544–1553
27. Ajami B, Bennett JL, Krieger C, McNagny KM, Rossi FM (2011) Infiltrating monocytes trigger EAE progression, but do not contribute to the resident microglia pool. Nat Neurosci 14(9):1142–1149
28. Glass CK, Saijo K, Winner B, Marchetto MC, Gage FH (2010) Mechanisms underlying inflammation in neurodegeneration. Cell 140(6):918–934
29. Fellner L, Jellinger KA, Wenning GK, Stefanova N (2011) Glial dysfunction in the pathogenesis of alpha-synucleinopathies: emerging concepts. Acta Neuropathol 121(6):675–693
30. Reitz C, Brayne C, Mayeux R (2011) Epidemiology of Alzheimer disease. Nat Rev Neurol 7(3):137–152
31. Lawson LJ, Perry VH, Gordon S (1992) Turnover of resident microglia in the normal adult mouse brain. Neuroscience 48(2):405–415
32. Lawson LJ, Perry VH, Dri P, Gordon S (1990) Heterogeneity in the distribution and morphology of microglia in the normal adult mouse brain. Neuroscience 39(1):151–170
33. Rivest S (2009) Regulation of innate immune responses in the brain. Nat Rev Immunol 9(6):429–439
34. Ransohoff RM, Perry VH (2009) Microglial physiology: unique stimuli, specialized responses. Annu Rev Immunol 27:119–145
35. Sierra A, Encinas JM, Deudero JJ, Chancey JH, Enikolopov G, Overstreet-Wadiche LS, Tsirka SE, Maletic-Savatic M (2010) Microglia shape adult hippocampal neurogenesis through apoptosis-coupled phagocytosis. Cell Stem Cell 7(4):483–495
36. Davalos D, Grutzendler J, Yang G, Kim JV, Zuo Y, Jung S, Littman DR, Dustin ML, Gan WB (2005) ATP mediates rapid microglial response to local brain injury in vivo. Nat Neurosci 8(6):752–758
37. Nimmerjahn A, Kirchhoff F, Helmchen F (2005) Resting microglial cells are highly dynamic surveillants of brain parenchyma in vivo. Science 308(5726):1314–1318
38. Tremblay ME, Lowery RL, Majewska AK (2010) Microglial interactions with synapses are modulated by visual experience. PLoS Biol 8(11):e1000527
39. Zettel M, Tremblay ME, Ison JR, Allen PD, Morrison WZ, Inglis A, Stanley HE, Rosene DL, Majewska AK (2010) Age-related loss of hearing or sight in mice yields greater apparent aging effects in corresponding sensory cortices. Soc Neurosci Abstr 36:765.3
40. Perry VH, Hume DA, Gordon S (1985) Immunohistochemical localization of macrophages and microglia in the adult and developing mouse brain. Neuroscience 15(2):313–326
41. Dalmau I, Finsen B, Zimmer J, Gonzalez B, Castellano B (1998) Development of microglia in the postnatal rat hippocampus. Hippocampus 8(5):458–474
42. Fiske BK, Brunjes PC (2000) Microglial activation in the developing rat olfactory bulb. Neuroscience 96(4):807–815
43. Wake H, Moorhouse AJ, Jinno S, Kohsaka S, Nabekura J (2009) Resting microglia directly monitor the functional state of synapses in vivo and determine the fate of ischemic terminals. J Neurosci 29(13):3974–3980
44. Schafer DP, Lehrman EK, Kautzman AG, Barres B, Stevens B (2010) Synaptic pruning in the CNS: the role of microglia and the complement system. Soc Neurosci Abstr 36:554.5
45. Stevens B, Allen NJ, Vazquez LE, Howell GR, Christopherson KS, Nouri N, Micheva KD, Mehalow AK, Huberman AD, Stafford B, Sher A, Litke AM, Lambris JD, Smith SJ, John

SW, Barres BA (2007) The classical complement cascade mediates CNS synapse elimination. Cell 131(6):1164–1178

46. Paolicelli RC, Bolasco G, Pagani F, Maggi L, Scianni M, Panzanelli P, Giustetto M, Ferreira TA, Guiducci E, Dumas L, Ragozzino D, Gross CT (2011) Synaptic pruning by microglia is necessary for normal brain development. Science 333(6048):1456–1458
47. Boulanger LM (2009) Immune proteins in brain development and synaptic plasticity. Neuron 64(1):93–109
48. Carson MJ, Doose JM, Melchior B, Schmid CD, Ploix CC (2006) CNS immune privilege: hiding in plain sight. Immunol Rev 213:48–65
49. Brockhaus J, Ilschner S, Banati RB, Kettenmann H (1993) Membrane properties of ameboid microglial cells in the corpus callosum slice from early postnatal mice. J Neurosci 13(10):4412–4421
50. Haas S, Brockhaus J, Verkhratsky A, Kettenmann H (1996) ATP-induced membrane currents in ameboid microglia acutely isolated from mouse brain slices. Neuroscience 75(1):257–261
51. Streit WJ, Xue QS (2009) Life and death of microglia. J Neuroimmune Pharmacol 4(4):371–379
52. Sievers J, Parwaresch R, Wottge HU (1994) Blood monocytes and spleen macrophages differentiate into microglia-like cells on monolayers of astrocytes: morphology. Glia 12(4):245–258
53. Wollmer MA, Lucius R, Wilms H, Held-Feindt J, Sievers J, Mentlein R (2001) ATP and adenosine induce ramification of microglia in vitro. J Neuroimmunol 115(1–2):19–27
54. Schilling T, Nitsch R, Heinemann U, Haas D, Eder C (2001) Astrocyte-released cytokines induce ramification and outward K+channel expression in microglia via distinct signalling pathways. Eur J Neurosci 14(3):463–473
55. Carson MJ, Bilousova TV, Puntambekar SS, Melchior B, Doose JM, Ethell IM (2007) A rose by any other name? The potential consequences of microglial heterogeneity during CNS health and disease. Neurotherapeutics 4(4):571–579
56. Colton CA (2009) Heterogeneity of microglial activation in the innate immune response in the brain. J Neuroimmune Pharmacol 4(4):399–418
57. Hanisch UK, Kettenmann H (2007) Microglia: active sensor and versatile effector cells in the normal and pathologic brain. Nat Neurosci 10(11):1387–1394
58. Perry VH, Cunningham C, Holmes C (2007) Systemic infections and inflammation affect chronic neurodegeneration. Nat Rev Immunol 7(2):161–167
59. Schwartz M, Butovsky O, Bruck W, Hanisch UK (2006) Microglial phenotype: is the commitment reversible? Trends Neurosci 29(2):68–74
60. Streit WJ, Walter SA, Pennell NA (1999) Reactive microgliosis. Prog Neurobiol 57(6):563–581
61. Meeuwsen S, Bsibsi M, Persoon-Deen C, Ravid R, van Noort JM (2005) Cultured human adult microglia from different donors display stable cytokine, chemokine and growth factor gene profiles but respond differently to a pro-inflammatory stimulus. Neuroimmunomodulation 12(4):235–245
62. Sierra A, Gottfried-Blackmore AC, McEwen BS, Bulloch K (2007) Microglia derived from aging mice exhibit an altered inflammatory profile. Glia 55(4):412–424
63. Hart AD, Wyttenbach A, Hugh PV, Teeling JL (2012) Age related changes in microglial phenotype vary between CNS regions: grey versus white matter differences. Brain Behav Immun 26:754–765
64. Choi JH, Won MH (2011) Microglia in the normally aged hippocampus. Lab Anim Res 27(3):181–187
65. Miller KR, Streit WJ (2007) The effects of aging, injury and disease on microglial function: a case for cellular senescence. Neuron Glia Biol 3(3):245–253
66. Streit WJ (2006) Microglial senescence: does the brain's immune system have an expiration date? Trends Neurosci 29(9):506–510
67. Zhang Y, Barres BA (2010) Astrocyte heterogeneity: an underappreciated topic in neurobiology. Curr Opin Neurobiol 20(5):588–594

68. Freeman MR (2010) Specification and morphogenesis of astrocytes. Science 330(6005): 774–778
69. Sofroniew MV, Vinters HV (2010) Astrocytes: biology and pathology. Acta Neuropathol 119(1):7–35
70. Oberheim NA, Goldman SA, Nedergaard M (2012) Heterogeneity of astrocytic form and function. Methods Mol Biol 814:23–45
71. Parpura V, Heneka MT, Montana V, Oliet SH, Schousboe A, Haydon PG, Stout RF Jr, Spray DC, Reichenbach A, Pannicke T, Pekny M, Pekna M, Zorec R, Verkhratsky A (2012) Glial cells in (patho)physiology. J Neurochem 121:4–27
72. Achur RN, Freeman WM, Vrana KE (2010) Circulating cytokines as biomarkers of alcohol abuse and alcoholism. J Neuroimmune Pharmacol 5(1):83–91
73. Laso FJ, Vaquero JM, Almeida J, Marcos M, Orfao A (2007) Production of inflammatory cytokines by peripheral blood monocytes in chronic alcoholism: relationship with ethanol intake and liver disease. Cytometry B Clin Cytom 72(5):408–415
74. Kiefer F, Jahn H, Schick M, Wiedemann K (2002) Alcohol intake, tumour necrosis factor-alpha, leptin and craving: factors of a possibly vicious circle? Alcohol Alcohol 37(4): 401–404
75. Kim DJ, Kim W, Yoon SJ, Choi BM, Kim JS, Go HJ, Kim YK, Jeong J (2003) Effects of alcohol hangover on cytokine production in healthy subjects. Alcohol 31(3):167–170
76. Liu J, Lewohl JM, Harris RA, Iyer VR, Dodd PR, Randall PK, Mayfield RD (2006) Patterns of gene expression in the frontal cortex discriminate alcoholic from nonalcoholic individuals. Neuropsychopharmacology 31(7):1574–1582
77. Okvist A, Johansson S, Kuzmin A, Bazov I, Merino-Martinez R, Ponomarev I, Mayfield RD, Harris RA, Sheedy D, Garrick T, Harper C, Hurd YL, Terenius L, Ekstrom TJ, Bakalkin G, Yakovleva T (2007) Neuroadaptations in human chronic alcoholics: dysregulation of the NF-kappaB system. PLoS One 2(9):e930
78. Yakovleva T, Bazov I, Watanabe H, Hauser KF, Bakalkin G (2011) Transcriptional control of maladaptive and protective responses in alcoholics: a role of the NF-kappaB system. Brain Behav Immun 25(Suppl 1):S29–S38
79. He J, Crews FT (2008) Increased MCP-1 and microglia in various regions of the human alcoholic brain. Exp Neurol 210(2):349–358
80. Blednov YA, Bergeson SE, Walker D, Ferreira VM, Kuziel WA, Harris RA (2005) Perturbation of chemokine networks by gene deletion alters the reinforcing actions of ethanol. Behav Brain Res 165(1):110–125
81. Nelson TE, Hao C, Manos J, Ransohoff RM, Gruol DL (2011) Altered hippocampal synaptic transmission in transgenic mice with astrocyte-targeted enhanced CCL2 expression. Brain Behav Immun 25(Suppl 1):S106–S119
82. Ahluwalia B, Wesley B, Adeyiga O, Smith DM, Da-Silva A, Rajguru S (2000) Alcohol modulates cytokine secretion and synthesis in human fetus: an in vivo and in vitro study. Alcohol 21(3):207–213
83. Guerri C, Bazinet A, Riley EP (2009) Foetal alcohol spectrum disorders and alterations in brain and behaviour. Alcohol Alcohol 44(2):108–114
84. Clancy B, Darlington RB, Finlay BL (2001) Translating developmental time across mammalian species. Neuroscience 105(1):7–17
85. Napper RM, West JR (1995) Permanent neuronal cell loss in the cerebellum of rats exposed to continuous low blood alcohol levels during the brain growth spurt: a stereological investigation. J Comp Neurol 362(2):283–292
86. Pierce DR, Kane CJ, Serbus DC, Light KE (1997) Microencephaly and selective decreases in cerebellar Purkinje cell numbers following combined exposure to ethanol and methadone during rat brain development. Dev Neurosci 19(5):438–445
87. Goodlett CR, Eilers AT (1997) Alcohol-induced Purkinje cell loss with a single binge exposure in neonatal rats: a stereological study of temporal windows of vulnerability. Alcohol Clin Exp Res 21(4):738–744

88. Pierce DR, Williams DK, Light KE (1999) Purkinje cell vulnerability to developmental ethanol exposure in the rat cerebellum. Alcohol Clin Exp Res 23(10):1650–1659
89. Light KE, Belcher SM, Pierce DR (2002) Time course and manner of Purkinje neuron death following a single ethanol exposure on postnatal day 4 in the developing rat. Neuroscience 114(2):327–337
90. Kane CJ, Phelan KD, Han L, Smith RR, Xie J, Douglas JC, Drew PD (2011) Protection of neurons and microglia against ethanol in a mouse model of fetal alcohol spectrum disorders by peroxisome proliferator-activated receptor-γ agonists. Brain Behav Immun 25(Suppl 1):S137–S145
91. Tiwari V, Chopra K (2011) Resveratrol prevents alcohol-induced cognitive deficits and brain damage by blocking inflammatory signaling and cell death cascade in neonatal rat brain. J Neurochem 117(4):678–690
92. Schneider ML, Moore CF, Adkins MM (2011) The effects of prenatal alcohol exposure on behavior: rodent and primate studies. Neuropsychol Rev 21(2):186–203
93. Kumar A, Singh CK, Lavoie HA, Dipette DJ, Singh US (2011) Resveratrol restores Nrf2 level and prevents ethanol-induced toxic effects in the cerebellum of a rodent model of fetal alcohol spectrum disorders. Mol Pharmacol 80(3):446–457
94. DeVito WJ, Stone S (2001) Prenatal exposure to ethanol alters the neuroimmune response to a central nervous system wound in the adult rat. Alcohol 25(1):39–47
95. Zhang X, Sliwowska JH, Weinberg J (2005) Prenatal alcohol exposure and fetal programming: effects on neuroendocrine and immune function. Exp Biol Med 230(6):376–388
96. Silveri MM, Spear LP (1998) Decreased sensitivity to the hypnotic effects of ethanol early in ontogeny. Alcohol Clin Exp Res 22(3):670–676
97. Van Skike CE, Botta P, Chin VS, Tokunaga S, McDaniel JM, Venard J, Diaz-Granados JL, Valenzuela CF, Matthews DB (2010) Behavioral effects of ethanol in cerebellum are age dependent: potential system and molecular mechanisms. Alcohol Clin Exp Res 34(12): 2070–2080
98. Monti PM, Miranda R Jr, Nixon K, Sher KJ, Swartzwelder HS, Tapert SF, White A, Crews FT (2005) Adolescence: booze, brains, and behavior. Alcohol Clin Exp Res 29(2):207–220
99. Crews F, He J, Hodge C (2007) Adolescent cortical development: a critical period of vulnerability for addiction. Pharmacol Biochem Behav 86(2):189–199
100. Peleg-Oren N, Saint-Jean G, Cardenas GA, Tammara H, Pierre C (2009) Drinking alcohol before age 13 and negative outcomes in late adolescence. Alcohol Clin Exp Res 33(11):1966–1972
101. Doremus-Fitzwater TL, Varlinskaya EI, Spear LP (2010) Motivational systems in adolescence: possible implications for age differences in substance abuse and other risk-taking behaviors. Brain Cogn 72(1):114–123
102. Nixon K, McClain JA (2010) Adolescence as a critical window for developing an alcohol use disorder: current findings in neuroscience. Curr Opin Psychiatry 23(3):227–232
103. Maldonado-Devincci AM, Alipour KK, Michael LA, Kirstein CL (2010) Repeated binge ethanol administration during adolescence enhances voluntary sweetened ethanol intake in young adulthood in male and female rats. Pharmacol Biochem Behav 96(4):476–487
104. Noonan MA, Bulin SE, Fuller DC, Eisch AJ (2010) Reduction of adult hippocampal neurogenesis confers vulnerability in an animal model of cocaine addiction. J Neurosci 30(1):304–315
105. He J, Crews FT (2007) Neurogenesis decreases during brain maturation from adolescence to adulthood. Pharmacol Biochem Behav 86(2):327–333
106. McClain JA, Hayes DM, Morris SA, Nixon K (2011) Adolescent binge alcohol exposure alters hippocampal progenitor cell proliferation in rats: effects on cell cycle kinetics. J Comp Neurol 519(13):2697–2710
107. Nixon K, Crews FT (2002) Binge ethanol exposure decreases neurogenesis in adult rat hippocampus. J Neurochem 83(5):1087–1093
108. Russo I, Barlati S, Bosetti F (2011) Effects of neuroinflammation on the regenerative capacity of brain stem cells. J Neurochem 116(6):947–956
109. Spulber S, Oprica M, Bartfai T, Winblad B, Schultzberg M (2008) Blunted neurogenesis and gliosis due to transgenic overexpression of human soluble IL-1ra in the mouse. Eur J Neurosci 27(3):549–558

110. Sun Y, Jin K, Childs JT, Xie L, Mao XO, Greenberg DA (2005) Neuronal nitric oxide synthase and ischemia-induced neurogenesis. J Cereb Blood Flow Metab 25(4):485–492
111. Vallieres L, Campbell IL, Gage FH, Sawchenko PE (2002) Reduced hippocampal neurogenesis in adult transgenic mice with chronic astrocytic production of interleukin-6. J Neurosci 22(2):486–492
112. Monje ML, Toda H, Palmer TD (2003) Inflammatory blockade restores adult hippocampal neurogenesis. Science 302(5651):1760–1765
113. Liu BF, Gao EJ, Zeng XZ, Ji M, Cai Q, Lu Q, Yang H, Xu QY (2006) Proliferation of neural precursors in the subventricular zone after chemical lesions of the nigrostriatal pathway in rat brain. Brain Res 1106(1):30–39
114. Goshen I, Kreisel T, Ben-Menachem-Zidon O, Licht T, Weidenfeld J, Ben-Hur T, Yirmiya R (2008) Brain interleukin-1 mediates chronic stress-induced depression in mice via adrenocortical activation and hippocampal neurogenesis suppression. Mol Psychiatry 13(7):717–728
115. Rolls A, Shechter R, London A, Ziv Y, Ronen A, Levy R, Schwartz M (2007) Toll-like receptors modulate adult hippocampal neurogenesis. Nat Cell Biol 9(9):1081–1088
116. Ekdahl CT, Claasen JH, Bonde S, Kokaia Z, Lindvall O (2003) Inflammation is detrimental for neurogenesis in adult brain. Proc Natl Acad Sci USA 100(23):13632–13637
117. Baron R, Nemirovsky A, Harpaz I, Cohen H, Owens T, Monsonego A (2008) IFN-γ enhances neurogenesis in wild-type mice and in a mouse model of Alzheimer's disease. FASEB J 22(8):2843–2852
118. Ziv Y, Ron N, Butovsky O, Landa G, Sudai E, Greenberg N, Cohen H, Kipnis J, Schwartz M (2006) Immune cells contribute to the maintenance of neurogenesis and spatial learning abilities in adulthood. Nat Neurosci 9(2):268–275
119. McClain JA, Morris SA, Deeny MA, Marshall SA, Hayes DM, Kiser ZM, Nixon K (2011) Adolescent binge alcohol exposure induces long-lasting partial activation of microglia. Brain Behav Immun 25(Suppl 1):S120–S128
120. Mandrekar P, Szabo G (2009) Signalling pathways in alcohol-induced liver inflammation. J Hepatol 50(6):1258–1266
121. Wang HJ, Zakhari S, Jung MK (2010) Alcohol, inflammation, and gut-liver-brain interactions in tissue damage and disease development. World J Gastroenterol 16(11):1304–1313
122. Rivest S, Lacroix S, Vallieres L, Nadeau S, Zhang J, Laflamme N (2000) How the blood talks to the brain parenchyma and the paraventricular nucleus of the hypothalamus during systemic inflammatory and infectious stimuli. Proc Soc Exp Biol Med 223(1):22–38
123. Qin L, Wu X, Block ML, Liu Y, Breese GR, Hong JS, Knapp DJ, Crews FT (2007) Systemic LPS causes chronic neuroinflammation and progressive neurodegeneration. Glia 55(5):453–462
124. Qin L, He J, Hanes RN, Pluzarev O, Hong JS, Crews FT (2008) Increased systemic and brain cytokine production and neuroinflammation by endotoxin following ethanol treatment. J Neuroinflammation 5:10
125. Zahr NM, Luong R, Sullivan EV, Pfefferbaum A (2010) Measurement of serum, liver, and brain cytokine induction, thiamine levels, and hepatopathology in rats exposed to a 4-day alcohol binge protocol. Alcohol Clin Exp Res 34(11):1858–1870
126. Riikonen J, Jaatinen P, Rintala J, Porsti I, Karjala K, Hervonen A (2002) Intermittent ethanol exposure increases the number of cerebellar microglia. Alcohol Alcohol 37(5):421–426
127. Nixon K, Kim DH, Potts EN, He J, Crews FT (2008) Distinct cell proliferation events during abstinence after alcohol dependence: microglia proliferation precedes neurogenesis. Neurobiol Dis 31(2):218–229
128. Mulligan MK, Ponomarev I, Hitzemann RJ, Belknap JK, Tabakoff B, Harris RA, Crabbe JC, Blednov YA, Grahame NJ, Phillips TJ, Finn DA, Hoffman PL, Iyer VR, Koob GF, Bergeson SE (2006) Toward understanding the genetics of alcohol drinking through transcriptome meta-analysis. Proc Natl Acad Sci USA 103(16):6368–6373
129. Breese GR, Knapp DJ, Overstreet DH, Navarro M, Wills TA, Angel RA (2008) Repeated lipopolysaccharide (LPS) or cytokine treatments sensitize ethanol withdrawal-induced anxiety-like behavior. Neuropsychopharmacology 33(4):867–876

130. Blednov YA, Ponomarev I, Geil C, Bergeson S, Koob GF, Harris RA (2012) Neuroimmune regulation of alcohol consumption: behavioral validation of genes obtained from genomic studies. Addict Biol 17(1):108–120
131. Kimpel MW, Strother WN, McClintick JN, Carr LG, Liang T, Edenberg HJ, McBride WJ (2007) Functional gene expression differences between inbred alcohol-preferring and -non-preferring rats in five brain regions. Alcohol 41(2):95–132
132. Liu J, Lewohl JM, Harris RA, Dodd PR, Mayfield RD (2007) Altered gene expression profiles in the frontal cortex of cirrhotic alcoholics. Alcohol Clin Exp Res 31(9):1460–1466
133. Brown DR (2009) Role of microglia in age-related changes to the nervous system. Scientific World Journal 9:1061–1071
134. Lucin KM, Wyss-Coray T (2009) Immune activation in brain aging and neurodegeneration: too much or too little? Neuron 64(1):110–122
135. Corona AW, Fenn AM, Godbout JP (2012) Cognitive and behavioral consequences of impaired immunoregulation in aging. J Neuroimmune Pharmacol 7:7–23
136. Dlugos CA, Pentney RJ (2001) Quantitative immunocytochemistry of glia in the cerebellar cortex of old ethanol-fed rats. Alcohol 23(2):63–69
137. Rintala J, Jaatinen P, Kiianmaa K, Riikonen J, Kemppainen O, Sarviharju M, Hervonen A (2001) Dose-dependent decrease in glial fibrillary acidic protein-immunoreactivity in rat cerebellum after lifelong ethanol consumption. Alcohol 23(1):1–8
138. Blanco AM, Valles SL, Pascual M, Guerri C (2005) Involvement of TLR4/type I IL-1 receptor signaling in the induction of inflammatory mediators and cell death induced by ethanol in cultured astrocytes. J Immunol 175(10):6893–6899
139. Floreani NA, Rump TJ, Abdul Muneer PM, Alikunju S, Morsey BM, Brodie MR, Persidsky Y, Haorah J (2010) Alcohol-induced interactive phosphorylation of Src and toll-like receptor regulates the secretion of inflammatory mediators by human astrocytes. J Neuroimmune Pharmacol 5(4):533–545
140. Blanco AM, Perez-Arago A, Fernandez-Lizarbe S, Guerri C (2008) Ethanol mimics ligand-mediated activation and endocytosis of IL-1RI/TLR4 receptors via lipid rafts caveolae in astroglial cells. J Neurochem 106(2):625–639
141. Fernandez-Lizarbe S, Pascual M, Guerri C (2009) Critical role of TLR4 response in the activation of microglia induced by ethanol. J Immunol 183(7):4733–4744
142. Alfonso-Loeches S, Pascual-Lucas M, Blanco AM, Sanchez-Vera I, Guerri C (2010) Pivotal role of TLR4 receptors in alcohol-induced neuroinflammation and brain damage. J Neurosci 30(24):8285–8295
143. Pascual M, Balino P, Alfonso-Loeches S, Aragon CM, Guerri C (2011) Impact of TLR4 on behavioral and cognitive dysfunctions associated with alcohol-induced neuroinflammatory damage. Brain Behav Immun 25(Suppl 1):S80–S91
144. Wu Y, Lousberg EL, Moldenhauer LM, Hayball JD, Coller JK, Rice KC, Watkins LR, Somogyi AA, Hutchinson MR (2012) Inhibiting the TLR4-MyD88 signalling cascade by genetic or pharmacologic strategies reduces acute alcohol dose-induced sedation and motor impairment in mice. Br J Pharmacol 165:1319–1329
145. Liu J, Yang AR, Kelly T, Puche A, Esoga C, June HL Jr, Elnabawi A, Merchenthaler I, Sieghart W, June HL Sr, Aurelian L (2011) Binge alcohol drinking is associated with GABAA alpha2-regulated Toll-like receptor 4 (TLR4) expression in the central amygdala. Proc Natl Acad Sci USA 108(11):4465–4470
146. Miller MW (1995) Generation of neurons in the rat dentate gyrus and hippocampus: effects of prenatal and postnatal treatment with ethanol. Alcohol Clin Exp Res 19(6):1500–1509
147. Moulder KL, Fu T, Melbostad H, Cormier RJ, Isenberg KE, Zorumski CF, Mennerick S (2002) Ethanol-induced death of postnatal hippocampal neurons. Neurobiol Dis 10(3):396–409
148. Jacobs JS, Miller MW (2001) Proliferation and death of cultured fetal neocortical neurons: effects of ethanol on the dynamics of cell growth. J Neurocytol 30(5):391–401
149. Maier SE, Cramer JA, West JR, Sohrabji F (1999) Alcohol exposure during the first two trimesters equivalent alters granule cell number and neurotrophin expression in the developing rat olfactory bulb. J Neurobiol 41(3):414–423

150. De A, Boyadjieva NI, Pastorcic M, Reddy BV, Sarkar DK (1994) Cyclic AMP and ethanol interact to control apoptosis and differentiation in hypothalamic beta-endorphin neurons. J Biol Chem 269(43):26697–26705
151. Sarkar DK, Kuhn P, Marano J, Chen C, Boyadjieva N (2007) Alcohol exposure during the developmental period induces beta-endorphin neuronal death and causes alteration in the opioid control of stress axis function. Endocrinology 148(6):2828–2834
152. Boyadjieva NI, Sarkar DK (2010) Role of microglia in ethanol's apoptotic action on hypothalamic neuronal cells in primary cultures. Alcohol Clin Exp Res 34(11):1835–1842
153. Drew PD, Xu J, Racke MK (2008) PPAR-γ: therapeutic potential for multiple sclerosis. PPAR Res 2008:627463
154. Stopponi S, Somaini L, Cippitelli A, Cannella N, Braconi S, Kallupi M, Ruggeri B, Heilig M, Demopulos G, Gaitanaris G, Massi M, Ciccocioppo R (2011) Activation of nuclear PPAR-γ receptors by the antidiabetic agent pioglitazone suppresses alcohol drinking and relapse to alcohol seeking. Biol Psychiatry 69(7):642–649
155. Pascual M, Blanco AM, Cauli O, Minarro J, Guerri C (2007) Intermittent ethanol exposure induces inflammatory brain damage and causes long-term behavioural alterations in adolescent rats. Eur J Neurosci 25(2):541–550
156. Agrawal RG, Hewetson A, George CM, Syapin PJ, Bergeson SE (2011) Minocycline reduces ethanol drinking. Brain Behav Immun 25(Suppl 1):S165–S169

Chapter 12
Alcohol–Chemokine Interaction and Neurotransmission

Donna L. Gruol

12.1 Introduction

Chemokines are a large family of small, structurally related proteins (molecular weights ranging from 8 to 14 kDa) that are members of the cytokine family of immune factors. Chemokines share gene and amino acid sequence homology including conserved amino acids that are important for establishing their 3-dimensional or tertiary structure. Chemokines were initially identified in the immune system where they play an important role as mediators of directed cell migration, referred to as chemotaxis [1]. Chemotaxis plays a critical role in host defense against injury and disease and in normal processes of tissue maintenance or development. Chemokines that act to recruit cells of the immune system to a site of injury or infection are considered to be proinflammatory chemokines, whereas chemokines that are involved in chemotaxis during normal processes of tissue maintenance or development are considered to be homeostatic in function.

In addition to their role in the immune system, emerging research indicates that chemokines mediate a wide range of biological activities in many different cell types including cells of the central nervous system (CNS), where recent studies show that chemokines are important signaling molecules [2, 3]. It is now recognized that within the CNS, chemokines and other immune factors are produced by an innate immune system that performs many of the same functions as the peripheral immune system and utilizes many of the same factors. The cells that comprise the innate immune system of the CNS are the glial cells. CNS-produced immune factors are referred to as neuroimmune factors to distinguish them from factors produced by cells of the peripheral immune system. The CNS neuroimmune system

D.L. Gruol (✉)
Molecular and Integrative Neuroscience Department (MIND), SP30-1522,
The Scripps Research Institute, 10550 North Torrey Pines Road, La Jolla, CA 92037, USA
e-mail: gruol@scripps.edu

C. Cui et al. (eds.), *Neural-Immune Interactions in Brain Function and Alcohol Related Disorders*, DOI 10.1007/978-1-4614-4729-0_12,

is now understood to participate in essential, normal CNS functions, including neuronal development and processes of learning and memory, in addition to roles in CNS injury and disease [4–6].

Until recently, neuroimmune factors such as chemokines and their actions in the CNS were studied primarily with respect to a role in CNS injury and infection, such as occurs in stroke, trauma, viral or bacterial infection, and in immunoinflammatory disorders (e.g., multiple sclerosis), conditions where elevated levels of neuroimmune factors in the CNS were initially documented [7–12]. However, emerging research shows that increased CNS expression of neuroimmune factors also occurs in other conditions associated with impaired CNS function such as aging, neurodegenerative disorders (e.g., Alzheimer's disease, Parkinson's disease), and psychiatric disorders (e.g., depression, schizophrenia) [13]. Importantly, there is now convincing evidence that alcohol (ethanol) exposure induces the production of neuroimmune factors including chemokines within the CNS and that these factors participate in or mediate the response of the CNS to alcohol [14–17]. This discovery has led to a new area of research that will undoubtedly provide new insights into the actions of alcohol on the CNS and the mechanisms mediating CNS dysfunction associated with alcohol use and abuse.

While the realization that chemokines are present in the CNS and can alter CNS function is relatively new, the ability of alcohol to alter CNS function has been known and investigated for decades. One of the most important effects of alcohol on the CNS is cognitive impairment including disruption of memory and learning [18, 19]. Based on the known involvement of the hippocampus in memory and learning, this CNS region has become a focus of studies to understand actions of alcohol on the CNS. A number of studies have shown that alcohol exposure alters synaptic function in the hippocampus, and emerging research shows that chemokines also regulate synaptic function in the hippocampus. This commonality of actions provides a substrate for alcohol–chemokine interactions that could play an important role in the effects of alcohol on the CNS. Few studies have investigated alcohol–chemokine interactions at the level of synaptic function in the hippocampus or other CNS region. Therefore, this chapter will focus on research that has revealed actions of chemokines or alcohol at common targets. Future studies on alcohol–chemokine interactions are needed and will be instrumental to an understanding of the role of chemokines in the effects of alcohol on the CNS.

12.1.1 Chemokine Nomenclature

Chemokines and their receptors comprise a large family of signaling proteins, with over 45 different chemokines and 19 different chemokine receptors identified to date (Table 12.1). Historically chemokines were known by a variety of names related to their biological function or other properties. This complex and often confusing nomenclature has now been replaced by a systematic nomenclature paralleling that developed for other receptor systems [20, 21]. Chemokines are now classified into

Table 12.1 Chemokine ligands and receptors [20, 21]

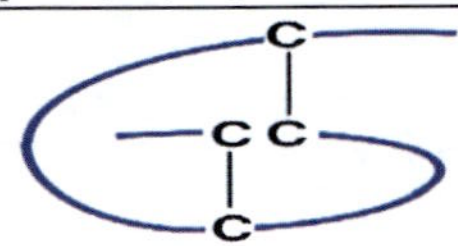

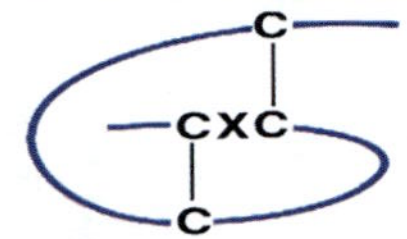

CC chemokines Ligand name	Other names for ligand	Receptor	CXC chemokines Ligand name	Other names for ligand	Receptor
CCL1	I-309, TCA-3	CCR8	CXCL1	Gro-α, GRO1, NAP-3, KC	CXCR2
CCL2	MCP-1	CCR2	CXCL2	Gro-β, GRO2, MIP2α	CXCR2
CCL3	MIP-1α	CCR1	CXCL3	GRO3, MIP2b	CXCR2
CCL4	MIP-1β	CCR1, CCR5	CXCL4	PF4	CXCR3B
CCL5	RANTES	CCR5	CXCL5	ENA-78	CXCR2
CCL6	C-10, MRP-2	CCR1	CXCL6	GCP-2	CXCR1, CXCR2
CCL7	MARC, MCP-3	CCR2	CXCL7	NAP-2, CTAPlll, b-Tα, PEP	–
CCL8	MCP-2	CCR1, CCR2B, CCR5	CXCL8	IL-8, NAP-1, MDNCF, GCP-1	CXCR1, CXCR2
CCL9/10	MRP-2, CCF18, MIP1	CCR1	CXCL9	MIG, CRG-10	CXCR3
CCL11	Eotaxin	CCR2, CCR3, CCR5	CXCL10	IP-10, CRG-2	CXCR3
CCL12	MCP-5	–	CXCL11	I-TAC, β-R!, IP-9	CXCR3, CXCR7
CCL13	MCP-4, NCC-1, Ckβ10	CCR2, CCR3, CCR5	CXCL12	SDF-1, PBSF	CXCR4, CXCR7
CCL14	Ckβ1, MC1F, NCC-2	CCR1	CXCL13	BCA-1, BCL	CXCR5
CCL15	Leukotactin-1, MIP-5, HCC-2, NCC-3	CCR1, CCR3	CXCL14	BRAK, bolekine	–
CCL16	LEC, NCC-4, LMC, Ckβ12	–	CXCL15	Lungkine, WECHE	–
CCL17	TARC, dendrokine, ABCD-2	CCR4	CXCL16	SRPSOX	CXCR6
CCL18	PARC, DC-CK-1, AMAC-1, MIP-4, Ckβ7	–	CXCL17	DMC, VCC-1	–

(continued)

Table 12.1 (continued)

CC chemokines Ligand name	Other names for ligand 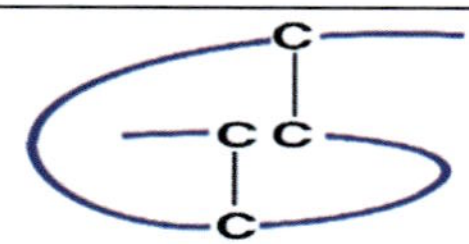	Receptor	CXC chemokines Ligand name	Other names for ligand 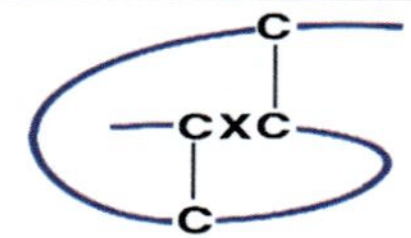	Receptor
CCL19	Exodus-3, ECL, Ckβ11	CCR7	*C-Chemokines*	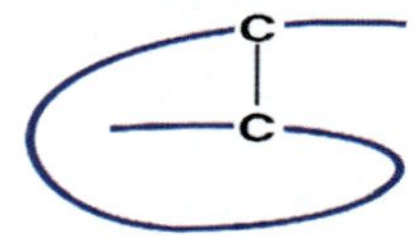	
CCL20	LARC, Exodus-1, Ckβ4	CCR6			
CCL21	SLC, 6Ckine, Exodus-2, TCA4, Ckβ9	CCR7	XCL1	Lymphotactin α, SCM-1α, ATAC	XCR1
CCL22	MDC,DC/β-CK	CCR4	XCL2	Lymphotactin β, SCM-1 β	XCR1
CCL23	MPIF-1, MIP-3, MPIF-1, Ckβ8	CCR1	*CXC-Chemokine*	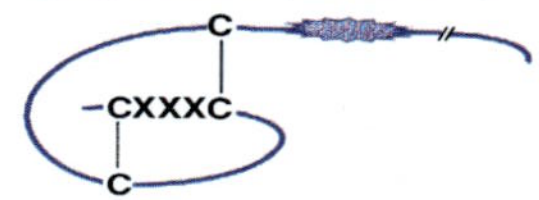	
CCL24	Exotaxin-2, MPIF-2, Ckβ6	CCR3			
CCL25	TECK, Ckβ15	CCR9	CX3CL1	Fractalkine, neurotactin, ABCD-3	CXCR3
CCL26	Exptaxin-3, IMAC, MIP-4α, TSC-1	CCR3			
CCL27	CTACK, ILC, ESkine, PESKY, skinkine	CCR10			
CCL28	MEC	CCR3, CCR10			

four basic subfamilies, based primarily on the position of specific conserved cysteine residues in the N-terminal structure: a-(or CXC) chemokines, b-(or CC) chemokines, g-(or C) chemokines, and δ-(or CX3C) chemokines. Chemokines and their receptors are named according to the subfamily that they belong to and are distinguished by the family acronym (i.e., CXC, CC, C, or CX3C), a letter and a number. The letter "L" is used to designate ligands, and the letter "R" is used to designate receptors (e.g., CCL2 and CCR2) (Table 12.1) [21]. The chemokine signaling system shows redundancy in that some chemokines can signal through several different chemokine receptor subtypes and several different chemokines can signal through the same chemokine receptor subtype. For example, both CXCL9 and CXCL10 can activate CXCR3, whereas CXCL12 can activate both CXCR4 and CXCR7 (Table 12.1).

12.1.2 Chemokine Receptors

Chemokines produce their biological effects in CNS cells and other cell types primarily by acting at cognate seven transmembrane G-protein (Gi/Go) coupled receptors (GPCRs). Much of our understanding about the function of chemokines and their receptors in CNS cells comes from concepts uncovered for other GPCRs and from studies of chemokines and their receptors in immune cells. As for other GPCRs, ligand binding to the receptor initiates dissociation of the G-protein heterotrimer coupled to the receptor into Ga and Gbg subunits (Fig. 12.1). The Ga and Gbg subunits then activate signal transduction pathways or directly act at downstream effectors such as ion channels [22, 23]. The parallel signaling pathways of Ga and Gbg engage several different signaling molecules and provide a variety of avenues for the regulation of cellular biology.

A number of studies have examined the cellular expression of chemokine receptors in the CNS from both humans and animal models. Results indicate that all cells of the CNS express functional chemokine receptors, including astrocytes, microglia, and neurons, although the particular chemokine receptors expressed depend on cell type [24–30].

12.1.3 Signal Transduction

The commonality of the signal transduction pathways linked to chemokine receptors and other GPCRs known to regulate neuronal function (e.g., metabotropic glutamate and g-aminobutyric acid (GABA) receptors) is consistent with the ability of chemokines to regulate neuronal function through receptor activation of signal transduction pathways. Because the physiological properties of CNS neurons vary across neuronal types, the effects of chemokines are also likely to vary across neuronal types and to depend on physiological or pathological context. Therefore, specific chemokines could selectively target certain neuronal functions through the

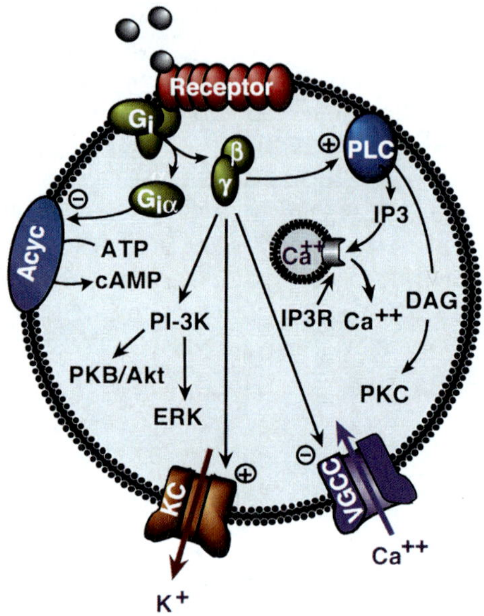

Fig. 12.1 Diagram showing some of the signal transduction molecules that mediate the effects of chemokines acting through their cognate GPCRs in the CNS. The *O plus* symbol indicates activation of the target molecule, and the *O minus* symbol indicates inhibition of the target molecule. *Acyc* adenylate cyclase, *ATP* adenosine-5′-triphosphate, *cAMP* cyclic adenosine monophosphate, *DAG* diacylglycerol, *IP3* inositol trisphosphate, *IP3R* inositol trisphosphate receptor, *KC* potassium channel, *PKC* protein kinase C, *PI-3K* phosphatidylinositol-3-OH kinase, *PKB/akt* protein kinase identified in the AKT virus (also known as protein kinase B), *PLC* phospholipase C, *VGCC* voltage-gated Ca^{2+} channel

signal transduction pathway that they activate. Limited information is available on the signal transduction pathway activated by chemokines in CNS neurons, and more research is needed in this area.

Activation of signal transduction molecules typically involves phosphorylation of the molecules at specific sites. Antibodies directed at the specific phosphorylation sites are available for many signal transduction molecules and are commonly used in Western blot studies to identify alterations in the level of the phosphorylation and therefore activation of the molecule. Such studies have identified some of the signal transduction pathways used by chemokines in CNS cells. Many of these studies were carried out in hippocampal cultures containing neurons and glial cells. Cultured cells are advantageous for such studies because both acute and chronic exposure paradigms can be used. Although cultured cells and culture environments diverge from the in vivo condition in many respects, over the years, culture models have provided a wealth of information about cellular and molecular mechanisms that would not be easy to access in more intact preparations. A limitation of culture preparations is that typically immature CNS tissue (embryonic or early postnatal) is used to prepare the cultures, and although maturation of the cells occurs during the

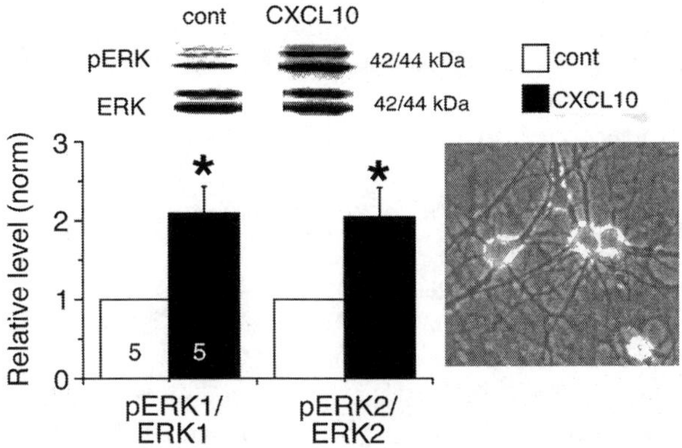

Fig. 12.2 Activation of ERK1/2 in hippocampal cultures by acute application of CXCL10. Graph shows relative level of phosphorylated ERK 1 and 2 normalized to total ERK 1 and 2 measured in Western blots of homogenates of hippocampal cultures at 14 days in vitro. The treated cultures were exposed to CXCL10 (100 nM in physiological saline) for 15 min. Saline-treated sibling cultures were used as controls. Representative Western blots are shown above the graph. A phase contrast image of hippocampal neurons in the cultures is shown to the right (Bajova, Nelson, and Gruol, unpublished; methods were similar to that reported previously [31])

culture period, the results obtained from culture preparations may not be directly translatable to in vivo conditions. Moreover, effects of chemokines are likely to depend on context, and the culture model may not sufficiently reflect the in vivo context, especially in the case of disease models. Culture characteristics depend on the culture method, but in general, by 10 days to 2 weeks in vitro, hippocampal neurons and glia are well developed, and the cultures have numerous functional excitatory and inhibitory synaptic connections.

In studies of rat hippocampal cultures, acute exposure to CXCL10 (interferon-γ protein-10, IP-10; 100 nM) was found to increase the level of activation (i.e., phosphorylation) of extracellular-signal-regulated kinase 1/2 (ERK1/2; also known as p42/p44 mitogen-activated protein kinase, MAPK) (Fig. 12.2). Chronic treatment of similar hippocampal cultures with CXCL10 (100–250 nM; from day 2 to day 11 in vitro) also increased the level of ERK1/2 activation [31]. In addition, chronic treatment with CXCL10 increased the level of activation of the transcription factors cAMP response element-binding protein (CREB) and nuclear factor kappa-light-chain-enhancer of activated B cells (NF-κB) in the hippocampal cultures. The levels of the antiapoptotic protein B-cell lymphoma 2 (Bcl-2) and the antioxidant enzyme manganese superoxide dismutase 2 (SOD2) were also increased [31]. ERK1/2, CREB, and NF-κB are known to have neuroprotective actions by inducing expression of Bcl-2 proteins and SOD2 [32–35]. Therefore, it is likely that these signaling molecules mediate the increased expression of Bcl-2 and SOD2 produced by CXCL10 in the hippocampal cultures. Taken together, these results indicate CXCL10 may function in neuroprotection or repair, at least under some conditions. Studies of other chemokines also indicate the ability for positive, neurotrophic effects on hippocampal cultures under some conditions including CXCL8 (IL-8; 500–1,000 ng/ml)

[36] and CX3CL1 (fractalkine; 100 nM) [37]. For example, chronic treatment of hippocampal cultures with CXCL8 (500–1,000 ng/ml) from culture days 1–7 produced neurotrophic effects that increased with exposure time during the treatment period. The neurotrophic effects included increased neuronal survival and increased immunostaining for neuron-specific enolase. Chronic treatment with CXCL8 also promoted growth of glia cells in the hippocampal cultures [36].

CX3CL1 is expressed in the CNS in two forms, a soluble form and a membrane-attached form. In rat hippocampal cultures, acute exposure to the soluble form of CX3CL1 (100 nM) was found to activate ERK1/2 and CREB in addition to the Ser/Thr kinase Akt (protein kinase identified in the AKT virus) but not p38 MAPK [25, 37, 38]. Acute exposure to CX3CL1 (100 nM) has also been reported to reduce forskolin-stimulated cAMP production in rat hippocampal cultures, consistent with an inhibitory effect of CX3CL1 signaling on adenylate cyclase [39]. CXCL12 (stromal cell-derived factor-1, SDF-1; 1–100 nM) was also shown to activate ERK1/2 and CREB when applied acutely to rat hippocampal cultures [40]. In contrast, no significant effect of acute CCL2 (monocyte chemotactic protein-1, MCP-1; 1–25 nM) on the level of activated ERK1/2 was observed in rat hippocampal cultures [41]. Instead, acute application of CCL2 (1–25 nM) activated p38 MAPK in a dose- and time-dependent manner [41]. The kinases (mitogen-activated protein kinase kinase 3/6; MKK3/6) that activate p38 MAPK were also activated by CCL2 in a corresponding time- and dose-dependent manner in the same study. Taken together, these results show that several chemokines activate signal transduction pathways linked to ERK1/2 and CREB in hippocampal cultures, whereas CCL2 preferentially activates the p38 MAPK pathway.

12.2 Chemokine Expression in the CNS

12.2.1 Astrocytes and Microglia

The primary cell types involved in production of chemokines in the CNS are glial cells, particularly astrocytes and microglia, although some neurons can also produce chemokines [13, 42, 43]. Astrocytes and microglia are important components of the normal CNS and are found throughout the gray and white matter where they form signaling systems that enable cell-to-cell communication over long distances. Widespread increases in the levels of chemokines and other neuroinflammatory factors can be induced in the CNS by these pathways. Astrocytes are the most abundant cell type in the CNS, whereas microglia comprise up to 20 % of the total glial population of the CNS [44]. Astrocytes are closely associated with neurons and play key roles in normal CNS function by regulating the levels of excitatory transmitters and ions at the synapse, controlling the flow of materials across the blood–brain barrier, providing metabolic support for neurons, and secreting signaling molecules that regulate the functions of other cell types [45–47]. Microglia provide a surveillance and scavenging function and play a major role in synaptic pruning and remodeling [48, 49].

Glial expression of chemokines in the CNS is generally low under normal conditions. Adverse conditions such as disease or injury result in activation of glia (astrocytes and/or microglia) and increased production of neuroimmune factors including chemokines. For example, in a recent study of patients with traumatic CNS injury, cerebral microdialysis was used to measure the levels of neuroimmune factors in the cortex [10]. A variety of cytokines and chemokines were found to be upregulated with the largest increases observed for CCL2 (median value of 2,562 pg/ml of dialysate) and CXCL10 (median value of 2,834 pg/ml of dialysate). The upregulated production of neuroimmune factors can impact the surrounding neural and nonneural cells either beneficially or detrimentally depending on the context. In general, high concentrations of chemokines that reflect dysregulated production are thought to lead to detrimental effects on the CNS, whereas lower concentrations are considered to mediate beneficial actions. However, the relationship between chemokine levels and their beneficial or detrimental effects has yet to be fully resolved.

12.2.2 Chemokine Expression in CNS Disease

Studies showing a correlation between elevated levels of chemokines in the CNS parenchyma or cerebrospinal fluid (CSF) and cognitive dysfunction in disease states were instrumental to the recognition that chemokines are important signaling molecules in the CNS. For example, in studies of human immunodeficiency virus (HIV)-infected patients, the levels of CCL2 in the CSF correlated with the severity of dementia and increased over time in patients who developed dementia [7]. CCL2 has also been implicated in the detrimental effects of aging on the CNS. CCL2 levels in the CSF significantly increase with the age of patients with and without neuropsychiatric disease [50]. In Alzheimer's patients, CCL2 and CXCL8 levels were significantly increased in individuals with neurocognitive impairment as compared with healthy controls [51]. In the same study, the level of CXCL10 was increased in the CSF of patients with mild neurocognitive impairment and mildly impaired Alzheimer's disease, but not in Alzheimer's patients with severe neurocognitive decline [51]. These results may reflect different roles of CXCL10 in different stages of a disease. CXCL10 expression has been described in a number of neurological disorders besides Alzheimer's disease [52] including HIV-associated dementia (HAD) [53] and multiple sclerosis/experimental autoimmune encephalomyelitis (EAE) [9, 54]. Elevated levels of CXCL10 in the CSF were found to be highly prevalent in HIV infection and to strongly correlate with the severity of HIV-associated neurologic disorders and the extent of lymphocyte chemotaxis [55].

Measurements of chemokine levels in the CNS parenchyma or CSF have contributed significantly to the identification of important neuroimmune players and their CNS levels during CNS disease and injury. However, the relationship between these measurements and chemokine levels at the most relevant CNS location, the interstitial space adjacent to cells expressing receptors, is unknown. Moreover, the lack of understanding of the cellular and molecular actions of chemokines in the CNS often limits the impact of the correlative information.

In addition to a role for chemokines in disease and injury, emerging research supports a physiological role for some chemokines. For example, selective deletion of either CXCL12 or its cognate receptor CXCR4 disrupts the normal migratory processes occurring during morphogenesis of neurons in the cerebellum and hippocampus and leads to abnormal formation of these CNS regions [56–60]. These results are consistent with a role for CXCL12 in cerebellar and hippocampal development. Evidence for a role of CXCL12/CXCR4 in neuronal migration and axonal pathfinding during development of other CNS regions has also appeared [61].

Research also indicates a role for some chemokines in normal cognitive function. For example, CXCR4 mRNA and protein expression were upregulated in rats trained in a behavioral paradigm, suggesting a role for CXCR4 in learning and memory [62]. In a mouse model of Alzheimer's disease, CXCL12 and its receptor CXCR4 were found to be downregulated coincident with cognitive deficits [63]. In the same study, CXCL12 levels were found to be decreased in Alzheimer patients as compared to controls. These results are consistent with a role for CXCL12 in cognitive function. Studies also indicate a role for CX3CR1 in cognitive function. In mice with selective deletion of CX3CR1 (CX3CR1$^{-/-}$ mice), cognitive function was significantly altered. In one study, CX3CR1$^{-/-}$ mice were shown to have deficiencies in associative and spatial memory and significantly reduced hippocampal long-term synaptic plasticity (LTP) compared with wild-type control mice [64]. Synaptic plasticity is a modification of synaptic strength produced by intense synaptic activity and is considered to be cellular mechanisms underlying learning and memory [65]. These effects of CX3CR1 deficiency were shown to involve actions of IL-1b, a cytokine that was upregulated in the CX3CR1-deficient mice [64]. In another study, CX3CR1$^{-/-}$ mice showed behavioral deficits and inability of hippocampal LTP to be enhanced by an enriched housing environment that produced enhanced hippocampal LTP in wild-type mice [66]. These results are consistent with an involvement of CX3CL1/CX3CR1 signaling in experience-dependent hippocampal processes that underlie learning and memory [66].

12.3 Chemokines and the Hippocampus

The association of altered levels of chemokines in CNS disease states with impaired cognitive function has stimulated interest in the actions of chemokines on hippocampal neurons and circuits that play a central role in learning and memory, processes that are essential for normal cognitive function [67, 68]. Many CNS regions participate in memory formation. However, the hippocampus is central to memory formation for a broad domain including memories that are generated within the hippocampus and memories that are generated in other CNS regions but depend on information that is processed in the hippocampus [69–71]. It is well documented that memories formed in the hippocampus are encoded by long-term changes in the strength of synaptic transmission, the primary mechanism through which neurons of the CNS communicate to produce behavior, and that the information inherent in these synaptic changes is later reorganized and stored in other CNS regions.

The hippocampus sends projections to a large number of CNS regions including the cortex, hypothalamus, amygdala, nucleus accumbens, and septum [72, 73]. Therefore, alterations in hippocampal synaptic transmission can have widespread effects on the function of other CNS regions. Studies show that astrocytes and microglia, the primary sources of chemokines in the CNS, are closely associated with synapses and dynamically interact with neurons to regulate synaptic transmission and plasticity [74–76]. This close association provides an avenue for chemokine signaling between glia and the pre- and postsynaptic mechanisms that mediate synaptic transmission [47, 77].

A variety of experimental approaches have been employed to assess the role of different chemokines and other biologically active molecules including alcohol on hippocampal synaptic function. One of the most frequently used approaches is electrophysiological recording in hippocampal slices. A brief description of this approach is included in the Appendix. Results from studies using this and other approaches are described below.

12.3.1 Acute Exposure to Chemokines

One of the first chemokines to be studied with respect to potential effects on synaptic transmission in the hippocampus was the CXC chemokine, IL-8 (CXCL8) [78]. Field potential recordings of excitatory postsynaptic potential (fEPSPs) evoked in the CA1 pyramidal neurons by electrical stimulation of the Schaffer collaterals in hippocampal slices (see Appendix) from rats (15–36 days old) showed that exogenous application of IL-8 did not alter basal synaptic transmission. However, LTP of the fEPSP was altered in a concentration-dependent manner. At the low concentration of IL-8 tested (5 ng/ml), LTP was enhanced, whereas at the high concentrations tested (50–100 ng/ml), LTP was depressed. A similar approach was used to determine if the CXC chemokine CXCL10 altered synaptic transmission and plasticity at the Schaffer collateral to CA1 pyramidal neuron synapse [79]. Exogenous application of CXCL10 (10 ng/ml) to hippocampal slices from 5- to 6-month-old mice did not significantly alter the fEPSP, indicating the lack of an effect of CXCL10 on synaptic transmission at this synapse. However, CXCL10 significantly depressed three forms of synaptic plasticity of the fEPSP produced by a high-frequency LTP induction stimulation protocol—posttetanic potentiation (PTP), short-term synaptic plasticity (STP), and LTP (see Appendix) [79]. Because both PTP and LTP were altered, these results suggest that CXCL10 acts at both presynaptic and postsynaptic mechanisms to reduce synaptic plasticity. The concentration of CXCL10 used in these studies (10 ng/ml) was considered by the authors to be a high concentration relative to the Ki for recombinant CXCL10 at CXCR3 receptors.

The effect of CCL2 on synaptic transmission was also investigated in hippocampal slices. In these studies, intracellular (whole-cell voltage clamp) recordings were used to measure current flow (excitatory postsynaptic current, EPSC) through the glutamate receptors activated by synaptically released glutamate [80]. EPSCs underlie the synaptic response (fEPSP) measured in the field potential recordings.

Hippocampal slices from 15- to 30-day-old rats were used for these studies. Application of 0.023–2.3 nM CCL2 enhanced the amplitude of the EPSC in the CA1 pyramidal neurons elicited by stimulation of the Schaffer collaterals in a dose-dependent manner. Further analysis of the effects of CCL2 on synaptic transmission at this synapse indicated that the enhanced synaptic response resulted from a presynaptic action of CCL2 that produced an increase in transmitter release [80].

In contrast to CCL2 and CXCL10, acute application of the chemokine CX3CL1 (0.2–20 nM) produced a depression of the fEPSP at the Schaffer collateral to CA1 pyramidal neuron synapse in hippocampal slices from 13- to 29-day-old mice and adult mice, indicating a reduction in synaptic transmission [81, 82] (Fig. 12.3a). CX3CL1 (2–5 nM) also inhibited LTP (Fig. 12.3b) and produced long-term depression (LTD) of synaptic transmission [81, 82]. LTD is a long-lasting reduction in synaptic strength and can be experimentally induced by low-frequency stimulation. Studies on the mechanisms mediating the depressive effects of CX3CL1 using intracellular recording methods revealed a complex mechanism involving adenosine receptors [82, 83].

In addition to these studies where effects of chemokine on synaptic transmission were directly tested, other studies have examined neuronal properties that are important for synaptic transmission. For example, in Cajal–Retzius cells, hippocampal interneurons that express CXCR4 of juvenile mice (postnatal day 12–24) exogenous application of CXCL12 (1–50 nM) reduced spontaneous spike activity measured in current clamp recordings [84]. In contrast, in current clamp recordings from CA1 pyramidal neurons of hippocampal slices from 15- to 30-day-old rats, exogenous application of CCL2 (2.3 nM) was reported to depolarize the membrane potential and increased spontaneous spike firing [80].

Taken together, these in vitro studies show that exogenous application of chemokines can regulate synaptic function in hippocampal circuits and that different chemokines produce different effects. However, an important aspect to note about these studies is the differences in the age of the animals studied, which could be an important factor relative to differences in chemokine actions. For example, typically rodents are not considered to be adults until around 2 months of age [85]. Therefore, studies of young animals may reflect chemokine actions specific to the CNS of juveniles (younger than postnatal day 28) or adolescents (postnatal day 28–42), whereas studies in older animals would be most relevant to chemokine actions in the adult CNS [86]. In addition to their actions on synaptic function, emerging research indicates that chemokines play an important role in neuronal development (e.g., [87–90]), a topic that is not addressed in this chapter.

12.3.2 Chronic Exposure to Chemokines

Studies using acute application of chemokines have provided convincing evidence that chemokines can alter neuronal and synaptic function and, in some cases, have identified potential mechanisms involved in these actions. In the diseased state,

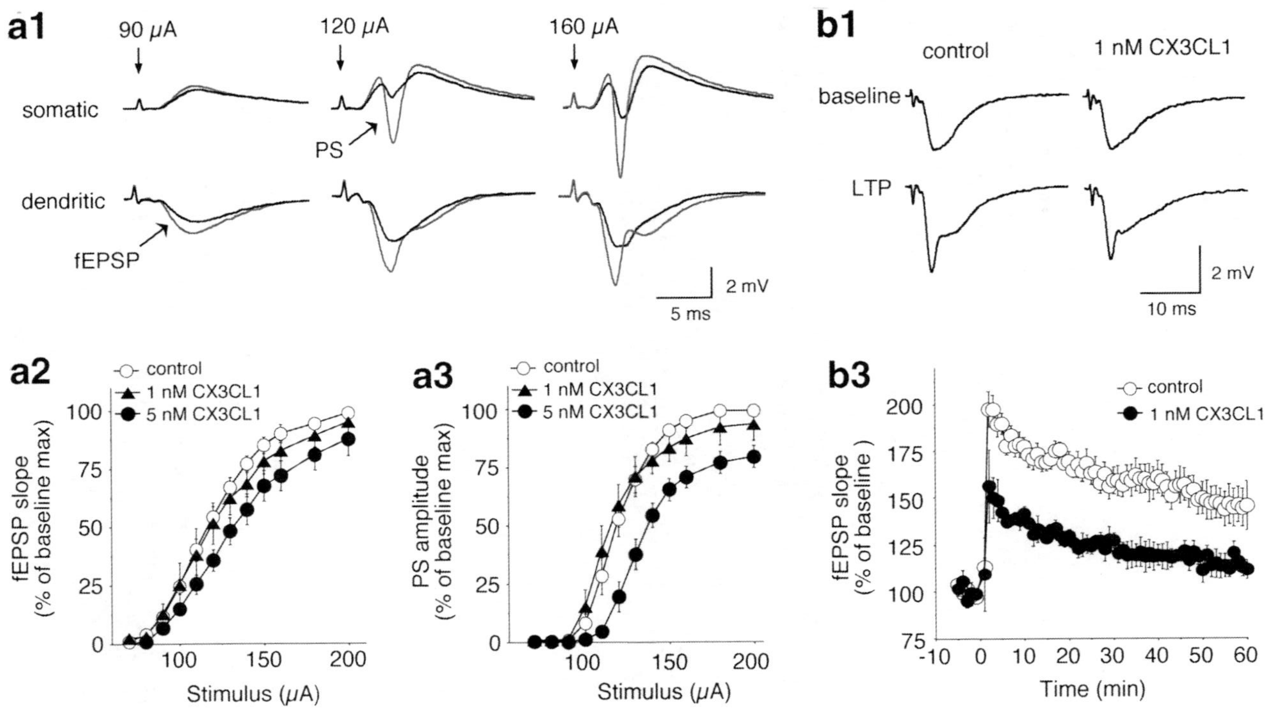

Fig. 12.3 Effect of exogenous application of CX3CL1 on synaptic transmission and LTP at the Schaffer collateral to CA1 pyramidal neuron synapse in a hippocampal slice from an adult mouse. (**A1**) Representative field potential recordings of simultaneous somatic (PS) and dendritic (fEPSP) responses evoked by Schaffer collateral stimulation under baseline conditions (*gray line*) and after superfusion of CX3CL1 (5 nM) (*black line*). CX3CL1 reduced the magnitude of the fEPSP with a corresponding reduction in the PS. (**A2**, **A3**) Graphs showing summarized results. Schaffer collaterals were stimulated at increasing intensity, and the magnitude of the resulting responses (fEPSP and PS) plotted as a function of stimulus intensity to show input/output relationships for synaptic transmission. (**B2**) Graph showing effect of CX3CL1 on synaptic plasticity evoked by high-frequency (theta burst) stimulation of Schaffer collaterals. Exogenous application of CX3CL1 significantly depressed PTP, STP, and LTP (Nelson, Huitron-Resendiz, and Gruol, unpublished)

chemokines are often chronically present at elevated levels for at least some period of time. To gain an understanding of how such long-term exposure can affect synaptic function, several investigators have constructed genetically modified mice that chronically express elevated levels of a specific chemokine in the CNS. These animals offer a model system where the actions of a chemokine can be investigated without the complexities that exist in a disease state that could confound analyses. One strategy that is particularly advantageous for studies of chemokine actions in the CNS is genetic manipulation through altered astrocyte expression. Two transgenic lines, CCL2 [91] and CXCL10 [92], have been developed with this strategy and utilized for studies of hippocampal synaptic function. Wild-type mice or non-transgenic littermates are used as controls. In both cases, elevated expression was accomplished by astrocyte-targeted gene expression under the transcriptional control of the GFAP promoter. Therefore, the source of the chemokine (i.e., astrocytes) in the CNS of the transgenic mice is consistent with an innate source in the CNS during physiological and pathological conditions.

Young CCL2 transgenic mice (>6 months of age) show little behavioral effects of the CCL2 overexpression in the CNS. However, progressive neurological impairment was observed in aged mice (e.g., 1 year and older) with symptoms beginning after 7 months of age. The impairment is characterized by postural changes, reduced grooming, difficulty in righting reflex, limb weakness, and hind limb paralysis [93]. There are no anatomical studies on the hippocampus of the CCL2 transgenic mice. However, anatomical studies have been carried out on the cervical region of the spinal cord. Results showed only minimal alterations of axons, neuronal somata, and synapses in the CCL2 transgenic mice compared to non-transgenic littermate controls, and these changes were observed only in the older mice with neurological impairment [93]. Huang et al. [93] reported that CCL2 levels in lysates of CNS tissue from the CCL2 transgenic mice averaged 100 ng/ml (7.14 nM, measured by ELISA), a level similar to that observed in CNS tissues from mice with EAE. In the same study, cultured astrocytes from the transgenic mice secreted an average of 3.5 ng/ml (0.25 nM) CCL2.

For the CXCL10 transgenic mice, other than leukocyte infiltration and accumulation, no apparent pathologic alterations in the CNS (up to 12 months of age) were observed in a histological analysis of hematoxylin and eosin (H&E)-stained paraffin sections (nuclei and cellular proteins are stained) [92]. Assessment of genes associated with inflammation showed similar levels in CXCL10 transgenic mice and their non-transgenic littermate controls, consistent with an absence of degenerative pathology in the CNS of the CXCL10 transgenic mice [92].

Field potential recordings in the CA1 region of synaptic transmission (fEPSPs) and synaptic plasticity (PTP, STP, LTP) produced by electrical stimulation of the Schaffer collaterals showed no significant differences between hippocampal slices from CXCL10 transgenic mice (5–6 months of age) and non-transgenic littermate controls [79]. However, when acute application of CXCL10 (10 ng/ml) was tested, PTP in the CXCL10 transgenic hippocampal slices was resistant to a depressing effect of acute CXCL10 observed in the non-transgenic hippocampal slices. These results indicate that chronic in vivo exposure to CXCL10 produces neuroadaptive changes that alter the effect of CXCL10 on presynaptic mechanisms involved in transmitter release induced by high-frequency synaptic activity, with little or no effect on mechanisms that mediate basic synaptic transmission or LTP.

In contrast, in hippocampal slices from CCL2 transgenic mice (7–9 months of age), synaptic transmission (fEPSP) at the Schaffer collateral to CA1 pyramidal neuron synapse was significantly depressed compared with synaptic transmission in hippocampal slices from non-transgenic mice [94]. Moreover, PTP and STP were significantly enhanced in hippocampal slices from the CCL2 transgenic mice compared with synaptic transmission in hippocampal slices from the non-transgenic mice, although there was no difference in LTP. These differences between CCL2 transgenic and non-transgenic hippocampus were not observed at 2–3 months of age [95]. Thus, chronic in vivo exposure to CCL2 produces in time neuroadaptive changes that affect presynaptic mechanisms involved in transmitter release induced by high-frequency synaptic activity.

12.3.3 Studies in Hippocampal Cultures

In addition to studies in hippocampal slice preparations, effects of chemokines on synaptic transmission have been investigated in rodent hippocampal culture preparations. Neurons in the hippocampal cultures are immature at plating but develop many characteristics of mature neurons during the culture period including the formation of synaptic networks. A population of the cultured hippocampal neurons is spontaneously active and can initiate spontaneous synaptic network activity in the cultures. The spontaneous synaptic activity is dependent on activation of NMDA receptors (NMDARs) and voltage-gated Ca^{2+} channels (VGCCs). Ca^{2+} influx through NMDARs and VGCCs during the spontaneous synaptic activity produces Ca^{2+} oscillations that are a reflection of the spontaneous synaptic activity [96]. A relatively noninvasive approach to studying this synaptic network activity is through the use of Ca^{2+} imaging techniques to measure the Ca^{2+} oscillations (Fig. 12.4). Recent studies using this approach showed that both acute and chronic applications of chemokines alter synaptic network activity in hippocampal cultures.

In studies using acute application, CCL5 (RANTES), CCL3 (MIP-1α), CXCL12 (SDF-1), and soluble CX3CL1 (fractalkine) (50–100 nM) decreased the amplitude of the spontaneous Ca^{2+} oscillations in cultured rat hippocampal neurons, reflecting a reduction in synaptic activity by these chemokines [25, 98]. In contrast, acute application of CXCL10 (100 nM) (Fig. 12.5) or CCL2 (25 nM) enhanced spontaneous Ca^{2+} oscillations in cultured rat hippocampal neurons, reflecting increased synaptic activity [97, 99].

In cultures chronically treated with CXCL10 (100–250 nM; from 2 to 11 days in vitro), the spontaneous Ca^{2+} oscillations were significantly larger in the CXCL10-treated neurons compared with neurons in untreated control cultures [100]. Measurement of protein levels in the CXCL10-treated and control cultures by Western blot showed

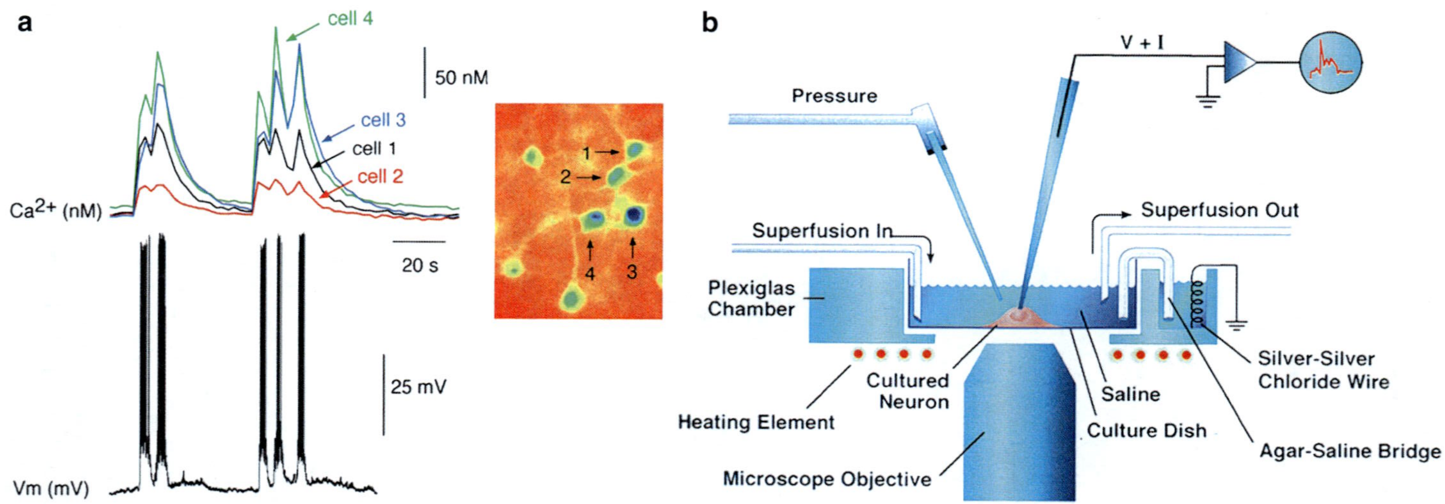

Fig. 12.4 Spontaneous synaptic network activity in hippocampal cultures. (**a**) Simultaneous Ca^{2+} imaging (*top trace*) and intracellular (current clamp; *bottom trace*) recording of spontaneous synaptic network activity in a rat hippocampal culture (14 days in vitro) using techniques previously described [97]. A fluorescent image of the fura-2-loaded neurons in the microscopic field used for these recordings is shown to the right. Cell number 1 is the neuron recorded with both electrophysiological and Ca^{2+} imaging techniques. Three other neurons in the field were recorded with just Ca^{2+} imaging techniques. Spontaneous, synaptically evoked burst discharges are evident in the electrophysiological recording. The associated Ca^{2+} oscillations produced by this activity are evident in the Ca^{2+} imaging recording. All recorded neurons show a similar pattern of Ca^{2+} oscillations, consistent with the involvement of synaptic network connections in the generation of this activity. (**b**) Diagram showing the setup for simultaneous Ca^{2+} imaging and electrophysiological recording from cultured neurons. The pressure pipette is used to apply drugs locally to neurons as was done in the recording shown in Fig. 12.5

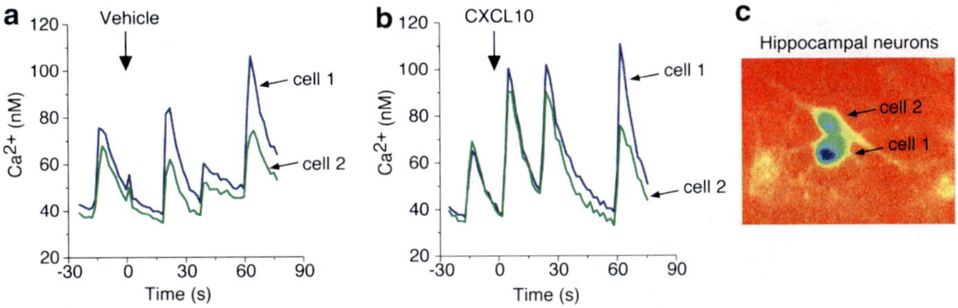

Fig. 12.5 CXCL10 enhances spontaneous network activity in hippocampal cultures. (**a**, **b**) Ca^{2+} imaging recording of Ca^{2+} oscillations produced by spontaneous synaptic network activity in two neurons in a rat hippocampal culture (14 days in vitro). CXCL10 was applied for 5 s by pressure application from a drug pipette placed near the neurons (**b**) and produced an increase in the amplitude of the Ca^{2+} oscillations. Pressure application of vehicle (saline, **a**) did not alter the network activity. The recorded fura-2-loaded neurons are shown in the fluorescent images in **c** (Nelson and Gruol, unpublished; methods were similar to that reported previously [97])

alterations in neuronal and glial proteins consistent with the ability of chronic CXCL10 to produce neuroadaptive changes that could affect neuronal and synaptic function [31, 100]. Thus, higher levels of synapsin I, a protein involved in transmitter release, the NR1 subunit of NMDAR, and group 2 metabotropic glutamate receptors (the antibody used detected both mGluR2 and mGluR3) were observed in the CXCL10-treated hippocampal cultures compared with control cultures. In contrast, reduced protein levels were observed for glutamic acid decarboxylase (GAD) 65/67, the synthetic enzyme for the inhibitory transmitter γ-aminobutyric acid (GABA), and the metabotropic GABA receptor ($GABA_BR$) in the CXCL10-treated cultures [100]. Higher levels of GFAP, an astrocyte-specific structural protein, neuron-specific enolase, a neuron-specific protein, and the housekeeping protein β-actin, were also observed in the CXCL10-treated hippocampal cultures.

The effect of chronic exposure to the chemokine CCL3 on neuronal function was also tested in hippocampal cultures [101]. The cultures were exposed to CCL3 (20–200 nM) for 3 days starting on culture day 9. Non-treated sibling cultures were used as controls. Chronic exposure to CCL3 (20 nM) enhanced spontaneous Ca^{2+} oscillations in the hippocampal cultures, indicating increased synaptic network activity. In addition, the peak amplitude of the Ca^{2+} signal evoked by brief application of NMDA to the neurons was enhanced. Western blot analysis showed that chronic exposure to CCL3 at low CCL3 concentration (20 nM) had no effect on the levels of b-actin but increased the levels of several proteins including GFAP, enolase, GAD65/67, NR1, and synapsin 1. The larger response to NMDA in the CCL3-treated neurons is consistent with the increased level of NR1 protein in the CCL2-treated hippocampal cultures [101]. At high concentrations (200 nM), chronic exposure to CCL3 produced no changes in the level of GFAP, enolase, GAD65/67, or NR1, but still decreased the level of synapsin 1. This negative concentration dependency may reflect dose-dependent actions of CCL3 at opposing sites. Taken together, these studies of acute and chronic chemokine exposure in hippocampal slices and cultures show that chemokines can alter hippocampal synaptic function. In addition, these studies show that different chemokines can have both overlapping and distinctive actions.

12.4 Alcohol and Chemokines

Recent studies provide strong support for a role of neuroimmune factors, and in particular the chemokine CCL2, in alcohol use and abuse disorders. Importantly, increased levels of CCL2 have been observed in the CNS of alcoholics [102], raising the possibility of a role for CCL2 in the actions of alcohol on the CNS. Consistent with these results, studies in experimental animals showed that levels of mRNA and protein for CCL2 were significantly increased in the CNS of mice (8 weeks old) subjected to either single or repetitive alcohol exposure (10 days; 5 g/kg body weight intragastrically). The increased levels of CCL2 persisted for days after the repetitive alcohol exposure, suggesting that mechanisms regulating levels of CCL2 in the CNS are particularly sensitive to alcohol [103]. In another study, intraperitoneal injections (i.p.) of alcohol (3 g/kg body weight) to mice (7–8 weeks old) increased the levels of CCL2 mRNA in the hippocampus but not in the corpus striatum, suggesting that CNS regions differ in the degree to which alcohol induces CCL2 expression [104]. In behavioral tests, both CCL2 and CCR2 null mice (10–12 weeks of age) showed lower preference for alcohol and lower alcohol consumption than wild-type mice, consistent with a role for CCL2 in the motivational aspects of alcohol consumption [105]. Mice (6 weeks of age) that were exposed to CCL2 at weekly intervals (two injections) by intracerebroventricular (i.c.v.) injection prior to being subjected to a 5-day chronic alcohol diet exhibited reduced social interaction during withdrawal, indicating an ability of CCL2 to sensitize the mice to withdrawal-induced anxiety-like behavior [106]. In addition to in vivo studies, studies in cultured preparations show that CCL2 and other neuroimmune factors are increased in CNS cells by alcohol exposure. For example, in rat hippocampal–entorhinal slice cultures, alcohol exposure (100 mM) increased mRNA levels for TNF-α and IL-1β in addition to mRNA levels for CCL2 [107]. In a study of cultured microglia, exposure to alcohol (50–100 mM) was shown to increase microglial secretion of TNF-α, IL-6, CCL3 (MIP-1a), and CXCL2 (MIP-2) [108].

12.4.1 Alcohol and Glia

Alcohol-induced alterations in the levels of neuroimmune factors are likely to relate to the actions of alcohol on astrocytes and microglia, the primary producers of neuroimmune factors in the CNS [13, 42]. Studies show that the structure, function, and number of glia are altered by alcohol exposure [109–112]. For example, studies in experimental animals and cultures of CNS neurons and glia showed that chronic alcohol exposure increases the level of glial fibrillary acidic protein (GFAP) in astrocytes, indicative of gliosis and glial activation [113–115]. Microglial activation is also produced by alcohol exposure [116]. In the activated state, astrocytes and microglia release a range of biologically active factors, such as pro- and anti-inflammatory cytokines, neurotrophic factors, chemokines, complement factors, and reactive oxygen species (ROS) [45]. These glial-produced factors can impact beneficially or detrimentally on the surrounding neural and nonneural cells [117]. Importantly, alcohol has been shown to activate toll-like receptor 4 (TLR4) signaling in CNS astrocytes and microglia. TLR4 is a receptor that recognizes the bacterial coat protein lipopolysaccharide (LPS) and is primarily expressed in astrocytes and microglia in the CNS, although some neurons also express TLR4 [118, 119]. Studies in vivo and in vitro show that activation of TLR4 by alcohol results in activation of kinases (e.g., ERK1/2, p38 MAPK) and downstream transcription factors (NF-κB and AP-1), which in turn leads to the production of inflammatory mediators including cytokines, chemokines, and generators of ROS (e.g., inducible nitric oxide synthase and cyclooxygenase-2) [120–122]. These effects of alcohol were not observed in TLR4-deficient astrocytes or microglia. In studies where wild-type and TLR4 knockout (KO) mice were subjected to a paradigm of chronic alcohol exposure (5 months) and withdrawal (15 days), behavioral and cognitive deficits and glial activation persisted 2 weeks after alcohol withdrawal in the wild-type mice but were not observed in the TLR4 KO mice [123]. These results provide support for the idea that alcohol interactions with the innate immune system of the CNS play an important role in the actions of alcohol on the CNS and downstream behavioral consequences of alcohol use [15, 123].

12.5 Alcohol and Hippocampal Synaptic Function

Considerable literature exists, showing that synaptic function is an important target of both acute and chronic alcohol in the CNS and that alcohol actions that affect synaptic function correlate with behavioral dysfunction produced by alcohol [124, 125]. An understanding of the mechanisms that underlie the effects of alcohol on synaptic function is an ongoing process, although much has already been learned. It is now widely accepted that alcohol alters synaptic function through a variety of mechanisms including alterations in transmitter release, the function of transmitter receptors and the function of signal transduction mechanisms that regulate receptor function or the mediate downstream effects of synaptic transmission. A number of excellent reviews are available that cover different aspects of the actions of alcohol on synaptic function [126–130]. A brief summary of studies at hippocampal synapses is provided below. Studies of chemokine actions on synaptic function, although limited, indicate that chemokines can alter some of the same synaptic functions as alcohol. However, it is unknown if alcohol-induced chemokine production plays a role in the actions of alcohol on synaptic function.

12.5.1 Acute Exposure to Alcohol and Synaptic Function

Early in vivo studies using field potential or single unit recordings showed that acute alcohol applied by i.p. injection altered synaptic transmission in the hippocampus [131, 132]. Early in vivo studies also showed that the altered synaptic activity in the dentate gyrus of the hippocampus involved an increase in recurrent inhibition mediated by GABAergic interneurons [133]. These effects were studied further using in vitro preparations and field potential or voltage clamp recordings.

Studies on the effects of acute alcohol on synaptic transmission at the Schaffer collateral to CA1 pyramidal neuron synapse in hippocampal slices from young rats (1–2 months old) revealed that at a low concentration (0.2 mM), acute alcohol enhanced the fEPSP, whereas at a high concentration (200 mM), acute alcohol reduced the fEPSP, indicating a dose-dependent sensitivity of excitatory synaptic transmission to alcohol. Excitatory synaptic transmission in the hippocampus is mediated by a variety of glutamate receptors including NMDA, α-amino-3-hydroxy-5-methyl-4-isoxazolepropionic acid (AMPA), and kainate receptors. Synaptic responses mediated by a specific receptor subtype can be isolated for study with pharmacological agents that block the action of other receptors. In studies using this approach, acute alcohol (25–100 mM) was found to inhibit NMDAR-mediated fEPSPs at the Schaffer collateral to CA1 pyramidal neuron synapse in field potential recordings from adult rat hippocampal slices [134]. In voltage clamp studies of cultured neurons from mouse hippocampus, acute alcohol exposure inhibited NMDAR-mediated ion currents produced by exogenous application of NMDA [135–137]. These results and others identified NMDAR as an important target of alcohol and indicated that effects of alcohol on NMDAR-mediated synaptic transmission involve a postsynaptic action of alcohol at NMDARs or a protein that regulates NMDAR function [138–142]. In other studies, a presynaptic effect of acute alcohol on NMDAR-mediated synaptic transmission was also identified, in addition to postsynaptic effects involving NMDARs [143]. In field potential studies, NMDAR-mediated fEPSPs recorded in the CA1 area and evoked by Schaffer collateral stimulation were found to be more sensitive to acute alcohol in hippocampal slices from young rats (20–30 days old) than in hippocampal slices from adult rats (80–100 days old) [144]. This age-dependent differential sensitivity of NMDAR-mediated responses to acute alcohol and differences in alcohol sensitivity of NMDAR-mediated responses across CNS regions is due in part to differences in the subunit composition of the receptor [140, 145, 146].

Acute alcohol also produced an inhibition of kainate receptor-mediated excitatory synaptic transmission. For example, acute alcohol (80 mM) reduced postsynaptic currents (EPSCs) elicited by stimulation of the mossy fiber pathway and mediated by kainate receptors in CA3 pyramidal neurons studied under voltage clamp in hippocampal slices from rats (20–40 days old) [147]. AMPA receptor (AMPAR)-mediated EPSCs were not altered by alcohol in these studies [147]. However, in studies of CA1 pyramidal neurons of the hippocampus, AMPAR-mediated synaptic responses were inhibited by alcohol (100 mM) but to a lesser extent than NMDAR- and kainate receptor-mediated synaptic responses [134, 148,

149]. Acute alcohol (5–20 mM) was also reported to inhibit responses mediated by kainate receptors activated either synaptically by local stimulation of fibers or by exogenous application of kainate receptor agonists to interneurons in the CA1 region of hippocampus from rats (21–40 days old and adult) [150, 151]. This effect of alcohol could serve to reduce GABAergic inhibitory synaptic transmission and may explain some of the excitatory actions of alcohol reported in studies of the hippocampus.

In voltage clamp studies of CA1 pyramidal neurons in hippocampal slices from rats (40–90 days old), acute alcohol (10–50 mM) potentiated pharmacologically isolated $GABA_A$ receptor ($GABA_AR$)-mediated inhibitory postsynaptic currents (IPSCs) evoked by stimulation of inhibitory interneurons in the stratum radiatum [152]. Similar results were found when alcohol (80 mM) was tested on $GABA_AR$-mediated IPSCs in CA1 pyramidal neurons in hippocampal slices from young adult mice (6–12 weeks old) [137]. Results from voltage clamp studies also showed that this effect of alcohol on the $GABA_AR$-mediated IPSC could be influenced by the activation of $GABA_B$ receptors ($GABA_BR$) such that simultaneous activation of $GABA_ARs$ and $GABA_BRs$ occluded the effects of alcohol on the $GABA_AR$-mediated IPSCs [153]. Similarly, in hippocampal slices from mice (6–12 weeks old), blocking $GABA_ARs$ caused $GABA_AR$-mediated IPSCs to become more sensitive to alcohol [137]. Consistent with these results, a facilitatory effect of alcohol (80 mM) on presynaptic $GABA_BR$ function was observed in voltage clamp studies of GABAergic synapses in CA1 pyramidal neurons in hippocampal slices from mice (4–6 weeks old). This action of alcohol limited the potentiating effect of alcohol on $GABA_AR$-mediated IPSCs [154]. Studies also showed that the effects of alcohol on $GABA_AR$-mediated responses were influenced by levels of protein kinase C (PKC) activation [152, 155].

12.5.2 Chronic Exposure to Alcohol

Acute exposure to alcohol typically produces alterations in CNS function that recover within a relatively short period of time. In contrast, chronic exposure to alcohol causes persistent neuroadaptations that can alter CNS function for a prolonged period. Moreover, withdrawal from alcohol produces additional neuroadaptive changes that can compound the effects of alcohol exposure. A variety of experimental paradigms for alcohol exposure in animal models have been developed to enable studies on the effects of prolonged alcohol exposure. In these paradigms, alcohol exposure can involve drinking, alcohol injection, forced alcohol consumption (gavage feeding), or exposure to alcohol vapor. The exposure protocols can vary with respect to alcohol concentration and the duration and pattern of alcohol exposure. In spite of the differences in alcohol administration, taken together, results show that chronic exposure to alcohol results in altered excitatory and inhibitory synaptic transmission and plasticity in the hippocampus. For example, the fEPSP at the Schaffer collateral to CA1 pyramidal neuron synapse was significantly smaller in slices from rats (young adults) treated by the chronic intermittent alcohol vapor method (blood alcohol levels, 150–200 mg%) for 12–14 days compared with slices from control rats [156]. Further analysis suggested that altered transmitter release contributed to the effects of chronic alcohol exposure [156]. In another study, the Schaffer collateral stimulation frequency threshold necessary for LTP induction in CA1 neurons was reduced in hippocampal slices from rats fed a Lieber–DeCarli liquid diet (ethanol 36 % of the total calories) for 12 weeks compared with hippocampal slices from control rats [157].

Withdrawal from chronic alcohol produces hyperexcitability in the hippocampus, an effect that studies show is due in part to increased NMDAR-mediated synaptic activity [143, 158–161]. For example, NMDAR-mediated fEPSPs evoked in the CA1 region by Schaffer collateral stimulation were unaltered in hippocampal slices from rats (young adults) subjected to chronic intermittent alcohol vapor for 2 weeks (blood alcohol levels, 150–200 mg%) and prepared for recording shortly after cessation of alcohol treatment compared with slices from control rats. However, at 7 days of withdrawal from alcohol, NMDAR-mediated fEPSPs in hippocampal slices were enhanced relative to NMDAR-mediated fEPSPs in hippocampal slices from age-matched controls [156]. The enhanced NMDAR-mediated fEPSPs correlated with elevated levels of protein for NR2A and NR2B subunits observed in Western blots of the CA1 region of hippocampal slices from alcohol-treated animals compared with hippocampal slices from age-matched control animals [156]. Similar results were obtained in a hippocampal explant culture model exposed chronically to alcohol (75 mM ethanol, 5–9 days) [162]. In contrast, when young adult rats were subjected to 2 weeks of chronic alcohol liquid-diet consumption, NMDAR-mediated function in CA1 pyramidal neurons in hippocampal slices was altered compared with control slices, but there was no associated upregulation of NMDAR subunit expression [136]. These results are consistent with the involvement of mechanisms other than alterations in expression levels in alcohol withdrawal-induced hyperexcitability. For example, studies show that chronic exposure to alcohol can alter the function of signal transduction pathways that regulate the function of NMDARs and ion channels that provide feedback control of the membrane depolarization produced by activation of NMDARs [163, 164]. Mechanisms involved in the effects of chronic alcohol exposure on synaptic function may be influenced by the alcohol exposure paradigm used [165].

In addition to effects on excitatory synaptic transmission, studies show that chronic exposure to alcohol followed by a period of withdrawal alters $GABA_A R$ receptor subunit composition, subunit expression, receptor pharmacology, and the relative function of synaptic vs. extrasynaptic $GABA_A Rs$ [126, 166–168]. One or more of these changes presumably underlie the attenuation in GABA-mediated inhibitory transmission in the CA1 region of the hippocampus that results from chronic exposure to alcohol [169, 170]. For example, in voltage clamp studies of spontaneous miniature IPSPs (mIPSPs) recorded in CA1 pyramidal neurons of hippocampal slices from control and chronic alcohol-treated rats (2–3 months of age), significant differences were observed for the amplitude and kinetics of the mIPSCs between the alcohol-treated and control hippocampal slices [171]. These alcohol-induced changes are consistent with reduced synaptic transmission resulting from alcohol-induced presynaptic and postsynaptic actions [171].

12.5.3 Synaptic Plasticity

LTP induction is mediated by activation of the NMDAR, and the resulting Ca^{2+} flux through the receptor channels is modulated by activation of GABAergic synaptic transmission. Therefore, it is not surprising that acute and chronic alcohol exposure can significantly alter LTP in rat hippocampus. A number of studies have shown that acute alcohol can reduce LTP at the Schaffer collateral to CA1 pyramidal neuron synapse [172–178]. The sensitivity of LTP to alcohol appears to be dependent on the age of the animal, alcohol concentration, and LTP induction protocol used. For example, in one study, LTP of the fEPSP was reduced by acute exposure to 60 mM alcohol in hippocampal slices from immature (15–25 days old) but not adult (70–100 days old) rats [179]. In another study, a different LTP induction stimulation protocol induces LTP in hippocampal slices from adult rats (2–3 months old) when a low concentration (4.3 mM) but not at higher concentration (8.6 or 17.1 mM) of acute alcohol was tested [157]. For acute alcohol to alter LTP, alcohol must be present during the LTP induction stimulation protocol [180]. Studies indicate that the effects of acute alcohol on LTP primarily involve mechanisms linked to NMDAR function, although effects on GABA receptor-mediated synaptic transmission also contribute [175, 180, 181]. Acute alcohol has also been reported to alter LTD [175, 182].

Chronic exposure to alcohol also alters LTP [183, 184]. Long periods of chronic alcohol treatment (e.g., several months) produce a lasting impairment of LTP in the hippocampus [185], whereas short-term chronic alcohol treatment (e.g., 2 weeks) produces an impairment of LTP that is reversible [184]. For example, in hippocampal slices from control rats (2–3 months old) and rats exposed to a chronic intermittent alcohol vapor protocol (blood alcohol levels ~180 mg/dl), an LTP induction protocol that induced LTP in slices from control rats was ineffective in inducing LTP slices from alcohol-exposed rats. However, after 5 days of withdrawal, recovery of LTP was observed [184]. A recent detailed review on alcohol effects on synaptic plasticity is available for more detailed information on this topic [186].

12.5.4 Signal Transduction

Alcohol exposure has been reported to activate or inhibit a variety of signal transduction pathways, many of which are used by GPCRs including chemokine receptors. For example, acute alcohol administered i.p. (3.5 g/kg; blood alcohol levels ~370 mg/dl) reduced the level of ERK1/2 and CREB activation in hippocampal neurons of postnatal day (PN) 5, PN 21, and adult rats [187]. In contrast to another study, acute alcohol administered i.p. (3.2 or 1.6 g/kg) increased CRE-mediated gene transcription in the CA1 region of the hippocampus and other CNS regions [188]. CREB is a downstream effector of ERK1/2 and cAMP-dependent protein kinase A (PKA). CREB phosphorylation on Ser-133 promotes transcription of genes with an upstream cAMP response element (CRE).

Other studies support a role for PKC in the actions of acute alcohol [189]. Acute alcohol administered to rats (58–64 days old) by gavage (0.5, 1.0, or 2 g/kg) was reported to increase the level of phosphorylated PKC immunoreactivity in the dentate granule cell layer and in the CA3 subregion of the hippocampus but not in the CA1 region of the hippocampus [190]. These results are consistent with differences in target sites of alcohol action in different CNS regions. Experiments in genetically modified mice also support a role for PKC in actions of alcohol in the CNS [191, 192] as well as a role for fyn kinase [142, 193].

Chronic exposure to alcohol also alters signal transduction pathways. For example, in young adult rats (6–8 weeks old) maintained for 2 weeks on a liquid alcohol diet, a significant increase in the levels of phosphorylated p38 MAPK was observed in the CA1 region of hippocampal slices [136]. Chronic intermittent alcohol exposure by the vapor method was also shown to reduce the level activated of ERK1/2 in the hippocampus [194, 195].

12.6 Summary and Conclusions

Effects of alcohol on the CNS have been intensively investigated both clinically and experimentally. In contrast, studies on the actions of chemokines in the CNS are limited. The discovery that alcohol induces neuroinflammation and the production of chemokines in the CNS is a significant finding. By analogy with other CNS conditions where elevated levels of chemokines have been shown to occur, we can expect that alcohol-induced production of chemokines in the CNS will play a role in the actions of alcohol on the CNS. The alcohol-induced chemokines could be helpful, for example, by providing neuroprotection, or could contribute to the impairment of CNS function produced by alcohol.

Emerging research shows that chemokines acting through chemokine receptors can interact with the same molecules and biological process as alcohol. This commonality provides a rational basis for the idea that alcohol-induced chemokines can mediate or modulate the effects of alcohol on the CNS. Our recent studies on synaptic function in hippocampal slices from CCL2 transgenic mice that express elevated levels of CCL2 in the CNS support this idea. The CCL2 transgenic mice model, one aspect of effects of chronic alcohol exposure on the CNS, elevated levels of CCL2 in the CNS. Our studies show that the actions of acute alcohol on hippocampal synaptic transmission and plasticity are altered in hippocampal slices from CCL2 transgenic mice compared with hippocampal slices from non-transgenic littermate controls [95]. These results are consistent with an important role for alcohol-induced chemokines in alcohol use disorders. Research in this challenging, interesting, and important area is needed and will undoubtedly bring new insight and understanding of the effects and mechanisms of alcohol action on the CNS and behavior.

Acknowledgments I thank Drs. Tom Nelson, Salvador Huitron-Resendiz, and Hilda Bajova for contributing data and figures; Dr. Jennifer Bray for helpful comments on the manuscript; and Floriska Chizer-Byas for administrative assistance. This work is supported by RC1 AA019261.

Appendix: Hippocampal Slice Recordings

A variety of experimental approaches have been employed to assess the role of different chemokines and other biologically active molecules including alcohol on hippocampal synaptic function. Perhaps the most common approaches are (a) the manipulation of extracellular levels through exogenous application to in vivo, ex vivo, or culture models; (b) the use of genetically modified animals; and (c) the use of animal models of a disease condition. An important issue inherent in such studies is how the experimental protocols and preparations relate to the in vivo conditions of interest. Typically, models do not exactly simulate an in vivo condition. Moreover, differences in concentration, timing of exposure, and age of the animal when exposure occurs are all likely to be important factors in the final consequences of chemokine action, as has been shown for the actions of alcohol. Therefore, integration of information from a variety of models and approaches will most likely be necessary for the most comprehensive understanding. Often, results from different models do not concur, a situation that reflects the limitations of models, complexity of the problem, and deficits in our knowledge base.

One of the most frequently used experimental approaches for studies on neuroactive molecules such as chemokines and alcohol is to test their ability to alter synaptic function in an in vitro slice preparation of the hippocampus. An important advantage of the hippocampal slice preparation is that it enables studies of synaptic function in hippocampus from both young and adult animals under conditions that retain many features of the in vivo tissue. The well-characterized lamellar structure of the hippocampus makes it well suited for measurement of synaptic function within well-defined circuits in a slice preparation, for example, in the CA1 pyramidal layer at the site of synaptic input from the Schaffer collaterals (Fig. 12.6). The Schaffer collateral to CA1 pyramidal neuron synapse is one of the most highly studied synapses in the CNS. Two electrophysiological recording techniques are commonly used for such studies, field potential recordings and intracellular recordings. Field potential recordings detect changes in electrical activity produced by synaptic transmission from an extracellular vantage point (Figs. 12.7 and 12.8). The recorded activity reflects the summed electrical events of a population of neurons and can provide a useful and sensitive measure of synaptic function and changes in synaptic function including an overall assessment of changes in pre- and postsynaptic components of synaptic transmission. In intracellular recordings, electrical activity is measured from an intracellular vantage point necessitating access to the intracellular compartment by the recording electrode. Intracellular recordings are used to measure changes in membrane potential (current clamp recordings) (Fig. 12.8) or ion flow through the membrane channels (voltage clamp) associated with synaptic transmission. Intracellular recordings are important for identification of cellular and molecular mechanisms mediating synaptic function and alterations in these cellular and molecular mechanisms produced by experimental manipulations such as exposure to chemokines or alcohol.

An assessment of synaptic function typically involves studies of synaptic transmission and synaptic plasticity. Synaptic transmission at the Schaffer collateral to

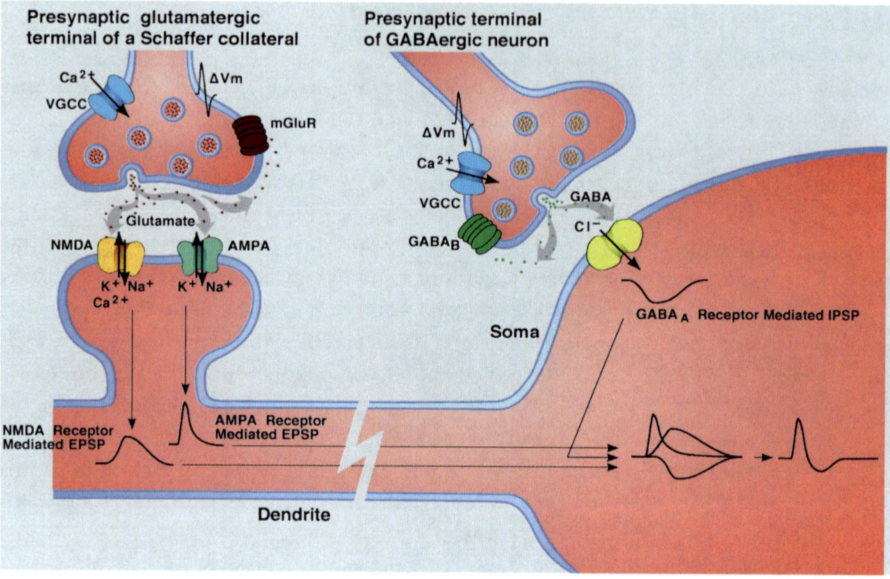

Fig. 12.6 Simplified diagram showing pre- and postsynaptic aspects of synaptic transmission at the Schaffer collateral to CA1 pyramidal neuron synapse. Activation of the Schaffer collaterals results in a presynaptic action potential and Ca^{2+} influx through voltage-gated Ca^{2+} channels (VGCC) in the axon terminals, which initiates glutamate release. Glutamate diffuses to the postsynaptic membrane and evokes a membrane depolarization (excitatory postsynaptic potential, EPSP) mediated by AMPAR and, at strong stimulations, NMDARs in the dendritic synapse of the CA1 pyramidal neurons. The membrane depolarization spreads to the somatic region where it can evoke an action potential (not shown). The action potential can spread to axon terminals of the CA1 neurons that provide synaptic input to GABAergic inhibitory interneurons responsible for feedback control of somatic excitability. The inhibitory postsynaptic potential (IPSP) produced by this pathway summates with the EPSP to form an overlapping EPSP/IPSP at the somatic region (see Fig. 12.8). Receptors on the presynaptic terminals can modulate transmitter release

CA1 pyramidal neuron synapse is elicited experimentally by electrical stimulation of the Schaffer collaterals, which induces transmitter (glutamate) release from the Schaffer collateral terminals. The released transmitter interacts with several glutamate receptors at dendritic synapses of the CA1 pyramidal neurons and evokes a membrane depolarization referred to as an excitatory postsynaptic potential (EPSP; fEPSP in field potential recording) (Figs. 12.6, 12.7, and 12.8). The dendritic membrane depolarization spreads to the axosomatic region of the CA1 neurons where, if large enough, it elicits a somatic action potential [referred to as a population spike (PS) in field potential recordings]. The action potential is conducted to axon terminals of CA1 pyramidal neurons that innervate other CNS regions and GABAergic inhibitory interneurons in the CA1 region responsible for feedback control of the excitability of the CA1 somatic region (Fig. 12.8).

Synaptic plasticity is induced experimentally by specific stimulation protocols designed to simulate activity patterns of neuronal circuits. Two forms of synaptic plasticity, LTP and LTD, can be established at the Schaffer collateral to CA1

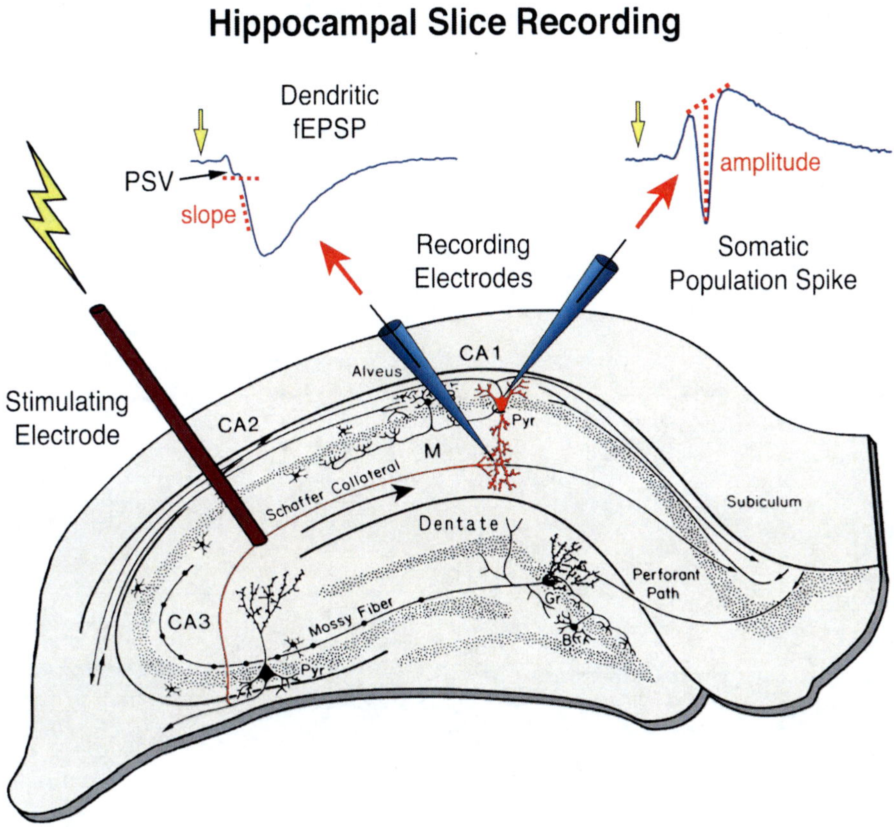

Fig. 12.7 Simplified diagram showing the placement of electrodes and recorded responses in a field potential recording of synaptic transmission at the Schaffer collateral to CA1 pyramidal neuron synapse in a hippocampal slice. Synaptic transmission is initiated experimentally by electrical stimulation of Schaffer collaterals, axons of the CA3 pyramidal neurons of the hippocampus. Stimulation of the Schaffer collateral elicits two electrical events at the CA1 synapse that are detected in field potential recordings as a presynaptic volley (PSV) and field excitatory postsynaptic potential (fEPSP). The PSV reflects a summation of the action potentials occurring in the stimulated Schaffer collateral afferent terminals, while the fEPSP reflects a summation of excitatory postsynaptic responses occurring in the dendrites of CA1 pyramidal neurons activated by the afferent fibers. A summated response of action potentials evoked by the synaptic responses and occurring in the somatic region of the pyramidal neurons can be recorded as a somatic population spike (PS) in the field potential recordings. Typically measurements are made of the slope of the fEPSP and amplitude of the PS. Field potential recordings are advantages for certain types of studies because the recording technique has little impact on the physiology of the neurons. In contrast, intracellular recordings, another important recording technique, can disturb the biochemistry of the cell and thereby limit the types of information that can be obtained from the recording

pyramidal neuron synapse by Schaffer collateral stimulation. Both LTP and LTD are thought to be cellular mechanisms underlying learning and memory [196, 197]. LTP and LTD are modifications in synaptic strength resulting from alterations in the function or number of postsynaptic glutamate receptors of AMPA subtype [198].

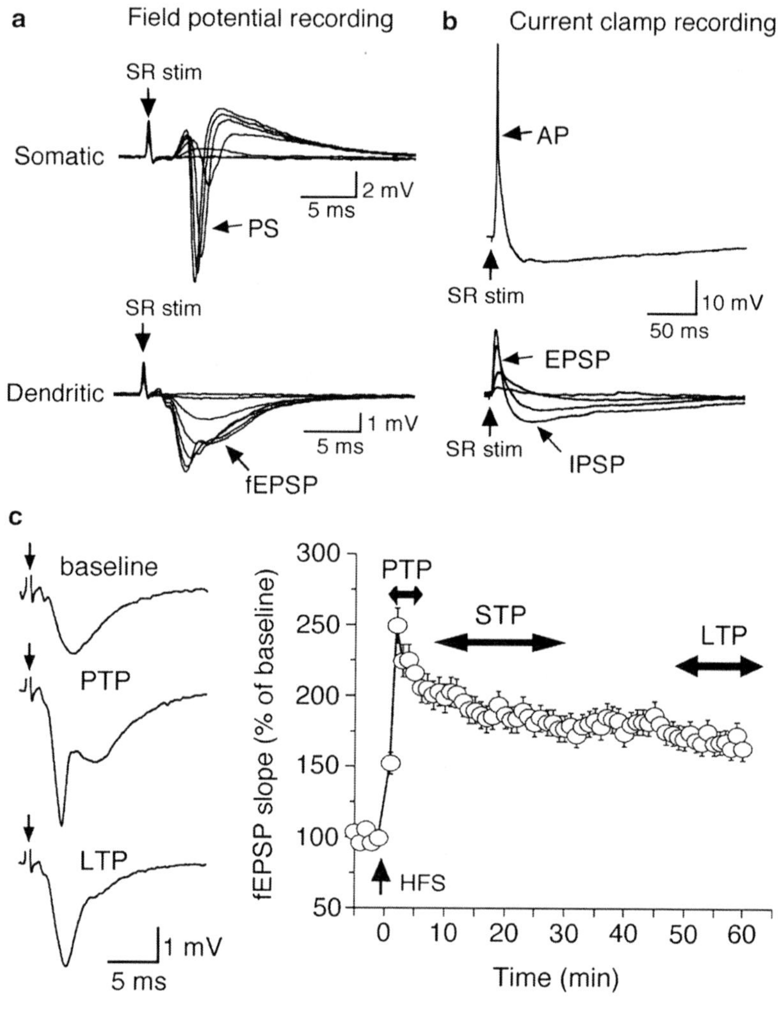

Fig. 12.8 Representative field potential and intracellular recordings of synaptic events elicited by activation of the Schaffer collateral to CA1 pyramidal neuron synapse in a hippocampal slice from adult mice. (**a**) Simultaneous recording of PS and fEPSP elicited by a series of electrical stimulations of increasing intensity applied to the Schaffer collaterals (SR stim). (**b**) Intracellular (current clamp) recording of synaptic responses in the somatic region of a CA1 pyramidal neuron elicited by series of electrical stimulations of increasing intensity applied to the Schaffer collaterals (SR stim). Stimulation of the Schaffer collaterals elicits an EPSP and feedback IPSP that summate. If the EPSP is large enough, an action potential (AP) is evoked (*top trace*). (**c**) Graph shows LTP of the fEPSP at the Schaffer collateral to CA1 pyramidal neuron synapse. At the left of the graph are shown representative field potential recordings of fEPSPs elicited by a standard test stimulation applied to the Schaffer collaterals during baseline recordings and at different times after application of the high-frequency stimulation protocol (HFS) used to induce LTP. Each point on the graph represents the slope of the fEPSP elicited by the standard test stimulus at the time point shown (Nelson and Gruol, unpublished; methods were similar to that reported previously [94])

LTP is induced by high-frequency stimulation that produces intense synaptic activity and results in a long-lasting enhancement of AMPAR-mediated synaptic strength (Fig. 12.8). LTD is induced by low-frequency stimulation and results in a long-lasting depression of AMPAR-mediated synaptic strength. In addition to LTP, two short-term forms of synaptic plasticity are produced by the LTP induction protocol, PTP and STP (Fig. 12.8c). PTP and STP are thought to be involved in short-term memory [199, 200]. PTP results from presynaptic mechanisms that produce an increase in transmitter release, whereas STP appears to result from both presynaptic and postsynaptic mechanisms [201–203].

Activity patterns that induce LTP or LTD also initiate downstream actions at signal transduction pathways leading to gene expression and protein synthesis, a process that is essential for encoding long-term memories. Numerous signaling molecules have been identified that mediate or regulate LTP or LTD including calcium/calmodulin-dependent protein kinase II (CaMKII), calcium/calmodulin-dependent protein kinase IV (CaMKIV), p42/p44 MAPK, p38 MAPK, PKA, PKC, protein phosphatases 1/2A, protein phosphatase 2B (calcineurin), and tyrosine kinase [204–207]. For example, increased activation of p42 MAP kinase (i.e., ERK2) is necessary for the induction of LTP [208]. In contrast, p38 MAPK is involved in LTD induction [209] as well as cytokine-induced inhibition of LTP (TNF-α and IL-1β) [210, 211]. CREB is also essential for the establishment of LTP and LTD by inducing change in gene expression required for long-term changes in hippocampal function [212].

References

1. Bennett LD, Fox JM, Signoret N (2011) Mechanisms regulating chemokine receptor activity. Immunology 134:246–256
2. Adler MW, Geller EB, Chen X, Rogers TJ (2005) Viewing chemokines as a third major system of communication in the brain. AAPS J 7:E865–E870
3. Rostene W, Kitabgi P, Parsadaniantz SM (2007) Chemokines: a new class of neuromodulator? Nat Rev Neurosci 8:895–903
4. Bilbo SD, Smith SH, Schwarz JM (2012) A lifespan approach to neuroinflammatory and cognitive disorders: a critical role for glia. J Neuroimmune Pharmacol 7:24–41
5. Mines M, Ding Y, Fan GH (2007) The many roles of chemokine receptors in neurodegenerative disorders: emerging new therapeutical strategies. Curr Med Chem 14:2456–2470
6. Melik-Parsadaniantz S, Rostene W (2008) Chemokines and neuromodulation. J Neuroimmunol 198:62–68
7. Kelder W, McArthur JC, Nance-Sproson T, McClernon D, Griffin DE (1998) Beta-chemokines MCP-1 and RANTES are selectively increased in cerebrospinal fluid of patients with human immunodeficiency virus-associated dementia. Ann Neurol 44:831–835
8. Rosler A et al (1998) Time course of chemokines in the cerebrospinal fluid and serum during herpes simplex type 1 encephalitis. J Neurol Sci 157:82–89
9. Sorensen TL et al (1999) Expression of specific chemokines and chemokine receptors in the central nervous system of multiple sclerosis patients. J Clin Invest 103:807–815
10. Helmy A, Carpenter KL, Menon DK, Pickard JD, Hutchinson PJ (2011) The cytokine response to human traumatic brain injury: temporal profiles and evidence for cerebral parenchymal production. J Cereb Blood Flow Metab 31:658–670

11. Asensio VC, Campbell IL (1999) Chemokines in the CNS: plurifunctional mediators in diverse states. Trends Neurosci 22:504–512
12. Mennicken F, Maki R, de Souza EB, Quirion R (1999) Chemokines and chemokine receptors in the CNS: a possible role in neuroinflammation and patterning. Trends Pharmacol Sci 20:73–78
13. Streit WJ, Mrak RE, Griffin WS (2004) Microglia and neuroinflammation: a pathological perspective. J Neuroinflammation 1:14
14. Rahman S, Miles MF (2001) Identification of novel ethanol-sensitive genes by expression profiling. Pharmacol Ther 92:123–134
15. Crews FT, Zou J, Qin L (2011) Induction of innate immune genes in brain create the neurobiology of addiction. Brain Behav Immun 25(Suppl 1):S4–S12
16. Blanco AM, Guerri C (2007) Ethanol intake enhances inflammatory mediators in brain: role of glial cells and TLR4/IL-1RI receptors. Front Biosci 12:2616–2630
17. Thibault C et al (2000) Expression profiling of neural cells reveals specific patterns of ethanol-responsive gene expression. Mol Pharmacol 58:1593–1600
18. Bates ME, Bowden SC, Barry D (2002) Neurocognitive impairment associated with alcohol use disorders: implications for treatment. Exp Clin Psychopharmacol 10:193–212
19. Matsumoto I, Alexander-Kaufman K, Iwazaki T, Kashem MA, Matsuda-Matsumoto H (2007) CNS proteomes in alcohol and drug abuse and dependence. Expert Rev Proteomics 4:539–552
20. IUIS/WHO Subcommittee on Chemokine Nomenclature (2002) Chemokine/chemokine receptor nomenclature. J Immunol Methods 262:1–3
21. Zlotnik A, Yoshie O (2000) Chemokines: a new classification system and their role in immunity. Immunity 12:121–127
22. Marinissen MJ, Gutkind JS (2001) G-protein-coupled receptors and signaling networks: emerging paradigms. Trends Pharmacol Sci 22:368–376
23. Luttrell LM (2006) Transmembrane signaling by G protein-coupled receptors. Methods Mol Biol 332:3–49
24. Nishiyori A et al (1998) Localization of fractalkine and CX3CR1 mRNAs in rat brain: does fractalkine play a role in signaling from neuron to microglia? FEBS Lett 429:167–172
25. Meucci O et al (1998) Chemokines regulate hippocampal neuronal signaling and gp120 neurotoxicity. Proc Natl Acad Sci USA 95:14500–14505
26. Coughlan CM et al (2000) Expression of multiple functional chemokine receptors and monocyte chemoattractant protein-1 in human neurons. Neuroscience 97:591–600
27. Klein RS et al (1999) Chemokine receptor expression and signaling in macaque and human fetal neurons and astrocytes: implications for the neuropathogenesis of AIDS. J Immunol 163:1636–1646
28. Horuk R et al (1997) Expression of chemokine receptors by subsets of neurons in the central nervous system. J Immunol 158:2882–2890
29. Lavi E, Kolson DL, Ulrich AM, Fu L, Gonzalez-Scarano F (1998) Chemokine receptors in the human brain and their relationship to HIV infection. J Neurovirol 4:301–311
30. van der Meer P et al (2000) Expression pattern of CXCR3, CXCR4, and CCR3 chemokine receptors in the developing human brain. J Neuropathol Exp Neurol 60:25–32
31. Bajova H, Nelson TE, Gruol DL (2008) Chronic CXCL10 alters the level of activated ERK1/2 and transcriptional factors CREB and NF-κB in hippocampal neuronal cell culture. J Neuroimmunol 195:36–46
32. Wang CX et al (2006) Cyclin-dependent kinase-5 prevents neuronal apoptosis through ERK-mediated upregulation of Bcl-2. Cell Death Differ 13:1203–1212
33. Hetman M, Gozdz A (2004) Role of extracellular signal regulated kinases 1 and 2 in neuronal survival. Eur J Biochem 271:2050–2055
34. Mattson MP, Goodman Y, Luo H, Fu W, Furukawa K (1997) Activation of NF-kappaB protects hippocampal neurons against oxidative stress-induced apoptosis: evidence for induction of manganese superoxide dismutase and suppression of peroxynitrite production and protein tyrosine nitration. J Neurosci Res 49:681–697
35. Mattson MP (2006) NF-kappaB in the survival and plasticity of neurons. Neurochem Res 30:883–893

36. Araujo DM, Cotman CW (1993) Trophic effects of interleukin-4, -7 and -8 on hippocampal neuronal cultures: potential involvement of glial-derived factors. Brain Res 600:49–55
37. Limatola C et al (2005) Chemokine CX3CL1 protects rat hippocampal neurons against glutamate-mediated excitotoxicity. J Neuroimmunol 166:19–28
38. Meucci O, Fatatis A, Simen AA, Miller RJ (2000) Expression of CX3CR1 chemokine receptors on neurons and their role in neuronal survival. Proc Natl Acad Sci USA 97:8075–8080
39. Ragozzino D et al (2006) Chemokine fractalkine/CX3CL1 negatively modulates active glutamatergic synapses in rat hippocampal neurons. J Neurosci 26:10488–10498
40. Luo Y, Lathia J, Mughal M, Mattson MP (2008) SDF1alpha/CXCR4 signaling, via ERKs and the transcription factor Egr1, induces expression of a 67-kDa form of glutamic acid decarboxylase in embryonic hippocampal neurons. J Biol Chem 283:24789–24800
41. Cho J, Gruol DL (2008) The chemokine CCL2 activates p38 mitogen-activated protein kinase pathway in cultured rat hippocampal cells. J Neuroimmunol 199:94–103
42. Farina C, Aloisi F, Meinl E (2007) Astrocytes are active players in cerebral innate immunity. Trends Immunol 28:138–145
43. Flugel A et al (2001) Neuronal MCP-1 expression in response to remote nerve injury. J Cereb Blood Flow Metab 21:69–76
44. Suk K (2007) Microglial signal transduction as a target of alcohol action in the brain. Curr Neurovasc Res 4:131–142
45. Allaman I, Belanger M, Magistretti PJ (2011) Astrocyte-neuron metabolic relationships: for better and for worse. Trends Neurosci 34:76–87
46. Ransom B, Behar T, Nedergaard M (2003) New roles for astrocytes (stars at last). Trends Neurosci 26:520–522
47. Halassa MM, Fellin T, Haydon PG (2007) The tripartite synapse: roles for gliotransmission in health and disease. Trends Mol Med 13:54–63
48. Paolicelli RC et al (2011) Synaptic pruning by microglia is necessary for normal brain development. Science 333:1456–1458
49. Tremblay ME et al (2011) The role of microglia in the healthy brain. J Neurosci 31:16064–16069
50. Blasko I et al (2006) Measurement of thirteen biological markers in CSF of patients with Alzheimer's disease and other dementias. Dement Geriatr Cogn Disord 21:9–15
51. Galimberti D et al (2006) Intrathecal chemokine synthesis in mild cognitive impairment and Alzheimer disease. Arch Neurol 63:538–543
52. Xia MQ, Bacskai BJ, Knowles RB, Qin SX, Hyman BT (2000) Expression of the chemokine receptor CXCR3 on neurons and the elevated expression of its ligand IP-10 in reactive astrocytes: in vitro ERK1/2 activation and role in Alzheimer's disease. J Neuroimmunol 108:227–235
53. Sanders VJ et al (1998) Chemokines and receptors in HIV encephalitis. AIDS 12: 1021–1026
54. Fife BT et al (2001) CXCL10 (IFN-gamma-inducible protein-10) control of encephalitogenic CD4+ T cell accumulation in the central nervous system during experimental autoimmune encephalomyelitis. J Immunol 166:7617–7624
55. Kolb SA et al (1999) Identification of a T cell chemotactic factor in the cerebrospinal fluid of HIV-1-infected individuals as interferon-gamma inducible protein 10. J Neuroimmunol 93:172–181
56. Tran PB, Miller RJ (2003) Chemokine receptors: signposts to brain development and disease. Nat Rev Neurosci 4:444–455
57. Ma Q et al (1998) Impaired B-lymphopoiesis, myelopoiesis, and derailed cerebellar neuron migration in CXCR4- and SDF-1-deficient mice. Proc Natl Acad Sci USA 95:9448–9453
58. Lu M, Grove EA, Miller RJ (2002) Abnormal development of the hippocampal dentate gyrus in mice lacking the CXCR4 chemokine receptor. Proc Natl Acad Sci USA 99:7090–7095
59. Bagri A et al (2002) The chemokine SDF1 regulates migration of dentate granule cells. Development 129:4249–4260
60. Zou YR, Kottmann AH, Kuroda M, Taniuchi I, Littman DR (1998) Function of the chemokine receptor CXCR4 in haematopoiesis and in cerebellar development. Nature 393:595–599

61. Stumm R, Hollt V (2007) CXC chemokine receptor 4 regulates neuronal migration and axonal pathfinding in the developing nervous system: implications for neuronal regeneration in the adult brain. J Mol Endocrinol 38:377–382
62. Su B, Pan S, He X, Li P, Liang Y (2012) Sprouting of nervous fibers and upregulation of C-X-C chemokine receptor type 4 expression in hippocampal formation of rats with enhanced spatial learning and memory. Anat Rec (Hoboken) 295:121–126
63. Parachikova A, Cotman CW (2007) Reduced CXCL12/CXCR4 results in impaired learning and is downregulated in a mouse model of Alzheimer disease. Neurobiol Dis 28:143–153
64. Rogers JT et al (2011) CX3CR1 deficiency leads to impairment of hippocampal cognitive function and synaptic plasticity. J Neurosci 31:16241–16250
65. Neves G, Cooke SF, Bliss TV (2008) Synaptic plasticity, memory and the hippocampus: a neural network approach to causality. Nat Rev Neurosci 9:65–75
66. Maggi L et al (2011) CX(3)CR1 deficiency alters hippocampal-dependent plasticity phenomena blunting the effects of enriched environment. Front Cell Neurosci 5:22
67. Ryabinin AE (1998) Role of hippocampus in alcohol-induced memory impairment: implications from behavioral and immediate early gene studies. Psychopharmacology (Berl) 139:34–43
68. Akhondzadeh S (1999) Hippocampal synaptic plasticity and cognition. J Clin Pharm Ther 24:241–248
69. Battaglia FP, Benchenane K, Sirota A, Pennartz CM, Wiener SI (2011) The hippocampus: hub of brain network communication for memory. Trends Cogn Sci 15:310–318
70. Wang SH, Morris RG (2010) Hippocampal-neocortical interactions in memory formation, consolidation, and reconsolidation. Annu Rev Psychol 61:49–79, C1–4
71. O'Reilly RC, Bhattacharyya R, Howard MD, Ketz N (2011) Complementary learning systems. Cogn Sci. 2011 Dec 5. doi: 10.1111/j.1551-6709.2011.01214.x. [Epub ahead of print]
72. van Groen T, Wyss JM (1990) Extrinsic projections from area CA1 of the rat hippocampus: olfactory, cortical, subcortical, and bilateral hippocampal formation projections. J Comp Neurol 302:515–528
73. Cenquizca LA, Swanson LW (2007) Spatial organization of direct hippocampal field CA1 axonal projections to the rest of the cerebral cortex. Brain Res Rev 56:1–26
74. Pannasch U et al (2011) Astroglial networks scale synaptic activity and plasticity. Proc Natl Acad Sci USA 108:8467–8472
75. Tremblay ME, Majewska AK (2011) A role for microglia in synaptic plasticity? Commun Integr Biol 4:220–222
76. Ben Achour S, Pascual O (2010) Glia: the many ways to modulate synaptic plasticity. Neurochem Int 57:440–445
77. Bessis A, Bechade C, Bernard D, Roumier A (2007) Microglial control of neuronal death and synaptic properties. Glia 55:233–238
78. Xiong H et al (2003) Inhibition of long-term potentiation by interleukin-8: implications for human immunodeficiency virus-1-associated dementia. J Neurosci Res 71:600–607
79. Vlkolinsky R, Siggins GR, Campbell IL, Krucker T (2004) Acute exposure to CXC chemokine ligand 10, but not its chronic astroglial production, alters synaptic plasticity in mouse hippocampal slices. J Neuroimmunol 150:37–47
80. Zhou Y, Tang H, Liu J, Dong J, Xiong H (2011) Chemokine CCL2 modulation of neuronal excitability and synaptic transmission in rat hippocampal slices. J Neurochem 116:406–414
81. Bertollini C, Ragozzino D, Gross C, Limatola C, Eusebi F (2006) Fractalkine/CX3CL1 depresses central synaptic transmission in mouse hippocampal slices. Neuropharmacology 51:816–821
82. Maggi L et al (2009) LTP impairment by fractalkine/CX3CL1 in mouse hippocampus is mediated through the activity of adenosine receptor type 3 (A3R). J Neuroimmunol 215:36–42
83. Piccinin S et al (2010) CX3CL1-induced modulation at CA1 synapses reveals multiple mechanisms of EPSC modulation involving adenosine receptor subtypes. J Neuroimmunol 224:85–92

84. Marchionni I et al (2010) Distinctive properties of CXC chemokine receptor 4-expressing Cajal–Retzius cells versus GABAergic interneurons of the postnatal hippocampus. J Physiol 588:2859–2878
85. Pautassi RM, Myers M, Spear LP, Molina JC, Spear NE (2008) Adolescent but not adult rats exhibit ethanol-mediated appetitive second-order conditioning. Alcohol Clin Exp Res 32:2016–2027
86. Spear LP (2000) The adolescent brain and age-related behavioral manifestations. Neurosci Biobehav Rev 24:417–463
87. Maccaferri G (2011) Modulation of hippocampal stratum lacunosum-moleculare microcircuits. J Physiol 589:1885–1891
88. Bhattacharyya BJ et al (2008) The chemokine stromal cell-derived factor-1 regulates GABAergic inputs to neural progenitors in the postnatal dentate gyrus. J Neurosci 28:6720–6730
89. Pujol F, Kitabgi P, Boudin H (2005) The chemokine SDF-1 differentially regulates axonal elongation and branching in hippocampal neurons. J Cell Sci 118:1071–1080
90. Tran PB, Ren D, Veldhouse TJ, Miller RJ (2004) Chemokine receptors are expressed widely by embryonic and adult neural progenitor cells. J Neurosci Res 76:20–34
91. Huang D et al (2002) Pertussis toxin-induced reversible encephalopathy dependent on monocyte chemoattractant protein-1 overexpression in mice. J Neurosci 22:10633–10642
92. Boztug K et al (2002) Leukocyte infiltration, but not neurodegeneration, in the CNS of transgenic mice with astrocyte production of the CXC chemokine ligand 10. J Immunol 169:1505–1515
93. Huang D et al (2005) Chronic expression of monocyte chemoattractant protein-1 in the central nervous system causes delayed encephalopathy and impaired microglial function in mice. FASEB J 19:761–772
94. Nelson TE, Hao C, Manos J, Ransohoff RM, Gruol DL (2011) Altered hippocampal synaptic transmission in transgenic mice with astrocyte-targeted enhanced CCL2 expression. Brain Behav Immun 25(Suppl 1):S106–S119
95. Bray JG, Reyes KC, Ransohoff RM, Gruol DL (2011) Synaptic plasticity in the hippocampus shows resistance to acute ethanol exposure in transgenic mice with astrocyte-targeted enhanced CCL2 expression. Alcohol Clin Exp Res 35(Suppl):61A
96. Przewlocki R et al (1999) Opioid enhancement of calcium oscillations and burst events involving NMDA receptors and L-type calcium channels in cultured hippocampal neurons. J Neurosci 19:9705–9715
97. Nelson TE, Gruol DL (2004) The chemokine CXCL10 modulates excitatory activity and intracellular calcium signaling in cultured hippocampal neurons. J Neuroimmunol 156: 74–87
98. Liu Z et al (2003) Frequency modulation of synchronized Ca2+ spikes in cultured hippocampal networks through G-protein-coupled receptors. J Neurosci 23:4156–4163
99. Cho JS, Gruol DL (2008) The chemokine CCL2 activates p38 mitogen-activated protein kinase pathway in cultured rat hippocampal cells, J Neuroimmunol 199:94–103
100. Cho J, Nelson TE, Bajova H, Gruol DL (2009) Chronic CXCL10 alters neuronal properties in rat hippocampal culture. J Neuroimmunol 207:92–100
101. Kuijpers M, van Gassen KL, de Graan PN, Gruol D (2010) Chronic exposure to the chemokine CCL3 enhances neuronal network activity in rat hippocampal cultures. J Neuroimmunol 229:73–80
102. He J, Crews FT (2008) Increased MCP-1 and microglia in various regions of the human alcoholic brain. Exp Neurol 210:349–358
103. Qin L et al (2008) Increased systemic and brain cytokine production and neuroinflammation by endotoxin following ethanol treatment. J Neuroinflammation 5:10
104. Flora G et al (2005) Proinflammatory synergism of ethanol and HIV-1 Tat protein in brain tissue. Exp Neurol 191:2–12
105. Blednov YA et al (2005) Perturbation of chemokine networks by gene deletion alters the reinforcing actions of ethanol. Behav Brain Res 165:110–125

106. Breese GR et al (2008) Repeated lipopolysaccharide (LPS) or cytokine treatments sensitize ethanol withdrawal-induced anxiety-like behavior. Neuropsychopharmacology 33:867–876
107. Zou J, Crews F (2010) Induction of innate immune gene expression cascades in brain slice cultures by ethanol: key role of NF-kappaB and proinflammatory cytokines. Alcohol Clin Exp Res 34:777–789
108. Boyadjieva NI, Sarkar DK (2010) Role of microglia in ethanol's apoptotic action on hypothalamic neuronal cells in primary cultures. Alcohol Clin Exp Res 34:1835–1842
109. Gonzalez A, Salido GM (2009) Ethanol alters the physiology of neuron-glia communication. Int Rev Neurobiol 88:167–198
110. Ward RJ et al (2009) Neuro-inflammation induced in the hippocampus of 'binge drinking' rats may be mediated by elevated extracellular glutamate content. J Neurochem 111: 1119–1128
111. Tomas M, Marin P, Megias L, Egea G, Renau-Piqueras J (2005) Ethanol perturbs the secretory pathway in astrocytes. Neurobiol Dis 20:773–784
112. Catlin MC, Guizzetti M, Costa LG (1999) Effects of ethanol on calcium homeostasis in the nervous system: implications for astrocytes. Mol Neurobiol 19:1–24
113. Satriotomo I et al (2000) Short-term ethanol exposure alters calbindin D28k and glial fibrillary acidic protein immunoreactivity in hippocampus of mice. Brain Res 879:55–64
114. Gonzalez A, Pariente JA, Salido GM (2007) Ethanol stimulates ROS generation by mitochondria through Ca2+ mobilization and increases GFAP content in rat hippocampal astrocytes. Brain Res 1178:28–37
115. Franke H (1995) Influence of chronic alcohol treatment on the GFAP-immunoreactivity in astrocytes of the hippocampus in rats. Acta Histochem 97:263–271
116. McClain JA et al (2011) Adolescent binge alcohol exposure induces long-lasting partial activation of microglia. Brain Behav Immun 25(Suppl 1):S120–S128
117. Sofroniew MV (2009) Molecular dissection of reactive astrogliosis and glial scar formation. Trends Neurosci 32:638–647
118. Okun E, Griffioen KJ, Mattson MP (2011) Toll-like receptor signaling in neural plasticity and disease. Trends Neurosci 34:269–281
119. Carty M, Bowie AG (2011) Evaluating the role of Toll-like receptors in diseases of the central nervous system. Biochem Pharmacol 81:825–837
120. Blanco AM, Valles SL, Pascual M, Guerri C (2005) Involvement of TLR4/type I IL-1 receptor signaling in the induction of inflammatory mediators and cell death induced by ethanol in cultured astrocytes. J Immunol 175:6893–6899
121. Fernandez-Lizarbe S, Pascual M, Guerri C (2009) Critical role of TLR4 response in the activation of microglia induced by ethanol. J Immunol 183:4733–4744
122. Alfonso-Loeches S, Pascual-Lucas M, Blanco AM, Sanchez-Vera I, Guerri C (2010) Pivotal role of TLR4 receptors in alcohol-induced neuroinflammation and brain damage. J Neurosci 30:8285–8295
123. Pascual M, Balino P, Alfonso-Loeches S, Aragon CM, Guerri C (2011) Impact of TLR4 on behavioral and cognitive dysfunctions associated with alcohol-induced neuroinflammatory damage. Brain Behav Immun 25(Suppl 1):S80–S91
124. White AM, Matthews DB, Best PJ (2000) Ethanol, memory, and hippocampal function: a review of recent findings. Hippocampus 10:88–93
125. Chin VS, Van Skike CE, Matthews DB (2010) Effects of ethanol on hippocampal function during adolescence: a look at the past and thoughts on the future. Alcohol 44:3–14
126. Weiner JL, Valenzuela CF (2006) Ethanol modulation of GABAergic transmission: the view from the slice. Pharmacol Ther 111:533–554
127. Kumar S et al (2009) The role of GABA(A) receptors in the acute and chronic effects of ethanol: a decade of progress. Psychopharmacology (Berl) 205:529–564
128. Tsai G, Coyle JT (1998) The role of glutamatergic neurotransmission in the pathophysiology of alcoholism. Annu Rev Med 49:173–184
129. Barclay JW et al (2010) Presynaptic targets for acute ethanol sensitivity. Biochem Soc Trans 38:172–176

130. Lovinger DM, Roberto M (2013) Synaptic effects induced by alcohol. Curr Top Behav Neurosci 13:31–86
131. Bloom FE, Siggins GR (1987) Electrophysiological action of ethanol at the cellular level. Alcohol 4:331–337
132. Newlin SA, Mancillas-Trevino J, Bloom FE (1981) Ethanol causes increases in excitation and inhibition in area CA3 of the dorsal hippocampus. Brain Res 209:113–128
133. Wiesner JB, Henriksen SJ, Bloom FE (1987) Ethanol enhances recurrent inhibition in the dentate gyrus of the hippocampus. Neurosci Lett 79:169–173
134. Lovinger DM, White G, Weight FF (1990) NMDA receptor-mediated synaptic excitation selectively inhibited by ethanol in hippocampal slice from adult rat. J Neurosci 10:1372–1379
135. Lovinger DM, White G, Weight FF (1990) Ethanol inhibition of neuronal glutamate receptor function. Ann Med 22:247–252
136. Wu PH, Coultrap S, Browning MD, Proctor WR (2010) Correlated changes in NMDA receptor phosphorylation, functional activity, and sedation by chronic ethanol consumption. J Neurochem 115:1112–1122
137. Proctor WR, Diao L, Freund RK, Browning MD, Wu PH (2006) Synaptic GABAergic and glutamatergic mechanisms underlying alcohol sensitivity in mouse hippocampal neurons. J Physiol 575:145–159
138. Roh MS, Cui FJ, Kim HK, Kang UG (2011) Regulation of NMDA receptor subunits after acute ethanol treatment in rat brain. Alcohol Alcohol 46:672–679
139. Hicklin TR et al (2011) Alcohol inhibition of the NMDA receptor function, long-term potentiation, and fear learning requires striatal-enriched protein tyrosine phosphatase. Proc Natl Acad Sci USA 108:6650–6655
140. Nagy J (2004) The NR2B subtype of NMDA receptor: a potential target for the treatment of alcohol dependence. Curr Drug Targets CNS Neurol Disord 3:169–179
141. Yaka R, Tang KC, Camarini R, Janak PH, Ron D (2003) Fyn kinase and NR2B-containing NMDA receptors regulate acute ethanol sensitivity but not ethanol intake or conditioned reward. Alcohol Clin Exp Res 27:1736–1742
142. Miyakawa T et al (1997) Fyn-kinase as a determinant of ethanol sensitivity: relation to NMDA-receptor function. Science 278:698–701
143. Hendricson AW, Sibbald JR, Morrisett RA (2004) Ethanol alters the frequency, amplitude, and decay kinetics of Sr2+-supported, asynchronous NMDAR mEPSCs in rat hippocampal slices. J Neurophysiol 91:2568–2577
144. Swartzwelder HS, Wilson WA, Tayyeb MI (1995) Differential sensitivity of NMDA receptor-mediated synaptic potentials to ethanol in immature versus mature hippocampus. Alcohol Clin Exp Res 19:320–323
145. Criswell HE, Ming Z, Griffith BL, Breese GR (2003) Comparison of effect of ethanol on N-methyl-D-aspartate- and GABA-gated currents from acutely dissociated neurons: absence of regional differences in sensitivity to ethanol. J Pharmacol Exp Ther 304:192–199
146. Gordey M, Mekmanee L, Mody I (2001) Altered effects of ethanol in NR2A(DeltaC/DeltaC) mice expressing C-terminally truncated NR2A subunit of NMDA receptor. Neuroscience 105:987–997
147. Weiner JL, Dunwiddie TV, Valenzuela CF (1999) Ethanol inhibition of synaptically evoked kainate responses in rat hippocampal CA3 pyramidal neurons. Mol Pharmacol 56:85–90
148. Martin D, Tayyeb MI, Swartzwelder HS (1995) Ethanol inhibition of AMPA and kainate receptor-mediated depolarizations of hippocampal area CA1. Alcohol Clin Exp Res 19:1312–1316
149. Moykkynen T, Korpi ER, Lovinger DM (2003) Ethanol inhibits alpha-amino-3-hydyroxy-5-methyl-4-isoxazolepropionic acid (AMPA) receptor function in central nervous system neurons by stabilizing desensitization. J Pharmacol Exp Ther 306:546–555
150. Carta M, Ariwodola OJ, Weiner JL, Valenzuela CF (2003) Alcohol potently inhibits the kainate receptor-dependent excitatory drive of hippocampal interneurons. Proc Natl Acad Sci USA 100:6813–6818

151. Crowder TL, Ariwodola OJ, Weiner JL (2002) Ethanol antagonizes kainate receptor-mediated inhibition of evoked GABA(A) inhibitory postsynaptic currents in the rat hippocampal CA1 region. J Pharmacol Exp Ther 303:937–944
152. Weiner JL, Zhang L, Carlen PL (1994) Potentiation of GABAA-mediated synaptic current by ethanol in hippocampal CA1 neurons: possible role of protein kinase C. J Pharmacol Exp Ther 268:1388–1395
153. Wan FJ, Berton F, Madamba SG, Francesconi W, Siggins GR (1996) Low ethanol concentrations enhance GABAergic inhibitory postsynaptic potentials in hippocampal pyramidal neurons only after block of GABAB receptors. Proc Natl Acad Sci USA 93:5049–5054
154. Ariwodola OJ, Weiner JL (2004) Ethanol potentiation of GABAergic synaptic transmission may be self-limiting: role of presynaptic GABA(B) receptors. J Neurosci 24: 10679–10686
155. Weiner JL, Valenzuela CF, Watson PL, Frazier CJ, Dunwiddie TV (1997) Elevation of basal protein kinase C activity increases ethanol sensitivity of GABA(A) receptors in rat hippocampal CA1 pyramidal neurons. J Neurochem 68:1949–1959
156. Nelson TE, Ur CL, Gruol DL (2005) Chronic intermittent ethanol exposure enhances NMDA-receptor-mediated synaptic responses and NMDA receptor expression in hippocampal CA1 region. Brain Res 1048:69–79
157. Fujii S, Yamazaki Y, Sugihara T, Wakabayashi I (2008) Acute and chronic ethanol exposure differentially affect induction of hippocampal LTP. Brain Res 1211:13–21
158. Whittington MA, Lambert JD, Little HJ (1995) Increased NMDA receptor and calcium channel activity underlying ethanol withdrawal hyperexcitability. Alcohol Alcohol 30:105–114
159. Thomas MP, Monaghan DT, Morrisett RA (1998) Evidence for a causative role of N-methyl-D-aspartate receptors in an in vitro model of alcohol withdrawal hyperexcitability. J Pharmacol Exp Ther 287:87–97
160. Molleman A, Little HJ (1995) Increases in non-N-methyl-D-aspartate glutamatergic transmission, but no change in gamma-aminobutyric acidB transmission, in CA1 neurons during withdrawal from in vivo chronic ethanol treatment. J Pharmacol Exp Ther 274:1035–1041
161. Gulya K, Grant KA, Valverius P, Hoffman PL, Tabakoff B (1991) Brain regional specificity and time-course of changes in the NMDA receptor-ionophore complex during ethanol withdrawal. Brain Res 547:129–134
162. Hendricson AW et al (2007) Aberrant synaptic activation of N-methyl-D-aspartate receptors underlies ethanol withdrawal hyperexcitability. J Pharmacol Exp Ther 321:60–72
163. Mulholland PJ, Becker HC, Woodward JJ, Chandler LJ (2011) Small conductance calcium-activated potassium type 2 channels regulate alcohol-associated plasticity of glutamatergic synapses. Biol Psychiatry 69:625–632
164. Nagy J (2008) Alcohol related changes in regulation of NMDA receptor functions. Curr Neuropharmacol 6:39–54
165. Rudolph JG, Walker DW, Iimuro Y, Thurman RG, Crews FT (1997) NMDA receptor binding in adult rat brain after several chronic ethanol treatment protocols. Alcohol Clin Exp Res 21:1508–1519
166. Fleming RL, Manis PB, Morrow AL (2009) The effects of acute and chronic ethanol exposure on presynaptic and postsynaptic gamma-aminobutyric acid (GABA) neurotransmission in cultured cortical and hippocampal neurons. Alcohol 43:603–618
167. Liang J et al (2006) Chronic intermittent ethanol-induced switch of ethanol actions from extrasynaptic to synaptic hippocampal GABAA receptors. J Neurosci 26:1749–1758
168. Olsen RW, Liang J, Cagetti E, Spigelman I (2005) Plasticity of GABAA receptors in brains of rats treated with chronic intermittent ethanol. Neurochem Res 30:1579–1588
169. Kang M, Spigelman I, Sapp DW, Olsen RW (1996) Persistent reduction of GABA(A) receptor-mediated inhibition in rat hippocampus after chronic intermittent ethanol treatment. Brain Res 709:221–228
170. Rogers CJ, Hunter BE (1992) Chronic ethanol treatment reduces inhibition in CA1 of the rat hippocampus. Brain Res Bull 28:587–592

171. Cagetti E, Liang J, Spigelman I, Olsen RW (2003) Withdrawal from chronic intermittent ethanol treatment changes subunit composition, reduces synaptic function, and decreases behavioral responses to positive allosteric modulators of GABAA receptors. Mol Pharmacol 63:53–64
172. Sinclair JG, Lo GF (1986) Ethanol blocks tetanic and calcium-induced long-term potentiation in the hippocampal slice. Gen Pharmacol 17:231–233
173. Puglia MP, Valenzuela CF (2010) Ethanol acutely inhibits ionotropic glutamate receptor-mediated responses and long-term potentiation in the developing CA1 hippocampus. Alcohol Clin Exp Res 34:594–606
174. Blitzer RD, Gil O, Landau EM (1990) Long-term potentiation in rat hippocampus is inhibited by low concentrations of ethanol. Brain Res 537:203–208
175. Izumi Y, Nagashima K, Murayama K, Zorumski CF (2005) Acute effects of ethanol on hippocampal long-term potentiation and long-term depression are mediated by different mechanisms. Neuroscience 136:509–517
176. Tokuda K, Zorumski CF, Izumi Y (2007) Modulation of hippocampal long-term potentiation by slow increases in ethanol concentration. Neuroscience 146:340–349
177. Mulkeen D, Anwyl R, Rowan MJ (1987) Enhancement of long-term potentiation by the calcium channel agonist Bayer K8644 in CA1 of the rat hippocampus in vitro. Neurosci Lett 80:351–355
178. Sugiura M, Shoyama Y, Saito H, Abe K (1995) The effects of ethanol and crocin on the induction of long-term potentiation in the CA1 region of rat hippocampal slices. Jpn J Pharmacol 67:395–397
179. Swartzwelder HS, Wilson WA, Tayyeb MI (1995) Age-dependent inhibition of long-term potentiation by ethanol in immature versus mature hippocampus. Alcohol Clin Exp Res 19:1480–1485
180. Schummers J, Bentz S, Browning MD (1997) Ethanol's inhibition of LTP may not be mediated solely via direct effects on the NMDA receptor. Alcohol Clin Exp Res 21:404–408
181. Morrisett RA, Swartzwelder HS (1993) Attenuation of hippocampal long-term potentiation by ethanol: a patch-clamp analysis of glutamatergic and GABAergic mechanisms. J Neurosci 13:2264–2272
182. Hendricson AW, Miao CL, Lippmann MJ, Morrisett RA (2002) Ifenprodil and ethanol enhance NMDA receptor-dependent long-term depression. J Pharmacol Exp Ther 301:938–944
183. Durand D, Carlen PL (1984) Impairment of long-term potentiation in rat hippocampus following chronic ethanol treatment. Brain Res 308:325–332
184. Roberto M, Nelson TE, Ur CL, Gruol DL (2002) Long-term potentiation in the rat hippocampus is reversibly depressed by chronic intermittent ethanol exposure. J Neurophysiol 87:2385–2397
185. Tremwel MF, Hunter BE (1994) Effects of chronic ethanol ingestion on long-term potentiation remain even after a prolonged recovery from ethanol exposure. Synapse 17:141–148
186. McCool BA (2011) Ethanol modulation of synaptic plasticity. Neuropharmacology 61:1097–1108
187. Chandler LJ, Sutton G (2005) Acute ethanol inhibits extracellular signal-regulated kinase, protein kinase B, and adenosine 3′:5′-cyclic monophosphate response element binding protein activity in an age- and brain region-specific manner. Alcohol Clin Exp Res 29: 672–682
188. Asyyed A, Storm D, Diamond I (2006) Ethanol activates cAMP response element-mediated gene expression in select regions of the mouse brain. Brain Res 1106:63–71
189. Kumar S, Lane BM, Morrow AL (2006) Differential effects of systemic ethanol administration on protein kinase cepsilon, gamma, and beta isoform expression, membrane translocation, and target phosphorylation: reversal by chronic ethanol exposure. J Pharmacol Exp Ther 319:1366–1375
190. Wilkie MB et al (2007) Acute ethanol administration rapidly increases phosphorylation of conventional protein kinase C in specific mammalian brain regions in vivo. Alcohol Clin Exp Res 31:1259–1267

191. Proctor WR et al (2003) Ethanol differentially enhances hippocampal GABA A receptor-mediated responses in protein kinase C gamma (PKC gamma) and PKC epsilon null mice. J Pharmacol Exp Ther 305:264–270
192. Choi DS et al (2008) Protein kinase Cdelta regulates ethanol intoxication and enhancement of GABA-stimulated tonic current. J Neurosci 28:11890–11899
193. Yaka R, Phamluong K, Ron D (2003) Scaffolding of Fyn kinase to the NMDA receptor determines brain region sensitivity to ethanol. J Neurosci 23:3623–3632
194. Roberto M et al (2003) The transient depression of hippocampal CA1 LTP induced by chronic intermittent ethanol exposure is associated with an inhibition of the MAP kinase pathway. Eur J Neurosci 17:1646–1654
195. Sanna PP, Simpson C, Lutjens R, Koob G (2002) ERK regulation in chronic ethanol exposure and withdrawal. Brain Res 948:186–191
196. Abraham WC, Williams JM (2008) LTP maintenance and its protein synthesis-dependence. Neurobiol Learn Mem 89:260–268
197. Kemp A, Manahan-Vaughan D (2007) Hippocampal long-term depression: master or minion in declarative memory processes? Trends Neurosci 30:111–118
198. Lee HK (2006) AMPA receptor phosphorylation in synaptic plasticity: insights from knockin mice. In: Kittler JT, Moss SJ (eds) The dynamic synapse: molecular methods in ionotropic receptor biology, Chap. 14. CRC Press, Boca Raton
199. Erickson MA, Maramara LA, Lisman J (2010) A single brief burst induces GluR1-dependent associative short-term potentiation: a potential mechanism for short-term memory. J Cogn Neurosci 22:2530–2540
200. Silva AJ et al (1996) Impaired learning in mice with abnormal short-lived plasticity. Curr Biol 6:1509–1518
201. Schulz PE, Fitzgibbons JC (1997) Differing mechanisms of expression for short- and long-term potentiation. J Neurophysiol 78:321–334
202. Lauri SE et al (2007) Presynaptic mechanisms involved in the expression of STP and LTP at CA1 synapses in the hippocampus. Neuropharmacology 52:1–11
203. Zucker RS, Regehr WG (2002) Short-term synaptic plasticity. Annu Rev Physiol 64:355–405
204. Miyamoto E (2006) Molecular mechanism of neuronal plasticity: induction and maintenance of long-term potentiation in the hippocampus. J Pharmacol Sci 100:433–442
205. Nguyen PV, Woo NH (2003) Regulation of hippocampal synaptic plasticity by cyclic AMP-dependent protein kinases. Prog Neurobiol 71:401–437
206. Malenka RC et al (1989) An essential role for postsynaptic calmodulin and protein kinase activity in long-term potentiation. Nature 340:554–557
207. Waltereit R, Weller M (2003) Signaling from cAMP/PKA to MAPK and synaptic plasticity. Mol Neurobiol 27:99–106
208. English JD, Sweatt JD (1997) A requirement for the mitogen-activated protein kinase cascade in hippocampal long term potentiation. J Biol Chem 272:19103–19106
209. Bolshakov VY, Carboni L, Cobb MH, Siegelbaum SA, Belardetti F (2000) Dual MAP kinase pathways mediate opposing forms of long-term plasticity at CA3-CA1 synapses. Nat Neurosci 3:1107–1112
210. Butler MP, O'Connor JJ, Moynagh PN (2004) Dissection of tumor-necrosis factor-alpha inhibition of long-term potentiation (LTP) reveals a p38 mitogen-activated protein kinase-dependent mechanism which maps to early-but not late-phase LTP. Neuroscience 124:319–326
211. Kelly A et al (2003) Activation of p38 plays a pivotal role in the inhibitory effect of lipopolysaccharide and interleukin-1 beta on long term potentiation in rat dentate gyrus. J Biol Chem 278:19453–19462
212. Josselyn SA, Nguyen PV (2005) CREB, synapses and memory disorders: past progress and future challenges. Curr Drug Targets CNS Neurol Disord 4:481–497

Chapter 13
Neuroimmune Genes and Alcohol Drinking Behavior

R. Adron Harris and Yuri A. Blednov

13.1 Introduction

Alcoholism (alcohol dependence) is one of the most expensive (more than $185 billion/year for the USA) and damaging chronic diseases. Treatment options are limited (there are only three FDA-approved drugs for alcohol dependence), and there is a high rate of relapse for all treatments. Emerging results suggest that the cytokine responses to alcohol (perhaps via endotoxins) promote persistent and excessive alcohol consumption, which may in turn promote further inflammatory responses, producing a positive feedback loop which that spirals out of control. Although there is considerable effort to develop pharmacotherapies to reduce alcohol consumption, craving, and relapse, most are directed at "traditional" targets involved in synaptic transmission. Neuroinflammatory pathways in brain (and other organs) may be unexplored targets for medication development to reduce excessive alcohol consumption and prevent relapse.

In the liver, alcohol activates inflammatory processes, and these signals are central to alcoholic liver disease [1]; however, their role in brain function and regulation of alcohol consumption has received little attention. It is well known that alcohol increases the permeability of the gut to endotoxins, and the action of compounds such as lipopolysaccharide (LPS) on the liver is a key factor in development of alcoholic cirrhosis [2]. LPS is a bacterial endotoxin normally confined to the gut, but it can leak from the gut as a result of chronic alcohol abuse [1]. LPS produces activation of the immune system when administered systemically by binding to Toll-like receptor 4 (TLR4) found on macrophages, Kupffer and stellate cells in the

R.A. Harris (✉) • Y.A. Blednov
Waggoner Center for Alcohol and Addiction Research, University of Texas at Austin, Austin, TX, USA
e-mail: harris@mail.utexas.edu

C. Cui et al. (eds.), *Neural-Immune Interactions in Brain Function and Alcohol Related Disorders*, DOI 10.1007/978-1-4614-4729-0_13,

liver, and endothelial cells [1, 3–5]. TLR4 is also found in the brain on neurons, astrocytes, microglia, and endothelial cells [6, 7]. Some publications suggest that LPS can directly activate brain TLR4 receptors [8–10]. However, LPS is a large molecule that would not be expected to cross the blood–brain barrier, and others have failed to find LPS within the brain parenchyma after systemic administration [11]. Activation of peripheral TLR4 by LPS produces the release of a number of immune mediators, including cytokines, and a "sickness" response characterized by decreased food and water intake, loss of weight, lethargy, and anhedonia [12]. As discussed above, these effects of LPS are likely due to indirect actions on the brain, apparently by cytokines that are released peripherally and that are transported across the blood–brain barrier. The increased levels of serum cytokines produced by LPS are transient (hours), and the sickness response lasts only a day or two. However, there are persistent actions of LPS exposure on brain neuroinflammatory signaling, including elevation of some brain cytokines, for up to 10 months [13, 14]. These studies showed that repeated injections of LPS or high doses of ethanol produce long-lasting increases in brain cytokine signaling and proliferation of brain microglia, even though these treatments produce only transient changes in liver cytokines [13, 14]. Interestingly, one of the cytokines increased in rat brain, monocyte chemotactic protein 1 [also known as chemokine (C–C motif) ligand 2] (MCP-1, Ccl2), was also increased in the brain of human alcoholics [15], and we found that deletion of Ccl2 decreased alcohol consumption in mice [16].

Neuroimmune factors are emerging as important regulators of brain function, particularly synaptic plasticity [17–20]. This role of neuroimmune signaling in normal, as well as pathological, brain function is a paradigm shift and may be a key to understanding the roles of these factors in regulating alcohol drinking. In this chapter, we will discuss the link between proinflammatory signaling in and changes brain function critical to alcohol dependence. Many novel anti-inflammatory drugs are currently under development and in clinical trials, and drugs developed for other purposes (e.g., some statins and antibiotics) unexpectedly produce some of their beneficial effects by anti-inflammatory actions. This provides the possibility of "off-the-shelf" compounds for future studies in human alcoholism. Uncovering the role for neuroimmunology in alcohol abuse opens the possibility that such processes are important for addiction to other drugs and that medications targeting these processes may be useful for treating multiple addictions. Although not a focus of this chapter, we are interested in the observations that stress can activate cytokine signaling and that proinflammatory events act as stressors [21]. Stress is known to increase alcohol intake in rodent models as well as humans, and this has been explored in the traditional framework of the hypothalamic–pituitary–adrenal axis. However, stress may instead regulate alcohol consumption through a cytokine network. This could have applications for conditions such as post-traumatic stress disorder, where alcohol abuse is a common comorbidity. Although beyond the scope of this chapter, studies showing that immune mediators produce long-lasting changes in brain function have important implications for other mental illnesses such as depression, Alzheimer's disease, autism, and schizophrenia, where there is increasing evidence for a role of inflammatory responses [22–27]. The role of inflammatory activation in depression may be particularly relevant to the comorbidity of depression and alcoholism and new treatments that could target the neuroinflammatory component of both diseases [23, 28]. Also relevant for alcohol dependence is the finding that activation of immune responses can lower seizure thresholds and trigger seizures [29], suggesting that inhibitors of neuroimmune signaling might be useful for prevention of alcohol withdrawal neurotoxicity. One hypothesis that we will consider is that compulsive, excessive alcohol consumption is mediated, at least in part, by signals from cytokines produced in brain as well as cytokines derived from the periphery that act on the brain.

13.2 Genetic and Genomic Evidence

Interest in the role of proinflammatory processes in alcoholism was initially prompted by comparisons of gene expression profiles in human brain tissue from alcoholic and nonalcoholic subjects. Genes classified as "immune/stress response" form one of the most prominent functional groups with differential expression in the frontal cortex of alcoholics in comparison with nonalcoholics [30]. Many of these changes in gene expression appear to be due to alterations in the coordinated function of the transcription factors, nuclear factor kappa-light-chain enhancer of activated B cells (NF-κB), and p50 homodimer in the alcoholic frontal cortex [31]. Human alcoholics also show altered levels of microRNAs (miR) compared to nonalcoholics [32], and six of these (miR-146, miR-203, miR-140, miR-194, miR-196, and miR-15) are known to regulate immune function [33].

In addition, analysis of brain gene expression in mice that are genetically predisposed to alcohol consumption indicates a key role for proinflammatory mediators in regulating alcohol intake [34, 35]. Recently, Kong et al. [36] found upregulation of genes in the Toll and immune deficiency (Imd) innate immune signaling pathways in Drosophila exposed to ethanol, extending the link between alcohol and immune mediators to invertebrates. Examples of gene changes from the human and mouse studies are shown in Table 13.1, and several of these genes (e.g., Cd14, Toll-interleukin 1 receptor domain-containing adaptor protein—Tirap, interleukins,

Table 13.1 Immune/inflammatory response genes in brain associated with increased alcohol consumption in mouse genetic models [34] and human alcoholics [30]

Functional group	Gene symbol	Species
Immune/inflammatory response	*Cd14, Csf2, Cxcl1, Fasl, Gbp2, Il10, Il1r1, Il1rn, Irak1, Lif, Pglyrp1, Ptx3, Tirap, Tlr1, Tlr6, Tlr7*	Mouse
	Ripk2	Mouse/human
	ZFP36, SMAD3	Human
NFkB pathway	*Casp8, Fadd, Ikbkb, Ikbkg, Map3k1, Map3k7, Tradd*	Mouse
	Slc20a1	Mouse/human
Cytokine activity	*Ebi3, Ifnb1, Irak4*	Mouse
Other immune system process	*Cd80, Cdkn1a*	Mouse

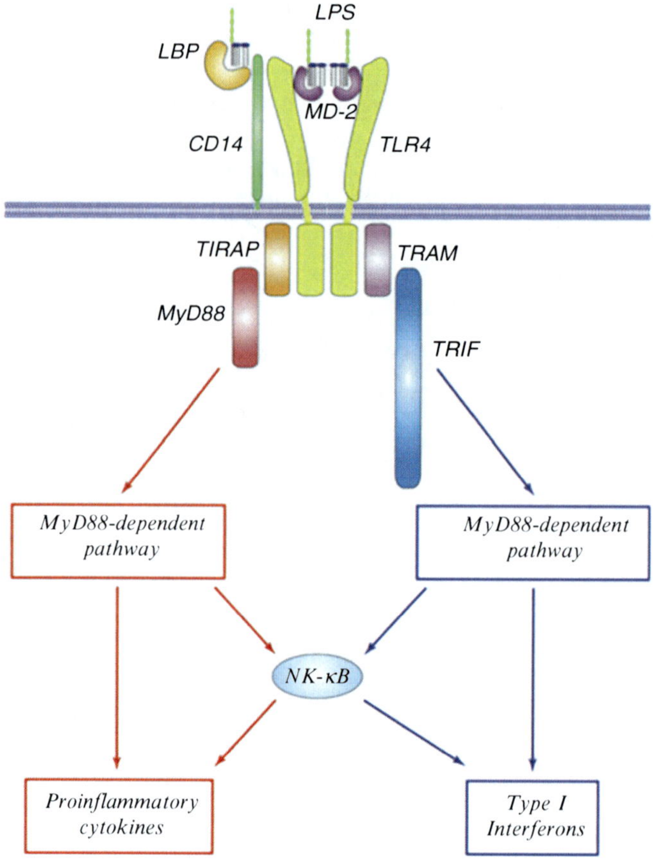

Fig. 13.1 Overview of LPS/TLR4 signaling. Lipopolysaccharide (LPS) recognition is facilitated by lipopolysaccharide-binding protein (LBP) and the cluster of differentiation 14 (CD14) protein and is mediated by Toll-like receptor 4 (TLR4/MD-2) receptor complex. LPS/TLR4 signaling can be separated into myeloid differentiation primary response gene (MyD), MyD88-dependent and MyD88-independent pathways, which mediate the activation of proinflammatory cytokine and type I interferon genes, respectively. Other adaptor proteins shown are Toll-interleukin 1 receptor adaptor protein (TIRAP), Toll-interleukin-1 receptor domain-containing adaptor-inducing interferon-beta (TRIF), and TRIF-related adaptor molecule (TRAM)

interferons) are part of the TLR4 signaling pathway shown in Fig. 13.1. A series of elegant studies from the Guerri group showed that TLR4 signaling is activated by alcohol treatments in cultured macrophages, glial cells, and animals consuming alcohol, and that it is required for alcohol-induced neuroinflammation [37–40]. This work, together with the gene expression data, suggests an intersection between human alcoholism and genetic animal models in regulation, or deregulation, of proinflammatory signaling in brain. It is interesting to note that there is extensive new literature suggesting endogenous ligands for TLR4 that can mimic the effects of LPS [41–43], and it is possible that alcohol consumption can promote this endogenous signaling.

13.3 Deletion of Neuroimmune Genes

The role of immune signaling in alcohol behaviors has been probed by testing of 13 different mutant mice with deletion or null mutations in immune-related genes and one with transgenic overexpression (Table 13.2). Based on the gene expression differences shown in Table 13.1, we studied six lines of mice with deletion of different candidate genes [44]. To select these genes, we began with a meta-analysis of gene expression changes associated with genetic predisposition to high alcohol consumption in mice (based on the 24-h, two-bottle choice test, or 2-BC) [34]. Initial examination of these data showed statistically significantly regulation of functional pathways related to immune/inflammatory responses, including interleukin (IL) IL-1, IL-2, IL-6, NFkB, Toll-like receptor, and tumor necrosis factor (TNF) receptor 1 signaling pathways. We then compared functional groups and individual genes identified in mice with expression studies in rats with genetic predisposition to high alcohol consumption [46] and in human alcoholics [30, 47, 48]. Examination of data from these four gene expression studies showed that a number of genes differentially expressed in each study could be classified into a functional category broadly defined as "immune/inflammatory/defense/stress response," and we chose this path-

Table 13.2 Summary of studies of behavioral effects of alcohol in immune-related mutant mice

				Two-bottle choice					
		LORR		EtOH (g/kg/day)		Preference		Total fluid intake	
Symbol	Reference	M	F	M	F	M	F	M	F
IL6	[44]			↓	↓	↓	↓	0	↑
IL6 (tg)	Not published	0	0	0	0	0	↑	↓	↓
Ccr2	[16]	0	0	↓	↓	↓	↓	↑	↑
Ccr5	[16]	0	0	0	↑	0	0	↑	↑
Ccl2	[16]	↑	↑	0	↓	0	↓	↑	0
Ccl3	[16]	↑	↑	↓	↓	↓	↓	0	↓
IL1rn	[44]	↑	↑	↓	↓	↓	↓	0	0
Cd14	[44]			0	↓	↓	↓	↑	↑
B2m	[44]			↓	↓	↓	↓	↓	↓
Ctss	[44]			↓	↓	↓	↓	↓	↓
Ctsf	[44]			↓	0	↓	0	0	0
Tlr4	[74], [45]	↓		0		0			
MyD88	[45]	↓							
C3H/HeJ	Not published			↓	↓	↓	↓	↓	0

All are null mutants (designated by gene symbols), except for the transgenic overexpression of IL6 [IL6(tg)]. C3H/HeJ is a naturally occurring null mutant for TLR4; C3H/HeOuJ was used as the control strain. LORR indicates duration of loss of righting reflex after ethanol injection. Alcohol consumption was measured in the continuous two-bottle choice drinking test, and results for amount of alcohol consumed (g/kg/day) and preference for alcohol compared to water and total fluid intake are given. Arrows indicate a change in mutant compared to wild type, 0 indicates no change, and a blank indicates not tested. Results are given for male (M) and female (F) mice

way for our validation study. We applied five criteria for selection of genes from this functional category: (1) statistical differences in gene expression (individual gene or overrepresented pathway) related to predisposition to alcohol consumption in mice [34], (2) statistical differences in gene expression in a rat or human study [30, 46, 47], (3) availability of knockout mice, (4) functional commonality, and (5) absence of publications linking the genes to alcohol consumption. This resulted in four genes that fulfilled all criteria and two genes that fulfilled three criteria and were found only in the Mulligan et al. [34] data (Table 13.1). Thus, we evaluated null mutant mice for six genes: B2m (beta-2 microglobulin), Cd14 (cluster of differentiation 14 or CD14 antigen), Il1rn (interleukin 1 receptor antagonist, also known as IL-1ra and IRAP), Il6 (interleukin 6), Ctss (cathepsin S), and Ctsf (cathepsin F). Deletion of any of these six genes resulted in decreased alcohol consumption in the two-bottle choice test, despite increasing total fluid consumption in most of the mutants (Table 13.2). It should be noted that several of the null mutations change alcohol consumption or preference in one sex but not the other (e.g., Ctsf, Cd14; Table 13.2). This is a common finding for effects of mutation on alcohol phenotypes [49], although the reasons are not known. Female mice consume more alcohol than males, and it is possible that levels of consumption may influence genetic effects.

Other genes related to immune function have also been deleted or mutated and evaluated for effects on alcohol consumption. Deletion of a chemokine receptor (Ccr2) or chemokines (Ccl2, Ccl3) reduced alcohol consumption. TLR4 is the site of action of lipopolysaccharide (LPS, discussed below) and is central to several aspects of immune signaling (see Fig. 13.1), and several naturally occurring null (or impaired) mutants of TLR4 are available [50, 51]. The TLR4-deficient C3H/HeJ mice showed, as expected, decreased alcohol consumption as compared to the control strain (C3H/HeOuJ) with normal TLR4 function (Table 13.2). The mutation did not alter total fluid consumption, but it should be noted that alcohol consumption is low in the control strain, due to the C3H genetic background. As discussed below, RNAi has been used to reduce TLR4 in amygdala, and this reduces operant self-administration of ethanol, consistent with the results from the C3H/HeJ mice. However, a null mutant for TLR4 has also been engineered, but this mouse did not show a change in alcohol consumption (Table 13.2). This is surprising, but may be due to differences in genetic background or differences in the consumption tests. Further studies are required to define the role of TLR4 in alcohol consumption. One transgenic overexpressing mutant (IL6 tg) has been tested and females (but not males) show increased preference for alcohol (Table 13.2). This is the opposite of the finding for the IL6 null mouse (and for most of the null mutants), suggesting that increased immune signaling can increase alcohol preference. It is important to note that the IL6 transgene is controlled by the glial fibrillary acidic protein (GFAP) promoter and is only expressed in brain astrocytes [52], suggesting a role for astrocytes in controlling alcohol preference. However, the IL6 transgenic mice also showed a decrease in total fluid consumption which resulted in no change in the amount of alcohol consumed despite an increase in preference (Table 13.2). It will be important to evaluate these mice with other tests of alcohol consumption.

The mutants have been tested primarily for alcohol consumption in the continuous two-bottle choice test, with only limited evaluation of other alcohol behaviors.

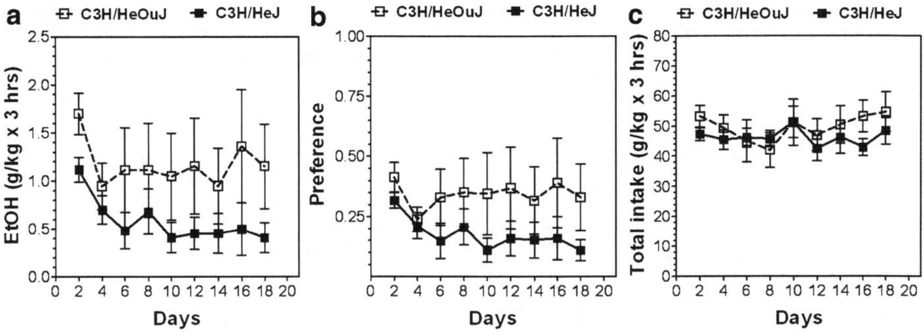

Fig. 13.2 A naturally occurring mutation in TLR4 reduces alcohol consumption in a two-bottle limited access test. C3H/HeJ inbred mice have defective TLR4 function and show decreased alcohol consumption when compared to a closely related strain of mice (C3H/HeOuJ) that have normal TLR4 signaling [51]. Data are from a two-bottle limited access (3 h) (modified "Drinking in the Dark" or DID) test and were obtained from female mice. The concentration of alcohol was 10 %. Values are mean ± SEM, $n=6$ for the two strains of mice. Significant differences were found in amount of alcohol consumed ($p<0.001$) (panel **a**) and preference ($p<0.01$) (panel **b**), but the total fluid intake was not different (panel **c**)

However, mice lacking *Ctss* and *Il1rn* genes also showed a reduction in ethanol intake in tests with limited access to alcohol ("Drinking in the Dark" or DID) [44]. In addition, the C3H/HeJ female mice which are deficient in TLR4 signaling show decreased alcohol consumption in the DID test (Fig. 13.2). A few neuroimmune mutants have been tested for the duration of loss of righting reflex (LORR) produced by ethanol with the TLR4 and MyD88 mutants showing decreases and several chemokine deletions as well as deletion of IL1rn producing increased LORR (Table 13.2). It should be noted that the deletion of CD14 increased the duration of LORR in contrast to the reduction of LORR observed in mice lacking Tlr4. Because CD14 is required for TLR4 function, deletion of either gene should produce the same effect on LORR (as it does on alcohol consumption). One possible explanation is that TLR2 has a role in LORR because CD14 is an adapter protein for TLR2 as well as TLR4. An alternative to manipulation is delivery of protein regulators, and mice treated with IL-1ra (IL1 receptor antagonist) showed reduced acute alcohol-induced sedation (faster recovery from acute alcohol-induced motor impairment and shorter duration of LORR than control animals) [45]. Another technique for manipulating the level of an mRNA, and thus the level of the protein, is direct injection of an siRNA vector into brain regions to produce a "knock down." This approach has been beautifully employed to answer several key questions about the role of TLR4 in alcohol self-administration. Injection of an siRNA for TLR4 into the central amygdala of alcohol-preferring (P) rats reduced operant self-administration of ethanol, but not sucrose [53]. This confirms and extends the mouse studies in several important ways: it extends the finding to another species and another behavioral test and shows that reduction of TLR4 in a single brain region is sufficient to reduce alcohol consumption. The latter point is particularly critical because the mouse studies used global knockouts which compromise peripheral immune signaling as well as central neuroimmune function.

Most of the mouse results discussed above were obtained from null mutants completely lacking the protein of interest, and it is important to ask if partial changes in function of immune proteins, as would be expected in a human population, are sufficient to alter consumption or behavioral actions of alcohol. These can be accomplished with remarkable ease and effectiveness in silico using GeneNetwork and WebQTL (http://webqtl.org/webqtl/main.py). This site brings together a huge range of phenotypes that have been measured by many different laboratories in identical strains of recombinant inbred (RI) mice and rats. This allows visualization of the variation of the phenotype in the RI population and correlations among phenotypes measured in the same RI strains. The BXD RI published phenotypes database listed 236 different "immune" measures and 78 "alcohol" measures as of October 25, 2011. There are also extensive gene expression databases for the RI mice and rats. As an example of the use of this resource to address a focused question, we asked if the BXD RI mice showed variation in TLR4 function and if this variation was related to any alcohol-related behaviors. One of the immune phenotypes is response of macrophages to stimulation by LPS (record 11001) which is a measure of the function of TLR4 receptors on the macrophages [50]. This function shows a wide range of variation and is correlated with several behavioral actions of alcohol such as preference in the two-bottle choice test, antianxiety actions of ethanol, and the severity of withdrawal signs (HIC) during acute alcohol withdrawal (Table 13.3). The possible role of TLR4 signaling in these behavioral effects of alcohol should be tested directly with tools such as RNAi, null mutants, and selective TLR4 antagonists. The importance of neuroinflammatory signaling in regulation of alcohol consumption is also supported by our data using lipopolysaccharide (LPS) to activate neuroinflammatory signaling (discussed in Sect. 1.5) [54].

13.4 Anti-inflammatory Treatments

Because genetic reduction of inflammatory signaling decreases alcohol consumption, it is possible that anti-inflammatory drug treatments might also reduce alcohol consumption. Many of these drugs are already used routinely in medicine and could easily be repurposed to treat alcoholism. This approach is complicated by the wide range of anti-inflammatory treatments with very different mechanisms of action. Many drugs including statins and antibiotics are anti-inflammatory, and even topiramate, a drug with efficacy in treating human alcoholism [55], also has anti-inflammatory effects [56].

The tetracycline derivative, minocycline, reduces alcohol consumption in the continuous two-bottle choice test [57]. Minocycline has potent anti-inflammatory effects independent from its microbiocidal properties and inhibits macrophage and microglial activation. It is probably the most commonly used drug to inhibit neuroimmune signaling, and there are over 300 publications on the neuroprotective effects of minocycline [58]. Of all the tetracyclines, it has the greatest permeability through the blood–brain barrier [59], and it confers therapeutic benefits in many

Table 13.3 Summary of WebQTL analysis of alcohol phenotypes correlated with LPS sensitivity of macrophages (WebQTL record 11001)

Phenotype	WebQTL record	Authors	Sample *r*	*n*	Sample *p*(*r*)
EtOH preference, residual mean, ♀	10142	[75]	0.745	23	1.7E–05
Ten-min anxiety post i.p. EtOH, ♀	12371	[77]	0.613	26	6.2E–04
Five-min anxiety post i.p. EtOH, ♀	12370	[77]	0.594	26	1.0E–03
BEC, ♀ and ♂	11967	[76]	0.582	27	1.1E–03
EtOH preference, 2-BC after 1 cycle CIE	12816	[78]	–0.798	11	2.0E–03
EtOH preference, raw mean, ♀	10140	[75]	0.512	23	1.1E–02
EtOH-induced HIC, 4 h post i.p., ♂ and ♀	11892	[76]	–0.836	7	1.6E–02

WebQTL was used to obtain data from BXD recombinant inbred mice for the response of macrophages to LPS [50], and these data were correlated with all alcohol phenotypes. Significant correlations, $p < 0.01$, are shown

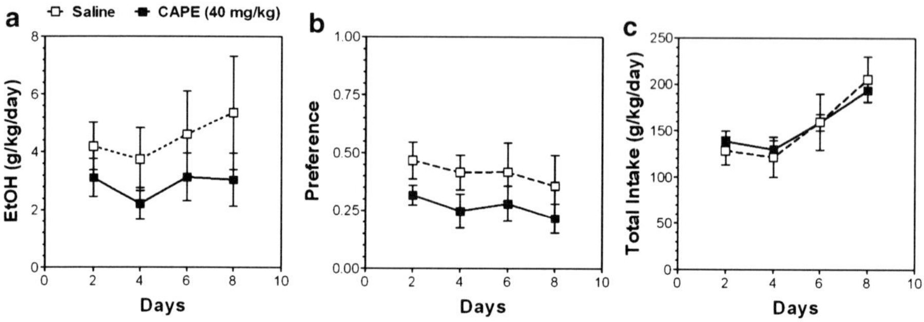

Fig. 13.3 Caffeic acid phenethyl ester (CAPE), an inhibitor of NFkappaB activation, reduces alcohol consumption. CAPE (40 mg/kg i.p. once daily just before the beginning of the dark cycle) reduced the amount of alcohol consumed ($p<0.05$) (panel **a**) and alcohol preference ($p<0.01$) (panel **b**) without changing total fluid intake (panel **c**). A continuous two-bottle choice test was used with a single concentration of alcohol (9 %) and daily (24 h) consumption is presented. Mixed genetic background (B6x129) female mice were used; values are mean ± SEM, $n=6$ for saline- and drug-treated groups

CNS disease models, including ischemia [60], Parkinson's disease [61], amyotrophic lateral sclerosis [62], multiple sclerosis [59, 62], and depressive-like behavior [63]. Documented actions of minocycline which may be responsible for the therapeutic benefits in neurodegenerative diseases include inhibition of microglial activation, attenuation of apoptosis, and suppression of reactive oxygen species production. In brain, minocycline treatment suppresses microglial proliferation/activation and subsequent release of cytokines such as IL 1β and 6 and TNF alpha and reduces expression of chemokines and their receptors [58].

A second drug, pioglitazone, was recently shown to reduce alcohol self-administration, consumption, and relapse in rats [64]. This drug was also found to protect from some of the toxic effects of fetal alcohol exposure [65] and to protect against alcohol liver injury [66]. Pioglitazone is an agonist of peroxisome proliferator-activated-type gamma receptors (PPAR-gamma) and is FDA approved as a treatment for diabetes because it potentiates the action of insulin. PPAR-gamma is found on most brain cell types, and activation of this receptor protects striatal neurons against LPS [67] and ameliorates symptoms of Alzheimer's disease and traumatic brain injury in animal models [68–70]. These effects are due, at least in part, to anti-inflammatory actions resulting from prevention of microglial activation and reduction of inflammatory cytokines and chemokines [71]. PPAR-gamma acts broadly to silence expression of a broad range of inflammatory genes in myeloid lineage cells, such as microglia and macrophages, and in the vasculature [70]. PPAR-gamma suppresses proinflammatory gene expression at the transcriptional level through antagonizing the actions of NFκB, AP-1, and STAT1 transcription factors; the main action of PPARγ in this context is thought to be inhibition of NFκB-dependent gene expression [70].

The importance of NFkB activity in regulation of voluntary alcohol intake is supported by our finding that daily injections of an inhibitor of NFkB activation (caffeic acid phenethyl ester, CAPE) [72] reduced ethanol intake and preference in the two-bottle choice paradigm without changes in total fluid intake (Fig. 13.3).

CAPE reduced alcohol consumption by about 50 % without altering total fluid intake in these studies, but it should be noted that alcohol consumption was relatively low (4–6 g/kg/day) because of the mixed genetic background of these mice. Testing of CAPE in other rodent models is justified by these initial results.

13.5 Neuroimmune Activation

In the liver, alcohol activates inflammatory processes, and these signals are central to alcoholic liver disease [1]; however, their role in brain function and regulation of alcohol consumption has received little attention. It is well known that alcohol increases the permeability of the gut to endotoxins, and the action of compounds such as LPS on the liver is a key factor in development of alcoholic cirrhosis [2]. LPS is a bacterial endotoxin normally confined to the gut, but it can leak from the gut as a result of chronic alcohol abuse [1]. LPS produces activation of the immune system when administered systemically by binding to TLR4 found on macrophages, Kupffer and stellate cells in the liver, and endothelial cells [1, 3–5]. TLR4 is also found in the brain on neurons, astrocytes, microglia, and endothelial cells [6, 7]. Some publications suggest that LPS can directly activate brain TLR4 receptors [8–10]. However, LPS is a large molecule that would not be expected to cross the blood–brain barrier, and others have failed to find LPS within the brain parenchyma after systemic administration [11]. Activation of peripheral TLR4 by LPS produces the release of a number of immune mediators, including cytokines, and produces a "sickness" response characterized by decreased food and water intake, loss of weight, lethargy, and anhedonia [12]. As discussed above, these effects of LPS are likely due to indirect actions on the brain, apparently by cytokines that are released peripherally and that are transported across the blood–brain barrier. The increased levels of serum cytokines produced by LPS are transient (hours), and the sickness response lasts only a day or two. However, there are persistent actions of LPS exposure on brain neuroinflammatory signaling, including elevation of some brain cytokines, for up to 10 months [13, 14]. These studies showed that repeated injections of LPS or high doses of ethanol produce long-lasting increases in brain cytokine signaling and proliferation of brain microglia, even though these treatments produce only transient changes in liver cytokines [13, 14]. This raises the possibility that injection of LPS may be a useful tool to produce long-lasting changes in behavior through persistent activation of brain cytokines. Indeed, we have discovered that one injection of LPS produces a long-lasting increase in alcohol consumption in mice that persists for up to 80 days after the last injection [54]. The action of LPS was absent in CD14 null mutant mice, indicating a requirement for TLR4/CD14 signaling, consistent with TLR4 as the site of action of LPS [51]. This LPS treatment also decreased ethanol conditioned taste aversion and decreased the firing rate of dopamine neurons in the ventral tegmental area [54]. Thus, activation of the TLR4 signaling pathway can produce long-lasting increases in alcohol consumption, decreased alcohol aversion, and changes in dopaminergic neuronal function.

13.6 Conclusions and Future Directions

Based on the findings presented above, we propose the following sequence of events: (1) excessive, long-term alcohol consumption enhances the entry of LPS from the gut, where it is continually produced by bacteria, and this LPS activates TLR4 in liver and brain; (2) alcohol also stimulates TLR4 signaling in brain by release of endogenous ligands and other actions on the signaling complex; (3) activation of proinflammatory cytokines and anti-inflammatory interferon pathways downstream from TLR4 (see Fig. 13.1) influences brain function and promotes alcohol consumption. The cascade given above, namely, LPS release, activation of TLR4 signaling, and stimulation of cytokine production, is well documented in human alcoholic liver disease [1] but is only beginning to be explored as a brain regulator of alcohol self-administration. It is important to note that most mouse models of alcohol consumption do not show the excessive increase in consumption that is characteristic of human alcoholism and do not show the liver damage seen in humans. It is possible that mouse models have less proinflammatory activation than human alcoholics and that LPS-enhanced alcohol consumption provides a better animal model for mechanistic studies as well as medication development for alcoholism.

There is considerable interest in the role of cytokines, chemokines, and other inflammatory factors in neurodegenerative diseases. In particular, activation of microglia and subsequent release of cytokines damage dopamine neurons, and this is thought to be important for Parkinson's disease [73]. Although this indicates an action of inflammatory factors on dopamine neurons, there is little information about possible regulation of function of these neurons by subtoxic levels of these factors. We found that a single injection of LPS decreases the firing of VTA dopamine neurons for at least 1 week after injection, suggesting that cytokine signaling can regulate reward circuitry [54]. A critical question that is difficult to address is the relative importance of the peripheral immune system in comparison to central neuroimmune signaling. The changes in brain gene expression discussed above as well as several experiments using brain delivery of RNAi or of transgenes discussed above indicate importance of central neuroimmune mechanisms, but do not rule out the participation of the peripheral immune system in the studies using global gene knockout or peripheral administration of LPS or anti-inflammatory drugs.

One hypothesis emerging from this field is that the normally functioning immune (or neuroimmune) system limits alcohol drinking, whereas overactivation of neuroimmune signaling promotes excessive alcohol consumption. This is supported by findings that null mutant mice with impaired immune signaling drink less alcohol (Table 13.2). In addition, activation of the immune system by injection of LPS increases alcohol consumption. It is possible that the well-known activation of inflammatory responses by alcohol is relayed to the brain by cytokines and other mediators. Actions of these immune mediators on the brain could promote excessive alcohol consumption, which in turn produces further activation of neuroimmune responses. It is important to note that alcoholism is likely a multifactorial (or multiple-hit) disorder, with access to alcohol playing a central role, but with the effects of alcohol on proinflammatory processes providing additional "hits," along with other genetic and environmental factors. Thus, inflammatory events (e.g., non-alcoholic steatohepatitis) alone would not be sufficient to produce alcohol dependence. The future goals of this research area should be to explore an innovative, neuroimmunological model of excessive alcohol consumption; to define molecular pathways, neural mechanisms, and gene networks critical for immune modulation of alcohol consumption; and to develop medications that normalize the dysregulations responsible for excessive alcohol consumption.

Acknowledgements This research was supported by grants from the National Institute of Alcohol Abuse and Alcoholism to the INIA West Consortium (AA13520, AA06399).

References

1. Mandrekar P, Szabo G (2009) Signalling pathways in alcohol-induced liver inflammation. J Hepatol 50:1258–1266
2. Jeong WI, Gao B (2008) Innate immunity and alcoholic liver fibrosis. J Gastroenterol Hepatol 23(Suppl 1):S112–S118
3. Andonegui G et al (2003) Endothelium-derived Toll-like receptor-4 is the key molecule in LPS-induced neutrophil sequestration into lungs. J Clin Invest 111:1011–1020
4. Suzumura A, Takeuchi H, Zhang G, Kuno R, Mizuno T (2006) Roles of glia-derived cytokines on neuronal degeneration and regeneration. Ann N Y Acad Sci 1088:219–229
5. Zanoni I, Granucci F (2010) Differences in lipopolysaccharide-induced signaling between conventional dendritic cells and macrophages. Immunobiology 215:709–712
6. Mallard C, Wang X, Hagberg H (2009) The role of Toll-like receptors in perinatal brain injury. Clin Perinatol 36:763–772, v–vi
7. Tang SC et al (2007) Pivotal role for neuronal Toll-like receptors in ischemic brain injury and functional deficits. Proc Natl Acad Sci U S A 104:13798–13803
8. Chakravarty S, Herkenham M (2005) Toll-like receptor 4 on nonhematopoietic cells sustains CNS inflammation during endotoxemia, independent of systemic cytokines. J Neurosci 25:1788–1796
9. Gosselin D, Rivest S (2008) MyD88 signaling in brain endothelial cells is essential for the neuronal activity and glucocorticoid release during systemic inflammation. Mol Psychiatry 13:480–497
10. Wisse BE et al (2007) Evidence that lipopolysaccharide-induced anorexia depends upon central, rather than peripheral, inflammatory signals. Endocrinology 148:5230–5237
11. Singh AK, Jiang Y (2004) How does peripheral lipopolysaccharide induce gene expression in the brain of rats? Toxicology 201:197–207
12. Dantzer R (2001) Cytokine-induced sickness behavior: where do we stand? Brain Behav Immun 15:7–24
13. Qin L et al (2008) Increased systemic and brain cytokine production and neuroinflammation by endotoxin following ethanol treatment. J Neuroinflammation 5:10
14. Qin L et al (2007) Systemic LPS causes chronic neuroinflammation and progressive neurodegeneration. Glia 55:453–462
15. He J, Crews FT (2008) Increased MCP-1 and microglia in various regions of the human alcoholic brain. Exp Neurol 210:349–358
16. Blednov YA et al (2005) Perturbation of chemokine networks by gene deletion alters the reinforcing actions of ethanol. Behav Brain Res 165:110–125

17. Di Filippo M, Sarchielli P, Picconi B, Calabresi P (2008) Neuroinflammation and synaptic plasticity: theoretical basis for a novel, immune-centred, therapeutic approach to neurological disorders. Trends Pharmacol Sci 29:402–412
18. Magistretti PJ (2011) Neuron-glia metabolic coupling and plasticity. *Exp Physiol* 96(4):407–410
19. Min SS et al (2009) Impairment of long-term depression induced by chronic brain inflammation in rats. Biochem Biophys Res Commun 383:93–97
20. Viviani B, Boraso M (2010) Cytokines and neuronal channels: a molecular basis for age-related decline of neuronal function? Exp Gerontol 46:199–206
21. Garcia-Bueno B, Caso JR, Leza JC (2008) Stress as a neuroinflammatory condition in brain: damaging and protective mechanisms. Neurosci Biobehav Rev 32:1136–1151
22. Crespi BJ, Thiselton DL (2011) Comparative immunogenetics of autism and schizophrenia. Genes Brain Behav 10:689–701
23. Dantzer R, O'Connor JC, Freund GG, Johnson RW, Kelley KW (2008) From inflammation to sickness and depression: when the immune system subjugates the brain. Nat Rev Neurosci 9:46–56
24. Hebb AL, Robertson HA (2008) PDEs as drug targets for CNS immune disorders. Curr Opin Investig Drugs 9:744–753
25. Hensley K (2010) Neuroinflammation in Alzheimer's disease: mechanisms, pathologic consequences, and potential for therapeutic manipulation. J Alzheimers Dis 21:1–14
26. Monji A, Kato T, Kanba S (2009) Cytokines and schizophrenia: microglia hypothesis of schizophrenia. Psychiatry Clin Neurosci 63:257–265
27. Wuwongse S, Chang RC, Law AC (2010) The putative neurodegenerative links between depression and Alzheimer's disease. Prog Neurobiol 91:362–375
28. Pettinati HM et al (2010) A double-blind, placebo-controlled trial combining sertraline and naltrexone for treating co-occurring depression and alcohol dependence. Am J Psychiatry 167:668–675
29. Rodgers KM et al (2009) The cortical innate immune response increases local neuronal excitability leading to seizures. Brain 132:2478–2486
30. Liu J et al (2006) Patterns of gene expression in the frontal cortex discriminate alcoholic from nonalcoholic individuals. Neuropsychopharmacology 31:1574–1582
31. Okvist A et al (2007) Neuroadaptations in human chronic alcoholics: dysregulation of the NF-kappaB system. PLoS One 2:e930
32. Lewohl JM et al (2011) Up-regulation of MicroRNAs in brain of human alcoholics. Alcohol Clin Exp Res 35:1928–1937
33. Sonkoly E, Stahle M, Pivarcsi A (2008) MicroRNAs and immunity: novel players in the regulation of normal immune function and inflammation. Semin Cancer Biol 18:131–140
34. Mulligan MK et al (2006) Toward understanding the genetics of alcohol drinking through transcriptome meta-analysis. Proc Natl Acad Sci U S A 103:6368–6373
35. Saba L et al (2006) Candidate genes and their regulatory elements: alcohol preference and tolerance. Mamm Genome 17:669–688
36. Kong EC et al (2010) Ethanol-regulated genes that contribute to ethanol sensitivity and rapid tolerance in Drosophila. Alcohol Clin Exp Res 34:302–316
37. Alfonso-Loeches S, Pascual-Lucas M, Blanco AM, Sanchez-Vera I, Guerri C (2010) Pivotal role of TLR4 receptors in alcohol-induced neuroinflammation and brain damage. J Neurosci 30:8285–8295
38. Blanco AM, Perez-Arago A, Fernandez-Lizarbe S, Guerri C (2008) Ethanol mimics ligand-mediated activation and endocytosis of IL-1RI/TLR4 receptors via lipid rafts caveolae in astroglial cells. J Neurochem 106:625–639
39. Fernandez-Lizarbe S, Pascual M, Gascon MS, Blanco A, Guerri C (2008) Lipid rafts regulate ethanol-induced activation of TLR4 signaling in murine macrophages. Mol Immunol 45:2007–2016
40. Fernandez-Lizarbe S, Pascual M, Guerri C (2009) Critical role of TLR4 response in the activation of microglia induced by ethanol. J Immunol 183:4733–4744

41. Brunn GJ, Platt JL (2006) The etiology of sepsis: turned inside out. Trends Mol Med 12:10–16
42. Erridge C (2010) Endogenous ligands of TLR2 and TLR4: agonists or assistants? J Leukoc Biol 87:989–999
43. Miyake K (2007) Innate immune sensing of pathogens and danger signals by cell surface Toll-like receptors. Semin Immunol 19:3–10
44. Blednov YA, Ponomarev I, Geil C, Bergeson S, Koob GF, Harris RA (2012) Neuroimmune regulation of alcohol consumption: behavioral validation of genes obtained from genomic studies. Addict Biol 17(1):108–120
45. Wu Y et al (2011) Attenuation of microglial and IL-1 signaling protects mice from acute alcohol-induced sedation and/or motor impairment. Brain Behav Immun 25(Suppl 1):S155–S164
46. Kimpel MW et al (2007) Functional gene expression differences between inbred alcohol-preferring and -non-preferring rats in five brain regions. Alcohol 41:95–132
47. Flatscher-Bader T, Zuvela N, Landis N, Wilce PA (2008) Smoking and alcoholism target genes associated with plasticity and glutamate transmission in the human ventral tegmental area. Hum Mol Genet 17:38–51
48. Liu J, Lewohl JM, Harris RA, Dodd PR, Mayfield RD (2007) Altered gene expression profiles in the frontal cortex of cirrhotic alcoholics. Alcohol Clin Exp Res 31:1460–1466
49. Crabbe JC, Phillips TJ, Harris RA, Arends MA, Koob GF (2006) Alcohol-related genes: contributions from studies with genetically engineered mice. Addict Biol 11:195–269
50. Cook DN et al (2006) Spontaneous mutations in recombinant inbred mice: mutant toll-like receptor 4 (Tlr4) in BXD29 mice. Genetics 172:1751–1755
51. Poltorak A et al (1998) Defective LPS signaling in C3H/HeJ and C57BL/10ScCr mice: mutations in Tlr4 gene. Science 282:2085–2088
52. Campbell IL et al (1993) Neurologic disease induced in transgenic mice by cerebral overexpression of interleukin 6. Proc Natl Acad Sci U S A 90:10061–10065
53. Liu J et al (2011) Binge alcohol drinking is associated with GABAA alpha2-regulated Toll-like receptor 4 (TLR4) expression in the central amygdala. Proc Natl Acad Sci U S A 108:4465–4470
54. Blednov YA, Benavidez JM, Geil C, Perra S, Morikawa H, Harris RA (2011) Activation of inflammatory signaling by lipopolysaccharide produces a prolonged increase of voluntary alcohol intake in mice. Brain Behav Immun 25 Suppl 1:S92–S105
55. Falk D et al (2010) Percentage of subjects with no heavy drinking days: evaluation as an efficacy endpoint for alcohol clinical trials. Alcohol Clin Exp Res 34:2022–2034
56. Dudley JT et al (2011) Computational repositioning of the anticonvulsant topiramate for inflammatory bowel disease. Sci Transl Med 3:96ra76
57. Agrawal RG, Hewetson A, George CM, Syapin PJ, Bergeson SE (2011) Minocycline reduces ethanol drinking. Brain Behav Immun 25(Suppl 1):S165–S169
58. Plane JM, Shen Y, Pleasure DE, Deng W (2010) Prospects for minocycline neuroprotection. Arch Neurol 67:1442–1448
59. Nikodemova M, Watters JJ, Jackson SJ, Yang SK, Duncan ID (2007) Minocycline down-regulates MHC II expression in microglia and macrophages through inhibition of IRF-1 and protein kinase C (PKC)alpha/betaII. J Biol Chem 282:15208–15216
60. Yrjanheikki J, Keinanen R, Pellikka M, Hokfelt T, Koistinaho J (1998) Tetracyclines inhibit microglial activation and are neuroprotective in global brain ischemia. Proc Natl Acad Sci U S A 95:15769–15774
61. Du Y et al (2001) Minocycline prevents nigrostriatal dopaminergic neurodegeneration in the MPTP model of Parkinson's disease. Proc Natl Acad Sci U S A 98:14669–14674
62. Zhu S et al (2002) Minocycline inhibits cytochrome c release and delays progression of amyotrophic lateral sclerosis in mice. Nature 417:74–78
63. O'Connor JC et al (2009) Lipopolysaccharide-induced depressive-like behavior is mediated by indoleamine 2,3-dioxygenase activation in mice. Mol Psychiatry 14:511–522
64. Stopponi S et al (2011) Activation of nuclear PPARgamma receptors by the antidiabetic agent pioglitazone suppresses alcohol drinking and relapse to alcohol seeking. Biol Psychiatry 69:642–649

65. Kane CJ et al (2011) Protection of neurons and microglia against ethanol in a mouse model of fetal alcohol spectrum disorders by peroxisome proliferator-activated receptor-gamma agonists. Brain Behav Immun 25(Suppl 1):S137–S145
66. Enomoto N et al (2003) Prevention of ethanol-induced liver injury in rats by an agonist of peroxisome proliferator-activated receptor-gamma, pioglitazone. J Pharmacol Exp Ther 306:846–854
67. Hunter RL, Choi DY, Ross SA, Bing G (2008) Protective properties afforded by pioglitazone against intrastriatal LPS in Sprague–Dawley rats. Neurosci Lett 432:198–201
68. Nicolakakis N, Hamel E (2010) The nuclear receptor PPARgamma as a therapeutic target for cerebrovascular and brain dysfunction in Alzheimer's disease. Front Aging Neurosci 2:1
69. Sauerbeck A et al (2011) Pioglitazone attenuates mitochondrial dysfunction, cognitive impairment, cortical tissue loss, and inflammation following traumatic brain injury. Exp Neurol 227:128–135
70. Sundararajan S, Jiang Q, Heneka M, Landreth G (2006) PPARgamma as a therapeutic target in central nervous system diseases. Neurochem Int 49:136–144
71. Kapadia R, Yi JH, Vemuganti R (2008) Mechanisms of anti-inflammatory and neuroprotective actions of PPAR-gamma agonists. Front Biosci 13:1813–1826
72. Natarajan K, Singh S, Burke TR Jr, Grunberger D, Aggarwal BB (1996) Caffeic acid phenethyl ester is a potent and specific inhibitor of activation of nuclear transcription factor NF-kappa B. Proc Natl Acad Sci U S A 93:9090–9095
73. Hernandez-Romero MC et al (2008) Simvastatin prevents the inflammatory process and the dopaminergic degeneration induced by the intranigral injection of lipopolysaccharide. J Neurochem 105:445–459
74. Pascual M, Balino P, Alfonso-Loeches S, Aragon CM, Guerri C (2011) Impact of TLR4 on behavioral and cognitive dysfunctions associated with alcohol-induced neuroinflammatory damage. Brain Behav Immun. 25 Suppl 1:S80–91
75. Fernandez JR, Vogler GP, Tarantino LM, Vignetti S, Plomin R, McClearn GE (1999) Sex-exclusive quantitative trait loci influences in alcohol-related phenotypes. Am J Med Genet 88(6):647–652
76. Philip VM, Duvvuru S, Gomero B, et al. (2010) High-throughput behavioral phenotyping in the expanded panel of BXD recombinant inbred strains. Genes Brain Behav 9(2):129–159
77. Cook et al. unpublished
78. Lopez et al. unpublished

Chapter 14
Immune Responses in HIV Infection, Alcoholism, and Aging: A Neuroimaging Perspective

N.M. Zahr, A. Manning-Bog, C. Alt, E.V. Sullivan, and A. Pfefferbaum

14.1 Introduction

Biological changes that accrue during normal, healthy aging contribute to behavioral (cognitive and motor) decline attributable to structural and metabolic alterations in the brain. Superimposed on the pattern of the healthy aging brain are the consequences of frequent concomitants of aging such as hypertension and hormone deficiency. Even more complex are the brain modifications that can occur in the presence of comorbidities such as infection with the human immunodeficiency virus (HIV) and alcohol dependence, each of which has independent deleterious effects on selective brain systems. Highly active antiretroviral therapy (HAART) has extended the average life span of individuals infected with HIV [1–3]. Accompanying longevity and improved health with treatment are increased opportunity to initiate or resume high-risk activities such as unsafe drinking (estimates of alcoholism among HIV patients range from 8 to 63%) [4–9], necessitating the consideration of the cumulative interactions of these disease processes on the aging brain [10].

Classically defined inflammation is a response to stimuli such as pathogens, damaged cells, or irritants. In acute inflammation, a cascade of biochemical events

N.M. Zahr, Ph.D. (✉) • A. Pfefferbaum
Policy Division, SRI International, Menlo Park, CA, USA

Psychiatry and Behavioral Sciences, Stanford University, School of Medicine, Stanford, CA, USA
e-mail: nzahr@stanford.edu

A. Manning-Bog
Policy Division, SRI International, Menlo Park, CA, USA

C. Alt
Biosciences Division, SRI International, Menlo Park, CA, USA

E.V. Sullivan
Psychiatry and Behavioral Sciences, Stanford University, School of Medicine, Stanford, CA, USA

C. Cui et al. (eds.), *Neural-Immune Interactions in Brain Function and Alcohol Related Disorders*, DOI 10.1007/978-1-4614-4729-0_14,

involves the local vascular system, the immune system, and various cells within the target tissue. A cardinal sign of classic inflammation is the movement of mobile cells of the innate immune system (i.e., leukocytes) from blood vessels to the site of the triggering stimulus. Acute inflammation can lead to chronic, persistent inflammation, distinguished by a progressive shift in the type of cells present at the site of inflammation and simultaneous healing and destruction of the tissue.

Examples of classical inflammation in the central nervous system (CNS) include the response to exogenous factors, such as bacteria, parasites, or viruses (e.g., HIV); "sterile" inflammation is the term used for the CNS inflammatory response to endogenous events, such as trauma or stroke [11]. A consequence of classical inflammation in the CNS is compromise of the blood–brain barrier (BBB). The innate immune system in CNS includes perivascular macrophages and microglia. Perivascular macrophages act as the first line of defense [12]. Microglia scavenge the CNS for cellular debris (e.g., damaged neurons) and pathogens. Upon activation, microglia undergo dramatic morphologic alterations; their surface molecules [e.g., complement receptors and major histocompatibility complex (MHC) molecules] are upregulated, and they release a variety of soluble factors [13]. They can remove waste by phagocytosis and act as antigen-presenting cells to T lymphocytes (i.e., T cells) [14]. Consequently, microglia can either promote a return to homeostasis (acute neuroinflammation) or incite inflammation (chronic neuroinflammation) by the combination of secreted signaling molecules (i.e., cytokines and chemokines) [15] (Fig. 14.1).

The term "neuroinflammation" is widely used to describe conditions that demonstrate activated microglia and the release of inflammatory mediators (e.g., cytokines) that potentially contribute to destructive processes [240, 16–21]). However, for conditions that do not include the presence of mobile, infiltrating cells or a breach of the BBB, a more appropriate term might be "microglial activation" (cf., [22]) (Fig. 14.1).

Thus, in this chapter, "neuroinflammation" will refer to processes where inflammation represents a primary or independent process, such as occurs in response to HIV. In HIV, significant evidence supports the view that the mechanisms contributing to brain alterations are due, at least in part, to neuroinflammatory processes [23]. In aging, the demonstration of "active" microglia, often interpreted as aging-related neuroinflammation, will instead be referred to as "microglial activation." Similarly, a provocative hypothesis in the alcohol literature suggests that activation of microglia may promote the CNS damage seen in alcoholism [24]. To establish a framework for the interpretation of cellular and molecular findings regarding CNS immune recruitment, in vivo brain imaging and postmortem findings will be presented for anatomical localization and characterization of pathology in each of these diseases. In vivo brain magnetic resonance (MR) techniques are unique in their ability to longitudinally track disease course. Longitudinal study can assess exacerbation and remission of the natural course of disease, response to treatment, and interaction of co-occurring conditions such as alcoholism, HIV infection, and advancing age. Quantitative assessment tools include structural MR imaging (MRI), which provides quantitative assessment of brain macrostructure; diffusion tensor

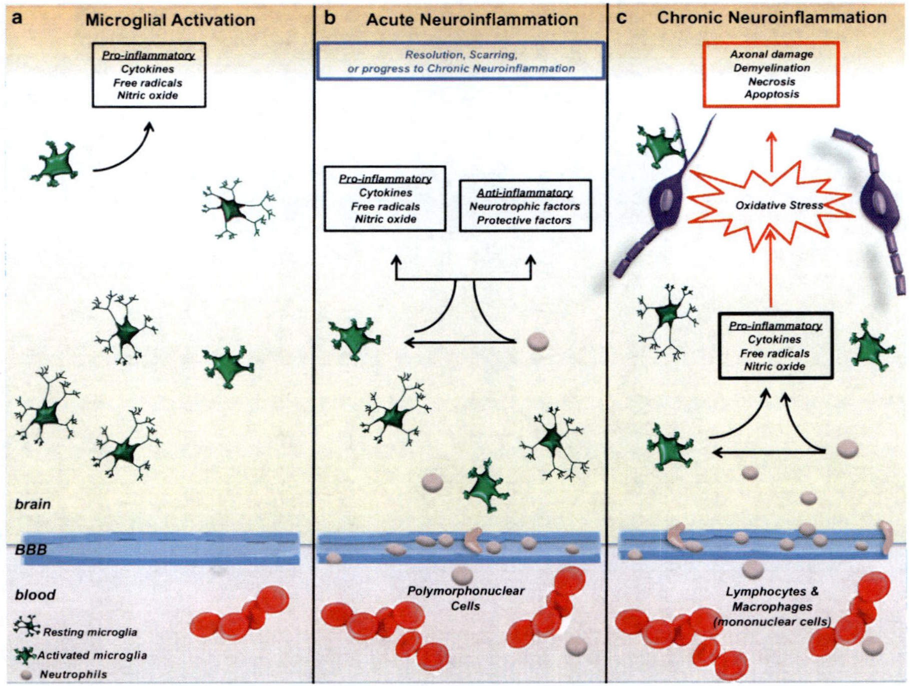

Fig. 14.1 Distinguishing features of (**a**) microglia activation, (**b**) acute neuroinflammation, and (**c**) chronic neuroinflammation. (**a**) In microglial activation, a stimulus (source unclear) may change the morphology of microglia so that they are in a state of increased alertness and are able to release a variety of factors (e.g., cytokines, chemokines). Such factors may have roles independent of immune or inflammatory responses. (**b**) Distinguishing features of acute neuroinflammation include a known stimulus (e.g., HIV), cellular (i.e., polymorphonuclear cells) trafficking into the brain, and the expression and release of various proinflammatory and toxic mediators. Acute inflammation can either be resolved or progress to chronic, persistent inflammation. (**c**) Chronic inflammation is distinguished by a progressive shift in the type of cells (e.g., lymphocytes, macrophages) present at the site of inflammation, compromise of blood–brain barrier (BBB), and possible destruction of tissue

imaging (DTI), which assesses the brain's microstructural integrity; and MR spectroscopy, which quantifies brain chemicals. Each level of analysis has its strengths and limitations: cellular and molecular studies can reveal mechanisms at a microscopic level but are largely limited to cross-sectional analysis; by contrast, neuroimaging studies can use in vivo paradigms in longitudinal design but have a highly restricted resolution. This chapter will survey primary literature regarding macrostructural, microstructural, and biochemical CNS changes in HIV, alcoholism, and aging and explore evidence for neuroimmune mechanisms in contributing to age- and disease-related changes.

14.2 HIV Infection

HIV, a ribonucleic acid (RNA) virus that primarily infects cells in the human immune system (e.g., T cells, macrophages), can lower the number of T cells expressing CD4 (a co-receptor that interacts directly with MHC class II molecules on antigen-presenting cells) and thereby increase the risk for opportunistic infections or malignancies. An "acquired immunodeficiency syndrome (AIDS)"-defining event refers to the occurrence of either a CD4 T-cell count below 200 cells/μL (i.e., below 15% of all lymphocytes) or one of the AIDS-defining illnesses (e.g., toxoplasmosis of the brain or Kaposi's sarcoma). Brain autopsy reveals in HIV with AIDS a characteristic neuropathology including macrophage infiltration, microglial nodules, and multinucleated giant cells [25–27]. Indeed, CNS HIV infection appears to begin with the transmigration of peripheral HIV-infected cells (e.g., monocytes or macrophages) across the BBB [28–36]. HIV primarily infects microglia, where CNS virus replication occurs [27, 37–42], via microglial chemokine receptors CXCR4 and CCR5 which serve as co-receptors for HIV entry into target cells [30, 40, 43–51]. This results in activation of microglia [52–56], as evidenced by microglial nodules and multinucleated giant cells, and the release of chemokine, cytokines, and neurotoxins that, in conjunction with secreted HIV protein, can damage the brain. Acute inflammation in HIV progresses to chronic inflammation because of the persistent presence of the noxious stimuli (i.e., HIV). Thus, the inflammatory process in HIV is composed of two essential components: (1) cellular trafficking into the brain and (2) the expression and release of various proinflammatory and toxic mediators.

Quantitative in vivo structural MRI studies report few abnormalities in asymptomatic HIV-infected individuals [57–59] but have noted escalating brain damage in symptomatic cases associated with advancing center for disease control (CDC) clinical stage [60, 61] or severity of cognitive deficits [41, 62, 63]. Gross brain abnormalities detected using MRI include global atrophy, cortical thinning, and ventricular enlargement [57, 60, 63–68]. More extensive ventricular enlargement is observed in HIV-infected individuals who have experienced an AIDS-defining event [61, 67]. Specific regions demonstrating significant volume loss as quantified with MR include frontal white matter, corpus callosum, caudate nucleus [61, 67, 69, 70], infratentorium [71], and cerebellar vermis [72]. Although volume shrinkage is often interpreted to reflect neuronal loss, it is a nonspecific indicator of neuronal integrity; volume loss may also reflect shrinkage of neuron cell bodies and their processes, shrinkage or loss of glial cells (i.e., astrocytes or microglia), alterations to oligodendrocytes or white matter, or other changes to the brain's neuropil [73].

In later-stage AIDS, neocortical gray matter is compromised, and there is evidence for frank neuronal loss, via apoptosis (but see [74, 75]), in putamen [76], substantia nigra [77], and cerebellum [78]. Such findings are supported by in vivo MRS results in AIDS dementia demonstrating decreased *N*-acetylaspartate (NAA) levels, assumed to reflect neuronal loss in frontal cortex, subjacent white matter, and basal ganglia [79–83]. In HIV without AIDS, a few postmortem studies have reported neuronal loss in the frontal cortex [84–86], but only rarely have apoptotic neurons been observed in non-AIDS, HIV brains. In agreement, MRS results do not observe NAA decreases in asymptomatic HIV-infected patients (but see, e.g., [87, 88]). MRS experiments also show, early in HIV, abnormally high myoinositol (mI) levels, possible evidence of gliosis, and abnormally high choline (Cho) levels, possible evidence for inflammation in the basal ganglia [28, 79, 89, 90].

Rather than compromising neurons, the preponderance of evidence suggests that HIV infection initially targets white matter [57, 91]. Postmortem histology reveals evidence for diffuse white matter degradation, widespread demyelination, and axonal injury [92–99]. HIV may directly injure white matter by damaging myelin via viral proteins (e.g., gp120 or Tat) or by infecting or damaging oligodendrocytes and brain endothelial cells [100]. Microglia-derived vehicles of white matter damage can include monocyte chemotactic protein-1 (CCL2/MCP-1), shown to be involved in demyelination in a variety of experimental animal models [97] and interleukin (IL)-1, which can lead to reversible demyelination of axons [93, 96].

Nonspecific white matter hyperintensities (WMHI) can be detected with MRI using the fluid-attenuated inversion recovery (FLAIR) acquisition protocol. FLAIR-identified WMHI occur in HIV patients [67, 69, 101], especially those with progressive multifocal leukoencephalopathy [102], and can predict length of survival [103]. WMHI in HIV likely reflects myelin pallor and rarefaction associated with presence of multinucleated giant cells [32, 92, 104]. Diffusion tensor MR imaging (DTI) reveals that white matter regions showing microstructural changes with HIV disease progression include the genu and splenium of the corpus callosum, frontal and parietal subcortical white matter, internal and external capsules, inferior longitudinal fascicules [105–109], superior cingulate bundles [110], and the optic radiations [111]. Pontocerebellar and commissural fibers coursing through premotor and sensorimotor callosal sectors appear resistant to HIV [110]. In unmedicated patients, particularly high transverse (i.e., radial) diffusivity, indicative of myelin compromise, is noted in superior longitudinal fasciculi, inferior cingulate bundles, and occipital forceps [110, 112]. The loss of microstructural integrity in normal-appearing periventricular white matter and corpus callosum of HIV-infected subjects correlates with high viral load and low CD4 count [107], while whole-brain microstructural integrity is predictive of cognitive compromise [113]. Thus, in HIV infection, it appears that CNS infection and the consequent neuroinflammation can contribute to CNS white matter damage and may underlie the macroscopic, microscopic, and biochemical abnormalities detected with in vivo neuroimaging methods (Fig. 14.2).

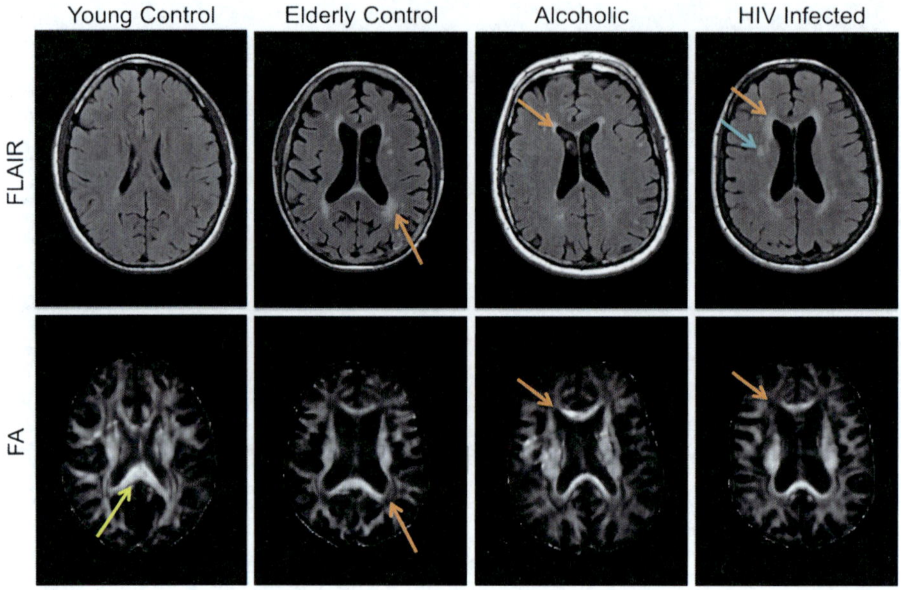

Fig. 14.2 *Top row*: axial FLAIR (fluid-attenuated inversion recovery) images at the level of the lateral ventricles of 4 individuals: a young healthy man (age 31 years), an elderly healthy women (age 84 years), an alcoholic man (age 59 years), and an HIV-infected woman (age 48 years). FLAIR images are used to detect *white matter* hyperintensities (WMHI). This figure demonstrates that aging, alcoholism, and HIV infection can result in similar phenotypes, that is, areas of periventricular (*orange arrows*) and deep WMHI (*blue arrows*). However, the mechanisms underlying the presence of WMHI may differ: in HIV, WMHI are thought to reflect myelin pallor and rarefaction associated with presence of multinucleated giant cells, whereas in alcoholism and aging, the available evidence suggests that WMHI reflect covert vascular brain injury or periventricular extravasation of CSF. *Bottom row*: axial diffusion tensor imaging (*DTI*) fractional anisotropy (*FA*) images at levels similar to those of the FLAIR images. FA provides an in vivo depiction and quantitative assessment of *white matter* microstructural integrity in terms of its linearity and orientational coherence of water diffusivity within voxels. FA ranges from 0 to 1; higher FA values are present in *white matter* regions and tracts that are highly linearly organized, such as the splenium of the corpus callosum (*yellow arrow*). Low FA values are present in *gray matter* and CSF-filled spaces, where diffusion is poorly organized. Low FA can also occur in areas of pathology or abnormality (*orange arrows*), such as WMHI, notable in the perisplenial region of the elderly control and the perigenual area of the alcoholic and HIV-infected individuals

14.3 Alcoholism

Alcohol use disorders, also known as alcoholism and alcohol dependence, are characterized by excessive alcohol use despite harmful effects on normal functioning and activities of daily life (Diagnostic and Statistical Manual, DSM-IV-TR). Alcoholism is the product of multiple interacting factors including complex genetics

[114, 115], the environment, predisposing personality characteristics, and a variety of psychiatric comorbidities that may be antecedent to or the consequence of the disease [116–118].

Similar to the distinction between HIV and AIDS, alcoholics may be distinguished by the absence (i.e., uncomplicated alcoholism) versus the presence of serious medical conditions such as cirrhosis, hepatic encephalopathy, alcoholic dementia, Wernicke's encephalopathy (WE), or Korsakoff's syndrome (KS). In KS, for example, there is evidence for frank neuronal loss in the thalamus [119], basal forebrain [120], and cerebellum [121]. In patients with uncomplicated alcoholism, neuropathological study provides little evidence for widespread cell loss [122–124]: no changes have been documented in the number of neurons in the cerebellum, basal ganglia [125], hippocampus [126, 127], or serotonergic raphe nuclei. Instead, available data suggest neuronal dysmorphology (e.g., reduction in dendritic length) in selective structures [119, 121, 128–133]. Postmortem evidence for loss of pyramidal neurons in the superior frontal and frontal association (dorsolateral portion) cortices [127, 134] has been questioned by recent techniques that suggest that rather than ostensible neuronal loss, frontal cortex changes may be accounted for by reversible dendritic shrinkage [135] or loss of astrocytes [136].

Cross-sectional MRI studies of brains of alcoholics compared with controls demonstrate a compromised cortical mantle and complementary ventriculomegaly [137–139]. Specific brain regions identified by structural MRI to be affected by chronic alcoholism corroborate postmortem findings and include cortical gray and white matter [140, 141], cerebellum [142, 143], hippocampus [144, 145], hypothalamus [146–148], thalamus [142, 149], and pons [150, 151].

Exemplary MRS studies in alcoholics report abnormally low levels of NAA in frontal [152–156] and cerebellar [152, 154, 157] regions compared with controls. Reversibility of NAA declines following abstinence from EtOH [152, 158], however, suggests temporary dysfunction and reversible injury rather than incontrovertible neuronal loss. Cho alterations vary across the course of alcoholism (e.g., [152, 158–160]), but in actively drinking alcohol dependents, Cho levels are elevated [155, 161]. Although in HIV Cho elevations are interpreted as evidence for inflammation, a number of processes can lead to Cho elevations including liver dysfunction [162], metabolic alterations [163–165], or demyelination [166].

Thus, as in HIV, rather than directly causing neuronal death, evidence also clearly suggests that alcoholism targets white matter. Brain white matter vulnerability to alcoholism is evidenced by postmortem observations for demyelination, loss of myelinated fibers, and axonal deletion [122, 123, 127, 167–173]. Structural MRI reveals pervasive white matter volume shrinkage [174–178]. Several studies have identified WMHI in alcoholism (but see [179–182]). In two of those studies, a U-shaped relationship was found between alcohol dose and WMHI such that moderate drinkers had the fewest number of WMHI relative to heavy drinkers and abstainers [180, 182]. Since some types of WMHI are thought to reflect vascular disease [183–185], such findings are consistent with studies regarding the beneficial effects of alcohol on the cardiovascular system [186–189] and suggest that low levels of alcohol consumption may be protective, while heavy alcohol consumption can increase the risk for cerebral ischemia [190]. Microstructural white matter abnormalities in alcoholism include degradation of callosal, pericallosal, and centrum semiovale tissue [191–194]. Quantitative fiber tracking in uncomplicated alcoholism reveals selective disruption of anterior and superior fiber systems and relative sparing of inferior and posterior fiber systems [195]. To the extent that DTI can discern the nature of the white matter degradation [196], decomposition of the diffusion metric indicates greater effect on the myelin sheath than the axon in alcoholics [195, 197]. Consistent with this possibility is a series of gene and protein expression studies showing in human alcoholic relative to control brains low expression of genes [198, 199] and proteins [200, 201] associated with myelin.

"Microglial activation" has been proposed as a mechanism underlying brain compromise in alcoholism. Indeed, a variety of authors have correlated induction of inflammatory mediators in alcoholism with neurodegeneration. However, as just reviewed, there is limited evidence for neurodegeneration in uncomplicated human alcoholism. Thus, caution must be used in the interpretation of cellular and molecular findings from in vitro studies. The process implicated is an alcohol-induced elevation of serum and liver proinflammatory cytokines, particularly TNF-α, that carries the inflammatory signal to the brain [202] where it activates microglia to increase the synthesis of additional cytokines and create a positive feedforward process of inflammation [203]. While the immune system is clearly responsive to alcohol, the direction of change (i.e., up- or downregulation, suppression or amplification of response) can depend on a variety of factors, including (a) length (e.g., acute vs chronic) [204] (b) schedule or pattern (e.g., chronic vs intermittent) [202] of the alcohol exposure protocol, (c) blood alcohol levels (BALs) attained [205], (d) sampling times following exposure, (e) the cell type (e.g., human monocytes vs rat- or mouse-mixed glial cultures) studied [206], (f) the species (e.g., rat vs mouse) or strains (Wistar vs Sprague-Dawley) examined, and (g) the sex of the animal. Further complexity is derived from the use of the bacterial toll-like receptor (TLR-4) ligand, lipopolysaccharide (LPS) [207], to promote robust immune responsiveness in experimental models [202, 208–210]. Finally, it is possible that disease concomitants to alcoholism must be present in order to observe immune recruitment. There is substantial support for the presence of peripheral inflammation and cytokine production in liver disease [211–213] and some experimental support for microglial activation and increased CNS proinflammatory cytokine production in animal models of thiamine deficiency [214–216]. The evidence for neuroinflammation in uncomplicated human alcoholism, however, is tentative.

Although it has been suggested that TNF-α transmits the peripheral alcohol-induced inflammatory signal to the brain [202], the evidence is at most preliminary. A critical gap in the alcohol literature is how the CNS immune system is recruited by exposure to chronic alcohol. There is currently no evidence for peripheral cell trafficking into the brain in alcoholism. By contrast, there is provisional evidence for microglial activation and expression and release of various proinflammatory and toxic mediators in the CNS in alcoholism. Significantly, human brain tissue recovered from chronic alcoholics compared with healthy donor tissue demonstrated increased levels of microglial markers Iba-1 and $GlutT_5$ in the cingulate cortex and midbrain and elevated levels of the chemokine CCL2/MCP-1 in homogenates of the VTA, substantia nigra, hippocampus, and amygdala [217]. Importantly, however, four of the eight alcoholic brains examined were harvested from alcoholics who died of complications related to liver disease, and within all brain regions examined, microglial morphology did not clearly distinguish control and alcoholic brains [217].

Ethanol exposure activates TLR-4 signaling ([218], but see [219]). The TLR-4 signaling cascade can initiate the transcription of proinflammatory cytokine genes in brain. In male BALB/c wild-type mice, prevention of TLR-4 signaling through either genetic knockout or pharmacological antagonism attenuated the acute sedative effects of ethanol as detected by sleep time and the motor effects of ethanol as detected by rotarod performance [220]. In male wild-type C57Bl mice, exposure to 10% ethanol in drinking water for 5 months to BALs of ~100 mg/dL followed by 15 days of ethanol withdrawal resulted in behavioral alterations, including decreased object recognition and an increase in astroglial (i.e., GFAP) and microglial (i.e., CD11b) markers in frontal cortex and striatum. In male TLR-4 knockout mice, however, behavioral responses and immunoreactivity changes following the ethanol exposure and withdrawal paradigm were attenuated [221]. In a similar study, female wild type compared with TLR-4 knockout C57Bl mice exposed to BALs of ~125 mg/dL showed downregulation of proteins involved in myelination [proteolipid protein (PLP), myelin basic protein (MBP), myelin-oligodendrocyte glycoprotein, 2,3-cyclic-nucleotide-3-phosphodiesterase, and myelin-associated glycoprotein], a reduced number of MBP-positive fibers, and an increase in caspase-3-positive oligodendrocytes indicating vulnerability of oligodendrocytes to chronic alcohol exposure [222]. Together, these data support a role for TLR-4 activation in the behavioral and structural changes in response to both ethanol exposure and withdrawal. However, TLR-4 signaling pathways are just beginning to be explored and may signal downstream events such as the regulation of cell proliferation and survival [223] independently of proinflammatory processes.

Ethanol via TLR-4 results in nuclear translocation of NF-κB, a ubiquitous transcription factor that is involved in many cellular processes, including regulation of innate immune gene expression [224], response to oxidative stress [225, 226], excitatory neurotransmission [227], synaptic plasticity, and pruning by apoptosis [228]. Ethanol has been shown to increase NF-κB–DNA binding following in vivo exposure in whole brain [229, 230] and in vitro exposure to hippocampal–entorhinal cortex brain slice cultures [231, 232]. However, by regulating transcription of a variety of targets, including glutamate receptors, growth factors, and kinases, NF-κB modulates synaptic plasticity and synaptic function and thus may play a role in plasticity, learning, and memory [227, 233–235].

In animal models, studies using various alcohol exposure protocols and several different markers suggest an increase in binding or staining of glial cells in response to alcohol. The hippocampus and entorhinal cortex of Sprague-Dawley rats exposed to binge alcohol treatment demonstrated increased [^{3}H]-PK-11195 binding [236], indicating increased, nonspecific binding to glial cells, including microglia, astrocytes, and infiltrating macrophages [237]. The dentate gyrus of the hippocampus of adolescent female Wistar rats exposed to alcohol intermittently showed increased

OX-6 immunohistochemical staining [238]. The molecular layer of the cerebellar vermis of adult male Wistar rats intermittently but not continuously exposed to an alcohol liquid diet for 5.5 months showed increased tomato lectin staining [239], and the cerebral cortex of female wild type but not TLR-4-deficient mice exposed to an alcohol liquid diet for 5 months showed upregulation of another microglial marker, CD11b [240]. The increased presence of such markers is believed to indicate enhanced microglial activation, but there is a question as to the quality and specificity of currently available microglial markers (e.g., [241–245]), and it is difficult to unambiguously identify activated microglia, especially in tissue from the human brain [246]. Inconsistency with respect to alcohol exposure protocols and types of microglial markers used emphasizes the limitation that only modest evidence for increased microglial activation in alcoholism is currently available.

Whole-brain homogenates from mice exposed to alcohol for 10 days showed induction of CCL2/MCP-1 and TNF-α proteins [202]. The cerebral cortex of adult female Wistar rats fed an alcohol liquid diet for 5 months demonstrated a significant increase in IL-1β protein [204]. Similarly, hippocampal–entorhinal cortex brain slice cultures exposed to alcohol showed increased expression of TNF-α, CCL2/MCP-1, and IL-1β [232]. Although microglia may be able to express such "immune molecules" (but see [247]), this expression is not synonymous with inflammation because these molecules can have CNS-specific roles independent of their roles in the immune or inflammatory response. Indeed, under physiological conditions, IL-1β can influence synaptic plasticity [248, 249]. CCL2/MCP-1 can modulate synaptic transmission through altering calcium channel excitability and neurotransmitter release [250]. Similarly, TNF-α has a neuromodulatory role in the brain [251]: it can modify glutamate homeostasis [252] by upregulating surface expression of AMPA receptors [253] and downregulating surface expression of GABA-A receptors [254], thus playing a critical role in homeostatic synaptic scaling [255]. Consequently, given the as yet limited understanding of the role of cytokines and chemokines in normal brain functioning, interpretation of the sparse data in the alcohol literature as evidence for immune activation is speculative.

14.4 Aging

A salient characteristic of normal aging is cognitive and motor slowing, which can attenuate response control for quick and accurate decisions, conflict resolution, and attention gating [256]. Successful aging includes functional neuro-adaptation to accommodate structural brain changes and the potential to enhance function with redistribution of resources, possibly at the expense of usurping functional reserve [257, 258].

Cross-sectional and longitudinal MRI studies in the adult age range provide consistent evidence for systematic age-related volume increases in CSF-filled spaces, including sulci, fissures, and ventricles, that occur at the expense of cortical gray matter (cross-sectional [143, 259–266] longitudinal [267–270]). As is observed in HIV and alcoholism, convergent data indicate an excessive vulnerability of the frontal cortex to aging [267, 268, 270–272]. Other gray matter structures demonstrating cross-sectional decline in volume with age and also observed in HIV infection and alcoholism include, but are not limited to, the thalamus [265] and cerebellum [143, 272–275]. The caudate, putamen, nucleus accumbens, medial septum/diagonal band ([276] but see [277, 278]), and the pons [265, 279–281] do not show substantial age-related volume decline.

Although controversial (cf., [282]), hippocampal volume shrinks little with age [283–285] and remains intact in size in healthy elderly adults without concurrent hypertension [275] or Alzheimer's disease [286]. While volume loss in Alzheimer's disease can reflect neuronal loss [287–290], volume loss in the various regions of the normal aging most likely reflects non-neuronal changes. Indeed, several MRS studies are consistent in reporting lack of age-related decline in NAA concentration per volume in healthy men and women [291–296] in various brain regions, including frontal white [297] and gray matter [298], parietal white matter [299, 300], and pons [301]. These findings suggest that the number of viable neurons per unit of gray matter is preserved with age.

Normal aging, like HIV infection and alcoholism, affects white matter integrity. Increasing age is a major risk factor for WMHI, and many studies have shown a positive correlation between age and WMHI volume [67, 302–304]. In aging, WMHI are associated with vascular risk factors and probably reveal evidence for cerebrovascular disease [183, 184, 305–308], even in apparently healthy elderly [309, 310]. These white matter markers of aging represent heterogeneous processes related to size and location of the hyperintensity, genetic contribution, and underlying pathology determined from postmortem correlation (for review [311]). With DTI, the patterns of white matter microstructural sparing and compromise differentiate regions, fiber type, and diffusivity characteristics [312–318]. Greater age-related compromise is observed in superior than inferior and anterior than posterior fiber bundles [319]. Higher Cho has been observed in elderly groups compared with younger groups [294, 320, 321] in frontal and parietal white matter [297–299, 322], frontal and occipital gray matter [298, 322], and in the pons [323] and may be interpreted as evidence for demyelination, among other explanations. Age-related changes in white matter may be due to perturbation of axonal cytoskeleton or myelin microstructure [324]. Postmortem studies observe trapping of fluid between thin or lysed myelin sheathes and bulbous swelling of oligodendrocytes [325–327]. Production of MBP is reduced in the aged brain [328].

While there are sufficient data to suggest that microglial function is impaired in the aged brain [329], the currently available evidence does not distinguish whether microglia secondarily react to age-related changes in their environment or whether the aging process directly affects microglia. There is no clear evidence for increased microglial numbers in brains from older compared with younger rats [330], but MHC class II, a purported marker for activated microglia, is increased in brains of aged humans [331, 332], monkeys [333, 334], rats [335, 336], and mice [337]. Microglia have the ability to respond to [338] and phagocytose myelin [339]; thus, age-related increases in myelin breakdown could account for the apparent increase in microglial activation. Increased MHC class II staining could also represent a phenotypic change in microglia, marking their senescence, rather than their activation [246].

In some studies, LPS [340] or *Escherichia coli* [341] administration has been shown to exaggerate the inflammatory cytokine response in the aged versus adult mouse brain. Thus, some have argued that the aging process serves as a "priming" stimulus for microglia, so that in the presence of a trigger, microglia are predisposed to activation and the excessive release of toxic mediators [342–344]. However, reports have indicated greater basal secretion of IL-6 from brain tissue from rats [345] and mice [346] but higher [345] or lower [347] secretion in response to stimulation (e.g., LPS) in older compared to younger animals. Similarly, while microglia from aged compared to young mice have been shown to constitutively secrete greater amounts of TNF-α ([348] but see [345, 347]), stimulated TNF-α secretion is variable [345, 348]. Variability in stimulated responsiveness may depend on the triggering stimulus [346], the animal model used, the anatomic region examined, or the marker evaluated. Altogether, current evidence does not support neuroinflammation as a primary contributor to age-related CNS decline; however, if microglia are typically protective, their dysfunction could have negative consequences for the aging brain.

14.5 Interactions

A variety of experimental models have been used to investigate the mechanisms of CNS toxicity in HIV, alcoholism, and aging. However, these entities do not exist independently, and these models do not recapitulate the full picture when studied separately. For example, it is widely accepted that aging is the only unequivocal risk factor for the development of Alzheimer's disease and Parkinson's disease, but few experimental studies examine these diseases in the context of aging. Similarly, there are limited animal models exploring the mechanisms of CNS injury in aging and its comorbidities such as HIV infection or alcohol dependence. Such studies are needed to fill critical gaps in knowledge and to provide a comprehensive account of the impact of such interactions on pathology.

14.6 HIV + Alcoholism

The rate of heavy drinking among HIV-infected individuals is almost twice that of the general population [5]. Studies based on HIV clinic attendees [4, 6–9] report significant alcohol use in upward of 63% of HIV patients, whereas an analysis of a national probability sample [5] reported that only 8% of patients with HIV infection reported heavy alcohol consumption. One study noted that of the 111 HIV-positive clinic patients 45 (41%) met the criteria for alcoholism [6]. Further, those who abuse alcohol have a 5–10% higher risk of acquiring HIV than the general population [81]. With extended life spans made possible by HAART, HIV-infected individuals remain at high risk for continuing or reverting to hazardous behavior, including excessive alcohol consumption [3], which can reduce medication compliance [1].

Mechanisms of alcohol's untoward effect on HIV disease include accelerated progression of the infection by contributing to immune suppression [349–352], diminished effectiveness of therapeutic regimens [7], and potentiation of the neurotoxicity of retroviral proteins shed by the HIV virus during glial infection [353, 354]. In neuron cultures [355], low to moderate alcohol concentrations potentiated gp120-induced neuronal apoptosis [353]. In hippocampal slice cultures, long-term exposure to alcohol (i.e., 12 months) enhanced hippocampal vulnerability to HIV toxicity through Tat- and NMDA-dependent mechanisms [354] independent of microglial activation as measured by benzodiazepine receptor binding [356]. In another in vitro study, moderate alcohol pretreatment [357, 358] was neuroprotective; higher levels potentiated damage [359]. Primate studies revealed that Rhesus monkeys that were pretreated with high doses of alcohol and then infected with the simian version of HIV (SIV) showed a greater presence of the virus than sucrose-exposed and SIV-infected controls [360, 361].

Elevations in HIV-positive blood-borne monocyte-derived macrophages in the alcoholic brain may be due to increased trafficking across the BBB [362]. HIV infection compromises the BBB and increases its permeability (for review, see [363]). Support for EtOH-induced BBB disruption comes predominantly from in vitro studies. In cultures of brain microvascular endothelial cells (BMVEC), alcohol metabolism in endothelial cells results in oxidative stress associated with a degradation of components of the BBB essential for its integrity, reduced barrier tightness, and enhanced permeability and monocyte migration [364, 365]. In animal models, long-term alcohol treatment markedly induced BBB leakage [366]. Other in vivo studies provide only provisional evidence for BBB disruption in alcoholism: rats exposed to 3 weeks of vaporized alcohol demonstrated increases in BBB permeability only if exposure was followed by starvation for 3 days [367]. Likewise, permeability was enhanced in alcohol-intoxicated relative to control rats only around small cerebral stab wounds in the frontal cortex [368]. Evidence for in vivo compromise of BBB in human alcoholics is largely inferential [369, 370].

The extent to which alcohol and HIV have synergistic deleterious effects on brain structure is seldom considered (e.g., [62, 81, 371–373]). In a structural MR study, the volume of the ventricular system and the cross-sectional area of the mid-sagittal corpus callosum were compared in four groups: patients with HIV infection, with and without alcoholism comorbidity, alcoholics, and controls. Disregarding HIV disease severity, a graded pattern of modest enlargement of the total ventricular system was observed. The pattern of callosal thinning showed a similar but small graded effect. A different pattern emerged, however, when HIV severity in the context of alcoholism comorbidity was factored into the analysis. Substantially greater volume abnormalities were present in individuals with a history of an AIDS-defining event or low CD4 T-cell counts irrespective of alcoholism comorbidity, and the effect of HIV severity was disproportionately exacerbated by alcoholism comorbidity, with a 1 standard deviation size deficit in the genu of corpus callosum and a nearly 2 standard deviation greater volume of the frontal and body regions of the ventricles for the AIDS + alcohol comorbid group. The differences in brain volumes between the AIDS groups with versus without alcoholism could not be attributed to differences in HIV disease severity, defined by CD4 T-cell count, viral load, or Karnofsky score. The substantial effect of the alcoholism–AIDS interaction on ventricular and callosal dysmorphology, in the context of the modest changes observed in non-AIDS, non-alcohol-abusing HIV-infected individuals, highlights the need to consider alcohol use disorders as a major risk factor for neuropathology among HIV-infected persons [67].

In a DTI study with the same four-group cohort, compared with controls, all patient groups had diffusivity changes in callosal regions and fiber bundles coursing through the genu and splenium, but these effects were only significant in the two groups with alcoholism. When the HIV-infected groups were divided by disease severity defined as an AIDS-defining event or low CD4 counts and alcoholism comorbidity, the HIV-infected subgroup with AIDS and alcoholism exhibited diffusivity abnormalities in the callosal sectors and fibers that were more than twice the effect sizes observed in the other HIV-infected subgroups. Degradation of callosal microstructure was consistently associated with alcoholism, with evidence for compounded alcoholism and HIV effects [112].

In a ^{31}phosphorus MRS study, the symptomatic HIV group and the alcoholic group both had low concentrations of brain energy metabolites in superior white matter, and patients with both conditions had augmented metabolite deficits [374]. In another MRS study including a four-group cohort, the two HIV groups were matched in T-cell count and were not demented; the two alcoholism groups were relatively matched in lifetime alcohol consumption. Significant group effects for NAA and Cr were observed only in the HIV + alcoholism group. The deficits were not related to HAART status. Neither HIV infection nor alcoholism independently resulted in parietal–occipital cortical metabolite abnormalities, yet each disease carried a liability that put affected individuals at a heightened risk for compromise when the diseases were compounded [375].

14.7 HIV + Aging

As of 2005, persons aged 50 and over accounted for 24% of individuals living with HIV/AIDS, increasing from 17% in 2001. This increase is due partially to the success of therapies, such as HAART, which has increased patient longevity, and partly due to new diagnoses in individuals over age 50 (15% of new HIV/AIDS diagnoses) [376]. Older adults have a higher risk of HIV-associated neurocognitive dysfunction [377–379].

Features that characterize CNS HIV infection and also present in normal aging include synaptodendritic damage and increased presence of reactive oxygen species [380–383]. Frontal cortex samples from HIV compared with age-matched controls reveal oxidative injury in the HIV-infected brain is worsened with aging [384]. Furthermore, HAART can promote formation of toxic free radicals [385]. Aged mice compared with their younger adult counterparts demonstrate enhanced hippocampal IL-1β and IL-6 production in response to gp120 [386]. Levels of spontaneous IL-6 secretion can be higher with age; in a microglia cell culture, IL-6 increased the cell surface expression of CCR5 [386] suggesting that HIV can be more infectious in the aged brain. More work is required to determine whether behavioral dysfunction in HIV and aging is due to a greater neuroinflammatory response.

With respect to imaging findings, some studies have found clear evidence for an interaction between age and HIV infection, while others have not. For example, while both HIV and aging show decreases in baseline cerebral blood flow and increases in functional blood flow, no interactions were observed [387]. In another study conducted within subjects at baseline and follow-up a year later, DTI measures were more changed in HIV individuals compared with their age-matched controls in the genu of the corpus callosum, evidence for an interaction, but not in other regions evaluated (i.e., frontal white matter, thalamus, caudate, globus pallidus) [388]. Similarly, the aging HIV-infected brain shows interactive increases in MRS markers for Cho and mI in frontal white matter and basal ganglia [389, 390] and interactive decreases in NAA in basal ganglia [390]. More imaging studies are required in order to clearly demonstrate the potential interactions of age and HIV on the brain in vivo.

14.8 Alcoholism + Aging

Alcohol abuse and alcoholism are common but under-recognized problems among older adults. One third of older alcoholic persons develop a problem with alcohol in later life, while the other two thirds grow older with the medical and psychosocial sequelae of early-onset alcoholism [391]. There are currently few studies focused on alcohol and aging effects with respect to CNS immune recruitment. Given the available evidence, however, if microglia are typically protective, their senescence could make the elderly brain more sensitive to the damaging effects of alcohol.

In brains from aging alcoholics compared with age-matched control cases, loss of cells and a reduction in the size of cell bodies were noted in superior frontal and motor cortical regions [124]. Stereological assessment revealed no significant change in granule layer volume or in the total number of cerebellar granule neurons in aged, ethanol-fed rats [132]. However, the number of Purkinje neuron synapses significantly declined [392], and terminal segments become abnormally elongated [393]. After a 20-week abstinence, the number of synapses and length of terminal segments returned to the levels observed in the pair-fed animals prior to the abstinence period

[393]. These data indicate that remodeling and compensation can occur even in aged rodents during recovery from alcohol-related structural injury [393].

Age–alcoholism interactions have been observed in callosal macrostructure, postmortem [394] and in vivo [176]. An age–alcoholism interaction was also observed in both FA and diffusivity, where older alcoholics had greater abnormalities for their age than younger ones [193]. Similarly, another MR study suggests that age-related gray matter shrinkage in treatment-naïve alcoholics compared with controls is a function of age and not a lifetime alcohol burden [395]. These data corroborate the premature aging hypothesis that suggests alcoholism accelerates aging such that brains of alcoholics are reminiscent of those of chronologically old, non-alcoholic brains [396].

14.9 Conclusion

MRI provides a powerful in vivo approach for anatomically localizing and characterizing CNS pathology caused by HIV infection, alcoholism, and aging separately and can also be used to assess the consequences of the combination of these disease states. From postmortem data, confirmed and extended by vivo data from DTI, it appears that at least initially, HIV infection, alcoholism, and aging compromise CNS white matter microstructural integrity. Macrostructural MRI data reveal similar extents of gray matter and white matter volume shrinkage in alcoholics and greater gray than white matter volume shrinkage in healthy men and women.

HIV infection has a complicated CNS pathology that includes direct effects of the virus and indirect effects of neuroinflammation. Incontrovertible support for classical neuroinflammation in HIV infection is derived from evidence for cellular trafficking into the brain and the expression and release of various proinflammatory and toxic mediators. There is also support for the idea that neuroinflammation in HIV can target white matter, which appears initially to be more vulnerable than gray matter. In AIDS, evidence for frank neuronal loss may be accounted for by inflammation but likely occurs as a result of a variety of converging mechanisms. In uncomplicated human alcoholism, current evidence only poorly supports immune recruitment as a pathological mechanism in alcoholism. More research will be necessary to determine whether "microglial activation" plays a primary role in the damaging effects of chronic alcohol consumption on the CNS. With aging, the senescence of microglia likely contributes to brain vulnerability, but again, there is as yet little support for "microglial activation" as a fundamental process contributing to age-related brain damage. Given that neuroimmune factors play multiple roles in the brain, more work is essential to determine how these molecules contribute to normal functioning, what in vivo neuroimaging signals specifically mark microglial activity, and pathological changes in the brain postmortem.

Funding NIAAA AA005965 A.P.; AA012388 A.P. & E.V.S.; AA0012388 R37; AA013521-INIA; AA017347

References

1. Braithwaite RS, McGinnis KA, Conigliaro J, Maisto SA, Crystal S, Day N, Cook RL, Gordon A, Bridges MW, Seiler JF, Justice AC (2005) A temporal and dose–response association between alcohol consumption and medication adherence among veterans in care. Alcohol Clin Exp Res 29(7):1190–1197
2. Everall IP, Hansen LA, Masliah E (2005) The shifting pattern of HIV encephalitis neuropathology. Neurotox Res 8:51–61
3. Stein M, Herman DS, Trisvan E, Pirraglia P, Engler P, Anderson BJ (2005) Alcohol use and sexual risk behavior among human immunodeficiency virus-positive persons. Alcohol Clin Exp Res 29(5):837–843
4. Cook RL, Sereika SM, Hunt SC, Woodward WC, Erlen JA, Conigliaro J (2001) Problem drinking and medication adherence among persons with HIV infection. J Gen Intern Med 16(2):83–88
5. Galvan FH, Bing EG, Fleishman JA, London AS, Caetano R, Burnam MA, Longshore D, Morton SC, Orlando M, Shapiro M (2002) The prevalence of alcohol consumption and heavy drinking among people with HIV in the United States: results from the HIV Cost and Services Utilization Study. J Stud Alcohol 63(2):179–186
6. LeFevre F, O'Leary B, Moran M, Mossar M, Yarnold PR, Martin GJ, Glassroth J (1995) Alcohol consumption among HIV-infected patients. J Gen Intern Med 10(8):458–460
7. Miguez MJ, Shor-Posner G, Morales G, Rodriguez A, Burbano X (2003) HIV treatment in drug abusers: impact of alcohol use. Addict Biol 8(1):33–37
8. Page-Shafer K, Delorenze GN, Satariano WA, Winkelstein W Jr (1996) Comorbidity and survival in HIV-infected men in the San Francisco Men's Health Survey. Ann Epidemiol 6(5):420–430
9. Samet JH, Horton NJ, Meli S, Freedberg KA, Palepu A (2004) Alcohol consumption and antiretroviral adherence among HIV-infected persons with alcohol problems. Alcohol Clin Exp Res 28(4):572–577
10. Justice AC, Lasky E, McGinnis KA, Skanderson M, Conigliaro J, Fultz SL, Crothers K, Rabeneck L, Rodriguez-Barradas M, Weissman SB, Bryant K (2006) Medical disease and alcohol use among veterans with human immunodeficiency infection: a comparison of disease measurement strategies. Med Care 44(8 Suppl 2):S52–S60
11. Chen GY, Nunez G (2010) Sterile inflammation: sensing and reacting to damage. Nat Rev Immunol 10(12):826–837
12. Ritter MR, Banin E, Moreno SK, Aguilar E, Dorrell MI, Friedlander M (2006) Myeloid progenitors differentiate into microglia and promote vascular repair in a model of ischemic retinopathy. J Clin Invest 116(12):3266–3276
13. Graeber MB, Streit WJ, Kreutzberg GW (1988) Axotomy of the rat facial nerve leads to increased CR3 complement receptor expression by activated microglial cells. J Neurosci Res 21(1):18–24
14. Gehrmann J, Matsumoto Y, Kreutzberg GW (1995) Microglia: intrinsic immuneffector cell of the brain. Brain Res Brain Res Rev 20(3):269–287
15. Aloisi F (2001) Immune function of microglia. Glia 36(2):165–179
16. Aschner M (1998) Astrocytic functions and physiological reactions to injury: the potential to induce and/or exacerbate neuronal dysfunction – a forum position paper. Neurotoxicology 19(1):7–17, discussion 37–38
17. Aschner M, Sonnewald U, Tan KH (2002) Astrocyte modulation of neurotoxic injury. Brain Pathol 12(4):475–481
18. Haorah J, Ramirez SH, Floreani N, Gorantla S, Morsey B, Persidsky Y (2008) Mechanism of alcohol-induced oxidative stress and neuronal injury. Free Radic Biol Med 45(11):1542–1550
19. Hirsch EC, Breidert T, Rousselet E, Hunot S, Hartmann A, Michel PP (2003) The role of glial reaction and inflammation in Parkinson's disease. Ann N Y Acad Sci 991:214–228

20. Kreutzberg GW (1996) Microglia: a sensor for pathological events in the CNS. Trends Neurosci 19(8):312–318
21. Rump TJ, Abdul Muneer PM, Szlachetka AM, Lamb A, Haorei C, Alikunju S, Xiong H, Keblesh J, Liu J, Zimmerman MC, Jones J, Donohue TM Jr, Persidsky Y, Haorah J (2010) Acetyl-L-carnitine protects neuronal function from alcohol-induced oxidative damage in the brain. Free Radic Biol Med 49(10):1494–1504
22. Graeber MB, Li W, Rodriguez ML (2011) Role of microglia in CNS inflammation. FEBS Lett 585(23):3798–3805
23. Block ML, Hong JS (2005) Microglia and inflammation-mediated neurodegeneration: multiple triggers with a common mechanism. Prog Neurobiol 76(2):77–98
24. Crews FT, Bechara R, Brown LA, Guidot DM, Mandrekar P, Oak S, Qin L, Szabo G, Wheeler M, Zou J (2006) Cytokines and alcohol. Alcohol Clin Exp Res 30(4):720–730
25. Gray F, Chretien F, Vallat-Decouvelaere AV, Scaravilli F (2003) The changing pattern of HIV neuropathology in the HAART era. J Neuropathol Exp Neurol 62(5):429–440
26. Langford TD, Letendre SL, Larrea GJ, Masliah E (2003) Changing patterns in the neuropathogenesis of HIV during the HAART era. Brain Pathol 13(2):195–210
27. Masliah E, DeTeresa RM, Mallory ME, Hansen LA (2000) Changes in pathological findings at autopsy in AIDS cases for the last 15 years. AIDS 14(1):69–74
28. Ghafouri M, Amini S, Khalili K, Sawaya BE (2006) HIV-1 associated dementia: symptoms and causes. Retrovirology 3:28
29. Grovit-Ferbas K, Harris-White ME (2010) Thinking about HIV: the intersection of virus, neuroinflammation and cognitive dysfunction. Immunol Res 48(1–3):40–58
30. Haase AT (1986) The AIDS lentivirus connection. Microb Pathog 1(1):1–4
31. Kaul M, Garden GA, Lipton SA (2001) Pathways to neuronal injury and apoptosis in HIV-associated dementia. Nature 410(6831):988–994
32. Langford TD, Letendre SL, Marcotte TD, Ellis RJ, McCutchan JA, Grant I, Mallory ME, Hansen LA, Archibald S, Jernigan T, Masliah E (2002) Severe, demyelinating leukoencephalopathy in AIDS patients on antiretroviral therapy. AIDS 16(7):1019–1029
33. Peluso R, Haase A, Stowring L, Edwards M, Ventura P (1985) A Trojan Horse mechanism for the spread of visna virus in monocytes. Virology 147(1):231–236
34. Rezai-Zadeh K, Gate D, Town T (2009) CNS infiltration of peripheral immune cells: D-Day for neurodegenerative disease? J Neuroimmune Pharmacol 4(4):462–475
35. Vazeux R, Brousse N, Jarry A, Henin D, Marche C, Vedrenne C, Mikol J, Wolff M, Michon C, Rozenbaum W et al (1987) AIDS subacute encephalitis. Identification of HIV-infected cells. Am J Pathol 126(3):403–410
36. Wiley CA, Schrier RD, Nelson JA, Lampert PW, Oldstone MB (1986) Cellular localization of human immunodeficiency virus infection within the brains of acquired immune deficiency syndrome patients. Proc Natl Acad Sci USA 83(18):7089–7093
37. Cosenza MA, Zhao ML, Si Q, Lee SC (2002) Human brain parenchymal microglia express CD14 and CD45 and are productively infected by HIV-1 in HIV-1 encephalitis. Brain Pathol 12(4):442–455
38. Fischer-Smith T, Croul S, Adeniyi A, Rybicka K, Morgello S, Khalili K, Rappaport J (2004) Macrophage/microglial accumulation and proliferating cell nuclear antigen expression in the central nervous system in human immunodeficiency virus encephalopathy. Am J Pathol 164(6):2089–2099
39. Gendelman HE, Persidsky Y, Ghorpade A, Limoges J, Stins M, Fiala M, Morrisett R (1997) The neuropathogenesis of the AIDS dementia complex. AIDS 11(Suppl A):S35–S45
40. Gonzalez-Scarano F, Martin-Garcia J (2005) The neuropathogenesis of AIDS. Nat Rev Immunol 5(1):69–81
41. Heindel WC, Jernigan TL, Archibald SL, Achim CL, Masliah E, Wiley CA (1994) The relationship of quantitative brain magnetic resonance imaging measures to neuropathologic indexes of human immunodeficiency virus infection. Arch Neurol 51(11):1129–1135
42. Kramer-Hammerle S, Rothenaigner I, Wolff H, Bell JE, Brack-Werner R (2005) Cells of the central nervous system as targets and reservoirs of the human immunodeficiency virus. Virus Res 111(2):194–213

43. Albright AV, Strizki J, Harouse JM, Lavi E, O'Connor M, Gonzalez-Scarano F (1996) HIV-1 infection of cultured human adult oligodendrocytes. Virology 217(1):211–219
44. Bagasra O, Lavi E, Bobroski L, Khalili K, Pestaner JP, Tawadros R, Pomerantz RJ (1996) Cellular reservoirs of HIV-1 in the central nervous system of infected individuals: identification by the combination of in situ polymerase chain reaction and immunohistochemistry. AIDS 10(6):573–585
45. Ensoli F, Cafaro A, Fiorelli V, Vannelli B, Ensoli B, Thiele CJ (1995) HIV-1 infection of primary human neuroblasts. Virology 210(1):221–225
46. Mizrachi Y, Ródriguez I, Sweetnam PM, Rubinstein A, Volsky DJ (1994) HIV type 1 infection of human cortical neuronal cells: enhancement by select neuronal growth factors. AIDS Res Hum Retroviruses 10(12):1593–1596
47. Nuovo GJ, Gallery F, MacConnell P, Braun A (1994) In situ detection of polymerase chain reaction-amplified HIV-1 nucleic acids and tumor necrosis factor-alpha RNA in the central nervous system. Am J Pathol 144(4):659–666
48. Obregon E, Punzon C, Fernandez-Cruz E, Fresno M, Munoz-Fernandez MA (1999) HIV-1 infection induces differentiation of immature neural cells through autocrine tumor necrosis factor and nitric oxide production. Virology 261(2):193–204
49. Radja F, Kay DG, Albrecht S, Jolicoeur P (2003) Oligodendrocyte-specific expression of human immunodeficiency virus type 1 Nef in transgenic mice leads to vacuolar myelopathy and alters oligodendrocyte phenotype in vitro. J Virol 77(21):11745–11753
50. Takahashi K, Wesselingh SL, Griffin DE, McArthur JC, Johnson RT, Glass JD (1996) Localization of HIV-1 in human brain using polymerase chain reaction/in situ hybridization and immunocytochemistry. Ann Neurol 39(6):705–711
51. Trillo-Pazos G, Diamanturos A, Rislove L, Menza T, Chao W, Belem P, Sadiq S, Morgello S, Sharer L, Volsky DJ (2003) Detection of HIV-1 DNA in microglia/macrophages, astrocytes and neurons isolated from brain tissue with HIV-1 encephalitis by laser capture microdissection. Brain Pathol 13(2):144–154
52. Gendelman HE, Baldwin T, Baca-Regen L, Swindells S, Loomis L, Skurkovich S (1994) Regulation of HIV1 replication by interferon alpha: from laboratory bench to bedside. Res Immunol 145(8–9):679–684, discussion 684–685
53. Langford D, Masliah E (2001) Crosstalk between components of the blood brain barrier and cells of the CNS in microglial activation in AIDS. Brain Pathol 11(3):306–312
54. Minagar A, Shapshak P, Fujimura R, Ownby R, Heyes M, Eisdorfer C (2002) The role of macrophage/microglia and astrocytes in the pathogenesis of three neurologic disorders: HIV-associated dementia, Alzheimer disease, and multiple sclerosis. J Neurol Sci 202(1–2):13–23
55. Pulliam L, Herndier BG, Tang NM, McGrath MS (1991) Human immunodeficiency virus-infected macrophages produce soluble factors that cause histological and neurochemical alterations in cultured human brains. J Clin Invest 87(2):503–512
56. Wiley CA (2003) Detection of HIV-1 DNA in microglia/macrophages, astrocytes and neurons isolated from brain tissue with HIV-1 encephalitis by laser capture microdissection. Brain Pathol 13(3):415, author reply 415–416
57. Aylward EH, Brettschneider PD, McArthur JC, Harris GJ, Schlaepfer TE, Henderer JD, Barta PE, Tien AY, Pearlson GD (1995) Magnetic resonance imaging measurement of gray matter volume reductions in HIV dementia. Am J Psychiatry 152:987–994
58. Jernigan TL, Archibald S, Hesselink JR, Atkinson JH, Velin RA, McCutchan JA, Chandler J, Grant I (1993) Magnetic resonance imaging morphometric analysis of cerebral volume loss in human immunodeficiency virus infection. Arch Neurol 50(3):250–255
59. Symonds LL, Archibald SL, Grant I, Zisook S, Jernigan TL (1999) Does an increase in sulcal or ventricular fluid predict where brain tissue is lost? J Neuroimaging 9(4):201–209
60. Di Sclafani V, Mackay RD, Meyerhoff DJ, Norman D, Weiner MW, Fein G (1997) Brain atrophy in HIV infection is more strongly associated with CDC clinical stage than with cognitive impairment. J Int Neuropsychol Soc 3(3):276–287
61. Thompson PM, Dutton RA, Hayashi KM, Lu A, Lee SE, Lee JY, Lopez OL, Aizenstein HJ, Toga AW, Becker JT (2006) 3D mapping of ventricular and corpus callosum abnormalities in HIV/AIDS. Neuroimage 31:12–23

62. Heaton RK, Grant I, Butters N, White DA, Kirson D, Atkinson JH, McCutchan JA, Taylor MJ, Kelly MD, Ellis RJ, Wolfson T, Velin R, Marcotte TD, Hesselink JR, Jernigan TL, Chandler J, Wallace M, Abramson I (1995) The HNRC 500 – neuropsychology of HIV infection at different disease stages. HIV Neurobehavioral Research Center. J Int Neuropsychol Soc 1(3):231–251
63. Stout J, Ellis R, Jernigan T, Archibald S, Abramson I, Wolfson T, McCutchan J, Wallace M, Atkinson J, Grant I (1998) Progressive cerebral volume loss in human immunodeficiency virus infection: a longitudinal volumetric magnetic resonance imaging study. Arch Neurol 55:161–168
64. Dal Pan GJ, McArthur JH, Aylward E, Selnes OA, Nance-Sproson TE, Kumar AJ, Mellits ED, McArthur JC (1992) Patterns of cerebral atrophy in HIV-1-infected individuals – results of a quantitative MRI analysis. Neurology 42(11):2125–2130
65. Manji H, Connolly S, McAllister R, Valentine AR, Kendall BE, Fell M, Durrance P, Thompson AJ, Newman S, Weller IVD, Harrison MJG (1994) Serial MRI of the brain in asymptomatic patients infected with HIV: results from the UCMSM/Medical Research Council neurology cohort. J Neurol Neurosurg Psychiatry 57(2):144–149
66. Pfefferbaum A, Rosenbloom M, Sassoon SA, Kemper CA, Deresinski S, Rohlfing T, Sullivan EV (2012) Regional brain structural dysmorphology in HIV infection: effects of AIDS, alcoholism, and age. Biol Psychiatry May 28. [Epub ahead of print]
67. Pfefferbaum A, Rosenbloom MJ, Rohlfing T, Adalsteinsson E, Kemper CA, Deresinski S, Sullivan EV (2006) Contribution of alcoholism to brain dysmorphology in HIV infection: effects on the ventricles and corpus callosum. Neuroimage 33(1):239–251
68. Thompson PM, Dutton RA, Hayashi KM, Toga AW, Lopez OL, Aizenstein HJ, Becker JT (2005) Thinning of the cerebral cortex visualized in HIV/AIDS reflects CD4+ T lymphocyte decline. Proc Natl Acad Sci USA 102(43):15647–15652
69. Archibald SL, Masliah E, Fennema-Notestine C, Marcotte TD, Ellis RJ, McCutchan JA, Heaton RK, Grant I, Mallory M, Miller A, Jernigan TL (2004) Correlation of in vivo neuroimaging abnormalities with postmortem human immunodeficiency virus encephalitis and dendritic loss. Arch Neurol 61(3):369–376
70. Chiang MC, Dutton RA, Hayashi KM, Lopez OL, Aizenstein HJ, Toga AW, Becker JT, Thompson PM (2007) 3D pattern of brain atrophy in HIV/AIDS visualized using tensor-based morphometry. Neuroimage 34(1):44–60
71. Sullivan EV, Rosenbloom MJ, Rohlfing T, Kemper CA, Deresinski S, Pfefferbaum A (2011) Pontocerebellar contribution to ataxia and psychomotor slowing in HIV infection without dementia. Brain Imaging Behav 5:12–24
72. Klunder AD, Chiang MC, Dutton RA, Lee SE, Toga AW, Lopez OL, Aizenstein HJ, Becker JT, Thompson PM (2008) Mapping cerebellar degeneration in HIV/AIDS. Neuroreport 19(17):1655–1659
73. Schuff N, Amend DL, Knowlton R, Norman D, Fein G, Weiner MW (1999) Age-related metabolite changes and volume loss in the hippocampus by magnetic resonance spectroscopy and imaging. Neurobiol Aging 20(3):279–285
74. Adle-Biassette H, Chretien F, Wingertsmann L, Hery C, Ereau T, Scaravilli F, Tardieu M, Gray F (1999) Neuronal apoptosis does not correlate with dementia in HIV infection but is related to microglial activation and axonal damage. Neuropathol Appl Neurobiol 25(2):123–133
75. An SF, Giometto B, Scaravilli T, Tavolato B, Gray F, Scaravilli F (1996) Programmed cell death in brains of HIV-1-positive AIDS and pre-AIDS patients. Acta Neuropathol 91(2):169–173
76. Everall I, Luthert P, Lantos P (1993) A review of neuronal damage in human immunodeficiency virus infection: its assessment, possible mechanism and relationship to dementia. J Neuropathol Exp Neurol 52(6):561–566
77. Reyes MG, Faraldi F, Senseng CS, Flowers C, Fariello R (1991) Nigral degeneration in acquired immune deficiency syndrome (AIDS). Acta Neuropathol 82(1):39–44

78. Graus F, Ribalta T, Abos J, Alom J, Cruz-Sanchez F, Mallolas J, Miro JM, Cardesa A, Tolosa E (1990) Subacute cerebellar syndrome as the first manifestation of AIDS dementia complex. Acta Neurol Scand 81(2):118–120
79. Chang L, Ernst T, Witt MD, Ames N, Gaiefsky M, Miller E (2002) Relationships among brain metabolites, cognitive function, and viral loads in antiretroviral-naive HIV patients. Neuroimage 17(3):1638–1648
80. Harezlak J, Buchthal S, Taylor M, Schifitto G, Zhong J, Daar E, Alger J, Singer E, Campbell T, Yiannoutsos C, Cohen R, Navia B (2011) Persistence of HIV-associated cognitive impairment, inflammation, and neuronal injury in era of highly active antiretroviral treatment. AIDS 25(5):625–633
81. Meyerhoff DJ (2001) Effects of alcohol and HIV infection on the central nervous system. Alcohol Res Health 25:288–298
82. Meyerhoff DJ, Bloomer C, Cardenas V, Norman D, Weiner MW, Fein G (1999) Elevated subcortical choline metabolites in cognitively and clinically asymptomatic HIV+ patients. Neurology 52(5):995–1003
83. Pfefferbaum A, Rosenbloom M, Sullivan EV (2002) Alcoholism and AIDS: magnetic resonance imaging approaches for detecting interactive neuropathology. Alcohol Clin Exp Res 26(7):1031–1046
84. Everall IP, Luthert PJ, Lantos PL (1991) Neuronal loss in the frontal cortex in HIV infection. Lancet 337(8750):1119–1121
85. Everall IP, Luthert PJ, Lantos PL (1993) Neuronal number and volume alterations in the neocortex of HIV infected individuals. J Neurol Neurosurg Psychiatry 56(5):481–486
86. Ketzler S, Weis S, Haug H, Budka H (1990) Loss of neurons in the frontal cortex in AIDS brains. Acta Neuropathol (Berl) 80(1):92–94
87. Pulliam L, Rempel H, Sun B, Abadjian L, Calosing C, Meyerhoff DJ (2011) A peripheral monocyte interferon phenotype in HIV infection correlates with a decrease in magnetic resonance spectroscopy metabolite concentrations. AIDS 25(14):1721–1726
88. Wilkinson I, Paley M, Miszkiel K, Hall-Craggs M, Kendall B, Chinn R, Harrison M (1997) Cerebral volumes and spectroscopic proton metabolites on MR: is sex important? Magn Reson Imaging 15:243–248
89. Chang L, Ernst T, St Hillaire C, Conant K (2004) Antiretroviral treatment alters relationship between MCP-1 and neurometabolites in HIV patients. Antivir Ther 9(3):431–440
90. Chang L, Ernst T, Witt MD, Ames N, Walot I, Jovicich J, DeSilva M, Trivedi N, Speck O, Miller EN (2003) Persistent brain abnormalities in antiretroviral-naive HIV patients 3 months after HAART. Antivir Ther 8:17–26
91. Budka H (1991) The definition of HIV-specific neuropathology. Acta Pathol Jpn 41(3):182–191
92. Budka H (1997) Neuropathology of myelitis, myelopathy, and spinal infections in AIDS. Neuroimaging Clin N Am (Neuroimaging of AIDS II) 7(3):639–650
93. Ferrari CC, Depino AM, Prada F, Muraro N, Campbell S, Podhajcer O, Perry VH, Anthony DC, Pitossi FJ (2004) Reversible demyelination, blood–brain barrier breakdown, and pronounced neutrophil recruitment induced by chronic IL-1 expression in the brain. Am J Pathol 165(5):1827–1837
94. Giometto B, An SF, Groves M, Scaravilli T, Geddes JF, Miller R, Tavolato B, Beckett AA, Scaravilli F (1997) Accumulation of beta-amyloid precursor protein in HIV encephalitis: relationship with neuropsychological abnormalities. Ann Neurol 42(1):34–40
95. Gray F, Belec L, Chretien F, Dubreuil-Lemaire ML, Ricolfi F, Wingertsmann L, Poron F, Gherardi R (1998) Acute, relapsing brain oedema with diffuse blood–brain barrier alteration and axonal damage in the acquired immunodeficiency syndrome. Neuropathol Appl Neurobiol 24(3):209–216
96. Hartung HP, Jung S, Stoll G, Zielasek J, Schmidt B, Archelos JJ, Toyka KV (1992) Inflammatory mediators in demyelinating disorders of the CNS and PNS. J Neuroimmunol 40(2–3):197–210

97. Kim TS, Perlman S (2005) Viral expression of CCL2 is sufficient to induce demyelination in RAG1–/– mice infected with a neurotropic coronavirus. J Virol 79(11):7113–7120
98. Medana IM, Esiri MM (2003) Axonal damage: a key predictor of outcome in human CNS diseases. Brain 126(Pt 3):515–530
99. Raja F, Sherriff FE, Morris CS, Bridges LR, Esiri MM (1997) Cerebral white matter damage in HIV infection demonstrated using beta-amyloid precursor protein immunoreactivity. Acta Neuropathol 93(2):184–189
100. Arese M, Ferrandi C, Primo L, Camussi G, Bussolino F (2001) HIV-1 Tat protein stimulates in vivo vascular permeability and lymphomononuclear cell recruitment. J Immunol 166(2):1380–1388
101. McMurtray A, Nakamoto B, Shikuma C, Valcour V (2008) Cortical atrophy and white matter hyperintensities in HIV: the Hawaii Aging with HIV Cohort Study. J Stroke Cerebrovasc Dis 17(4):212–217
102. Ernst T, Chang L, Witt M, Walot I, Aronow H, Leonido-Yee M, Singer E (1999) Progressive multifocal leukoencephalopathy and human immunodeficiency virus-associated white matter lesions in AIDS: magnetization transfer MR imaging. Radiology 210(2):539–543
103. Thurnher MM, Post MJ, Rieger A, Kleibl-Popov C, Loewe C, Schindler E (2001) Initial and follow-up MR imaging findings in AIDS-related progressive multifocal leukoencephalopathy treated with highly active antiretroviral therapy. AJNR Am J Neuroradiol 22(5):977–984
104. Navia BA, Gonzalez RG (1997) Functional imaging of the AIDS dementia complex and the metabolic pathology of the HIV-1 infected brain. Neuroimaging Clin N Am 7(3):431–445
105. Chen Y, An H, Zhu H, Stone T, Smith JK, Hall C, Bullitt E, Shen D, Lin W (2009) White matter abnormalities revealed by diffusion tensor imaging in non-demented and demented HIV+ patients. Neuroimage 47(4):1154–1162
106. Cloak CC, Chang L, Ernst T (2004) Increased frontal white matter diffusion is associated with glial metabolites and psychomotor slowing in HIV. J Neuroimmunol 157(1–2):147–152
107. Filippi CG, Ulug AM, Ryan E, Ferrando SJ, van Gorp W (2001) Diffusion tensor imaging of patients with HIV and normal-appearing white matter on MR images of the brain. AJNR Am J Neuroradiol 22(2):277–283
108. Pomara N, Crandall DT, Choi SJ, Johnson G, Lim KO (2001) White matter abnormalities in HIV-1 infection: a diffusion tensor imaging study. Psychiatry Res 106:15–24
109. Thurnher MM, Castillo M, Stadler A, Rieger A, Schmid B, Sundgren PC (2005) Diffusion-tensor MR imaging of the brain in human immunodeficiency virus-positive patients. AJNR Am J Neuroradiol 26(9):2275–2281
110. Pfefferbaum A, Rosenbloom MJ, Rohlfing T, Kemper CA, Deresinski S, Sullivan EV (2009) Frontostriatal fiber bundle compromise in HIV infection without dementia. AIDS 23(15):1977–1985
111. Gongvatana A, Schweinsburg BC, Taylor MJ, Theilmann RJ, Letendre SL, Alhassoon OM, Jacobus J, Woods SP, Jernigan TL, Ellis RJ, Frank LR, Grant I (2009) White matter tract injury and cognitive impairment in human immunodeficiency virus-infected individuals. J Neurovirol 15(2):187–195
112. Pfefferbaum A, Rosenbloom MJ, Adalsteinsson E, Sullivan EV (2007) Diffusion tensor imaging with quantitative fiber tracking in HIV infection and alcoholism comorbidity: synergistic white matter damage. Brain 130:48–64
113. Ragin AB, Storey P, Cohen BA, Epstein LG, Edelman RR (2004) Whole brain diffusion tensor imaging in HIV-associated cognitive impairment. AJNR Am J Neuroradiol 25(2):195–200
114. Crabbe JC (2002) Alcohol and genetics: new models. Am J Med Genet 114(8):969–974
115. Porjesz B, Begleiter H, Reich T, Vaneerdewegh P, Edenberg H, Foroud T, Goate A, Litke A, Chorlian D, Stimus A, Rice J, Blangero J, Almasy L, Sorbell J, Bauer L, Kuperman S, O'Connor S, Rohrbaugh J (1998) Amplitude of visual P3 event-related potential as a phenotypic marker for a predisposition to alcoholism: preliminary-results from the COGA project. Alcohol Clin Exp Res 22(6):1317–1323
116. Ceballos NA, Nixon SJ, Phillips JA, Tivis R (2003) Semantic processing in alcoholics with and without antisocial symptomatology. J Stud Alcohol 64(2):286–291

117. Grant BF, Stinson FS, Dawson DA, Chou SP, Ruan WJ, Pickering RP (2004) Co-occurrence of 12-month alcohol and drug use disorders and personality disorders in the United States: results from the National Epidemiologic Survey on alcohol and related conditions. Arch Gen Psychiatry 61(4):361–368
118. Kessler RC, Chiu WT, Demler O, Walters EE (2005) Prevalence, severity, and comorbidity of 12-month DSM-IV disorders in the National Comorbidity Survey Replication. Arch Gen Psychiatry 62(6):617–627
119. Harding A, Halliday G, Caine D, Kril J (2000) Degeneration of anterior thalamic nuclei differentiates alcoholics with amnesia. Brain 123(Pt 1):141–154
120. Cullen KM, Halliday GM, Caine D, Kril JJ (1997) The nucleus basalis (Ch4) in the alcoholic Wernicke–Korsakoff syndrome: reduced cell number in both amnesic and non-amnesic patients. J Neurol Neurosurg Psychiatry 63(3):315–320
121. Baker K, Harding A, Halliday G, Kril J, Harper C (1999) Neuronal loss in functional zones of the cerebellum of chronic alcoholics with and without Wernicke's encephalopathy. Neuroscience 91(2):429–438
122. Harper C (1998) The neuropathology of alcohol-specific brain damage, or does alcohol damage the brain? J Neuropathol Exp Neurol 57(2):101–110
123. Harper CG, Kril JJ, Daly JM (1987) Are we drinking our neurones away? Br Med J 294:534–536
124. Kril JJ, Harper CG (1989) Neuronal counts from four cortical regions of the alcoholic brain. Acta Neuropathol (Berl) 79:200–204
125. Harper C, Dixon G, Sheedy D, Garrick T (2003) Neuropathological alterations in alcoholic brains. Studies arising from the New South Wales Tissue Resource Centre. Prog Neuropsychopharmacol Biol Psychiatry 27(6):951–961
126. Harding AJ, Wong A, Svoboda M, Kril JJ, Halliday GM (1997) Chronic alcohol consumption does not cause hippocampal neuron loss in humans. Hippocampus 7(1):78–87
127. Kril JJ, Halliday GM, Svoboda MD, Cartwright H (1997) The cerebral cortex is damaged in chronic alcoholics. Neuroscience 79(4):983–998
128. McMullen PA, Saint-Cyr JA, Carlen PL (1984) Morphological alterations in the rat CA1 hippocampal pyramidal cell dendrites resulting from chronic ethanol consumption and withdrawal. J Comp Neurol 225:111–118
129. Pentney R, Mullan B, Felong A, Dlugos C (2002) The total numbers of cerebellar granule neurons in young and aged Fischer 344 and Wistar-Kyoto rats do not change as a result of lengthy ethanol treatment. Cerebellum 1:79–89
130. Pentney RJ (1993) Alterations in the structure of the cerebellum after long-term ethanol consumption. In: Hunt WA, Nixon SJ (eds) Alcohol-induced brain damage: NIAAA Research Monograph No. 22. National Institute of Health, Rockville, pp 249–276
131. Phillips SC, Harper CG, Kril J (1987) A quantitative histological study of the cerebellar vermis in alcoholic patients. Brain 110:301–314
132. Tabbaa S, Dlugos C, Pentney R (1999) The number of granule cells and spine density on Purkinje cells in aged, ethanol-fed rats. Alcohol 17(3):253–260
133. Torvik A, Torp S (1986) The prevalence of alcoholic cerebellar atrophy: a morphometric and histological study of an autopsy material. J Neurol Sci 75:43–51
134. Harper C, Kril J (1994) An introduction to alcohol-induced brain damage and its causes. Alcohol Alcohol Suppl 2:237–243
135. Harper C, Corbett D (1990) Changes in the basal dendrites of cortical pyramidal cells from alcoholic patients – a quantitative golgi study. J Neurol Neurosurg Psychiatry 53(10):856–861
136. Ikegami Y, Goodenough S, Inoue Y, Dodd PR, Wilce PA, Matsumoto I (2003) Increased TUNEL positive cells in human alcoholic brains. Neurosci Lett 349(3):201–205
137. Oscar-Berman M, Marinkovic K (2007) Alcohol: effects on neurobehavioral functions and the brain. Neuropsychol Rev 17(3):239–257
138. Sullivan EV, Pfefferbaum A (2005) Neurocircuitry in alcoholism: a substrate of disruption and repair. Psychopharmacology (Berl) 180:583–594
139. Zahr NM, Kaufman KL, Harper CG (2011) Clinical and pathological features of alcohol-related brain damage. Nat Rev Neurol 7(5):284–294

140. Jernigan TL, Butters N, DiTraglia G, Schafer K, Smith T, Irwin M, Grant I, Schuckit M, Cermak L (1991) Reduced cerebral grey matter observed in alcoholics using magnetic resonance imaging. Alcohol Clin Exp Res 15(3):418–427
141. Pfefferbaum A, Lim KO, Zipursky RB, Mathalon DH, Rosenbloom MJ, Lane B, Ha CN, Sullivan EV (1992) Brain gray and white matter volume loss accelerates with aging in chronic alcoholics: a quantitative MRI study. Alcohol Clin Exp Res 16(6):1078–1089
142. De Bellis MD, Narasimhan A, Thatcher DL, Keshavan MS, Soloff P, Clark DB (2005) Prefrontal cortex, thalamus, and cerebellar volumes in adolescents and young adults with adolescent-onset alcohol use disorders and comorbid mental disorders. Alcohol Clin Exp Res 29(9):1590–1600
143. Sullivan EV, Deshmukh A, Desmond JE, Lim KO, Pfefferbaum A (2000) Cerebellar volume decline in normal aging, alcoholism, and Korsakoff's syndrome: relation to ataxia. Neuropsychology 14(3):341–352
144. Agartz I, Momenan R, Rawlings RR, Kerich MJ, Hommer DW (1999) Hippocampal volume in patients with alcohol dependence. Arch Gen Psychiatry 56:356–363
145. Sullivan EV, Marsh L, Mathalon DH, Lim KO, Pfefferbaum A (1995) Anterior hippocampal volume deficits in nonamnesic, aging chronic alcoholics. Alcohol Clin Exp Res 19:110–122
146. Davila MD, Shear PK, Lane B, Sullivan EV, Pfefferbaum A (1994) Mammillary body and cerebellar shrinkage in chronic alcoholics: an MRI and neuropsychological study. Neuropsychology 8:433–444
147. Shear PK, Sullivan EV, Lane B, Pfefferbaum A (1996) Mammillary body and cerebellar shrinkage in chronic alcoholics with and without amnesia. Alcohol Clin Exp Res 20(8):1489–1495
148. Sullivan EV, Deshmukh A, Desmond JE, Mathalon DH, Rosenbloom MJ, Lim KO, Pfefferbaum A (2000) Contribution of alcohol abuse to cerebellar volume deficits in men with schizophrenia. Arch Gen Psychiatry 57(9):894–902
149. Sullivan EV, Rosenbloom MJ, Serventi KL, Deshmukh A, Pfefferbaum A (2003) Effects of alcohol dependence comorbidity and anti-psychotic medication on volumes of the thalamus and pons in schizophrenia. Am J Psychiatry 160:1110–1116
150. Pfefferbaum A, Rosenbloom MJ, Serventi K, Sullivan EV (2002) Corpus callosum, pons and cortical white matter in alcoholic women. Alcohol Clin Exp Res 26:400–405
151. Sullivan EV (2003) Compromised pontocerebellar and cerebellothalamocortical systems: speculations on their contributions to cognitive and motor impairment in nonamnesic alcoholism. Alcohol Clin Exp Res 27(9):1409–1419
152. Bendszus M, Weijers HG, Wiesbeck G, Warmuth-Metz M, Bartsch AJ, Engels S, Boning J, Solymosi L (2001) Sequential MR imaging and proton MR spectroscopy in patients who underwent recent detoxification for chronic alcoholism: correlation with clinical and neuropsychological data. AJNR Am J Neuroradiol 22(10):1926–1932
153. Fein G, Meyerhoff D, Di Sclafani V, Ezekiel F, Poole N, MacKay S, Dillon WP, Constans J-M, Weiner MW (1994) 1H magnetic resonance spectroscopic imaging separates neuronal from glial changes in alcohol-related brain atrophy. In: Lancaster F (ed) Alcohol and glial cells, NIAAA Research Monograph # 27. US Government Printing Office, Bethesda, pp 227–241
154. Jagannathan NR, Desai NG, Raghunathan P (1996) Brain metabolite changes in alcoholism: an in vivo proton magnetic resonance spectroscopy (MRS) study. Magn Reson Imaging 14:553–557
155. Meyerhoff DJ, Blumenfeld R, Truran D, Lindgren J, Flenniken D, Cardenas V, Chao LL, Rothlind J, Studholme C, Weiner MW (2004) Effects of heavy drinking, binge drinking, and family history of alcoholism on regional brain metabolites. Alcohol Clin Exp Res 28(4):650–661
156. Schweinsburg B, Taylor M, Videen J, Alhassoon O, Patterson T, Grant I (2000) Elevated myo-inositol in gray matter of recently detoxified but not long-term abstinent alcoholics: a preliminary MR spectroscopy study. Alcohol Clin Exp Res 24:699–705
157. Seitz D, Widmann U, Seeger U, Nagele T, Klose U, Mann K, Grodd W (1999) Localized proton magnetic resonance spectroscopy of the cerebellum in detoxifying alcoholics. Alcohol Clin Exp Res 23(1):158–163

158. Parks MH, Dawant BM, Riddle WR, Hartmann SL, Dietrich MS, Nickel MK, Price RR, Martin PR (2002) Longitudinal brain metabolic characterization of chronic alcoholics with proton magnetic resonance spectroscopy. Alcohol Clin Exp Res 26(9):1368–1380
159. Biller A, Bartsch AJ, Homola G, Solymosi L, Bendszus M (2009) The effect of ethanol on human brain metabolites longitudinally characterized by proton MR spectroscopy. J Cereb Blood Flow Metab 29:891–902
160. Durazzo TC, Gazdzinski S, Banys P, Meyerhoff DJ (2004) Cigarette smoking exacerbates chronic alcohol-induced brain damage: a preliminary metabolite imaging study. Alcohol Clin Exp Res 28(12):1849–1860
161. Ende G, Walter S, Welzel H, Demirakca T, Wokrina T, Ruf M, Ulrich M, Diehl A, Henn FA, Mann K (2006) Alcohol consumption significantly influences the MR signal of frontal choline-containing compounds. Neuroimage 32(2):740–746
162. Li Z, Vance DE (2008) Phosphatidylcholine and choline homeostasis. J Lipid Res 49(6):1187–1194
163. Djuricic B, Olson SR, Assaf HM, Whittingham TS, Lust WD, Drewes LR (1991) Formation of free choline in brain tissue during in vitro energy deprivation. J Cereb Blood Flow Metab 11(2):308–313
164. Klein J, Koppen A, Loffelholz K, Schmitthenner J (1992) Uptake and metabolism of choline by rat brain after acute choline administration. J Neurochem 58(3):870–876
165. Scremin OU, Jenden DJ (1989) Focal ischemia enhances choline output and decreases acetylcholine output from rat cerebral cortex. Stroke 20(1):92–95
166. De Stefano N, Bartolozzi ML, Guidi L, Stromillo ML, Federico A (2005) Magnetic resonance spectroscopy as a measure of brain damage in multiple sclerosis. J Neurol Sci 233(1–2):203–208
167. Alling C, Bostrom K (1980) Demyelination of the mamillary bodies in alcoholism. A combined morphological and biochemical study. Acta Neuropathol (Berl) 50:77–80
168. Badsberg-Jensen G, Pakkenberg B (1993) Do alcoholics drink their neurons away? Lancet 342(8881):1201–1204
169. De la Monte SM (1988) Disproportionate atrophy of cerebral white matter in chronic alcoholics. Arch Neurol 45:990–992
170. Harper C, Kril JJ (1989) Patterns of neuronal loss in the cerebral cortex in chronic alcoholic patients. J Neurol Sci 92:81–89
171. Harper CG, Kril JJ (1988) Corpus callosal thickness in alcoholics. Br J Addict 83:577–580
172. Harper CG, Kril JJ, Holloway RL (1985) Brain shrinkage in chronic alcoholics: a pathological study. Br Med J 290:501–504
173. Hommer D, Momenan R, Rawlings R, Ragan P, Williams W, Rio D, Eckardt M (1996) Decreased corpus callosum size among alcoholic women. Arch Neurol 53(4):359–363
174. Estruch R, Nicolas JM, Salamero M, Aragon C, Sacanella E, Fernandez-Sola J, Urbano-Marquez A (1997) Atrophy of the corpus callosum in chronic alcoholism. J Neurol Sci 146(2):145–151
175. Hommer DW, Momenan R, Kaiser E, Rawlings RR (2001) Evidence for a gender-related effect of alcoholism on brain volumes. Am J Psychiatry 158(2):198–204
176. Pfefferbaum A, Lim KO, Desmond JE, Sullivan EV (1996) Thinning of the corpus callosum in older alcoholic men: a magnetic resonance imaging study. Alcohol Clin Exp Res 20(4):752–757
177. Pfefferbaum A, Lim KO, Rosenbloom MJ (1992) Structural imaging of the brain in chronic alcoholism. In: Zakhari S, Witt E (eds) Imaging in alcohol research: NIAAA Research Monograph No. 21. U.S. Dept Health and Human Services, Rockville, pp 99–120
178. Pfefferbaum A, Sullivan EV, Mathalon DH, Lim KO (1997) Frontal lobe volume loss observed with magnetic resonance imaging in older chronic alcoholics. Alcohol Clin Exp Res 21(3):521–529
179. Anstey KJ, Jorm AF, Reglade-Meslin C, Maller J, Kumar R, von Sanden C, Windsor TD, Rodgers B, Wen W, Sachdev P (2006) Weekly alcohol consumption, brain atrophy, and white matter hyperintensities in a community-based sample aged 60 to 64 years. Psychosom Med 68(5):778–785

180. den Heijer T, Vermeer SE, van Dijk EJ, Prins ND, Koudstaal PJ, van Duijn CM, Hofman A, Breteler MM (2004) Alcohol intake in relation to brain magnetic resonance imaging findings in older persons without dementia. Am J Clin Nutr 80(4):992–997
181. Fein G, Shimotsu R, Di Sclafani V, Barakos J, Harper C (2009) Increased white matter signal hyperintensities in long-term abstinent alcoholics compared with nonalcoholic controls. Alcohol Clin Exp Res 33(1):70–78
182. Mukamal KJ, Longstreth WT Jr, Mittleman MA, Crum RM, Siscovick DS (2001) Alcohol consumption and subclinical findings on magnetic resonance imaging of the brain in older audlts: the Cardiovascular Health Study. Stroke 32:1939–1946
183. Basile AM, Pantoni L, Pracucci G, Asplund K, Chabriat H, Erkinjuntti T, Fazekas F, Ferro JM, Hennerici M, O'Brien J, Scheltens P, Visser MC, Wahlund LO, Waldemar G, Wallin A, Inzitari D (2006) Age, hypertension, and lacunar stroke are the major determinants of the severity of age-related white matter changes. The LADIS (Leukoaraiosis and Disability in the Elderly) Study. Cerebrovasc Dis 21(5–6):315–322
184. Ovbiagele B, Saver JL (2006) Cerebral white matter hyperintensities on MRI: current concepts and therapeutic implications. Cerebrovasc Dis 22(2–3):83–90
185. Rossi R, Boccardi M, Sabattoli F, Galluzzi S, Alaimo G, Testa C, Frisoni GB (2006) Topographic correspondence between white matter hyperintensities and brain atrophy. J Neurol 253(7):919–927
186. Kiechl S, Willeit J, Rungger G, Egger G, Oberhollenzer F, Bonora E (1998) Alcohol consumption and atherosclerosis: what is the relation? Prospective results from the Bruneck Study. Stroke 29(5):900–907
187. Mukamal KJ, Kuller LH, Fitzpatrick AL, Longstreth WT Jr, Mittleman MA, Siscovick DS (2003) Prospective study of alcohol consumption and risk of dementia in older adults. JAMA 289(11):1405–1413
188. Rimm EB, Williams P, Fosher K, Criqui M, Stampfer MJ (1999) Moderate alcohol intake and lower risk of coronary heart disease: meta-analysis of effects on lipids and haemostatic factors. BMJ 319(7224):1523–1528
189. Ruitenberg A, van Swieten JC, Witteman JC, Mehta KM, van Duijn CM, Hofman A, Breteler MM (2002) Alcohol consumption and risk of dementia: the Rotterdam Study. Lancet 359(9303):281–286
190. de Leeuw FE, de Groot JC, Bots ML, Witteman JC, Oudkerk M, Hofman A, van Gijn J, Breteler MM (2000) Carotid atherosclerosis and cerebral white matter lesions in a population based magnetic resonance imaging study. J Neurol 247(4):291–296
191. Pfefferbaum A, Adalsteinsson E, Sullivan EV (2006) Supratentorial profile of white matter microstructural integrity in recovering alcoholic men and women. Biol Psychiatry 59(4):364–372
192. Pfefferbaum A, Sullivan EV (2002) Microstructural but not macrostructural disruption of white matter in women with chronic alcoholism. Neuroimage 15:708–718
193. Pfefferbaum A, Sullivan EV (2005) Disruption of brain white matter microstructure by excessive intracellular and extracellular fluid in alcoholism: evidence from diffusion tensor imaging. Neuropsychopharmacology 30(2):423–432
194. Pfefferbaum A, Sullivan EV, Hedehus M, Adalsteinsson E, Lim KO, Moseley M (2000) In vivo detection and functional correlates of white matter microstructural disruption in chronic alcoholism. Alcohol Clin Exp Res 24(8):1214–1221
195. Pfefferbaum A, Rosenbloom MJ, Fama R, Sassoon SA, Sullivan EV (2010) Transcallosal white matter degradation detected with quantitative fiber tracking in alcoholic men and women: selective relations to dissociable functions. Alcohol Clin Exp Res 34(7):1201–1211
196. Song SK, Yoshino J, Le TQ, Lin SJ, Sun SW, Cross AH, Armstrong RC (2005) Demyelination increases radial diffusivity in corpus callosum of mouse brain. Neuroimage 26(1):132–140
197. Pfefferbaum A, Rosenbloom M, Rohlfing T, Sullivan EV (2009) Degradation of association and projection white matter systems in alcoholism detected with quantitative fiber tracking. Biol Psychiatry 65(8):680–690
198. Lewohl J, Wang L, Miles M, Zhang L, Dodd P, Harris R (2000) Gene expression in human alcoholism: microarray analysis of frontal cortex. Alcohol Clin Exp Res 24:1873–1882

199. Mayfield RD, Lewohl JM, Dodd PR, Herlihy A, Liu J, Harris RA (2002) Patterns of gene expression are altered in the frontal and motor cortices of human alcoholics. J Neurochem 81(4):802–813
200. Hasin DS, Liu X, Alderson D, Grant BF (2006) DSM-IV alcohol dependence: a categorical or dimensional phenotype? Psychol Med 36(12):1695–16705
201. Lewohl JM, Wixey J, Harper CG, Dodd PR (2005) Expression of MBP, PLP, MAG, CNP, and GFAP in the human alcoholic brain. Alcohol Clin Exp Res 29(9):1698–1705
202. Qin L, He J, Hanes RN, Pluzarev O, Hong JS, Crews FT (2008) Increased systemic and brain cytokine production and neuroinflammation by endotoxin following ethanol treatment. J Neuroinflammation 5:10
203. Qin L, Liu Y, Wang T, Wei SJ, Block ML, Wilson B, Liu B, Hong JS (2004) NADPH oxidase mediates lipopolysaccharide-induced neurotoxicity and proinflammatory gene expression in activated microglia. J Biol Chem 279(2):1415–1421
204. Valles SL, Blanco AM, Pascual M, Guerri C (2004) Chronic ethanol treatment enhances inflammatory mediators and cell death in the brain and in astrocytes. Brain Pathol 14(4):365–371
205. Knapp DJ, Crews FT (1999) Induction of cyclooxygenase-2 in brain during acute and chronic ethanol treatment and ethanol withdrawal. Alcohol Clin Exp Res 23(4):633–643
206. Choi DK, Lee H, Jeong J, Lim B, Suk K (2005) Differential effects of ethanol on glial signal transduction initiated by lipopolysaccharide and interferon-gamma. J Neurosci Res 82(2):225–231
207. Deaciuc IV (1997) Alcohol and cytokine networks. Alcohol 14(5):421–430
208. Fleming S, Toratani S, Shea-Donohue T, Kashiwabara Y, Vogel SN, Metcalf ES (2001) Pro- and anti-inflammatory gene expression in the murine small intestine and liver after chronic exposure to alcohol. Alcohol Clin Exp Res 25(4):579–589
209. Frost RA, Nystrom G, Burrows PV, Lang CH (2005) Temporal differences in the ability of ethanol to modulate endotoxin-induced increases in inflammatory cytokines in muscle under in vivo conditions. Alcohol Clin Exp Res 29(7):1247–1256
210. Jarvelainen HA, Fang C, Ingelman-Sundberg M, Lindros KO (1999) Effect of chronic coadministration of endotoxin and ethanol on rat liver pathology and proinflammatory and anti-inflammatory cytokines. Hepatology 29(5):1503–1510
211. Leevy CB, Elbeshbeshy HA (2005) Immunology of alcoholic liver disease. Clin Liver Dis 9(1):55–66
212. McClain CJ, Song Z, Barve SS, Hill DB, Deaciuc I (2004) Recent advances in alcoholic liver disease. IV. Dysregulated cytokine metabolism in alcoholic liver disease. Am J Physiol Gastrointest Liver Physiol 287(3):G497–G502
213. Tilg H, Jalan R, Kaser A, Davies NA, Offner FA, Hodges SJ, Ludwiczek O, Shawcross D, Zoller H, Alisa A, Mookerjee RP, Graziadei I, Datz C, Trauner M, Schuppan D, Obrist P, Vogel W, Williams R (2003) Anti-tumor necrosis factor-alpha monoclonal antibody therapy in severe alcoholic hepatitis. J Hepatol 38(4):419–425
214. Karuppagounder SS, Shi Q, Xu H, Gibson GE (2007) Changes in inflammatory processes associated with selective vulnerability following mild impairment of oxidative metabolism. Neurobiol Dis 26(2):353–362
215. Todd KG, Butterworth RF (1999) Early microglial response in experimental thiamine deficiency: an immunohistochemical analysis. Glia 25(2):190–198
216. Vemuganti R, Kalluri H, Yi JH, Bowen KK, Hazell AS (2006) Gene expression changes in thalamus and inferior colliculus associated with inflammation, cellular stress, metabolism and structural damage in thiamine deficiency. Eur J Neurosci 23(5):1172–1188
217. He J, Crews FT (2008) Increased MCP-1 and microglia in various regions of the human alcoholic brain. Exp Neurol 210(2):349–358
218. Fernandez-Lizarbe S, Pascual M, Guerri C (2009) Critical role of TLR4 response in the activation of microglia induced by ethanol. J Immunol 183(7):4733–4744
219. Mandrekar P, Dolganiuc A, Bellerose G, Kodys K, Romics L, Nizamani R, Szabo G (2002) Acute alcohol inhibits the induction of nuclear regulatory factor kappa B activation through CD14/toll-like receptor 4, interleukin-1, and tumor necrosis factor receptors: a common

mechanism independent of inhibitory kappa B alpha degradation? Alcohol Clin Exp Res 26(11):1609–1614
220. Wu Y, Lousberg EL, Moldenhauer LM, Hayball JD, Coller JK, Rice KC, Watkins LR, Somogyi AA, Hutchinson MR (2012) Inhibiting the TLR4-MyD88 signalling cascade by genetic or pharmacological strategies reduces acute alcohol-induced sedation and motor impairment in mice. Br J Pharmacol 165(5):1319–1329
221. Pascual M, Balino P, Alfonso-Loeches S, Aragon CM, Guerri C (2011) Impact of TLR4 on behavioral and cognitive dysfunctions associated with alcohol-induced neuroinflammatory damage. Brain Behav Immun 25(Suppl 1):S80–S91
222. Alfonso-Loeches S, Pascual M, Gomez-Pinedo U, Pascual-Lucas M, Renau-Piqueras J, Guerri C (2012) Toll-like receptor 4 participates in the myelin disruptions associated with chronic alcohol abuse. Glia 60(6):948–964
223. Li X, Jiang S, Tapping RI (2010) Toll-like receptor signaling in cell proliferation and survival. Cytokine 49(1):1–9
224. Mao XR, Moerman-Herzog AM, Chen Y, Barger SW (2009) Unique aspects of transcriptional regulation in neurons – nuances in NFkappaB and Sp1-related factors. J Neuroinflammation 6:16
225. Ghosh S, May MJ, Kopp EB (1998) NF-kappa B and Rel proteins: evolutionarily conserved mediators of immune responses. Annu Rev Immunol 16:225–260
226. Pahl HL (1999) Activators and target genes of Rel/NF-kappaB transcription factors. Oncogene 18(49):6853–6866
227. Meffert MK, Chang JM, Wiltgen BJ, Fanselow MS, Baltimore D (2003) NF-kappa B functions in synaptic signaling and behavior. Nat Neurosci 6(10):1072–1078
228. Mattson MP, Culmsee C, Yu Z, Camandola S (2000) Roles of nuclear factor kappaB in neuronal survival and plasticity. J Neurochem 74(2):443–456
229. Crews F, Nixon K, Kim D, Joseph J, Shukitt-Hale B, Qin L, Zou J (2006) BHT blocks NF-kappaB activation and ethanol-induced brain damage. Alcohol Clin Exp Res 30(11):1938–1949
230. Ward RJ, Zhang Y, Crichton RR, Piret B, Piette J, de Witte P (1996) Identification of the nuclear transcription factor NFkappaB in rat after in vivo ethanol administration. FEBS Lett 389(2):119–122
231. Zou J, Crews F (2006) CREB and NF-kappaB transcription factors regulate sensitivity to excitotoxic and oxidative stress induced neuronal cell death. Cell Mol Neurobiol 26:385–405
232. Zou J, Crews F (2010) Induction of innate immune gene expression cascades in brain slice cultures by ethanol: key role of NF-kappaB and proinflammatory cytokines. Alcohol Clin Exp Res 34(5):777–789
233. Albensi BC, Mattson MP (2000) Evidence for the involvement of TNF and NF-kappaB in hippocampal synaptic plasticity. Synapse 35(2):151–159
234. Kaltschmidt B, Ndiaye D, Korte M, Pothion S, Arbibe L, Prullage M, Pfeiffer J, Lindecke A, Staiger V, Israel A, Kaltschmidt C, Memet S (2006) NF-kappaB regulates spatial memory formation and synaptic plasticity through protein kinase A/CREB signaling. Mol Cell Biol 26(8):2936–2946
235. Wang J, Fu XQ, Lei WL, Wang T, Sheng AL, Luo ZG (2010) Nuclear factor kappaB controls acetylcholine receptor clustering at the neuromuscular junction. J Neurosci 30(33):11104–11113
236. Obernier JA, White AM, Swartzwelder HS, Crews FT (2002) Cognitive deficits and CNS damage after a 4-day binge ethanol exposure in rats. Pharmacol Biochem Behav 72(3):521–532
237. Cagnin A, Gerhard A, Banati RB (2002) In vivo imaging of neuroinflammation. Eur Neuropsychopharmacol 12(6):581–586
238. Ward RJ, Colivicchi MA, Allen R, Schol F, Lallemand F, de Witte P, Ballini C, Corte LD, Dexter D (2009) Neuro-inflammation induced in the hippocampus of 'binge drinking' rats may be mediated by elevated extracellular glutamate content. J Neurochem 111(5):1119–1128

239. Riikonen J, Jaatinen P, Rintala J, Porsti I, Karjala K, Hervonen A (2002) Intermittent ethanol exposure increases the number of cerebellar microglia. Alcohol Alcohol 37(5):421–426
240. Alfonso-Loeches S, Pascual-Lucas M, Blanco AM, Sanchez-Vera I, Guerri C (2010) Pivotal role of TLR4 receptors in alcohol-induced neuroinflammation and brain damage. J Neurosci 30(24):8285–8295
241. Ito D, Imai Y, Ohsawa K, Nakajima K, Fukuuchi Y, Kohsaka S (1998) Microglia-specific localisation of a novel calcium binding protein, Iba1. Brain Res Mol Brain Res 57(1):1–9
242. Lawson LJ, Perry VH, Dri P, Gordon S (1990) Heterogeneity in the distribution and morphology of microglia in the normal adult mouse brain. Neuroscience 39(1):151–170
243. Milligan CE, Cunningham TJ, Levitt P (1991) Differential immunochemical markers reveal the normal distribution of brain macrophages and microglia in the developing rat brain. J Comp Neurol 314(1):125–135
244. Savchenko VL, McKanna JA, Nikonenko IR, Skibo GG (2000) Microglia and astrocytes in the adult rat brain: comparative immunocytochemical analysis demonstrates the efficacy of lipocortin 1 immunoreactivity. Neuroscience 96(1):195–203
245. Shapiro LA, Wang L, Ribak CE (2008) Rapid astrocyte and microglial activation following pilocarpine-induced seizures in rats. Epilepsia 49(Suppl 2):33–41
246. Conde JR, Streit WJ (2006) Microglia in the aging brain. J Neuropathol Exp Neurol 65(3):199–203
247. Zahr NM, Luong R, Sullivan EV, Pfefferbaum A (2010) Measurement of serum, liver, and brain cytokine induction, thiamine levels, and hepatopathology in rats exposed to a 4-day alcohol binge protocol. Alcohol Clin Exp Res 34(11):1858–1870
248. O'Connor JJ, Coogan AN (1999) Actions of the pro-inflammatory cytokine IL-1 beta on central synaptic transmission. Exp Physiol 84(4):601–614
249. Ross FM, Allan SM, Rothwell NJ, Verkhratsky A (2003) A dual role for interleukin-1 in LTP in mouse hippocampal slices. J Neuroimmunol 144(1–2):61–67
250. Semple BD, Kossmann T, Morganti-Kossmann MC (2010) Role of chemokines in CNS health and pathology: a focus on the CCL2/CCR2 and CXCL8/CXCR2 networks. J Cereb Blood Flow Metab 30(3):459–473
251. Szelenyi J, Kiss JP, Puskas E, Szelenyi M, Vizi ES (2000) Contribution of differently localized alpha 2- and beta-adrenoceptors in the modulation of TNF-alpha and IL-10 production in endotoxemic mice. Ann N Y Acad Sci 917:145–153
252. Pickering M, Cumiskey D, O'Connor JJ (2005) Actions of TNF-alpha on glutamatergic synaptic transmission in the central nervous system. Exp Physiol 90(5):663–670
253. Beattie EC, Stellwagen D, Morishita W, Bresnahan JC, Ha BK, Von Zastrow M, Beattie MS, Malenka RC (2002) Control of synaptic strength by glial TNFalpha. Science 295(5563):2282–2285
254. Stellwagen D, Beattie EC, Seo JY, Malenka RC (2005) Differential regulation of AMPA receptor and GABA receptor trafficking by tumor necrosis factor-alpha. J Neurosci 25(12):3219–3228
255. Stellwagen D, Malenka RC (2006) Synaptic scaling mediated by glial TNF-alpha. Nature 440(7087):1054–1059
256. Rabbitt P, Scott M, Lunn M, Thacker N, Lowe C, Pendleton N, Horan M, Jackson A (2007) White matter lesions account for all age-related declines in speed but not in intelligence. Neuropsychology 21(3):363–370
257. Kramer JH, Blusewicz MJ, Preston KA (1989) The premature aging hypothesis: old before its time? J Consult Clin Psychol 57(2):257–262
258. Reuter-Lorenz PA, Stanczak L, Miller AC (1999) Neural recruitment and cognitive aging: two hemispheres are better than one, especially as you age. Psychol Sci 10:494–500
259. Blatter DD, Bigler ED, Gale SD, Johnson SC, Anderson C, Burnett BM, Parker N, Kurth S, Horn S (1995) Quantitative volumetric analysis of brain MRI: normative database spanning five decades of life. AJNR Am J Neuroradiol 16(2):241–245
260. Courchesne E, Chisum HJ, Townsend J, Cowles A, Covington J, Egaas B, Harwood M, Hinds S, Press GA (2000) Normal brain development and aging: quantitative analysis at in vivo MR imaging in healthy volunteers. Radiology 216(3):672–682

261. Good CD, Johnsrude IS, Ashburner J, Henson RN, Friston KJ, Frackowiak RS (2001) A voxel-based morphometric study of ageing in 465 normal adult brains. Neuroimage 14:21–36
262. Pfefferbaum A, Mathalon DH, Sullivan EV, Rawles JM, Zipursky RB, Lim KO (1994) A quantitative magnetic resonance imaging study of changes in brain morphology from infancy to late adulthood. Arch Neurol 51:874–887
263. Raz N, Gunning FM, Head D, Dupuis JH, McQuain J, Briggs SD, Loken WJ, Thornton AE, Acker JD (1997) Selective aging of the human cerebral cortex observed in vivo: differential vulnerability of the prefrontal gray matter. Cereb Cortex 7(3):268–282
264. Sowell ER, Thompson PM, Toga AW (2004) Mapping changes in the human cortex throughout the span of life. Neuroscientist 10(4):372–392
265. Sullivan EV, Rosenbloom MJ, Serventi KL, Pfefferbaum A (2004) Effects of age and sex on volumes of the thalamus, pons, and cortex. Neurobiol Aging 25:185–192
266. Taki Y, Kinomura S, Sato K, Goto R, Inoue K, Okada K, Ono S, Kawashima R, Fukuda H (2006) Both global gray matter volume and regional gray matter volume negatively correlate with lifetime alcohol intake in non-alcohol-dependent Japanese men: a volumetric analysis and a voxel-based morphometry. Alcohol Clin Exp Res 30(6):1045–1050
267. Liu RS, Lemieux L, Bell GS, Sisodiya SM, Shorvon SD, Sander JW, Duncan JS (2003) A longitudinal study of brain morphometrics using quantitative magnetic resonance imaging and difference image analysis. Neuroimage 20(1):22–33
268. Pfefferbaum A, Sullivan EV, Rosenbloom MJ, Mathalon DH, Lim KO (1998) A controlled study of cortical gray matter and ventricular changes in alcoholic men over a five year interval. Arch Gen Psychiatry 55(10):905–912
269. Raz N, Rodrigue KM, Kennedy KM, Acker JD (2007) Vascular health and longitudinal changes in brain and cognition in middle-aged and older adults. Neuropsychology 21(2):149–157
270. Resnick SM, Pham DL, Kraut MA, Zonderman AB, Davatzikos C (2003) Longitudinal magnetic resonance imaging studies of older adults: a shrinking brain. J Neurosci 23(8):3295–3301
271. Raz N, Gunning-Dixon F, Head D, Rodrigue K, Williamson A, Acker J (2004) Aging, sexual dimorphism, and hemispheric asymmetry of the cerebral cortex: replicability of regional differences in volume. Neurobiol Aging 25(3):377–396
272. Tang Y, Whitman GT, Lopez I, Baloh RW (2001) Brain volume changes on longitudinal magnetic resonance imaging in normal older people. J Neuroimaging 11(4):393–400
273. Luft AR, Skalej M, Schulz JB, Welte D, Kolb R, Burk K, Klockgether T, Voight K (1999) Patterns of age-related shrinkage in cerebellum and brainstem observed in vivo using three-dimensional MRI volumetry. Cereb Cortex 9(7):712–721
274. Oguro H, Okada K, Yamaguchi S, Kobayashi S (1998) Sex differences in morphology of the brain stem and cerebellum with normal ageing. Neuroradiology 40(12):788–792
275. Raz N, Lindenberger U, Rodrigue KM, Kennedy KM, Head D, Williamson A, Dahle C, Gerstorf D, Acker JD (2005) Regional brain changes in aging healthy adults: general trends, individual differences, and modifiers. Cereb Cortex 15:1676–1689
276. Sullivan EV, Deshmukh A, De Rosa E, Rosenbloom MJ, Pfefferbaum A (2005) Striatal and forebrain nuclei volumes: contribution to motor function and working memory deficits in alcoholism. Biol Psychiatry 57(7):768–776
277. Jernigan TL, Archibald SL, Fennema-Notestine C, Gamst AC, Stout JC, Bonner J, Hesselink JR (2001) Effects of age on tissues and regions of the cerebrum and cerebellum. Neurobiol Aging 22(4):581–594
278. Walhovd KB, Fjell AM, Reinvang I, Lundervold A, Dale AM, Eilertsen DE, Quinn BT, Salat D, Makris N, Fischl B (2005) Effects of age on volumes of cortex, white matter and subcortical structures. Neurobiol Aging 26(9):1261–1270, discussion 1275–1278
279. Driesen NR, Raz N (1995) The influence of sex, age, and handedness on corpus callosum morphology: a meta-analysis. Psychobiology 23(3):240–247

280. Raz N, Gunning-Dixon F, Head D, Williamson A, Acker JD (2001) Age and sex differences in the cerebellum and the ventral pons: a prospective MR study of healthy adults. AJNR Am J Neuroradiol 22(6):1161–1167
281. Sullivan EV, Pfefferbaum A, Adalsteinsson E, Swan GE, Carmelli D (2002) Differential rates of regional change in callosal and ventricular size: a 4-year longitudinal MRI study of elderly men. Cereb Cortex 12:438–445
282. Van Petten C (2004) Relationship between hippocampal volume and memory ability in healthy individuals across the lifespan: review and meta-analysis. Neuropsychologia 42:1394–1413
283. Sullivan EV, Marsh L, Mathalon DH, Lim KO, Pfefferbaum A (1995) Age-related decline in MRI volumes of temporal lobe gray matter but not hippocampus. Neurobiol Aging 16(4):591–606
284. Sullivan EV, Marsh L, Pfefferbaum A (2005) Preservation of hippocampal volume through adulthood in healthy men and women. Neurobiol Aging 26:1093–1098
285. Sullivan EV, Pfefferbaum A, Swan GE, Carmelli D (2001) Heritability of hippocampal size in elderly twin men: equivalent influence from genes and environment. Hippocampus 11:754–762
286. Golomb J, De Leon MJ, Kluger A, George AE, Tarshish C, Ferris SH (1993) Hippocampal atrophy in normal aging: an association with recent memory impairment. Arch Neurol 50(9):967–973
287. Braak H, Braak E (1994) Morphological criteria for the recognition of Alzheimer's disease and the distribution pattern of cortical changes related to this disorder. Neurobiol Aging 15:355–356
288. Hyman BT, Gomezisla T (1994) Alzheimer's disease is a laminar, regional, and neural system specific disease, not a global brain disease. Neurobiol Aging 15(3):353–354
289. Lippa CF, Hamos JE, Pulaski-Salo D, DeGennaro LJ, Drachman DA (1992) Alzheimer's disease and aging: effects on perforant pathway perikarya and synapses. Neurobiol Aging 13(3):405–411
290. West MJ, Coleman PD, Flood DG, Troncoso JC (1994) Differences in the pattern of hippocampal neuronal loss in normal ageing and Alzheimer's disease. Lancet 344(8925):769–772
291. Adalsteinsson E, Sullivan EV, Kleinhans N, Spielman DM, Pfefferbaum A (2000) Longitudinal decline of the neuronal marker N-acetyl aspartate in Alzheimer's disease. Lancet 355:1696–1697
292. Grachev ID, Apkarian AV (2000) Chemical heterogeneity of the living human brain: a proton MR spectroscopy study on the effects of sex, age, and brain region. Neuroimage 11(5 Pt 1): 554–563
293. Kwo-On-Yuen PF, Newmark RD, Budinger TF, Kaye JA, Ball MJ, Jagust WJ (1994) Brain N-acetyl-L-aspartic acid in Alzheimer's disease: a proton magnetic resonance spectroscopy study. Brain Res 667(2):167–174
294. Pfefferbaum A, Adalsteinsson E, Spielman D, Sullivan EV, Lim KO (1999) In vivo spectroscopic quantification of the N-acetyl moiety, creatine, and choline from large volumes of brain gray and white matter: effects of normal aging. Magn Reson Med 41(2):276–284
295. Saunders D, Howe F, van den Boogaart A, Griffiths J, Brown M (1999) Aging of the adult human brain: in vivo quantitation of metabolite content with proton magnetic resonance spectroscopy. J Magn Reson Imaging 9(5):711–716
296. Sijens PE, den Heijer T, Origgi D, Vermeer SE, Breteler MM, Hofman A, Oudkerk M (2003) Brain changes with aging: MR spectroscopy at supraventricular plane shows differences between women and men. Radiology 226(3):889–896
297. Kaiser LG, Schuff N, Cashdollar N, Weiner MW (2005) Age-related glutamate and glutamine concentration changes in normal human brain: 1H MR spectroscopy study at 4T. Neurobiol Aging 26(5):665–672
298. Chang L, Ernst T, Poland RE, Jenden DJ (1996) In vivo proton magnetic resonance spectroscopy of the normal aging human brain. Life Sci 58(22):2049–2056

299. Leary SM, Brex PA, MacManus DG, Parker GJ, Barker GJ, Miller DH, Thompson AJ (2000) A (1)H magnetic resonance spectroscopy study of aging in parietal white matter: implications for trials in multiple sclerosis. Magn Reson Imaging 18(4):455–459
300. Meyerhoff DJ, MacKay S, Constans JM, Norman D, Van Dyke C, Fein G, Weiner MW (1994) Axonal injury and membrane alterations in Alzheimer's disease suggested by in vivo proton magnetic resonance spectroscopic imaging. Ann Neurol 36(1):40–47
301. Costa MO, Lacerda MT, Garcia Otaduy MC, Cerri GG, Da Costa LC (2002) Proton magnetic resonance spectroscopy: normal findings in the cerebellar hemisphere in childhood. Pediatr Radiol 32(11):787–792
302. Fazekas F, Niederkorn K, Schmidt R, Offenbacher H, Horner S, Bertha G, Lechner H (1988) White matter signal abnormalities in normal individuals: correlation with carotid ultrasonography, cerebral blood flow measurements, and cerebrovascular risk factors. Stroke 19(10):1285–1288
303. Kuller LH, Longstreth WT Jr, Arnold AM, Bernick C, Bryan RN, Beauchamp NJ Jr (2004) White matter hyperintensity on cranial magnetic resonance imaging: a predictor of stroke. Stroke 35(8):1821–1825
304. Longstreth WT, Manolio TA, Arnold A, Burke GL, Bryan N, Jungreis CA, Enright PL, O'Leary D, Fried L (1996) Clinical correlates of white matter findings on cranial magnetic resonance imaging of 3301 elderly people: the cardiovascular health study. Stroke 27(8):1274–1282
305. Hachinski VC, Potter P, Merskey H (1987) Leuko-araiosis. Arch Neurol 44(1):21–23
306. Inzitari D (2003) Leukoaraiosis: an independent risk factor for stroke? Stroke 34(8):2067–2071
307. Roman GC, Erkinjuntti T, Wallin A, Pantoni L, Chui HC (2002) Subcortical ischaemic vascular dementia. Lancet Neurol 1(7):426–436
308. Young VG, Halliday GM, Kril JJ (2008) Neuropathologic correlates of white matter hyperintensities. Neurology 71(11):804–811
309. Marcus J, Gardener H, Rundek T, Elkind MS, Sacco RL, Decarli C, Wright CB (2011) Baseline and longitudinal increases in diastolic blood pressure are associated with greater white matter hyperintensity volume: the northern Manhattan study. Stroke 42(9):2639–2641
310. Raz N, Yang Y, Dahle CL, Land S (2012) Volume of white matter hyperintensities in healthy adults: contribution of age, vascular risk factors, and inflammation-related genetic variants. Biochim Biophys Acta 1822(3):361–369
311. Schmidt R, Grazer A, Enzinger C, Ropele S, Homayoon N, Pluta-Fuerst A, Schwingenschuh P, Katschnig P, Cavalieri M, Schmidt H, Langkammer C, Ebner F, Fazekas F (2011) MRI-detected white matter lesions: do they really matter? J Neural Transm 118(5):673–681
312. Bhagat YA, Beaulieu C (2004) Diffusion anisotropy in subcortical white matter and cortical gray matter: changes with aging and the role of CSF-suppression. J Magn Reson Imaging 20(2):216–227
313. Head D, Buckner RL, Shimony JS, Williams LE, Akbudak E, Conturo TE, McAvoy M, Morris JC, Snyder AZ (2004) Differential vulnerability of anterior white matter in nondemented aging with minimal acceleration in dementia of the Alzheimer type: evidence from diffusion tensor imaging. Cereb Cortex 14(4):410–423
314. Pfefferbaum A, Adalsteinsson E, Sullivan EV (2005) Frontal circuitry degradation marks healthy adult aging: evidence from diffusion tensor imaging. Neuroimage 26(3):891–899
315. Pfefferbaum A, Sullivan EV (2003) Increased brain white matter diffusivity in normal adult aging: relationship to anisotropy and partial voluming. Magn Reson Med 49:953–961
316. Pfefferbaum A, Sullivan EV, Hedehus M, Lim KO, Adalsteinsson E, Moseley M (2000) Age-related decline in brain white matter anisotropy measured with spatially corrected echo-planar diffusion tensor imaging. Magn Reson Med 44(2):259–268
317. Salat DH, Tuch DS, Greve DN, van der Kouwe AJ, Hevelone ND, Zaleta AK, Rosen BR, Fischl B, Corkin S, Rosas HD, Dale AM (2005) Age-related alterations in white matter microstructure measured by diffusion tensor imaging. Neurobiol Aging 26(8):1215–1227

318. Sullivan EV, Adalsteinsson E, Hedehus M, Ju C, Moseley M, Lim KO, Pfefferbaum A (2001) Equivalent disruption of regional white matter microstructure in aging healthy men and women. Neuroreport 12(22):99–104
319. Sullivan EV, Rohlfing T, Pfefferbaum A (2010) Quantitative fiber tracking of lateral and interhemispheric white matter systems in normal aging: relations to timed performance. Neurobiol Aging 31:464–481
320. Angelie E, Bonmartin A, Boudraa A, Gonnaud PM, Mallet JJ, Sappey-Marinier D (2001) Regional differences and metabolic changes in normal aging of the human brain: proton MR spectroscopic imaging study. AJNR Am J Neuroradiol 22(1):119–127
321. Soher BJ, Hurd RE, Sailasuta N, Barker PB (1996) Quantitation of automated single-voxel proton MRS using cerebral water as an internal reference. Magn Reson Med 36(3):335–339
322. Moats RA, Ernst T, Shonk TK, Ross BD (1994) Abnormal cerebral metabolite concentrations in patients with probable Alzheimer's disease. Magn Reson Med 32:110–115
323. Moreno-Torres A, Pujol J, Soriano-Mas C, Deus J, Iranzo A, Santamaria J (2005) Age-related metabolic changes in the upper brainstem tegmentum by MR spectroscopy. Neurobiol Aging 26(7):1051–1059
324. Arfanakis K, Haughton VM, Carew JD, Rogers BP, Dempsey RJ, Meyerand ME (2002) Diffusion tensor MR imaging in diffuse axonal injury. AJNR Am J Neuroradiol 23:794–802
325. Peters A, Sethares C (2002) Aging and the myelinated fibers in prefrontal cortex and corpus callosum of the monkey. J Comp Neurol 442(3):277–291
326. Peters A, Sethares C (2003) Is there remyelination during aging of the primate central nervous system? J Comp Neurol 460(2):238–254
327. Peters A, Sethares C, Killiany RJ (2001) Effects of age on the thickness of myelin sheaths in monkey primary visual cortex. J Comp Neurol 435(2):241–248
328. Wang DS, Bennett DA, Mufson EJ, Mattila P, Cochran E, Dickson DW (2004) Contribution of changes in ubiquitin and myelin basic protein to age-related cognitive decline. Neurosci Res 48(1):93–100
329. Dilger RN, Johnson RW (2008) Aging, microglial cell priming, and the discordant central inflammatory response to signals from the peripheral immune system. J Leukoc Biol 84(4):932–939
330. Long JM, Kalehua AN, Muth NJ, Calhoun ME, Jucker M, Hengemihle JM, Ingram DK, Mouton PR (1998) Stereological analysis of astrocyte and microglia in aging mouse hippocampus. Neurobiol Aging 19(5):497–503
331. Rogers J, Luber-Narod J, Styren SD, Civin WH (1988) Expression of immune system-associated antigens by cells of the human central nervous system: relationship to the pathology of Alzheimer's disease. Neurobiol Aging 9(4):339–349
332. Streit WJ, Sparks DL (1997) Activation of microglia in the brains of humans with heart disease and hypercholesterolemic rabbits. J Mol Med (Berl) 75(2):130–138
333. Sheffield LG, Berman NE (1998) Microglial expression of MHC class II increases in normal aging of nonhuman primates. Neurobiol Aging 19(1):47–55
334. Sloane JA, Hollander W, Moss MB, Rosene DL, Abraham CR (1999) Increased microglial activation and protein nitration in white matter of the aging monkey. Neurobiol Aging 20(4):395–405
335. Ogura K, Ogawa M, Yoshida M (1994) Effects of ageing on microglia in the normal rat brain: immunohistochemical observations. Neuroreport 5(10):1224–1226
336. Perry VH, Matyszak MK, Fearn S (1993) Altered antigen expression of microglia in the aged rodent CNS. Glia 7(1):60–67
337. Godbout JP, Chen J, Abraham J, Richwine AF, Berg BM, Kelley KW, Johnson RW (2005) Exaggerated neuroinflammation and sickness behavior in aged mice following activation of the peripheral innate immune system. FASEB J 19(10):1329–1331
338. Williams K, Ulvestad E, Waage A, Antel JP, McLaurin J (1994) Activation of adult human derived microglia by myelin phagocytosis in vitro. J Neurosci Res 38(4):433–443

339. Mosley K, Cuzner ML (1996) Receptor-mediated phagocytosis of myelin by macrophages and microglia: effect of opsonization and receptor blocking agents. Neurochem Res 21(4):481–487
340. Chen J, Buchanan JB, Sparkman NL, Godbout JP, Freund GG, Johnson RW (2008) Neuroinflammation and disruption in working memory in aged mice after acute stimulation of the peripheral innate immune system. Brain Behav Immun 22(3):301–311
341. Barrientos RM, Higgins EA, Biedenkapp JC, Sprunger DB, Wright-Hardesty KJ, Watkins LR, Rudy JW, Maier SF (2006) Peripheral infection and aging interact to impair hippocampal memory consolidation. Neurobiol Aging 27(5):723–732
342. DiPatre PL, Gelman BB (1997) Microglial cell activation in aging and Alzheimer disease: partial linkage with neurofibrillary tangle burden in the hippocampus. J Neuropathol Exp Neurol 56(2):143–149
343. Godbout JP, Johnson RW (2006) Age and neuroinflammation: a lifetime of psychoneuroimmune consequences. Neurol Clin 24(3):521–538
344. Perry VH (2010) Contribution of systemic inflammation to chronic neurodegeneration. Acta Neuropathol 120(3):277–286
345. Xie Z, Morgan TE, Rozovsky I, Finch CE (2003) Aging and glial responses to lipopolysaccharide in vitro: greater induction of IL-1 and IL-6, but smaller induction of neurotoxicity. Exp Neurol 182(1):135–141
346. Njie EG, Boelen E, Stassen FR, Steinbusch HW, Borchelt DR, Streit WJ (2012) Ex vivo cultures of microglia from young and aged rodent brain reveal age-related changes in microglial function. Neurobiol Aging 33(1):195.e1–12
347. Ye SM, Johnson RW (2001) An age-related decline in interleukin-10 may contribute to the increased expression of interleukin-6 in brain of aged mice. Neuroimmunomodulation 9(4):183–192
348. Sierra A, Gottfried-Blackmore AC, McEwen BS, Bulloch K (2007) Microglia derived from aging mice exhibit an altered inflammatory profile. Glia 55(4):412–424
349. Potula R, Haorah J, Knipe B, Leibhart J, Chrastil J, Heilman D, Dou H, Reddy R, Ghorpade A, Persidsky Y (2006) Alcohol abuse enhances neuroinflammation and impairs immune responses in an animal model of human immunodeficiency virus-1 encephalitis. Am J Pathol 168(4):1335–1344
350. Szabo G (1999) Consequences of alcohol consumption on host defence. Alcohol Alcohol 34(6):830–841
351. Wang X, Douglas SD, Metzger DS, Guo CJ, Li Y, O'Brien CP, Song L, Davis-Vogal A, Ho WZ (2002) Alcohol potentiates HIV-1 infection of human blood mononuclear phagocytes. Alcohol Clin Exp Res 26(12):1880–1886
352. Wang Y, Watson RR (1995) Is alcohol consumption a cofactor in the development of acquired immunodeficiency syndrome? Alcohol 12(2):105–109
353. Chen W, Tang Z, Fortina P, Patel P, Addya S, Surrey S, Acheampong EA, Mukhtar M, Pomerantz RJ (2005) Ethanol potentiates HIV-1 gp120-induced apoptosis in human neurons via both the death receptor and NMDA receptor pathways. Virology 334(1):59–73
354. Self RL, Mulholland PJ, Harris BR, Nath A, Prendergast MA (2004) Cytotoxic effects of exposure to the human immunodeficiency virus type 1 protein Tat in the hippocampus are enhanced by prior ethanol treatment. Alcohol Clin Exp Res 28(12):1916–1924
355. Wu P, Price P, Du B, Hatch WC, Terwilliger EF (1996) Direct cytotoxicity of HIV-1 envelope protein gp120 on human NT neurons. Neuroreport 7(5):1045–1049
356. Self RL, Smith KJ, Butler TR, Pauly JR, Prendergast MA (2009) Intra-cornu ammonis 1 administration of the human immunodeficiency virus-1 protein trans-activator of transcription exacerbates the ethanol withdrawal syndrome in rodents and activates N-methyl-D-aspartate glutamate receptors to produce persisting spatial learning deficits. Neuroscience 163(3):868–876
357. Belmadani A, Zou JY, Schipma MJ, Neafsey EJ, Collins MA (2001) Ethanol pre-exposure suppresses HIV-1 glycoprotein 120-induced neuronal degeneration by abrogating endogenous glutamate/Ca2+–mediated neurotoxicity. Neuroscience 104(3):769–781

358. Collins MA, Neafsey EJ, Zou JY (2000) HIV-I gpI20 neurotoxicity in brain cultures is prevented by moderate ethanol pretreatment. Neuroreport 11(6):1219–1222
359. Belmadani A, Neafsey EJ, Collins MA (2003) Human immunodeficiency virus type 1 gp120 and ethanol coexposure in rat organotypic brain slice cultures: curtailment of gp120-induced neurotoxicity and neurotoxic mediators by moderate but not high ethanol concentrations. J Neurovirol 9(1):45–54
360. Bagby GJ, Stoltz DA, Zhang P, Kolls JK, Brown J, Bohm RP Jr, Rockar R, Purcell J, Murphey-Corb M, Nelson S (2003) The effect of chronic binge ethanol consumption on the primary stage of SIV infection in *Rhesus macaques*. Alcohol Clin Exp Res 27(3): 495–502
361. Poonia B, Nelson S, Bagby GJ, Zhang P, Quniton L, Veazey RS (2006) Chronic alcohol consumption results in higher simian immunodeficiency virus replication in mucosally inoculated *Rhesus macaques*. AIDS Res Hum Retroviruses 22(6):589–594
362. Hudson LC, Colby BA, Meeker RB (2010) Ethanol suppression of peripheral blood mononuclear cell trafficking across brain endothelial cells in immunodeficiency virus infection. HIV AIDS (Auckl) 2:7–18
363. Strazza M, Pirrone V, Wigdahl B, Nonnemacher MR (2011) Breaking down the barrier: the effects of HIV-1 on the blood–brain barrier. Brain Res 1399:96–115
364. Haorah J, Knipe B, Leibhart J, Ghorpade A, Persidsky Y (2005) Alcohol-induced oxidative stress in brain endothelial cells causes blood–brain barrier dysfunction. J Leukoc Biol 78(6):1223–1232
365. Haorah J, Ramirez SH, Schall K, Smith D, Pandya R, Persidsky Y (2007) Oxidative stress activates protein tyrosine kinase and matrix metalloproteinases leading to blood–brain barrier dysfunction. J Neurochem 101(2):566–576
366. Ehrlich D, Pirchl M, Humpel C (2012) Effects of long-term moderate ethanol and cholesterol on cognition, cholinergic neurons, inflammation, and vascular impairment in rats. Neuroscience 205:154–166
367. Phillips SC, Cragg BG (1982) Weakening of the blood–brain barrier by alcohol-related stresses in the rat. J Neurol Sci 54(2):271–278
368. Rosengren LE, Persson LI (1979) Chlorpromazine treatment of blood–brain barrier dysfunction. A quantitative and fluorescence microscopical study on small cerebral stab wounds in the rat. Acta Neuropathol 46(1–2):145–150
369. Pratt OE, Rooprai HK, Shaw GK, Thomson AD (1990) The genesis of alcoholic brain tissue injury. Alcohol Alcohol 25(2–3):217–230
370. Thomsen H, Kaatsch HJ, Asmus R (1994) Magnetic resonance imaging of the brain during alcohol absorption and elimination – a study of the "rising tide phenomenon". Blutalkohol 31(3):178–185
371. Fein G, Biggins CA, MacKay S (1995) Delayed latency of the event-related brain potential P3A component in HIV disease: progressive effects with increasing cognitive impairment. Arch Neurol 52(11):1109–1118
372. Green JE, Saveanu RV, Bornstein RA (2004) The effect of previous alcohol abuse on cognitive function in HIV infection. Am J Psychiatry 161(2):249–254
373. Rothlind JC, Greenfield TM, Bruce AV, Meyerhoff DJ, Flenniken DL, Lindgren JA, Weiner MW (2005) Heavy alcohol consumption in individuals with HIV infection: effects on neuropsychological performance. J Int Neuropsychol Soc 11(1):70–83
374. Meyerhoff DJ, MacKay S, Sappey-Marinier D, Deicken R, Calabrese G, Dillon WP, Weiner MW, Fein G (1995) Effects of chronic alcohol abuse and HIV infection on brain phosphorus metabolites. Alcohol Clin Exp Res 19(3):685–692
375. Pfefferbaum A, Adalsteinsson E, Sullivan EV (2005) Cortical NAA deficits in HIV infection without dementia: influence of alcoholism comorbidity. Neuropsychopharmacology 30(7):1392–1399
376. CDC (2005) HIV/AIDS surveillance report, vol 17, Atlanta
377. Becker JT, Lopez OL, Dew MA, Aizenstein HJ (2004) Prevalence of cognitive disorders differs as a function of age in HIV virus infection. AIDS 18(Suppl 1):S11–S18

378. Valcour V, Shikuma C, Shiramizu B, Watters M, Poff P, Selnes O, Holck P, Grove J, Sacktor N (2004) Higher frequency of dementia in older HIV-1 individuals: the Hawaii Aging with HIV-1 Cohort. Neurology 63(5):822–827
379. Valcour V, Shiramizu B (2004) HIV-associated dementia, mitochondrial dysfunction, and oxidative stress. Mitochondrion 4(2–3):119–129
380. Agrawal L, Lu X, Jin Q, Alkhatib G (2006) Anti-HIV therapy: current and future directions. Curr Pharm Des 12(16):2031–2055
381. Price TO, Uras F, Banks WA, Ercal N (2006) A novel antioxidant N-acetylcysteine amide prevents gp120- and Tat-induced oxidative stress in brain endothelial cells. Exp Neurol 201(1):193–202
382. Reynolds A, Laurie C, Mosley RL, Gendelman HE (2007) Oxidative stress and the pathogenesis of neurodegenerative disorders. Int Rev Neurobiol 82:297–325
383. Visalli V, Muscoli C, Sacco I, Sculco F, Palma E, Costa N, Colica C, Rotiroti D, Mollace V (2007) N-acetylcysteine prevents HIV gp 120-related damage of human cultured astrocytes: correlation with glutamine synthase dysfunction. BMC Neurosci 8:106
384. Achim CL, Adame A, Dumaop W, Everall IP, Masliah E (2009) Increased accumulation of intraneuronal amyloid beta in HIV-infected patients. J Neuroimmune Pharmacol 4(2):190–199
385. Caron M, Auclairt M, Vissian A, Vigouroux C, Capeau J (2008) Contribution of mitochondrial dysfunction and oxidative stress to cellular premature senescence induced by antiretroviral thymidine analogues. Antivir Ther 13(1):27–38
386. Abraham J, Jang S, Godbout JP, Chen J, Kelley KW, Dantzer R, Johnson RW (2008) Aging sensitizes mice to behavioral deficits induced by central HIV-1 gp120. Neurobiol Aging 29(4):614–621
387. Ances BM, Vaida F, Yeh MJ, Liang CL, Buxton RB, Letendre S, McCutchan JA, Ellis RJ (2010) HIV infection and aging independently affect brain function as measured by functional magnetic resonance imaging. J Infect Dis 201(3):336–340
388. Chang L, Wong V, Nakama H, Watters M, Ramones D, Miller EN, Cloak C, Ernst T (2008) Greater than age-related changes in brain diffusion of HIV patients after 1 year. J Neuroimmune Pharmacol 3(4):265–274
389. Chang L, Lee PL, Yiannoutsos CT, Ernst T, Marra CM, Richards T, Kolson D, Schifitto G, Jarvik JG, Miller EN, Lenkinski R, Gonzalez G, Navia BA (2004) A multicenter in vivo proton-MRS study of HIV-associated dementia and its relationship to age. Neuroimage 23(4):1336–1347
390. Ernst T, Chang L (2004) Effect of aging on brain metabolism in antiretroviral-naive HIV patients. AIDS 18(Suppl 1):S61–S67
391. Rigler SK (2000) Alcoholism in the elderly. Am Fam Physician 61(6):1710–1716, 1883–1884, 1887–1888 passim
392. Dlugos CA, Pentney RJ (1997) Morphometric evidence that the total number of synapses on Purkinje neurons of old F344 rats is reduced after long-term ethanol treatment and restored to control levels after recovery. Alcohol Alcohol 32(2):161–172
393. Pentney RJ, Quackenbush LJ (1990) Dendritic hypertrophy in Purkinje neurons of old Fischer 344 rats after long-term ethanol treatment. Alcohol Clin Exp Res 14(6):878–886
394. Wiggins RC, Gorman A, Rolsten C, Samorajski T, Ballinger WE, Freund G (1988) Effects of aging and alcohol on the biochemical composition of histologically normal human brain. Metab Brain Dis 3(1):67–80
395. Fein G, Shimotsu R, Barakos J (2010) Age-related gray matter shrinkage in a treatment naive actively drinking alcohol-dependent sample. Alcohol Clin Exp Res 34(1):175–182
396. Oscar-Berman M (2000) Neuropsychological vulnerabilities in chronic alcoholism. In: Noronha A, Eckardt M, Warren K (eds) Review of NIAAA's neuroscience and behavioral research portfolio, NIAAA Research Monograph No. 34. National Institutes of Health, Bethesda, pp 437–472

Chapter 15
Effects of Morphine and Alcohol on the Hypothalamic–Pituitary–Adrenal Axis, Immunity, and Cognitive Behavior

Sulie L. Chang and Xiangqian Liu

15.1 Introduction

The hypothalamic–pituitary–adrenal (HPA) axis is a feedback loop that includes the hypothalamus, pituitary, and adrenal glands. The HPA axis has been shown to regulate immune responses. The main hormones involved in the HPA axis are corticotropin-releasing hormone (CRH), adrenocorticotropic hormone (ACTH), and glucocorticoids (GC), which are also called stress hormones. These hormones contribute to the regulation of immune responses and can also affect neuronal survival, neurogenesis, synaptic plasticity, and behavioral responses [1, 2]. The HPA axis is a three-tiered biological system that begins at the highest level with the release of CRH from the hypothalamic paraventricular nucleus (PVN). CRH-expressing neurons located in the PVN of the hypothalamus play a pivotal role in orchestrating the central stress response. CRH stimulates the release of ACTH from the anterior pituitary gland. In turn, ACTH acts on the adrenal cortex to increase the production and release of GC hormones. Proper functioning of all of these neurons is essential for maintaining a homeostatic state following a stressful event. Several neuronal pathways modulate HPA axis activity. For example, the hippocampus and prefrontal cortex inhibit the HPA axis, and the amygdala and monoaminergic input from the brainstem stimulate CRH production by PVN neurons. GC hormones exert negative

S.L. Chang (✉)
Institute of NeuroImmune Pharmacology, Seton Hall University, South Orange, NJ, USA

Department of Biological Sciences, Seton Hall University, South Orange, NJ, USA
e-mail: sulie.chang@shu.edu

X. Liu
Institute of Neuroimmune Pharmacology, Seton Hall University, South Orange, NJ, USA

Department of Histology and Embryology, Tongji Medical College,
Huazhong University of Science and Technology, Wuhan, P.R. China

C. Cui et al. (eds.), *Neural-Immune Interactions in Brain Function and Alcohol Related Disorders*, DOI 10.1007/978-1-4614-4729-0_15,

feedback control of the HPA axis by binding to GC receptors on hippocampal and hypothalamic PVN neurons [3–5].

Bidirectional interactions exist between the nervous and immune systems. The brain can modulate immune functions by controlling the HPA axis and autonomic nervous system; the immune system can affect brain functions through the release of cytokines [6–8]. CNS-mediated regulation of immunity occurs through local, regional, and systemic routes. The peripheral nervous system provides the first line of defense at local sites of inflammation through the release of neuropeptides that generally increase local inflammatory responses. The sympathetic (adrenergic) nervous system and the parasympathetic (cholinergic) nervous system generally inhibit inflammation on a regional level through innervation of immune organs. Neuroendocrine responses control inflammation at a systemic level via the HPA axis through the anti-inflammatory effects of GC hormones released from the adrenal cortex, via the hypothalamic–pituitary–gonadal (HPG) axis through sex hormones released from the ovaries and testes, and via the hypothalamic–pituitary–thyroid hormone (HPT) axis through hormones released from the thyroid gland [9].

Alcohol, one of the oldest known psychoactive substances, can have a wide range of effects on the human body, including suppression of immune function and impairment of cognitive behavior. Alcohol abuse has been closely associated with a number of infectious diseases and is a major factor in many neuropsychiatric disorders. Similarly, morphine, one of the most effective known analgesic substances, has been shown to induce immune dysfunction as well as physical and psychological dependence and tolerance, with a cluster of cognitive, behavioral, and physiological symptoms. Research has shown that morphine's dysregulation of the HPA axis is one of the mechanisms by which morphine modulates immune responses. There is less known about the involvement of the HPA axis in alcohol's actions on the immune system. This chapter will survey the literature regarding various mechanisms by which these two substances of abuse, morphine and alcohol, modulate the HPA axis, affect cognition, and dysregulate the immune system.

15.2 Effects of Morphine on the HPA Axis, Immunity, and Cognitive Behavior

Morphine is the principal constituent of opium obtained from the opium poppy (*Papaver somniferum*). It is one of the most effective analgesics for pain management. It has been found so effective that, in the past two decades, there has been a threefold increase in the use of morphine products for therapeutic purposes in the United States [10]. Because of their euphoric effects, morphine and other morphine-derived substances such as heroin have a high abuse liability. Whether used as part of a long-term morphine therapy regimen or illegally in a recreational fashion, chronic morphine use results in tolerance and psychological and physical dependence.

There are three primary opioid receptors, mu or μ (MOR), delta or δ (DOR), and kappa or κ (KOR). These receptors belong to the G protein-coupled receptor (GPCR) superfamily. They are widely distributed in the nervous system and have been identified in various immune cells. They play an important role in mediating various opioid functions, including analgesia, reward, withdrawal, and modulation of immune responses [11–16]. Morphine exerts its pharmacological effects primarily through the MOR [17] and multiple signal transduction pathways, including activation of inhibitory GTP-binding protein (G_i) and β-arrestin recruitment-mediated signaling cascades [18]. The binding of morphine to G_i-coupled MOR inhibits adenylyl cyclase activity and the cAMP-dependent pathway, inhibits voltage-dependent Ca^{2+} channels, and activates inwardly rectifying K^+ channels. The modulation of these signal transduction pathways is one of the molecular mechanisms by which morphine suppresses cellular activity [19]. The activation of G_i protein also results in MOR phosphorylation, followed by the recruitment of GPCR-associated β-arrestin to the receptor. Recruitment of GPCR-associated β-arrestin further recruits intracellular proteins involving receptor internalization and functions as a signaling intermediate to activate other signaling pathways such as the mitogen-activated protein kinase (MAPK)-dependent pathway [20, 21]. In addition, nuclear factor-κB (NF-κB), one of the most diverse and essential transcription factors, is involved in the biological actions of morphine. Morphine also modulates gene expression of a number of cytokines and chemokines through NF-κB action [11].

15.3 Morphine's Effects on the HPA Axis

Studies have demonstrated that chronic morphine exposure alters stress-response systems, indicating that alterations in the HPA axis are directly or indirectly involved [22]. Most of the studies investigating the effects of morphine on the HPA axis have been carried out using animal models; few studies have been performed using human subjects. Morphine's activation of the HPA axis is mediated by both catecholaminergic neurons in the hypothalamus and noradrenergic neurons in the nucleus tractus solitarius-A2 and ventrolateral medulla-A1 (NTS-A2/VLM-A1) that project to the PVN [23–27]. Clinical research shows that pain patients receiving both short- and long-term morphine therapies have a decrease in blood cortisol concentrations, but a functional HPA axis, or no significant change in blood ACTH and cortisol levels [28, 29]. In other studies performed with healthy subjects, acute morphine treatment suppressed basal hormone levels and/or blunted the ACTH and cortisol responses to CRH [30, 31].

In a rodent model, the HPA axis is affected differently by acute and chronic morphine treatments. Acute treatment with morphine activates the HPA axis by increasing both ACTH and corticosterone secretion and exaggerating the HPA response to stress. However, chronic morphine treatment (using morphine pellet implantation) either does not activate the HPA axis or attenuates the morphine- or stress-induced

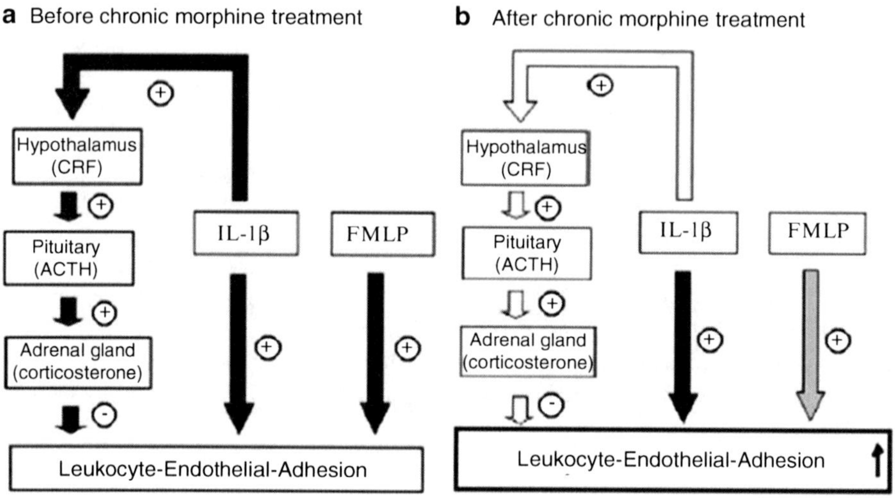

Fig. 15.1 Chronic exposure to morphine may potentiate the inflammatory response by altering IL-1β's modulatory actions on the HPA axis. (**a**) In the opiate-naive rat, IL-1β stimulates (+) the production of CRF, ACTH, and corticosterone by the HPA axis, which inhibits (−) LEA. Both IL-1β and FMLP also have direct local stimulatory effects on LEA. (**b**) After chronic exposure to morphine, IL-1β is less effective in stimulating production of the HPA hormones (indicated by the *open arrows*), thereby reducing the inhibitory effects of these products on LEA. Although the local stimulatory effects of n-formyl-methionyl-leucyl-phenylalanine (FMLP) may be weakened by morphine (indicated by the *gray arrow*), the loss of corticosterone inhibition disrupts the balance between IL-1β-mediated proinflammatory and anti-inflammatory effects, leading to a net increase in LEA (indicated by the *bold* and *bigger font* with an *arrow*)

ACTH and corticosterone responses [27, 32, 33]. Further, IL-1β-induced neuronal activation of the hypothalamic PVN, CRH mRNA expression, and plasma corticosterone are attenuated in animals in a state of morphine tolerance [34]. House et al. [34] conducted a series of studies using a rat model of morphine tolerance to determine possible mechanisms by which chronic exposure to morphine modulates the HPA axis. More specifically, they looked at differences in leukocyte–endothelial adhesion (LEA), CRF expression in the hypothalamic PVN, and serum levels of corticosterone in morphine-naïve rats and rats made morphine-tolerant by exposure to a 2+4 regimen of chronic morphine. LEA is the initial step in the immune response cascade; the PVN is the site from which CRF initiates activation of the HPA axis, and corticosterone, an anti-inflammatory hormone, is the final product of the HPA axis. The data show that, while basal levels between groups are comparable, after treatment with IL-1β, morphine-treated rats have a greater increase in LEA levels, attenuation of hypothalamic CRF production, and adrenal corticosterone production. In morphine-naïve rats, IL-1β injection initiates the HPA axis cascade by stimulating CRF expression in the hypothalamus, resulting in an increase in corticosterone via the HPA axis. In the case of chronic morphine exposure, IL-1β-induced HPA activation is desensitized, and the entire cascade is weakened, resulting in a decrease in production of the HPA stress hormones (Fig. 15.1). In addition, rats rendered morphine-tolerant and administered a non-pyrogenic dosage of LPS have significantly enhanced LEA as well as reduced corticosterone secretion [15]. Attenuation of these anti-inflammatory stress hormones reflects a decrease in the inhibitory effects of GC on immune responses. Consequently, LPS-induced production of proinflammatory cytokines, such as TNF-a, IL-1β, and IL-6, is potentiated in the morphine-tolerant state.

In addition to morphine tolerance, the severe stress associated with morphine withdrawal also results in immune system dysregulation, and many serious affective and somatic symptoms [22, 35] as well as activation of hypothalamic nuclei that control the HPA axis [22, 35]. Intermittent treatment with morphine, which reflects repeated cycles of partial morphine withdrawal, has been shown to elevate basal ACTH and corticosterone levels. In addition, repeated cycles of partial morphine withdrawal enhance the pituitary–adrenal response to a novel restraint when the rat is tested immediately prior to its expected next injection [36]. Both PKC-dependent pathways and L-type Ca^{2+} channels appear to be involved in mediating morphine withdrawal-induced hyperactivity of the HPA axis [37–39].

15.4 Morphine's Effects on Immunity

The immune system has two major components—innate or nonspecific immunity and adaptive or specific immunity. Cells of the innate immune system, such as monocytes, macrophages, dendritic cells, and neutrophils, can recognize pathogens by pattern recognition receptors (PRRs). Intracellular signaling cascades triggered by these PRRs, especially Toll-like receptors (TLRs), lead to transcriptional expression of various inflammatory cytokines, chemokines, and type I interferons that coordinate the elimination of pathogens and infected cells [40–42]. The adaptive immune system acts as a second line of defense against invading organisms. The priming of adaptive immunity depends on both direct antigen recognition by the antigen receptors and essential signals delivered by the antigen-presenting cells (such as macrophages and dendritic cells) of the innate immune system and leads to T-cell-mediated cellular immune responses and B-cell-mediated humoral immune responses [43, 44].

15.5 Morphine's Effects on Innate Immunity

Recruitment of immune cells is an essential first step in a host's defense against invading pathogens. Morphine inhibits recruitment of immune cells by altering the expression of cytokines and chemokines. For example, Breslow et al. [45] found the percentage and total number of neutrophils and macrophages within the peritoneal cavity of mice implanted with morphine pellets 48 h before intraperitoneal injection (i.p.) with *Acinetobacter baumannii* to be significantly lower than those of control animals [45]. They also found that morphine reduces the peritoneal exudate fluid levels of KC/CXCL1, a major chemokine for neutrophil recruitment, and IL-17A, which induces the expression of KC/CXCL1 [45]. Martin et al. [46] demonstrated that chronic morphine exposure decreases keratinocyte-derived cytokine (KC) and monocyte chemotactic protein-1 (MCP-1) levels, thus inhibiting recruitment of neutrophils and macrophages to injured tissues [46]. Treatment with morphine delays neutrophil migration and decreases the clearance of *Streptococcus pneumoniae* during lung infection [47]. These effects are associated with impairment in production of IL-23 by alveolar macrophages and dendritic cells and IL-17 by γδT lymphocytes [47] (Fig. 15.2).

Phagocytosis of invading pathogens by immune cells is the next important step in the defense against infection. Phagocytosis is suppressed by morphine via the MOR and DOR-2 receptors, but not KOR receptors [48, 49]. Phagocytosis by murine peritoneal macrophages and human monocyte-derived macrophages is inhibited following both acute morphine exposure and withdrawal from chronic morphine [50, 51]. Bosshart [233] reported that prior treatment with morphine (10^{-7} M, 60 min) can inhibit the phagocytic activity of human monocytic THP-1 cells stably transfected with human CD14 following exposure to *Escherichia coli* for 90 min. Morphine also inhibits *S. pneumoniae* clearance by alveolar macrophages through impaired TLR9-mediated NF-κB signaling [52]. Further, chronic morphine exposure modulates LPS-induced dendritic cell maturation and enhances the antigen-presenting function of dendritic cells to activate T cells [53].

Other immune cells, such as NK cells and mast cells, are also affected by morphine. Chronic morphine treatment decreases the percentage of NK cells in the peripheral blood [54] and suppresses NK cell activity in surgical patients [55]. Morphine inhibits TLR-4-mediated TNF-α release from intraperitoneal mast cells in response to LPS [56], indicating crosstalk between opioid receptor- and TLR-4 receptor-mediated signaling pathways [56].

Cytokines mediate a wide range of immune responses, and morphine has been shown to modulate the expression of cytokines by immune cells. In monocyte-derived human dendritic cells, chronic morphine exposure increases LPS-induced production of IL-12 (p70), a proinflammatory cytokine, and decreases LPS-induced production of IL-10, an anti-inflammatory cytokine [53]. In a mouse model, morphine enhances expression of TNF-α, IL-12 p40, and p70 expression [57] and inhibits *S. pneumonia*-induced IL-23 expression in bone marrow-derived dendritic cells through TLR2 and Nod2 signaling [58]. However, morphine inhibits the production of cytokines, including TNF-α and IL-6, in TLR2 agonist-stimulated human peripheral blood mononuclear cells and monocytes [59].

15.6 Morphine's Effects on Adaptive Immunity

In addition to its affects on innate immunity, morphine also affects adaptive immunity. It does so by inhibiting IL-12 production, thus causing Th cell differentiation to be biased toward the Th2 lineage [60, 61]. Mice administered morphine show a

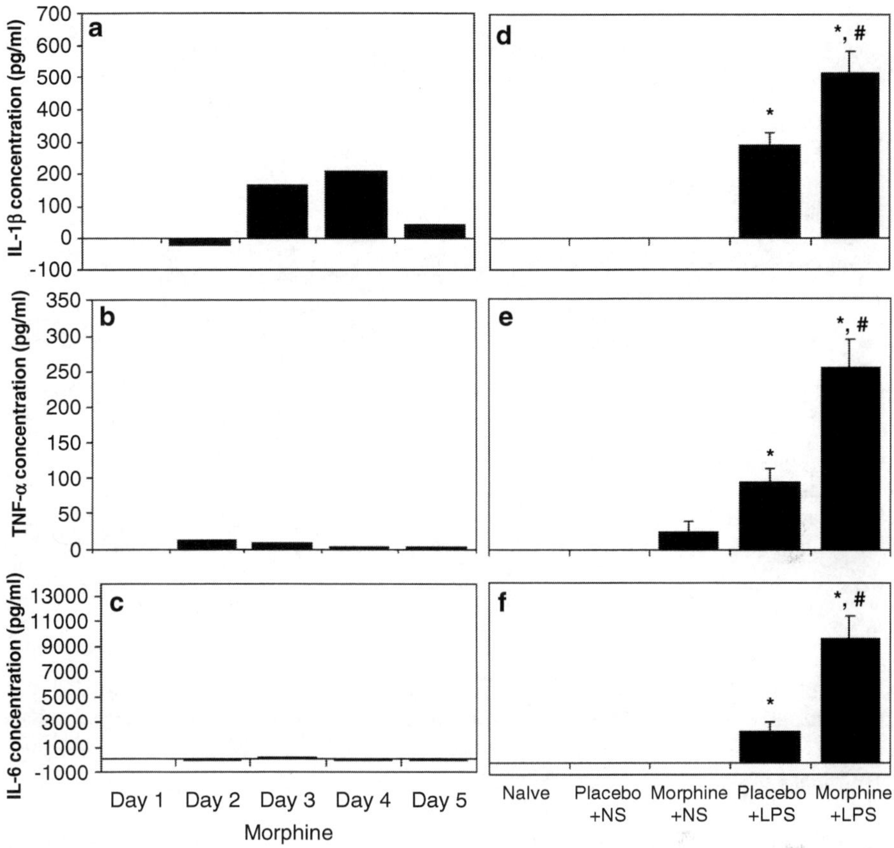

Fig. 15.2 The effect of chronic morphine treatment on the levels of the proinflammatory cytokines, IL-1β, TNF-α, and IL-6. (**a–c**) Serum cytokine levels were measured before and 5 days after the initiation of chronic morphine treatment. (**a**) IL-1β levels in chronic morphine-treated rats gradually increased until reaching a peak concentration on day 4. (**b**) The serum TNF-α concentration peaked on day 2 in chronic morphine-treated rats and declined on subsequent days. (**c**) IL-6 levels peaked on day 3, but otherwise remained low during the other 4 days. (**d–f**) Serum cytokine levels in naïve and placebo- and chronic morphine-treated rats given an i.p. injection of either 250 mg/kg LPS or normal saline (NS) were measured. There was no significant change in cytokine levels in naïve and placebo- and chronic morphine-treated rats given normal saline. LPS significantly increased serum concentrations of all three cytokines in both placebo- and chronic morphine-treated rats compared to NS treatment ($^{*}p<0.05$). Serum IL-1β, TNF-α, and IL-6 levels were significantly higher in the chronic morphine-treated rats given LPS compared to LPS-treated placebo rats ($^{\#}p<0.05$) ($n=5$ rats per group)

reduction in total number of B lymphocytes, $CD4^{+}$ T cells, and $CD8^{+}$ T cells in the spleen and lymph nodes [62]. While acute morphine treatment alone does not induce caspase activity, it does potentiate LPS-induced caspase activity in the spleen and thymus as well as bone marrow-derived immune cells [63]. This suggests a possible mechanism by which morphine induces loss of immune cells because activation of the caspase could lead to programmed cell death.

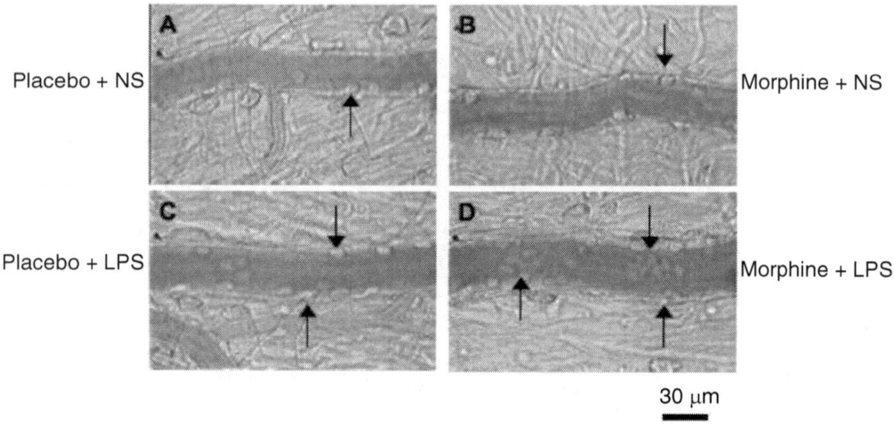

Fig. 15.3 Representative photomicrographs showing LEA in chronic morphine-treated rats given LPS. (**a**, **b**) LEA is minimal (*arrows*) following normal saline injection in placebo- and morphine-treated rats. (**c**, **d**) Injection with LPS led to an increase in LEA (*arrows*) in both placebo- and chronic morphine-treated groups. (**d**) Chronic morphine-treated rats demonstrated the greatest potentiation in the LEA response. Scale bar = 20 μm

Moreover, morphine can alter T-cell function. Acute morphine treatment reduces cell-mediated immunity in the spleen of mice infected with herpes simplex virus type 1 (HSV-1), reduces cytolytic lymphocyte responses, and inhibits proliferation of lymphocytes as well as the production of IFN-γ. Morphine-induced suppression of these cellular immune responses may explain why reactivation of HSV-1 often occurs in opioid addicts [64]. MOR-mediated inhibition of IL-2 expression in activated T lymphocytes and attenuation of the activation of transcription factors, including AP-1 and NF-κB, may be one of the molecular mechanisms by which morphine suppresses immunity [65] (Fig. 15.3).

15.7 Morphine's Effects on the Neuroimmune Axis

Recent research has found that morphine can affect the central nervous system by multiple immunological mechanisms. For example, morphine promotes splenocyte infiltration into the central nervous system [66] and can affect the permeability of the blood–brain barrier, which makes the brain more likely to be influenced by peripheral immune cells or cytokines [67–69]. Morphine also upregulates the production of the chemokine, CCL2, in cultured neurons [70] and alters the expression of alpha- and beta-chemokines and their receptors in astroglial cells [71]. Induction of proinflammatory cytokines by LPS is increased in the brain of rats in the morphine-tolerant state [72].

Morphine can also activate glial cells, which have been implicated in the development of morphine tolerance and dependence [73–79]. TLR2 plays an important role in morphine-induced microglial activation and dependence [80]. Unlike its actions at classical opioid receptors, morphine can selectively influence TLR4 signaling in a non-stereotypic manner, leading to glial activation and production of cytokines [81]. Morphine induces NF-κB-dependent TNF-α production in microglia [82] and increases microglial migration through interaction between the MOR and P2X4 receptors, which is dependent on the activation of the PI3K/Akt pathway [83].

We now know that morphine directly affects both innate immunity and adaptive immunity by acting on opioid receptors in immune cells. Chronic morphine promotes peripheral inflammatory reactions, in part, by disrupting the anti-inflammatory effects of the HPA axis [34, 84]. Morphine also induces apoptosis of thymocytes [85]; suppresses splenic NK cell activity, which is dependent on activation of the dopamine D1 receptor in the nucleus accumbens; activates neuropeptide Y (NPY) Y1 receptors in the spleen [86–88]; and suppresses circulating B-lymphocyte MHC-II expression, which is important in antigen presentation [89].

15.8 Morphine's Effects on the Brain

Investigation of the effects of morphine on the brain focus on two distinct aspects: (1) analgesia and tolerance and (2) abuse and dependence.

15.8.1 Analgesia and Morphine Tolerance

Opioids have long been used to treat moderate to severe pain [90]. Morphine is the most widely used opioid analgesic and the standard by which other new analgesics are judged [91]. The analgesic effects of opioids are mediated through binding to three primary opioid receptors distributed within the neuronal nociceptive system. These include moderate-density MOR and KOR binding in the periaqueductal gray, locus coeruleus, substantia nigra, ventral tegmental area, raphe nuclei, and nucleus tractus solitarii, and low-density DOR binding in the substantia nigra and nucleus tractus solitarii [12]. The MOR can be found in many pain-controlling brain areas [92, 93], and morphine exerts its analgesic effects mainly by binding to the MOR on both presynaptic and postsynaptic cells. Morphine binding on presynaptic cells causes the inhibition of presynaptic neurotransmitter release by blocking voltage-gated Ca^{2+} channels. Binding on postsynaptic cells promotes postsynaptic K^+ efflux via K^+ channels, neuronal hyperpolarization, and firing inhibition [94]. Further, chronic morphine treatment can activate the MAPK signaling pathway cascades, and the use of a MAPK inhibitor reduces morphine tolerance [95].

The use of morphine for chronic pain is often limited by the need for dose escalation due to the development of tolerance to the drug's analgesic effects. In high doses, morphine can cause excessive and unwanted side effects [96]. Neuronal apoptosis has been associated with the development of morphine tolerance in a rodent model [97]. However, the mechanisms underlying morphine tolerance are complex and also involve adaptive changes in the MOR and post-receptor signaling elements [98, 99]. Research on opioid desensitization and tolerance has focused largely on adaptations which occur at the level of the MOR itself. These include, but are not limited to, MOR phosphorylation and ensuing trafficking events. Additional adaptations, such as covalent modification of signaling molecules and altered associations occurring downstream from MOR sites, also affect morphine tolerance. More specifically, augmented isoform-specific synthesis of adenylyl cyclase (AC) and its phosphorylation, as well as augmented phosphorylation of the Gβ subunit of Gβγ, contribute to the development of opioid tolerance and desensitization. The aggregate effect of these changes shifts MOR-coupled signaling from predominantly Giα inhibitory to (Gi-derived) Gβγ stimulatory AC signaling [100].

Adaptive changes which take place in the MOR over a period of prolonged opioid exposure include receptor desensitization and downregulation of MOR expression [101]. Rapid receptor desensitization, endocytosis, and recycling following MOR activation are essential for maintaining MOR-mediated analgesic effects. Failure to promote MOR desensitization and endocytosis mediated, in part, by β-arrestin 2 may be one of the mechanisms involved in morphine tolerance [102, 103]. In addition, it has been proposed that one of the key mechanisms by which the development of morphine tolerance occurs is heterodimerization of the MOR and DOR. Use of a DOR antagonist together with morphine can produce analgesia with reduced morphine tolerance [101, 104].

Nitric oxide (NO) may also play a central role in the development of tolerance and desensitization. NO helps control physiological functions and is involved in mechanisms underlying both beneficial and detrimental actions of therapeutic agents [105]. It functions in the brain mainly as a neuromodulator and has been implicated in modulating the analgesic effects of morphine and potentiating morphine tolerance [105] by either enhancing or reducing anti-nociception [105].

15.8.2 Morphine Abuse and Dependence

Despite numerous and varied attempts to control drug abuse, drug addiction, a state of compulsive drug use, persists in our society [106, 107]. Addiction involves the pathological usurpation of neural processes that normally serve reward-related learning. Persistent, compulsive drug use is related to molecular and cellular mechanisms that underlie long-term associative memory in several forebrain circuits that receive input from midbrain dopamine neurons. These circuits include the ventral and dorsal striatum and prefrontal cortex. The hippocampus is a central brain region involved in learning and memory [108, 109] and has been implicated in morphine-induced cognitive deficits with altered brain-derived neurotrophic factor (BDNF) signaling, or decreased neurogenesis [110, 111]. In rodents, morphine impairs acquisition of radial maze and Y-maze choice escape, radial maze performance, acquisition and retrieval of spatial recognition memory in the Y-maze, acquisition and performance of avoidance, and spatial memory acquisition in the Morris Water Maze task [112–116]. Transient anterograde and retrograde memory impairments are observed in patients receiving chronic opioid therapy in palliative care [117].

15.9 Effects of Alcohol on the HPA Axis, Immunity, and Cognitive Behavior

Like morphine, alcohol, also referred to as ethanol, can have a wide range of molecular and systemic effects on the human body. Alcohol abuse is the world's third largest risk factor for disease burden and results in approximately 2.5 million deaths each year. The abuse of alcohol is associated with several infectious diseases and is a major factor in many neuropsychiatric disorders [118].

15.10 Alcohol's Effects on Regions of the Brain

Acute exposure to ethanol activates specific areas in the rat brain, including the bed nucleus of the stria terminalis, PVN, the central nucleus of the amygdala, Edinger–Westphal nucleus, locus coeruleus, and the parabrachial nucleus in a dose- and time-dependent manner [119]. Researchers now know that chronic alcohol abuse and alcohol dependence can affect both gray and white matters in more widespread brain regions, such as cortical and subcortical regions, the cerebellum, and brainstem [120–124].

15.11 Effects of Alcohol on the HPA Axis

Alcohol affects the activity of the HPA axis primarily by acting on the hypothalamus. Prolonged alcohol exposure can cause degeneration of hypothalamic neurons, resulting in HPA axis dysfunction. Silva et al. [125] provided rats with an ethanol solution as their only available liquid source for either 6 or 10 months and found that chronic alcohol consumption leads to a loss of 35 % of the neurons located in the posterior magnocellular division of the PVN, the site where activation of the HPA axis is initiated [125]. In a human study, neuron loss and abnormal neuron morphology was noted in the hypothalamic supraoptic nucleus (SON) and PVN in postmortem brains of patients with chronic alcoholism [126, 127]. Acute treatment of rats with 16 % (w/v) ethanol activates the PVN in a dose-dependent manner; however, repeated administration of ethanol (16 %, w/v) results in desensitization of the PVN [119]. Alcohol can alter gene transcription in the PVN and, like morphine, affects the release of CRH [128]. It enhances G protein-coupled CRH receptor 1α-mediated signal transduction, which activates adenylyl cyclase, leading to ACTH release [129]. Similar to morphine, alcohol's effects on the HPA axis are dependent on dosage, frequency, and duration of alcohol administration. Rivier et al. [131] showed that acute alcohol treatment (3 g/kg, i.p.) stimulates activity of hypothalamic neurons that express CRH and produces a rapid increase in plasma ACTH in the rat, whereas chronic intermittent exposure to alcohol vapors results in decreased PVN neuronal activity and decreases the ACTH response to certain stimuli [130–132]. In addition, Sinha et al. [133] reported that alcohol-dependent individuals have an abnormal HPA axis, including a higher basal ACTH level and a lack of stress- and cue-induced ACTH and cortisol responses [133].

15.12 Alcohol's Effects on Immunity

Alcohol is a potent immunomodulator [134] that interferes with the host's immune surveillance system. Research findings, using both animal models and human subjects, point to a whole spectrum of alcohol-induced immune dysfunctions. Alcohol has been implicated in atrophy of the thymus and spleen, loss or redistribution of peripheral blood leukocytes, and diminished humoral and cell-mediated immune responses [135]. Alcohol causes abnormalities in the function and/or structure of a broad array of cells involved in humoral and cellular immunity, including lymphocytes, Kupffer cells, and other macrophages, as well as the endothelium of blood vessels and lymphatics. Regulatory cytokines and neuroendocrine factors mediate some of these immunomodulatory effects [136]. The severity of the alterations in the immune system caused by alcohol depends on different factors including age, race, gender, body composition, environmental factors, time of consumption, and amount and type of alcoholic beverage [137].

15.13 Alcohol's Effects on Innate Immunity

Challenges to innate immunity may directly or indirectly affect the body's overall immune responses. Alcohol consumption has been shown to alter innate immunity in a biphasic manner. Acute alcohol exposure inhibits inflammatory cell activation and decreases production of inflammatory mediators, whereas chronic alcohol use leads to an increase in inflammatory cell responses to various microbial compounds, particularly lipopolysaccharide (LPS) [134, 138]. Szabo et al. [139–141] showed that, in vitro, acute treatment with alcohol inhibits induction of proinflammatory mediators, including IL-1β, IL-6, and monocyte chemotactic protein-1 (MCP-1) in blood monocytes [139–141]. In an in vivo binge drinking model, expression of the proinflammatory cytokines, IL-1β and IL-6, and the chemokine, CCL2 (MCP-1), is significantly decreased in the spleen of adolescent rats administered a 52 % (4.8 g/kg/day) ethanol solution for 3 days [142].

Interestingly, females appear to have a greater sensitivity to the effects of alcohol than males. Acute ethanol intoxication in female rats is associated with lower phagocytic activity by liver-recruited polymorphonuclear leukocytes and Kupffer cells than in their male counterparts, and an LPS challenge elicits a lesser phagocytic response in liver-recruited polymorphonuclear leukocytes and Kupffer cells of acutely ethanol-intoxicated female rats than in males, which indicates a smaller phagocytic response to fight infection in the liver of acutely alcohol-intoxicated female rats [143]. This could involve an indirect mechanism by which heavy drinking decreases the production of estrogen and nullifies the stimulatory effects of estrogen on female immune responses [144].

The liver plays an important role in innate immunity and is particularly important in the body's response to alcohol. The liver receives blood from both the systemic circulation and the intestines. Cells involved in innate immunity are unusually abundant in the liver in order to effectively and quickly defend against potentially toxic agents. The liver is also involved in controlling systemic innate immunity through the biosynthesis of numerous soluble pathogen-recognition receptors and complement [145]. Chronic exposure to alcohol damages liver tissue and can have devastating effects on liver function. In the short-term, alcohol exposure enhances endotoxin hepatotoxicity and inactivates Kupffer cells [146]. These effects contribute to the impairment of innate immunity in alcoholism and the progression of alcoholic liver injury. Moderate alcohol exposure increases TNF-α and decreases NF-κB activity, leading to cellular apoptosis [147]. During acute alcohol intoxication, oxidative stress mediates endotoxin-induced hepatic TNF-α production [148]. Chronic alcohol intoxication is associated with increased basal H_2O_2 formation, enhanced nuclear translocation, and binding of NF-κB, AP-1, and MNP-1 in Kupffer cells. Further, upregulation of chemokines during alcohol consumption is selective. During the early phase of alcoholic liver disease, alpha-chemokines predominate and are associated with neutrophil infiltration of the liver, whereas in the later stages of alcoholic liver disease, beta-chemokines predominate [149]. A deficiency of MCP-1 can protect against alcoholic liver injury independent of inhibition of proinflammatory cytokines and induction of genes related to fatty acid oxidation, thus linking chemokines to hepatic lipid metabolism [150].

Examination of the effects of alcohol during HIV-1 infection is helpful in highlighting the importance of liver function in innate immunity. The SIV/nonhuman primate model has been used to examine the relationship between chronic alcohol exposure and SIV infection on chemokine secretion and innate immune functions in the liver. Excessive production of free radicals and chemokines suppresses chemotaxis and phagocytosis by hepatic non-parenchymal cells. As a result, susceptible individuals become highly vulnerable to hepatitis and reduced microbial clearance by the liver [151]. Acute ethanol administration also attenuates the release of HIV-1 gp120-induced chemokines from splenocytes and Kupffer cells, which could compromise inhibition of HIV-1 entry into target cells [152].

15.14 Alcohol's Effects on Adaptive Immunity

15.14.1 Dendritic Cells

Alcohol use may affect adaptive immunity either directly or indirectly by inhibiting the antigen-presenting function of innate immune cells, including macrophages and dendritic cells (DCs) [134, 153]. DCs play a key role in initiation of adaptive immune responses. Alcohol exposure interferes with DC functions by inhibiting the generation of bone marrow-derived myeloid and plasmacytoid DC subsets, interfering with production of interferon-α by plasmacytoid DCs, affecting the expression of B7 family co-signaling molecules on resting DC subsets following CpG oligodeoxynucleotide (CpG-ODN) stimulation, and inhibiting splenic DC maturation and function [154].

15.14.2 Natural Killer Cells

Natural killer (NK) cells are involved not only in innate immune responses against viral infection and tumor development but also participate in a complex network of cell-to-cell interaction that leads to the development of adaptive immune responses [155, 156]. Chronic alcohol consumption perturbs the balance between thymus-derived and bone marrow-derived NK cells and decreases the percentage and number of NK cells in the spleen and peripheral lymph nodes [157, 158].

15.14.3 Lymphocytes

Alcohol also has profound effects on lymphocyte production and function. As many as 13.3 % of alcohol-dependent male inpatients at the Wuhan Mental Health Center (China) with a history of alcohol use for at least 10 years had a low white blood cell count, and 26.7 % of this population had a low percentage of circulating lymphocytes (Chang et al. unpublished data). While B-cell functions appear to be relatively unaffected by acute alcohol exposure [134], a number of investigators have shown diminished numbers of peripheral blood B cells (relative and absolute) after long-term alcohol consumption. Alcoholics also exhibit poor antibody levels upon immunization with the hepatitis B vaccine, a response that requires the participation of T-helper cells. These findings suggest not only attrition of B cells but diminished ability to generate protective antibodies. In contrast, it has been commonly reported that alcoholic patients have elevated concentration of serum IgA and IgG. A number of investigators have further demonstrated the presence of autoantibodies reactive against various self-tissues. Finally, in vitro studies have shown that peripheral blood B cells from alcoholics secrete higher levels of immunoglobulins compared to normal controls [159].

Many reports have documented chronic alcohol-induced T-cell subset abnormalities. Heavy alcohol use by alcohol-dependent adolescents is associated with significantly lower mean numbers of circulating total T lymphocytes ($CD3^+$) and T-lymphocyte subsets $CD4^+$ (T-helper) and $CD8^+$ (T-cytotoxic) compared with light/nondrinking control groups [160]. This reduction in $CD3^+$, $CD4^+$, and $CD8^+$ cell numbers has also been found in the isolated splenocytes of male Fischer 344 rats fed an ethanol-containing liquid diet for 3 and 4 weeks [161]. Using an attenuated *Listeria monocytogenes* strain to inoculate mice, Gurung et al. [162] found chronic ethanol consumption reduces antigen-specific $CD8^+$ T-cell responses, but not antigen-specific $CD4^+$ T-cell responses [162]. In addition, Zhang et al. [163] reported that chronic alcohol intake, with 20 % (w/v) alcohol as the sole drinking source, inhibits tumor-induced memory $CD8^+$ T-cell proliferation in the spleen and accelerates the decay of interferon-γ-producing cells within the splenic $CD8^+$ T-cell population in melanoma-bearing female mice [163].

15.15 Alcohol's Effects on the Neuroimmune Axis

Alcohol abuse can cause brain damage by multiple mechanisms. Robust immune responses occur in the brain even though the brain lacks major histocompatibility complex (MHC) molecules and normal lymphatic drainage and reduced immune surveillance [164]. Both neurons and glial cells in the brain express genes encoding several cytokines, chemokines, and their receptors. Microglial cells constitute the resident macrophage population in the CNS, and they play a key role in the inflammatory response of the brain [165, 166]. Microglial cells are sensitive to addictive drugs and stress, which can lead to progressive adaptation that can alter neurobiology and behavior through innate immune gene induction [167]. Astrocytes are the most abundant glial cell in the CNS and have various biological activities. The activation of astrocytes is associated with several neurodegenerative disorders [168, 169].

Alcohol can stimulate both microglia and astrocytes to release inflammatory mediators via TLR4/NF-κB-activated signaling and induce neuronal cell death [170–173]. NF-κB plays a key role in immune and inflammatory responses and is abundant in the brain. NF-κB regulates various functions in the brain, including synaptic plasticity and neuroinflammation [174]. Acute alcohol exposure activates NF-κB in the brain, which induces neuroinflammatory responses and leads to neuronal damage [175, 176]. Postmortem studies on the brains of alcoholics suggest that repeated cycles of alcohol intoxication and withdrawal may cause adaptive changes in the NF-κB system, which alters the expression of multiple genes contributing to cognitive impairment and alcohol dependence [174, 177]. In addition, alcohol abuse may lead to blood–brain barrier disruption, enhancing leukocyte migration from blood to brain and promoting neuroinflammatory processes [178, 179].

Besides the direct effects on the brain, alcohol can indirectly affect the brain through GC hormones produced through HPA axis activation. The level of circulating

GC hormones is one of the most powerful and well-controlled feedback mechanisms of proinflammatory signal transduction [3, 180]. However, controversy still exists regarding whether the HPA axis itself mediates the effects of alcohol on the immune system. Some studies have reported that the immunological effects of ethanol are mediated through the immunosuppressive actions of GC hormones [181–183], and ethanol-induced release of corticosteroids is suggested to be responsible for most of the cell loss in the thymus, spleen, mesenteric lymph nodes, and Peyer's patches [184]. Excessive alcohol consumption, as well as withdrawal from chronic heavy alcohol consumption, reportedly increases circulating GC levels, causing adverse effects on neuronal functions and cognition [185, 186]. The immunosuppressive and anti-inflammatory activity of GCs, along with their cost-effectiveness, makes these compounds the treatment of choice for the majority of autoimmune and inflammatory diseases, despite serious side effects that frequently accompany GC therapy. The activated GC receptor (GR) engages a number of pathways to ultimately suppress autoimmune responses. The primary mechanisms responsible for the anti-inflammatory actions of GC hormones are GC-mediated apoptosis of numerous hematopoietic cell types and suppression of proinflammatory cytokine gene expression. There are unexpected pathways, factors, and mechanisms that have recently begun to emerge as novel targets for GC-mediated immunosuppression. GC hormones are widely known for their ability to suppress, directly or indirectly, the activation of proinflammatory cytokine genes. GC-mediated suppression of TNF-α and IL-1β production has long been considered the basis for their efficacy in relieving symptoms of rheumatoid arthritis, inflammatory bowel disease, and psoriasis [187]. Conversely, other studies suggest that GC hormones do not mediate the immunosuppressive effects of acute or chronic alcohol exposure [188, 189]. Repeated ethanol treatment desensitizes the HPA axis, and can lead to unexpected potentiation of LEA, the initial step in the immune response cascade [15, 119]. In addition, it has been suggested that stress and GC hormones may sensitize glial neuroinflammatory responses to drugs of abuse [190].

Research indicates that cytokines and the HPA axis work together to regulate the neuroimmune axis. Alcohol can alter cytokine levels in a variety of tissues, including plasma, lung, liver, and brain (see review: [191]). High concentrations of ethanol have recently been shown to decrease the expression of splenic inflammatory mediators, such as IL-1β, IL-6, and CCL2, in the rat independent of GC hormone levels [142]. Interestingly, alterations were found in neurotransmitter receptor expression, including a decrease in the $GABA_A$ receptor $\alpha 2$ subunit and an increase in mGluR5 and 5-HT3A receptors [142].

15.16 Alcohol's Effects on Cognitive Behavior

Cognition is considered a higher-level function of the brain and is responsible for processes such as perception, attention, working memory, long-term memory, executive functioning, language, and social cognition [192]. Circulating immune cells play a key role in supporting brain plasticity manifested in normal behavior, cognition, and mental activity [193]. Because mental behavior cannot be studied directly, outward behavior is recorded in order to determine possible changes in cognitive function. Neuroinflammation is associated with a variety of neurologic and psychological disorders, including addictive behaviors [194–196]. This suggests a connection between the immune system and cognitive function. In the case of alcohol, the neuroimmune axis may be at least partially responsible for changes in performance and cognitive function. Alcohol can activate glial TLR4/NF-κB signaling to release inflammatory mediators. The release of these mediators has been shown to result in behavioral and cognitive changes [197]. In addition, heavy alcohol use and acute phases of alcohol withdrawal can activate the HPA axis and increase circulating concentration of GCs. High concentrations of circulating GC have also been associated with cognitive deficits and addictive behavior [198].

Chronic alcohol abuse affects several brain structures and results in various neuropsychological impairments in terms of cognition, emotion, and behavior [199–203]. As many as 50–80 % of individuals with alcohol use disorders experience mild to severe neurocognitive impairment [204]. The effects of alcohol on cognitive processes depend, in part, on the quantity and pattern of consumption [205]. In comparison to their moderate drinking counterparts, young people who consume large quantities of alcohol performed significantly worse on tests of cognitive function [205]. Neafsey et al. [206] reviewed 143 papers and found that, although earlier studies indicated that moderate drinking impairs cognition, light to moderate drinking overall does not appear to impair cognition in younger subjects and actually seems to reduce the risk of dementia and cognitive decline in older subjects [206].

Prolonged heavy alcohol consumption can result in the development of alcohol dependence, commonly known as alcoholism. Despite the vast research on alcoholism, the complex mechanisms underlying alcoholism are not entirely understood. It is believed that neuroadaptation to an altered neurotransmitter balance is responsible for the development of this disorder [207]. Alcohol has been shown to alter brain function by interacting with multiple neurotransmitter systems, such as glutamate, gamma-aminobutyric acid (GABA), and dopamine [208–211]. Repeated alcohol exposure can cause neuronal adaptations in various areas of the brain, including the prefrontal cortex and the hippocampus [177, 212]. Using a rat model of adolescent binge drinking, alcohol was found to alter the expression of multiple neurotransmitter system components in different areas of the brain, particularly the dopamine receptors in the striatum (Chang et al. unpublished data). Molecular and cellular adaptations in the nucleus accumbens, ventral tegmental area, amygdala, and prefrontal cortex have all been implicated in behavioral changes induced by chronic alcohol consumption, including craving and relapse [177].

The developing brain is particularly vulnerable to the effects of alcohol. Alcohol exposure during adolescence is more likely to interfere with normal brain functions, such as the development of self-regulation, judgment, reasoning, problem solving, and impulse control, and can increase the likelihood of alcohol dependence in adulthood. A number of studies have reported altered brain structure and function in alcohol-dependent or abusing adolescents and young adults compared to healthy individuals. These studies show smaller frontal and hippocampal volumes, altered white matter microstructure, and poorer memory [213–217].

While high alcohol consumption suppresses a wide range of immune responses, leading to an increased incidence in a number of infectious diseases, moderate alcohol consumption may have a beneficial impact on the immune system as compared to alcohol abuse or even abstinence. These seemingly contradictory effects are most likely due to the protective effects of the multiple components of polyphenol-rich alcoholic beverages on the immune system with moderate alcohol consumption [218]. However, it is clear that excessive alcohol drinking can cause cognitive and neuropathological changes. For example, acute alcohol intoxication can be followed by episodes of "blackout," a period of alcohol-induced amnesia [219]. Further, "binge" drinking, a pattern of rapid alcohol consumption which results in a blood alcohol concentration (BAC) of 0.08 gram percent or above, can lead to increased impulsivity, impaired spatial working memory, impaired reversal and emotional learning, facilitated excitatory neurotransmission, reduced synaptic plasticity in the amygdala and hippocampus, and neurodegeneration in the olfactory bulb, piriform cortex, perirhinal cortex, entorhinal cortex, and dentate gyrus [220–222]. For the typical male adult, a binge consists of consuming five or more drinks in a 2-h period, and for a female, four or more drinks within the same timeframe. Binge drinking is dangerous for both the drinker and society [207].

Although alcoholics show neurodegeneration after decades of drinking, recent studies using an animal model of binge drinking have found corticolimbic damage after as few as 4 days of binge exposure. Neurodegeneration can occur through either apoptotic or necrotic mechanisms. Binge drinking causes necrotic neurodegeneration after 2 days and increased damage after 4 days, but neurodegeneration does not increase during withdrawal [221]. Studies on the relationship between binge drinking and cognitive and emotional functioning in young adults found evidence for increased impulsivity, impairment in spatial working memory, and impaired emotional learning [220]. Even repetitive administration of a moderate dose of ethanol transiently impairs spatial learning and memory and promotes cognitive inflexibility.

15.17 Convergence of the Effects of Morphine and Alcohol on the HPA Axis

The endogenous opioid system is made up of three distinct families of opioid peptides processed from their precursor molecules, proopiomelanocortin, proenkephalin, and prodynorphin, and their opioid receptors. The endogenous opioid peptides, β-endorphins, enkephalins, and dynorphins, and three major classes of opioid receptor types, μ, δ, and κ, have been well characterized. Opioid peptides and receptors are broadly expressed throughout the peripheral and central nervous systems and are involved in many functions, such as nociception and analgesia; mood and well-being; addictive behaviors; and regulation of numerous physiological functions, including responses to stress, and endocrine and immune functions [223]. Morphine acts directly on opioid receptors, whereas alcohol interacts indirectly with opioid receptors by increasing endogenous levels of beta-endorphin and enkephalin, whose actions are mediated through the MOR [224]. The circulation level of corticosterone, the final product of the HPA axis, is not increased in MOR knockout mice in response to morphine injection [225]. Thus, it is evident that the MOR is at least one of the mechanisms underlying morphine and alcohol's actions on the HPA axis.

Alcohol promotes the release of several endogenous opioid peptides (such as β-endorphins), modifies their synthesis rate, and alters the binding properties of the opioid receptors [226]. Alterations in the activity of the brain and pituitary endogenous opioid system have been proposed to mediate or modulate, at least in part, many of the behavioral and neuroendocrine effects of alcohol. Acute ethanol exposure stimulates the release of brain β-endorphin. This enhanced β-endorphin release may interact with specific μ- and δ-opioid receptors in distinct regions of the brain and mediate, at least in part, many of the neurobehavioral effects of ethanol, including that of reinforcement and acquisition of ethanol drinking behavior. Conversely, the decrease in β-endorphin activity following prolonged exposure to ethanol can promote and maintain ethanol consumption through the mechanisms of negative rather than positive reinforcement. Alcohol administration, acute or chronic, can alter not only the activity of the endogenous opioid peptide systems but also the density or affinity of specific classes of opioid receptors in distinct regions of the brain. Such changes would alter the interactions of the opioid receptors with their respective endogenous ligands and, as a result, alter the functional activity of the endogenous opioid system. These ethanol-induced changes can be important in mediating some of the neurobehavioral effects of ethanol (including ethanol reinforcement) and can play a role in controlling ethanol consumption [227]. Acute or light alcohol consumption stimulates the release of opioid peptides in brain regions that are associated with reward and reinforcement and that mediate, at least in part, the reinforcing effects of ethanol. However, chronic heavy alcohol consumption induces a central opioid deficiency, which can be perceived as opioid withdrawal and can promote alcohol consumption through the mechanisms of negative reinforcement [228].

15.18 Conclusion

Table 15.1 summarizes and compares the effects of acute and chronic alcohol and morphine use on the HPA axis, innate immunity, adaptive immunity, neuroimmune axis, and cognition in order to give a comprehensive overview of the research discussed in this chapter.

Table 15.1 Effects of acute and chronic alcohol and morphine use on the HPA axis, immunity, neuroimmune axis, and cognition

	Alcohol		Morphine	
	Acute	Chronic	Acute	Chronic
HPA axis				
Causes neuron loss in the posterior magnocellular division of the PVN	–	Yes [125]	–	–
Activates the PVN in a dose-dependent manner	Yes [119]	–	–	–
Stimulates CRH-expressing hypothalamic neurons	Yes [131]	–	–	–
Neuronal activation in the PVN	Yes [119]	–	Yes [119, 229]	–
Desensitizes neuronal activation of the PVN; decreases ACTH response	–	Yes [119, 130, 132, 230]	Yes [31, 231]	Yes [34, 229, 230]
Desensitization of HPA axis and potentiation of immune responses	–	–	–	Yes [15, 34, 229]
Innate immunity				
Inhibits inflammatory cell activation	Yes [134, 138]	No [134, 138]	–	Yes [55]
Inhibits induction of proinflammatory mediator	Yes [139–141]	–	Yes [59]	–
Inhibits phagocytosis	Yes [232]	–	Yes [50, 51, 233]	Yes [50, 51]
Adaptive immunity				
Effects B-cell presence and/or function	Yes [134]	Yes [159]	Yes [62]	–
Causes T-cell abnormalities	–	Yes [160–163]	Yes [62, 64]	–
Enhances LPS-induced caspase activity	–	–	Yes [63]	–
Neuroimmune axis				
Impacts the immunosuppressive effects of GC hormones	Yes [181–183] No [234]	No [189]	Yes [89]	–
Activates NF-κB in the brain	Yes [175, 176]	–	Yes [82]	–
Induces neuronal activation via the MOR	–	–	Yes [235–237]	Yes [238]
Cognition				
Affects brain structures leading to cognitive impairment	Yes [221]	Yes [201–203]	–	–
Impacts molecular and/or cellular adaptations in PFC and amygdala	–	Yes [177]	–	–
Results in amnesia	Yes [219]	–	–	Yes [117]
Results in blackouts	Yes [219]	–	–	–

15.19 Summary

In summary, both alcohol and morphine cause neuroadaptative changes and immunomodulation. Alcohol interacts with multiple neurotransmitter systems, with inhibitory effects occurring after a short-term alcohol exposure and the development of tolerance, craving, alcohol-seeking behaviors, and alcohol withdrawal syndrome after long-term alcohol intake [239]. Multiple signal transduction systems are engaged once the MOR is activated by morphine. Thus, the behavioral manifestation of morphine tolerance and dependence is likely caused by a complex interplay of several events or adaptations occurring in multiple neurotransmitter systems [240]. Interestingly, chronic morphine treatment results in desensitization of the HPA axis and, consequently, the potentiation of the immune response. There have been conflicting results reported regarding the effects of exposure to alcohol on production of GC, the final product of the HPA axis. Many similarities and differences in the mechanisms by which morphine and alcohol modulate immune responses have been presented; however, the question remains if the discrepancy in results reported in the production of GC following treatment with alcohol is due to desensitization of the HPA axis following chronic treatment with alcohol. As shown in the studies surveyed, the effects of both alcohol and morphine on the neuroimmune axis are complex, and their modulation of the HPA axis may be the key point underlying the complexity of their effects on immunity and cognitive behavior [241].

Acknowledgments This work was supported, in part, by the National Institutes of Health/National Institute on Drug Abuse and National Institute on Alcohol Abuse and Alcoholism (R01 DA007058, R01 DA 026356, K02 DA016149, and RC2 AA019415 to SLC). The authors would also like to thank Dr. Louaine Spriggs and Ms. Kaitlyn P. Connaghan for the significant editorial support.

References

1. Erdmann G, Berger S, Schutz G (2008) Genetic dissection of glucocorticoid receptor function in the mouse brain. J Neuroendocrinol 20(6):655–659
2. Joels M (2007) Role of corticosteroid hormones in the dentate gyrus. Prog Brain Res 163:355–370
3. Itoi K, Jiang YQ, Iwasaki Y, Watson SJ (2004) Regulatory mechanisms of corticotropin-releasing hormone and vasopressin gene expression in the hypothalamus. J Neuroendocrinol 16(4):348–355
4. Swaab DF, Bao AM, Lucassen PJ (2005) The stress system in the human brain in depression and neurodegeneration. Ageing Res Rev 4(2):141–194
5. Reyes G, Elhai JD, Ford JD (2008) The Encyclopedia of Psychological Trauma. In: The Encyclopedia of Psychological Trauma Hoboken, NJ: John Wiley & Sons Inc.
6. Maier SF (2003) Bi-directional immune-brain communication: Implications for understanding stress, pain, and cognition. Brain Behav Immun 17(2):69–85
7. Haddad JJ (2008) On the mechanisms and putative pathways involving neuroimmune interactions. Biochem Biophys Res Commun 370(4):531–535
8. Wrona D (2006) Neural-immune interactions: an integrative view of the bidirectional relationship between the brain and immune systems. J Neuroimmunol 172(1–2):38–58

9. Sternberg EM (2006) Neural regulation of innate immunity: a coordinated nonspecific host response to pathogens. Nat Rev Immunol 6(4):318–328
10. U.S. Department of Justice DEA (2011) Drugs of Abuse- A DEA Resource Guide. (Administration, DE ed): U.S. Department of Justice, Drug Enforcement Administration http://www.justice.gov/dea/pubs/drugs_of_abuse.pdf Accessed: Dec 2011
11. Chen YL, Law PY, Loh HH (2006) Nuclear factor kappaB signaling in opioid functions and receptor gene expression. J Neuroimmune Pharmacol 1(3):270–279
12. Przewlocki R, Przewlocka B (2001) Opioids in chronic pain. Eur J Pharmacol 429(1–3):79–91
13. Langsdorf EF, Chang SL (2011) Methamphetamine-mediated modulation of MOR expression in the SH-SY5Y neuroblastoma cell line. Synapse 65(9):858–865
14. Chang SL, Beltran JA, Swarup S (2007) Expression of the mu opioid receptor in the human immunodeficiency virus type 1 transgenic rat model. J Virol 81(16):8406–8411
15. Ocasio FM, Jiang Y, House SD, Chang SL (2004) Chronic morphine accelerates the progression of lipopolysaccharide-induced sepsis to septic shock. J Neuroimmunol 149(1–2):90–100
16. Langsdorf EF, Mao X, Chang SL (2011) A role for reactive oxygen species in endotoxin-induced elevation of MOR expression in the nervous and immune systems. J Neuroimmunol 236(1–2):57–64
17. Law PY, Loh HH, Wei LN (2004) Insights into the receptor transcription and signaling: implications in opioid tolerance and dependence. Neuropharmacology 47(Suppl 1):300–311
18. Frolich N, Dees C, Paetz C, Ren X, Lohse MJ, Nikolaev VO, Zenk MH (2011) Distinct pharmacological properties of morphine metabolites at G(i)-protein and beta-arrestin signaling pathways activated by the human mu-opioid receptor. Biochem Pharmacol 81(10):1248–1254
19. Gris P, Cheng P, Pierson J, Maixner W, Diatchenko L (2010) Molecular assays for characterization of alternatively spliced isoforms of the u opioid receptor (MOR). Methods Mol Biol 617:421–435
20. Lefkowitz RJ, Shenoy SK (2005) Transduction of receptor signals by beta-arrestins. Science 308(5721):512–517
21. Zheng H, Loh HH, Law PY (2010) Agonist-selective signaling of G protein-coupled receptor: mechanisms and implications. IUBMB Life 62(2):112–119
22. Zhou Y, Proudnikov D, Yuferov V, Kreek MJ (2010) Drug-induced and genetic alterations in stress-responsive systems: implications for specific addictive diseases. Brain Res 1314:235–252
23. Benavides M, Laorden ML, Garcia-Borron JC, Milanes MV (2003) Regulation of tyrosine hydroxylase levels and activity and Fos expression during opioid withdrawal in the hypothalamic PVN and medulla oblongata catecholaminergic cell groups innervating the PVN. Eur J Neurosci 17(1):103–112
24. Laorden ML, Nunez C, Almela P, Milanes MV (2002) Morphine withdrawal-induced c-fos expression in the hypothalamic paraventricular nucleus is dependent on the activation of catecholaminergic neurones. J Neurochem 83(1):132–140
25. Laorden ML, Castells MT, Milanes MV (2002) Effects of morphine and morphine withdrawal on brainstem neurons innervating hypothalamic nuclei that control the pituitary-adrenocortical axis in rats. Br J Pharmacol 136(1):67–75
26. Laorden ML, Fuertes G, Gonzalez-Cuello A, Milanes MV (2000) Changes in catecholaminergic pathways innervating paraventricular nucleus and pituitary-adrenal axis response during morphine dependence: implication of alpha(1)- and alpha(2)-adrenoceptors. J Pharmacol Exp Ther 293(2):578–584
27. Milanes MV, Puig MM, Vargas ML (1993) Simultaneous changes in hypothalamic catecholamine levels and plasma corticosterone concentration in the rat after acute morphine and during tolerance. Neuropeptides 24(5):279–284
28. Palm S, Moenig H, Maier C (1997) Effects of oral treatment with sustained release morphine tablets on hypothalamic-pituitary-adrenal axis. Methods Find Exp Clin Pharmacol 19(4):269–273

29. Provinciali M, Di Stefano G, Stronati S, Raffaeli W, Pari G, Fabris N (1996) Role of prolactin in the modulation of NK and LAK cell activity after short- or long-term morphine administration in neoplastic patients. Int J Immunopharmacol 18(10):577–586
30. Allolio B, Schulte HM, Deuss U, Kallabis D, Hamel E, Winkelman W (1987) Effect of oral morphine and naloxone on pituitary-adrenal response in man induced by human corticotropin-releasing hormone. Acta Endocrinol (Copenh) 114(4):509–514
31. Rittmaster RS, Cutler GB Jr, Sobel DO, Goldstein DS, Koppelman MC, Loriaux DL, Chrousos GP (1985) Morphine inhibits the pituitary-adrenal response to ovine corticotropin-releasing hormone in normal subjects. J Clin Endocrinol Metab 60(5):891–895
32. Esmaeili Mahani S, Motamedi F, Javan M, Ahmadiani A (2005) Involvement of hypothalamic pituitary adrenal axis on the effects of nifedipine in the development of morphine tolerance in rats. Pharmacol Biochem Behav 81(1):152–157
33. Buckingham JC, Cooper TA (1984) Differences in hypothalamo-pituitary-adrenocortical activity in the rat after acute and prolonged treatment with morphine. Neuroendocrinology 38(5):411–417
34. House SD, Mao X, Wu G, Espinelli D, Li WX, Chang SL (2001) Chronic morphine potentiates the inflammatory response by disrupting interleukin-1beta modulation of the hypothalamic-pituitary-adrenal axis. J Neuroimmunol 118(2):277–285
35. Nunez C, Foldes A, Laorden ML, Milanes MV, Kovacs KJ (2007) Activation of stress-related hypothalamic neuropeptide gene expression during morphine withdrawal. J Neurochem 101(4):1060–1071
36. Houshyar H, Gomez F, Manalo S, Bhargava A, Dallman MF (2003) Intermittent morphine administration induces dependence and is a chronic stressor in rats. Neuropsychopharmacology 28(11):1960–1972
37. Martin F, Mora L, Laorden M, Milanes M (2011) Protein kinase C phosphorylates the cAMP response element binding protein in the hypothalamic paraventricular nucleus during morphine withdrawal. Br J Pharmacol 163(4):857–875
38. Martin F, Laorden ML, Milanes MV (2009) Morphine withdrawal regulates phosphorylation of cAMP response element binding protein (CREB) through PKC in the nucleus tractus solitarius-A2 catecholaminergic neurons. J Neurochem 110(5):1422–1432
39. Esmaeili-Mahani S, Fathi Y, Motamedi F, Hosseinpanah F, Ahmadiani A (2008) L-type calcium channel blockade attenuates morphine withdrawal: in vivo interaction between L-type calcium channels and corticosterone. Horm Behav 53(2):351–357
40. Takeuchi O, Akira S (2010) Pattern recognition receptors and inflammation. Cell 140(6):805–820
41. Kawai T, Akira S (2010) The role of pattern-recognition receptors in innate immunity: update on Toll-like receptors. Nat Immunol 11(5):373–384
42. Kumar H, Kawai T, Akira S (2011) Pathogen recognition by the innate immune system. Int Rev Immunol 30(1):16–34
43. Schenten D, Medzhitov R (2011) The control of adaptive immune responses by the innate immune system. Adv Immunol 109:87–124
44. Iwasaki A, Medzhitov R (2010) Regulation of adaptive immunity by the innate immune system. Science 327(5963):291–295
45. Breslow JM, Monroy MA, Daly JM, Meissler JJ, Gaughan J, Adler MW, Eisenstein TK (2011) Morphine, but not trauma, sensitizes to systemic *Acinetobacter baumannii* infection. J Neuroimmune Pharmacol 6(4):551–565
46. Martin JL, Koodie L, Krishnan AG, Charboneau R, Barke RA, Roy S (2010) Chronic morphine administration delays wound healing by inhibiting immune cell recruitment to the wound site. Am J Pathol 176(2):786–799
47. Ma J, Wang J, Wan J, Charboneau R, Chang Y, Barke RA, Roy S (2010) Morphine disrupts interleukin-23 (IL-23)/IL-17-mediated pulmonary mucosal host defense against *Streptococcus pneumoniae* infection. Infect Immun 78(2):830–837
48. Tomassini N, Renaud F, Roy S, Loh HH (2004) Morphine inhibits Fc-mediated phagocytosis through mu and delta opioid receptors. J Neuroimmunol 147(1–2):131–133

49. Tomassini N, Renaud FL, Roy S, Loh HH (2003) Mu and delta receptors mediate morphine effects on phagocytosis by murine peritoneal macrophages. J Neuroimmunol 136(1–2):9–16
50. Delgado-Velez M, Lugo-Chinchilla A, Lizardo L, Morales I, Robles Y, Bruno N, Rodriguez JW, Rios-Olivares E, Correa M, Renaud FL (2008) Chronic exposure of human macrophages in vitro to morphine and methadone induces a putative tolerant/dependent state. J Neuroimmunol 196(1–2):94–100
51. Lazaro MI, Tomassini N, Gonzalez I, Renaud FL (2000) Reversibility of morphine effects on phagocytosis by murine macrophages. Drug Alcohol Depend 58(1–2):159–164
52. Wang J, Barke RA, Charboneau R, Schwendener R, Roy S (2008) Morphine induces defects in early response of alveolar macrophages to *Streptococcus pneumoniae* by modulating TLR9-NF-kappa B signaling. J Immunol 180(5):3594–3600
53. Messmer D, Hatsukari I, Hitosugi N, Schmidt-Wolf IG, Singhal PC (2006) Morphine reciprocally regulates IL-10 and IL-12 production by monocyte-derived human dendritic cells and enhances T cell activation. Mol Med 12(11–12):284–290
54. Campana G, Sarti D, Spampinato S, Raffaeli W (2010) Long-term intrathecal morphine and bupivacaine upregulate MOR gene expression in lymphocytes. Int Immunopharmacol 10(9):1149–1152
55. Yokota T, Uehara K, Nomoto Y (2000) Intrathecal morphine suppresses NK cell activity following abdominal surgery. Can J Anaesth 47(4):303–308
56. Madera-Salcedo IK, Cruz SL, Gonzalez-Espinosa C (2011) Morphine decreases early peritoneal innate immunity responses in Swiss-Webster and C57BL6/J mice through the inhibition of mast cell TNF-alpha release. J Neuroimmunol 232(1–2):101–107
57. Peng X, Mosser DM, Adler MW, Rogers TJ, Meissler JJ Jr, Eisenstein TK (2000) Morphine enhances interleukin-12 and the production of other pro-inflammatory cytokines in mouse peritoneal macrophages. J Leukoc Biol 68(5):723–728
58. Wang J, Ma J, Charboneau R, Barke R, Roy S (2011) Morphine inhibits murine dendritic cell IL-23 production by modulating Toll-like receptor 2 and Nod2 signaling. J Biol Chem 286(12):10225–10232
59. Bonnet MP, Beloeil H, Benhamou D, Mazoit JX, Asehnoune K (2008) The mu opioid receptor mediates morphine-induced tumor necrosis factor and interleukin-6 inhibition in toll-like receptor 2-stimulated monocytes. Anesth Analg 106(4):1142–1149 (table of contents)
60. Kelschenbach J, Ninkovic J, Wang J, Krishnan A, Charboneau R, Barke RA, Roy S (2008) Morphine withdrawal inhibits IL-12 induction in a macrophage cell line through a mechanism that involves cAMP. J Immunol 180(6):3670–3679
61. Kelschenbach J, Barke RA, Roy S (2005) Morphine withdrawal contributes to Th cell differentiation by biasing cells toward the Th2 lineage. J Immunol 175(4):2655–2665
62. Zhang E, Xiong J, Parker B, Chen A, Fields P, Ma X, Qiu J, Yankee T (2011) Depletion and recovery of lymphoid subsets following morphine administration. Br J Pharmacol 164(7):1829–1844
63. Olin MR, Roy S, Molitor T (2010) In vivo morphine treatment synergistically increases LPS-induced caspase activity in immune organs. J Neuroimmune Pharmacol 5(4):546–552
64. Mojadadi S, Jamali A, Khansarinejad B, Soleimanjahi H, Bamdad T (2009) Acute morphine administration reduces cell-mediated immunity and induces reactivation of latent herpes simplex virus type 1 in BALB/c mice. Cell Mol Immunol 6(2):111–116
65. Borner C, Warnick B, Smida M, Hartig R, Lindquist JA, Schraven B, Hollt V, Kraus J (2009) Mechanisms of opioid-mediated inhibition of human T cell receptor signaling. J Immunol 183(2):882–889
66. Olin M, Oh S, Roy S, Peterson P, Molitor T (2011) Morphine induces splenocyte trafficking into the CNS. J Neuroimmune Pharmacol 7:436–443
67. Sharma HS, Sjoquist PO, Ali SF (2010) Alterations in blood–brain barrier function and brain pathology by morphine in the rat. Neuroprotective effects of antioxidant H-290/51. Acta Neurochir Suppl 106:61–66
68. Mahajan SD, Aalinkeel R, Sykes DE, Reynolds JL, Bindukumar B, Fernandez SF, Chawda R, Shanahan TC, Schwartz SA (2008) Tight junction regulation by morphine and HIV-1 tat modulates blood–brain barrier permeability. J Clin Immunol 28(5):528–541

69. Wen H, Lu Y, Yao H, Buch S (2011) Morphine induces expression of platelet-derived growth factor in human brain microvascular endothelial cells: implication for vascular permeability. PLoS One 6(6):e21707
70. Rock RB, Hu S, Sheng WS, Peterson PK (2006) Morphine stimulates CCL2 production by human neurons. J Neuroinflammation 3:32
71. Mahajan SD, Schwartz SA, Shanahan TC, Chawda RP, Nair MP (2002) Morphine regulates gene expression of alpha- and beta-chemokines and their receptors on astroglial cells via the opioid mu receptor. J Immunol 169(7):3589–3599
72. Staikos L, Malellari L, Chang SL (2008) Lipopolysaccharide-induced pro-inflammatory cytokines in the brain of rats in the morphine-tolerant state. J Neuroimmune Pharmacol 3(4):236–240
73. Suder P, Bodzon-Kulakowska A, Mak P, Bierczynska-Krzysik A, Daszykowski M, Walczak B, Lubec G, Kotlinska JH, Silberring J (2009) The proteomic analysis of primary cortical astrocyte cell culture after morphine administration. J Proteome Res 8(10):4633–4640
74. Cottingham K (2009) Astrocyte proteins and morphine addiction. J Proteome Res 8(10):4394
75. Narita M, Suzuki M, Kuzumaki N, Miyatake M, Suzuki T (2008) Implication of activated astrocytes in the development of drug dependence: differences between methamphetamine and morphine. Ann N Y Acad Sci 1141:96–104
76. Narita M, Miyatake M, Shibasaki M, Shindo K, Nakamura A, Kuzumaki N, Nagumo Y, Suzuki T (2006) Direct evidence of astrocytic modulation in the development of rewarding effects induced by drugs of abuse. Neuropsychopharmacology 31(11):2476–2488
77. Lazriev IL, Kiknadze GI, Kutateladze II, Nebieridze MI (2001) Effect of morphine on the number and branching of astrocytes in various regions of rat brain. Bull Exp Biol Med 131(3):248–250
78. Mika J (2008) Modulation of microglia can attenuate neuropathic pain symptoms and enhance morphine effectiveness. Pharmacol Rep 60(3):297–307
79. Lipovsky MM, Gekker G, Hu S, Hoepelman AI, Peterson PK (1998) Morphine enhances complement receptor-mediated phagocytosis of *Cryptococcus neoformans* by human microglia. Clin Immunol Immunopathol 87(2):163–167
80. Zhang Y, Li H, Li Y, Sun X, Zhu M, Hanley G, Lesage G, Yin D (2011) Essential role of toll-like receptor 2 in morphine-induced microglia activation in mice. Neurosci Lett 489(1):43–47
81. Hutchinson MR, Zhang Y, Shridhar M, Evans JH, Buchanan MM, Zhao TX, Slivka PF, Coats BD, Rezvani N, Wieseler J, Hughes TS, Landgraf KE, Chan S, Fong S, Phipps S, Falke JJ, Leinwand LA, Maier SF, Yin H, Rice KC, Watkins LR (2010) Evidence that opioids may have toll-like receptor 4 and MD-2 effects. Brain Behav Immun 24(1):83–95
82. Sawaya BE, Deshmane SL, Mukerjee R, Fan S, Khalili K (2009) TNF alpha production in morphine-treated human neural cells is NF-kappaB-dependent. J Neuroimmune Pharmacol 4(1):140–149
83. Horvath RJ, DeLeo JA (2009) Morphine enhances microglial migration through modulation of P2X4 receptor signaling. J Neurosci 29(4):998–1005
84. Chang SL, Wu GD, Patel NA, Vidal EL, Fiala M (1998) The effects of interaction between morphine and interleukin-1 on the immune response. Adv Exp Med Biol 437:67–72
85. Zhang Y, Wu GC, He QZ, Cao XD (2000) Effect of morphine and electro-acupuncture (EA) on apoptosis of thymocytes. Acupunct Electrother Res 25(1):17–26
86. Saurer TB, Carrigan KA, Ijames SG, Lysle DT (2006) Suppression of natural killer cell activity by morphine is mediated by the nucleus accumbens shell. J Neuroimmunol 173(1–2):3–11
87. Saurer TB, Ijames SG, Lysle DT (2006) Neuropeptide Y Y1 receptors mediate morphine-induced reductions of natural killer cell activity. J Neuroimmunol 177(1–2):18–26
88. Saurer TB, Ijames SG, Carrigan KA, Lysle DT (2008) Neuroimmune mechanisms of opioid-mediated conditioned immunomodulation. Brain Behav Immun 22(1):89–97
89. Nugent AL, Houghtling RA, Bayer BM (2011) Morphine suppresses MHC-II expression on circulating B lymphocytes via activation of the HPA. J Neuroimmune Pharmacol 6(1):130–141

90. Sinatra R (2006) Opioid analgesics in primary care: challenges and new advances in the management of noncancer pain. J Am Board Fam Med 19(2):165–177
91. Hamilton GR, Baskett TF (2000) In the arms of Morpheus the development of morphine for postoperative pain relief. Can J Anaesth 47(4):367–374
92. Schulz S, Schreff M, Koch T, Zimprich A, Gramsch C, Elde R, Hollt V (1998) Immunolocalization of two mu-opioid receptor isoforms (MOR1 and MOR1B) in the rat central nervous system. Neuroscience 82(2):613–622
93. Ding YQ, Kaneko T, Nomura S, Mizuno N (1996) Immunohistochemical localization of mu-opioid receptors in the central nervous system of the rat. J Comp Neurol 367(3):375–402
94. Ninkovic J, Roy S (2011) Role of the mu-opioid receptor in opioid modulation of immune function. Amino Acids [2011 Dec 15 E-pub ahead of print]
95. Chen Y, Sommer C (2009) The role of mitogen-activated protein kinase (MAPK) in morphine tolerance and dependence. Mol Neurobiol 40(2):101–107
96. Cherny N, Ripamonti C, Pereira J, Davis C, Fallon M, McQuay H, Mercadante S, Pasternak G, Ventafridda V (2001) Strategies to manage the adverse effects of oral morphine: an evidence-based report. J Clin Oncol 19(9):2542–2554
97. Hassanzadeh K, Roshangar L, Habibi-asl B, Farajnia S, Izadpanah E, Nemati M, Arasteh M, Mohammadi S (2011) Riluzole prevents morphine-induced apoptosis in rat cerebral cortex. Pharmacol Rep 63(3):697–707
98. Garzon J, Rodriguez-Munoz M, Sanchez-Blazquez P (2008) Do pharmacological approaches that prevent opioid tolerance target different elements in the same regulatory machinery? Curr Drug Abuse Rev 1(2):222–238
99. Gintzler AR, Chakrabarti S (2006) Post-opioid receptor adaptations to chronic morphine; altered functionality and associations of signaling molecules. Life Sci 79(8):717–722
100. Chakrabarti S, Regec A, Gintzler AR (2005) Chronic morphine acts via a protein kinase Cgamma-G(beta)-adenylyl cyclase complex to augment phosphorylation of G(beta) and G(betagamma) stimulatory adenylyl cyclase signaling. Brain Res Mol Brain Res 138(1):94–103
101. Rozenfeld R, Abul-Husn NS, Gomez I, Devi LA (2007) An emerging role for the delta opioid receptor in the regulation of mu opioid receptor function. Sci World J 7:64–73
102. Martini L, Whistler JL (2007) The role of mu opioid receptor desensitization and endocytosis in morphine tolerance and dependence. Curr Opin Neurobiol 17(5):556–564
103. Zuo Z (2005) The role of opioid receptor internalization and beta-arrestins in the development of opioid tolerance. Anesth Analg 101(3):728–734 (table of contents)
104. Dietis N, Guerrini R, Calo G, Salvadori S, Rowbotham DJ, Lambert DG (2009) Simultaneous targeting of multiple opioid receptors: a strategy to improve side-effect profile. Br J Anaesth 103(1):38–49
105. Toda N, Kishioka S, Hatano Y, Toda H (2009) Modulation of opioid actions by nitric oxide signaling. Anesthesiology 110(1):166–181
106. Hyman SE (2005) Addiction: a disease of learning and memory. Am J Psychiatry 162(8):1414–1422
107. Hyman SE, Malenka RC, Nestler EJ (2006) Neural mechanisms of addiction: the role of reward-related learning and memory. Annu Rev Neurosci 29:565–598
108. Battaglia FP, Benchenane K, Sirota A, Pennartz CM, Wiener SI (2011) The hippocampus: hub of brain network communication for memory. Trends Cogn Sci 15(7):310–318
109. Bird CM, Burgess N (2008) The hippocampus and memory: insights from spatial processing. Nat Rev Neurosci 9(3):182–194
110. Miladi-Gorji H, Rashidy-Pour A, Fathollahi Y, Akhavan MM, Semnanian S, Safari M (2011) Voluntary exercise ameliorates cognitive deficits in morphine dependent rats: the role of hippocampal brain-derived neurotrophic factor. Neurobiol Learn Mem 96(3):479–491
111. Eisch AJ, Barrot M, Schad CA, Self DW, Nestler EJ (2000) Opiates inhibit neurogenesis in the adult rat hippocampus. Proc Natl Acad Sci USA 97(13):7579–7584
112. Sala M, Braida D, Leone MP, Calcaterra P, Frattola D, Gori E (1994) Chronic morphine affects working memory during treatment and withdrawal in rats: possible residual long-term impairment. Behav Pharmacol 5(6):570–580

113. Spain JW, Newsom GC (1991) Chronic opioids impair acquisition of both radial maze and Y-maze choice escape. Psychopharmacology (Berl) 105(1):101–106
114. Ma MX, Chen YM, He J, Zeng T, Wang JH (2007) Effects of morphine and its withdrawal on Y-maze spatial recognition memory in mice. Neuroscience 147(4):1059–1065
115. Aguilar MA, Minarro J, Simon VM (1998) Dose-dependent impairing effects of morphine on avoidance acquisition and performance in male mice. Neurobiol Learn Mem 69(2): 92–105
116. Farahmandfar M, Karimian SM, Naghdi N, Zarrindast MR, Kadivar M (2010) Morphine-induced impairment of spatial memory acquisition reversed by morphine sensitization in rats. Behav Brain Res 211(2):156–163
117. Kamboj SK, Tookman A, Jones L, Curran HV (2005) The effects of immediate-release morphine on cognitive functioning in patients receiving chronic opioid therapy in palliative care. Pain 117(3):388–395
118. World Health Organization, Global Status Report on Alcohol and Health (2011) http://www.who.int/substance_abuse/publications/global_alcohol_report/en/ Accessed: Dec 2011
119. Chang SL, Patel NA, Romero AA (1995) Activation and desensitization of Fos immunoreactivity in the rat brain following ethanol administration. Brain Res 679(1):89–98
120. Mechtcheriakov S, Brenneis C, Egger K, Koppelstaetter F, Schocke M, Marksteiner J (2007) A widespread distinct pattern of cerebral atrophy in patients with alcohol addiction revealed by voxel-based morphometry. J Neurol Neurosurg Psychiatry 78(6):610–614
121. Chanraud S, Martelli C, Delain F, Kostogianni N, Douaud G, Aubin HJ, Reynaud M, Martinot JL (2007) Brain morphometry and cognitive performance in detoxified alcohol-dependents with preserved psychosocial functioning. Neuropsychopharmacology 32(2):429–438
122. Makris N, Oscar-Berman M, Jaffin SK, Hodge SM, Kennedy DN, Caviness VS, Marinkovic K, Breiter HC, Gasic GP, Harris GJ (2008) Decreased volume of the brain reward system in alcoholism. Biol Psychiatry 64(3):192–202
123. Fortier CB, Leritz EC, Salat DH, Venne JR, Maksimovskiy AL, Williams V, Milberg WP, McGlinchey RE (2011) Reduced cortical thickness in abstinent alcoholics and association with alcoholic behavior. Alcohol Clin Exp Res 35(12):2193–2201
124. Zahr NM, Kaufman KL, Harper CG (2011) Clinical and pathological features of alcohol-related brain damage. Nat Rev Neurol 7(5):284–294
125. Silva SM, Madeira MD, Ruela C, Paula-Barbosa MM (2002) Prolonged alcohol intake leads to irreversible loss of vasopressin and oxytocin neurons in the paraventricular nucleus of the hypothalamus. Brain Res 925(1):76–88
126. Harding AJ, Halliday GM, Ng JL, Harper CG, Kril JJ (1996) Loss of vasopressin-immunoreactive neurons in alcoholics is dose-related and time-dependent. Neuroscience 72(3):699–708
127. Sivukhina EV, Dolzhikov AA, Morozov Iu E, Jirikowski GF, Grinevich V (2006) Effects of chronic alcoholic disease on magnocellular and parvocellular hypothalamic neurons in men. Horm Metab Res 38(6):382–390
128. Allen CD, Lee S, Koob GF, Rivier C (2011) Immediate and prolonged effects of alcohol exposure on the activity of the hypothalamic-pituitary-adrenal axis in adult and adolescent rats. Brain Behav Immun 25(Suppl 1):S50–S60
129. Pronko SP, Saba LM, Hoffman PL, Tabakoff B (2010) Type 7 adenylyl cyclase-mediated hypothalamic-pituitary-adrenal axis responsiveness: influence of ethanol and sex. J Pharmacol Exp Ther 334(1):44–52
130. Seo DO, Lee S, Rivier C (2004) Prolonged exposure to intermittent alcohol vapors decreases the ACTH as well as hypothalamic nitric oxide and cytokine responses to endotoxemia. Alcohol Clin Exp Res 28(6):848–854
131. Rivier C, Lee S (1996) Acute alcohol administration stimulates the activity of hypothalamic neurons that express corticotropin-releasing factor and vasopressin. Brain Res 726(1–2):1–10
132. Lee S, Schmidt D, Tilders F, Cole M, Smith A, Rivier C (2000) Prolonged exposure to intermittent alcohol vapors blunts hypothalamic responsiveness to immune and non-immune signals. Alcohol Clin Exp Res 24(1):110–122

133. Sinha R, Fox HC, Hong KI, Hansen J, Tuit K, Kreek MJ (2011) Effects of adrenal sensitivity, stress- and cue-induced craving, and anxiety on subsequent alcohol relapse and treatment outcomes. Arch Gen Psychiatry 68(9):942–952
134. Szabo G, Mandrekar P (2009) A recent perspective on alcohol, immunity, and host defense. Alcohol Clin Exp Res 33(2):220–232
135. Cook RT (1998) Alcohol abuse, alcoholism, and damage to the immune system – a review. Alcohol Clin Exp Res 22(9):1927–1942
136. Watson RR, Borgs P, Witte M, McCuskey RS, Lantz C, Johnson MI, Mufti SI, Earnest DL (1994) Alcohol, immunomodulation, and disease. Alcohol Alcohol 29(2):131–139
137. Diaz LE, Montero A, Gonzalez-Gross M, Vallejo AI, Romeo J, Marcos A (2002) Influence of alcohol consumption on immunological status: a review. Eur J Clin Nutr 56(Suppl 3):S50–S53
138. Goral J, Karavitis J, Kovacs EJ (2008) Exposure-dependent effects of ethanol on the innate immune system. Alcohol 42(4):237–247
139. Szabo G, Mandrekar P, Catalano D (1995) Inhibition of superantigen-induced T cell proliferation and monocyte IL-1 beta, TNF-alpha, and IL-6 production by acute ethanol treatment. J Leukoc Biol 58(3):342–350
140. Szabo G, Mandrekar P, Girouard L, Catalano D (1996) Regulation of human monocyte functions by acute ethanol treatment: decreased tumor necrosis factor-alpha, interleukin-1 beta and elevated interleukin-10, and transforming growth factor-beta production. Alcohol Clin Exp Res 20(5):900–907
141. Szabo G, Chavan S, Mandrekar P, Catalano D (1999) Acute alcohol consumption attenuates interleukin-8 (IL-8) and monocyte chemoattractant peptide-1 (MCP-1) induction in response to ex vivo stimulation. J Clin Immunol 19(1):67–76
142. Liu X, Mao X, Chang L (2011) Altered gene expression in the spleen of adolescent rats following high ethanol concentration binge drinking. Int J Clin Exp Med 4(4):252–257
143. Spitzer JA, Zhang P (1996) Gender differences in phagocytic responses in the blood and liver, and the generation of cytokine-induced neutrophil chemoattractant in the liver of acutely ethanol-intoxicated rats. Alcohol Clin Exp Res 20(5):914–920
144. Kovacs EJ, Messingham KA (2002) Influence of alcohol and gender on immune response. Alcohol Res Health 26(4):257–263
145. Gao B, Jeong WI, Tian Z (2008) Liver: an organ with predominant innate immunity. Hepatology 47(2):729–736
146. Yamashina S, Ikejima K, Enomoto N, Takei Y, Sato N (2003) Ethanol changes sensitivity of Kupffer cells to endotoxin. Nihon Arukoru Yakubutsu Igakkai Zasshi 38(5):415–424
147. Robin MA, Demeilliers C, Sutton A, Paradis V, Maisonneuve C, Dubois S, Poirel O, Letteron P, Pessayre D, Fromenty B (2005) Alcohol increases tumor necrosis factor alpha and decreases nuclear factor-kappab to activate hepatic apoptosis in genetically obese mice. Hepatology 42(6):1280–1290
148. Zhou Z, Wang L, Song Z, Lambert JC, McClain CJ, Kang YJ (2003) A critical involvement of oxidative stress in acute alcohol-induced hepatic TNF-alpha production. Am J Pathol 163(3):1137–1146
149. Bautista AP (2000) Impact of alcohol on the ability of Kupffer cells to produce chemokines and its role in alcoholic liver disease. J Gastroenterol Hepatol 15(4):349–356
150. Mandrekar P, Ambade A, Lim A, Szabo G, Catalano D (2011) An essential role for monocyte chemoattractant protein-1 in alcoholic liver injury: regulation of proinflammatory cytokines and hepatic steatosis in mice. Hepatology 54(6):2185–2197
151. Thiele GM, Szabo G, Kovacs EJ, Bautista AP, Sosa L, Jerrells TR (2002) Modulation of immunity and viral-host interactions by alcohol. Alcohol Clin Exp Res 26(12):1897–1908
152. Bautista AP, Wang E (2002) Acute ethanol administration downregulates human immunodeficiency virus-1 glycoprotein 120-induced KC and RANTES production by murine Kupffer cells and splenocytes. Life Sci 71(4):371–382
153. Szabo G, Catalano D, White B, Mandrekar P (2004) Acute alcohol consumption inhibits accessory cell function of monocytes and dendritic cells. Alcohol Clin Exp Res 28(5):824–828

154. Lau AH, Abe M, Thomson AW (2006) Ethanol affects the generation, cosignaling molecule expression, and function of plasmacytoid and myeloid dendritic cell subsets in vitro and in vivo. J Leukoc Biol 79(5):941–953
155. Vivier E, Raulet DH, Moretta A, Caligiuri MA, Zitvogel L, Lanier LL, Yokoyama WM, Ugolini S (2011) Innate or adaptive immunity? The example of natural killer cells. Science 331(6013):44–49
156. Marcenaro E, Carlomagno S, Pesce S, Moretta A, Sivori S (2011) Bridging innate NK cell functions with adaptive immunity. Adv Exp Med Biol 780:45–55
157. Zhang H, Zhu Z, Meadows GG (2011) Chronic alcohol consumption decreases the percentage and number of NK cells in the peripheral lymph nodes and exacerbates B16BL6 melanoma metastasis into the draining lymph nodes. Cell Immunol 266(2):172–179
158. Zhang H, Meadows GG (2008) Chronic alcohol consumption perturbs the balance between thymus-derived and bone marrow-derived natural killer cells in the spleen. J Leukoc Biol 83(1):41–47
159. Verma S, Alexander CM, Carlson MJ, Tygrett LT, Waldschmidt TJ (2008) B-cell studies in chronic ethanol mice. Methods Mol Biol 447:295–323
160. Naude CE, Bouic P, Senekal M, Kidd M, Ferrett HL, Fein G, Carey PD (2011) Lymphocyte measures in treatment-naive 13-15-year old adolescents with alcohol use disorders. Alcohol 45(5):507–514
161. Boyadjieva NI, Dokur M, Advis JP, Meadows GG, Sarkar DK (2002) Beta-endorphin modulation of lymphocyte proliferation: effects of ethanol. Alcohol Clin Exp Res 26(11):1719–1727
162. Gurung P, Young BM, Coleman RA, Wiechert S, Turner LE, Ray NB, Waldschmidt TJ, Legge KL, Cook RT (2009) Chronic ethanol induces inhibition of antigen-specific CD8+ but not CD4+ immunodominant T cell responses following *Listeria monocytogenes* inoculation. J Leukoc Biol 85(1):34–43
163. Zhang H, Meadows GG (2010) Chronic alcohol consumption enhances myeloid-derived suppressor cells in B16BL6 melanoma-bearing mice. Cancer Immunol Immunother 59(8):1151–1159
164. Bailey SL, Carpentier PA, McMahon EJ, Begolka WS, Miller SD (2006) Innate and adaptive immune responses of the central nervous system. Crit Rev Immunol 26(2):149–188
165. Hanisch UK, Kettenmann H (2007) Microglia: active sensor and versatile effector cells in the normal and pathologic brain. Nat Neurosci 10(11):1387–1394
166. Walter L, Neumann H (2009) Role of microglia in neuronal degeneration and regeneration. Semin Immunopathol 31(4):513–525
167. Crews FT, Zou J, Qin L (2011) Induction of innate immune genes in brain create the neurobiology of addiction. Brain Behav Immun 25(Suppl 1):S4–S12
168. Pekny M, Nilsson M (2005) Astrocyte activation and reactive gliosis. Glia 50(4):427–434
169. Pekny M, Wilhelmsson U, Bogestal YR, Pekna M (2007) The role of astrocytes and complement system in neural plasticity. Int Rev Neurobiol 82:95–111
170. Fernandez-Lizarbe S, Pascual M, Guerri C (2009) Critical role of TLR4 response in the activation of microglia induced by ethanol. J Immunol 183(7):4733–4744
171. Blanco AM, Pascual M, Valles SL, Guerri C (2004) Ethanol-induced iNOS and COX-2 expression in cultured astrocytes via NF-kappa B. Neuroreport 15(4):681–685
172. Valles SL, Blanco AM, Pascual M, Guerri C (2004) Chronic ethanol treatment enhances inflammatory mediators and cell death in the brain and in astrocytes. Brain Pathol 14(4):365–371
173. Alfonso-Loeches S, Pascual-Lucas M, Blanco AM, Sanchez-Vera I, Guerri C (2010) Pivotal role of TLR4 receptors in alcohol-induced neuroinflammation and brain damage. J Neurosci 30(24):8285–8295
174. Yakovleva T, Bazov I, Watanabe H, Hauser KF, Bakalkin G (2011) Transcriptional control of maladaptive and protective responses in alcoholics: a role of the NF-kappaB system. Brain Behav Immun 25(Suppl 1):S29–S38
175. Rulten SL, Ripley TL, Hunt CL, Stephens DN, Mayne LV (2006) Sp1 and NFkappaB pathways are regulated in brain in response to acute and chronic ethanol. Genes Brain Behav 5(3):257–273

176. Crews F, Nixon K, Kim D, Joseph J, Shukitt-Hale B, Qin L, Zou J (2006) BHT blocks NF-kappaB activation and ethanol-induced brain damage. Alcohol Clin Exp Res 30(11):1938–1949
177. Okvist A, Johansson S, Kuzmin A, Bazov I, Merino-Martinez R, Ponomarev I, Mayfield RD, Harris RA, Sheedy D, Garrick T, Harper C, Hurd YL, Terenius L, Ekstrom TJ, Bakalkin G, Yakovleva T (2007) Neuroadaptations in human chronic alcoholics: dysregulation of the NF-kappaB system. PLoS One 2(9):e930
178. Haorah J, Knipe B, Gorantla S, Zheng J, Persidsky Y (2007) Alcohol-induced blood–brain barrier dysfunction is mediated via inositol 1,4,5-triphosphate receptor (IP3R)-gated intracellular calcium release. J Neurochem 100(2):324–336
179. Haorah J, Knipe B, Leibhart J, Ghorpade A, Persidsky Y (2005) Alcohol-induced oxidative stress in brain endothelial cells causes blood–brain barrier dysfunction. J Leukoc Biol 78(6):1223–1232
180. Rivest S (2010) Interactions between the immune and neuroendocrine systems. Prog Brain Res 181:43–53
181. Glover M, Cheng B, Deng X, Pruett S (2011) The role of glucocorticoids in the immediate vs. delayed effects of acute ethanol exposure on cytokine production in a binge drinking model. Int Immunopharmacol 11(6):755–761
182. Glover M, Pruett SB (2006) Role of corticosterone in immunosuppressive effects of acute ethanol exposure on Toll-like receptor mediated cytokine production. J Neuroimmune Pharmacol 1(4):435–442
183. Collier SD, Wu WJ, Pruett SB (2000) Ethanol suppresses NK cell activation by polyinosinic-polycytidylic acid (poly I:C) in female B6C3F1 mice: role of endogenous corticosterone. Alcohol Clin Exp Res 24(3):291–299
184. Padgett EL, Sibley DA, Jerrells TR (2000) Effect of adrenalectomy on ethanol-associated changes in lymphocyte cell numbers and subpopulations in thymus, spleen, and gut-associated lymphoid tissues. Int J Immunopharmacol 22(4):285–298
185. Rose AK, Shaw SG, Prendergast MA, Little HJ (2010) The importance of glucocorticoids in alcohol dependence and neurotoxicity. Alcohol Clin Exp Res 34(12):2011–2018
186. Spencer RL, Hutchison KE (1999) Alcohol, aging, and the stress response. Alcohol Res Health 23(4):272–283
187. Flammer JR, Rogatsky I (2011) Minireview: glucocorticoids in autoimmunity: unexpected targets and mechanisms. Mol Endocrinol 25(7):1075–1086
188. Pruett SB, Fan R, Zheng Q (2003) Acute ethanol administration profoundly alters poly I:C-induced cytokine expression in mice by a mechanism that is not dependent on corticosterone. Life Sci 72(16):1825–1839
189. Cook RT, Schlueter AJ, Coleman RA, Tygrett L, Ballas ZK, Jerrells TR, Nashelsky MB, Ray NB, Haugen TH, Waldschmidt TJ (2007) Thymocytes, pre-B cells, and organ changes in a mouse model of chronic ethanol ingestion – absence of subset-specific glucocorticoid-induced immune cell loss. Alcohol Clin Exp Res 31(10):1746–1758
190. Frank MG, Watkins LR, Maier SF (2011) Stress- and glucocorticoid-induced priming of neuroinflammatory responses: potential mechanisms of stress-induced vulnerability to drugs of abuse. Brain Behav Immun 25(Suppl 1):S21–S28
191. Crews FT, Bechara R, Brown LA, Guidot DM, Mandrekar P, Oak S, Qin L, Szabo G, Wheeler M, Zou J (2006) Cytokines and alcohol. Alcohol Clin Exp Res 30(4):720–730
192. Keeler JF, Robbins TW (2011) Translating cognition from animals to humans. Biochem Pharmacol 81(12):1356–1366
193. Schwartz M (2010) The emergence of a new science of the mind: immunology benefits the mind. Mol Psychiatry 15(4):337–338
194. Dirnagl U, Schwab JM (2009) Brain-immune interactions in acute and chronic brain disorders. Neuroscience 158(3):969–971
195. Irwin MR, Miller AH (2007) Depressive disorders and immunity: 20 years of progress and discovery. Brain Behav Immun 21(4):374–383
196. Petitto JM (2006) A shared neuroimmunological mechanism for drugs of addiction? J Neuroimmunol 173(1–2):1–2

197. Pascual M, Balino P, Alfonso-Loeches S, Aragon CM, Guerri C (2011) Impact of TLR4 on behavioral and cognitive dysfunctions associated with alcohol-induced neuroinflammatory damage. Brain Behav Immun 25(Suppl 1):S80–S91
198. Prendergast MA, Little HJ (2007) Adolescence, glucocorticoids and alcohol. Pharmacol Biochem Behav 86(2):234–245
199. Matsumoto I (2009) Proteomics approach in the study of the pathophysiology of alcohol-related brain damage. Alcohol Alcohol 44(2):171–176
200. Pfefferbaum A, Rosenbloom M, Rohlfing T, Sullivan EV (2009) Degradation of association and projection white matter systems in alcoholism detected with quantitative fiber tracking. Biol Psychiatry 65(8):680–690
201. Fitzpatrick LE, Jackson M, Crowe SF (2008) The relationship between alcoholic cerebellar degeneration and cognitive and emotional functioning. Neurosci Biobehav Rev 32(3):466–485
202. Oscar-Berman M, Marinkovic K (2007) Alcohol: effects on neurobehavioral functions and the brain. Neuropsychol Rev 17(3):239–257
203. Harper C, Matsumoto I (2005) Ethanol and brain damage. Curr Opin Pharmacol 5(1):73–78
204. Bates ME, Bowden SC, Barry D (2002) Neurocognitive impairment associated with alcohol use disorders: implications for treatment. Exp Clin Psychopharmacol 10(3):193–212
205. Sanhueza C, Garcia-Moreno LM, Exposito J (2011) Weekend alcoholism in youth and neurocognitive aging. Psicothema 23(2):209–214
206. Neafsey EJ, Collins MA (2011) Moderate alcohol consumption and cognitive risk. Neuropsychiatr Dis Treat 7:465–484
207. NIAAA (2009) Neuroscience: pathways to alcohol dependence. Alcohol Alert No. 77
208. Mukherjee S, Das SK, Vaidyanathan K, Vasudevan DM (2008) Consequences of alcohol consumption on neurotransmitters – an overview. Curr Neurovasc Res 5(4):266–272
209. Spanagel R (2009) Alcoholism: a systems approach from molecular physiology to addictive behavior. Physiol Rev 89(2):649–705
210. Ward RJ, Lallemand F, de Witte P (2009) Biochemical and neurotransmitter changes implicated in alcohol-induced brain damage in chronic or 'binge drinking' alcohol abuse. Alcohol Alcohol 44(2):128–135
211. Heinz A, Beck A, Wrase J, Mohr J, Obermayer K, Gallinat J, Puls I (2009) Neurotransmitter systems in alcohol dependence. Pharmacopsychiatry 42(Suppl 1):S95–S101
212. Shaham Y, Hope BT (2005) The role of neuroadaptations in relapse to drug seeking. Nat Neurosci 8(11):1437–1439
213. Alfonso-Loeches S, Guerri C (2011) Molecular and behavioral aspects of the actions of alcohol on the adult and developing brain. Crit Rev Clin Lab Sci 48(1):19–47
214. Committee on Substance Abuse (2010) From the American Academy of Pediatrics: Policy Statement – Alcohol use by youth and adolescents: a pediatric concern. Pediatrics 125(5):1078–1087
215. U.S. Department of Health and Human Services (2007) The surgeon General's Call to action to prevent and reduce underage drinking. U.S. Department of Health and human Services, Office of the Surgeon General. http://www.surgeongeneral.gov/library/calls/underagedrinking/calltoaction.pdf. Accessed: Dec 2011
216. Liu X, Chang SL (2011) Development and aging of the mesolimbic dopamine system and alcohol use. J Med Sci 31(2):061–066
217. Casey BJ, Jones RM (2010) Neurobiology of the adolescent brain and behavior: implications for substance use disorders. J Am Acad Child Adolesc Psychiatry 49(12):1189–1201 (quiz 1285)
218. Romeo J, Warnberg J, Marcos A (2010) Drinking pattern and socio-cultural aspects on immune response: an overview. Proc Nutr Soc 69(3):341–346
219. Lee H, Roh S, Kim DJ (2009) Alcohol-induced blackout. Int J Environ Res Public Health 6(11):2783–2792
220. Stephens DN, Duka T (2008) Review. Cognitive and emotional consequences of binge drinking: role of amygdala and prefrontal cortex. Philos Trans R Soc Lond B Biol Sci 363(1507):3169–3179
221. Obernier JA, White AM, Swartzwelder HS, Crews FT (2002) Cognitive deficits and CNS damage after a 4-day binge ethanol exposure in rats. Pharmacol Biochem Behav 72(3):521–532

222. Kuzmin A, Liljequist S, Meis J, Chefer V, Shippenberg T, Bakalkin G (2012) Repeated moderate-dose ethanol bouts impair cognitive function in Wistar rats. Addict Biol 17:132–140
223. Le Merrer J, Becker JA, Befort K, Kieffer BL (2009) Reward processing by the opioid system in the brain. Physiol Rev 89(4):1379–1412
224. Cosgrove KP (2010) Imaging receptor changes in human drug abusers. Curr Top Behav Neurosci 3:199–217
225. Roy S, Wang JH, Balasubramanian S, Sumandeep, Charboneau R, Barke R, Loh HH (2001) Role of hypothalamic-pituitary axis in morphine-induced alteration in thymic cell distribution using mu-opioid receptor knockout mice. J Neuroimmunol 116(2):147–155
226. Sanchis-Segura C, Grisel JE, Olive MF, Ghozland S, Koob GF, Roberts AJ, Cowen MS (2005) Role of the endogenous opioid system on the neuropsychopharmacological effects of ethanol: new insights about an old question. Alcohol Clin Exp Res 29(8):1522–1527
227. Gianoulakis C (2009) Endogenous opioids and addiction to alcohol and other drugs of abuse. Curr Top Med Chem 9(11):999–1015
228. Gianoulakis C (2001) Influence of the endogenous opioid system on high alcohol consumption and genetic predisposition to alcoholism. J Psychiatry Neurosci 26(4):304–318
229. Chang SL, Patel NA, Romero AA, Thompson J, Zadina JE (1996) FOS expression induced by interleukin-1 or acute morphine treatment in the rat hypothalamus is attenuated by chronic exposure to morphine. Brain Res 736(1–2):227–236
230. Chang SL, Kenigs V, Moldow RL, Zadina JE (1995) Chronic treatment with morphine and ethanol, but not cocaine, attenuates IL-1 beta activation of FOS expression in the rat hypothalamic paraventricular nucleus. Adv Exp Med Biol 373:201–208
231. Allolio B, Schulte HM, Deuss U, Kallabis D, Hamel E, Winkelman W (1987) Effect of oral morphine and naloxone on pituitary-adrenal response in man induced by human corticotropin-releasing hormone. Acta Endocrinol 114(4):509–514
232. Pruett SB, Fan R, Cheng B, Glover M, Tan W, Deng X (2010) Innate immunity and inflammation in sepsis: mechanisms of suppressed host resistance in mice treated with ethanol in a binge-drinking model. Toxicol Sci 117(2):314–324
233. Bosshart H (2010) Morphine-mediated suppression of phagocytosis. Int Immunopharmacol 10(2):264–265
234. Pruett SB, Fan R, Zheng Q (2003) Characterization of glucocorticoid receptor translocation, cytoplasmic IkappaB, nuclear NFkappaB, and activation of NFkappaB in T lymphocytes exposed to stress-inducible concentrations of corticosterone in vivo. Int Immunopharmacol 3(1):1–16
235. Sharp FR, Liu J, Nickolenko J, Bontempi B (1995) NMDA and D1 receptors mediate induction of c-fos and junB genes in striatum following morphine administration: implications for studies of memory. Behav Brain Res 66(1–2):225–230
236. Bontempi B, Sharp FR (1997) Systemic morphine-induced Fos protein in the rat striatum and nucleus accumbens is regulated by mu opioid receptors in the substantia nigra and ventral tegmental area. J Neurosci 17(21):8596–8612
237. Garcia MM, Anderson AT, Edwards R, Harlan RE (2003) Morphine induction of c-fos expression in the rat forebrain through glutamatergic mechanisms: role of non-n-methyl-D-aspartate receptors. Neuroscience 119(3):787–794
238. Frankel PS, Harlan RE, Garcia MM (1999) Chronic administration of morphine alters immediate-early gene expression in the forebrain of post-dependent rats. Brain Res 835(2):204–212
239. Valenzuela CF (1997) Alcohol and neurotransmitter interactions. Alcohol Health Res World 21(2):144–148
240. He L, Kim JA, Whistler JL (2009) Biomarkers of morphine tolerance and dependence are prevented by morphine-induced endocytosis of a mutant mu-opioid receptor. FASEB J 23(12):4327–4334
241. Lovallo WR (2006) The hypothalamic-pituitary-adrenocortical axis in addiction. Int J Psychophysiol 59(3):193–194

Chapter 16
Alcohol Abuse, HIV-1, and Hepatitis C Infection

Yuri Persidsky, Wen-Zhe Ho, Xu Wang, and Raghava Potula

16.1 Introduction: Alcohol and HIV-1 Infection

Alcohol is the most commonly used and abused drug in the United States. Approximately 14 million Americans meet the criteria for alcohol abuse or dependence [1]. Alcohol abuse significantly affects morbidity and mortality from infectious diseases [2] and contributes to 85,000 deaths each year in the United States [3]. Alcohol use is common among HIV-infected individuals, and its impact on HIV disease progression has been suggested. Since alcohol use is widespread and it is often heavy among HIV-infected individuals and among the most sexually active age groups at risk of HIV infection, it has been of a great interest to investigate the role of alcohol abuse in promoting HIV transmission and infection. It was suggested that acute and chronic alcohol use might increase host susceptibility to HIV infection [4, 5]. Heavy alcohol consumption was strongly associated with incidence of HIV infection [6]. Alcohol use and abuse have been identified as potential behavioral risk factors for the transmission of HIV, in the form of drinking before risky sexual events or frequent binge drinking as associated with HIV incidence [7–9]. Alcohol consumption, particularly at high levels, has a negative impact on adherence to medications in general [10] and specifically to combined antiretroviral therapy, cART [11–14]. Chander et al. found that hazardous levels of alcohol use were associated with decreased cART utilization, adherence, and viral suppression, independent of active drug use. Combined alcohol and drug use was associated with lower odds of adherence and viral suppression than either drugs or alcohol alone [15].

Y. Persidsky, M.D., Ph.D. (✉) • W.-Z. Ho • X. Wang • R. Potula
Department of Pathology and Laboratory Medicine, Temple University School of Medicine, Philadelphia, PA, USA
e-mail: yuri.persidsky@tuhs.temple.edu

C. Cui et al. (eds.), *Neural-Immune Interactions in Brain Function and Alcohol Related Disorders*, DOI 10.1007/978-1-4614-4729-0_16,

Table 16.1 In vitro impact of alcohol on HIV/SIV and HCV

Virus	In vitro effect	Target cells	Host	References
HIV	↑	Human macrophages	Human	[24]
	↑	Human PBMCs	Human	[25, 26]
		Human PBMCs		[27, 28]
	↑	Human oral keratinocytes	Human	[29]
	↑	Human T lymphocytes	Human	[30, 31]
	No effect (ex vivo)	Human lymphocytes	Human	[201]
SIV	↑	Monkey PBMC and alveolar macrophages	Monkey	[32]
HCV	↑	Human HCV replicon hepatoma cell line (Huh8, Huh3, FCA-1)	Human	[33, 34]
	↑	Huh7 HCV replicon cells	Human	[35]
	↑	Human hepatocyte derived cellular carcinoma cell line (Huh7)	Human	[36, 37]
	↑	Human primary hepatocytes	Human	[36]
	↓	Human HCV replicon hepatoma cell line (BB7, FL-Neo)	Human	[202]
	↓	Human hepatocyte derived cellular carcinoma cell line (Huh7)	Human	[202]

It remains unclear whether alcohol intake increases HIV replication and facilitates HIV disease progression. Although earlier studies [16, 17] did not find an association between alcohol consumption and markers of HIV disease progression, several lines of evidence pointed to a strong association between alcohol consumption and worsening of HIV infection. In a cross-sectional study of HIV disease in intravenous drug users, the relative risk of AIDS was 3.8 times higher in heavier drinkers than in moderate drinkers [18]. Pol et al. showed that HIV-infected alcohol abusers had a 41% increase in the number of $CD4^{+}$ T cells after cessation of alcohol use, whereas only a 15% increase was seen in uninfected control subjects who stopped drinking [19]. Among patients who have a history of alcohol problems and who are receiving antiretroviral treatment, alcohol consumption was associated with higher HIV RNA levels and lower $CD4^{+}$ T cell counts [20]. In a prospective study of HIV-infected drug abusers, alcohol abuse did not correlate with changes in the percentage of $CD4^{+}$ T cells; however, the percentage of $CD8^{+}$ T cells significantly increased among the heaviest drinkers [21]. Recently, data from several groups indicated a significant association between heavy alcohol consumption and lower levels of $CD4^{+}$ T cells in HIV-infected alcoholic patients not receiving cART [20, 22]. Baum et al. demonstrated that frequent alcohol users receiving cART [23] showed higher HIV viral load [22]. Frequent alcohol intake and the combination of frequent alcohol and crack-cocaine accelerated progression of HIV infection [22].

The findings from epidemiological studies are supported by several in vitro studies, indicating that alcohol augmented HIV replication (Table 16.1). Bagasra et al. [25–28] demonstrated that alcohol enhanced HIV replication in peripheral blood

Table 16.2 In vivo impact of alcohol on HIV/SIV and HCV

Virus	In vivo effect	Study materials	Host	References
HIV	↑	Plasma, PBMC	Human	[5, 8]
	↑	Plasma, PBMC	Human	[9, 203]
HIV	↑	Macrophages in HIVE/ SCID mice brain	Mice	[38]
SIV	↑	Plasma	Monkey	[39, 40]
	↑	Plasma	Monkey	[41, 42]
	↑	Plasma, CSF	Monkey	[43]
	No effect	Plasma	Monkey	[44]
	↓	CSF	Monkey	[44]
HCV	↑	PBMC, plasma	Human	[182, 187]
	↑	Plasma	Human	[182]
	↑	Serum	Human	[204]

mononuclear cells (PBMC). Alcohol also enhanced HIV infection in human blood mononuclear phagocytes [24] and T lymphocytes [30]. Alcohol exposure facilitated the entry of CXCR4-tropic HIV in peripheral blood lymphocytes tenfold when compared to untreated cells [31]. In addition, alcohol promoted HIV infection of normal human oral keratinocytes by upregulating CXCR4 expression [29]. These in vitro findings are supported by animal studies (Table 16.2). Bagby et al. found that alcohol could promote simian immunodeficiency virus (SIV) infection and replication in both PBMC and alveolar macrophages of rhesus monkeys [32]. Several recent studies using nonhuman primates (Table 16.2) also showed that, under physiologic conditions, chronic alcohol exposure accelerated progression of SIV disease [39–44]. Another aspect of alcohol exposure is accelerated AIDS wasting that has been suggested as a mechanism by which alcohol might result in more rapid HIV disease progression [41].

Taken together, while it is still debatable, studies that have been reviewed indicate that alcohol consumption acts as a cofactor that facilitates the severity of HIV-1 infection [45]. This impact of alcohol on HIV is likely to be a consequence of its immune suppression in $CD4^+$ T cells and macrophages, the targets for HIV infection. Alcohol diminished exogenous and allogeneic antigen presentation and altered the formation of peptide–MHCII complexes and co-stimulatory molecule expression on the cell surface [46]. Haorah et al. showed alcohol and HIV decrease proteasome and immunoproteasome function in macrophages and potentially affect antigen presentation, further dampening immune responses against HIV-1 [47]. cART, including NRTIs and protease inhibitors (PIs), is metabolized via the cytochrome P450 system. It is known that alcohol consumption significantly alters the metabolism of cART by two different mechanisms, enzymatic induction [48], associated with chronic alcohol use, and enzymatic inhibition, due to competition of ethanol for various cytochrome P450 isozymes related to acute alcohol exposure.

Because P450 is involved in metabolism of multiple drugs, chronic alcohol users may be affected by altered drug concentrations in plasma. Due to the complexity of alcohol interactions with multiple organ systems involving in many types of cells, it is challenging to determine the precise mechanisms by which alcohol enhances susceptibility of immune cells to HIV-1 infection and promotes HIV-1 disease progression.

16.2 Alcohol and HIV-1 Infection: Effects on Neurocognitive Performance

Despite introduction of cART achieving significant control of HIV-1 replication, prevalence of HIV-1-associated neurocognitive disorders (HAND) remains the same or increases [49]. It has been increasingly recognized that augmented levels of proinflammatory molecules [50, 51] and an increase in specific subsets of circulating blood monocytes [52] are associated with cognitive impairment. Interestingly, both are linked to blood–brain barrier (BBB) compromise [53], acknowledging their role in HAND with additional confirmation derived from well-controlled animal studies [54]. Such changes can be exacerbated by other conditions, like alcohol use disorders, that affect the nervous system.

While cognitive deficits associated with HIV-1 infection and chronic alcoholism have been well documented separately, few studies have addressed the combined effects of these conditions. Impaired verbal working memory has been reported [55]. Martin et al. [56] documented enhanced cognitive impulsivity in studies of individuals with HIV-1 infection and alcoholism or drug use disorders when compared with seronegative drug users. HIV-1 infection and concurrent heavy alcohol consumption had synergistic negative effects on motor and visual spatial tasks [57]. Additive and interactive effects of previous alcohol abuse and HIV-1 infection on verbal reasoning were found in HIV-1-infected individuals with past alcohol use disorders [58]. Schulte and colleagues [59] introduced a novel test, the Match-to-Sample Stroop Task and showed that HIV-1-infected individuals with alcoholism had normally reduced reaction times when a valid match cue introduced a Stroop stimulus but were disproportionately slow when the cue was invalid, indicating impairment in attention disengagement. A number of reports indicated significantly impaired memory and executive functions both in HIV-1 infection and chronic alcoholism [60–62]. Pitel and colleagues [61, 62] examined both memory and executive function abilities in abstinent alcoholics and demonstrated that executive function impairments did not account for the majority of variance in memory scores, concluding the memory impairment to be "genuine" rather than solely the result of executive function deficits. Winsauer and colleagues demonstrated that alcohol administration in SIV-infected monkeys produced greater behavioral and cognitive deficits than either alcohol or SIV infection alone [63].

There is an important link between high levels of alcohol use and progression of HIV-1 infection and its potential contribution to HAND. Increased monocyte activation markers and proinflammatory molecules (CD14, CCL2, and IL-6), together with higher-plasma LPS levels, were associated with HAND [64]. LPS levels predicted HIV-1-associated dementia (HAD), independent of viral load and CD4 counts. Increased circulating LPS, a consequence of microbial translocation due to increased permeability of the gut barrier, can contribute to monocyte activation in chronic HIV-1 infection and AIDS [65] and supports the idea that circulating LPS may be linked to development of HIV-1-associated cognitive decline via increased trafficking of activated monocytes into the brain. This study also indicated that both virologic and immune controls in the periphery are linked to key events involved in HIV-1 neuropathogenesis [66]. Importantly, it was shown that alcohol consumption increased permeability of the gut barrier, leading to increased LPS concentrations in plasma; elimination of LPS due to treatment with nonabsorbable antibiotics prevented most of alcohol's effects in the liver [65]. Furthermore, depletion of Kupffer cells (resident liver macrophages) resulted in similar effects, indicating the important contribution of these cells to a state of chronic inflammation associated with alcohol exposure [67]. Chronic alcohol exposure results in significant changes in gut microbiota that could further promote endotoxemia and an inflammatory state in alcoholics [68]. It has been acknowledged that the combination of alcoholism and HIV-1 further lessened barrier tightness [69] via induction of inducible nitric oxide synthase [70]. Therefore, immune dysfunction associated with alcohol abuse could further compromise antiviral responses and promote chronic neuroinflammation [45]. Validity of these observations was underscored by the finding that immediate episodic memory was impaired in individuals comorbid with HIV-1 infection and alcoholism [71]. Imaging studies show faster brain atrophy in HIV-1-infected alcoholics as compared to uninfected heavy drinkers [72, 73].

16.3 Mechanisms of HIV-1 and Alcohol-Related Associated Neurodegeneration

Despite seemingly different effects, HIV-1 infection of the central nervous system (CNS) and chronic alcohol exposure appear to have similar targets, resulting in synergistic neurodegenerative effects. Indeed, oxidative stress, high levels of glutamate, chronic inflammation, dysfunction of astrocytes and microglia, and blood–brain barrier (BBB) compromise have been reported in both HIV-1 infection and alcohol-associated injury [74–80]. All cell types of the CNS cells are affected by alcohol exposure. A wealth of evidence underscores the role of microglia cells in shaping neural networks. Microglia support neuronal function via production of neurotrophins and growth factors [like nerve growth factor (NGF), brain-derived neurotrophic factor (BDNF), neurotrophin (NT-4/5), glia-derived nerve factor (GDNF), and insulin-like growth factor (IGF)] [81–90]. Through the expression of immune molecules and other neurotrophic factors, microglia guide neurodevelopment, aid in the formation of new synapses, and help strengthen existing neuronal connections. Numerous studies [91, 92] have demonstrated the role of microglia in immune surveillance in the brain, noting that microglia, while highly active under normal conditions, rapidly deploy to sites of injury. Under physiologic conditions, microglia secrete neurotrophic factors, which support neuronal function, plasticity, survival, and differentiation [93]. However, as immune effector cells in the CNS [94], microglia are activated in pathological conditions [including Alzheimer's disease, Parkinson's disease, multiple sclerosis, and HIV-1 encephalitis (HIVE)] undergoing transformation from the resting to the reactive state accompanied by changes in cell shape [95], enhanced expression of macrophage-specific activation markers, and secretion of cytokines and reactive oxygen species [96], all contributing to neurodegeneration [97].

Inflammation is an underlying component of a diverse group of neurodegenerative diseases and their associated neuropathology, and increasing evidence suggests that microglia are a key causative factor in this process. Microglial activation is an important pathogenic component of alcohol-mediated neurodegenerative disease [45]. Chronic alcohol use causes structural and functional abnormalities of the brain and has a profound impact on the neuroimmune system [73]. Several lines of evidence both from humans and animal models suggest that alcohol induces inflammatory processes in the brain leading to neurodegeneration. Recently published data demonstrate an increased amount of microglia in chronic alcoholics and animals exposed to alcohol [98] and an overexpression of inflammatory markers in these cells [45]. Alcohol abuse and HIV-1 infection of the CNS result in combined toxic effects leading to neuronal injury and cognitive dysfunction (Fig. 16.1).

Previous studies clearly indicated that long-term alcohol abuse produced profound functional or morphological changes in the CNS regardless of nutritional status. Neuropathologic examination of brain tissue from chronic alcoholics suggested that alcohol abuse results in neuronal degeneration, ranging from minor dendritic structural changes and synaptic changes to neuronal cell death in the CNS [99]. Multiple lines of evidence suggest that chronic and excessive ethanol consumption may enhance oxidative injury of neurons and result in cell death. By mechanisms not well understood, ethanol induces activity of CYP2E1 (a major ethanol-metabolizing enzyme), enhances ROS generation, changes cytokine signaling pathways [resulting in upregulation of iNOS and phospholipase A_2 (PLA_2)], and produces prostanoids through the PLA_2/cyclooxygenase (COX) pathways (for review, see [100]). Knapp and Crews [101] showed that COX-2 immunoreactivity was progressively increased in rat brains during chronic treatment. Expression of CYP2E1 (metabolizing ethanol to ROS and acetaldehyde) was detected in neuronal cell bodies and astroglia. Interestingly, staining of blood vessels was prominent in the white matter, and immunoreactive astrocytes were seen to have end-feet on the microvessels [102]. Ethanol could increase inflammation caused by other factors (e.g., HIV-1 CNS infection). Davis and Syapin demonstrated that exposure of an astrocytic cell line to ethanol (0.5–6 h, 50 mM) enhanced iNOS activity induced by a cytokine cocktail (IL-1β, TNF-α, and IFNγ) via NF-κB activation) [103, 104]. Interestingly, chronic ethanol exposure resulted in an increased number of microglia in vivo before any ethanol-induced brain atrophy was detected [105].

Our previous work uncovered a proinflammatory pathway of neurodegeneration associated with alcohol abuse [106]. We found that the ethanol-metabolizing

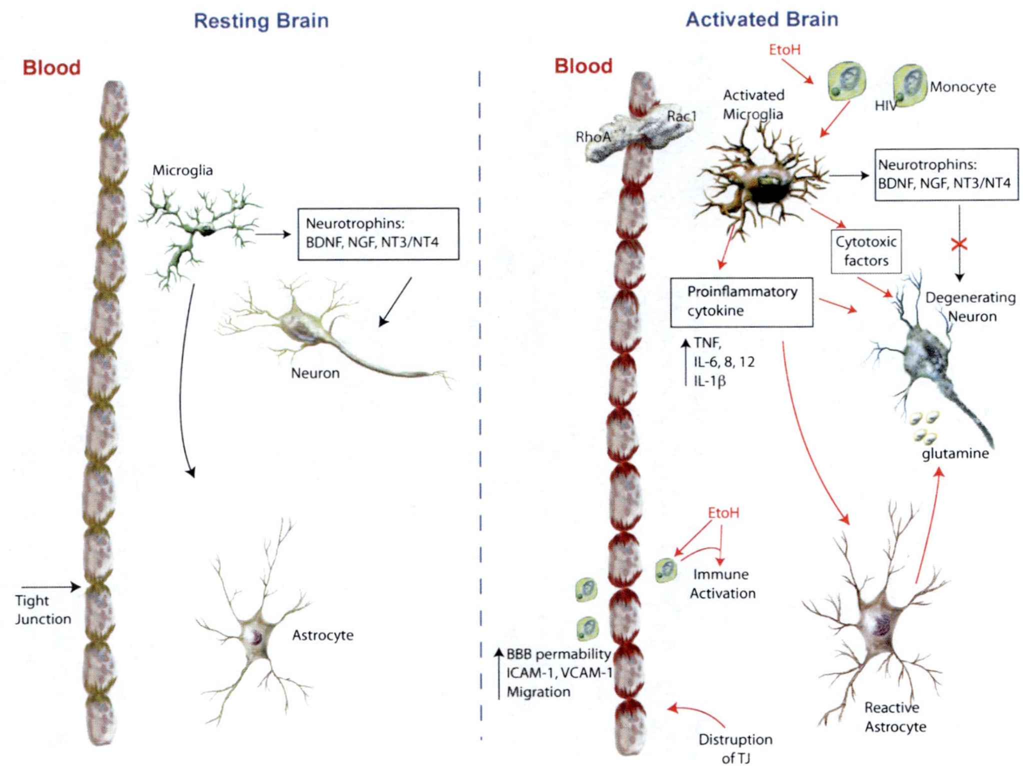

Fig. 16.1 Microglias, the sentinels of CNS in normal resting brain, perform homeostatic activity and mediate their beneficial effect (tropic support factors) via several effector mechanisms. The combined toxic effects of alcohol abuse and HIV leading to neuronal injury and cognitive dysfunction are not only a consequence of altered (functional and morphological) proinflammatory microglia phenotype but also a resultant of deregulation of protective/trophic microglia phenotype. Reduction in microglia effector function and insufficient homeostatic activity of microglia can contribute to neuronal degeneration. Insights into understanding the significance of microglia neuroprotective/immunoregulatory function deregulation may add to our current understanding of the underlying neurodegenerative mechanisms in the context of alcohol abuse and HIV infection

enzymes, CYP2E1/ADH, metabolized ethanol in primary human neurons and correlated with ROS production. We detected ROS levels in cultured neurons after ethanol/acetaldehyde (AcH) exposure. The increase in ROS production (by 68%) after ethanol treatment was inhibited by 4-MP (an inhibitor of CY2E1 and ADH), suggesting that alcohol metabolism resulted in ROS generation. Inhibition of SOD further enhanced the ethanol-induced ROS level. AcH exposure also led to a 68% increase in ROS levels, suggesting the potential involvement of AcH in ROS generation. AcH is not an ROS donor, nor is AcH metabolism capable of producing ROS. Therefore, increased levels of reactive species might be due to AcH-mediated activation of the major ROS generating enzymes, NADPH/xanthine oxidase (NOX/XOX). In order to address whether AcH increases ROS level via NOX/XOX activation, we studied the dose-dependent effects of AcH on ROS production in cultured neurons using the NOX inhibitor, apocynin (APC), or the XOX inhibitor, allopurinol (AP). There were dose-dependent increments of 38–69% in ROS production after AcH application. AP/APC individually or in combination inhibited (27–43%) the increased ROS production caused by AcH. These findings suggest that alcohol metabolism by ADH/CYP2E1 in human neurons generated ROS and AcH, and AcH subsequently activated NOX and XOX, exacerbating oxidative stress in the CNS associated with chronic alcohol exposure. Neuroimaging studies demonstrated more significant white matter abnormalities in HIV-1-positive alcohol abusers as compared to noninfected alcoholics [107]. Diffuse microglial activation is thought to be an underlying process in HIV-1-mediated neurodegeneration [108]. Using a small animal model of human immunodeficiency viral encephalitis, we demonstrated that alcohol abuse and HIV-1 infection synergistically enhance microglial activation [38]. These changes were associated with increased viral load and inefficient immune responses due to impairment of immunoproteasomes.

Ethanol exposure activates NF-κB transcription and increases proinflammatory cytokines and oxidative stress, thus promoting neurodegeneration. Furthermore, by reducing pCREB transcription, ethanol sensitizes neurons to insults (like oxidative stress) contributing to neuronal injury and impaired neurogenesis. It has been suggested that alcohol-induced neurodegeneration is primarily due to increased oxidative stress and neurotoxic proinflammatory proteins [109]. Although enhanced microglial reaction in alcohol exposure in vivo has been interpreted as a sign of neurotoxic activity of resident brain macrophages, it is plausible that the initiation and progression of alcohol-mediated neurodegeneration is driven in part by release of cytotoxic molecules from activated microglia such as proinflammatory cytokines, ROS, proteinases, and complement proteins [110]. Alternatively, it could be hypothesized that alcohol-induced neuroinflammation and associated neurodegeneration are in part due to deregulation of microglial neuroprotective factors. One could consider that the microglial reaction is an indicator of a compensatory reaction attempting to prevent neurodegeneration and that the imbalance between secreted molecules with pro- and anti-inflammatory effects contributes to the neuropathology of alcohol-related brain damage. A key issue, therefore, is to elucidate the novel mechanism of alcohol-induced neurodegeneration and identification of signaling pathways responsible for deregulation of microglial neuroprotective factors.

16.4 BBB Dysfunction in HIV-1 Infection and Alcohol Abuse

The role of BBB dysfunction is being increasingly recognized in various neurodegenerative disorders, including HIV-1 infection and alcohol abuse [45]. For example, the pathogenesis of HAND involves activation of monocytes and their subsequent recruitment into the CNS altering BBB function [111]. The molecules and mechanisms regulating monocyte migration across the BBB remain poorly defined. Enhanced expression of adhesion molecules on brain microvascular endothelial cells (BMVEC) triggered by inflammatory mediators [112, 113] control leukocyte trafficking into the CNS. Increased expression of adhesion molecules and BBB permeability has been demonstrated in HAND patients [114]. A disrupted BBB allows accumulation of toxic serum proteins and increased infiltration of monocytes and lymphocytes, thereby accelerating inflammation and viral entry into the CNS. cART failed to control BBB leakage and inflammation in HAND patients [115, 116], in part because it does not reduce the high levels of CD40 ligand (CD40L) found in the plasma and cerebrospinal fluid (CSF) of HIV-1-infected patients [117, 118]. CD40L, a 33-kDa type II membrane glycoprotein, is expressed predominantly by activated leukocytes and platelets [119]. 31- and 18-kDa forms of soluble sCD40L are secreted or shed from activated cells. CD40L activates CD40, a 45- to 50-kDa type I membrane glycoprotein expressed at a low level in resting cells of myeloid and vascular origin [118, 120, 121]. As shown in other organs [122–124], sCD40L can promote CNS inflammation at the level of the BBB. Upregulation of CD40 expression has been detected on microglia of HIV-1-infected brain tissues [125], indicating its role in HIV-1 neuropathogenesis. CD40L was shown to potentiate the ability of HIV-1 Tat protein to activate monocytes and microglia leading to the secretion of neurotoxic inflammatory mediators [118].

CD40/CD40L interactions in brain endothelium have not been explored in detail; therefore, we used a combination of ex vivo and in vitro assays to investigate the role of CD40 activation by sCD40L in brain endothelium. We found that CD40 is upregulated in brain endothelial cells of patients affected by HIVE as compared to seronegative controls. Interestingly, CD40 expression on brain microvessels was also increased in HIV-1-infected patients without HIVE. These observations suggested an important role for the CD40/CD40L dyad in regulating BBB functions. In vitro experiments further indicated validity of this idea. CD40L upregulated the expression of the adhesion molecules, ICAM-1 and VCAM-1, in primary human BMVEC causing a fourfold increase in monocyte adhesion to brain endothelium and stimulated migration across an in vitro BBB model. We studied the intracellular signaling pathways that govern these events and found that c-JUN N-terminal kinase (JNK) is critical to CD40 activation in the BMVEC. CD40L induced activation of mixed lineage kinase-3 and JNK leading to the subsequent activation of c-JUN/activating protein-1. JNK inhibition in the BMVEC suppressed CD40L-mediated induction of adhesion molecules, monocyte adhesion, and transendothelial migration [53]. These findings support the concept that the CD40/CD40L dyad plays a major role in HAND pathogenesis. Importantly, levels of CD40L are increased in a variety of inflammatory conditions [e.g., hepatitis C virus (HCV)] that could lead to subclinical impairment of the BBB, further exacerbated by alcohol exposure, a phenomenon that has not been recognized until now.

Our previous work indicated that BBB dysfunction in HIV-1 CNS infection was associated with activation of small GTPases, RhoA, and Rac1 in brain endothelium after engagement by monocytes [126]. Activation of RhoA and its downstream enzyme, Rho kinase, led to phosphorylation of tight junction (TJ) proteins and BBB loosening [127]. Diminished BBB tightness and migration of HIV-1-infected monocytes across the BBB can be diminished via RhoA, Rac1, and Rho kinase suppressions [113, 126]. Importantly, TJ phosphorylation in brain endothelium was also found after ethanol exposure indicating potential synergistic mechanisms in BBB impairment by HIV-1 infection and alcohol abuse. We demonstrated molecular mechanisms of BBB disruption by alcohol in pathophysiologically relevant concentrations (25–50 mM). Such mechanisms include alcohol metabolism in primary human BMVEC (via CYP2E1) causing oxidative stress, leading to Ca^{2+} release via IP_3R stimulation, activation of myosin light-chain kinase, and phosphorylation of myosin light chain (MLC) and TJ proteins [78, 79, 128]. These changes decreased structural integrity of BMVEC monolayers, increased BBB permeability in vitro and in vivo, and enhanced monocyte migration across the BBB, events potentially enhancing neurotoxicity associated with HIV-1 CNS infection. We next established that chronic exposure to ethanol (24–72 h) resulted in activation of metalloprotease (MMP)-2 and -9 via activation of protein tyrosine kinase, degradation of basement membrane components, increased barrier "leakiness," and monocyte migration across human brain endothelial monolayers [129]. Impairment of antioxidant systems, such as a reduction in catalase and SOD activity, by ethanol exposure may elevate the levels of ROS/NO in endothelium, resulting in BBB damage. We studied whether stabilization of antioxidant enzyme activity results in suppression of ROS levels by anti-inflammatory agents [130]. To explore this idea, we determined the effects of ethanol on the kinetic profile of SOD and catalase activity and ROS/NO generation in primary human BMVEC. We found an enhanced production of ROS and NO levels due to the metabolism of ethanol in brain endothelium. Similar changes were observed after BMVEC exposure to AcH. Ethanol simultaneously augmented ROS generation and the activity of anti-oxidative enzymes. SOD activity was increased for a much longer period of time than catalase activity. A decline in SOD activity and protein levels preceded elevation of oxidant levels. SOD stabilization by the antioxidant and mitochondria-protecting agent acetyl-L-carnitine (ALC) and the anti-inflammatory agent rosiglitazone suppressed ROS levels, with a marginal increase in NO levels. Mitochondrial membrane protein damage and decreased membrane potential after ethanol exposure indicated mitochondrial injury that was prevented by ALC. All functional alterations and signal transduction events caused by ethanol exposure can be reproduced by application of AcH or exogenous ROS donors [128, 131]; therefore, other conditions associated with oxidative stress (like HIV-1 infection) may worsen the BBB impairment seen in alcohol exposure.

16.5 Astrocyte Dysfunction in HIV-1 Infection and Alcohol Abuse

The induction of both CYP2E1 and ROS was also noted in astrocytes exposed to ethanol [132], suggesting that glial activation might contribute to ethanol induction of oxidative stress in the brain. Blanco and colleagues demonstrated that chronic ethanol treatment activated iNOS/COX [133] and increased the levels of IL-1β and TNF-α in both rat brain and cultured astrocytes, activating signaling pathways that are usually associated with inflammation (MAPKs, NF-κB, AP-1). Such events coincided with an increase in cell death [134] via RhoA activation [135]. These data suggest that ethanol can stimulate glial cells by triggering the production of toxic compounds, such as ROS or NO [134], cytokines, and glutamate contributing to ethanol-induced brain damage, similar to what occurs in several brain disorders and neurodegenerative diseases including HAND [136–138]. Therefore, alcohol abuse contributes to brain damage and neurodegeneration by increasing oxidative stress and exacerbating neuroinflammation causing astrocyte dysfunction. We investigated the idea that alcohol could increase the production of reactive metabolites (ROS, AcH) due to ethanol metabolism by CYP2E1 in astrocytes. Pathophysiologically relevant concentration of 20 mM ethanol increased CYP2E1 activity, paralleling enhanced ROS production, similar to the findings in rat and human astrocytes [80, 132], suggesting that CYP2E1 indeed has a prominent role in ROS generation in human astrocytes. Our results suggested that activation of NOX appeared to be the main source of ROS production because NOX inhibitor suppressed the ethanol/AcH-induced ROS generation. Toll-like receptors (TLRs) could mediate alcohol-induced astrocytic responses because TLRs activation by alcohol in glial cells promoted secretion of proinflammatory molecules [139]. Ethanol exhibited biphasic effects on glial cells, where 10–50 mM ethanol activates TLR4/IL-1 receptor and 100–300 mM ethanol inhibits TLR4/IL-1 receptor [133, 139]. Reactive metabolites (ROS) could activate (phosphorylate) Src (a non-receptor tyrosine kinase) through TLR4 recruitment. Indeed, ethanol exposure elevated secretion of prostaglandin E2 (PGE2) in human astrocytes that was associated with induction of $cPLA_2$ activity and protein content and COX-2 protein level in a Src phosphorylation-dependent manner [80]. The interactive tyrosine phosphorylation of TLR4–Src complex at the cell membrane triggered the activation of $cPLA_2$ and COX-2 in the cytoplasm through Src signaling. Inhibition of ethanol metabolism, blockage of Src activity, or inactivation of TLR4 prevented the activation of $cPLA_2$ and COX-2 as well as diminished PGE2 production [80], suggesting that interactive phosphorylation of TLR4–Src regulated the proinflammatory response in astrocytes.

Another aspect of astrocyte dysfunction seen in alcohol exposure and HIV-1 infection is associated with the role astrocytes play in glutamate-mediated excitotoxicity. Glutamate, being the main excitatory neurotransmitter [140], is released from glutamatergic neuronal vesicles through a calcium-dependent mechanism (for review, see [141]). The amount of glutamate is in the micromolar range in the synaptic cleft under physiologic conditions [141] because sustained glutamate receptor hyperstimulation induces neuronal death [142] via calcium and sodium deregulation (so-called excitotoxicity). Extracellular glutamate concentration is controlled by EAATs that include five cloned subtypes [143, 144]. EAAT-1 and EAAT-2 are primarily detected in astrocytes [145]. The Na^+ and K^+ electrochemical gradients are used as a driving force by EAAT to take up extracellular glutamate against a several thousand-fold concentration gradient [146]. The astrocytic transporters, EAAT-1 and EAAT–2, are essential for protection against excitotoxicity as was shown in EAAT-gene knockout experiments in mice [147]. Astrocytes protect from excitotoxicity via clearance of extracellular glutamate, and glutamate metabolism within glutamate-scavenging cells is vital to prevent excitotoxicity. This is accomplished by astrocytic glutamine synthase (GS) rapidly converting glutamate into glutamine [148].

Excitotoxicity is seen in many other acute and chronic neurological diseases, including HIV-1 CNS infection [149, 150]. Among other mechanisms, HIV-1-associated neurotoxicity involves glutamate-related excitotoxicity and oxidative stress, resulting in neuronal injury and apoptosis [149]. In vitro experiments demonstrated that EAAT expression and function in astrocytes were diminished by HIV-1, presumably due to the effects of inflammatory mediators and viral proteins [151–153]. TNF-α decreased astrocyte clearance of glutamate [151] and decreased EAAT-2 expression in these cells [154]. Oxidative stress (present during HIV-1 brain infection [155]) also results in decreased EAAT-2 expression and function [156]. Since oxidative stress plays an important role in ethanol-mediated neurotoxicity, EAAT dysfunction seen in HIV-1 CNS infection could be further exacerbated by alcohol abuse leading to increased levels of extracellular glutamate. EAAT-2 could be modulated by RhoA in astrocytes [157]. As RhoA activation by alcohol exposure was described recently in astrocytes [135], this could be one putative pathway controlling EAAT-2 in alcohol abuse.

While glutamate imbalance plays an important role in alcohol-associated brain effects, available data in the literature are somewhat conflicting. Chronic exposure to ethanol (at high doses of 100 mM) increased EAAT-2 expression in organotypic cortical slice brain cultures [158], did not change in brain tissues of animals exposed to ethanol [75], or reduced (both density and function) in alcohol-preferring rats [159] or in vitro systems [160]. In cortical astroglial cell cultures exposed to ethanol, uptake of radioactively labeled glutamate was increased, and this was interpreted as a maladaptive process [158]. Concentrations of extracellular glutamate were increased in animals exposed to ethanol for 4–8 days suggesting deficits in glutamate transport [75]. Increased gene expression for EAAT-1 was shown in the brains (frontal cortex) of alcoholics, while no results were presented for EAAT-2 [161]. In contrast to discrepant experimental results, a number of clinical studies showed the efficacy of anti-glutamatergic approaches for treating alcohol withdrawal symptoms [162] and dependence [163]. Furthermore, increased glutamate levels in animals with defective glutamate transporters enhanced their alcohol consumption [164], pointing to an association between dysfunction of glutamate transporters and enhanced alcohol intake. Therefore, chronic neuroinflammation (like in HIV-1 CNS infection) could cause glutamate transporter dysfunction promoting an alcohol-craving phenotype and alcoholism. How the combination of chronic inflammatory process (like HIVE) and alcohol abuse affects glutamate transport and resulting neurotoxicity is currently unknown.

16.6 Alcohol Derails Microglia-Mediated Neuroprotection

Neurotrophic factors play a vital role in neuronal survival and maturation and are important in regulating naturally occurring cell death through apoptosis. Also, neurotrophins play a role in regulating the function of microglia through tropomyosin-receptor-kinase (Trk) and/or LNGFRp75-mediated signal transduction. Reduced levels of trophic factors from glial cells can induce alterations in the pattern of neuronal synaptic connections or cause cell death. Recent studies suggest that chronic exposure to ethanol can reduce the availability of brain-derived neurotrophic factors and alter its receptor TrkB function [165]. Receptors for nerve growth factor (TrkA) were also diminished after alcohol exposure [166]. Therefore, such changes may impede neuronal function, increase susceptibility to ethanol-induced cell injury, and neuronal death. Additionally, another group of receptors that are highly expressed on microglia and affected by alcohol are the purinergic receptors [167, 168], which can contribute to alcohol-caused dysfunction of resident brain macrophages. Change in expression of receptors important for modulating inflammatory function, including P_2 purinergic receptors (P_2Y and P_2X), is required for microglial process movement [169]. Downregulation of CX3CR1, receptor for fractalkine (neurotactin) [170], and TREM2 (triggering receptor expressed on myeloid cells-2) [171] has been shown to increase proinflammatory cytokine expression promoting neuroinflammation. Effects of alcohol abuse on neurotrophin secretion by microglia and identification of mechanisms deregulating neurotrophin secretion may uncover novel pathways of alcohol-induced neurodegeneration. Interestingly, such events are operative in acute nervous system injury [89, 172]. In essence, we suggest that the proinflammatory microglia phenotype is an indicator of microglia compensatory reaction, attempting to prevent neurodegeneration secondary to diminished secretion of neurotrophins.

Understanding the temporal expression of proinflammatory cytokines, neurotrophins, and growth factors will elucidate the unique profile of secreted microglia proteins that play a role in neuroregulatory functions. Downregulation of microglia-mediated neuroprotection by alcohol abuse would tilt the balance in priming microglia to a proinflammatory phenotype and exacerbate alcohol-induced neurodegeneration. The yin and yang between various microglial factors would determine a protective/trophic or destructive/toxic role of microglia in a given disease or injury condition. Given the scenario, analysis of secretory profiles and pathways by which alcohol exposure changes the phenotypic and functional properties of microglia allows deciphering the cellular mechanism processes that contribute to alcohol-mediated neurodegeneration. Furthermore, illustrating the genes associated with neurotrophic signaling molecules, neuronal cell growth and differentiation, neuronal regeneration and survival, cytokines and receptors involved in neuronal signaling, and genes involved in neuronal apoptosis in response to neurotrophic factors, transcription factors, and regulators indicative of the activation pathways downstream of the neuronal system will aid in this process. Insights into the effects of alcohol abuse on deregulation of microglia neuroprotective functions, as a potential cause of neurodegeneration, not only would add to our understanding but also would shed light on similar mechanisms that might play a role in other neurodegenerative diseases.

16.7 Alcohol and HCV in Combination with HIV-1 Infection

Alcohol consumption and viral hepatitis infection, both recognized as major causes of liver disease worldwide, frequently coexist in patients with chronic liver disease [173–175] and often coincide with HIV-1 infection leading to progression of liver injury [176]. Furthermore, HIV-1-infected adults coinfected with HCV are more likely to experience pain that interferes with daily living, muscle or joint pain, and headaches compared to those not coinfected [177]. It has been suggested that HIV-1 infection per se without coinfection can cause liver disease as well [178], a notion that requires further investigation. Alcohol abuse has become a major social as well as a clinical problem with nearly 20% of alcoholics developing fibrosis and subsequent cirrhosis, neither of which has an acceptable cure outside of liver transplantation [179]. Almost one-third of alcoholics with clinical symptoms of liver disease have been infected with HCV, which is four times the rate of HCV infection found in alcoholics who do not have liver disease [3, 179]. When the etiology of liver disease in the United States is categorized by single cause, chronic alcohol use and HCV infection account for 70–90% of cases, with roughly equal prevalence of each [180]. Alcohol and HCV most likely act synergistically to promote the development and progression of liver damage [174, 175]. Multiple studies have confirmed this finding and demonstrated that there are associations between heavy alcohol consumption and increased rates of cirrhosis and risk of death in chronic HCV-infected patients [174, 181, 182]. Patients who drink more than 50 g of alcohol each day—an amount that is equivalent to approximately four to five alcoholic drinks (330 ml of beer, 120 ml of wine, or 40 ml of liquor)—have an increased rate of progression of liver fibrosis [3].

Clinical trials indicate a therapeutic benefit of IFNα treatment in chronic HCV infection. Although a history of alcohol abuse is not a contraindication to clinical therapy of HCV infection, continued alcohol use during therapy adversely affects the response to HCV treatment. Heavy alcohol consumption reduces the efficacy of IFNα therapy for chronic HCV infection [23], and this adverse effect of alcohol drinking on efficacy can be partly reversed by abstinence from alcohol. Alcohol consumption accelerates liver damage, diminishes therapeutic response to IFNα, and increases the rate of hepatocellular carcinoma in patients with chronic HCV infection [182–187]. Sustained virological response to IFNα treatment is lower in

HCV-infected subjects with a recent history of alcohol use [3]; the mechanism for this IFNα resistance is unknown. These studies are supported by in vitro investigations, showing that alcohol treatment of HCV-infected hepatocytes compromised the anti-HCV effect of IFNα [33, 36].

Although we do not fully understand how alcohol consumption accelerates liver injury in patients with HCV infection, it is likely that alcohol impairs immune function, thereby enhancing HCV replication in the liver. Alcohol ingestion has an additional effect of diminishing immune clearance and increasing viral burden to hasten the onset of cirrhosis and hepatocellular carcinoma [188]. High levels of alcohol intake have been associated with an accelerated course of chronic HCV infection. Several lines of evidence showed that greater alcohol consumption is related to higher HCV RNA load [188]. Oshita et al. [188] reported that serum HCV RNA levels were significantly higher in habitual alcohol drinkers with chronic HCV infection than those in infrequent alcohol drinkers with chronic HCV infection. In addition, HCV RNA levels were significantly higher in alcohol drinkers than in abstainers [188]. These in vivo findings were supported by in vitro studies (Table 16.2) showing that alcohol potentiates both HCV replicon expression and full-cycle HCV replication in hepatic cells [34–37, 39]. Although the underlying mechanisms responsible for the effect of alcohol on HCV replication remain to be determined, several mechanisms have been suggested, including alcohol-induced changes in oxidative stress and mitochondrial function [34, 189, 190] and alcohol-induced immune dysregulation [191]. McCartney et al. [34] reported that CYP2E1 metabolism of alcohol, which stimulates microsomal ROS production, may play a key role in the increase of HCV replication by alcohol. We showed that the expression of endogenous type I IFNs (both IFNα and IFNβ) in human hepatocytes was significantly inhibited by alcohol treatment [36]. Such negative impact of alcohol on IFNα/β was due to its modulating effect on both positive (IRF5 and IRF7) and negative (SOCS-2 and SOCS-3) regulators of the JAK-STAT pathway that is crucial for IFN-mediated signaling and activation of gene expression [36]. Because type I IFNs and their inducible antiviral factors can directly inhibit HCV at different steps of the viral replication cycle, alcohol-mediated suppression of these innate antiviral factors provides a favorable microenvironment for HCV growth in hepatocytes. These data in conjunction with studies by others [33, 34, 189] indicate that alcohol abusers may have a compromised innate immunity within the liver, which may contribute to the chronicity of HCV infection and the poor efficacy of IFNα-based therapy.

The impact of HIV-1–HCV coinfection on the CNS remains largely undetermined. A number of studies have shown the impact of interactions between HCV and HIV-1 on the CNS and neuropsychological function [56, 192–196]. It remains to be determined that HCV has an additive effect on HIV-1-associated cognitive impairment. It has been reported there is more severe neuropsychological impairment among coinfected patients than mono-infected individuals [192–194, 197], as coinfected subjects had decreased processing speed and psychomotor speed. One study observed significantly slower reaction times among coinfected patients compared to individuals with either HIV-1 or HCV mono-infection [198]. This finding is supported by the report showing that coinfected individuals had slower reaction times on the task compared to mono-infected subjects [56]. It was reported that 80% of a coinfected group met criteria for cognitive impairment compared to 69% of HIV-1-mono-infected subjects that met criteria for impairment [199]. Further, it has also been reported that the primary cognitive domain affected in the coinfected subjects was psychomotor slowing, with 84% of the coinfected individuals impaired in this domain compared to 56% of the mono-infected individuals [193]. However, the finding that there is greater cognitive impairment among coinfected subjects has not been confirmed in other studies. One study indicated that there was no significant difference between coinfected patients and a comparison group of HCV mono-infected patients [195]. Several other studies also failed to find significant differences between coinfected and mono-infected persons on individual neuropsychological measures, although trends for greater impairment in coinfected persons were suggested [196]. Interestingly, although there was no difference in neuropsychological measures, one report demonstrated a differential pattern of impairment on a cognitive task of reaction time and response inhibition, with HCV-infected patients exhibiting overall slowed processing speed and HIV-1-infected patients showing impaired executive ability [56]. Winston et al. [200] documented the synergistic effects of HIV-1 and acute HCV infection on cognitive performance associated with functional changes found by proton magnetic resonance spectroscopy. Apparently, more studies are needed in order to determine how HIV-1 and/or HCV contribute to psychological and psychiatric signs.

16.8 Summary

Alcohol abuse affects numerous aspects of HIV infection including transmission, cell susceptibility (receptor expression), virus replication, innate and acquired immune responses to virus, compromise of the gut–blood and blood–brain barriers, poor adherence to cART, and neurocognitive performance. Accelerated progression of HIV infection in heavy drinkers might be associated with microbial translocation contributing to the state of immune activation, a defining feature of HIV infection. Chronic inflammatory responses appear to play a significant role in alcohol-associated neurodegeneration due to dysfunction of microglia and astrocytes. Furthermore, new data suggest that neurotrophic factors are diminished by chronic alcohol exposure, and this is further exacerbated in HIV infection. HCV infection is also affected by alcohol exposure, leading to progressive liver injury. HCV replication is enhanced in HIV coinfected individuals, and the combination with alcohol use further compromises immune responses (thereby controlling both viruses) and liver tissue injury. Better understanding of the complex interactions between alcohol, HIV, and HCV infection allows the design of therapeutic approaches protecting end organ injury and cognitive impairment in affected patients.

Acknowledgments This work was supported in part by NIH grants R37 AA015913, RO1 AA017398, RO1 MH65151 and RO1 DA025566 (to YP), NIH RO1 DA12815, RO1 DA27550, RO1 DA25477, and RO1DA-22177 (to WZH).

References

1. Bryant K (2010) Alcohol and AIDS: a guide to research issues and opportunities. National Institute on alcohol abuse and Alcoholism 33:167–178
2. Cook RT (1998) Alcohol abuse, alcoholism, and damage to the immune system—a review. Alcohol Clin Exp Res 22:1927–1942
3. Saitz R (2005) Clinical practice. Unhealthy alcohol use. N Engl J Med 352:596–607
4. Szabo G (1997) Alcohol's contribution to compromised immunity. Alcohol Health Res World 21:30–41
5. Szabo G (1999) Consequences of alcohol consumption on host defence. Alcohol Alcohol 34:830–841
6. O'Leary A, Hatzenbuehler M (2009) Alcohol and AIDS. In: Korsmeyer P, Kranzler HR (eds) Encyclopedia of drugs, alcohol, & addictive behavior, 3rd edn. Macmillian Reference, Detroit
7. Erbelding EJ, Hutton HE, Zenilman JM, Hunt WP, Lyketsos CG (2004) The prevalence of psychiatric disorders in sexually transmitted disease clinic patients and their association with sexually transmitted disease risk. Sex Transm Dis 31:8–12
8. Fisher JC, Bang H, Kapiga SH (2007) The association between HIV infection and alcohol use: a systematic review and meta-analysis of African studies. Sex Transm Dis 34:856–863
9. Kalichman SC, Simbayi LC, Kaufman M, Cain D, Jooste S (2007) Alcohol use and sexual risks for HIV/AIDS in sub-Saharan Africa: systematic review of empirical findings. Prev Sci 8:141–151
10. Weiss RD (2004) Adherence to pharmacotherapy in patients with alcohol and opioid dependence. Addiction 99:1382–1392
11. Cook RL et al (2001) Problem drinking and medication adherence among persons with HIV infection. J Gen Intern Med 16:83–88
12. Hendershot CS, Stoner SA, Pantalone DW, Simoni JM (2009) Alcohol use and antiretroviral adherence: review and meta-analysis. J Acquir Immune Defic Syndr 52:180–202
13. Meyerhoff DJ (2001) Effects of alcohol and HIV infection on the central nervous system. Alcohol Res Health 25:288–298
14. Petry NM (1999) Alcohol use in HIV patients: what we don't know may hurt us. Int J STD AIDS 10:561–570
15. Chander G, Lau B, Moore RD (2006) Hazardous alcohol use: a risk factor for non-adherence and lack of suppression in HIV infection. J Acquir Immune Defic Syndr 43:411–417
16. Coates RA et al (1990) Cofactors of progression to acquired immunodeficiency syndrome in a cohort of male sexual contacts of men with human immunodeficiency virus disease. Am J Epidemiol 132:717–722
17. Kaslow RA et al (1989) No evidence for a role of alcohol or other psychoactive drugs in accelerating immunodeficiency in HIV-1-positive individuals. A report from the Multicenter AIDS Cohort Study. JAMA 261:3424–3429
18. Lake-Bakaar G, Grimson R (1996) Alcohol abuse and stage of HIV disease in intravenous drug abusers. J R Soc Med 89:389–392
19. Pol S et al (1998) Retrospective analysis of the impact of HIV infection and alcohol use on chronic hepatitis C in a large cohort of drug users. J Hepatol 28:945–950
20. Samet JH et al (2007) Alcohol consumption and HIV disease progression. J Acquir Immune Defic Syndr 46:194–199
21. Crum RM, Galai N, Cohn S, Celentano DD, Vlahov D (1996) Alcohol use and T-lymphocyte subsets among injection drug users with HIV-1 infection: a prospective analysis. Alcohol Clin Exp Res 20:364–371
22. Baum MK et al (2010) Alcohol use accelerates HIV disease progression. AIDS Res Hum Retroviruses 26:511–518
23. Ohnishi K et al (1996) Interferon therapy for chronic hepatitis C in habitual drinkers: comparison with chronic hepatitis C in infrequent drinkers. Am J Gastroenterol 91:1374–1379

24. Wang X et al (2002) Alcohol potentiates HIV-1 infection of human blood mononuclear phagocytes. Alcohol Clin Exp Res 26:1880–1886
25. Bagasra O, Kajdacsy-Balla A, Lischner HW (1989) Effects of alcohol ingestion on in vitro susceptibility of peripheral blood mononuclear cells to infection with HIV and of selected T-cell functions. Alcohol Clin Exp Res 13:636–643
26. Saravolatz LD, Cerra RF, Pohlod DJ, Smereck S (1990) The effect of alcohol on HIV infection in vitro. Prog Clin Biol Res 325:267–271
27. Bagasra O et al (1996) Increased human immunodeficiency virus type 1 replication in human peripheral blood mononuclear cells induced by ethanol: potential immunopathogenic mechanisms. J Infect Dis 173:550–558
28. Bagasra O, Kajdacsy-Balla A, Lischner HW, Pomerantz RJ (1993) Alcohol intake increases human immunodeficiency virus type 1 replication in human peripheral blood mononuclear cells. J Infect Dis 167:789–797
29. Chen H et al (2004) Alcohol enhances HIV type 1 infection in normal human oral keratinocytes by up-regulating cell-surface CXCR4 coreceptor. AIDS Res Hum Retroviruses 20:513–519
30. Wang X et al (2006) Naltrexone inhibits alcohol-mediated enhancement of HIV infection of T lymphocytes. J Leukoc Biol 30:30
31. Liu X, Zha J, Nishitani J, Chen H, Zack JA (2003) HIV-1 infection in peripheral blood lymphocytes (PBLs) exposed to alcohol. Virology 307:37–44
32. Bagby GJ, Stoltz DA, Zhang P, Bohm RP Jr, Nelson S (1998) Simian immunodeficiency virus, infection, alcohol, and host defense. Alcohol Clin Exp Res 22:193S–195S
33. Zhang T et al (2003) Alcohol potentiates hepatitis C virus replicon expression. Hepatology 38:57–65
34. McCartney EM et al (2008) Alcohol metabolism increases the replication of hepatitis C virus and attenuates the antiviral action of interferon. J Infect Dis 198:1766–1775
35. Trujillo-Murillo K et al (2007) Additive effect of ethanol and HCV subgenomic replicon expression on COX-2 protein levels and activity. J Viral Hepat 14:608–617
36. Ye L et al (2010) Alcohol impairs interferon signaling and enhances full cycle hepatitis C virus JFH-1 infection of human hepatocytes. Drug Alcohol Depend 112:107–116
37. Seronello S, Ito C, Wakita T, Choi J (2010) Ethanol enhances hepatitis C virus replication through lipid metabolism and elevated NADH/NAD+. J Biol Chem 285:845–854
38. Potula R et al (2006) Alcohol abuse enhances neuroinflammation and impairs immune responses in an animal model of human immunodeficiency virus-1 encephalitis. Am J Pathol 168:1335–1344
39. Bagby GJ et al (2003) The effect of chronic binge ethanol consumption on the primary stage of SIV infection in rhesus macaques. Alcohol Clin Exp Res 27:495–502
40. Poonia B et al (2006) Chronic alcohol consumption results in higher simian immunodeficiency virus replication in mucosally inoculated rhesus macaques. AIDS Res Hum Retroviruses 22:589–594
41. Molina PE, Lang CH, McNurlan M, Bagby GJ, Nelson S (2008) Chronic alcohol accentuates simian acquired immunodeficiency syndrome-associated wasting. Alcohol Clin Exp Res 32:138–147
42. Molina PE et al (2006) Chronic alcohol accentuates nutritional, metabolic, and immune alterations during asymptomatic simian immunodeficiency virus infection. Alcohol Clin Exp Res 30:2065–2078
43. Kumar R et al (2005) Increased viral replication in simian immunodeficiency virus/simian-HIV-infected macaques with self-administering model of chronic alcohol consumption. J Acquir Immune Defic Syndr 39:386–390
44. Marcondes MC et al (2008) Chronic alcohol consumption generates a vulnerable immune environment during early SIV infection in rhesus macaques. Alcohol Clin Exp Res 32:1583–1592

45. Persidsky Y et al (2011) HIV-1 infection and alcohol abuse: neurocognitive impairment, mechanisms of neurodegeneration and therapeutic interventions. Brain Behav Immun 25(Suppl 1):S61–S70
46. Eken A, Ortiz V, Wands JR (2011) Ethanol inhibits antigen presentation by dendritic cells. Clin Vaccine Immunol 18:1157–1166
47. Haorah J et al (2004) Alcohol and HIV decrease proteasome and immunoproteasome function in macrophages: implications for impaired immune function during disease. Cell Immunol 229:139–148
48. Lieber CS, DeCarli LM (1970) Hepatic microsomal ethanol-oxidizing system. In vitro characteristics and adaptive properties in vivo. J Biol Chem 245:2505–2512
49. McArthur JC, Steiner J, Sacktor N, Nath A (2010) Human immunodeficiency virus-associated neurocognitive disorders: mind the gap. Ann Neurol 67:699–714
50. Lyons JL et al (2011) Plasma sCD14 is a biomarker associated with impaired neurocognitive test performance in attention and learning domains in HIV infection. J Acquir Immune Defic Syndr 57:371–379
51. Letendre SL et al (2011) Chemokines in cerebrospinal fluid correlate with cerebral metabolite patterns in HIV-infected individuals. J Neurovirol 17:63–69
52. Burdo TH et al (2011) Soluble CD163 made by monocyte/macrophages is a novel marker of HIV activity in early and chronic infection prior to and after anti-retroviral therapy. J Infect Dis 204:154–163
53. Ramirez SH et al (2010) Dyad of CD40/CD40 ligand fosters neuroinflammation at the blood–brain barrier and is regulated via JNK signaling: implications for HIV-1 encephalitis. J Neurosci 30:9454–9464
54. Renner NA et al (2012) S100beta as a novel and accessible indicator for the presence of monocyte-driven encephalitis in AIDS. Neuropathol Appl Neurobiol 38(2):162–174
55. Farinpour R et al (2000) Verbal working memory in HIV-seropositive drug users. J Int Neuropsychol Soc 6:548–555
56. Martin EM et al (2004) Cognitive impulsivity and HIV serostatus in substance dependent males. J Int Neuropsychol Soc 10:931–938
57. Rothlind JC et al (2005) Heavy alcohol consumption in individuals with HIV infection: effects on neuropsychological performance. J Int Neuropsychol Soc 11:70–83
58. Green JE, Saveanu RV, Bornstein RA (2004) The effect of previous alcohol abuse on cognitive function in HIV infection. Am J Psychiatry 161:249–254
59. Schulte T, Mueller-Oehring EM, Rosenbloom MJ, Pfefferbaum A, Sullivan EV (2005) Differential effect of HIV infection and alcoholism on conflict processing, attentional allocation, and perceptual load: evidence from a Stroop Match-to-Sample task. Biol Psychiatry 57:67–75
60. Fein G, Torres J, Price LJ, Di Sclafani V (2006) Cognitive performance in long-term abstinent alcoholic individuals. Alcohol Clin Exp Res 30:1538–1544
61. Pitel AL et al (2007) Genuine episodic memory deficits and executive dysfunctions in alcoholic subjects early in abstinence. Alcohol Clin Exp Res 31:1169–1178
62. Pitel AL et al (2007) Effect of episodic and working memory impairments on semantic and cognitive procedural learning at alcohol treatment entry. Alcohol Clin Exp Res 31: 238–248
63. Winsauer PJ et al (2002) Alcohol unmasks simian immunodeficiency virus-induced cognitive impairments in rhesus monkeys. Alcohol Clin Exp Res 26:1846–1857
64. Ancuta P et al (2008) Microbial translocation is associated with increased monocyte activation and dementia in AIDS patients. PLoS One 3:e2516
65. Brenchley JM et al (2006) Microbial translocation is a cause of systemic immune activation in chronic HIV infection. Nat Med 12:1365–1371
66. Persidsky Y, Poluektova L (2006) Immune privilege and HIV-1 persistence in the CNS. Immunol Rev 213:180–194
67. Szabo G, Bala S, Petrasek J, Gattu A (2010) Gut-liver axis and sensing microbes. Dig Dis 28:737–744
68. Mutlu EA et al (2012) Colonic microbiome is altered in alcoholism. Am J Physiol Gastrointest Liver Physiol 302:G966–G978

69. Szabo G, Zakhari S (2011) Mechanisms of alcohol-mediated hepatotoxicity in human-immunodeficiency-virus-infected patients. World J Gastroenterol 17:2500–2506
70. Forsyth CB, Tang Y, Shaikh M, Zhang L, Keshavarzian A (2011) Role of snail activation in alcohol-induced iNOS-mediated disruption of intestinal epithelial cell permeability. Alcohol Clin Exp Res 35:1635–1643
71. Fama R, Rosenbloom MJ, Nichols BN, Pfefferbaum A, Sullivan EV (2009) Working and episodic memory in HIV infection, alcoholism, and their comorbidity: baseline and 1-year follow-up examinations. Alcohol Clin Exp Res 33:1815–1824
72. Schulte T, Muller-Oehring EM, Javitz H, Pfefferbaum A, Sullivan EV (2008) Callosal compromise differentially affects conflict processing and attentional allocation in alcoholism, HIV, and their comorbidity. Brain Imaging Behav 2:27–38
73. Sullivan EV, Zahr NM (2008) Neuroinflammation as a neurotoxic mechanism in alcoholism: commentary on "Increased MCP-1 and microglia in various regions of human alcoholic brain". Exp Neurol 213:10–17
74. Blanco AM, Guerri C (2007) Ethanol intake enhances inflammatory mediators in brain: role of glial cells and TLR4/IL-1RI receptors. Front Biosci 12:2616–2630
75. Melendez RI, Hicks MP, Cagle SS, Kalivas PW (2005) Ethanol exposure decreases glutamate uptake in the nucleus accumbens. Alcohol Clin Exp Res 29:326–333
76. Zhao J et al (2004) Mitochondrial glutaminase enhances extracellular glutamate production in HIV-1-infected macrophages: linkage to HIV-1 associated dementia. J Neurochem 88:169–180
77. Ramirez SH et al (2010) Inhibition of glycogen synthase kinase 3{beta} (GSK3{beta}) decreases inflammatory responses in brain endothelial cells. Am J Pathol 176:881–892
78. Haorah J et al (2005) Ethanol-induced activation of myosin light chain kinase leads to dysfunction of tight junctions and blood–brain barrier compromise. Alcohol Clin Exp Res 29:999–1009
79. Haorah J, Knipe B, Gorantla S, Zheng J, Persidsky Y (2007) Alcohol-induced blood–brain barrier dysfunction is mediated via inositol 1,4,5-triphosphate receptor (IP3R)-gated intracellular calcium release. J Neurochem 100:324–336
80. Floreani NA et al (2010) Alcohol-induced interactive phosphorylation of Src and toll-like receptor regulates the secretion of inflammatory mediators by human astrocytes. J Neuroimmune Pharmacol 5:533–545
81. Arkins S, Rebeiz N, Biragyn A, Reese DL, Kelley KW (1993) Murine macrophages express abundant insulin-like growth factor-I class I Ea and Eb transcripts. Endocrinology 133:2334–2343
82. Elkabes S, DiCicco-Bloom EM, Black IB (1996) Brain microglia/macrophages express neurotrophins that selectively regulate microglial proliferation and function. J Neurosci 16:2508–2521
83. Batchelor PE et al (1999) Activated macrophages and microglia induce dopaminergic sprouting in the injured striatum and express brain-derived neurotrophic factor and glial cell line-derived neurotrophic factor. J Neurosci 19:1708–1716
84. Batchelor PE et al (2002) Macrophages and microglia produce local trophic gradients that stimulate axonal sprouting toward but not beyond the wound edge. Mol Cell Neurosci 21:436–453
85. Dougherty KD, Dreyfus CF, Black IB (2000) Brain-derived neurotrophic factor in astrocytes, oligodendrocytes, and microglia/macrophages after spinal cord injury. Neurobiol Dis 7:574–585
86. Satake K et al (2000) Up-regulation of glial cell line-derived neurotrophic factor (GDNF) following traumatic spinal cord injury. Neuroreport 11:3877–3881
87. Caroleo MC, Costa N, Bracci-Laudiero L, Aloe L (2001) Human monocyte/macrophages activate by exposure to LPS overexpress NGF and NGF receptors. J Neuroimmunol 113:193–201
88. Nakajima K et al (1998) Neurotrophins regulate the function of cultured microglia. Glia 24:272–289

89. Nakajima K et al (2001) Neurotrophin secretion from cultured microglia. J Neurosci Res 65:322–331
90. Shibata A et al (2003) Peripheral nerve induces macrophage neurotrophic activities: regulation of neuronal process outgrowth, intracellular signaling and synaptic function. J Neuroimmunol 142:112–129
91. Davalos D et al (2005) ATP mediates rapid microglial response to local brain injury in vivo. Nat Neurosci 8:752–758
92. Nimmerjahn A, Kirchhoff F, Helmchen F (2005) Resting microglial cells are highly dynamic surveillants of brain parenchyma in vivo. Science 308:1314–1318
93. Canals JM et al (1999) The neurotrophin receptors trkA, trkB and trkC are differentially regulated after excitotoxic lesion in rat striatum. Brain Res Mol Brain Res 69:242–248
94. Nakajima K, Kohsaka S (2004) Microglia: neuroprotective and neurotrophic cells in the central nervous system. Curr Drug Targets Cardiovasc Haematol Disord 4:65–84
95. Bechmann I, Nitsch R (1997) Astrocytes and microglial cells incorporate degenerating fibers following entorhinal lesion: a light, confocal, and electron microscopical study using a phagocytosis-dependent labeling technique. Glia 20:145–154
96. Gao HM, Liu B, Hong JS (2003) Critical role for microglial NADPH oxidase in rotenone-induced degeneration of dopaminergic neurons. J Neurosci 23:6181–6187
97. Streit WJ, Walter SA, Pennell NA (1999) Reactive microgliosis. Prog Neurobiol 57:563–581
98. He J, Crews FT (2008) Increased MCP-1 and microglia in various regions of the human alcoholic brain. Exp Neurol 210:349–358
99. Harper C (1998) The neuropathology of alcohol-specific brain damage, or does alcohol damage the brain? J Neuropathol Exp Neurol 57:101–110
100. Sun AY, Sun GY (2001) Ethanol and oxidative mechanisms in the brain. J Biomed Sci 8:37–43
101. Knapp DJ, Crews FT (1999) Induction of cyclooxygenase-2 in brain during acute and chronic ethanol treatment and ethanol withdrawal. Alcohol Clin Exp Res 23:633–643
102. Hansson T, Tindberg N, Ingelman-Sundberg M, Kohler C (1990) Regional distribution of ethanol-inducible cytochrome P450 IIE1 in the rat central nervous system. Neuroscience 34:451–463
103. Davis RL, Syapin PJ (2004) Acute ethanol exposure modulates expression of inducible nitric-oxide synthase in human astroglia: evidence for a transcriptional mechanism. Alcohol 32:195–202
104. Davis RL, Syapin PJ (2004) Ethanol increases nuclear factor-kappa B activity in human astroglial cells. Neurosci Lett 371:128–132
105. Riikonen J et al (2002) Intermittent ethanol exposure increases the number of cerebellar microglia. Alcohol Alcohol 37:421–426
106. Haorah J et al (2008) Mechanism of alcohol-induced oxidative stress and neuronal injury. Free Radic Biol Med 45:1542–1550
107. Mann K et al (2001) Neuroimaging in alcoholism: ethanol and brain damage. Alcohol Clin Exp Res 25:104S–109S
108. Persidsky Y et al (1999) Microglial and astrocyte chemokines regulate monocyte migration through the blood–brain barrier in human immunodeficiency virus-1 encephalitis. Am J Pathol 155:1599–1611
109. Crews FT, Nixon K (2009) Mechanisms of neurodegeneration and regeneration in alcoholism. Alcohol Alcohol 44:115–127
110. Chana G et al (2006) Cognitive deficits and degeneration of interneurons in HIV + methamphetamine users. Neurology 67:1486–1489
111. Persidsky Y, Ramirez SH, Haorah J, Kanmogne GD (2006) Blood–brain barrier: structural components and function under physiologic and pathologic conditions. J Neuroimmune Pharmacol 1:223–236
112. Mondal D, Pradhan L, Ali M, Agrawal KC (2004) HAART drugs induce oxidative stress in human endothelial cells and increase endothelial recruitment of mononuclear cells: exacerbation by inflammatory cytokines and amelioration by antioxidants. Cardiovasc Toxicol 4:287–302

113. Ramirez SH et al (2008) Activation of peroxisome proliferator-activated receptor gamma (PPARgamma) suppresses Rho GTPases in human brain microvascular endothelial cells and inhibits adhesion and transendothelial migration of HIV-1 infected monocytes. J Immunol 180:1854–1865
114. Eugenin EA et al (2006) Shedding of PECAM-1 during HIV infection: a potential role for soluble PECAM-1 in the pathogenesis of NeuroAIDS. J Leukoc Biol 79:444–452
115. Avison MJ et al (2004) Neuroimaging correlates of HIV-associated BBB compromise. J Neuroimmunol 157:140–146
116. Eilers M, Roy U, Mondal D (2008) MRP (ABCC) transporters-mediated efflux of anti-HIV drugs, saquinavir and zidovudine, from human endothelial cells. Exp Biol Med (Maywood) 233:1149–1160
117. Sipsas NV, Sfikakis PP, Kontos A, Kordossis T (2002) Levels of soluble CD40 ligand (CD154) in serum are increased in human immunodeficiency virus type 1-infected patients and correlate with CD4(+) T-cell counts. Clin Diagn Lab Immunol 9:558–561
118. Sui Z et al (2007) Functional synergy between CD40 ligand and HIV-1 Tat contributes to inflammation: implications in HIV type 1 dementia. J Immunol 178:3226–3236
119. Li H, Nord EP (2009) IL-8 amplifies CD40/CD154-mediated ICAM-1 production via the CXCR-1 receptor and p38-MAPK pathway in human renal proximal tubule cells. Am J Physiol Renal Physiol 296:F438–F445
120. Mancino A et al (2008) Divergent effects of hypoxia on dendritic cell functions. Blood 112:3723–3734
121. Pluvinet R et al (2008) CD40: an upstream master switch for endothelial cell activation uncovered by RNAi-coupled transcriptional profiling. Blood 112:3624–3637
122. Ishikawa M et al (2005) CD40/CD40 ligand signaling in mouse cerebral microvasculature after focal ischemia/reperfusion. Circulation 111:1690–1696
123. Piguet PF et al (2001) Role of CD40-CVD40L in mouse severe malaria. Am J Pathol 159:733–742
124. Sitati E, McCandless EE, Klein RS, Diamond MS (2007) CD40-CD40 ligand interactions promote trafficking of CD8+ T cells into the brain and protection against West Nile virus encephalitis. J Virol 81:9801–9811
125. D'Aversa TG, Eugenin EA, Berman JW (2005) NeuroAIDS: contributions of the human immunodeficiency virus-1 proteins Tat and gp120 as well as CD40 to microglial activation. J Neurosci Res 81:436–446
126. Persidsky Y et al (2006) Rho-mediated regulation of tight junctions during monocyte migration across the blood–brain barrier in HIV-1 encephalitis (HIVE). Blood 107:4770–4780
127. Yamamoto M et al (2008) Phosphorylation of claudin-5 and occludin by rho kinase in brain endothelial cells. Am J Pathol 172:521–533
128. Haorah J, Knipe B, Leibhart J, Ghorpade A, Persidsky Y (2005) Alcohol-induced oxidative stress in brain endothelial cells causes blood–brain barrier dysfunction. J Leukoc Biol 78:1223–1232
129. Haorah J, Schall K, Ramirez SH, Persidsky Y (2008) Activation of protein tyrosine kinases and matrix metalloproteinases causes blood–brain barrier injury: novel mechanism for neurodegeneration associated with alcohol abuse. Glia 56:78–88
130. Haorah J, Floreani NA, Knipe B, Persidsky Y (2011) Stabilization of superoxide dismutase by acetyl-l-carnitine in human brain endothelium during alcohol exposure: novel protective approach. Free Radic Biol Med 51:1601–1609
131. Haorah J et al (2007) Oxidative stress activates protein tyrosine kinase and matrix metalloproteinases leading to blood–brain barrier dysfunction. J Neurochem 101:566–576
132. Montoliu C et al (1995) Ethanol increases cytochrome P4502E1 and induces oxidative stress in astrocytes. J Neurochem 65:2561–2570
133. Blanco AM, Pascual M, Valles SL, Guerri C (2004) Ethanol-induced iNOS and COX-2 expression in cultured astrocytes via NF-kappa B. Neuroreport 15:681–685

134. Valles SL, Blanco AM, Pascual M, Guerri C (2004) Chronic ethanol treatment enhances inflammatory mediators and cell death in the brain and in astrocytes. Brain Pathol 14:365–371
135. Minambres R, Guasch RM, Perez-Arago A, Guerri C (2006) The RhoA/ROCK-I/MLC pathway is involved in the ethanol-induced apoptosis by anoikis in astrocytes. J Cell Sci 119:271–282
136. Jiang ZG et al (2001) Glutamate is a mediator of neurotoxicity in secretions of activated HIV-1-infected macrophages. J Neuroimmunol 117:97–107
137. Heyes M et al (1991) Quinolinic acid in cerebrospinal fluid and serum in HIV-1 infection: relationship to clinical and neurological status. Ann Neurol 29:202–209
138. Boven LA et al (1999) Increased peroxynitrite activity in AIDS dementia complex: implications for the neuropathogenesis of HIV-1 infection. J Immunol 162:4319–4327
139. Blanco AM, Valles SL, Pascual M, Guerri C (2005) Involvement of TLR4/type I IL-1 receptor signaling in the induction of inflammatory mediators and cell death induced by ethanol in cultured astrocytes. J Immunol 175:6893–6899
140. Fonnum F (1984) Glutamate: a neurotransmitter in mammalian brain. J Neurochem 42:1–11
141. Fillenz M (1995) Physiological release of excitatory amino acids. Behav Brain Res 71:51–67
142. Rothman SM (1985) The neurotoxicity of excitatory amino acids is produced by passive chloride influx. J Neurosci 5:1483–1489
143. Tanaka K (1994) Pharmacological characterization of a cloned rat glutamate transporter (GluT-1). Brain Res Mol Brain Res 21:167–170
144. Storck T, Schulte S, Hofmann K, Stoffel W (1992) Structure, expression, and functional analysis of a Na(+)-dependent glutamate/aspartate transporter from rat brain. Proc Natl Acad Sci U S A 89:10955–10959
145. Rothstein JD et al (1994) Localization of neuronal and glial glutamate transporters. Neuron 13:713–725
146. Gegelashvili G, Schousboe A (1997) High affinity glutamate transporters: regulation of expression and activity. Mol Pharmacol 52:6–15
147. Rothstein JD et al (1996) Knockout of glutamate transporters reveals a major role for astroglial transport in excitotoxicity and clearance of glutamate. Neuron 16:675–686
148. Martinez-Hernandez A, Bell KP, Norenberg MD (1977) Glutamine synthetase: glial localization in brain. Science 195:1356–1358
149. Kaul M, Zheng J, Okamoto S, Gendelman HE, Lipton SA (2005) HIV-1 infection and AIDS: consequences for the central nervous system. Cell Death Differ 12(Suppl 1):878–892
150. Gras G, Porcheray F, Samah B, Leone C (2006) The glutamate-glutamine cycle as an inducible, protective face of macrophage activation. J Leukoc Biol 80:1067–1075
151. Fine SM et al (1996) Tumor necrosis factor alpha inhibits glutamate uptake by primary human astrocytes. Implications for pathogenesis of HIV-1 dementia. J Biol Chem 271: 15303–15306
152. Patton HK, Zhou ZH, Bubien JK, Benveniste EN, Benos DJ (2000) gp120-induced alterations of human astrocyte function: Na(+)/H(+) exchange, K(+) conductance, and glutamate flux. Am J Physiol Cell Physiol 279:C700–C708
153. Rozyczka J, Figiel M, Engele J (2004) Endothelins negatively regulate glial glutamate transporter expression. Brain Pathol 14:406–414
154. Sitcheran R, Gupta P, Fisher PB, Baldwin AS (2005) Positive and negative regulation of EAAT2 by NF-kappaB: a role for N-myc in TNFalpha-controlled repression. EMBO J 24:510–520
155. Turchan J et al (2003) Oxidative stress in HIV demented patients and protection ex vivo with novel antioxidants. Neurology 60:307–314
156. Trotti D et al (1997) Neuronal and glial glutamate transporters possess an SH-based redox regulatory mechanism. Eur J Neurosci 9:1236–1243

157. Chen CJ et al (2006) L-Glutamate activates RhoA GTPase leading to suppression of astrocyte stellation. Eur J Neurosci 23:1977–1987
158. Zink M, Schmitt A, Vengeliene V, Henn FA, Spanagel R (2004) Ethanol induces expression of the glutamate transporters EAAT1 and EAAT2 in organotypic cortical slice cultures. Alcohol Clin Exp Res 28:1752–1757
159. Schreiber R, Freund WD (2000) Glutamate transport is downregulated in the cerebral cortex of alcohol-preferring rats. Med Sci Monit 6:649–652
160. Kim JH, Do SH, Kim YL, Zuo Z (2005) Effects of chronic exposure to ethanol on glutamate transporter EAAT3 expressed in Xenopus oocytes: evidence for protein kinase C involvement. Alcohol Clin Exp Res 29:2046–2052
161. Flatscher-Bader T et al (2005) Alcohol-responsive genes in the frontal cortex and nucleus accumbens of human alcoholics. J Neurochem 93:359–370
162. Krupitsky EM et al (2007) Antiglutamatergic strategies for ethanol detoxification: comparison with placebo and diazepam. Alcohol Clin Exp Res 31:604–611
163. Ma JZ, Ait-Daoud N, Johnson BA (2006) Topiramate reduces the harm of excessive drinking: implications for public health and primary care. Addiction 101:1561–1568
164. Spanagel R et al (2005) The clock gene Per2 influences the glutamatergic system and modulates alcohol consumption. Nat Med 11:35–42
165. Climent E, Pascual M, Renau-Piqueras J, Guerri C (2002) Ethanol exposure enhances cell death in the developing cerebral cortex: role of brain-derived neurotrophic factor and its signaling pathways. J Neurosci Res 68:213–225
166. Miller R, King MA, Heaton MB, Walker DW (2002) The effects of chronic ethanol consumption on neurotrophins and their receptors in the rat hippocampus and basal forebrain. Brain Res 950:137–147
167. Xiao C, Zhou C, Li K, Davies DL, Ye JH (2008) Purinergic type 2 receptors at GABAergic synapses on ventral tegmental area dopamine neurons are targets for ethanol action. J Pharmacol Exp Ther 327:196–205
168. Calvert RC, Banks FC, Thompson CS, Mikhailidis DP, Morgan RJ (2002) Chronic ethanol consumption impairs adrenoceptor- and purinoceptor-mediated relaxations in isolated rat detrusor smooth muscle. BJU Int 89:793–794
169. Burnstock G (2008) Purinergic signalling and disorders of the central nervous system. Nat Rev Drug Discov 7:575–590
170. Cardona AE, Huang D, Sasse ME, Ransohoff RM (2006) Isolation of murine microglial cells for RNA analysis or flow cytometry. Nat Protoc 1:1947–1951
171. Takahashi K, Rochford CD, Neumann H (2005) Clearance of apoptotic neurons without inflammation by microglial triggering receptor expressed on myeloid cells-2. J Exp Med 201:647–657
172. Streit WJ (2002) Microglia as neuroprotective, immunocompetent cells of the CNS. Glia 40:133–139
173. Corrao G, Arico S (1998) Independent and combined action of hepatitis C virus infection and alcohol consumption on the risk of symptomatic liver cirrhosis. Hepatology 27:914–919
174. Regev A, Jeffers LJ (1999) Hepatitis C and alcohol. Alcohol Clin Exp Res 23:1543–1551
175. Wiley TE, McCarthy M, Breidi L, Layden TJ (1998) Impact of alcohol on the histological and clinical progression of hepatitis C infection. Hepatology 28:805–809
176. Bonacini M (2011) Alcohol use among patients with HIV infection. Ann Hepatol 10:502–507
177. Tsui JI, Cheng DM, Libman H, Bridden C, Samet J (2012) Hepatitis C virus infection is associated with painful symptoms in HIV-infected adults. AIDS Care
178. Gotti D et al (2011) Cryptogenic liver diseases: sailing by sight from HIV co-infection with hepatitis viruses to HIV mono-infection through the Pillars of Hercules. Curr HIV Res 9:61–69
179. Hines IN, Wheeler MD (2004) Recent advances in alcoholic liver disease III. Role of the innate immune response in alcoholic hepatitis. Am J Physiol Gastrointest Liver Physiol 287:G310–G314

180. Biron CA (1998) Role of early cytokines, including alpha and beta interferons (IFN-alpha/beta), in innate and adaptive immune responses to viral infections. Semin Immunol 10:383–390
181. Bhattacharya R, Shuhart MC (2003) Hepatitis C and alcohol: interactions, outcomes, and implications. J Clin Gastroenterol 36:242–252
182. Safdar K, Schiff ER (2004) Alcohol and hepatitis C. Semin Liver Dis 24:305–315
183. Backmund M, Meyer K, Wachtler M, Eichenlaub D (2003) Hepatitis C virus infection in injection drug users in Bavaria: risk factors for seropositivity. Eur J Epidemiol 18:563–568
184. Garten RJ et al (2004) Rapid transmission of hepatitis C virus among young injecting heroin users in Southern China. Int J Epidemiol 33:182–188
185. Quaglio G et al (2003) Factors associated with hepatitis C virus infection in injection and noninjection drug users in Italy. Clin Infect Dis 37:33–40
186. Schaefer M, Heinz A, Backmund M (2004) Treatment of chronic hepatitis C in patients with drug dependence: time to change the rules? Addiction 99:1167–1175
187. McCartney EM, Beard MR (2010) Impact of alcohol on hepatitis C virus replication and interferon signaling. World J Gastroenterol 16:1337–1343
188. Oshita M et al (1994) Increased serum hepatitis C virus RNA levels among alcoholic patients with chronic hepatitis C. Hepatology 20:1115–1120
189. Otani K et al (2005) Hepatitis C virus core protein, cytochrome P450 2E1, and alcohol produce combined mitochondrial injury and cytotoxicity in hepatoma cells. Gastroenterology 128:96–107
190. Rigamonti C et al (2003) Moderate alcohol consumption increases oxidative stress in patients with chronic hepatitis C. Hepatology 38:42–49
191. Szabo G (2003) Pathogenic interactions between alcohol and hepatitis C. Curr Gastroenterol Rep 5:86–92
192. Cherner M et al (2005) Hepatitis C augments cognitive deficits associated with HIV infection and methamphetamine. Neurology 64:1343–1347
193. Clifford DB, Yang Y, Evans S (2005) Neurologic consequences of hepatitis C and human immunodeficiency virus coinfection. J Neurovirol 11(Suppl 3):67–71
194. Letendre SL et al (2005) The effects of hepatitis C, HIV, and methamphetamine dependence on neuropsychological performance: biological correlates of disease. AIDS 19(Suppl 3):S72–S78
195. Perry W et al (2005) Neuropsychological test performance in patients co-infected with hepatitis C virus and HIV. AIDS 19(Suppl 3):S79–S84
196. Ryan EL, Morgello S, Isaacs K, Naseer M, Gerits P (2004) Neuropsychiatric impact of hepatitis C on advanced HIV. Neurology 62:957–962
197. Richardson JL et al (2005) Neuropsychological functioning in a cohort of HIV- and hepatitis C virus-infected women. AIDS 19:1659–1667
198. von Giesen HJ et al (2004) Psychomotor slowing in hepatitis C and HIV infection. J Acquir Immune Defic Syndr 35:131–137
199. Hilsabeck RC, Hassanein TI, Ziegler EA, Carlson MD, Perry W (2005) Effect of interferon-alpha on cognitive functioning in patients with chronic hepatitis C. J Int Neuropsychol Soc 11:16–22
200. Winston A et al (2010) Does acute hepatitis C infection affect the central nervous system in HIV-1 infected individuals? J Viral Hepat 17:419–426
201. Cook RT, Stapleton JT, Ballas ZK, Klinzman D (1997) Effect of a single ethanol exposure on HIV replication in human lymphocytes. J Investig Med 45:265–271
202. Plumlee CR, Lazaro CA, Fausto N, Polyak SJ (2005) Effect of ethanol on innate antiviral pathways and HCV replication in human liver cells. Virol J 2:89
203. Baliunas D, Rehm J, Irving H, Shuper P (2010) Alcohol consumption and risk of incident human immunodeficiency virus infection: a meta-analysis. Int J Public Health 55:59–166
204. Pessione F et al (1998) Effect of alcohol consumption on serum hepatitis C virus RNA and histological lesions in chronic hepatitis C. Hepatology 27:1717–1722

Chapter 17
Therapeutic Strategies to Treat Alcohol-Related Disorders Targeting Central Immune Signaling

Yue Wu and Mark R. Hutchinson

Abbreviations

ATP	Adenosine triphosphate
B2M	β-2-Microglobulin
BBB	Blood–brain barrier
CCL2	Chemokine (C–C motif) ligand receptor 2
CCR2	Chemokine (C–C) motif ligand 2
CNS	Central nervous system
COX-2	Cyclooxygenase-2
CTSF	Cathepsin F
CTSS	Cathepsin S
CXCL12	Chemokine (C–X–C motif) ligand 12
DAMP	Danger-associated molecular pattern
ERK	Extracellular regulated kinase
GABA	Gamma-aminobutyric acid
GDNF	Glial cell line-derived neurotrophic factor
GFAP	Glial fibrillary acidic protein
HIV	Human immunodeficiency virus
IFN-γ	Interferon-γ
IL	Interleukin
IL-1Ra	Interleukin-1 receptor antagonist
iNOS	Inducible nitric oxide synthase

Y. Wu
Discipline of Pharmacology, School of Medical Sciences, University of Adelaide, Adelaide, SA, Australia

M.R. Hutchinson, Ph.D. (✉)
Discipline of Physiology, School of Medical Sciences, University of Adelaide, Adelaide, SA, Australia
e-mail: mark.hutchinson@adelaide.edu.au

C. Cui et al. (eds.), *Neural-Immune Interactions in Brain Function and Alcohol Related Disorders*, DOI 10.1007/978-1-4614-4729-0_17,

IRAK	Serine/threonine kinase interleukin-1 receptor-associated kinase 4
IRF3	IFN regulatory factor 3
IκBα	NFκB inhibitor α
JNK	c-Jun N-terminal kinase
LPS	Lipopolysaccharide
MAL	Myelin and lymphocyte protein
MAPK	Mitogen-activated protein kinase
MCP-1	Monocyte chemoattractant protein-1
MD-2	Myeloid differentiation factor 2
MHC	Major histocompatibility complex
MOR	μ-Opioid receptor
MyD88	Myeloid differentiation primary response gene 88
NAc	Nucleus accumbens
NFκB	Nuclear factor κ-light-chain-enhancer of activated B cells
NMDA	Neuronal *N*-methyl-D-aspartate
NO	Nitric oxide
PAMP	Pathogen-associated molecular pattern
PI3K	Phosphoinositide 3-kinase
SN	Substantia nigra
TLR	Toll-like receptor
TNF-α	Tumor necrosis factor-α
TRAM	TRIF-related adaptor molecule
TRIF	Toll/IL-1R domain containing adaptor inducing IFN-β
VTA	Ventral tegmental area

17.1 Introduction

With its consumption believed to have begun more than 10,000 years ago, alcohol continues to feature prominently in many societies and cultures worldwide. Unfortunately among the two billion annual consumers, alcohol abuse poses a significant issue, with over 76 million people diagnosed with an alcohol abuse disorder [1]. Related to more than 60 types of disease and injury, alcohol consumption is the third biggest risk factor for disease burden in developed countries, such as the United States, Canada, Germany, France, the United Kingdom, Australia, and Japan, and is the largest risk factor in developing countries, such as China, the Philippines, Vietnam, Indonesia, Thailand, Brazil, and Mexico [1]. As alcohol dependence and abuse remain continuing and important health problems, intervention strategies to overcome this drug addiction are urgently required to reduce the burden of their related diseases in society.

Understanding the mechanisms underpinning the effects of alcohol would have wide-reaching implications in developing targeted pharmacological intervention therapies for alcohol dependence. Much research has focused on how the nervous system adapts to alcohol exposure. However, despite several decades of investigation on the neuronal mechanisms of alcohol's effects, the number of "unknowns" in this research area has increased significantly. Apart from the neuronal consequences of alcohol exposure, recent studies have implicated alcohol-induced glial activation and Toll-like receptor (TLR)-dependent proinflammatory signaling in the central nervous system (CNS) in the neurotoxic effects and behavioral responses of alcohol. This chapter will focus on the new central immune signaling mechanisms of alcohol's effects and the potential treatment and implications which can be developed from these mechanisms.

17.2 Alcohol and Central Immune Signaling

17.2.1 Glia: The Immunocompetent Cells in the CNS

Glia and immune factors within the CNS have been considered to be involved in the effects of alcohol since the 1980s [2–4]. The defining characteristic of neurons is their ability to fire action potentials. All the other cells in the brain that lack this property are classified into a broad class termed glia [5]. Glia, making up approximately 90% of the human brain and roughly 65% of the mouse brain, were thought by many that their purpose was to solely provide physical support and housekeeping for neurons [6]. Being called "brain glue" until the last two decades, glial cells are in fact not all the same and have different subtypes. They are pivotal to maintaining a healthy brain yet are also related to a variety of diseases. Glia are classified into microglia, astrocytes, and oligodendrocytes on the basis of morphology, function, and location in the CNS. Microglia are recognized as the prime component of the immune system in the brain [7]. They are quickly activated in acute pathological states such as those observed during infection, inflammation, and neurodegeneration, and contribute to altered CNS function [8]. In contrast, astrocytes play an important role in the CNS through the "tripartite synapse" [9] or the "neural threesome" hypothesis [10], which is a bidirectional communication system between astrocytes, presynaptic neurons, and postsynaptic neurons [11]. Oligodendrocytes produce a lipid-rich membrane called myelin, which enwraps axons, thereby speeding up the conduction of electrical impulses [5, 12].

Generally, as immunocompetent cells in the CNS, glial cells can be activated by different stimuli, including viruses and bacteria which contain factors termed pathogen-associated molecular patterns (PAMPs). The PAMPs are a variety of microbial molecules which share a number of different recognizable biochemical features which alert the organism via innate immune mechanisms to intruding pathogens [13]. In addition, endogenous signals can also drive central immune signaling, resulting from damage-associated molecular patterns (DAMPs). The PAMPs and DAMPs are recognized by hereditable innate immune system receptors, primarily the TLRs, and subsequently activating several different signaling pathways, in which the nuclear factor κ-light-chain-enhancer of activated B cells (NFκB) signaling is the most distinctive [14]. Such signaling could in turn activate other glial cells to enhance this signal [15].

17.2.2 Alcohol and Proinflammatory Glial Activation

So, what does alcohol do to glial cells? Alcohol exposure can activate microglia and astrocytes in a proinflammatory fashion. It has been demonstrated that rodents chronically treated with alcohol display increased levels of glial fibrillary acidic protein (GFAP, proinflammatory astrocyte marker) in the ventral tegmental area (VTA) and cerebral cortex [16, 17], as well as CD11b (proinflammatory microglial marker) within the cerebral cortex [18], implicating the activation of glial cells by alcohol exposure. The activation of microglia, as determined by the increased expression of activation markers, in the rat hippocampus post-binge alcohol exposure lasted up to 28 days [19]. This microglial activation contributes to the behavioral consequences of alcohol, including sedation, motor impairment [20], and alcohol consumption [21], in mouse models owing to the minocycline sensitivity of the response [20, 21]. Increased microglial activation was also found in the human postmortem brains from alcoholics compared to those from moderate drinking controls [22]. However, the direct mechanisms by which alcohol acts on glial cells and produces this proinflammatory reactivity remains unclear.

Recently, Blanco et al. [23] showed that alcohol directly acted on glia by interacting with lipid rafts caveolae, which are the plasma membrane microdomains. Rafts/caveolae, which have been identified as platforms for receptor signaling, constitute an important integrator of signal events and trafficking. The regulation of lipid rafts by alcohol promotes interleukin-1β receptor type I and TLR4 recruitment, triggering their endocytosis via caveosomes and downstream signaling stimulation [23]. Interestingly, Lee et al. [24] demonstrated that astrocytes were capable of expressing gamma-aminobutyric acid (GABA) receptors (both class A and class B), and GABA could in turn suppress the reactive response of both astrocytes and microglia to lipopolysaccharide (LPS, the classical ligand for TLR4) and interferon-γ (IFN-γ) by inhibiting induction of inflammatory pathways mediated by NFκB and p38/mitogen-activated protein kinase (MAPK). Thus, the suppressed GABA receptor signaling by acute alcohol exposure may lead to glial activation. Although many researchers have observed the central immune signaling induced by alcohol exposure [18, 19, 25], it is still debatable if alcohol directly or indirectly acts on glial cells, which requires further research.

Although both astrocytes and microglia can be activated by alcohol, microglia are possibly the initial cells where this neuroinflammation occurs, considering the pivotal role of TLR2 and TLR4 in alcohol-induced neuroinflammation (see Sect. 2). Human microglia express TLRs 1–9 mRNA, whereas human astrocytes express robust level of TLR3, low-level TLR1, 4, 5, and 9, and rare TLR2, 6, 7, 8, and 10 mRNA [26]. In mice, microglia have been demonstrated to be the major non-neuronal cell type in the CNS that express TLR4 [27]. As such, pharmacological treatment targeting microglial activation may inhibit the initial effect of alcohol on central immune signaling and therefore prevent the robust neuroimmune activation that follows. Microglial inhibitors may prove beneficial in the conditions related to acute alcohol exposure, such as binge drinking-induced brain damage.

17.2.3 Glial–Neuronal Communication

Generally, activated glial cells release a variety of active substances, including proinflammatory cytokines, prostaglandins, brain-derived neurotrophic factor, adenosine triphosphate (ATP), nitric oxide (NO), D-serine, and glutamate [15], which in turn modulate neuronal activity, although the mechanisms by which such modulation occurs are only beginning to be understood [28]. There is a profound relationship between the activation of TLR4 and cell-signaling proteins involved in innate immunity and altered neuronal receptor functions post-alcohol exposure. Neuronal *N*-methyl-D-aspartate (NMDA) and GABA receptors are critical for the neuronal activities of alcohol [29, 30]. However, inhibition of $GABA_A\alpha2$ attenuated TLR4 expression in the central nucleus of the amygdale of rats and also inhibited binge drinking [31]. This finding provides solid evidence for the relationship between GABA receptors, TLRs, and alcohol-drinking behavior. In addition, the NMDA receptor was resistant to its inhibition by acute alcohol exposure, following chronic alcohol consumption [32]. This resistance occurred with enhanced levels of p38 phosphorylation and acquisition of tolerance to alcohol-induced sedation [32], suggesting the role of MAPK pathways in glial–neuronal crosstalk induced by alcohol exposure.

In neurons, it has been demonstrated that chronic alcohol exposure activated extracellular-regulated kinase (ERK)/p44 and c-Jun N-terminal kinase (JNK)/p46 [33]. This chronic alcohol treatment also upregulated the effects of LPS on ERK, p38, and JNK phosphorylation, enhanced LPS-induced activation of NFκB/RelA-p50, but blocked NFκB/p50–p50 activation by LPS [33]. These changes in signaling proteins in MAPK and NFκB pathways may be possible consequences of glial–neuronal communication following alcohol exposure. The activation of MAPK and NFκB signaling may accelerate the glial–neuronal crosstalk and extend the neuroimmune activation in the CNS. Pharmacological strategies attenuating these pathways could not only inhibit the activation of glia and neurons but also block the communication between those cells.

Alcohol may also modulate glutamate and GABA signaling in glia directly and therefore alter proinflammatory signaling. The expression of NMDA and GABA receptors on glia was not reported until recently [24, 34, 35]. Alcohol has been found to reduce kainate-induced glutamate secretion in rat hippocampal astrocytes [36], and glutamate can mediate the signaling transmission between neurons and astrocytes [37]. GABA suppressed the LPS- or IFN-γ-induced immune response in both astrocytes and microglia by inhibiting induction of NFκB and MAPK signaling pathways, and this led to a reduction in tumor necrosis factor-α (TNF-α) and IL-6 release [24].

The proinflammatory cytokines released by activated microglia and astrocytes are generally considered to be the mediators underpinning alcohol-induced neuronal death caused as a result of neuroinflammatory damage [38–40] and the behavioral effects of alcohol [20]. Interestingly, IL-1β signaling, which can be activated by alcohol exposure [40], also reduces astrocytic glutamate transporter-1 expression and subsequently drives excitotoxic motor neuron injury [41]. In addition, binding of glutamate to the NMDA receptor modulates LPS-induced innate immune activation, including TNF-α signaling, in a TLR4-dependent manner [42]. With regard to GABA signaling, its synaptic activity could be enhanced by chemokine (C–X–C motif) ligand 12 (CXCL12) on serotonergic neurons in rats [43]. The above findings suggest that proinflammatory cytokines and chemokines are capable of altering the downstream signaling of NMDA and GABA receptors in neurons. Furthermore, in another type of central glial cells, oligodendrocytes, cytokines could reduce the toxic effects of alcohol [44]. As such, alcohol could activate microglia and astrocytes and subsequently induce the release of cytokines and chemokines, which could in turn modify the function of neurons through neuronal targets and oligodendrocytes.

In summary, the glial–neuronal crosstalk post-alcohol exposure is mediated by neuronal transmitters, TLRs, cytokines, and intracellular signaling proteins in MAPK and NFκB pathways. Alcohol could act on neuronal receptors, mainly GABA and NMDA receptors, on neurons and astrocytes, which further alters the release of neuronal transmitters. This in turn modifies the activation and release of cytokines and chemokines by glia. In contrast, alcohol activates microglia and astrocytes through TLRs, and the cytokines and chemokines released from activated glia could modify the function of neuronal targets. Thus, the receptors, neuronal transmitters, cytokines, and chemokines that are involved in this neuroimmune communication could be potential therapeutic targets for alcohol dependence. Due to the interaction between glia and neurons, adding glial signaling inhibitors to the traditional treatment for alcohol dependence targeting neuronal receptors (see Sect. 3) may increase the therapeutic effects and decrease the side effects of these medications.

17.2.4 Behavioral Consequences of Alcohol-Induced Neuroimmune Activation

As discussed above, alcohol acts on both neurons and glia, and this neuroimmune modification in the CNS will subsequently alter behaviors. Recent study showed that minocycline (a microglial attenuator) administration decreased the acute sedative effect of alcohol [20] and reduced alcohol drinking [21] in mice, and these were independent of minocycline-induced changes of body weight. These findings suggest that microglial activation is related to the initial effect of alcohol and possibly in the development of alcohol dependence.

Multiple cytokines, including chemokines, have been demonstrated to play a role in the behavioral effects of alcohol in both animal alcohol models and human alcoholic experiments (Table 17.1). Animal evidence showed that deletion of chemokine (C–C motif) ligand receptor 2 (CCR2), chemokine (C–C) motif ligand 2 (CCL2)/monocyte chemoattractant protein-1 (MCP-1) (females), or CCL3 in mice resulted in lower preference for alcohol, and CCL2- or CCL3-deficient mice showed longer duration of alcohol-induced loss of righting reflex than wild-type

Table 17.1 Evidence of alcohol-induced neuroimmune activation

Target	Description	Material	Treatment	Animal/ human	Effect	TLR4-dependent	Reference
In vitro cell culture study							
MyD88	MyD88-dependent pathway	Astrocytes	Acute alcohol[a]	Mice	↑	Yes	[18, 39, 40]
IRF-1	TRIF-dependent pathway	Microglia	Acute alcohol	Mice	↑	Yes	[39]
ERK	MAPK signaling	Astrocytes, microglia	Acute alcohol	Mice	↑ phosphorylation	Yes	[18, 39, 40]
IRAK	MyD88-dependent pathway	Astrocytes, microglia	Acute alcohol	Mice	↑ phosphorylation	Yes	[18, 39, 40]
p38	MAPK signaling	Astrocytes, microglia	Acute alcohol	Mice	↑ phosphorylation	Yes	[18, 39, 40]
JNK	MAPK signaling	Astrocytes, microglia	Acute alcohol	Mice	↑ phosphorylation	Yes	[18, 39, 40]
IκBα	NFκB signaling	Astrocytes	Acute alcohol	Mice	↑ phosphorylation	Yes	[18, 39, 40]
COX-2	Inflammatory mediator	Astrocytes	Acute alcohol	Mice	↑	Yes	[18, 39, 40]
iNOS	Inflammatory mediator	Astrocytes	Acute alcohol	Mice	↑	Yes	[18, 39, 40]
TNF-α	Cytokine	Microglia	Acute alcohol	Mice	↑	Yes	[39]
IFN-β	Cytokine	Microglia	Acute alcohol	Mice	↑	Yes	[39]
Ex vivo study							
CD11b	Microglial marker	Cerebral cortex	Chronic alcohol	Mice	↑	Yes	[18]
Iba-1	Microglial marker	Cingulate cortex	Chronic alcohol	Humans	↑	No	[22]
GluT5	Microglial marker	Cingulate cortex, VTA, midbrain	Chronic alcohol	Humans	↑	No	[22]
GFAP	Astrocyte marker	Cerebral cortex, VTA	Chronic alcohol	Mice	↑	Yes	[16, 17]
CD14	TLR4 signaling	Cerebral cortex	Chronic alcohol	Mice	↑	Yes	[18]
MyD88	MyD88-dependent pathway	Cerebral cortex	Chronic alcohol	Mice	↑	Yes	[18]
ERK	MAPK signaling	Cerebral cortex	Chronic alcohol	Mice	↑ phosphorylation	Yes	[18]
AKT	Non-MAPK signaling	Striatum	Acute alcohol	Mice	↑ phosphorylation	No	[16]
NFκB-p65	NFκB signaling	Cerebral cortex	Chronic alcohol	Mice	↑	Yes	[18]
NFκB	NFκB signaling	Corticolimbic brain	Chronic alcohol	Rats	↑ DNA binding	No	[45]
iNOS	Inflammatory mediator	Cerebral cortex	Chronic alcohol	Mice	↑	Yes	[18]

(continued)

Table 17.1 (continued)

Target	Description	Material	Treatment	Animal/ human	Effect	TLR4-dependent	Reference
COX-2	Inflammatory mediator	Corticolimbic brain, cerebral cortex	Chronic alcohol	Rats, mice	↑	Yes	[45]
TNF-α	Cytokine	Cerebral cortex	Chronic alcohol	Mice	↑	Yes	[18]
IL-1β	Cytokine	Cerebral cortex	Chronic alcohol	Mice	↑	Yes	[18]
IL-6	Cytokine	Cerebral cortex	Chronic alcohol	Mice	↑	Yes	[18]
CCL2	Chemokine	Superior frontal cortex, VTA, SN, hippocampus, amygdala	Chronic alcohol	Humans	↑	No	[22, 46]
Behavioral study							
TLR4	TLR4 signaling	Central nucleus of the amygdala	Blockade	Rats	↓ binge drinking	Yes	[31]
TLR4	TLR4 signaling	–	Genetic deficiency	Mice	Withdrawal induced by chronic alcohol consumption	Yes	[47]
TLR4	TLR4 signaling	–	Genetic deficiency or pharmacological blockade	Mice	↓ sedative and motor effects of alcohol	Yes	[48]
TLR2	TLR2 signaling	–	Genetic deficiency	Mice	↓ sedative effects of alcohol	No	[49]
MyD88	MyD88-dependent pathway		Genetic deficiency	Mice	↓ sedative and motor effects of alcohol	No	[48]
CD14	TLR4 signaling	–	Genetic deficiency	Mice	↓ alcohol consumption	No	[50]
MAPK	MAPK signaling	–	Pharmacological activation	Rats	↓ alcohol consumption and seeking	No	[51]

CCL2	Chemokine	–	Genetic deficiency	Mice	↓ preference for alcohol (females), ↑ alcohol-induced sedation	No	[50]
CCR2	Chemokine receptor	–	Genetic deficiency	Mice	↓ preference for alcohol	No	[50]
CCL3	Chemokine	–	Genetic deficiency	Mice	↓ preference for alcohol, ↑ alcohol-induced sedation	No	[50]
IL-1Ra	Cytokine receptor signaling	–	Genetic deficiency	Mice	↓ alcohol consumption	No	[50]
IL-1 receptor	Cytokine receptor	–	Pharmacological blockade	Mice	↓ sedative and motor effects of alcohol	No	[51]
IL-6	Cytokine	–	Genetic deficiency	Mice	↓ alcohol consumption	No	[50]
B2M	MHC molecule	–	Genetic deficiency	Mice	↓ alcohol consumption	No	[50]
CTSS	MHC molecule	–	Genetic deficiency	Mice	↓ alcohol consumption	No	[50]
CTSF	MHC molecule	–	Genetic deficiency	Mice	↓ alcohol consumption	No	[50]
Microglia	Microglia	–	Pharmacological blockade	Mice	↓ sedative and ↑ motor effects of alcohol	No	[51]
Pharmacogenetic study							
NFκB	NFκB signaling	–	–	Humans	–	No	[51]
IL-10	Cytokine	–	–	Humans	–	No	[52]
IL-1Ra	Cytokine receptor signaling	–	–	Humans	–	No	[53, 54]
IL-1β	Cytokine	–	–	Humans	–	No	[53, 55]

SN substantia nigra, *VTA* ventral tegmental area, *B2M* β-2-microglobulin, *CTSS* cathepsin S, *CTSF* cathepsin F, *MHC* major histocompatibility complex

[a]Acute alcohol, less than 24 h

mice [56], which suggested the involvement of these chemokines and their receptors. This is further supported by work demonstrating elevated CCL2 expression in human brains of alcoholics [22, 46]. Mice deficient in interleukin-1 (IL-1) receptor antagonist (IL-1Ra, encoded by *IL1RN* gene) or IL-6 had reduced alcohol consumption in the 24-h two-bottle choice test [50], suggesting a role for IL-1Ra and IL-6 in alcohol-drinking behavior. Moreover, IL-1 receptor signaling could be activated acutely by alcohol administration, and this subsequently modulates the sedative and motor effects of alcohol [20]. Human immunogenetic studies have demonstrated associations between *IL1B* [53, 55], *IL1RN* [53, 54], *IL10* [52], and *NFKB1* [57] genetic polymorphisms and alcohol dependence, thereby providing indirect evidence of a proinflammatory contribution to alcohol dependence.

In animal preclinical studies, LPS has been used as an immune stressor leading to proinflammatory signaling activation. The systemic administration of LPS enhanced alcohol-induced motor impairment 24 h later via the rotarod test [58]. A prolonged increase of voluntary alcohol intake in wild-type mice induced by LPS administration was lacking in mice deficient in CD14 (a key component of TLR4 signaling; see the discussion in Sect. 2.1) [59]. Although LPS unlikely passed the blood–brain barrier (BBB) to activate brain TLR4 signaling directly, the above evidence still highlights the fact that immune activation facilitated the behavioral effects of alcohol. Together, the neuroimmune activation contributes to the behavioral effects of alcohol, and therefore glia and immune factors could be the research targets for treating alcohol dependence. The potential pharmacotherapy based on alcohol-induced glial activation will be discussed in greater detail in Sect. 3.

17.3 Alcohol and TLRs

How do glia become activated in a proinflammatory fashion? As many studies have indicated, TLR4 appears to be a key receptor in the mechanisms of alcohol's effects on microglia and astrocytes [18, 39, 40, 48]. The activation of TLR4 signaling enhances proinflammatory mediators through NFκB [60, 61]. TLR2 can also be activated by alcohol administration, as the TLR2-deficient mice show reduced sedative effects of alcohol [49], although the cell type where this activation occurs remains to be determined. Therefore, pharmacotherapies inhibiting TLR signaling may have beneficial effects in the treatment of alcohol abuse.

17.3.1 Similarities and Differences Between TLR2 and TLR4 Pathways

The well-understood function of TLR4, a homodimer [62, 63], is to recognize LPS from gram-negative bacteria, cooperating with myeloid differentiation factor 2 (MD-2) and CD14 [62]. TLR4 signaling cascades are classified into the myeloid

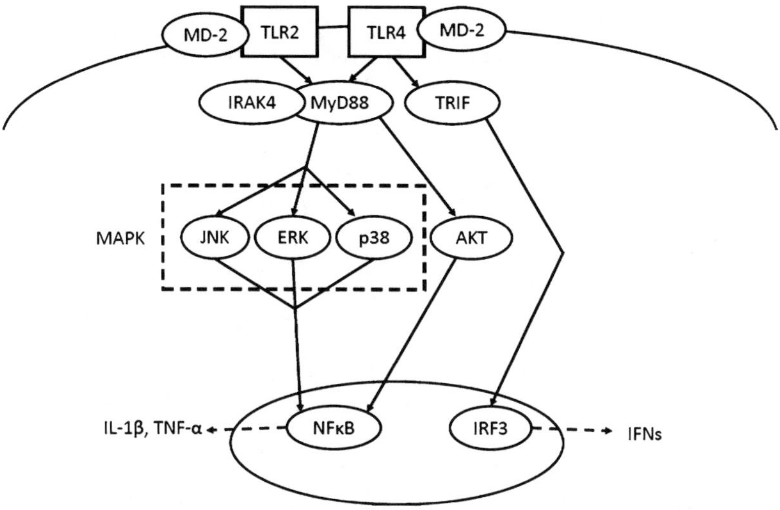

Fig. 17.1 Schematic representation of the TLR2 and TLR4 pathways. The TLR4 downstream pathways are classified into the MyD88-dependent and MyD88-independent (TRIF-dependent) pathways. The activation of the MyD88-dependent pathway induces NFκB activation via the MAPK (JNK, ERK, and p38) or non-MAPK (AKT) pathways. The NFκB signaling activation triggers an increase in the expression of proinflammatory cytokines, such as IL-1β and TNF-α. The activation of the TRIF-dependent pathway leads to the activation of the transcription factor IRF3 and the induction of genes such as those encoding type I IFNs. In contrast, TLR2 signaling activation only leads to MyD88-dependent signaling

differentiation primary response gene 88 (MyD88)-dependent and MyD88-independent pathways. For the MyD88-dependent pathway, myelin and lymphocyte protein (MAL) and MyD88 lead to the activation of the serine/threonine kinase interleukin-1 receptor-associated kinase 4 (IRAK4), which engages with MAPK cascades (JNK, ERK, and p38) and leads to NFκB activation and the induction of proinflammatory cytokines such as IL-1β and TNF-α. Alternatively, downstream of TLR4, phosphoinositide 3-kinase (PI3K)/AKT pathways could also induce NFκB activation [64]. With regard to the MyD88-independent pathway, TLR4 recruits Toll/IL-1R domain-containing adaptor inducing IFN-β (TRIF) and TRIF-related adaptor molecule (TRAM) in endosomes, which leads to the activation of the transcription factor IFN regulatory factor 3 (IRF3) and the induction of genes such as those encoding Type I IFNs [65, 66]. In contrast, TLR2 heterodimerizes with TLR1 or TLR6 [67, 68]. After binding to lipopeptides, TLR2 is activated and subsequently leads to MyD88-dependent signaling [69, 70] (Fig. 17.1). As at various conditions, TLR2 and TLR4 may be differently activated [49], the activation of TLR2 and TLR4 at different alcohol concentrations, durations of alcohol exposure, cell types, and brain regions are worthy of investigating and comparing. The similarities and differences between TLR2 and TLR4 pathways provide potential therapeutic targets, acting on TLR2 or TLR4 selectively, for alcohol abuse. For example, TLR2-deficient, but not TLR4-deficient, mice were lacking of the alcohol–morphine interaction as assessed by sedation [49], and therefore, TLR2 signaling inhibitors may be suitable for this drug interaction. In contrast, alcohol exposure activates both TLR2 and TLR4 signaling acutely [48, 49], and pharmacological blockade of the function of both TLR2 and TLR4 or their shared downstream signaling could be a candidate treatment. Due to the limited knowledge of the involvement of TLR2 in the effect of alcohol, the following section will focus on TLR4 signaling.

17.3.2 TLR4 Downstream Signaling Induced by Alcohol Exposure

The effect of alcohol on the activation of glial TLR4 signaling depends on the dose and time-course of alcohol exposure, as well as the specific pathways involved [71]. It has been demonstrated that alcohol activates glial MAPK and NFκB pathways within minutes of stimulation in vitro, which is TLR4 dependent [18, 39, 40]. In wild-type astrocytic culture, 10–50 mM of alcohol increased ERK, IRAK, p38, and JNK phosphorylation and MyD88 protein levels, with peak levels at 10 min post-alcohol exposure. Moreover, the enhanced phosphorylation of NFκB inhibitor α (IκBα), cyclooxygenase-2 (COX-2), and inducible nitric oxide synthase (iNOS) were also observed, with peak concentrations at 30 min post-alcohol treatment [18, 40]. The increase of both COX-2 and iNOS in astrocytes by 75 mM alcohol even lasted for 7 days [61]. However, in the presence of TLR4 or IL-1 receptor antibodies, neither wild-type [40] nor TLR4-deficient astrocytes [18] were activated by alcohol treatment, which implies the involvement of TLR4 and IL-1 pathways and the possibility that these pathways could be pharmacologically targeted. Similarly, in wild-type cortical microglial cells, 50 mM of alcohol induced the phosphorylation of ERK, JNK, p38, and IRAK, which peaked at 30 min to 1h post-alcohol exposure then decreased to baseline after exposure to alcohol for 24 h [39]. Furthermore, mRNA or protein expression of TNF-α, IFN-β, and IRF-1 were increased after wild-type microglial cells being treated with alcohol for 3 h [39]. This result suggested the involvement of microglial MyD88-dependent and MyD88-independent pathways in the effect of alcohol. However, this quick proinflammatory activation by alcohol was not found in TLR4-deficient microglia, which indicated the key role of TLR4 in the microglial activation induced by acute alcohol exposure. The above evidence implies that alcohol exposure is capable of rapidly activating TLR4-dependent MAPK and NFκB pathways in both astrocytes and microglia in vitro. The receptors, intracellular signaling proteins, and cytokines which are influenced by alcohol exposure have been summarized in Table 17.1. While the contributions of each of these pathways to the behavioral consequences of alcohol-induced TLR signaling have yet to be determined, pharmacological targeting of the key mediators may prove beneficial, as will be reviewed in Sect. 3.

During conditions of chronic alcohol treatment, MAPK, NFκB, and proinflammatory cytokine signaling are also found to be activated. Utilizing a chronic ex vivo animal model, 5 months of alcohol treatment elevated CD14,

MyD88, phosphorylated ERK, NFκB-p65, TNF-α, IL-1β, IL-6, iNOS, and COX-2 in the brains of female wild-type mice. However, these changes were lacking in the brains of female TLR4-deficient mice [18]. In male animals, on the other hand, four days of alcohol treatment triggered microglial activation, induction of NFκB-DNA binding, and enhancement of COX-2 expression in rat brain [45] (Table 17.1).

Importantly, the role of alternative non-MAPK pathways has been highlighted. It has been demonstrated that alcohol challenge induces a robust AKT phosphorylation in mouse striatum ex vivo [72] (Table 17.1), which suggests the involvement of AKT pathways (Fig. 17.1). This AKT signaling activation was blocked by (−)-naltrexone [72], a TLR4 signaling inhibitor and μ-opioid receptor (MOR) antagonist [73], which implies that such activation may be TLR4- or MOR-dependent or a combination of the two. In another study, the activation of AKT signaling occurred at 15 min post-alcohol administration in the nucleus accumbens (NAc) of rats, and inhibition of the AKT signaling within the NAc attenuated binge drinking of alcohol [74]. Chronic alcohol exposure also induced AKT signaling activation in neurons [75]. However, if the AKT signaling induced by alcohol occurs in glia and if it is TLR4-dependent have not been examined to date.

Collectively, first of all, alcohol activated glial TLR4 signaling cascades in 10 min post-alcohol exposure in vitro and in a long period or even 5 months of alcohol treatment ex vivo. As such, there is a lack of evidence at present showing glial activation by acute alcohol exposure ex vivo. Secondly, the role of glial AKT signaling is worthy of investigation. Finally, the activation of TLR2-dependent signaling and the communication between TLR2 and TLR4 pathways following alcohol exposure are to be examined. With a clear understanding of the effect of alcohol on TLR signaling, the new therapy targeting TLR cascade may be developed for treating different alcohol abuse-related conditions.

17.3.3 Behavioral Consequences of Alcohol-Induced TLR2 and TLR4 Activation

TLR2 and TLR4 contribute significantly to the acute sedation and motor impairment induced by alcohol and alcohol-drinking behavior. The acute sedative and motor effects of alcohol [48] and chronic alcohol consumption [47] were attenuated by inhibition of TLR4 activity using genetic-deficient mice models (TLR4-deficient and MyD88-deficient mice) or pharmacological blockade using (+)-naloxone (a TLR4 signaling inhibitor). Specifically, selective knockdown of TLR4 using inhibitory RNAs in the central nucleus of amygdala of alcohol-preferring rats inhibited binge drinking [31]. Furthermore, TLR2-deficient mice showed reduced acute sedative effects of alcohol compared to their wild-type counterparts [49]. With regard to the downstream signaling of TLRs, the MAPK cascade activation may be related to alcohol consumption. The motivation to both consume and seek alcohol was reduced by activation of MAPK pathways by the glial cell line-derived neurotrophic factor (GDNF) administration [51]. In addition, several genes involved in the MAPK pathways, including *Mapk3*, *p38*, *Map2k1*, and *Map2k2*, were found to be upregulated in the NAc of the high-alcohol-consuming rat strain [76]. These results provide solid evidence that TLR2 and TLR4 signaling activation by alcohol exposure is related to alcohol behaviors. Therefore, pharmacological treatment targeting TLR2 and TLR4 pathways could be potential drug therapies for alcohol dependence.

17.4 Potential Drug Treatment for Alcohol Dependence

The pharmacotherapy of alcohol dependence has significantly changed during the past few years. Disulfiram and lithium have been studied and used in alcohol dependence for more than two decades [77, 78]. More recently, the pharmacotherapy for treating alcohol dependence has been switched to opioid receptor antagonists [(−)-isomers of naltrexone, naloxone, and nalmefene], acamprosate, and various serotonergic agents. Among these pharmacological agents, as several systematic reviews indicated [77, 79], (−)-naltrexone and acamprosate improved therapeutic outcomes by decreasing alcohol consumption theoretically via opioid receptor antagonism and beneficial NMDA receptor modulation, respectively. However, most of the subjects did resume drinking alcohol during such treatments [77, 79], which indicates the problematic efficacy of these drugs.

Considering the emerging evidence of glia-mediated modulation of alcohol's effects by TLR4 and IL-1β and the pivotal neuroinflammatory effects of alcohol, central immune signaling inhibitors could potentially be an additional treatment for alcohol abuse-related conditions. Alcohol and other drugs of abuse, such as morphine [80] and methamphetamine [81], activate glial TLR4-IL-1 signaling and share similar reward pathways [82]. Therefore, central immune signaling inhibitors which have been studied to alter the effect of other drugs of abuse may modulate the effects of alcohol. These central immune signaling inhibitors and their effects on alcohol and other drugs of abuse are discussed below.

17.4.1 Glial Attenuators

Glial attenuators, including minocycline, ibudilast (AV411/MN-166), and propentofylline (HWA 285), have been reported to modulate the effects of drugs of abuse in animal studies. Minocycline, a putative microglial attenuator, was found to suppress the acute sedative effect of alcohol [20] and alcohol consumption [21] but facilitated alcohol-induced motor impairment [20]. In addition, minocycline decreased morphine-induced respiratory depression and reward and enhanced systemic morphine administration-induced analgesia [80, 83]; inhibited the development of cocaine-induced locomotor sensitization [84, 85]; and attenuated methamphetamine-induced hyperlocomotion, behavioral sensitization, and impairment of recognition memory [81, 86]. Similar to minocycline, administration of ibudilast attenuated several effects of morphine, including its activation of brain microglia and astrocytes and induction of elevated dopamine levels [87], reward, tolerance, and withdrawal [88, 89]. Furthermore, ibudilast inhibited prime- and stress-induced methamphetamine relapse [90]. Propentofylline, another glial attenuator, diminished the rewarding effects induced by morphine and methamphetamine [91], morphine tolerance, and withdrawal-induced hyperalgesia [92]. Since alcohol, similar to other drugs of abuse, activates proinflammatory central glial signaling [18, 93], ibudilast and propentofylline may modulate the behavioral effects of alcohol.

As a semisynthetic tetracycline that has been used for more than 30 years, minocycline has an overall good safety record for its clinical use [94]. Minocycline is a small (457 g/mol) and highly lipophilic molecule at physiological pH (apparent partition coefficients with octanol–aqueous buffer at pH 6.6: tetracycline, 0.052; doxycycline, 0.92; minocycline, 0.48 [95]), which is capable of crossing the BBB better than other tetracyclines, such as doxycycline [96, 97]. The neuroprotective effect of minocycline has been studied and established in numerous disease studies using animal models, such as ischemic and hemorrhagic stroke, multiple sclerosis, spinal cord injury, Parkinson's disease, Huntington's disease, and amyotrophic lateral sclerosis models [98]. Although the mechanisms of minocycline's effects are still unclear, minocycline is considered to be a potent inhibitor of microglial activation and of apoptotic pathways, because microglial activation and the attendant neurotoxic products generated by persistently activated microglia are common features of these diseases summarized above [98]. Collectively, since minocycline is safe to use in humans and it modulates drinking, sedative, and motor behaviors of alcohol in mice, it is worthy of investigating the effect of minocycline on the behavioral actions of alcohol in human clinical studies.

Ibudilast (AV411/MN-166, molecular weight: 230 g/mol), a nonselective phosphodiesterase (PDE) inhibitor [99], attenuates the activation of microglia and astrocytes, proinflammatory cytokine release, TLR4 signaling, migration inhibitory factor (MIF), and the release of leukotrienes and reactive oxygen species [100]. It has been generally prescribed for the indication of asthma since the 1980s and is considered to be well tolerated by humans at the regular as well as higher doses used in asthma. Importantly, ibudilast shows CNS activities since it is able to partition into the CNS [101, 102]. In animal models, ibudilast beneficially altered the behavioral effects of methamphetamine and opioids [89, 90]. Preliminary clinical studies of ibudilast have been conducted in patients of poststroke dizziness, neuropathic pain, and multiple sclerosis, and more recently ibudilast was studied for treating opioid withdrawal symptoms [100]. Considering the CNS activity of ibudilast and its effects on drugs of abuse, it could also be a good candidate in alcohol abuse-related studies.

Propentofylline (HWA 285, molecular weight: 306 g/mol), a novel xanthine derivative, is known to depress the activation of microglia and astrocytes under pathological conditions [103]. The mechanism of glial modulation and proinflammatory cytokine suppression [104] by propentofylline is considered to be via its inhibitory action on PDE and subsequent augmentation of cAMP signaling [103, 105]. Although propentofylline has been shown in human preclinical studies to have its effects on the processes of Alzheimer's disease and vascular dementia [106], it has not been used clinically because of some negative results [107]. The effect of propentofylline on the actions of alcohol in both preclinical and clinical settings remains to be uncovered.

17.4.2 TLRs Signaling Inhibitors

As TLR4 signaling has only been considered to be involved in the abuse mechanisms of drugs since the 2000s, only a few of these signaling inhibitors, including (+)-naloxone and (+)-naltrexone [108], have been examined in this area. Other TLR4 signaling inhibitors, such as eritoran (E5564), resatorvid (TAK-242), and anti-TLR4 antibodies (NI-0101 and 1A6), and anti-TLR2 antibodies (OPN-305) [63, 65, 109, 110] have been tested in clinical studies for treating severe sepsis and inflammation. It is possible that these TLRs signaling inhibitors could modulate the effect of drugs of abuse, including alcohol.

Unlike their (−)-isomers, (+)-naloxone and (+)-naltrexone (molecular weight: 327 and 341 g/mol) are inactive at classical neuronal opioid receptors [73], and the result of in silico docking analysis indicated that they were human MD-2 (signaling protein facilitating TLR4 responses) sensitive [108]. In mice, (+)-naloxone inhibited acute sedative and motor effects of alcohol [48]. Regarding other drugs of abuse, (+)-isomers of naloxone and naltrexone blocked morphine-induced hyperalgesia [108, 111], and (+)-naloxone reduced amphetamine- or cocaine-induced hyperactivity [112, 113] in rodents. Importantly, (+)-naloxone and (+)-naltrexone are capable of passing the BBB of rodents and have actions in the CNS, unlike many other available TLR4 inhibitors. Since the TLR4 signaling activities of (+)-naloxone and (+)-naltrexone were discovered only in 2008 [73], human clinical data of these two drugs are still not available.

Eritoran (E5564, molecular weight: 1,314 g/mol), an analog of LPS, antagonizes the activity of LPS by binding to the TLR4-MD-2 complex [63]. It is shown to be well tolerated with a comparatively long half-life in healthy volunteers [109, 114]. However, in phase II clinical studies, administration of eritoran did not appear to confer any clear benefit to patients of severe sepsis [115] and elective cardiac surgery [116]. The CNS permeability and activity of eritoran and its effects on alcohol action should be conducted.

Resatorvid (TAK-242, molecular weight: 362 g/mol), a small-molecule inhibitor of TLR4-mediated signaling, suppresses the production of multiple cytokines induced by LPS exposure [110, 117]. It selectively binds to TLR4 among TLR2, 3, 4, 5, and 9 [118, 119] by directly binding to Cys747 of TLR4 [118]. Acting by inhibiting the association between TLR4 and TIRAP or TRAM, resatorvid disrupts TLR4 signal transduction and downstream signaling [119]. Although promising results have been drawn from animal studies [120], the human clinical study showed that resatorvid failed to suppress the increased cytokine levels in severe sepsis patients [121]. Again, the activity of resatorvid in the CNS is yet to be determined.

Compared to glial attenuators, the TLR inhibitors may have significant advantages in their selectivity. Since both glial and neuronal TLRs are involved in the mechanisms of drugs of abuse [39, 122], selectively blocking TLRs may increase the efficiency of treatment. We have demonstrated that minocycline-treated mice had decreased sedative effect but increased motor dysfunction induced by alcohol administration. In contrast, (+)-naloxone-treated mice were protected from both sedation and motor incoordination following alcohol exposure [20, 48]. The differential effects of minocycline are possibly due to its nonselective blockade of immune signaling in microglia. Together, (+)-isomers of naloxone and naltrexone as well as other TLR4 signaling inhibitors that have been investigated in clinical studies might offer further possibilities of therapeutic manipulation for alcohol abuse-related conditions.

17.4.3 Other Anti-inflammatory Therapies

Numerous drugs of abuse, including alcohol, increase the production of proinflammatory cytokines in the CNS [18, 93]. Treatments blocking these cytokine signaling pathways could potentially be therapies for dependence and other conditions related to alcohol abuse. Among anti-inflammatory drugs, human recombinant IL-1Ra is capable of inhibiting acute sedative and motor effects of alcohol [20] and preventing the development of morphine tolerance [123] in rodents via systemic injection.

Human recombinant IL-1Ra (Anakinra), consisting of 153 amino acids, blocks the biological activity of IL-1, including IL-1α and IL-1β, by competitively inhibiting both IL-1α and IL-1β binding to IL-1 receptor I [124]. It is indicated for the treatment of active adult rheumatoid arthritis and has shown modest benefits in clinical trials [125, 126]. Human recombinant IL-1Ra not only is capable of crossing the BBB of rodents [123] but also reduces CNS inflammation in humans [127]. Considering its effects on the behavioral actions of alcohol in mice [20], it is possible that it could prove beneficial for alcohol dependence in clinical studies. However, as a recombinant protein, anakinra is administered by subcutaneous injection. Therefore, it is not as convenient as drugs with oral formulations.

The various drugs summarized above, all acting on glial or TLRs signaling cascades, have shown their effectiveness on altering the behavioral effects of drugs of abuse, including alcohol, in animals. In addition, some of them have been demonstrated to be well tolerated in human clinical studies [65]. Thus, these medications may provide new targeted therapies for alcohol and possibly other drug dependence. Notably, alcohol also triggers peripheral TLR4 activation and subsequently induces liver diseases [128]. On the other hand, recovery from alcoholic liver diseases may also help the treatment for alcohol dependence. Other key processes in alcohol-induced neuroimmune activation, such as MAPK and NFκB cascades, present new opportunities for developing therapeutic drugs for alcohol dependence. Notably, these immune signaling inhibitors may influence the immune profile of patients, and therefore benefit or risk the patients on immune therapies. These new questions raised by the current findings will be discussed below.

17.5 Speculative Mechanistic Contributors to the Propensity for Alcoholism

The modification of central immune signaling by alcohol may be influenced by preexisted immune activation or underlying propensity for proinflammatory central immune signaling, which leads to various questions about alcohol abuse detailed below.

17.5.1 Alcohol and Other Drugs of Abuse

What is the mechanism of the cross-tolerance and synergistic interactions between alcohol and other drugs of abuse [129]? Why does alcohol consumption increase the risk of heroin-related deaths [130, 131]? Since alcohol and other drugs of abuse share similar neuronal and immune signaling cascades, this may be a possible explanation for the above questions owing to a possible common site of action. The sedative interaction between alcohol and morphine has been demonstrated to be TLR2 dependent [49], which indicates the role of innate immune signaling in the interaction between alcohol and opioids. The neuroimmune activation by drugs of abuse may mediate multidrug abuse-induced toxicity, synergism, and cross-tolerance, thereby providing a novel pharmacological intervention target. Future studies are required to test this hypothesis.

17.5.2 Alcohol and Immunosuppressants

As alcohol activates both peripheral [132] and central immune signaling [133], alcohol consumption may potentially decrease the efficacy and increase the side effects of immunosuppressants. For example, the variation of *TLR4* Asp299Gly single nucleotide polymorphism (SNP) which leads to diminished TLR4-NFκB signaling [134, 135] is correlated with increased development of adverse drug events of methotrexate [136]. This may serve as one of the mechanisms of alcohol-elicited methotrexate toxicity; however, studies are required to test this hypothesis. In addition, FK506 (tacrolimus), used for the graft versus host disease after transplantation, is found to impact microglia and astrocyte function, providing neuroprotection [137]. As such, alcohol exposure may modify the immune function complicating the primary indication of immunosuppressant usage by patients and therefore may be a risk factor for patients on immunosuppressants. Further research is required to test this hypothesis.

17.5.3 Alcohol and Liver Disease

Liver transplantation is becoming a more frequent procedure for the patients of end-stage alcoholic liver disease. The risk of relapse to alcohol drinking after liver transplantation contributes substantially to the selection criteria for liver transplantation [138]. However, the risk factor for relapse is poorly understood [139]. An average alcohol relapse rate of 5.6% per year and a relapse rate of 2.5% per year for heavy alcohol use have been reported in transplant patients [139]. In contrast, in randomized clinical trials for the treatment of alcoholism in patients with alcohol dependence, 37% of (−)-naltrexone-treated and 48% of placebo-treated patients relapsed drinking in 12 weeks [140]. Clearly there are significant discrepancies between the two populations and transplantation is a special case, with significant motivation and support mechanisms put in place to support the individual. It is nonetheless surprising that the transplant recipients achieved a very low relapse rate for drinking alcohol. Is it possible that the removal of the diseased liver or the immunosuppressants prescribed to the patients prevent them from drinking again? Is it possible therefore, that the chronic inflammatory consequences of liver disease contribute to alcohol abuse via a central immune signaling mechanism? Such hypotheses raise the question of whether treatment of liver disease should be a higher priority in alcohol abuse settings. Further research is required to test these hypotheses.

17.5.4 Alcohol and Other Inflammatory Conditions

Other preexisting conditions, such as human immunodeficiency virus (HIV) infection [141], stress [142], and Alzheimer's disease [143], may be influenced by the neuroimmune activation of alcohol as discussed in previous literatures. Alcohol consumption may contribute to more severe neuroinflammation in these conditions, and these conditions may lead to increased alcohol consumption owing to existing proinflammatory central immune signaling. Therefore, many more studies are urgently required in these new research areas.

17.6 Conclusion

Overall, there is significant evidence that alcohol activates proinflammatory central immune signaling, microglia, and astrocytes through TLRs, which contributes significantly to alcohol actions. This neuroimmune mechanism of alcohol offers novel insights and approaches that form the foundation of a new era of research in the pursuit of more effective pharmacotherapies for alcohol abuse and dependence.

References

1. WHO (2004) WHO global status report on alcohol 2004. WHO, Geneva
2. Hansson E et al (1987) Astroglial primary cultures: a model to study ethanol effects on the cell membrane lipid composition. Alcohol Alcohol Suppl 1:679–683
3. Ronnback L, Hansson E, Alling C (1988) Primary astroglial cultures in alcohol and drug research. Alcohol Alcohol 23(6):465–475
4. Alling C et al (1986) Lipids and fatty acids in membranes from astroglial cells cultured in ethanol-containing media. Drug Alcohol Depend 18(2):115–126
5. Allen NJ, Barres BA (2009) Neuroscience: glia—more than just brain glue. Nature 457(7230):675–677
6. Miller G (2005) Neuroscience. The dark side of glia. Science 308(5723):778–781
7. Streit WJ, Mrak RE, Griffin WS (2004) Microglia and neuroinflammation: a pathological perspective. J Neuroinflammation 1(1):14
8. Kreutzberg GW (1996) Microglia: a sensor for pathological events in the CNS. Trends Neurosci 19(8):312–318
9. Araque A et al (1999) Tripartite synapses: glia, the unacknowledged partner. Trends Neurosci 22(5):208–215
10. Smith K (2010) Neuroscience: settling the great glia debate. Nature 468(7321):160–162
11. Perea G, Araque A (2002) Communication between astrocytes and neurons: a complex language. J Physiol Paris 96(3–4):199–207
12. Fields RD, Stevens-Graham B (2002) New insights into neuron-glia communication. Science 298(5593):556–562
13. Janeway CA Jr, Medzhitov R (2002) Innate immune recognition. Annu Rev Immunol 20:197–216
14. Bianchi ME (2007) DAMPs, PAMPs and alarmins: all we need to know about danger. J Leukoc Biol 81(1):1–5
15. Watkins LR, Maier SF (2005) Immune regulation of central nervous system functions: from sickness responses to pathological pain. J Intern Med 257(2):139–155
16. Ortiz J et al (1995) Biochemical actions of chronic ethanol exposure in the mesolimbic dopamine system. Synapse 21(4):289–298
17. Udomuksorn W et al (2011) Effects of alcohol administration during adulthood on parvalbumin and glial fibrillary acidic protein immunoreactivity in the rat cerebral cortex. Acta Histochem 113(3):283–289
18. Alfonso-Loeches S et al (2010) Pivotal role of TLR4 receptors in alcohol-induced neuroinflammation and brain damage. J Neurosci 30(24):8285–8295
19. McClain JA et al (2011) Adolescent binge alcohol exposure induces long-lasting partial activation of microglia. Brain Behav Immun 25(Suppl 1):S120–S128
20. Wu Y et al (2011) Attenuation of microglial and IL-1 signaling protects mice from acute alcohol-induced sedation and/or motor impairment. Brain Behav Immun 25(Suppl 1): S155–S164
21. Agrawal RG et al (2011) Minocycline reduces ethanol drinking. Brain Behav Immun 25(Suppl 1):S165–S169
22. He J, Crews FT (2008) Increased MCP-1 and microglia in various regions of the human alcoholic brain. Exp Neurol 210(2):349–358
23. Blanco AM et al (2008) Ethanol mimics ligand-mediated activation and endocytosis of IL-1RI/TLR4 receptors via lipid rafts caveolae in astroglial cells. J Neurochem 106(2):625–639
24. Lee M, Schwab C, McGeer PL (2011) Astrocytes are GABAergic cells that modulate microglial activity. Glia 59(1):152–165
25. Zou J, Crews F (2010) Induction of innate immune gene expression cascades in brain slice cultures by ethanol: key role of NF-kappaB and proinflammatory cytokines. Alcohol Clin Exp Res 34(5):777–789

26. Jack CS et al (2005) TLR signaling tailors innate immune responses in human microglia and astrocytes. J Immunol 175(7):4320–4330
27. Lehnardt S et al (2002) The toll-like receptor TLR4 is necessary for lipopolysaccharide-induced oligodendrocyte injury in the CNS. J Neurosci 22(7):2478–2486
28. Ren K, Dubner R (2008) Neuron-glia crosstalk gets serious: role in pain hypersensitivity. Curr Opin Anaesthesiol 21(5):570–579
29. Ikonomidou C et al (2000) Ethanol-induced apoptotic neurodegeneration and fetal alcohol syndrome. Science 287(5455):1056–1060
30. Vengeliene V et al (2008) Neuropharmacology of alcohol addiction. Br J Pharmacol 154(2):299–315
31. Liu J et al (2011) Binge alcohol drinking is associated with GABAA alpha2-regulated Toll-like receptor 4 (TLR4) expression in the central amygdala. Proc Natl Acad Sci U S A 108(11):4465–4470
32. Wu PH et al (2010) Correlated changes in NMDA receptor phosphorylation, functional activity, and sedation by chronic ethanol consumption. J Neurochem 115(5):1112–1122
33. Singh AK et al (2007) Effects of chronic ethanol drinking on the blood brain barrier and ensuing neuronal toxicity in alcohol-preferring rats subjected to intraperitoneal LPS injection. Alcohol Alcohol 42(5):385–399
34. Palygin O, Lalo U, Pankratov Y (2011) Distinct pharmacological and functional properties of NMDA receptors in mouse cortical astrocytes. Br J Pharmacol 163(8):1755–1766
35. Lee MC et al (2010) Characterisation of the expression of NMDA receptors in human astrocytes. PLoS One 5(11):e14123
36. Santofimia-Castano P, Salido GM, Gonzalez A (2011) Ethanol reduces kainate-evoked glutamate secretion in rat hippocampal astrocytes. Brain Res 1402:1–8
37. Bardoni R et al (2010) Glutamate-mediated astrocyte-to-neuron signalling in the rat dorsal horn. J Physiol 588(Pt 5):831–846
38. Boyadjieva NI, Sarkar DK (2010) Role of microglia in ethanol's apoptotic action on hypothalamic neuronal cells in primary cultures. Alcohol Clin Exp Res 34(11):1835–1842
39. Fernandez-Lizarbe S, Pascual M, Guerri C (2009) Critical role of TLR4 response in the activation of microglia induced by ethanol. J Immunol 183(7):4733–4744
40. Blanco AM et al (2005) Involvement of TLR4/type I IL-1 receptor signaling in the induction of inflammatory mediators and cell death induced by ethanol in cultured astrocytes. J Immunol 175(10):6893–6899
41. Prow NA, Irani DN (2008) The inflammatory cytokine, interleukin-1 beta, mediates loss of astroglial glutamate transport and drives excitotoxic motor neuron injury in the spinal cord during acute viral encephalomyelitis. J Neurochem 105(4):1276–1286
42. Glezer I et al (2003) Modulation of the innate immune response by NMDA receptors has neuropathological consequences. J Neurosci 23(35):11094–11103
43. Heinisch S, Kirby LG (2010) SDF-1alpha/CXCL12 enhances GABA and glutamate synaptic activity at serotonin neurons in the rat dorsal raphe nucleus. Neuropharmacology 58(2):501–514
44. Benjamins JA et al (2011) Cytokines reduce toxic effects of ethanol on oligodendroglia. Neurochem Res 36(9):1677–1686
45. Crews F et al (2006) BHT blocks NF-kappaB activation and ethanol-induced brain damage. Alcohol Clin Exp Res 30(11):1938–1949
46. Liu J et al (2006) Patterns of gene expression in the frontal cortex discriminate alcoholic from nonalcoholic individuals. Neuropsychopharmacology 31(7):1574–1582
47. Pascual M et al (2011) Impact of TLR4 on behavioral and cognitive dysfunctions associated with alcohol-induced neuroinflammatory damage. Brain Behav Immun 25(Suppl 1):S80–S91
48. Wu Y et al (2011) Inhibiting the TLR4-MyD88 signalling cascade by genetic or pharmacologic strategies reduces acute alcohol dose-induced sedation and motor impairment in mice. Brit J Pharmacology 165(5):1319–1329
49. Hutchinson MR et al (2011) Neuroimmunopharmacology of opioids & alcohol: behavioral, cellular signaling and binding evidence of a role for Toll like receptors. In: Society for Neuroscience. Neuroscience Meeting Planner, Washington, DC

50. Blednov YA et al (2011) Neuroimmune regulation of alcohol consumption: behavioral validation of genes obtained from genomic studies. Addict Biol 17(1):108–120
51. Carnicella S et al (2008) GDNF is a fast-acting potent inhibitor of alcohol consumption and relapse. Proc Natl Acad Sci U S A 105(23):8114–8119
52. Marcos M et al (2008) Interleukin-10 gene polymorphism is associated with alcoholism but not with alcoholic liver disease. Alcohol Alcohol 43(5):523–528
53. Pastor IJ et al (2005) Interleukin-1 gene cluster polymorphisms and alcoholism in Spanish men. Alcohol Alcohol 40(3):181–186
54. Pastor IJ et al (2000) Polymorphism in the interleukin-1 receptor antagonist gene is associated with alcoholism in Spanish men. Alcohol Clin Exp Res 24(10):1479–1482
55. Liu L et al (2009) Association of IL-1B genetic polymorphisms with an increased risk of opioid and alcohol dependence. Pharmacogenet Genomics 19:869–876
56. Blednov YA et al (2005) Perturbation of chemokine networks by gene deletion alters the reinforcing actions of ethanol. Behav Brain Res 165(1):110–125
57. Edenberg HJ et al (2008) Association of NFKB1, which encodes a subunit of the transcription factor NF-kappaB, with alcohol dependence. Hum Mol Genet 17(7):963–970
58. Drugan RC et al (2007) Environmental and immune stressors enhance alcohol-induced motor ataxia in rat. Pharmacol Biochem Behav 86(1):125–131
59. Blednov YA et al (2011) Activation of inflammatory signaling by lipopolysaccharide produces a prolonged increase of voluntary alcohol intake in mice. Brain Behav Immun 25(Suppl 1):S92–S105
60. Blanco AM et al (2004) Ethanol-induced iNOS and COX-2 expression in cultured astrocytes via NF-kappa B. Neuroreport 15(4):681–685
61. Valles SL et al (2004) Chronic ethanol treatment enhances inflammatory mediators and cell death in the brain and in astrocytes. Brain Pathol 14(4):365–371
62. Park BS et al (2009) The structural basis of lipopolysaccharide recognition by the TLR4-MD-2 complex. Nature 458(7242):1191–1195
63. Kim HM et al (2007) Crystal structure of the TLR4-MD-2 complex with bound endotoxin antagonist Eritoran. Cell 130(5):906–917
64. Hua F et al (2007) Activation of Toll-like receptor 4 signaling contributes to hippocampal neuronal death following global cerebral ischemia/reperfusion. J Neuroimmunol 190(1–2):101–111
65. Hennessy EJ, Parker AE, O'Neill LA (2010) Targeting Toll-like receptors: emerging therapeutics? Nat Rev Drug Discov 9(4):293–307
66. Akira S, Takeda K (2004) Functions of toll-like receptors: lessons from KO mice. C R Biol 327(6):581–589
67. Jin MS et al (2007) Crystal structure of the TLR1-TLR2 heterodimer induced by binding of a tri-acylated lipopeptide. Cell 130(6):1071–1082
68. Kang JY et al (2009) Recognition of lipopeptide patterns by Toll-like receptor 2-Toll-like receptor 6 heterodimer. Immunity 31(6):873–884
69. Sheedy FJ, O'Neill LA (2007) The Troll in Toll: Mal and Tram as bridges for TLR2 and TLR4 signaling. J Leukoc Biol 82(2):196–203
70. Rakoff-Nahoum S, Medzhitov R (2009) Toll-like receptors and cancer. Nat Rev Cancer 9(1):57–63
71. Suk K (2007) Microglial signal transduction as a target of alcohol action in the brain. Curr Neurovasc Res 4(2):131–142
72. Bjork K et al (2010) Ethanol-induced activation of AKT and DARPP-32 in the mouse striatum mediated by opioid receptors. Addict Biol 15(3):299–303
73. Hutchinson MR et al (2008) Non-stereoselective reversal of neuropathic pain by naloxone and naltrexone: involvement of toll-like receptor 4 (TLR4). Eur J Neurosci 28(1):20–29
74. Neasta J et al (2011) AKT signaling pathway in the nucleus accumbens mediates excessive alcohol drinking behaviors. Biol Psychiatry 70(6):575–582
75. Liu L et al (2010) Mesenchymal stem cells inhibition of chronic ethanol-induced oxidative damage via upregulation of phosphatidylinositol-3-kinase/Akt and modulation of extracellular signal-regulated kinase 1/2 activation in PC12 cells and neurons. Neuroscience 167(4):1115–1124

76. Arlinde C et al (2004) A cluster of differentially expressed signal transduction genes identified by microarray analysis in a rat genetic model of alcoholism. Pharmacogenomics J 4(3):208–218
77. Garbutt JC et al (1999) Pharmacological treatment of alcohol dependence: a review of the evidence. JAMA 281(14):1318–1325
78. Lejoyeux M, Ades J (1993) Evaluation of lithium treatment in alcoholism. Alcohol Alcohol 28(3):273–279
79. Srisurapanont M, Jarusuraisin N (2005) Opioid antagonists for alcohol dependence. Cochrane Database Syst Rev 1:CD001867
80. Hutchinson MR et al (2008) Minocycline suppresses morphine-induced respiratory depression, suppresses morphine-induced reward, and enhances systemic morphine-induced analgesia. Brain Behav Immun 22:1248–1256
81. Zhang L et al (2006) Protective effects of minocycline on behavioral changes and neurotoxicity in mice after administration of methamphetamine. Prog Neuropsychopharmacol Biol Psychiatry 30(8):1381–1393
82. Hyman SE, Malenka RC, Nestler EJ (2006) Neural mechanisms of addiction: the role of reward-related learning and memory. Annu Rev Neurosci 29:565–598
83. Hutchinson MR et al (2011) Exploring the neuroimmunopharmacology of opioids: an integrative review of mechanisms of central immune signaling and their implications for opioid analgesia. Pharmacol Rev 63(3):772–810
84. Chen H, Manev H (2011) Effects of minocycline on cocaine sensitization and phosphorylation of GluR1 receptors in 5-lipoxygenase deficient mice. Neuropharmacology 60(7–8):1058–1063
85. Chen H, Uz T, Manev H (2009) Minocycline affects cocaine sensitization in mice. Neurosci Lett 452(3):258–261
86. Mizoguchi H et al (2008) Improvement by minocycline of methamphetamine-induced impairment of recognition memory in mice. Psychopharmacology (Berl) 196(2):233–241
87. Bland ST et al (2009) The glial activation inhibitor AV411 reduces morphine-induced nucleus accumbens dopamine release. Brain Behav Immun 23(4):492–497
88. Ledeboer A et al (2007) Ibudilast (AV-411). A new class therapeutic candidate for neuropathic pain and opioid withdrawal syndromes. Expert Opin Investig Drugs 16(7): 935–950
89. Hutchinson MR et al (2009) Reduction of opioid withdrawal and potentiation of acute opioid analgesia by systemic AV411 (ibudilast). Brain Behav Immun 23(2):240–250
90. Beardsley PM et al (2011) The glial cell modulator and phosphodiesterase inhibitor, AV411 (ibudilast), attenuates prime- and stress-induced methamphetamine relapse. Eur J Pharmacol 637(1–3):102–108
91. Narita M et al (2006) Direct evidence of astrocytic modulation in the development of rewarding effects induced by drugs of abuse. Neuropsychopharmacology 31(11):2476–2488
92. Raghavendra V, Tanga FY, DeLeo JA (2004) Attenuation of morphine tolerance, withdrawal-induced hyperalgesia, and associated spinal inflammatory immune responses by propentofylline in rats. Neuropsychopharmacology 29(2):327–334
93. Watkins LR et al (2009) The "toll" of opioid-induced glial activation: improving the clinical efficacy of opioids by targeting glia. Trends Pharmacol Sci 30(11):581–591
94. Kircik LH (2010) Doxycycline and minocycline for the management of acne: a review of efficacy and safety with emphasis on clinical implications. J Drugs Dermatol 9(11):1407–1411
95. Colaizzi JL, Klink PR (1969) pH-Partition behavior of tetracyclines. J Pharm Sci 58(10):1184–1189
96. Carney S et al (1974) Minocycline excretion and distribution in relation to renal function in man. Clin Exp Pharmacol Physiol 1(4):299–308
97. Saivin S, Houin G (1988) Clinical pharmacokinetics of doxycycline and minocycline. Clin Pharmacokinet 15(6):355–366
98. Yong VW et al (2004) The promise of minocycline in neurology. Lancet Neurol 3(12):744–751

99. Gibson LC et al (2006) The inhibitory profile of Ibudilast against the human phosphodiesterase enzyme family. Eur J Pharmacol 538(1–3):39–42
100. Rolan P, Hutchinson M, Johnson K (2009) Ibudilast: a review of its pharmacology, efficacy and safety in respiratory and neurological disease. Expert Opin Pharmacother 10(17):2897–2904
101. Kishi Y et al (2001) Ibudilast: a non-selective PDE inhibitor with multiple actions on blood cells and the vascular wall. Cardiovasc Drug Rev 19(3):215–225
102. Ledeboer A et al (2006) The glial modulatory drug AV411 attenuates mechanical allodynia in rat models of neuropathic pain. Neuron Glia Biol 2(4):279–291
103. Schubert P et al (1997) Support of homeostatic glial cell signaling: a novel therapeutic approach by propentofylline. Ann N Y Acad Sci 826:337–347
104. Sweitzer SM, Schubert P, DeLeo JA (2001) Propentofylline, a glial modulating agent, exhibits antiallodynic properties in a rat model of neuropathic pain. J Pharmacol Exp Ther 297(3):1210–1217
105. Tawfik VL et al (2007) Reprint of "efficacy of propentofylline, a glial modulating agent, on existing mechanical allodynia following peripheral nerve injury" [Brain Behav. Immun. 21 (2007) 238–246]. Brain Behav Immun 21(5):677–685
106. Marcusson J et al (1997) A 12-month, randomized, placebo-controlled trial of propentofylline (HWA 285) in patients with dementia according to DSM III-R. The European Propentofylline Study Group. Dement Geriatr Cogn Disord 8(5):320–328
107. Salimi S et al (2008) A placebo controlled study of the propentofylline added to risperidone in chronic schizophrenia. Prog Neuropsychopharmacol Biol Psychiatry 32(3):726–732
108. Hutchinson MR et al (2010) Possible involvement of toll-like receptor 4/myeloid differentiation factor-2 activity of opioid inactive isomers causes spinal proinflammation and related behavioral consequences. Neuroscience 167(3):880–893
109. Rossignol DP et al (2008) Continuous pharmacodynamic activity of eritoran tetrasodium, a TLR4 antagonist, during intermittent intravenous infusion into normal volunteers. Innate Immun 14(6):383–394
110. Yamada M et al (2005) Discovery of novel and potent small-molecule inhibitors of NO and cytokine production as antisepsis agents: synthesis and biological activity of alkyl 6-(N-substituted sulfamoyl)cyclohex-1-ene-1-carboxylate. J Med Chem 48(23):7457–7467
111. Hutchinson MR et al (2010) Evidence that opioids may have toll-like receptor 4 and MD-2 effects. Brain Behav Immun 24(1):83–95
112. Chatterjie N et al (1996) Prevention of cocaine-induced hyperactivity by a naloxone isomer with no opiate antagonist activity. Neurochem Res 21(6):691–693
113. Chatterjie N et al (1998) Dextro-naloxone counteracts amphetamine-induced hyperactivity. Pharmacol Biochem Behav 59(2):271–274
114. Rossignol DP et al (2004) Safety, pharmacokinetics, pharmacodynamics, and plasma lipoprotein distribution of eritoran (E5564) during continuous intravenous infusion into healthy volunteers. Antimicrob Agents Chemother 48(9):3233–3240
115. Tidswell M et al (2010) Phase 2 trial of eritoran tetrasodium (E5564), a toll-like receptor 4 antagonist, in patients with severe sepsis. Crit Care Med 38(1):72–83
116. Bennett-Guerrero E et al (2007) A phase II, double-blind, placebo-controlled, ascending-dose study of Eritoran (E5564), a lipid A antagonist, in patients undergoing cardiac surgery with cardiopulmonary bypass. Anesth Analg 104(2):378–383
117. Yamada M et al (2006) Optically active cyclohexene derivative as a new antisepsis agent: an efficient synthesis of ethyl (6R)-6-[N-(2-chloro-4-fluorophenyl)sulfamoyl]cyclohex-1-ene-1-carboxylate (TAK-242). Chem Pharm Bull (Tokyo) 54(1):58–62
118. Takashima K et al (2009) Analysis of binding site for the novel small-molecule TLR4 signal transduction inhibitor TAK-242 and its therapeutic effect on mouse sepsis model. Br J Pharmacol 157(7):1250–1262
119. Matsunaga N et al (2011) TAK-242 (resatorvid), a small-molecule inhibitor of toll-like receptor (TLR) 4 signaling, binds selectively to TLR4 and interferes with interactions between TLR4 and its adaptor molecules. Mol Pharmacol 79(1):34–41

120. Sha T et al (2007) Therapeutic effects of TAK-242, a novel selective Toll-like receptor 4 signal transduction inhibitor, in mouse endotoxin shock model. Eur J Pharmacol 571(2–3):231–239
121. Rice TW et al (2010) A randomized, double-blind, placebo-controlled trial of TAK-242 for the treatment of severe sepsis. Crit Care Med 38(8):1685–1694
122. Li Y et al (2010) Toll-like receptor 2 is required for opioids-induced neuronal apoptosis. Biochem Biophys Res Commun 391(1):426–430
123. Shavit Y et al (2005) Interleukin-1 antagonizes morphine analgesia and underlies morphine tolerance. Pain 115(1–2):50–59
124. Granowitz EV et al (1992) Pharmacokinetics, safety and immunomodulatory effects of human recombinant interleukin-1 receptor antagonist in healthy humans. Cytokine 4(5):353–360
125. Gabay C, Lamacchia C, Palmer G (2010) IL-1 pathways in inflammation and human diseases. Nat Rev Rheumatol 6(4):232–241
126. Cohen S et al (2002) Treatment of rheumatoid arthritis with anakinra, a recombinant human interleukin-1 receptor antagonist, in combination with methotrexate: results of a twenty-four-week, multicenter, randomized, double-blind, placebo-controlled trial. Arthritis Rheum 46(3):614–624
127. Goldbach-Mansky R et al (2006) Neonatal-onset multisystem inflammatory disease responsive to interleukin-1beta inhibition. N Engl J Med 355(6):581–592
128. Miller AM et al (2011) Molecular mechanisms of alcoholic liver disease: innate immunity and cytokines. Alcohol Clin Exp Res 35(5):787–793
129. Taslim N, Soderstrom K, Dar MS (2011) Role of mouse cerebellar nicotinic acetylcholine receptor (nAChR) alpha(4)beta(2)- and alpha(7) subtypes in the behavioral cross-tolerance between nicotine and ethanol-induced ataxia. Behav Brain Res 217(2):282–292
130. Hickman M et al (2008) Does alcohol increase the risk of overdose death: the need for a translational approach. Addiction 103(7):1060–1062
131. Levine B, Green D, Smialek JE (1995) The role of ethanol in heroin deaths. J Forensic Sci 40(5):808–810
132. Szabo G, Bala S (2010) Alcoholic liver disease and the gut-liver axis. World J Gastroenterol 16(11):1321–1329
133. Crews FT, Zou J, Qin L (2011) Induction of innate immune genes in brain create the neurobiology of addiction. Brain Behav Immun 25(Suppl 1):S4–S12
134. Rallabhandi P et al (2006) Analysis of TLR4 polymorphic variants: new insights into TLR4/MD-2/CD14 stoichiometry, structure, and signaling. J Immunol 177(1):322–332
135. Arbour NC et al (2000) TLR4 mutations are associated with endotoxin hyporesponsiveness in humans. Nat Genet 25(2):187–191
136. Kooloos WM et al (2010) Functional polymorphisms and methotrexate treatment outcome in recent-onset rheumatoid arthritis. Pharmacogenomics 11(2):163–175
137. Kaminska B, Gaweda-Walerych K, Zawadzka M (2004) Molecular mechanisms of neuroprotective action of immunosuppressants–facts and hypotheses. J Cell Mol Med 8(1):45–58
138. McCallum S, Masterton G (2006) Liver transplantation for alcoholic liver disease: a systematic review of psychosocial selection criteria. Alcohol Alcohol 41(4):358–363
139. Dew MA et al (2008) Meta-analysis of risk for relapse to substance use after transplantation of the liver or other solid organs. Liver Transpl 14(2):159–172
140. Bouza C et al (2004) Efficacy and safety of naltrexone and acamprosate in the treatment of alcohol dependence: a systematic review. Addiction 99(7):811–828
141. Persidsky Y et al (2011) HIV-1 infection and alcohol abuse: neurocognitive impairment, mechanisms of neurodegeneration and therapeutic interventions. Brain Behav Immun 25(Suppl 1):S61–S70
142. Frank MG, Watkins LR, Maier SF (2011) Stress- and glucocorticoid-induced priming of neuroinflammatory responses: potential mechanisms of stress-induced vulnerability to drugs of abuse. Brain Behav Immun 25(Suppl 1):S21–S28
143. Lampron A, Gosselin D, Rivest S (2011) Targeting the hematopoietic system for the treatment of Alzheimer's disease. Brain Behav Immun 25(Suppl 1):S71–S79

Section III
Summary

Chapter 18
Alcohol and the Neuroimmune Response: Current Status and Future Directions

Lindsey Grandison, Changhai Cui, and Antonio Noronha

18.1 Introduction

The recent progress in the understanding of the immune response within the CNS and the novel findings that neuroimmune factors, such as the cytokines and the chemokines, have a role beyond immune response including developmental changes and modulation of behavior add new perspective to interpreting brain function. These advances also invite reexamination of mechanisms involved in brain dysfunction and diseases. This book attempts to bring forth the most recent advances in the neuroimmune field and to summarize the recent findings implicating neuroimmune factors in the response to alcohol and in the modulation of drinking behaviors. This chapter tries to integrate these recent observations linking neuroimmune response and neuroimmune modulators to the mechanisms producing alcoholism and the neuroadaptations responsible for alcohol dependence. Since this area of alcohol research is at its beginning, significant advance is expected in the near future. The recent findings in neuroimmune research may identify new processes through which alcohol impacts health. Several of the novel neuroimmune findings with potential relevance for future alcohol research are noted here.

18.2 Interaction Between Alcohol Consumption and Neuroimmune Function

In the interaction between alcohol use and the neuroimmune system, there are two aspects with critical relevance to alcoholism. First is the effect of alcohol exposure on neuroimmune activity. Alcohol exposure can alter production of cytokines,

L. Grandison, Ph.D. (✉) • C. Cui • A. Noronha
Division of Neuroscience and Behavior, National Institute on Alcohol Abuse and Alcoholism, National Institutes of Health, Bethesda, MD, USA
e-mail: lgrandis@mail.nih.gov

C. Cui et al. (eds.), *Neural-Immune Interactions in Brain Function and Alcohol Related Disorders*, DOI 10.1007/978-1-4614-4729-0_18,

chemokines, and other neuroimmune factors leading to a proinflammatory state within the CNS and, with chronic exposure, contribute to neurodegeneration. The second aspect, only recently appreciated, is the effect of enhanced neuroimmune activity on alcohol consumption including heavy drinking, the transition from social drinking to uncontrolled drinking, and the affective states associated with alcohol dependence. This summary chapter will focus on those findings described in earlier chapters that directly examine these two issues.

18.3 Effects of Alcohol on Immune Response and CNS Neuroimmune Activity

It has been well established that alcohol consumption will diminish the peripheral immune response as evidenced by higher incidences of infections and decreased indicators of immune reaction following acute alcohol exposure. Alcohol affects both adaptive and innate immune responses in the periphery. Within the CNS, it has its action primarily on the innate immune cells since there are few, if any, resident cells within the brain that have the capacity for adaptive immune response. The immune response to alcohol is dependent on the dose and duration of exposure, and recently, it has been appreciated that the immunological response in the periphery and in the CNS can be desynchronized [2]. Within the CNS, alcohol can enhance neuroimmune response, activate neuroinflammatory signal, and as a long-term consequence, these actions may contribute to the neurodegenerative effects of chronic alcohol use.

18.4 Alcohol-Induced Proinflammatory Response Within the CNS

Alcohol consumption can lead to a proinflammatory response within the CNS. This effect is complex involving several mechanisms both indirect and direct that lead to activation of a neuroimmune response. Alcohol induces the breakdown of gut epithelium and entry of endotoxins into the blood. Once in the blood, endotoxins activate liver Kupffer cells with the associated release of cytokines such as TNF-alpha. Serum-borne TNF-alpha can be transported to the brain and act on perivascular cells to trigger CNS reactivity or be translocated across the blood–brain barrier into the CNS where it directly activates glia to release proinflammatory cytokines such as TNF-alpha, interleukin-1beta, and monocyte chemotactic protein-1 (MCP-1). In addition, metabolism of alcohol leads to reactive oxidative species and damage to the perivascular cells of the brain. These stressed cells are the locus of peripheral immune cell adhesion and migration across the blood–brain barrier into the neuronal environment [1]. This indirect effect of alcohol has been well characterized and recognized earlier; however, alcohol does act directly on the CNS to influence neuroimmune activity [2].

In the absence of an immunological challenge "resting" microglia are dynamic and have a role in synaptic pruning during development, in neuroplasticity and in experience dependent remodeling [32]. However microglia can be induced through several levels of activation. Alcohol exposure depending on dose and duration can produce microglial activation and even cytokine release. Acute alcohol exposure can activate microglia in vitro and induce the release of IL-1beta as well as the production of other neuroimmune factors such as MCP-1 [3–4]. Within the CNS, the alcohol-induced activation of microglia produces a feed forward response in which cytokines can act on surrounding microglia to induce further cytokine release. In addition to inducing cytokines, alcohol can also increase expression of the receptors for the cytokines (Qin et al. in press). With chronic, intermittent exposure, alcohol has been shown to induce microgliosis [5]. The direct effect of alcohol on the neuroimmune system is further discussed next.

The interconnected yet separable control of the peripheral versus central immune response is demonstrated by the time course of cytokine production following systemic immunological challenge. In response to peripheral endotoxin, blood levels of cytokines increase acutely, returning to base line within 7 days. However, from this same stimulus, cytokine production within the CNS can persist for upward of 10 months [6].

18.5 Direct Effects of Alcohol on the CNS Neuroimmune System

Ethanol can act on microglia, the primary immune-responsive cells of the CNS. Recent investigation has indicated that ethanol activation of cultured microglia involves Toll-like receptor 4 (TLR-4) [7]. Exposure of microglia cells to ethanol was associated with induction of the same second messengers and the release of the same immunomodulators as occurs after TLR4 activation. These responses to ethanol were absent in microglia lacking TLR4 [7]. By acting through TLR4 on the microglia, ethanol can induce release of cytokines which in turn can induce inflammation and, if sufficiently prolonged, neurodegenerative responses.

However, the action of ethanol on TLR4 appears to be dependent on dose and duration of exposure. Acute exposure to ethanol was found to downregulate TLR4 response in monocytes and macrophage [8], while chronic exposure to ethanol was associated with increased cytokine production in response to the TLR4 agonist, Lipopolysaccharide (LPS), a bacterial membrane molecule [9, 10].

It is proposed that ethanol may interact within the microdomain of lipid rafts on the plasma membrane to either activate or inhibit the TLR4 response [11]. Exposure of macrophage to moderate doses of ethanol (10–50 mM) initiates within 10 min the phosphorylation of ERK, JNK, and p38 and by 30 min the recruitment into lipid rafts of second messenger signaling molecules and the translocation of NF-KB to the nucleus. NF-KB is a transcription factor that regulates the expression of many neuroimmune factors. After 18 h of exposure to a low concentration of ethanol (10 mM), TNF-alpha was increased. These responses were absent when the concentration of ethanol was increased to 100 mM [7]. While these low concentrations of ethanol activated TLR4, these same low concentrations interfered with LPS activation of TLR4.

In microglia, 100 mM ethanol but not 200 mM induced activation through TRL4 and involved both My88- and MyD88-independent pathways [7]. These studies have been extended to indicate that TLR4 has an essential role in mediating the neuroinflammation associated with chronic ethanol exposure [12, 13].

In addition, other signal transduction pathways have been proposed for alcohol-induced neuroinflammation. Using cultured human astrocytes, Florean et al. have implicated ethanol in the initiation of a proinflammatory response [14]. Evidence was provided to show that ethanol was metabolized by P450-2E1 generating acetyaldehyde and reactive oxygen species (ROS) primarily through NADPH oxidase. As a result of elevated ROS, the TLR4 recruited and interacted with the cytosolic tyrosine kinase Src through autophosphorylation. This led to phosphorylation and activation of $cPLA_2$ increasing free arachidonate. COX2 was also increased and converted the arachidonate to the proinflammatory mediator prostaglandin E2.

With chronic exposure, alcohol can induce neural degeneration and promote cell death. During cell necrosis, HMGB1 (high-mobility group box 1) protein is released. It can bind to the receptor for advanced glycation end products (RAGE) on healthy cells to activate the transcription factor NF-KB. This alarmin protein, HMGB1, is increased in alcoholics and in mice exposed to alcohol [33].

18.6 Fetal Alcohol Exposure

Chronic alcohol consumption during pregnancy is associated with elevated cord blood levels of cytokines including TNF-alpha, IL-1beta, and IL-6 [15]. Elevations in maternal blood cytokines, induced by other stimuli, have been implicated in increased fetal brain concentration of these cytokines [16]. However, the ability of a stimulus to induce increased cytokines within the fetus may be restricted to a window of time within the period of pregnancy. Further, it appears that only a subset of maternally produced cytokines may be transferred to the fetus. Elevations in fetal cytokines are attributed to both transfer across the placental barrier and induction of cytokine production in the fetus. The mechanism of transfer across the placental barrier remains to be resolved. Fetal cytokines can impact CNS development and, later in the life of the offspring, vulnerability to infections and increased cytokine production. Furthermore, the early exposure to cytokines has the potential to set the sensitivity of the hypothalamic–pituitary–adrenal axis making it more responsive to stress in adulthood and slower to deactivate following a stress-induced increase in glucocorticoid. The effects of alcohol exposure in utero at critical developmental stages may potentially involve the action of alcohol on maternal and fetal cytokine production. The role of neuroimmune perturbations in fetal alcohol spectrum disorders is in the early stages of investigation.

18.7 Effects of Neuroimmune Activation on Behavior

Molecules released from activated immune cells of the CNS regulate the immune response but also have been implicated in a broader range of CNS functioning. Cytokines, chemokines, and signal transduction pathways that were previously thought to have an exclusive neuroimmune role have recently been appreciated to have an expanded role [17, 18]. Some of these neuroimmune modulators may influence alcohol intake, promote neuroadaptation between controlled drinking and compulsive drinking, perpetuate alcohol dependence, and lastly, mediate alcohol-induced neurocognitive impairments.

Analysis of gene expression has been done in an attempt to identify the factors which contribute to alcohol intake in lines of mice exhibiting high alcohol consumption [19, 20], in rats selectively bred for high alcohol intake [21], and in human alcoholics [22]. An analysis of these studies indicated that one of the common features associated with enhanced alcohol intake was an alteration in expression of genes related to the neuroimmune response [23].

The relevance for alcohol consumption of the neuroimmune genes identified in animal lines with a trait of high alcohol intake has been further examined by determining alcohol consumption in the null mutants of these neuroimmune genes. Deletion of beta-2-microglobulin, cathepsin S, cathepsin F, interleukin-1 receptor antagonist, CD14, or interleukin-6 was associated with a decrease in alcohol consumption. However, the effect varied dependent on gender and the protocol for assessing alcohol self-administration [23].

Reduction in the TLR4 in central amygdala resulting from local injection of a siRNA vector was associated with decreased drinking in P rats [22, 24]. GABA has been implicated in modulation of alcohol consumption, and changes in GABA receptor expression have been shown to affect TLR4 expression.

Consistent with the hypothesis that immunomodulators contribute to increased alcohol intake is the observation that LPS, a potent activator of immune response, was found to increase alcohol consumption [25]. Conversely, minocycline which disrupts the activation of microglia [26] and pioglitazone which prevents microglial activation and reduces proinflammatory cytokines and chemokines [27] have been found to reduce alcohol drinking [24, 28]. Another molecule found to disrupt activation of microglia by blocking TLR4 signal transduction is the opioid receptor antagonist naloxone. Both stereoisomers of naloxone (– and + naloxone) are effective in blocking the TLR4 receptor whereas blockade of opioid receptors is produced only by the [-] stereoisomer of naloxone. Naltrexone, a long-acting, modified naloxone molecule, is one of the three clinically approved treatments for alcohol dependence. It is not known whether naltrexone blockade of TLR4 has a role in its reduction of alcohol consumption. Within the microglia, activation of the transcription factor NF-kappa B leads to increased expression of many cytokines. Blockade of NF-kappa B with caffeic acid phenethyl ester reduced ethanol intake (see Harris and Blednov).

Alcohol induces sedation and motor impairment, and these effects can be reduced by blocking TLR4 receptor signaling [29]; (+) naloxone reduced alcohol sedative effects and motor impairment. Similarly, these effects of alcohol were reduced when the TLR signaling was eliminated by a null mutation of TLR4 gene or a null mutation of the MyD88 gene, which is an essential component of the TLR4 signaling cascade.

18.8 Effects of Cytokines on Neurons and Alcohol-Related Responses

LPS binds to the TLR4 receptors on microglia and astrocytes and activates these neuroimmune cells to produce cytokines. When LPS was superfused onto a brain slice of the central amygdala, a brain site implicated in regulation of alcohol-drinking behavior, the amplitude of IPSPs was increased. The effects resulted predominately from postsynaptic mechanisms. This action of LPS is similar to the effect of ethanol on these same neurons. Knocking out CD14 (an essential accessory molecule for initiation of signal transduction through TLR4) reduced the effect of both LPS and ethanol on IPSPs [30]. In contrast to LPS which activates microglia, interleukin-1β (IL-1β), which is released from microglia, produced the opposite effects. IL-1β superfused onto a slice of the central nucleus of the amygdala decreased IPSP amplitude.

Ethanol exposure increased production of Interleukin 10 (IL-10), an anti-inflammatory cytokine. Interleukin-10 (IL-10) has been found to decrease mIPSCs of both cultured cortical neurons and dentate granule cells in brain slices [31]. IL-10 was also found to decrease expression of the alpha-1 subunit of the GABAA receptor. These observations suggest a potential role for IL-10 in the neuroadaptation of GABA receptors following ethanol exposure.

18.9 Conclusion

These results have extended the findings that alcohol can modulate peripheral immune function to now include its action on neuroimmune reactivity in the CNS. They demonstrate the separate effects of alcohol on the peripheral and the CNS immune response. These observations provide evidence for a role of alcohol-induced neuroimmune response in the immediate reaction but potentially, more importantly, in the long-term adverse consequences of chronic alcohol consumption. More intriguing is the findings implicating a role for alcohol-induced neuroimmune response in regulation of alcohol consumption. These observations expand the role of neuroimmune signaling to include a contribution to reward and addiction behaviors. While the evidence implicating neuroimmune signaling in alcohol response and alcohol consumption is compelling, the relative role of neuroimmunce activity in the overall response to alcohol is poorly defined. However, these observations do provide new target candidates for treatment intervention for alcoholism, a disease that currently has few effective treatment options.

18.10 Opportunities for Future Advances in Alcohol Research

The rapid advance in neuroimmune research and the recognition only recently of the involvement of neuroimmune molecules in alcohol response and in alcohol consumption have provided opportunities for further understanding the mechanisms of alcoholism. The neuroimmune advances described here could have direct relevance for alcohol research. Several of these new findings raise interest because alcohol has been implicated with a process that is now observed to be influenced by neuroimmune factors. While links exist, large gaps remain in understanding. Filling these gaps could provide new medication targets or identify other routes of intervention or repair.

There is much promise for continuing the investigations that have been noted in earlier chapters on the effects of neuroimmune and alcohol interactions. Interventions targeting neuroimmune signaling may yield new medications for treatment, relapse prevention, or reduction of alcohol effects. In addition, there are areas of neuroimmune function that have not yet been examined or exploited in the effort to advance alcohol research. Further understanding is needed on how alcohol impacts microglia activities in the healthy brain and development (see Davis and Carson). The role of immune molecules in synaptic function including both development and plasticity has been reviewed (see Pribiag and Stellwagen; Groul; McIver et al.). While alcohol is known to act within this compartment, the mechanism and full extent of response are poorly defined. Early alcohol exposure produces fetal alcohol spectrum disorders (FASD). There are effects of early alcohol exposure on neuroimmune function. The role for a disrupted neuroimmune function in mediating some of the FASD consequences requires further examination (see Drew and Kane; Bodnar and Weinberg). One of the hypotheses of alcohol dependence suggests that chronic alcohol produces an allostatic load involving a negative affective state and that this altered state motivates further alcohol consumption as a way to relieve the negative state. Stress, mood disorders, and drugs of abuse are associated with the dysregulation of the neuroimmune and neuroendocrine networks (see Deak et al.; Walker et al.; Chang and Liu). The effect of the immune activator LPS to enhance anxiety and potentiate alcohol withdrawal anxiety (Breese et al.) suggests there could be a connection. Yet much further examination would be necessary to link these processes. The importance of alcohol use to the progression and treatment of HIV-associated neurocognitive disorders (McMillian and Gendelman; Zahr et al.; Persidsky et al.) has immediate relevance. The use of neuroimaging approaches to define the involvement of neuroimmune processes in alcohol response (see Hannestad; Zahr et al.) may provide new insights. While many topics have been noted, still others remain to be developed, e.g., potential role of alcohol and neuroimmune response in stem cell development, sleep disorders, and pain. The seminal observations of these investigators will provide the opportunities and starting points for future advances.

References

1. Alikunju S, Abdul Muneer PM et al (2011) The inflammatory footprints of alcohol-induced oxidative damage in neurovascular components. Brain Behav Immun 25 Suppl 1, S129–136
2. Qin L et al (2008) Increased systemic and brain cytokine production and neuroinflammation by endotoxin following ethanol treatment. J Neuroinflammation 5:10
3. He J, Crews FT (2008) Increased MCP-1 and microglia in various regions of the human alcoholic brain. Exp Neurol 210(2):349–358
4. Wu Y et al (2011) Attenuation of microglial and IL-1 signaling protects mice from acute alcohol-induced sedation and/or motor impairment. Brain Behav Immun 25(Suppl 1):S155–S164
5. Riikonen J et al (2002) Intermittent ethanol exposure increases the number of cerebellar microglia. Alcohol Alcohol 37(5):421–426
6. Qin, L. et al. Systemic LPS causes chronic neuroinflammation and progressive neurodegeneration. Glia 55, 453-462 (2007)
7. Fernandez-Lizarbe S, Pascual M, Guerri C (2009) Critical role of TLR4 response in the activation of microglia induced by ethanol. J Immunol 183(7):4733–4744
8. Goral J, Kovacs EJ (2005) In vivo ethanol exposure down-regulates TLR2-, TLR4-, and TLR9-mediated macrophage inflammatory response by limiting p38 and ERK1/2 activation. J Immunol 174(1):456–463
9. Shi L et al (2002) Chronic ethanol increases lipopolysaccharide-stimulated Egr-1 expression in RAW 264.7 macrophages: contribution to enhanced tumor necrosis factor alpha production. J Biol Chem 277(17):14777–14785
10. Zhang Z et al (2001) Prolonged ethanol treatment enhances lipopolysaccharide/phorbol myristate acetate-induced tumor necrosis factor-alpha production in human monocytic cells. Alcohol Clin Exp Res 25(3):444–449
11. Blanco AM, Guerri C (2007) Ethanol intake enhances inflammatory mediators in brain: role of glial cells and TLR4/IL-1RI receptors. Front Biosci 12:2616–2630
12. Alfonso-Loeches S, Guerri C (2011) Molecular and behavioral aspects of the actions of alcohol on the adult and developing brain. Crit Rev Clin Lab Sci 48(1):19–47
13. Alfonso-Loeches S et al (2010) Pivotal role of TLR4 receptors in alcohol-induced neuroinflammation and brain damage. J Neurosci 30(24):8285–8295
14. Floreani NA et al (2010) Alcohol-induced interactive phosphorylation of Src and toll-like receptor regulates the secretion of inflammatory mediators by human astrocytes. J Neuroimmune Pharmacol 5(4):533–545
15. Ahluwalia B et al (2000) Alcohol modulates cytokine secretion and synthesis in human fetus: an in vivo and in vitro study. Alcohol 21(3):207–213
16. Bilbo SD and Schwarz JM (2009) Early-life programming of later-life brain and behavior: a critical role for the immune system. *Front Behav Neurosci* 3, 14, doi:10.3389/neuro.08.014.2009
17. Fourgeaud L et al (2010) MHC class I modulates NMDA receptor function and AMPA receptor trafficking. Proc Natl Acad Sci USA 107(51):22278–22283
18. Boulanger LM (2004) MHC class I in activity-dependent structural and functional plasticity. Neuron Glia Biol. 1, 283–289
19. Mulligan MK et al (2006) Toward understanding the genetics of alcohol drinking through transcriptome meta-analysis. Proc Natl Acad Sci USA 103(16):6368–6373

20. Tabakoff B et al (2008) The genomic determinants of alcohol preference in mice. Mamm Genome 19(5):352–365
21. Kimpel MW et al (2007) Functional gene expression differences between inbred alcohol-preferring and -non-preferring rats in five brain regions. Alcohol 41(2):95–132
22. Liu J et al (2011) Binge alcohol drinking is associated with GABAA alpha2-regulated Toll-like receptor 4 (TLR4) expression in the central amygdala. Proc Natl Acad Sci USA 108(11):4465–4470
23. Blednov YA et al (2005) Perturbation of chemokine networks by gene deletion alters the reinforcing actions of ethanol. Behav.Brain Res. 165, 110–125
24. Agrawal RG et al (2011) Minocycline reduces ethanol drinking. Brain Behav Immun 25 (Suppl 1): S165–S169
25. Blednov YA et al (2011) Activation of inflammatory signaling by lipopolysaccharide produces a prolonged increase of voluntary alcohol intake in mice. *Brain Behav. Immun.* 25 Suppl 1, S92–S105
26. Plane JM et al (2010) Prospects for minocycline neuroprotection. Arch Neurol 67(12): 1442–1448
27. Kapadia R Yi JH Vemuganti R (2008) Mechanisms of anti-inflammatory and neuroprotective actions of PPAR-gamma agonists. Front Biosci 13:1813–1826
28. Stopponi S et al (2011) Activation of nuclear PPAR-gamma receptors by the antidiabetic agent pioglitazone suppresses alcohol drinking and relapse to alcohol seeking. Biol Psychiatry 69(7):642–649
29. Wu Y et al (2012) Inhibiting the TLR4-MyD88 signalling cascade by genetic or pharmacologic strategies reduces acute alcohol dose-induced sedation and motor impairment in mice. Br.J.Pharmacol
30. Bajo M et al. (2010) Activation of TLR4 and IL-1 receptors modulates GABAergic transmission in the mouse central amygdala. Neuroscience Meeting Planner 103.10
31. Suryanarayanan A et al. (2011) Ethanol modulation of interleukon-10 (IL-10) and inhibitory effects of IL-10 on GABAergic transmission in rat cortical and hippocampal neurons. www.sfn.org/am2011/pdf/final_program/final_program_b2.pdf
32. Tremblay M-E et al. (2011) The role of microglia in the healthy brain. J Neurosci 31(45): 16064–16069
33. Qin 1 Crews FT (2012) Chronic ethanol increases systemic TLR3 agonist-induced neuroinflammation and neurodegeneration. J Neuroinflamm 9:130–148

Index

C. Cui et al. (eds.), *Neural-Immune Interactions in Brain Function and Alcohol Related Disorders*, DOI 10.1007/978-1-4614-4729-0,

B

C

I

Printed by Publishers' Graphics LLC
MRO20121221.19.32.18